Contents

v

Feedback Control of Dynamic Systems

Sixth Edition

Gene F. Franklin
Stanford University

J. David Powell
Stanford University

Abbas Emami-Naeini
SC Solutions, Inc.

Upper Saddle River   Boston   Columbus   San Francisco   New York
Indianapolis   London   Toronto   Sydney   Singapore   Tokyo   Montreal
Dubai   Madrid   Hong Kong   Mexico City   Munich   Paris   Amsterdam   Cape Town

Vice President and Editorial Director,
  ECS: *Marcia J. Horton*
Acquisition Editor: *Andrew Gilfillan*
Editorial Assistant: *William Opaluch*
Director of Team-Based Project
  Management: *Vince O'Brien*
Marketing Manager: *Tim Galligan*
Marketing Assistant: *Mack Patterson*
Permissions Project Manager: *Vicki Menanteaux*
Senior Managing Editor: *Scott Disanno*
Production Project Manager: *Clare Romeo*
Senior Operations Specialist: *Alan Fischer*

Operations Specialist: *Lisa McDowell*
Art Director: *Kenny Beck*
Cover Designer: *Kristine Carney*
Manager, Rights and Permissions: *Zina Arabia*
Manager, Visual Research: *Beth Brenzel*
Image Permission Coordinator: *Debbie Latronica*
Manager, Cover Photo Permissions:
  Karan Sanatar
Composition: *TexTech International*
Full-Service Project Management: *TexTech International*
Printer/Binder: *Courier Westford*
Typeface: 10/12 Times

The author and publisher of this book have used their best efforts in preparing this book. These efforts include the development, research, and testing of theories and programs to determine their effectiveness. The author and publisher make no warranty of any kind, expressed or implied, with regard to these programs or the documentation contained in this book. The author and publisher shall not be liable in any event for incidental or consequential damages with, or arising out of, the furnishing, performance, or use of these programs.

Pearson Education Ltd., London
Pearson Education Singapore, Pte. Ltd
Pearson Education Canada, Inc.
Pearson Education—Japan
Pearson Education Australia PTY, Limited
Pearson Education North Asia, Ltd., Hong Kong
Pearson Educación de Mexico, S.A. de C.V.
Pearson Education Malaysia, Pte. Ltd.
Pearson Education, Inc., Upper Saddle River, New Jersey

Library of Congress Cataloging-in-Publication Data

Franklin, Gene F.
  Feedback control of dynamic systems / Gene Franklin, J. David Powell,
Abbas Emami-Naeini. — 6th ed.
    p. cm.
  Includes bibliographical references and index.
  ISBN-13: 978-0-13-601969-5
  ISBN-10: 0-13-601969-2

1. Feedback control systems.   I. Powell, J. David, 1938–   II. Emami-Naeini, Abbas.
III. Title.
TJ216.F723   2009
629.8′3—dc22   2009025662

Prentice Hall
is an imprint of

www.pearsonhighered.com

10 9 8 7 6 5 4 3 2 1
ISBN-13: 978-0-13-601969-5
ISBN-10:   0-13-601969-2

*To Gertrude, David, Carole,
Valerie, Daisy, Annika, Davenport,
Malahat, Sheila, and Nima*

Preface

In this Sixth Edition we again present a text in support of a first course in control and have retained the best features of our earlier editions. For this edition, we have substantially rewritten Chapter 4 on the Basic Properties of Feedback, placing the material in a more logical order and presenting it in a much more effective (bottom up?) manner. We have also updated the text throughout on how computer-aided design is utilized to more fully reflect how design is carried out today. At the same time, we also strive to equip control system engineers with a basic understanding so that computer results can be guided and verified. In support of this updating, MATLAB® referrals have been updated and include some of the latest capabilities in that software. The case studies in Chapter 10 have been retained and a new case study of the emerging Bioengineering field has been added. A Historical Perspective section has been added at the end of each chapter in order to add to the knowledge of how all these concepts came into being. Finally, in order to guide the reader in finding specific topics, we have expanded the Table of Contents to include subsections.

The basic structure of the book is unchanged and we continue to combine analysis with design using the three approaches of the root locus, frequency response, and state-variable equations. The text continues to include many carefully worked out examples to illustrate the material. As before, we provide a set of review questions at the end of each chapter with answers in the back of the book to assist the students in verifying that they have learned the material.

In the three central chapters on design methods we continue to expect the students to learn how to perform the very basic calculations by hand and make a rough sketch of a root locus or Bode plot as a sanity check on the computer results and as an aid to design. However, we introduce the use of MATLAB early on in recognition of the universal use of software tools in control analysis and design. Furthermore, in recognition of the fact that increasingly controllers are implemented in imbedded computers, we again introduce digital control in Chapter 4 and in a number of cases compare the responses of feedback systems using analog controllers with those having a digital "equivalent" controller. As before, we have prepared a collection of all the MATLAB files (both "m" files and SIMULINK® files) used to produce the figures in the book. These are available at the following Web site:

www.FPE6e.com

For the SIMULINK files, there are equivalent LabView files that can be obtained by a link from the same web site.

We have removed some material that was judged to be less useful for the teaching of a first course in controls. However, recognizing that there may still be some instructors who choose to teach the material, or students who want to refer to enrichment and/or review material, we have moved the material to the website above for access by anyone. The Table of Contents provides a guide as to where the various topics are located.

We feel that this Sixth Edition presents the material with good pedagogical support, provides strong motivation for the study of control, and represents a solid foundation for meeting the educational challenges. We introduce the study of feedback control, both as a specialty of itself and as support for many other fields.

Addressing the Educational Challenges

Some of the educational challenges facing students of feedback control are long-standing; others have emerged in recent years. Some of the challenges remain for students across their entire engineering education; others are unique to this relatively sophisticated course. Whether they are old or new, general or particular, the educational challenges we perceived were critical to the evolution of this text. Here we will state several educational challenges and describe our approaches to each of them.

- CHALLENGE   *Students must master design as well as analysis techniques.*

Design is central to all of engineering and especially so to control systems. Students find that design issues, with their corresponding opportunities to tackle practical applications, particularly motivating. But students also find design problems difficult because design problem statements are usually poorly posed and lack unique solutions. Because of both its inherent importance for and its motivational effect on students, design is emphasized throughout this text so that confidence in solving design problems is developed from the start.

The emphasis on design begins in Chapter 4 following the development of modeling and dynamic response. The basic idea of feedback is introduced first, showing its influence on disturbance rejection, tracking accuracy, and robustness to parameter changes. The design orientation continues with uniform treatments of the root locus, frequency response, and state variable feedback techniques. All the treatments are aimed at providing the knowledge necessary to find a good feedback control design with no more complex mathematical development than is essential to clear understanding.

Throughout the text, examples are used to compare and contrast the design techniques afforded by the different design methods and, in the capstone case studies of Chapter 10, complex real-world design problems are attacked using all the methods in a unified way.

- CHALLENGE   *New ideas continue to be introduced into control.*

Control is an active field of research and hence there is a steady influx of new concepts, ideas, and techniques. In time, some of these elements develop to the point where they join the list of things every control engineer must know. This text is devoted to supporting students equally in their need to grasp both traditional and more modern topics.

In each of our editions we have tried to give equal importance to root locus, frequency response, and state-variable methods for design. In this edition we continue to emphasize solid mastery of the underlying techniques. coupled with computer based methods for detailed calculation. We also provide an early introduction to data sampling and discrete controllers in recognition of the major role played by digital controllers in our field. While this material can be skipped to save time without harm to the flow of the text, we feel that it is very important for students to understand that computer control is widely used and that the most basic techniques of computer control are easily mastered.

• CHALLENGE *Students need to manage a great deal of information.*

The vast array of systems to which feedback control is applied and the growing variety of techniques available for the solution of control problems means that today's student of feedback control must learn many new ideas. How do students keep their perspective as they plow through lengthy and complex textual passages? How do they identify highlights and draw appropriate conclusions? How do they review for exams? Helping students with these tasks was a criterion for the Fourth and Fifth Editions and continues to be addressed in this Sixth Edition. We outline these features below.

FEATURE

1. *Chapter openers* offer perspective and overview. They place the specific chapter topic in the context of the discipline as a whole and they briefly overview the chapter sections.
2. *Margin notes* help students scan for chapter highlights. They point to important definitions, equations, and concepts.
3. *Boxed highlights* identify key concepts within the running text. They also function to summarize important design procedures.
4. *Bulleted chapter summaries* help with student review and prioritization. These summaries briefly reiterate the key concepts and conclusions of the chapter.
5. *Synopsis of design aids.* Relationships used in design and throughout the book are collected inside the back cover for easy reference.
6. *The color blue* is used (1) to highlight useful pedagogical features, (2) to highlight components under particular scrutiny within block diagrams, (3) to distinguish curves on graphs, and (4) to lend a more realistic look to figures of physical systems.
7. *Review questions* at the end of each chapter with solutions in the back to guide the student in self-study

• CHALLENGE *Students of feedback control come from a wide range of disciplines.*

Feedback control is an interdisciplinary field in that control is applied to systems in every conceivable area of engineering. Consequently, some schools have separate introductory courses for control within the standard disciplines and some, like Stanford, have a single set of courses taken by students from many disciplines. However, to restrict the examples to one field is to miss much of the range and power of feedback but to cover the whole range of applications is overwhelming. In this book we develop the interdisciplinary nature of the field and provide review material for several of the most common technologies so that students from many disciplines will be comfortable with the presentation. For Electrical Engineering students who typically have a good background in transform analysis, we include in Chapter 2 an introduction to writing equations of motion for mechanical mechanisms. For mechanical engineers, we include in Chapter 3 a review of the Laplace Transform and dynamic response as needed in control. In addition, we introduce other technologies briefly and, from time to time, we present the equations of motion of a physical system without derivation but with enough physical description to be understood from a response point of view. Examples of some of the physical systems represented in the text include the read–write head for a computer disk drive, a satellite tracking system, the fuel–air ratio in an automobile engine, and an airplane automatic pilot system.

Outline of the Book

The contents of the book are organized into ten chapters and three appendixes. Optional sections of advanced or enrichment material marked with a triangle ($\triangle$) are included at the end of some chapters. There is additional enrichment material on the website. Examples and problems based on this material are also marked with a triangle ($\triangle$). The appendices include background and reference material. The appendices in the book include Laplace transform tables, answers to the end-of-chapter review questions, and a list of MATLAB commands. The appendixes on the website include a review of complex variables, a review of matrix theory, some important results related to State-Space design, a tutorial on RLTOOL for MATLAB, and optional material supporting or extending several of the chapters.

In Chapter 1, the essential ideas of feedback and some of the key design issues are introduced. This chapter also contains a brief history of control, from the ancient beginnings of process control to flight control and electronic feedback amplifiers. It is hoped that this brief history will give a context for the field, introduce some of the key figures who contributed to its development, and provide motivation to the student for the studies to come.

Chapter 2 is a short presentation of dynamic modeling and includes mechanical, electrical, electromechanical, fluid, and thermodynamic devices. This material can be omitted, used as the basis of review homework to smooth out the usual nonuniform preparation of students, or covered in-depth depending on the needs of the students.

Chapter 3 covers dynamic response as used in control. Again, much of this material may have been covered previously, especially by electrical engineering students. For many students, the correlation between pole locations and transient response and the effects of extra zeros and poles on dynamic response represent new material. Stability of dynamic systems is also introduced in this Chapter. This material needs to be covered carefully.

Chapter 4 presents the basic equations and transfer functions of feedback along with the definitions of the sensitivity function. With these tools, open-loop and closed-loop control are compared with respect to disturbance rejection, tracking accuracy, and sensitivity to model errors. Classification of systems according to their ability to track polynomial reference signals or to reject polynomial disturbances is described with the concept of system type. Finally, the classical proportional, integral, and derivative (PID) control structure is introduced and the influence of the controller parameters on a system's characteristic equation is explored along with PID tuning methods. The end-of-chapter optional section treats digital control.

Following the overview of feedback in Chapter 4, the core of the book presents the design methods based on root locus, frequency response, and state-variable feedback in Chapters 5, 6, and 7, respectively.

Chapter 8 develops in more detail the tools needed to design feedback control for implementation in a digital computer. However, for a complete treatment of feedback control using digital computers, the reader is referred to the companion text, *Digital Control of Dynamic Systems*, by Franklin, Powell, and Workman; Ellis-Kagle Press, 1998.

In Chapter 9 the nonlinear material includes techniques for the linearization of equations of motion, analysis of zero memory nonlinearity as a variable gain, frequency response as a describing function, the phase plane, Lyapunov stability theory, and the circle stability criterion.

In Chapter 10 the three primary approaches are integrated in several case studies and a framework for design is described that includes a touch of the real-world context of practical control design.

Course Configurations

The material in this text can be covered flexibly. Most first-course students in controls will have some dynamics and Laplace transforms. Therefore, Chapter 2 and most of Chapter 3 would be a review for those students. In a ten-week quarter, it is possible to review Chapter 3, and all of Chapters 1, 4, 5, and 6. Most boxed sections should be omitted. In the second quarter, Chapters 7, and 9 can be covered comfortably including the boxed sections. Alternatively, some boxed sections could be omitted and selected portions of Chapter 8 included. A semester course should comfortably accommodate Chapters 1–7, including the review material of Chapters 2 and 3, if needed. If time remains after this core coverage, some introduction of digital control from Chapter 8, selected nonlinear issues from Chapter 9 and some of the case studies from Chapter 10 may be added.

The entire book can also be used for a three-quarter sequence of courses consisting of modeling and dynamic response (Chapters 2 and 3), classical control (Chapters 4–6), and modern control (Chapters 7–10).

Two basic 10-week courses are offered at Stanford and are taken by seniors and first-year graduate students who have not had a course in control, mostly in the departments of Aeronautics and Astronautics, Mechanical Engineering, and Electrical Engineering. The first course reviews Chapters 2 and 3 and covers Chapters 4–6. The more advanced course is intended for graduate students and reviews Chapters 4–6 and covers Chapters 7–10. This sequence complements a graduate course in linear systems and is the prerequisite to courses in digital control, nonlinear control, optimal control, flight control, and smart product design. Several of the subsequent courses include extensive laboratory experiments. Prerequisites for the course sequence include dynamics or circuit analysis and Laplace transforms.

Prerequisites to This Feedback Control Course

This book is for a first course at the senior level for all engineering majors. For the core topics in Chapters 4–7, prerequisite understanding of modeling and dynamic response is necessary. Many students will come into the course with sufficient background in those concepts from previous courses in physics, circuits, and dynamic response. For those needing review, Chapters 2 and 3 should fill in the gaps.

An elementary understanding of matrix algebra is necessary to understand the state-space material. While all students will have much of this in prerequisite math courses, a review of the basic relations is given in Appendix WE and a brief treatment of particular material needed in control is given at the start of Chapter 7. The emphasis is on the relations between linear dynamic systems and linear algebra.

Supplements

The Web site mentioned above includes the dot-m and dot-mdl files used to generate all the MATLAB figures in the book and these may be copied and distributed to the students as desired. An instructor's manual with complete solutions to all homework problems is available. The Web site also includes advanced material and appendixes.

Acknowledgments

Finally, we wish to acknowledge our great debt to all those who have contributed to the development of feedback control into the exciting field it is today and specifically to the considerable help and education we have received from our students and our colleagues. In particular, we have benefited in this effort by many discussions with the following who taught introductory control at Stanford: A. E. Bryson, Jr., R. H. Cannon, Jr., D. B. DeBra, S. Rock, S. Boyd, C. Tomlin, P. Enge, and C. Gerdes. Other colleagues who have helped us include D. Fraser, N. C. Emami, B. Silver, M. Dorfman, D. Brennan, K. Rudie, L. Pao, F. Khorrami, K. Lorell, and P. D. Mathur.

Special thanks go to the many students who have provided almost all the solutions to the problems in the book.

G.F.F.
J.D.P.
A.E.-N.
Stanford, California

An Overview and Brief History of Feedback Control

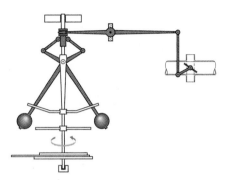

A Perspective on Feedback Control

Feedback control of dynamic systems is a very old concept with many characteristics that have evolved over time. The central idea is that a system's output can be measured and fed back to a controller of some kind and used to effect the control. It has been shown that signal feedback can be used to control a vast array of dynamic systems including, for example, airplanes and hard-disk data storage devices. To achieve good control there are four basic requirements.

- The system must be stable at all times
- The system output must track the command input signal
- The system output must be prevented from responding too much to disturbance inputs
- These goals must be met even if the model used in the design is not completely accurate or if the dynamics of the physical system change over time or with environmental changes.

The requirement of stability is basic and may have two causes. In the first place, the system may be unstable. This is illustrated by the Segway vehicle, which will simply fall over if the control is turned off. On the other hand, adding feedback may itself drive the system unstable. In ordinary experience

1

such an instability is called a "vicious circle," where the feedback signal that is circled back makes the situation worse rather than better.

There are many examples of the requirement of having the system's output track a command signal. For example, driving a car so that the vehicle stays in its lane is command tracking. Similarly, flying an airplane in the approach to a landing strip requires that a glide path be accurately tracked.

Disturbance rejection is one of the very oldest applications of feedback control. In this case, the "command" is simply a constant set point to which the output is to be held as the environment changes. A very common example of this is the room thermostat whose job it is to hold the room temperature close to the set point as outside temperature and wind change, and as doors and windows are opened and closed.

Finally, to design a controller for a dynamic system, it is necessary to have a mathematical model of the dynamic response of the system in all but the simplest cases. Unfortunately, almost all physical systems are very complex and often nonlinear. As a result, the design will usually be based on a simplified model and must be robust enough that the system meets its performance requirements when applied to the real device. Furthermore, again in almost all cases, as time and the environment change, even the best of models will be in error because the system dynamics have changed. Again, the design must not be too sensitive to these inevitable changes and it must work well enough regardless.

The tools available to control engineers to solve these problems have evolved over time as well. Especially important has been the development of digital computers both as computation aids and as embedded control devices. As computation devices, computers have permitted identification of increasingly complex models and the application of very sophisticated control design methods. Also, as embedded devices, digital devices have permitted the implementation of very complex control laws. Control engineers must not only be skilled in manipulating these design tools but also need to understand the concepts behind these tools to be able to make the best use of them. Also important is that the control engineer understand both the capabilities and the limitations of the controller devices available.

Chapter Overview

In this chapter we begin our exploration of feedback control using a simple familiar example: a household furnace controlled by a thermostat. The generic components of a control system are identified within the context of this example. In another example—an automobile cruise control—we develop the elementary static equations and assign numerical values to elements of the system model in order to compare the performance of open-loop control to that of feedback control when dynamics are ignored. In order to provide a context for our studies and to give you a glimpse of how the field has evolved, Section 1.3 provides a brief history of control theory and design. In addition, later chapters have brief sections of additional historical notes on the topics covered there. Finally, Section 1.4 provides a brief overview of the contents and organization of the entire book.

1.1   A Simple Feedback System

In feedback systems the variable being controlled—such as temperature or speed—is measured by a sensor and the measured information is fed back to the controller to influence the controlled variable. The principle is readily illustrated by a very common system, the household furnace controlled by a thermostat. The components of this system and their interconnections are shown in Fig. 1.1. Such a picture identifies the major parts of the system and shows the directions of information flow from one component to another.

We can easily analyze the operation of this system qualitatively from the graph. Suppose both the temperature in the room where the thermostat is located and the

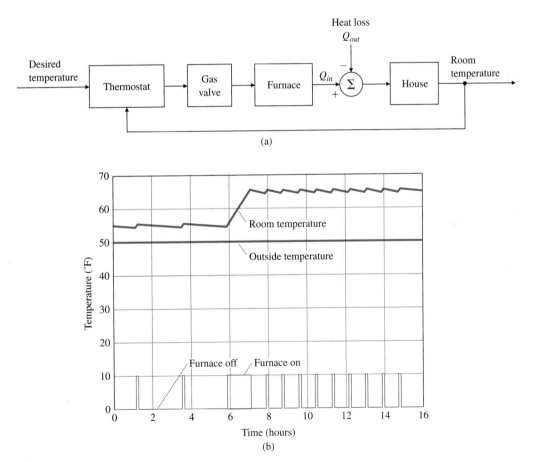

Figure 1.1

(a) Component block diagram of a room temperature control system; (b) plot of room temperature and furnace action

outside temperature are significantly below the reference temperature (also called the set point) when power is applied. The thermostat will be *on* and the control logic will open the furnace gas valve and light the fire box. This will cause heat Q_{in} to be supplied to the house at a rate that will be significantly larger than the heat loss Q_{out}. As a result, the room temperature will rise until it exceeds the thermostat reference setting by a small amount. At this time the furnace will be turned off and the room temperature will start to fall toward the outside value. When it falls a small amount below the set point, the thermostat will come on again and the cycle will repeat. Typical plots of room temperature along with the furnace cycles of on and off are shown in Fig. 1.1. The outside temperature is held at 50°F and the thermostat is initially set at 55°F. At 6 a.m., the thermostat is stepped to 65°F and the furnace brings it to that level and cycles the temperature around that figure thereafter.[1] Notice that the house is well insulated, so that the fall of temperature with the furnace off is significantly slower than the rise with the furnace on. From this example, we can identify the generic components of the elementary feedback control system as shown in Fig. 1.2.

The central component of this feedback system is the **process** whose output is to be controlled. In our example the process would be the house whose output is the room temperature and the **disturbance** to the process is the flow of heat from the house due to conduction through the walls and roof to the lower outside temperature. (The outward flow of heat also depends on other factors such as wind, open doors, etc.) The design of the process can obviously have a major impact on the effectiveness of the controls. The temperature of a well-insulated house with thermopane windows is clearly easier to control than otherwise. Similarly, the design of aircraft with control in mind makes a world of difference to the final performance. In every case, the earlier the issues of control are introduced into the process design, the better. The **actuator** is the device that can influence the controlled variable of the process and in our case,

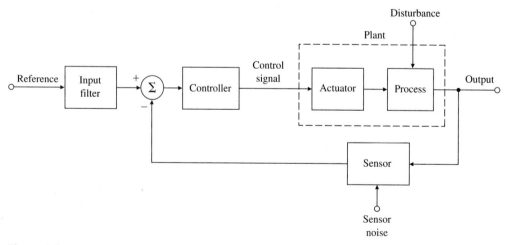

Figure 1.2

Component block diagram of an elementary feedback control

---

[1] Notice that the furnace had come on a few minutes before 6 a.m. on its regular nighttime schedule.

the actuator is a gas furnace. Actually, the furnace usually has a pilot light or striking mechanism, a gas valve, and a blower fan, which turns on or off depending on the air temperature in the furnace. These details illustrate the fact that many feedback systems contain components that themselves form other feedback systems.[2] The central issue with the actuator is its ability to move the process output with adequate speed and range. The furnace must produce more heat than the house loses on the worst day and must distribute it quickly if the house temperature is to be kept in a narrow range. Power, speed, and reliability are usually more important than accuracy. Generally, the process and the actuator are intimately connected and the control design centers on finding a suitable input or control signal to send to the actuator. The combination of process and actuator is called the **plant** and the component that actually computes the desired control signal is the **controller**. Because of the flexibility of electrical signal processing, the controller typically works on electrical signals although the use of pneumatic controllers based on compressed air has a long and important place in process control. With the development of digital technology, cost-effectiveness and flexibility have led to the use of digital signal processors as the controller in an increasing number of cases. The component labeled **thermostat** in Fig. 1.1 measures the room temperature and is called the **sensor** in Fig. 1.2, a device whose output inevitably contains sensor noise. Sensor selection and placement are very important in control design, for it is sometimes not possible for the true controlled variable and the sensed variable to be the same. For example, although we may really wish to control the house temperature as a whole, the thermostat is in one particular room, which may or may not be at the same temperature as the rest of the house. For instance, if the thermostat is set to 68°F but is placed in the living room near a roaring fireplace, a person working in the study could still feel uncomfortably cold.[3,4] As we will see, in addition to placement, important properties of a sensor are the accuracy of the measurements as well as low noise, reliability, and linearity. The sensor will typically convert the physical variable into an electrical signal for use by the controller. Our general system also includes an **input filter** whose role is to convert the reference signal to electrical form for later manipulation by the controller. In some cases the input filter can modify the reference command input in ways that improve the system response. Finally, there is a **comparator** to compute the difference between the reference signal and the sensor output to give the controller a measure of the system error.

This text will present methods for analyzing feedback control systems and their components and will describe the most important design techniques engineers can use with confidence in applying feedback to solve control problems. We will also study

---

[2] Jonathan Swift (1733) said it this way: "So, Naturalists observe, a flea Hath smaller fleas that on him prey; And these have smaller still to bite 'em; And so proceed, *ad infinitum*."

[3] In the renovations of the kitchen in the house of one of the authors, the new ovens were placed against the wall where the thermostat was mounted on the other side. Now when dinner is baked in the kitchen on a cold day, the author freezes in his study unless the thermostat is reset.

[4] The story is told of the new employee at the nitroglycerin factory who was to control the temperature of a critical part of the process manually. He was told to "keep that reading below 300°." On a routine inspection tour, the supervisor realized that the batch was dangerously hot and found the worker holding the thermometer under cold water tap to bring it down to 300°. They got out just before the explosion. Moral: sometimes automatic control is better than manual.

the specific advantages of feedback that compensate for the additional complexity it demands. However, although the temperature control system is easy to understand, it is nonlinear as seen by the fact that the furnace is either on or off, and to introduce linear controls we need another example.

1.2 A First Analysis of Feedback

The value of feedback can be readily demonstrated by quantitative analysis of a simplified model of a familiar system, the cruise control of an automobile (Fig. 1.3). To study this situation analytically, we need a mathematical **model** of our system in the form of a set of quantitative relationships among the variables. For this example, we ignore the dynamic response of the car and consider only the steady behavior. (Dynamics will, of course, play a major role in later chapters.) Furthermore, we assume that for the range of speeds to be used by the system, we can approximate the relations as linear. After measuring the speed of the vehicle on a level road at 65 mph, we find that a 1° change in the throttle angle (our control variable) causes a 10-mph change in speed. From observations while driving up and down hills it is found that when the grade changes by 1%, we measure a speed change of 5 mph. The speedometer is found to be accurate to a fraction of 1 mph and will be considered exact. With these relations, we can draw the **block diagram** of the plant (Fig. 1.4), which shows these mathematical relationships in graphical form. In this diagram the connecting lines carry signals and a block is like an ideal amplifier which multiplies the signal at its input by the value marked in the block to give the output signal. To sum two or more signals, we show lines for the signals coming into a summer, a circle with the summation sign Σ inside. An algebraic sign (plus or minus) beside each arrow head indicates whether the input adds to or subtracts from the total output of the summer. For this analysis, we wish to compare the effects of a 1% grade on the output speed when the reference speed is set for 65 with and without feedback to the controller.

Figure 1.3

Component block diagram of automobile cruise control

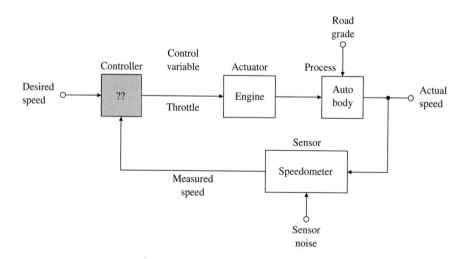

Figure 1.4

Block diagram of the cruise control plant

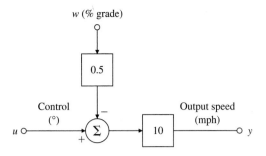

Figure 1.5

Open-loop cruise control

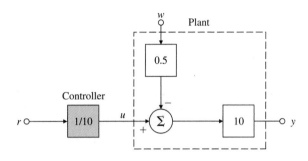

Open-loop control

In the first case, shown in Fig. 1.5, the controller does not use the speedometer reading but sets $u = r/10$. This is an example of an **open-loop control system.** The term *open loop* refers to the fact that there is no closed path or loop around which the signals go in the block diagram. In our simple example the open-loop output speed, y_{ol}, is given by the equations

$$y_{ol} = 10(u - 0.5w)$$

$$= 10\left(\frac{r}{10} - 0.5w\right)$$

$$= r - 5w.$$

The error in output speed is

$$e_{ol} = r - y_{ol} \qquad (1.1)$$

$$= 5w \qquad (1.2)$$

and the percent error is

$$\% \, error = 500\frac{w}{r}. \qquad (1.3)$$

If $r = 65$ and the road is level then $w = 0$ and the speed will be 65 with no error. However, if $w = 1$ corresponding to a 1% grade, then the speed will be 60 and we have a 5-mph error, which is a 7.69% error in the speed. For a grade of 2%, the speed error would be 10 mph, which is an error of 15.38%, and so on. The example shows that there would be no error when $w = 0$ but this result depends on the controller gain being the exact inverse of the plant gain of 10. In practice, the plant gain is subject to change and if it does, errors are introduced by this means also. If there is an error in the plant gain in open-loop control, the percent speed error would be the same as the percent plant-gain error.

The block diagram of a feedback scheme is shown in Fig. 1.6, where the controller gain has been set to 10. Recall that in this simple example, we have assumed that we have an ideal sensor whose block is not shown. In this case the equations are

$$y_{cl} = 10u - 5w,$$
$$u = 10(r - y_{cl}).$$

Combining them yields

$$y_{cl} = 100r - 100y_{cl} - 5w,$$
$$101y_{cl} = 100r - 5w,$$
$$y_{cl} = \frac{100}{101}r - \frac{5}{101}w,$$
$$e_{cl} = \frac{r}{101} + \frac{5w}{101}.$$

Thus the feedback has reduced the sensitivity of the speed error to the grade by a factor of 101 when compared with the open-loop system. Note, however, that there is now a small speed error on level ground because even when $w = 0$,

$$y_{cl} = \frac{100}{101}r = 0.99r \text{ mph.}$$

This error will be small as long as the loop gain (product of plant and controller gains) is large.[5] If we again consider a reference speed of 65 mph and compare speeds with a 1% grade, the percent error in the output speed is

$$\% \ error = 100\frac{\frac{65 \times 100}{101} - \left(\frac{65 \times 100}{101} - \frac{5}{101}\right)}{\frac{65 \times 100}{101}} \tag{1.4}$$

$$= 100\frac{5 \times 101}{101 \times 65 \times 100} \tag{1.5}$$

$$= 0.0769\%. \tag{1.6}$$

Figure 1.6

Closed-loop cruise control

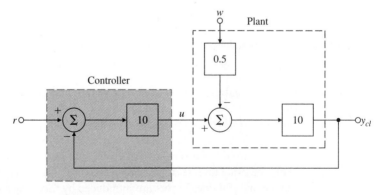

The reduction of the speed sensitivity to grade disturbances and plant gain in our example is due to the loop gain of 100 in the feedback case. Unfortunately, there are limits to how high this gain can be made; when dynamics are introduced, the feedback can make the response worse than before, or even cause the system to become unstable. The dilemma is illustrated by another familiar situation where it is easy to change a feedback gain. If one tries to raise the gain of a public-address amplifier too much, the sound system will squeal in a most unpleasant way. This is a situation where the gain in the feedback loop—from the speakers to the microphone through the amplifier back to the speakers—is too much. The issue of how to get the gain as large as possible to reduce the errors without making the system become unstable and squeal is what much of feedback control design is all about.

The design trade-off

1.3 A Brief History

An interesting history of early work on feedback control has been written by O. Mayr (1970), who traces the control of mechanisms to antiquity. Two of the earliest examples are the control of flow rate to regulate a water clock and the control of liquid level in a wine vessel, which is thereby kept full regardless of how many cups are dipped from it. The control of fluid flow rate is reduced to the control of fluid level, since a small orifice will produce constant flow if the pressure is constant, which is the case if the level of the liquid above the orifice is constant. The mechanism of the liquid-level control invented in antiquity and still used today (for example, in the water tank of the ordinary flush toilet) is the **float valve**. As the liquid level falls, so does the float, allowing the flow into the tank to increase; as the level rises, the flow is reduced and if necessary cut off. Figure 1.7 shows how a float valve operates. Notice here that sensor and actuator are not separate devices but are contained in the carefully shaped float-and-supply-tube combination.

Liquid-level control

A more recent invention described by Mayr (1970) is a system, designed by Cornelis Drebbel in about 1620, to control the temperature of a furnace used to heat an incubator[6] (Fig. 1.8). The furnace consists of a box to contain the fire, with a flue at the top fitted with a damper. Inside the fire box is the double-walled incubator box, the hollow walls of which are filled with water to transfer the heat evenly to the incubator. The temperature sensor is a glass vessel filled with alcohol and mercury and placed in the water jacket around the incubator box. As the fire heats the box and

Drebbel's incubator

Figure 1.7
Early historical control of liquid level and flow

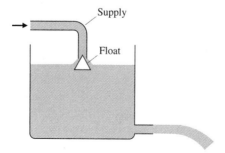

---

[6]French doctors introduced incubators into the care of premature babies over 100 years ago.

Figure 1.8

Drebbel's incubator for hatching chicken eggs

Source: Adapted from Mayr, 1970

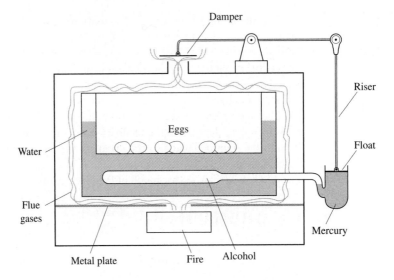

water, the alcohol expands and the riser floats up, lowering the damper on the flue. If the box is too cold, the alcohol contracts, the damper is opened, and the fire burns hotter. The desired temperature is set by the length of the riser, which sets the opening of the damper for a given expansion of the alcohol.

A famous problem in the chronicles of control systems was the search for a means to control the rotation speed of a shaft. Much early work (Fuller, 1976) seems to have been motivated by the desire to automatically control the speed of the grinding stone in a wind-driven flour mill. Of various methods attempted, the one with the most promise used a conical pendulum, or **fly-ball governor**, to measure the speed of the mill. The sails of the driving windmill were rolled up or let out with ropes and pulleys, much like a window shade, to maintain fixed speed. However, it was adaptation of these principles to the steam engine in the laboratories of James Watt around 1788 that made the fly-ball governor famous. An early version is shown in Fig. 1.9, while Figs. 1.10 and 1.11 show a close-up of a fly-ball governor and a sketch of its components.

The action of the fly-ball governor (also called a centrifugal governor) is simple to describe. Suppose the engine is operating in equilibrium. Two weighted balls spinning around a central shaft can be seen to describe a cone of a given angle with the shaft. When a load is suddenly applied to the engine, its speed will slow, and the balls of the governor will drop to a smaller cone. Thus the ball angle is used to sense the output speed. This action, through the levers, will open the main valve to the steam chest (which is the actuator) and admit more steam to the engine, restoring most of the lost speed. To hold the steam valve at a new position it is necessary for the fly balls to rotate at a different angle, implying that the speed under load is not exactly the same as before. We saw this effect earlier with cruise control, where feedback control gave a very small error. To recover the exact same speed in the system, it would require resetting the desired speed setting by changing the length of the rod from the lever to the valve. Subsequent inventors introduced mechanisms that integrated the speed error to provide automatic reset. In Chapter 4 we will analyze these systems to show

Fly-ball governor

Figure 1.9

Photograph of an early Watt steam engine

Source: British Crown Copyright, Science Museum, London

Figure 1.10

Close-up of the fly-ball governor

Source: British Crown Copyright, Science Museum, London

Beginnings of control theory

that such integration can result in feedback systems with zero steady-state error to constant disturbances.

Because Watt was a practical man, like the millwrights before him, he did not engage in theoretical analysis of the governor. Fuller (1976) has traced the early development of control theory to a period of studies from Christian Huygens in 1673 to James Clerk Maxwell in 1868. Fuller gives particular credit to the contributions of G. B. Airy, professor of mathematics and astronomy at Cambridge University from 1826 to 1835 and Astronomer Royal at Greenwich Observatory from 1835 to 1881. Airy was concerned with speed control; if his telescopes could be rotated counter to the rotation of the earth, a fixed star could be observed for extended periods. Using the centrifugal-pendulum governor he discovered that it was capable of unstable motion— "and the machine (if I may so express myself) became perfectly wild" (Airy, 1840; quoted in Fuller, 1976). According to Fuller, Airy was the first worker to discuss

Figure 1.11

Operating parts of a fly-ball governor

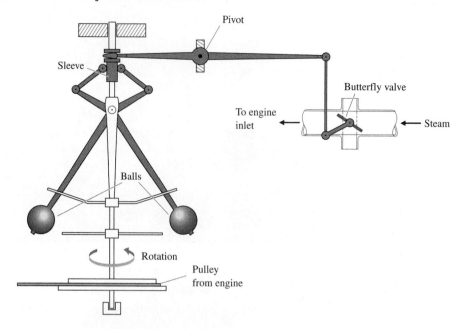

Stability analysis

instability in a feedback control system and the first to analyze such a system using differential equations. These attributes signal the beginnings of the study of feedback control dynamics.

The first systematic study of the stability of feedback control was apparently given in the paper "On Governors" by J. C. Maxwell (1868).[7] In this paper, Maxwell developed the differential equations of the governor, linearized them about equilibrium, and stated that stability depends on the roots of a certain (characteristic) equation having negative real parts. Maxwell attempted to derive conditions on the coefficients of a polynomial that would hold if all the roots had negative real parts. He was successful only for second- and third-order cases. Determining criteria for stability was the problem for the Adams Prize of 1877, which was won by E. J. Routh.[8] His criterion, developed in his essay, remains of sufficient interest that control engineers are still learning how to apply his simple technique. Analysis of the characteristic equation remained the foundation of control theory until the invention of the electronic feedback amplifier by H. S. Black in 1927 at Bell Telephone Laboratories.

Shortly after publication of Routh's work, the Russian mathematician A. M. Lyapunov (1893) began studying the question of stability of motion. His studies were based on the nonlinear differential equations of motion and also included results for linear equations that are equivalent to Routh's criterion. His work was fundamental to what is now called the state-variable approach to control theory, but was not introduced into the control literature until about 1958.

---

[7]An exposition of Maxwell's contribution is given in Fuller (1976).

[8]E. J. Routh was first academically in his class at Cambridge University in 1854, while J. C. Maxwell was second. In 1877 Maxwell was on the Adams Prize Committee that chose the problem of stability as the topic for the year.

Frequency response

The development of the feedback amplifier is briefly described in an interesting article based on a talk by H. W. Bode (1960) reproduced in Bellman and Kalaba (1964). With the introduction of electronic amplifiers, long-distance telephoning became possible in the decades following World War I. However, as distances increased, so did the loss of electrical energy; in spite of using larger-diameter wires, increasing numbers of amplifiers were needed to replace the lost energy. Unfortunately, large numbers of amplifiers resulted in much distortion since the small nonlinearity of the vacuum tubes then used in electronic amplifiers were multiplied many times. To solve the problem of reducing distortion, Black proposed the feedback amplifier. As mentioned earlier in connection with the automobile cruise control, the more we wish to reduce errors (or distortion), the more feedback we need to apply. The loop gain from actuator to plant to sensor to actuator must be made very large. With high gain the feedback loop begins to squeal and is unstable. Here was Maxwell's and Routh's stability problem again, except that in this technology the dynamics were so complex (with differential equations of order 50 being common) that Routh's criterion was not very helpful. So the communications engineers at Bell Telephone Laboratories, familiar with the concept of frequency response and the mathematics of complex variables, turned to complex analysis. In 1932 H. Nyquist published a paper describing how to determine stability from a graphical plot of the loop frequency response. From this theory there developed an extensive methodology of feedback-amplifier design described by Bode (1945) and extensively used still in the design of feedback controls. Nyquist and Bode plots are discussed in more detail in Chapter 6.

PID control

Simultaneous with the development of the feedback amplifier, feedback control of industrial processes was becoming standard. This field, characterized by processes that are not only highly complex but also nonlinear and subject to relatively long time delays between actuator and sensor, developed **proportional-integral-derivative (PID) control**. The PID controller was first described by Callender et al. (1936). This technology was based on extensive experimental work and simple linearized approximations to the system dynamics. It led to standard experiments suitable to application in the field and eventually to satisfactory "tuning" of the coefficients of the PID controller. (PID controllers are covered in Chapter 4.) Also under development at this time were devices for guiding and controlling aircraft; especially important was the development of sensors for measuring aircraft altitude and speed. An interesting account of this branch of control theory is given in McRuer (1973).

An enormous impulse was given to the field of feedback control during World War II. In the United States engineers and mathematicians at the MIT Radiation Laboratory combined their knowledge to bring together not only Bode's feedback amplifier theory and the PID control of processes but also the theory of stochastic processes developed by N. Wiener (1930). The result was the development of a comprehensive set of techniques for the design of **servomechanisms**, as control mechanisms came to be called. Much of this work was collected and published in the records of the Radiation Laboratory by James et al. (1947).

Another approach to control systems design was introduced in 1948 by W. R. Evans, who was working in the field of guidance and control of aircraft. Many of his problems involved unstable or neutrally stable dynamics, which made the frequency methods difficult, so he suggested returning to the study of the characteristic equation that had been the basis of the work of Maxwell and Routh nearly 70 years

Root locus

State-variable design

Modern control
Classical control

earlier. However, Evans developed techniques and rules allowing one to follow graph-ically the paths of the roots of the characteristic equation as a parameter was changed. His method, the **root locus**, is suitable for design as well as for stability analysis and remains an important technique today. The root-locus method developed by Evans is covered in Chapter 5.

During the 1950s several authors, including R. Bellman and R. E. Kalman in the United States and L. S. Pontryagin in the U.S.S.R., began again to consider the ordinary differential equation (ODE) as a model for control systems. Much of this work was stimulated by the new field of control of artificial earth satellites, in which the ODE is a natural form for writing the model. Supporting this endeavor were digital computers, which could be used to carry out calculations unthinkable 10 years before. (Now, of course, these calculations can be done by any engineering student with a desktop computer.) The work of Lyapunov was translated into the language of control at about this time, and the study of optimal controls, begun by Wiener and Phillips during World War II, was extended to optimizing trajectories of nonlinear systems based on the calculus of variations. Much of this work was presented at the first conference of the newly formed International Federation of Automatic Control held in Moscow in 1960.[9] This work did not use the frequency response or the characteristic equation but worked directly with the ODE in "normal" or "state" form and typically called for extensive use of computers. Even though the foundations of the study of ODEs were laid in the late 19th century, this approach is now often called **modern control** to distinguish it from **classical control**, which uses the complex variable methods of Bode and others. In the period from the 1970s continuing through the present, we find a growing body of work that seeks to use the best features of each technique.

Thus we come to the current state of affairs where the principles of control are applied in a wide range of disciplines, including every branch of engineering. The well-prepared control engineer needs to understand the basic mathematical theory that underlies the field and must be able to select the best design technique suited to the problem at hand. With the ubiquitous use of computers it is especially important that the engineer is able to use his or her knowledge to guide and verify calculations done on the computer.[10]

1.4   An Overview of the Book

The central purpose of this book is to introduce the most important techniques for single-input–single-output control systems design. **Chapter 2** will review the tech-niques necessary to obtain models of the dynamic systems that we wish to control. These include model making for mechanical, electric, electromechanical, and a few other physical systems. Chapter 2 also describes briefly the linearization of nonlinear models, although this will be discussed more thoroughly in **Chapter 9**.

---

[9]Optimal control gained a large boost when Bryson and Denham (1962) showed that the path of a supersonic aircraft should actually dive at one point in order to reach a given altitude in minimum time. This nonintuitive result was later demonstrated to skeptical fighter pilots in flight tests.

[10]For more background on the history of control, see the survey papers appearing in the *IEEE Control Systems Magazine* of November 1984 and June 1996.

In **Chapter 3** and **Appendix A** we will discuss the analysis of dynamic response using Laplace transforms along with the relationship between time response and the poles and zeros of a transfer function. The chapter also includes a discussion of the critical issue of system stability, including the Routh test.

In **Chapter 4** we will cover the basic equations and features of feedback. An analysis of the effects of feedback on disturbance rejection, tracking accuracy, sensitivity to parameter changes, and dynamic response will be given. The idea of elementary PID control is discussed. Also in this chapter a brief introduction is given to the digital implementation of transfer functions and thus of linear time-invariant controllers so that the effects of digital control can be compared with analog controllers as these are designed.

In **Chapters 5, 6, and 7** we introduce the techniques for realizing the control objectives first identified in Chapter 4 in more complex dynamic systems. These methods include the root locus, frequency response, and state-variable techniques. These are alternative means to the same end and have different advantages and disadvantages as guides to design of controls. The methods are fundamentally complementary, and each needs to be understood to achieve the most effective control systems design.

In **Chapter 8** we develop further the ideas of implementing controllers in a digital computer that were introduced in Chapter 4. The chapter addresses how one "digitizes" the control equations developed in Chapters 5 through 7, how the sampling introduces a delay that tends to destabilize the system, and how the sample rate needs to be a certain multiple of the system frequencies for good performance. The analysis of sampled systems requires another analysis tool—the z-transform—and that tool is described and its use is illustrated.

Most real systems are nonlinear to some extent. However, the analyses and design methods in most of the book up to here are for linear systems. In **Chapter 9** we explain why the study of linear systems is pertinent, why it is useful for design even though most systems are nonlinear, and how designs for linear systems can be modified to handle most common nonlinearities in the systems being controlled. The chapter covers saturation, describing functions and the anti windup controller, and contains a brief introduction to Lyapunov stability theory.

Application of all the techniques to problems of substantial complexity are discussed in **Chapter 10**, in which the design methods are brought to bear simultaneously on specific case studies.

Computer aids

Control designers today make extensive use of computer-aided control systems design software that is commercially available. Furthermore, most instructional programs in control systems design make software tools available to the students. The most widely used software for the purpose are MATLAB® and SIMULINK® from The Mathworks. MATLAB routines have been included throughout the text to help illustrate this method of solution and many problems require computer aids for solution. Many of the figures in the book were created using MATLAB and the files for their creation are available free of charge on the web at the site: **http://www.FPE6e.com**. Students and instructors are invited to use these files as it is believed that they should be helpful in learning how to use computer methods to solve control problems.

Needless to say, many topics are not treated in the book. We do not extend the methods to multivariable controls, which are systems with more than one input and/or output, except as part of the case study of the rapid thermal processor in Chapter 10. Nor is optimal control treated in more than a very introductory manner in Chapter 7.

Also beyond the scope of this text is a detailed treatment of the experimental testing and modeling of real hardware, which is the ultimate test of whether any design really works. The book concentrates on analysis and design of linear controllers for linear plant models—not because we think that is the final test of a design, but because that is the best way to grasp the basic ideas of feedback and is usually the first step in arriving at a satisfactory design. We believe that mastery of the material here will provide a foundation of understanding on which to build knowledge of these more advanced and realistic topics—a foundation strong enough to allow one to build a personal design method in the tradition of all those who worked to give us the knowledge we present here.

SUMMARY

- **Control** is the process of making a system variable adhere to a particular value, called the **reference value.** A system designed to follow a changing reference is called **tracking control** or a **servo**. A system designed to maintain an output fixed regardless of the disturbances present is called a **regulating control** or a **regulator**.
- Two kinds of control were defined and illustrated based on the information used in control and named by the resulting structure. In **open-loop control** the system does *not* measure the output and there is no correction of the actuating signal to make that output conform to the reference signal. In **closed-loop control** the system includes a sensor to measure the output and uses **feedback** of the sensed value to influence the control variable.
- A simple feedback system consists of the **process** whose output is to be controlled, the **actuator** whose output causes the process output to change, **reference** and **output sensors** that measure these signals, and the **controller** that implements the logic by which the control signal that commands the actuator is calculated.
- **Block diagrams** are helpful for visualizing system structure and the flow of information in control systems. The most common block diagrams represent the mathematical relationships among the signals in a control system.
- The theory and design techniques of control have come to be divided into two categories: **classical control** methods use the Laplace or Fourier Transforms and were the dominant methods for control design until about 1960 while **modern control** methods are based on ODEs in state form and were introduced into the field starting in the1960s. Many connections have been discovered between the two categories and well prepared engineers must be familiar with both techniques.

REVIEW QUESTIONS

1. What are the main components of a feedback control system?
2. What is the purpose of the sensor?
3. Give three important properties of a good sensor.

4. What is the purpose of the actuator?
5. Give three important properties of a good actuator.
6. What is the purpose of the controller? Give the input(s) and output(s) of the controller.
7. What physical variable(s) of a process can be directly measured by a Hall effect sensor?
8. What physical variable is measured by a tachometer?
9. Describe three different techniques for measuring temperature.
10. Why do most sensors have an electrical output, regardless of the physical nature of the variable being measured?

PROBLEMS

1.1 Draw a component block diagram for each of the following feedback control systems.

(a) The manual steering system of an automobile

(b) Drebbel's incubator

(c) The water level controlled by a float and valve

(d) Watt's steam engine with fly-ball governor

In each case, indicate the location of the elements listed below and give the units associated with each signal.

- the process
- the process desired output signal
- the sensor
- the actuator
- the actuator output signal
- the controller
- the controller output signal
- the reference signal
- the error signal

Notice that in a number of cases the same physical device may perform more than one of these functions.

1.2 Identify the physical principles and describe the operation of the thermostat in your home or office.

1.3 A machine for making paper is diagrammed in Fig. 1.12. There are two main parameters under feedback control: the density of fibers as controlled by the consistency of the thick stock that flows from the headbox onto the wire, and the moisture content of the final product that comes out of the dryers. Stock from the machine chest is diluted by white water returning from under the wire as controlled by a control valve (CV). A meter supplies a reading of the consistency. At the "dry end" of the machine, there is a moisture sensor. Draw a signal graph and identify the nine components listed in Problem 1.1 part (d) for

(a) control of consistency

(b) control of moisture

1.4 Many variables in the human body are under feedback control. For each of the following controlled variables, draw a graph showing the process being controlled, the sensor that measures the variable, the actuator that causes it to increase and/or decrease, the

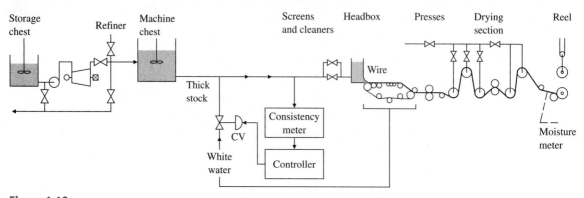

Figure 1.12

A papermaking machine

Source: From Åström (1970, p. 192); reprinted with permission

information path that completes the feedback path, and the disturbances that upset the variable. You may need to consult an encyclopedia or textbook on human physiology for information on this problem.

(a) blood pressure

(b) blood sugar concentration

(c) heart rate

(d) eye-pointing angle

(e) eye-pupil diameter

1.5 Draw a graph of the components for temperature control in a refrigerator or automobile air-conditioning system.

1.6 Draw a graph of the components for an elevator-position control. Indicate how you would measure the position of the elevator car. Consider a combined coarse and fine measurement system. What accuracies do you suggest for each sensor? Your system should be able to correct for the fact that in elevators for tall buildings there is significant cable stretch as a function of cab load.

1.7 Feedback control requires being able to sense the variable being controlled. Because electrical signals can be transmitted, amplified, and processed easily, often we want to have a sensor whose output is a voltage or current proportional to the variable being measured. Describe a sensor that would give an electrical output proportional to:

(a) temperature

(b) pressure

(c) liquid level

(d) flow of liquid along a pipe (or blood along an artery) force

(e) linear position

(f) rotational position

(g) linear velocity

(h) rotational speed

(i) translational acceleration

(j) torque

1.8 Each of the variables listed in Problem 1.7 can be brought under feedback control. Describe an actuator that could accept an electrical input and be used to control the variables listed. Give the units of the actuator output signal.

2

Dynamic Models

A Perspective on Dynamic Models

The overall goal of feedback control is to use the principle of feedback to cause the output variable of a dynamic process to follow a desired reference variable accurately, regardless of the reference variable's path and regardless of any external disturbances or any changes in the dynamics of the process. This complex goal is met as the result of a number of simple, distinct steps. The first of these is to develop a mathematical description (called a **dynamic model**) of the process to be controlled. The term **model**, as it is used and understood by control engineers, means a set of differential equations that describe the dynamic behavior of the process. A model can be obtained using principles of the underlying physics or by testing a prototype of the device, measuring its response to inputs, and using the data to construct an analytical model. We will focus only on using physics in this chapter. There are entire books written on experimentally determining models, sometimes called System Identification, and these techniques are described very briefly in Chapter 3. A careful control system designer will typically rely on at least some experiments to verify the accuracy of the model when it is derived from physical principles.

In many cases the modeling of complex processes is difficult and expensive, especially when the important steps of building and testing prototypes are included. However, in this introductory text, we will focus on the most basic principles of modeling for the most common physical systems. More comprehensive sources and specialized texts will be referenced throughout the text where appropriate for those wishing more detail.

In later chapters we will explore a variety of analysis methods for dealing with the equations of motion and their solution for purposes of designing feedback control systems.

Chapter Overview

The fundamental step in building a dynamic model is writing the equations of motion for the system. Through discussion and a variety of examples, Section 2.1 demonstrates how to write the equations of motion for a variety of mechanical systems. In addition, the section demonstrates the use of MATLAB® to find the time response of a simple system to a step input. Furthermore, the ideas of transfer functions and block diagrams are introduced, along with the idea that problems can also be solved via SIMULINK®.

Electric circuits and electromechanical systems are modeled in Sections 2.2 and 2.3, respectively.

For those wanting modeling examples for more diverse dynamic systems, Section 2.4, which is optional, extends the discussion to heat and fluid-flow systems.

The chapter concludes with Section 2.5, a discussion of the history behind the discoveries that led to the knowledge that we take for granted today.

The differential equations developed in modeling are often nonlinear. Because nonlinear systems are significantly more challenging to solve than linear ones and because linear models are usually adequate, the emphasis in the early chapters is primarily on linear systems. However, we do show how to linearize simple nonlinearities here in Chapter 2 and show how to use SIMULINK to numerically solve for the motion of a nonlinear system. A much more extensive discussion of linearization and analysis of nonlinear systems is contained in Chapter 9.

In order to focus on the important first step of developing mathematical models, we will defer explanation of the computational methods used to solve the equations of motion developed in this chapter until Chapter 3.

2.1 Dynamics of Mechanical Systems

2.1.1 Translational Motion

Newton's law for translational motion

The cornerstone for obtaining a mathematical model, or the **equations of motion**, for any mechanical system is Newton's law,

$$\mathbf{F} = m\mathbf{a}, \tag{2.1}$$

where

$\mathbf{F}$ = the vector sum of all forces applied to each body in a system, newtons (N) or pounds (lb),

$\mathbf{a}$ = the vector acceleration of each body with respect to an inertial reference frame (i.e., one that is neither accelerating nor rotating with respect to the stars); often called **inertial acceleration**, m/sec^2 or ft/sec^2,

m = mass of the body, kg or slug.

Note that here in Eq. (2.1), as throughout the text, we use the convention of boldfacing the type to indicate that the quantity is a matrix or vector, possibly a vector function.

Figure 2.1

Cruise control model

$\longrightarrow u$

In SI units a force of 1 N will impart an acceleration of 1 m/sec^2 to a mass of 1 kg. In English units a force of 1 lb will impart an acceleration of 1 ft/sec^2 to a mass of 1 slug. The "weight" of an object is mg, where g is the acceleration of gravity ($= 9.81$ m/sec$^2 = 32.2$ ft/sec^2). In English units it is common usage to refer to the mass of an object in terms of its weight in pounds, which is the quantity measured on scales. To obtain the mass in slugs for use in Newton's law, divide the weight by g. Therefore, an object weighing 1 lb has a mass of $1/32.2$ slugs. A slug has units lb·sec^2/ft. In metric units, scales are typically calibrated in kilograms, which is a direct measure of mass.

Use of free-body diagram in applying Newton's law

Application of this law typically involves defining convenient coordinates to account for the body's motion (position, velocity, and acceleration), determining the forces on the body using a free-body diagram, and then writing the equations of motion from Eq. (2.1). The procedure is simplest when the coordinates chosen express the position with respect to an inertial frame because, in this case, the accelerations needed for Newton's law are simply the second derivatives of the position coordinates.

EXAMPLE 2.1

A Simple System; Cruise Control Model

1. Write the equations of motion for the speed and forward motion of the car shown in Fig. 2.1 assuming that the engine imparts a force u as shown. Take the Laplace transform of the resulting differential equation and find the transfer function between the input u and the output v.
2. Use MATLAB to find the response of the velocity of the car for the case in which the input jumps from being $u = 0$ at time $t = 0$ to a constant $u = 500$ N thereafter. Assume that the car mass m is 1000 kg and viscous drag coefficient, $b = 50$ N·sec/m.

Solution

1. **Equations of motion:** For simplicity we assume that the rotational inertia of the wheels is negligible and that there is friction retarding the motion of the car that is proportional to the car's speed with a proportionality constant, b.[1] The car can

---

[1] If the speed is v, the aerodynamic friction force is proportional to v^2. In this simple model we have taken a linear approximation.

Figure 2.2

Free-body diagram for cruise control

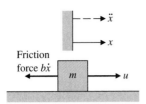

then be approximated for modeling purposes using the free-body diagram seen in Fig. 2.2, which defines coordinates, shows all forces acting on the body (heavy lines), and indicates the acceleration (dashed line). The coordinate of the car's position x is the distance from the reference line shown and is chosen so that positive is to the right. Note that in this case the inertial acceleration is simply the second derivative of x (i.e., $\mathbf{a} = \ddot{x}$) because the car position is measured with respect to an inertial reference. The equation of motion is found using Eq. (2.1). The friction force acts opposite to the direction of motion; therefore it is drawn opposite the direction of positive motion and entered as a negative force in Eq. (2.1). The result is

$$u - b\dot{x} = m\ddot{x}, \tag{2.2}$$

or

$$\ddot{x} + \frac{b}{m}\dot{x} = \frac{u}{m}. \tag{2.3}$$

For the case of the automotive cruise control where the variable of interest is the speed, $v\ (=\dot{x})$, the equation of motion becomes

$$\dot{v} + \frac{b}{m}v = \frac{u}{m}. \tag{2.4}$$

The solution of such an equation will be covered in detail in Chapter 3; however, the essence is that you assume a solution of the form $v = V_o e^{st}$ given an input of the form $u = U_o e^{st}$. Then, since $\dot{v} = sV_o e^{st}$, the differential equation can be written as

$$\left(s + \frac{b}{m}\right)V_o e^{st} = \frac{1}{m}U_o e^{st}. \tag{2.5}$$

The e^{st} term cancels out, and we find that

$$\frac{V_o}{U_o} = \frac{\frac{1}{m}}{s + \frac{b}{m}}. \tag{2.6}$$

For reasons that will become clear in Chapter 3, this is usually written as

$$\frac{V(s)}{U(s)} = \frac{\frac{1}{m}}{s + \frac{b}{m}}. \tag{2.7}$$

Transfer function

This expression of the differential equation (2.4) is called the **transfer function** and will be used extensively in later chapters. Note that, in essence, we have substituted s for d/dt in Eq. (2.4).[2]

---

[2]The use of an operator for differentiation was developed by Cauchy about 1820 based on the Laplace transform, which was developed about 1780. In Chapter 3 we will show how to derive transfer functions using the Laplace transform. Reference: Gardner and Barnes, 1942.

2. **Time response:** The dynamics of a system can be prescribed to MATLAB in terms of row vectors containing the coefficients of the polynomials describing the numerator and denominator of its transfer function. The transfer function for this problem is that given in part (a). In this case, the numerator (called num) is simply one number since there are no powers of s, so that $num = 1/m = 1/1000$. The denominator (called den) contains the coefficients of the polynomial $s + b/m$, which are

$$den = \begin{bmatrix} 1 & \dfrac{b}{m} \end{bmatrix} = \begin{bmatrix} 1 & \dfrac{50}{1000} \end{bmatrix}.$$

The step function in MATLAB calculates the time response of a linear system to a unit step input. Because the system is linear, the output for this case can be multiplied by the magnitude of the input step to derive a step response of any amplitude. Equivalently, num can be multiplied by the magnitude of the input step.

Step response with MATLAB

The statements

```
num = 1/1000;            %   1/m
den = [1  50/1000];      %   s  +  b/m
sys = tf(num*500, den);  % step gives unit step response, so num*500
                             gives u = 500.
step(sys);               % plots the step response
```

calculate and plot the time response for an input step with a 500-N magnitude. The step response is shown in Fig. 2.3.

Newton's law also can be applied to systems with more than one mass. In this case it is particularly important to draw the free-body diagram of each mass, showing the applied external forces as well as the equal and opposite internal forces that act from each mass on the other.

Figure 2.3

Response of the car velocity to a step in u

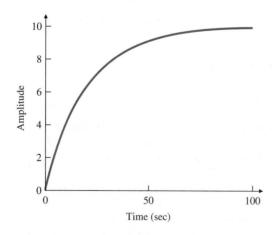

EXAMPLE 2.2

A Two-Mass System: Suspension Model

Figure 2.4 shows an automobile suspension system. Write the equations of motion for the automobile and wheel motion assuming one-dimensional vertical motion of one quarter of the car mass above one wheel. A system consisting of one of the four wheel suspensions is usually referred to as a quarter-car model. Assume that the model is for a car with a mass of 1580 kg, including the four wheels, which have a mass of 20 kg each. By placing a known weight (an author) directly over a wheel and measuring the car's deflection, we find that $k_s = 130{,}000$ N/m. Measuring the wheel's deflection for the same applied weight, we find that $k_w \simeq 1{,}000{,}000$ N/m. By using the results in Section 3.3, Fig. 3.18(b), and qualitatively observing that the car's response as the author jumps off matches the $\zeta = 0.7$ curve, we conclude that $b = 9800$ N·sec/m.

Solution. The system can be approximated by the simplified system shown in Fig. 2.5. The coordinates of the two masses, x and y, with the reference directions as shown, are the displacements of the masses from their equilibrium conditions. The equilibrium positions are offset from the springs' unstretched positions because of the force of gravity. The shock absorber is represented in the schematic diagram by a dashpot symbol with friction constant b. The magnitude of the force from the shock

Figure 2.4

Automobile suspension

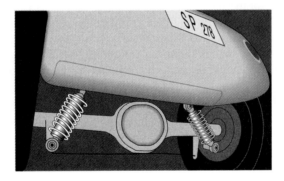

Figure 2.5

The quarter-car model

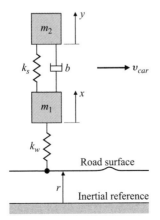

Figure 2.6

Free-body diagrams for suspension system

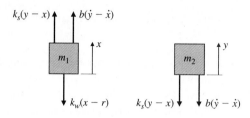

absorber is assumed to be proportional to the rate of change of the relative displacement of the two masses—that is, the force $= b(\dot{y} - \dot{x})$. The force of gravity could be included in the free-body diagram; however, its effect is to produce a constant offset of x and y. By defining x and y to be the distance from the equilibrium position, the need to include the gravity forces is eliminated.

The force from the car suspension acts on both masses in proportion to their relative displacement with spring constant k_s. Figure 2.6 shows the free-body diagram of each mass. Note that the forces from the spring on the two masses are equal in magnitude but act in opposite directions, which is also the case for the damper. A positive displacement y of mass m_2 will result in a force from the spring on m_2 in the direction shown and a force from the spring on m_1 in the direction shown. However, a positive displacement x of mass m_1 will result in a force from the spring k_s on m_1 in the opposite direction to that drawn in Fig. 2.6, as indicated by the *minus* x term for the spring force.

The lower spring k_w represents the tire compressibility, for which there is insufficient damping (velocity-dependent force) to warrant including a dashpot in the model. The force from this spring is proportional to the distance the tire is compressed and the nominal equilibrium force would be that required to support m_1 and m_2 against gravity. By defining x to be the distance from equilibrium, a force will result if either the road surface has a bump (r changes from its equilibrium value of zero) or the wheel bounces (x changes). The motion of the simplified car over a bumpy road will result in a value of $r(t)$ that is not constant.

As previously noted, there is a constant force of gravity acting on each mass; however, this force has been omitted, as have the equal and opposite forces from the springs. Gravitational forces can always be omitted from vertical-spring mass systems (1) if the position coordinates are defined from the equilibrium position that results when gravity is acting, and (2) if the spring forces used in the analysis are actually the perturbation in spring forces from those forces acting at equilibrium.

Applying Eq. (2.1) to each mass and noting that some forces on each mass are in the negative (down) direction yields the system of equations

$$b(\dot{y} - \dot{x}) + k_s(y - x) - k_w(x - r) = m_1\ddot{x}, \qquad (2.8)$$

$$-k_s(y - x) - b(\dot{y} - \dot{x}) = m_2\ddot{y}. \qquad (2.9)$$

Some rearranging results in

$$\ddot{x} + \frac{b}{m_1}(\dot{x} - \dot{y}) + \frac{k_s}{m_1}(x - y) + \frac{k_w}{m_1}x = \frac{k_w}{m_1}r, \qquad (2.10)$$

$$\ddot{y} + \frac{b}{m_2}(\dot{y} - \dot{x}) + \frac{k_s}{m_2}(y - x) = 0. \qquad (2.11)$$

Check for sign errors

The most common source of error in writing equations for systems like these are sign errors. The method for keeping the signs straight in the preceding development entailed mentally picturing the displacement of the masses and drawing the resulting force in the direction that the displacement would produce. Once you have obtained the equations for a system, a check on the signs for systems that are obviously stable from physical reasoning can be quickly carried out. As we will see when we study stability in Section 3.6, a stable system always has the same signs on similar variables. For this system, Eq. (2.10) shows that the signs on the $\ddot{x}$, $\dot{x}$, and x terms are all positive, as they must be for stability. Likewise, the signs on the $\ddot{y}$, $\dot{y}$, and y terms are all positive in Eq. (2.11).

The transfer function is obtained in a similar manner as before. Substituting s for d/dt in the differential equations yields

$$s^2 X(s) + s\frac{b}{m_1}(X(s) - Y(s)) + \frac{k_s}{m_1}(X(s) - Y(s)) + \frac{k_w}{m_1}X(s) = \frac{k_w}{m_1}R(s),$$

$$s^2 Y(s) + s\frac{b}{m_2}(Y(s) - X(s)) + \frac{k_s}{m_2}(Y(s) - X(s)) = 0,$$

which, after some algebra and rearranging, yields the transfer function

$$\frac{Y(s)}{R(s)} = \frac{\dfrac{k_w b}{m_1 m_2}\left(s + \dfrac{k_s}{b}\right)}{s^4 + \left(\dfrac{b}{m_1} + \dfrac{b}{m_2}\right)s^3 + \left(\dfrac{k_s}{m_1} + \dfrac{k_s}{m_2} + \dfrac{k_w}{m_1}\right)s^2 + \left(\dfrac{k_w b}{m_1 m_2}\right)s + \dfrac{k_w k_s}{m_1 m_2}}.$$

$$(2.12)$$

To determine numerical values, we subtract the mass of the four wheels from the total car mass of 1580 kg and divide by 4 to find that $m_2 = 375$ kg. The wheel mass was measured directly to be $m_1 = 20$ kg. Therefore, the transfer function with the numerical values is

$$\frac{Y(s)}{R(s)} = \frac{1.31e06(s + 13.3)}{s^4 + (516.1)s^3 + (5.685e04)s^2 + (1.307e06)s + 1.733e07}. \qquad (2.13)$$

2.1.2 Rotational Motion

Newton's law for rotational motion

Application of Newton's law to one-dimensional rotational systems requires that Eq. (2.1) be modified to

$$M = I\alpha, \qquad (2.14)$$

where

$M =$ the sum of all external moments about the center of mass of a body, N·m or lb·ft,

$I =$ the body's mass moment of inertia about its center of mass, kg·m^2 or slug·ft^2,

$\alpha =$ the angular acceleration of the body, rad/sec^2.

EXAMPLE 2.3   *Rotational Motion: Satellite Attitude Control Model*

Satellites, as shown in Fig. 2.7, usually require attitude control so that antennas, sensors, and solar panels are properly oriented. Antennas are usually pointed toward a particular location on earth, while solar panels need to be oriented toward the sun for maximum power generation. To gain insight into the full three-axis attitude control system, it is helpful to consider one axis at a time. Write the equations of motion for one axis of this system and show how they would be depicted in a block diagram. In addition, determine the transfer function of this system and construct the system as if it were to be evaluated via MATLAB's SIMULINK.

Solution.   Figure 2.8 depicts this case, where motion is allowed only about the axis perpendicular to the page. The angle θ that describes the satellite orientation must be measured with respect to an inertial reference—that is, a reference that has no angular acceleration. The control force comes from reaction jets that produce a moment of $F_c d$ about the mass center. There may also be small disturbance moments M_D on the

Figure 2.7

Communications satellite

Source: Courtesy Space Systems/Loral

Figure 2.8

Satellite control
schematic

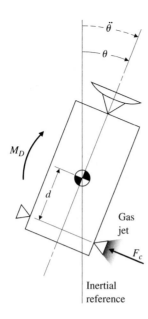

Figure 2.8

Satellite control
schematic

Figure 2.9

Block diagrams
representing Eq. (2.15)
in the upper half and Eq.
(2.16) in the lower half

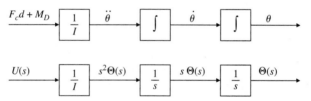

satellite, which arise primarily from solar pressure acting on any asymmetry in the
solar panels. Applying Eq. (2.14) yields the equation of motion

$$F_c d + M_D = I\ddot{\theta}. \tag{2.15}$$

Double-integrator plant

The output of this system, θ, results from integrating the sum of the input torques
twice; hence this type of system is often referred to as the **double-integrator plant**.
The transfer function can be obtained as described for Eq. (2.7) and is

$$\frac{\Theta(s)}{U(s)} = \frac{1}{I}\frac{1}{s^2}, \tag{2.16}$$

$1/s^2$ plant

where $U = F_c d + M_D$. In this form, the system is often referred to as the $1/s^2$ **plant**.

Figure 2.9 shows a block diagram representing Eq. (2.15) in the upper half and
a block diagram representing Eq. (2.16) in the lower half. This simple system can
be analyzed using the linear analysis techniques that are described in later chapters,
or via MATLAB as we saw in Example 2.1. It can also be numerically evaluated for
an arbitrary input time history using SIMULINK. SIMULINK is a sister software
package to MATLAB for interactive, nonlinear simulation and has a graphical user
interface with drag and drop properties. Figure 2.10 shows a block diagram of the
system as depicted by SIMULINK.

Figure 2.10
SIMULINK block diagram of the double-integrator plant

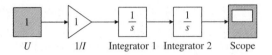

Figure 2.11
Disk read/write mechanism

Source: Courtesy of Hewlett-Packard Company

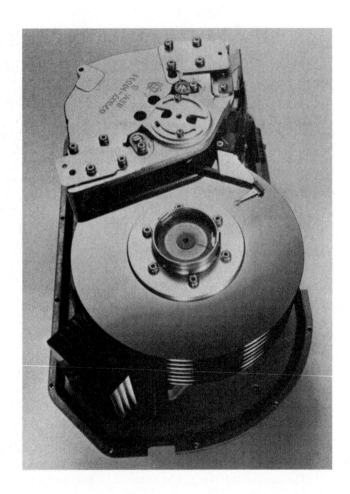

In many cases a system, such as the disk-drive read/write head shown in Fig. 2.11, in reality has some flexibility, which can cause problems in the design of a control system. Particular difficulty arises when there is flexibility, as in this case, between the sensor and actuator locations. Therefore, it is often important to include this flexibility in the model even when the system seems to be quite rigid.

EXAMPLE 2.4   *Flexibility: Flexible Read/Write for a Disk Drive*

Assume that there is some flexibility between the read head and the drive motor in Fig. 2.11. Find the equations of motion relating the motion of the read head to a torque applied to the base.

Figure 2.12

Disk read/write head schematic for modeling

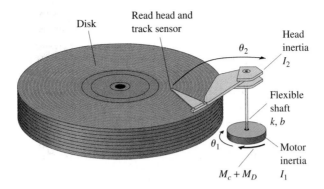

Figure 2.13

Free-body diagrams of the disk read/write head

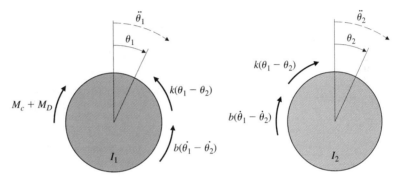

Solution. The dynamic model for this situation is shown schematically in Fig. 2.12. This model is dynamically similar to the resonant system shown in Fig. 2.5 and results in equations of motion that are similar in form to Eqs. (2.10) and (2.11). The moments on each body are shown in the free-body diagrams in Fig. 2.13. The discussion of the moments on each body is essentially the same as the discussion for Example 2.2, except that the springs and damper in that case produced forces, instead of moments that act on each inertia, as in this case. When the moments are summed, equated to the accelerations according to Eq. (2.14), and rearranged, the result is

$$I_1 \ddot{\theta}_1 + b(\dot{\theta}_1 - \dot{\theta}_2) + k(\theta_1 - \theta_2) = M_c + M_D, \qquad (2.17)$$

$$I_2 \ddot{\theta}_2 + b(\dot{\theta}_2 - \dot{\theta}_1) + k(\theta_2 - \theta_1) = 0. \qquad (2.18)$$

Ignoring the disturbance torque M_D and the damping b for simplicity, we find the transfer function from the applied torque M_c to the read head motion to be

$$\frac{\Theta_2(s)}{M_c(s)} = \frac{k}{I_1 I_2 s^2 \left(s^2 + \dfrac{k}{I_1} + \dfrac{k}{I_2} \right)}. \qquad (2.19)$$

It might also be possible to sense the motion of the inertia where the torque is applied, θ_1, in which case the transfer function with the same simplifications would be

$$\frac{\Theta_1(s)}{M_c(s)} = \frac{I_2 s^2 + k}{I_1 I_2 s^2 \left(s^2 + \dfrac{k}{I_1} + \dfrac{k}{I_2} \right)}. \qquad (2.20)$$

Figure 2.14

Pendulum

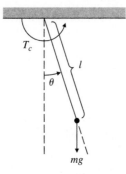

These two cases are typical of many situations in which the sensor and actuator may or may not be placed in the same location in a flexible body. We refer to the situation between sensor and actuator in Eq. (2.19) as the "noncollocated" case, whereas Eq. (2.20) describes the "collocated" case. You will see in Chapter 5 that it is far more difficult to control a system when there is flexibility between the sensor and actuator (noncollocated case) than when the sensor and actuator are rigidly attached to one another (the collocated case).

Collocated sensor and actuator

In the special case in which a point in a rotating body is fixed with respect to an inertial reference, as is the case with a pendulum, Eq. (2.14) can be applied such that M is the sum of all moments about the *fixed* point and I is the moment of inertia about the fixed point.

EXAMPLE 2.5   *Rotational Motion: Pendulum*

1. Write the equations of motion for the simple pendulum shown in Fig. 2.14, where all the mass is concentrated at the end point and there is a torque, T_c, applied at the pivot.
2. Use MATLAB to determine the time history of θ to a step input in T_c of 1 N·m. Assume $l = 1$ m, $m = 1$ kg, and $g = 9.81$ m/sec^2.

Solution

1. **Equations of motion:** The moment of inertia about the pivot point is $I = ml^2$. The sum of moments about the pivot point contains a term from gravity as well as the applied torque T_c. The equation of motion, obtained from Eq. (2.14), is

$$T_c - mgl \sin \theta = I\ddot{\theta}, \qquad (2.21)$$

which is usually written in the form

$$\ddot{\theta} + \frac{g}{l} \sin \theta = \frac{T_c}{ml^2}. \qquad (2.22)$$

This equation is nonlinear due to the $\sin \theta$ term. A general discussion of nonlinear equations is contained in Chapter 9; however, we can proceed with a linearization of this case by assuming the motion is small enough that $\sin \theta \cong \theta$. Then Eq. (2.22) becomes the linear equation

$$\ddot{\theta} + \frac{g}{l} \theta = \frac{T_c}{ml^2}. \qquad (2.23)$$

With no applied torque, the natural motion is that of a harmonic oscillator with a natural frequency of [3]

$$\omega_n = \sqrt{\frac{g}{l}}. \tag{2.24}$$

The transfer function can be obtained as described for Eq. (2.7), yielding

$$\frac{\Theta(s)}{T_c(s)} = \frac{\frac{1}{ml^2}}{s^2 + \frac{g}{l}}. \tag{2.25}$$

2. **Time history:** The dynamics of a system can be prescribed to MATLAB in terms of row vectors containing the coefficients of the polynomials describing the numerator and denominator of its transfer function. In this case, the numerator (called num) is simply one number, since there are no powers of s, so that

$$\text{num} = \frac{1}{ml^2} = \frac{1}{(1)(1)^2} = [1],$$

and the denominator (called den) contains the coefficients of the descending powers of s in $(s^2 + g/l)$ and is a row vector with three elements:

$$\text{den} = \begin{bmatrix} 1 & 0 & \frac{g}{l} \end{bmatrix} = \begin{bmatrix} 1 & 0 & 9.81 \end{bmatrix}.$$

The desired response of the system can be obtained by using the MATLAB step response function, called step. The MATLAB statements

```
t = 0:0.02:10;        % vector of times for output, 0 to 10 at 0.02 increments
num = 1;
den = [1 0 9.81];
sys = tf(num,den);    % defines the system by its numerator and denominator
y = step(sys,t);      % computes step responses at times given by t for step
                      %   at t = 0
plot(t, 57.3*y)       % converts radians to degrees and plots step response
```

will produce the desired time history shown in Fig. 2.15.

As we saw in this example, the resulting equations of motion are often nonlinear. Such equations are much more difficult to solve than linear ones, and the kinds of possible motions resulting from a nonlinear model are much more difficult to categorize than those resulting from a linear model. It is therefore useful to linearize models in order to gain access to linear analysis methods. It may be that the linear models and linear analysis are used only for the design of the control system (whose function may be to maintain the system in the linear region). Once a control system is synthesized and shown to have desirable performance based on linear analysis, it is then prudent to carry out further analysis or an accurate numerical simulation of the system with the significant nonlinearities in order to validate that performance. SIMULINK is an expedient way to carry out these simulations and can handle most

SIMULINK

---

[3] In a grandfather clock it is desired to have a pendulum period of exactly 2 sec. Show that the pendulum should be approximately 1 m in length.

Figure 2.15

Response of the
pendulum to a step
input of 1 N-m in the
applied torque

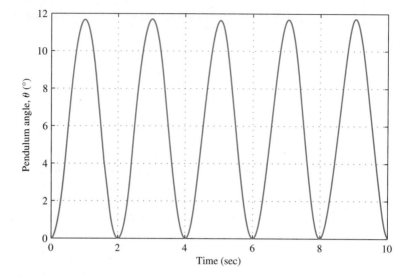

Figure 2.16

The SIMULINK block
diagram representing
the linear
equation (2.26)

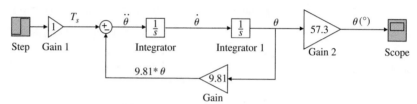

nonlinearities. Use of this simulation tool is carried out by constructing a block diagram[4] that represents the equations of motion. The linear equation of motion for the pendulum with the parameters as specified in Example 2.5 can be seen from Eq. (2.23) to be

$$\ddot{\theta} = -9.81 * \theta + 1, \tag{2.26}$$

and this is represented in SIMULINK by the block diagram in Fig. 2.16. Note that the circle on the left side of the figure with the $+$ and $-$ signs indicating addition and subtraction implements the equation above.

The result of running this numerical simulation will be essentially identical to the linear solution shown in Fig. 2.15 because the solution is for relatively small angles where $\sin \theta \cong \theta$. However, using SIMULINK to solve for the response enables us to simulate the nonlinear equation so that we could analyze the system for larger motions. In this case, Eq. (2.26) becomes

$$\ddot{\theta} = -9.81 * \sin \theta + 1, \tag{2.27}$$

---

[4]A more extensive discussion of block diagrams is contained in Section 3.2.1.

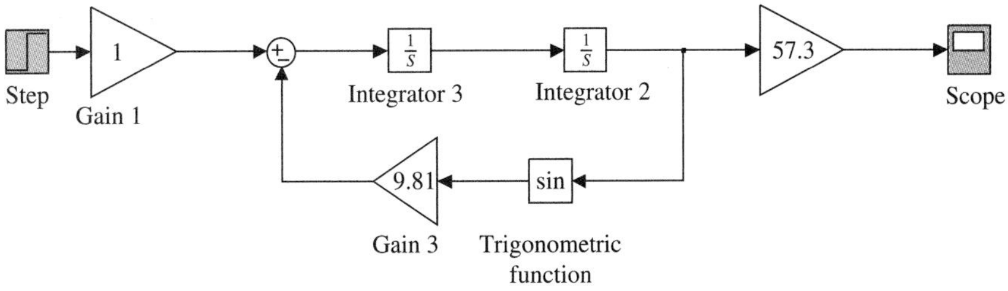

Figure 2.17
The SIMULINK block diagram representing the nonlinear equation (2.27)

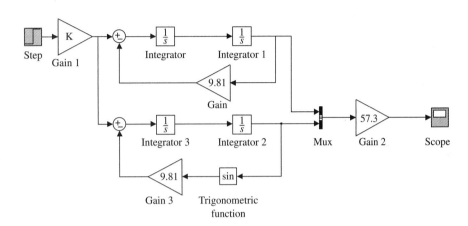

Figure 2.18
Block diagram of the pendulum for both the linear and nonlinear models

and the SIMULINK block diagram shown in Fig. 2.17 implements this nonlinear equation.

SIMULINK is capable of simulating all commonly encountered nonlinearities, including deadzones, on–off functions, stiction, hysteresis, aerodynamic drag (a function of v^2), and trigonometric functions. All real systems have one or more of these characteristics in varying degrees.

EXAMPLE 2.6

Use of SIMULINK for Nonlinear Motion: Pendulum

Use SIMULINK to determine the time history of θ for the pendulum in Example 2.5. Compare it against the linear solution for T_c values of 1 N·m and 4 N·m.

Solution. Time history: The SIMULINK block diagrams for the two cases discussed above are combined and both outputs in Fig. 2.16 and 2.17 are sent via a "multiplexer block (Mux)" to the "scope" so they can be plotted on the same graph. Fig. 2.18 shows the combined block diagram where the gain, K, represents the values of T_c. The outputs of this system for T_c values of 1 N · m and 4 N· m are shown in Fig. 2.19. Note that for $T_c = 1$ N·m, the outputs at the top of the figure remain at 12° or less and the linear approximation is extremely close to the nonlinear output. For $T_c = 4$ N·m, the output angle grows to near 50° and a substantial difference in the response magnitude and frequency is apparent due to θ being a poor approximation to $\sin \theta$ at these magnitudes.

Figure 2.19

Response of the
pendulum SIMULINK
numerical simulation
for the linear and
nonlinear models. (a)
for $T_c = 1$ N·m and (b)
$T_c = 4$ N·m

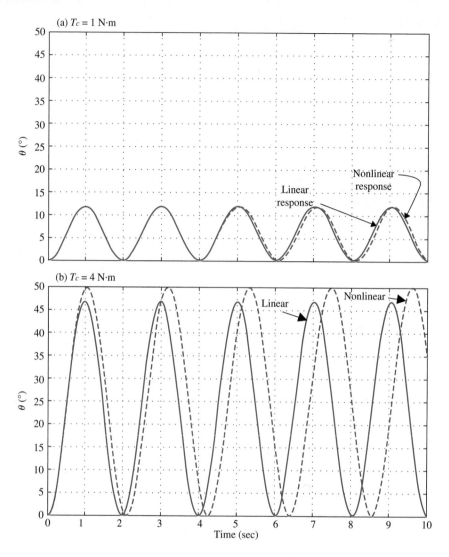

Chapter 9 is devoted to the analysis of nonlinear systems and greatly expands on
these ideas.

2.1.3 Combined Rotation and Translation

In some cases, mechanical systems contain both translational and rotational portions.
The procedure is the same as that described in Sections 2.1.1 and 2.1.2: sketch the
free-body diagrams, define coordinates and positive directions, determine all forces
and moments acting, and apply Eqs. (2.1) and/or (2.14). An exact derivation of the
equations for these systems can become quite involved; therefore, the complete analysis for the following examples are contained in Appendix W2 and only the linearized
equations of motion and their transfer functions are given here.

Figure 2.20

Schematic of the crane
with hanging load

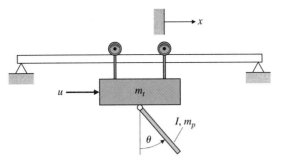

Figure 2.21

Inverted pendulum

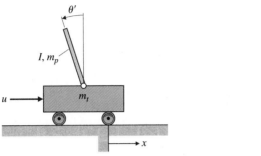

EXAMPLE 2.7   *Rotational and Translational Motion: Hanging Crane*

Write the equations of motion for the hanging crane shown schematically in Fig. 2.20. Linearize the equations about $\theta = 0$, which would typically be valid for the hanging crane. Also linearize the equations for $\theta = \pi$, which represents the situation for the inverted pendulum shown in Fig. 2.21. The trolley has mass, m_t, and the hanging crane (or pendulum) has mass, m_p, and inertia about its mass center of I. The distance from the pivot to the mass center of the pendulum is l; therefore, the moment of inertia of the pendulum about the pivot point is $(I + m_p l^2)$.

Solution. Free-body diagrams need to be drawn for the trolley and the pendulum and the reaction forces considered where the two attach to one another. We carry out this process in Appendix W2. After Newton's Laws are applied for the translational motion of the trolley and the rotational motion of the pendulum, it will be found that the reaction forces between the two bodies can be eliminated, and the only unknowns will be θ and x. The results are two coupled second-order nonlinear differential equations in θ and x with the input being the force applied to the trolley, u. They can be linearized in a similar manner that was done for the simple pendulum by assuming small angles. For small motions about $\theta = 0$, we let $\cos\theta \cong 1$, $\sin\theta \cong \theta$, and $\dot{\theta}^2 \cong 0$; thus the equations are approximated by

$$(I + m_p l^2)\ddot{\theta} + m_p g l\theta = -m_p l\ddot{x},$$

$$(m_t + m_p)\ddot{x} + b\dot{x} + m_p l\ddot{\theta} = u. \qquad (2.28)$$

Note that the first equation is very similar to the simple pendulum, Eq. (2.21), where the applied torque arises from the trolley accelerations. Likewise, the second equation representing the trolley motion, x, is very similar to the car translation in

Eq. (2.3) where the forcing term arises from the angular acceleration of the pendulum. Neglecting the friction term b leads to the transfer function from the control input u to hanging crane angle θ:

$$\frac{\theta(s)}{U(s)} = \frac{-m_p l}{((I + m_p l^2)(m_t + m_p) - m_p^2 l^2)s^2 + m_p g l(m_t + m_p)}. \tag{2.29}$$

For the inverted pendulum in Fig. 2.21, where $\theta \cong \pi$, assume $\theta = \pi + \theta'$, where θ' represents motion from the vertical *upward* direction. In this case, $\sin\theta \cong -\theta'$, $\cos\theta \cong -1$, and the nonlinear equations become[5]

Inverted pendulum equations

$$(I + m_p l^2)\ddot{\theta}' - m_p g l \theta' = m_p l \ddot{x},$$

$$(m_t + m_p)\ddot{x} + b\dot{x} - m_p l \ddot{\theta}' = u. \tag{2.30}$$

As noted in Example 2.2, a stable system will always have the same signs on each variable, which is the case for the stable hanging crane modeled by Eqs. (2.28). However, the signs on θ and $\ddot{\theta}$ in the top Eq. (2.30) are opposite, thus indicating instability, which is the characteristic of the inverted pendulum.

The transfer function, again without friction, is

$$\frac{\theta'(s)}{U(s)} = \frac{m_p l}{((I + m_p l^2) - m_p^2 l^2)s^2 - m_p g l(m_t + m_p)}. \tag{2.31}$$

In Chapter 5 you will learn how to stabilize systems using feedback and will see that even unstable systems like an inverted pendulum can be stabilized providing there is a sensor that measures the output quantity and a control input. For the case of the inverted pendulum perched on a trolley, it would be required to measure the pendulum angle, θ', and provide a control input, u, that accelerated the trolley in such a way that the pendulum remained pointing straight up. In years past, this system existed primarily in university control system laboratories as an educational tool. However, more recently, there is a practical device in production and being sold that employs essentially this same dynamic system: The Segway. It uses a gyroscope so that the angle of the device is known with respect to vertical, and electric motors provide a torque on the wheels so that it balances the device and provides the desired forward or backward motion. It is shown in Fig. 2.22.

2.1.4 Distributed Parameter Systems

All the preceding examples contained one or more rigid bodies, although some were connected to others by springs. Actual structures—for example, satellite solar panels, airplane wings, or robot arms—usually bend, as shown by the flexible beam in Fig. 2.23(a). The equation describing its motion is a fourth-order *partial* differential equation that arises because the mass elements are continuously distributed along the beam with a small amount of flexibility between elements. This type of system is called a **distributed parameter system.** The dynamic analysis methods presented

---

[5]The inverted pendulum is often described with the angle of the pendulum being positive for *clockwise* motion. If defined that way, then reverse the sign on all terms in Eqs. (2.30) in θ' or $\ddot{\theta}'$.

in this section are not sufficient to analyze this case; however, more advanced texts (Thomson and Dahleh, 1998) show that the result is

$$EI\frac{\partial^4 w}{\partial x^4} + \rho\frac{\partial^2 w}{\partial t^2} = 0, \tag{2.32}$$

where

$E =$ Young's modulus,

$I =$ beam area moment of inertia,

$\rho =$ beam density,

$w =$ beam deflection at length x along the beam.

The exact solution to Eq. (2.32) is too cumbersome to use in designing control systems, but it is often important to account for the gross effects of bending in control systems design.

The continuous beam in Fig. 2.23(b) has an infinite number of vibration-mode shapes, all with different frequencies. Typically, the lowest-frequency modes have the largest amplitude and are the most important to approximate well. The simplified model in Fig. 2.23(c) can be made to duplicate the essential behavior of the first

Figure 2.23

(a) Flexible robot arm
used for research at
Stanford University;
(b) model for a
continuous flexible
beam; (c) simplified
model for the first
bending mode;
(d) model for the first
and second bending
modes

*Source: Photo courtesy of
E. Schmitz*

(a)

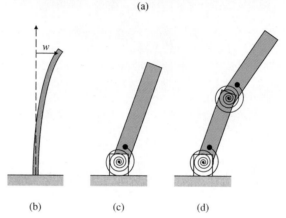

(b)   (c)   (d)

bending mode shape and frequency and would usually be adequate for controller
design. If frequencies higher than the first bending mode are anticipated in the control
system operation, it may be necessary to model the beam as shown in Fig. 2.23(d),
which can be made to approximate the first two bending modes and frequencies.
Likewise, higher-order models can be used if such accuracy and complexity are
deemed necessary (Thomson and Dahleh, 1998; Schmitz, 1985). When a continuously
bending object is approximated as two or more rigid bodies connected by springs, the
resulting model is sometimes referred to as a **lumped parameter model**.

*A flexible structure can be
approximated by a lumped
parameter model*

2.1.5   Summary: Developing Equations of Motion for Rigid Bodies

The physics necessary to write the equations of motion of a rigid body is entirely
given by Newton's laws of motion. The method is as follows:

1. Assign variables such as x and θ that are both necessary and sufficient to describe
   an *arbitrary* position of the object.
2. Draw a free-body diagram of each component. Indicate *all* forces acting on each
   body and their reference directions. Also indicate the accelerations of the center
   of mass with respect to an inertial reference for each body.

3. Apply Newton's law in translation [Eq. (2.1)] and/or rotation [Eq. (2.14)] form.
4. Combine the equations to eliminate internal forces.
5. The number of independent equations should equal the number of unknowns.

2.2 Models of Electric Circuits

Electric circuits are frequently used in control systems largely because of the ease of manipulation and processing of electric signals. Although controllers are increasingly implemented with digital logic, many functions are still performed with analog circuits. Analog circuits are faster than digital and, for very simple controllers, an analog circuit would be less expensive than a digital implementation. Furthermore, the power amplifier for electromechanical control and the anti-alias prefilters for digital control must be analog circuits.

Electric circuits consist of interconnections of sources of electric voltage and current, and other electronic elements such as resistors, capacitors, and transistors. An important building block for circuits is an operational amplifier (or op-amp),[6] which is also an example of a complex feedback system. Some of the most important methods of feedback system design were developed by the designers of high-gain, wide-bandwidth feedback amplifiers, mainly at the Bell Telephone Laboratories between 1925 and 1940. Electric and electronic components also play a central role in electromechanical energy conversion devices such as electric motors, generators, and electrical sensors. In this brief survey we cannot derive the physics of electricity or give a comprehensive review of all the important analysis techniques. We will define the variables, describe the relations imposed on them by typical elements and circuits, and describe a few of the most effective methods available for solving the resulting equations.

Symbols for some linear circuit elements and their current–voltage relations are given in Fig. 2.24. Passive circuits consist of interconnections of resistors, capacitors, and inductors. With electronics, we increase the set of electrical elements by adding active devices, including diodes, transistors, and amplifiers.

Kirchhoff's laws

The basic equations of electric circuits, called Kirchhoff's laws, are as follows:

1. **Kirchhoff's current law (KCL):** The algebraic sum of currents leaving a junction or node equals the algebraic sum of currents entering that node.
2. **Kirchhoff's voltage law (KVL):** The algebraic sum of all voltages taken around a closed path in a circuit is zero.

With complex circuits of many elements, it is essential to write the equations in a careful, well organized way. Of the numerous methods for doing this, we choose for description and illustration the popular and powerful scheme known as **node analysis**. One node is selected as a reference and we assume the voltages of all other nodes to be unknowns. The choice of reference is arbitrary in theory, but in actual electronic

---

[6]Oliver Heaviside introduced the mathematical operation p to signify differentiation so that $pv = dv/dt$. The Laplace transform incorporates this idea, using the complex variable s. Ragazzini et al. (1947) demonstrated that an ideal, high-gain electronic amplifier permitted one to realize arbitrary "operations" in the Laplace transform variable s, so they named it the operational amplifier, commonly abbreviated to op-amp.

Figure 2.24

Elements of electric circuits

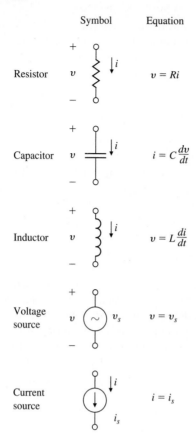

	Symbol	Equation
Resistor	v	$v = Ri$
Capacitor	v	$i = C\dfrac{dv}{dt}$
Inductor	v	$v = L\dfrac{di}{dt}$
Voltage source	v v_s	$v = v_s$
Current source	i_s	$i = i_s$

circuits the common, or ground, terminal is the obvious and standard choice. Next, we write equations for the selected unknowns using the current law (KCL) at each node. We express these currents in terms of the selected unknowns by using the element equations in Fig. 2.24. If the circuit contains voltage sources, we must substitute a voltage law (KVL) for such sources. Example 2.8 illustrates how node analysis works.

EXAMPLE 2.8

Equations for the Bridged Tee Circuit

Determine the differential equations for the circuit shown in Fig. 2.25.

Solution. We select node 4 as the reference and the voltages v_1, v_2, and v_3 at nodes 1, 2, and 3 as the unknowns. We start with the degenerate KVL relationship

$$v_1 = v_i. \tag{2.33}$$

At node 2 the KCL is

$$-\frac{v_1 - v_2}{R_1} + \frac{v_2 - v_3}{R_2} + C_1 \frac{dv_2}{dt} = 0, \tag{2.34}$$

Figure 2.25

Bridged tee circuit

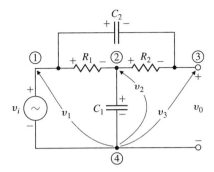

and at node 3 the KCL is

$$\frac{v_3 - v_2}{R_2} + C_2 \frac{d(v_3 - v_1)}{dt} = 0. \tag{2.35}$$

These three equations describe the circuit.

Operational amplifier

Kirchhoff's laws can also be applied to circuits that contain an **operational amplifier**. The simplified circuit of the op-amp is shown in Fig. 2.26(a) and the schematic symbol is drawn in Fig. 2.26(b). If the positive terminal is not shown, it is assumed to be connected to ground, $v_+ = 0$, and the reduced symbol of Fig. 2.26(c) is used. For use in control circuits, it is usually assumed that the op-amp is *ideal* with the values $R_1 = \infty$, $R_0 = 0$, and $A = \infty$. The equations of the ideal op-amp are extremely simple, being

$$i_+ = i_- = 0, \tag{2.36}$$

$$v_+ - v_- = 0. \tag{2.37}$$

The gain of the amplifier is assumed to be so high that the output voltage becomes $v_{out} = $ *whatever it takes* to satisfy these equations. Of course, a real amplifier only

Figure 2.26

(a) Op-amp simplified circuit; (b) op-amp schematic symbol; (c) reduced symbol for $v_+ = 0$

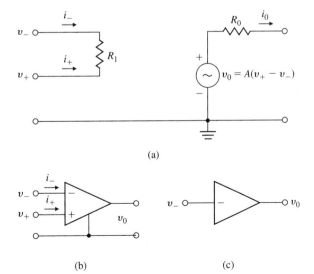

Figure 2.27

The op-amp summer

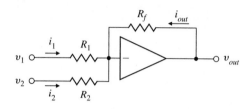

approximates these equations, but unless they are specifically described, we will assume all op-amps are ideal. More realistic models are the subject of several problems given at the end of the chapter.

EXAMPLE 2.9   *Op-Amp Summer*

Find the equations and transfer functions of the circuit shown in Fig. 2.27.

Solution. Equation (2.37) requires that $v_- = 0$, and thus the currents are $i_1 = v_1/R_1$, $i_2 = v_2/R_2$, and $i_{out} = v_{out}/R_f$. To satisfy Eq. (2.36), $i_1 + i_2 + i_{out} = 0$, from which it follows that $v_1/R_1 + v_2/R_2 + v_{out}/R_f = 0$, and we have

$$v_{out} = -\left[\frac{R_f}{R_1}v_1 + \frac{R_f}{R_2}v_2\right]. \qquad (2.38)$$

The op-amp summer

From this equation we see that the circuit output is a weighted sum of the input voltages with a sign change. The circuit is called a **summer**.

A second important example for control is given by the op-amp integrator.

EXAMPLE 2.10   *Integrator*

Op-amp as integrator

Find the transfer function for the circuit shown in Fig. 2.28.

Solution. In this case the equations are differential and Eqs. (2.36) and (2.37) require

$$i_{in} + i_{out} = 0, \qquad (2.39)$$

so that

$$\frac{v_{in}}{R_{in}} + C\frac{dv_{out}}{dt} = 0. \qquad (2.40)$$

Figure 2.28

The op-amp integrator

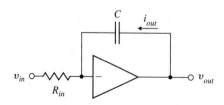

Eq. (2.40) can be written in integral form as

$$v_{out} = -\frac{1}{R_{in}C} \int_0^t v_{in}(\tau)\, d\tau + v_{out}(0). \qquad (2.41)$$

Using the operational notation that $d/dt = s$ in Eq. (2.40), the transfer function (which assumes zero initial conditions) can be written as

$$V_{out}(s) = -\frac{1}{s}\frac{V_{in}(s)}{R_{in}C}. \qquad (2.42)$$

Thus the ideal op-amp in this circuit performs the operation of integration and the circuit is simply referred to as an **integrator**.

2.3 Models of Electromechanical Systems

Electric current and magnetic fields interact in two ways that are particularly important to an understanding of the operation of most electromechanical actuators and sensors. If a current of i amperes in a conductor of length l meters is arranged at right angles in a magnetic field of B tesls, then there is a force on the conductor at right angles to the plane of i and B, with magnitude

Law of motors

$$F = Bli \text{ newtons.} \qquad (2.43)$$

This equation is the basis of conversion of electric energy to mechanical work and is called the **law of motors**.

EXAMPLE 2.11   *Modeling a Loudspeaker*

A typical geometry for a loudspeaker for producing sound is sketched in Fig. 2.29. The permanent magnet establishes a radial field in the cylindrical gap between the poles of the magnet. The force on the conductor wound on the bobbin causes the

Figure 2.29

Geometry of a loudspeaker: (a) overall configuration; (b) the electromagnet and voice coil

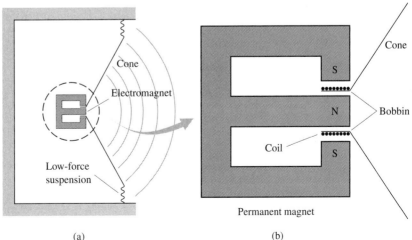

(a)   (b)

voice coil to move, producing sound.[7] The effects of the air can be modeled as if the cone had equivalent mass M and viscous friction coefficient b. Assume that the magnet establishes a uniform field B of 0.5 tesla and the bobbin has 20 turns at a 2-cm diameter. Write the equations of motion of the device.

Solution. The current is at right angles to the field, and the force of interest is at right angles to the plane of i and B, so Eq. (2.43) applies. In this case the field strength is $B = 0.5$ tesla and the conductor length is

$$l = 20 \times \frac{2\pi}{100} m = 1.26 \ m.$$

Thus, the force is

$$F = 0.5 \times 1.26 \times i = 0.63i \ N.$$

The mechanical equation follows from Newton's laws, and for a mass M and friction coefficient b, the equation is

$$M\ddot{x} + b\dot{x} = 0.63i. \tag{2.44}$$

This second-order differential equation describes the motion of the loudspeaker cone as a function of the input current i driving the system. Substituting s for d/dt in Eq. (2.44) as before, the transfer function is easily found to be

$$\frac{X(s)}{I(s)} = \frac{0.63/M}{s(s + b/M)}. \tag{2.45}$$

The second important electromechanical relationship is the effect of mechanical motion on electric voltage. If a conductor of length l meters is moving in a magnetic field of B teslas at a velocity of v meters per second at mutually right angles, an electric voltage is established across the conductor with magnitude

Law of generators

$$e(t) = Blv \ V. \tag{2.46}$$

This expression is called the **law of generators.**

EXAMPLE 2.12   *Loudspeaker with Circuit*

For the loudspeaker in Fig. 2.29 and the circuit driving it in Fig. 2.30, find the differential equations relating the input voltage v_a to the output cone displacement x. Assume the effective circuit resistance is R and the inductance is L.

Solution. The loudspeaker motion satisfies Eq. (2.44), and the motion results in a voltage across the coil as given by Eq. (2.46), with the velocity $\dot{x}$. The resulting voltage is

$$e_{coil} = Bl\dot{x} = 0.63\dot{x}. \tag{2.47}$$

This induced voltage effect needs to be added to the analysis of the circuit. The equation of motion for the electric circuit is

$$L\frac{di}{dt} + Ri = v_a - 0.63\dot{x}. \tag{2.48}$$

---

[7] Similar voice-coil motors are commonly used as the actuator for the read/write head assembly of computer hard-disk data access devices.

Figure 2.30

A loudspeaker showing
the electric circuit

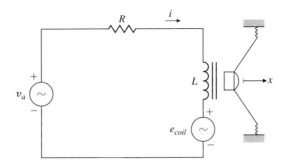

These two coupled equations, (2.44) and (2.48), constitute the dynamic model for the loudspeaker.

Again substituting s for d/dt in these equations, the transfer function between the applied voltage and the loudspeaker displacement is found to be

$$\frac{X(s)}{V_a(s)} = \frac{0.63}{s\left[(Ms+b)(Ls+R)+(0.63)^2\right]}. \tag{2.49}$$

DC motor actuators

A common actuator based on these principles and used in control systems is the DC motor to provide rotary motion. A sketch of the basic components of a DC motor is given in Fig. 2.31. In addition to housing and bearings, the nonturning part (stator) has magnets, which establish a field across the rotor. The magnets may be electromagnets or, for small motors, permanent magnets. The brushes contact the rotating commutator, which causes the current always to be in the proper conductor windings so as to produce maximum torque. If the direction of the current is reversed, the direction of the torque is reversed.

Back emf

The motor equations give the torque T on the rotor in terms of the armature current i_a and express the back emf voltage in terms of the shaft's rotational velocity $\dot{\theta}_m$.[8]

Figure 2.31

Sketch of a DC motor

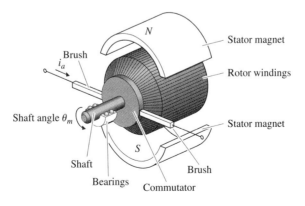

---

[8] Because the generated electromotive force (emf) works against the applied armature voltage, it is called the **back emf**.

Thus

$$T = K_t\, i_a, \tag{2.50}$$

$$e = K_e\, \dot{\theta}_m. \tag{2.51}$$

Torque

In consistent units, the torque constant K_t equals the electric constant K_e, but in some cases the torque constant will be given in other units, such as ounce-inches per ampere, and the electric constant may be expressed in units of volts per 1000 rpm. In such cases the engineer must make the necessary translations to be certain the equations are correct.

EXAMPLE 2.13

Modeling a DC Motor

Find the equations for a DC motor with the equivalent electric circuit shown in Fig. 2.32(a). Assume that the rotor has inertia J_m and viscous friction coefficient b.

Solution. The free-body diagram for the rotor, shown in Fig. 2.32(b), defines the positive direction and shows the two applied torques, T and $b\dot{\theta}_m$. Application of Newton's laws yields

$$J_m\ddot{\theta}_m + b\dot{\theta}_m = K_t i_a. \tag{2.52}$$

Analysis of the electric circuit, including the back emf voltage, shows the electrical equation to be

$$L_a\frac{di_a}{dt} + R_a i_a = v_a - K_e\dot{\theta}_m. \tag{2.53}$$

With s substituted for d/dt in Eqs. (2.52) and (2.53), the transfer function for the motor is readily found to be

$$\frac{\Theta_m(s)}{V_a(s)} = \frac{K_t}{s[(J_m s + b)(L_a s + R_a) + K_t K_e]}. \tag{2.54}$$

In many cases the relative effect of the inductance is negligible compared with the mechanical motion and can be neglected in Eq. (2.53). If so, we can combine Eqs. (2.52) and (2.53) into one equation to get

$$J_m\ddot{\theta}_m + \left(b + \frac{K_t K_e}{R_a}\right)\dot{\theta}_m = \frac{K_t}{R_a}v_a. \tag{2.55}$$

Figure 2.32

DC motor: (a) electric circuit of the armature; (b) free-body diagram of the rotor

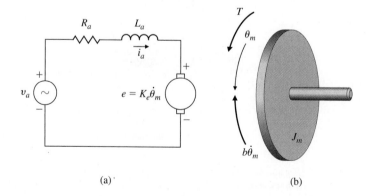

(a)   (b)

From Eq. (2.55) it is clear that in this case the effect of the back emf is indistinguishable from the friction, and the transfer function is

$$\frac{\Theta_m(s)}{V_a(s)} = \frac{\dfrac{K_t}{R_a}}{J_m s^2 + \left(b + \dfrac{K_t K_e}{R_a}\right)s} \qquad (2.56)$$

$$= \frac{K}{s(\tau s + 1)}, \qquad (2.57)$$

where

$$K = \frac{K_t}{b R_a + K_t K_e}, \qquad (2.58)$$

$$\tau = \frac{R_a J_m}{b R_a + K_t K_e}. \qquad (2.59)$$

In many cases, a transfer function between the motor input and the output speed ($\omega = \dot{\theta}_m$) is required. In such cases, the transfer function would be

$$\frac{\Omega(s)}{V_a(s)} = s \frac{\Theta_m(s)}{V_a(s)} = \frac{K}{\tau s + 1}. \qquad (2.60)$$

AC motor actuators

Another device used for electromechanical energy conversion is the alternating current (AC) induction motor invented by N. Tesla. Elementary analysis of the AC motor is more complex than that of the DC motor. A typical experimental set of curves of torque versus speed for fixed frequency and varying amplitude of applied (sinusoidal) voltage is given in Fig. 2.33. Although the data in the figure are for a constant engine speed, they can be used to extract the motor constants that will provide a dynamic model for the motor. For analysis of a control problem involving an AC

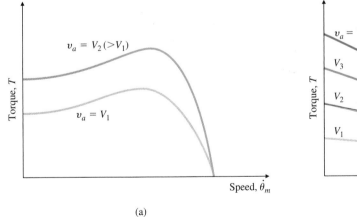

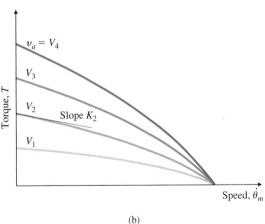

(a) (b)

Figure 2.33

Torque–speed curves for a servo motor showing four amplitudes of armature voltage: (a) low-rotor-resistance machine; (b) high-rotor-resistance machine showing four values of armature voltage, v_a

motor such as that described by Fig. 2.33, we make a linear approximation to the curves for speed near zero and at a midrange voltage to obtain the expression

$$T = K_1 v_a - K_2 \dot{\theta}_m. \tag{2.61}$$

The constant K_1 represents the ratio of a change in torque to a change in voltage at zero speed and is proportional to the distance between the curves at zero speed. The constant K_2 represents the ratio of a change in torque to a change in speed at zero speed and a midrange voltage; therefore, it is the slope of a curve at zero speed as shown by the line at V_2. For the electrical portion, values for the armature resistance R_a and inductance L_a are also determined by experiment. Once we have values for K_1, K_2, R_a, and L_a, the analysis proceeds as the analysis in Example 2.13 for the DC motor. For the case in which the inductor can be neglected, we can substitute K_1 and K_2 into Eq. (2.55) in place of K_t/R_a and $K_t K_e/R_a$, respectively.

In addition to the DC and AC motors mentioned here, control systems use brushless DC motors (Reliance Motion Control Corp., 1980) and stepping motors (Kuo, 1980). Models for these machines, developed in the works just cited, do not differ in principle from the motors considered in this section. In general, the analysis, supported by experiment, develops the torque as a function of voltage and speed similar to the AC motor torque–speed curves given in Fig. 2.33. From such curves one can obtain a linearized formula such as Eq. (2.61) to use in the mechanical part of the system and an equivalent circuit consisting of a resistance and an inductance to use in the electrical part.

△ 2.4   Heat and Fluid-Flow Models

Thermodynamics, heat transfer, and fluid dynamics are each the subject of complete textbooks. For purposes of generating dynamic models for use in control systems, the most important aspect of the physics is to represent the dynamic interaction between the variables. Experiments are usually required to determine the actual values of the parameters and thus to complete the dynamic model for purposes of control systems design.

2.4.1   Heat Flow

Some control systems involve regulation of temperature for portions of the system. The dynamic models of temperature control systems involve the flow and storage of heat energy. Heat energy flows through substances at a rate proportional to the temperature difference across the substance; that is,

$$q = \frac{1}{R}(T_1 - T_2), \tag{2.62}$$

where

$q =$ heat energy flow, joules per second (J/sec), or British Thermal Unit/sec (BTU/sec),

$R =$ thermal resistance, °C/J·sec or °F/BTU·sec,

$T =$ temperature, °C or °F.

The net heat-energy flow into a substance affects the temperature of the substance according to the relation

$$\dot{T} = \frac{1}{C}q,\tag{2.63}$$

where C is the thermal capacity. Typically, there are several paths for heat to flow into or out of a substance, and q in Eq. (2.63) is the sum of heat flows obeying Eq. (2.62).

EXAMPLE 2.14

Equations for Heat Flow

A room with all but two sides insulated ($1/R = 0$) is shown in Fig. 2.34. Find the differential equations that determine the temperature in the room.

Solution. Application of Eqs. (2.62) and (2.63) yields

$$\dot{T}_I = \frac{1}{C_I}\left(\frac{1}{R_1} + \frac{1}{R_2}\right)(T_O - T_I),$$

where

C_I = thermal capacity of air within the room,

T_O = temperature outside,

T_I = temperature inside,

R_2 = thermal resistance of the room ceiling,

R_1 = thermal resistance of the room wall.

Normally the material properties are given in tables as follows:

Specific heat

1. The specific heat at constant volume c_v, which is converted to heat capacity by

$$C = mc_v,\tag{2.64}$$

where m is the mass of the substance;

Thermal conductivity

2. The thermal conductivity[9] k, which is related to thermal resistance R by

$$\frac{1}{R} = \frac{kA}{l},$$

where A is the cross-sectional area and l is the length of the heat-flow path.

Figure 2.34

Dynamic model for room temperature

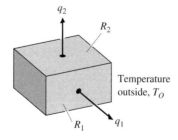

In addition to flow due to transfer, as expressed by Eq. (2.62), heat can also flow when a warmer mass flows into a cooler mass, or vice versa. In this case,

$$q = wc_v(T_1 - T_2), \tag{2.65}$$

where w is the mass flow rate of the fluid at T_1 flowing into the reservoir at T_2. For a more complete discussion of dynamic models for temperature control systems, see Cannon (1967) or textbooks on heat transfer.

EXAMPLE 2.15   *Equations for Modeling a Heat Exchanger*

A heat exchanger is shown in Fig. 2.35. Steam enters the chamber through the controllable valve at the top, and cooler steam leaves at the bottom. There is a constant flow of water through the pipe that winds through the middle of the chamber so that it picks up heat from the steam. Find the differential equations that describe the dynamics of the measured water outflow temperature as a function of the area A_s of the steam-inlet control valve when open. The sensor that measures the water outflow temperature, being downstream from the exit temperature in the pipe, lags the temperature by t_d seconds.

Solution.   The temperature of the water in the pipe will vary continuously along the pipe as the heat flows from the steam to the water. The temperature of the steam will also reduce in the chamber as it passes over the maze of pipes. An accurate thermal model of this process is therefore quite involved because the actual heat transfer from the steam to the water will be proportional to the local temperatures of each fluid. For many control applications it is not necessary to have great accuracy because the feedback will correct for a considerable amount of error in the model. Therefore, it makes sense to combine the spatially varying temperatures into single temperatures T_s and T_w for the outflow steam and water temperatures, respectively. We then assume that the heat transfer from steam to water is proportional to the difference in these temperatures, as given by Eq. (2.62). There is also a flow of heat into the chamber

Figure 2.35

Heat exchanger

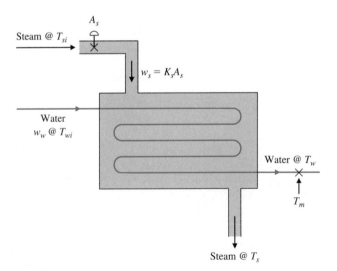

from the inlet steam that depends on the steam flow rate and its temperature according to Eq. (2.65),

$$q_{in} = w_s c_{vs}(T_{si} - T_s),$$

where

$$w_s = K_s A_s, \text{ mass flow rate of the steam,}$$
$$A_s = \text{area of the steam inlet valve,}$$
$$K_s = \text{flow coefficient of the inlet valve,}$$
$$c_{vs} = \text{specific heat of the steam,}$$
$$T_{si} = \text{temperature of the inflow steam,}$$
$$T_s = \text{temperature of the outflow steam.}$$

The net heat flow into the chamber is the difference between the heat from the hot incoming steam and the heat flowing out to the water. This net flow determines the rate of temperature change of the steam according to Eq. (2.63),

$$C_s \dot{T}_s = A_s K_s c_{vs}(T_{si} - T_s) - \frac{1}{R}(T_s - T_w), \tag{2.66}$$

where

$C_s = m_s c_{vs}$ is the thermal capacity of the steam in the chamber with mass m_s,

$R = $ the thermal resistance of the heat flow averaged over the entire exchanger.

Likewise, the differential equation describing the water temperature is

$$C_w \dot{T}_w = w_w c_{cw}(T_{wi} - T_w) + \frac{1}{R}(T_s - T_w), \tag{2.67}$$

where

$$w_w = \text{mass flow rate of the water,}$$
$$c_{cw} = \text{specific heat of the water,}$$
$$T_{wi} = \text{temperature of the incoming water,}$$
$$T_w = \text{temperature of the outflowing water.}$$

To complete the dynamics, the time delay between the measurement and the exit flow is described by the relation

$$T_m(t) = T_w(t - t_d),$$

where T_m is the measured downstream temperature of the water and t_d is the time delay. There may also be a delay in the measurement of the steam temperature T_s, which would be modeled in the same manner.

Equation (2.66) is nonlinear because the quantity T_s is multiplied by the control input A_s. The equation can be linearized about T_{so} (a specific value of T_s) so that $T_{si} - T_s$ is assumed constant for purposes of approximating the nonlinear term, which we will define as ΔT_s. In order to eliminate the T_{wi} term in Eq. (2.67), it is convenient to measure all temperatures in terms of deviation in degrees from T_{wi}. The resulting equations are then

$$C_s \dot{T}_s = -\frac{1}{R} T_s + \frac{1}{R} T_w + K_s c_{vs} \Delta T_s A_s,$$

$$C_w \dot{T}_w = -\left(\frac{1}{R} + w_w c_{vw}\right) T_w + \frac{1}{R} T_s,$$

$$T_m = T_w(t - t_d).$$

Although the time delay is not a nonlinearity, we will see in Chapter 3 that operationally, $T_m = e^{-t_d s} T_w$. Therefore, the transfer function of the heat exchanger has the form

$$\frac{T_m(s)}{A_s(s)} = \frac{K e^{-t_d s}}{(\tau_1 s + 1)(\tau_2 s + 1)}. \tag{2.68}$$

2.4.2 Incompressible Fluid Flow

Fluid flows are common in many control systems components. One example is the hydraulic actuator, which is used extensively in control systems because it can supply a large force with low inertia and low weight. They are often used to move the aero-dynamic control surfaces of airplanes, to gimbal rocket nozzles, to move the linkages in earth-moving equipment, farm tractor implements, snow-grooming machines, and to move robot arms.

The physical relations governing fluid flow are continuity, force equilibrium, and

The continuity relation

resistance. The continuity relation is simply a statement of the conservation of matter; that is,

$$\dot{m} = w_{in} - w_{out}, \tag{2.69}$$

where

$$m = \text{fluid mass within a prescribed portion of the system,}$$

$$w_{in} = \text{mass flow rate into the prescribed portion of the system,}$$

$$w_{out} = \text{mass flow rate out of the prescribed portion of the system.}$$

EXAMPLE 2.16   *Equations for Describing Water Tank Height*

Determine the differential equation describing the height of the water in the tank in Fig. 2.36.

Figure 2.36

Water tank example

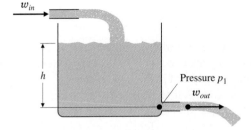

Solution. Application of Eq. (2.69) yields

$$\dot{h} = \frac{1}{A\rho}\left(w_{in} - w_{out}\right),$$ (2.70)

where

$$A = \text{area of the tank,}$$

$$\rho = \text{density of water,}$$

$$h = m/A\rho = \text{height of water,}$$

$$m = \text{mass of water in the tank.}$$

Force equilibrium must apply exactly as described by Eq. (2.1) for mechanical systems. Sometimes in fluid-flow systems some forces result from fluid pressure acting on a piston. In this case the force from the fluid is

$$f = pA,$$ (2.71)

where

$$f = \text{force,}$$

$$p = \text{pressure in the fluid,}$$

$$A = \text{area on which the fluid acts.}$$

EXAMPLE 2.17   *Modeling a Hydraulic Piston*

Determine the differential equation describing the motion of the piston actuator shown in Fig. 2.37, given that there is a force F_D acting on it and a pressure p in the chamber.

Solution. Equations (2.1) and (2.71) apply directly, where the forces include the fluid pressure as well as the applied force. The result is

$$M\ddot{x} = Ap - F_D,$$

where

$$A = \text{area of the piston,}$$

$$p = \text{pressure in the chamber,}$$

Figure 2.37
Hydraulic piston
actuator

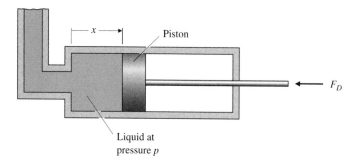

Piston

x

F_D

Liquid at
pressure p

$$M = \text{mass of the piston,}$$

$$x = \text{position of the piston.}$$

In many cases of fluid-flow problems the flow is resisted either by a constriction in the path or by friction. The general form of the effect of resistance is given by

$$w = \frac{1}{R}(p_1 - p_2)^{1/\alpha}, \tag{2.72}$$

where

$$w = \text{mass flow rate,}$$

$$p_1, p_2 = \text{pressures at ends of the path through which flow is occurring,}$$

$$R, \alpha = \text{constants whose values depend on the type of restriction.}$$

Or, as is more commonly used in hydraulics,

$$Q = \frac{1}{\rho R}(p_1 - p_2)^{1/\alpha}, \tag{2.73}$$

where

$$Q = \text{volume flow rate, where } Q = {}^{w}/_{\rho},$$

$$\rho = \text{fluid density.}$$

The constant α takes on values between 1 and 2. The most common value is approximately 2 for high flow rates (those having a Reynolds number Re $>$ 10^5) through pipes or through short constrictions or nozzles. For very slow flows through long pipes or porous plugs wherein the flow remains laminar (Re $\lesssim$ 1000), $\alpha = 1$. Flow rates between these extremes can yield intermediate values of α. The Reynolds number indicates the relative importance of inertial forces and viscous forces in the flow. It is proportional to a material's velocity and density and to the size of the restriction, and it is inversely proportional to the viscosity. When Re is small, the viscous forces predominate and the flow is laminar. When Re is large, the inertial forces predominate and the flow is turbulent.

Note that a value of $\alpha = 2$ indicates that the flow is proportional to the square root of the pressure difference and therefore will produce a nonlinear differential equation. For the initial stages of control systems analysis and design, it is typically very useful to linearize these equations so that the design techniques described in this book can be applied. Linearization involves selecting an operating point and expanding the nonlinear term to be a small perturbation from that point.

EXAMPLE 2.18   *Linearization of Water Tank Height and Outflow*

Find the nonlinear differential equation describing the height of the water in the tank in Fig. 2.36. Assume that there is a relatively short restriction at the outlet and that $\alpha = 2$. Also linearize your equation about the operating point h_o.

Solution. Applying Eq. (2.72) yields the flow out of the tank as a function of the height of the water in the tank:

$$w_{out} = \frac{1}{R}(p_1 - p_a)^{1/2}. \tag{2.74}$$

Here,

$$p_1 = \rho g h + p_a, \text{the hydrostatic pressure,}$$

$$p_a = \text{ambient pressure outside the restriction.}$$

Substituting Eq. (2.74) into Eq. (2.70) yields the nonlinear differential equation for the height:

$$\dot{h} = \frac{1}{A\rho}\left(w_{in} - \frac{1}{R}\sqrt{p_1 - p_a}\right). \tag{2.75}$$

Linearization involves selecting the operating point $p_o = \rho g h_o + p_a$ and substituting $p_1 = p_o + \Delta p$ into Eq. (2.74). Then we expand the nonlinear term according to the relation

$$(1 + \varepsilon)^\beta \cong 1 + \beta\varepsilon, \tag{2.76}$$

where $\varepsilon \ll 1$. Equation (2.74) can thus be written as

$$w_{out} = \frac{\sqrt{p_o - p_a}}{R}\left(1 + \frac{\Delta p}{p_o - p_a}\right)^{1/2}$$

$$\cong \frac{\sqrt{p_o - p_a}}{R}\left(1 + \frac{1}{2}\frac{\Delta p}{p_o - p_a}\right). \tag{2.77}$$

The linearizing approximation made in Eq. (2.77) is valid as long as $\Delta p \ll p_o - p_a$; that is, as long as the deviations of the system pressure from the chosen operating point are relatively small.

Combining Eqs. (2.70) and (2.77) yields the following linearized equation of motion for the water tank level:

$$\Delta \dot{h} = \frac{1}{A\rho}\left[w_{in} - \frac{\sqrt{p_o - p_a}}{R}\left(1 + \frac{1}{2}\frac{\Delta p}{p_o - p_a}\right)\right].$$

Because $\Delta p = \rho g \Delta h$, this equation reduces to

$$\Delta \dot{h} = -\frac{g}{2AR\sqrt{p_o - p_a}}\Delta h + \frac{w_{in}}{A\rho} - \frac{\sqrt{p_o - p_a}}{\rho AR}, \tag{2.78}$$

which is a linear differential equation for $\Delta \dot{h}$. The operating point is not an equilibrium point because some control input is required to maintain it. In other words, when the system is at the operating point ($\Delta h = 0$) with no input ($w_{in} = 0$), it will move from that point because $\Delta \dot{h} \neq 0$. So if no water is flowing into the tank, the tank will drain, thus moving it from the reference point. To define an operating point that is also an equilibrium point, we need to require that there be a nominal flow rate,

$$\frac{w_{in_o}}{A\rho} = \frac{\sqrt{p_o - p_a}}{\rho AR},$$

and define the linearized input flow to be a perturbation from that value.

Hydraulic actuators

Hydraulic actuators obey the same fundamental relationships we saw in the water tank: continuity [Eq. (2.69)], force balance [Eq. (2.71)], and flow resistance [Eq. (2.72)]. Although the development here assumes the fluid is perfectly incompressible, in fact, hydraulic fluid has some compressibility due primarily to entrained air. This feature causes hydraulic actuators to have some resonance because the compressibility of the fluid acts like a stiff spring. This resonance limits their speed of response.

EXAMPLE 2.19

Modeling a Hydraulic Actuator

1. Find the nonlinear differential equations relating the movement θ of the control surface to the input displacement x of the valve for the hydraulic actuator shown in Fig. 2.38.
2. Find the linear approximation to the equations of motion when $\dot{y} = $ constant, with and without an applied load—that is, when $F \neq 0$ and when $F = 0$. Assume that θ motion is small.

Solution

1. **Equations of motion:** When the valve is at $x = 0$, both passages are closed and no motion results. When $x > 0$, as shown in Fig. 2.38, the oil flows clockwise as shown and the piston is forced to the left. When $x < 0$, the fluid flows counterclockwise. The oil supply at high pressure p_s enters the *left* side of the large piston chamber, forcing the piston to the right. This causes the oil to flow out of the valve chamber from the rightmost channel instead of the left.

   We assume that the flow through the orifice formed by the valve is proportional to x; that is,

$$Q_1 = \frac{1}{\rho R_1}(p_s - p_1)^{1/2}x. \tag{2.79}$$

   Similarly,

$$Q_2 = \frac{1}{\rho R_2}(p_2 - p_e)^{1/2}x. \tag{2.80}$$

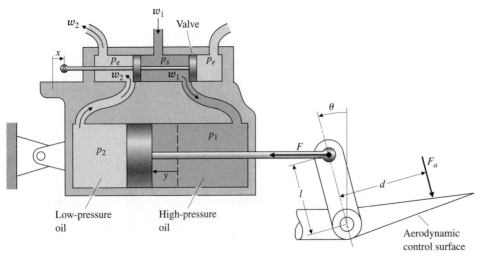

Figure 2.38
Hydraulic actuator with valve

The continuity relation yields

$$A\dot{y} = Q_1 = Q_2, \tag{2.81}$$

where

$$A = \text{piston area.}$$

The force balance on the piston yields

$$A(p_1 - p_2) - F = m\ddot{y}, \tag{2.82}$$

where

$m = $ mass of the piston and the attached rod,

$F = $ force applied by the piston rod to the control surface attachment point.

Furthermore, the moment balance of the control surface using Eq. (2.14) yields

$$I\ddot{\theta} = Fl\cos\theta - F_a d, \tag{2.83}$$

where

$I = $ moment of inertia of the control surface and attachment about the hinge,

$F_a = $ applied aerodynamic load.

To solve this set of five equations, we require the following additional kinematic relationship between θ and y:

$$y = l\sin\theta. \tag{2.84}$$

The actuator is usually constructed so that the valve exposes the two passages equally; therefore, $R_1 = R_2$, and we can infer from Eqs. (2.79) to (2.81) that

$$p_s - p_1 = p_2 - p_e. \tag{2.85}$$

These relations complete the nonlinear differential equations of motion; they are formidable and difficult to solve.

2. **Linearization and simplification:** For the case in which $\dot{y} = $ a constant ($\ddot{y} = 0$) and there is no applied load ($F = 0$), Eqs. (2.82) and (2.85) indicate that

$$p_1 = p_2 = \frac{p_s + p_e}{2}. \tag{2.86}$$

Therefore, using Eq. (2.81) and letting $\sin\theta = \theta$ (since θ is assumed to be small), we get

$$\dot{\theta} = \frac{\sqrt{p_s - p_e}}{\sqrt{2A\rho Rl}}x. \tag{2.87}$$

This represents a single integration between the input x and the output θ, where the proportionality constant is a function only of the supply pressure and the fixed parameters of the actuator. For the case $\dot{y} = $ constant but $F \neq 0$, Eqs. (2.82) and (2.85) indicate that

$$p_1 = \frac{p_s + p_e + F/A}{2}$$

and

$$\dot{\theta} = \frac{\sqrt{p_s - p_e - F/A}}{\sqrt{2}A_\rho Rl}x. \qquad (2.88)$$

This result is also a single integration between the input x and the output θ, but the proportionality constant now depends on the applied load F.

As long as the commanded values of x produce θ motion that has a sufficiently small value of $\ddot{\theta}$, the approximation given by Eqs. (2.87) or (2.88) is valid and no other linearized dynamic relationships are necessary. However, as soon as the commanded values of x produce accelerations in which the inertial forces ($m\ddot{y}$ and the reaction to $I\ddot{\theta}$) are a significant fraction of $p_s - p_e$, the approximations are no longer valid. We must then incorporate these forces into the equations, thus obtaining a dynamic relationship between x and θ that is much more involved than the pure integration implied by Eqs. (2.87) or (2.88). Typically, for initial control system designs, hydraulic actuators are assumed to obey the simple relationship of Eqs. (2.87) or (2.88). When hydraulic actuators are used in feedback control systems, resonances have been encountered that are not explained by using the approximation that the device is a simple integrator as in Eqs. (2.87) or (2.88). The source of the resonance is the neglected accelerations discussed above along with the additional feature that the oil is slightly compressible due to small quantities of entrained air. This phenomenon is called the "oil-mass resonance."

2.5   Historical Perspective

Newton's second law of motion (Eq. 2.1) was first published in his *Philosophiae Naturalis Principia Mathematica* in 1686 along with his two other famous laws of motion. The first: A body will continue with the same uniform motion unless acted on by an external unbalanced force, and the third: To every action there is an equal and opposite reaction. Isaac Newton also published his law of gravitation in this same publication, which stated that every mass particle attracts all other particles by a force proportional to the inverse of the square of the distance between them and the product of their two masses. His basis for developing these laws was the work of several other early scientists, combined with his own development of the calculus in order to reconcile all the observations. It is amazing that these laws still stand today as the basis for almost all dynamic analysis with the exception of Einstein's additions in the early 1900s for relativistic effects. It is also amazing that Newton's development of calculus formed the foundation of our mathematics that enable dynamic modeling. In addition to being brilliant, he was also very eccentric. As Brennan writes in *Heisenberg Probably Slept Here*, "He was seen about campus in his disheveled clothes, his wig askew, wearing run-down shoes and a soiled neckpiece. He seemed to care about nothing but his work. He was so absorbed in his studies that he forgot to eat." Another interesting aspect of Newton is that he initially developed the calculus and the now famous laws of physics about 20 years prior to publishing them! The incentive to publish them arose from a bet between three men having lunch at a pub in 1684: Edmond Halley, Christopher Wren, and Robert Hooke. They all had the opinion that

Kepler's elliptical characterization of planetary motion could be explained by the inverse square law, but nobody had ever proved it, so they "placed a bet as to who could first prove the conjecture."[10] Halley went to Newton for help due to his fame as a mathematician, who responded he had already done it many years ago and would forward the papers to him. He not only did that shortly afterwards, but followed it up with the *Principia* with all the details two years later.

The basis for Newton's work started with the astronomer Nicholas Copernicus more than a hundred years before the *Principia* was published. He was the first to speculate that the planets revolved around the sun, rather than everything in the skies revolving around the earth. But Copernicus' heretical notion was largely ignored at the time, except by the church who banned his publication. However, two scientists did take note of his work: Galileo Galilei in Italy and Johannes Kepler in Austria. Kepler relied on a large collection of astronomical data taken by a Danish astronomer, Tycho Brahe, and concluded that the planetary orbits were ellipses rather than the circles that Copernicus had postulated. Galileo was an expert telescope builder and was able to clearly establish that the earth was not the center of all motion, partly because he was able to see moons revolving around other planets. He also did experiments with rolling balls down inclined planes that strongly suggested that $F = ma$ (alas, it's a myth that he did his experiments by dropping objects out of the Leaning Tower of Pisa). Galileo published his work in 1632, which raised the ire of the church who then later banned him to house arrest until he died.[11] It was not until 1985 that the church recognized the important contributions of Galileo! These men laid the groundwork for Newton to put it all together with his laws of motion and the inverse square gravitational law. With these two physical principles, all the observations fit together with a theoretical framework that today forms the basis for the modeling of dynamic systems.

The sequence of discoveries that ultimately led to the laws of dynamics that we take for granted today were especially remarkable when we stop to think that they were all carried out without a computer, a calculator, or even a slide rule. On top of that, Newton had to invent calculus in order to reconcile the data.

After publishing the *Principia*, Newton went on to be elected to Parliament and was given high honors, including being the first man of science to be knighted by the Queen. He also got into fights with other scientists fairly regularly and used his powerful positions to get what he wanted. In one instance, he wanted data from the Royal Observatory that was not forthcoming fast enough. So he created a new board with authority over the Observatory and had the Astronomer Royal expelled from the Royal Society. Newton also had other less scientific interests. Many years after his death, John Maynard Keynes found that Newton had been spending as much of his time on metaphysical occult, alchemy, and biblical works as he had been on physics.

More than a hundred years after Newton's *Principia*, Michael Faraday performed a multitude of experiments and postulated the notion of electromagnetic lines of force in free space. He also discovered induction (Faraday's Law), which led to the

---

[10]Much of the background on Newton was taken from *Heisenberg Probably Slept Here*, by Richard P. Brennan, 1997. The book discusses his work and the other early scientists that laid the groundwork for Newton.

[11]Galileo's life, accomplishments, and house arrest are very well described in Dava Sobel's book, *Galileo's Daughter.*

electric motor and the laws of electrolysis. Faraday was born into a poor family, had virtually no schooling, and became an apprentice to a bookbinder at age 14. There he read many of the books being bound and became fascinated by science articles. Enthralled by these, he maneuvered to get a job as a bottle washer for a famous scientist, eventually learned enough to be a competitor to him, and ultimately became a professor at the Royal Institution in London. But lacking a formal education, he had no mathematical skills, and lacked the ability to create a theoretical framework for his discoveries. Faraday became a famous scientist in spite of his humble origins. After he had achieved fame for his discoveries and was made a Fellow of the Royal Society, the prime minister asked him what good his inventions could be.[12] Faraday's answer was, "Why Prime Minister, someday you can tax it." But in those days, scientists were almost exclusively men born into privilege; so Faraday had been treated like a second-class citizen by some of the other scientists. As a result, he rejected knighthood as well as burial at Westminster Abbey. Faraday's observations, along with those by Coulomb and Ampere, led James Clerk Maxwell to integrate all their knowledge on magnetism and electricity into Maxwell's Equations. Against the beliefs of most prominent scientists of the day (Faraday being an exception), Maxwell invented the concepts of fields and waves that explained magnetic and electrostatic forces and was the key to creating the unifying theory. Although Newton had discovered the spectrum of light, Maxwell was also the first to realize that light was one type of the same electromagnetic waves, and its behavior was explained as well by Maxwell's Equations. In fact, the only constant in his equations are μ and ε. The constant speed of light is $c = 1/\sqrt{\mu\varepsilon}$.

Maxwell was a Scottish mathematician and theoretical physicist. His work has been called the second great unification in physics, the first being that due to Newton. Maxwell was born into the privileged class and was given the benefits of an excellent education and he excelled at it. In fact, he was an extremely gifted theoretical and experimental scientist as well as a very generous and kind man with many friends and little vanity. In addition to unifying the observations of electromagnetics into a theory that still governs our engineering analyses today, he was the first to present an explanation of how light travels, the primary colors, the kinetic theory of gases, the stability of Saturn's rings, and the stability of feedback control systems! His discovery of the three primary colors (red, green, and blue) forms the basis of our color television to this day. His theory showing the speed of light is a constant was difficult to reconcile with Newton's laws and led Albert Einstein to create the special theory of relativity in the early 1900s. This led Einstein to say, "One scientific epoch ended and another began with James Clerk Maxwell."[13]

SUMMARY

Mathematical modeling of the system to be controlled is the first step in analyzing and designing the required system controls. In this chapter we developed models

---

[12] $E = MC^2$, *A Biography of the World's Most Famous Equation*, by David Bodanis, Walker and Co., New York, 2000.

[13] *The Man Who Changed Everything: The Life of James Clerk Maxwell*, Basil Mahon, Wiley, 2003.

TABLE 2.1

Key Equations for Dynamic Models

System	Important Laws or Relationships	Associated Equations	Equation Number
Mechanical	Translation motion (Newton's law)	$F = ma$	(2.1)
	Rotational motion	$M = I\alpha$	(2.14)
Electrical	Operational amplifier		(2.36), (2.37)
Electromechanical	Law of motors	$F = Bli$	(2.43)
	Law of the generator	$e(t) = Blv$	(2.46)
	Torque developed in a rotor	$T = K_t i_a$	(2.50)
Back emf	Voltage generated as a result of rotation of a rotor	$e = K_e \dot{\theta}_m$	(2.51)
Heat flow	Heat-energy flow	$q = {}^1/_R(T_1 - T_2)$	(2.62)
	Temperature as a function of heat-energy flow	$\dot{T} = {}^1/_C q$	(2.63)
	Specific heat	$C = mc_v$	(2.64)
Fluid flow	Continuity relation (conservation of matter)	$\dot{m} = w_{in} - w_{out}$	(2.69)
	Force of a fluid acting on a piston	$f = pA$	(2.71)
	Effect of resistance to fluid flow	$w = {}^1/_R(p_1 - p_2)^{1/\alpha}$	(2.72)

for representative systems. Important equations for each category of system are summarized in Table 2.1.

REVIEW QUESTIONS

1. What is a "free-body diagram"?
2. What are the two forms of Newton's law?
3. For a structural process to be controlled, such as a robot arm, what is the meaning of "collocated control"? "Noncollocated control"?
4. State Kirchhoff's current law.
5. State Kirchhoff's voltage law.
6. When, why, and by whom was the device named an "operational amplifier"?
7. What is the major benefit of having zero input current to an operational amplifier?
8. Why is it important to have a small value for the armature resistance R_a of an electric motor?
9. What are the definition and units of the electric constant of a motor?
10. What are the definition and units of the torque constant of an electric motor?

11. Why do we approximate a physical model of the plant (which is *always* nonlinear) with a linear model?

△ 12. Give the relationships for

(a) heat flow across a substance, and

(b) heat storage in a substance.

△ 13. Name and give the equations for the three relationships governing fluid flow.

PROBLEMS

Problems for Section 2.1: Dynamics of Mechanical Systems

2.1 Write the differential equations for the mechanical systems shown in Fig. 2.39. For (a) and (b), state whether you think the system will eventually decay so that it has no motion at all, given that there are non-zero initial conditions for both masses, and give a reason for your answer.

Figure 2.39
Mechanical systems

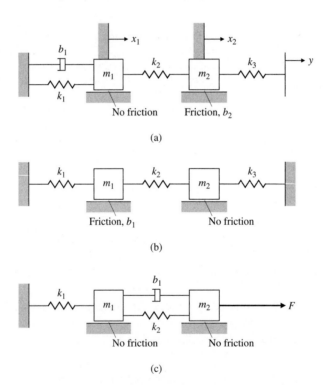

(a)

(b)

(c)

2.2 Write the differential equation for the mechanical system shown in Fig. 2.40. State whether you think the system will eventually decay so that it has no motion at all, given that there are non-zero initial conditions for both masses, and give a reason for your answer.

2.3 Write the equations of motion for the double-pendulum system shown in Fig. 2.41. Assume that the displacement angles of the pendulums are small enough to ensure that the spring is always horizontal. The pendulum rods are taken to be massless, of length l, and the springs are attached three-fourths of the way down.

Figure 2.40

Mechanical system for
Problem 2.2

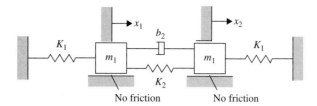

No friction   No friction

Figure 2.41

Double pendulum

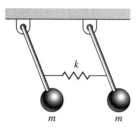

2.4 Write the equations of motion of a pendulum consisting of a thin, 4 kg stick of length l suspended from a pivot. How long should the rod be in order for the period to be exactly 2 sec? (The inertia I of a thin stick about an end point is $\frac{1}{3}ml^2$. Assume that θ is small enough that $\sin\theta \cong \theta$.) Why do you think grandfather clocks are typically about 6 ft high?

2.5 For the car suspension discussed in Example 2.2, plot the position of the car and the wheel after the car hits a "unit bump"(i.e., r is a unit step) using MATLAB. Assume that $m_1 = 10$ kg, $m_2 = 350$ kg, $K_w = 500{,}000$ N/m, $K_s = 10{,}000$ N/m. Find the value of b that you would prefer if you were a passenger in the car.

2.6 Write the equations of motion for a body of mass M suspended from a fixed point by a spring with a constant k. Carefully define where the body's displacement is zero.

2.7 Automobile manufacturers are contemplating building active suspension systems. The simplest change is to make shock absorbers with a changeable damping, $b(u_1)$. It is also possible to make a device to be placed in parallel with the springs that has the ability to supply an equal force, u_2, in opposite directions on the wheel axle and the car body.

(a) Modify the equations of motion in Example 2.2 to include such control inputs.

(b) Is the resulting system linear?

(c) Is it possible to use the forcer u_2 to completely replace the springs and shock absorber? Is this a good idea?

2.8 Modify the equation of motion for the cruise control in Example 2.1, Eq. (2.4), so that it has a control law; that is, let

$$u = K(v_r - v), \qquad (2.89)$$

where

$$v_r = \text{reference speed}, \qquad (2.90)$$

$$K = \text{constant}. \qquad (2.91)$$

This is a "proportional"control law in which the difference between v_r and the actual speed is used as a signal to speed the engine up or slow it down. Revise the equations of motion with v_r as the input and v as the output and find the transfer function. Assume that $m = 1000$ kg and $b = 50$ N·sec/m, and find the response for a unit step in v_r using

MATLAB. Using trial and error, find a value of K that you think would result in a control system in which the actual speed converges as quickly as possible to the reference speed with no objectionable behavior.

2.9 In many mechanical positioning systems there is flexibility between one part of the system and another. An example is shown in Fig. 2.7 where there is flexibility of the solar panels. Fig. 2.42 depicts such a situation, where a force u is applied to the mass M and another mass m is connected to it. The coupling between the objects is often modeled by a spring constant k with a damping coefficient b, although the actual situation is usually much more complicated than this.

   (a) Write the equations of motion governing this system.

   (b) Find the transfer function between the control input u and the output y.

Figure 2.42

Schematic of a system with flexibility

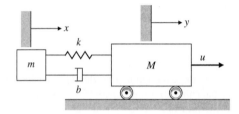

Problems for Section 2.2: Models of Electric Circuits

2.10 A first step toward a realistic model of an op-amp is given by the following equations and is shown in Fig. 2.43:

$$V_{out} = \frac{10^7}{s+1}[V_+ - V_-],$$

$$i_+ = i_- = 0.$$

Find the transfer function of the simple amplification circuit shown using this model.

Figure 2.43

Circuit for Problem 2.10

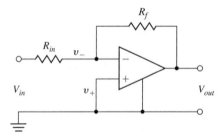

2.11 Show that the op-amp connection shown in Fig. 2.44 results in $V_{out} = V_{in}$ if the op-amp is ideal. Give the transfer function if the op-amp has the nonideal transfer function of Problem 2.10.

2.12 Show that, with the nonideal transfer function of Problem 2.10, the op-amp connection shown in Fig. 2.45 is unstable.

Figure 2.44
Circuit for Problem 2.11

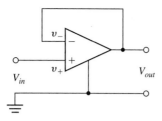

Figure 2.45
Circuit for Problem 2.12

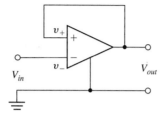

2.13 A common connection for a motor power amplifier is shown in Fig. 2.46. The idea is to have the motor current follow the input voltage, and the connection is called a current amplifier. Assume that the sense resistor r_s is very small compared with the feedback resistor R, and find the transfer function from V_{in} to I_a. Also show the transfer function when $R_f = \infty$.

Figure 2.46
Op-amp circuit for
Problem 2.13

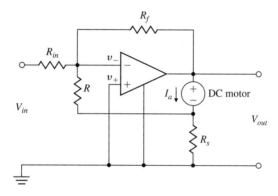

2.14 An op-amp connection with feedback to both the negative and the positive terminals is shown in Fig. 2.47. If the op-amp has the nonideal transfer function given in Problem 2.10, give the maximum value possible for the positive feedback ratio, $P = \frac{r}{r+R}$, in terms of the negative feedback ratio, $N = \frac{R_{in}}{R_{in}+R_f}$, for the circuit to remain stable.

2.15 Write the dynamic equations and find the transfer functions for the circuits shown in Fig. 2.48.

    (a) passive lead circuit

    (b) active lead circuit

    (c) active lag circuit

    (d) passive notch circuit

Figure 2.47

Op-amp circuit for
Problem 2.14

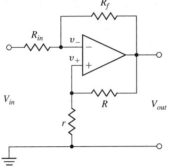

Figure 2.48

(a) Passive lead; (b)
active lead; (c) active
lag; and (d) passive
notch circuits

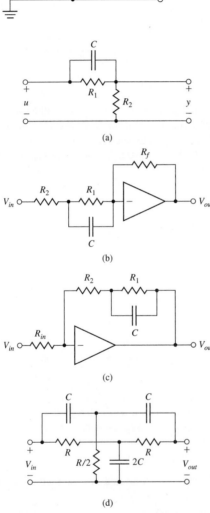

2.16 The very flexible circuit shown in Fig. 2.49 is called a biquad because its transfer function
can be made to be the ratio of two second-order or quadratic polynomials. By selecting
different values for R_a, R_b, R_c, and R_d, the circuit can realize a low-pass, band-pass,
high-pass, or band-reject (notch) filter.

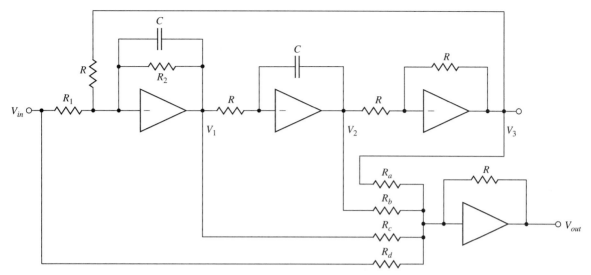

Figure 2.49
Op-amp biquad

(a) Show that if $R_a = R$, and $R_b = R_c = R_d = \infty$, the transfer function from V_{in} to V_{out} can be written as the low-pass filter

$$\frac{V_{out}}{V_{in}} = \frac{A}{\dfrac{s^2}{\omega_n^2} + 2\zeta\dfrac{s}{\omega_n} + 1}, \tag{2.92}$$

where

$$A = \frac{R}{R_1},$$

$$\omega_n = \frac{1}{RC},$$

$$\zeta = \frac{R}{2R_2}.$$

(b) Using the MATLAB command step, compute and plot on the same graph the step responses for the biquad of Fig. 2.49 for $A = 1$, $\omega_n = 1$, and $\zeta = 0.1, 0.5$, and 1.0.

2.17 Find the equations and transfer function for the biquad circuit of Fig. 2.49 if $R_a = R$, $R_d = R_1$, and $R_b = R_c = \infty$.

Problems for Section 2.3: Models of Electromechanical Systems

2.18 The torque constant of a motor is the ratio of torque to current and is often given in ounce-inches per ampere. (Ounce-inches have dimension force × distance, where an ounce is $1/16$ of a pound.) The electric constant of a motor is the ratio of back emf to speed and is often given in volts per 1000 rpm. In consistent units, the two constants are the same for a given motor.

(a) Show that the units ounce-inches per ampere are proportional to volts per 1000 rpm by reducing both to MKS (SI) units.

(b) A certain motor has a back emf of 25 V at 1000 rpm. What is its torque constant in ounce-inches per ampere?

(c) What is the torque constant of the motor of part (b) in newton-meters per ampere?

2.19 The electromechanical system shown in Fig. 2.50 represents a simplified model of a capacitor microphone. The system consists in part of a parallel plate capacitor connected into an electric circuit. Capacitor plate a is rigidly fastened to the microphone frame. Sound waves pass through the mouthpiece and exert a force $f_s(t)$ on plate b, which has mass M and is connected to the frame by a set of springs and dampers. The capacitance C is a function of the distance x between the plates, as follows:

$$C(x) = \frac{\varepsilon A}{x},$$

where

$$\varepsilon = \text{dielectric constant of the material between the plates,}$$

$$A = \text{surface area of the plates.}$$

The charge q and the voltage e across the plates are related by

$$q = C(x)e.$$

The electric field in turn produces the following force f_e on the movable plate that opposes its motion:

$$f_e = \frac{q^2}{2\varepsilon A}.$$

(a) Write differential equations that describe the operation of this system. (It is acceptable to leave in nonlinear form.)

(b) Can one get a linear model?

(c) What is the output of the system?

Figure 2.50

Simplified model for capacitor microphone

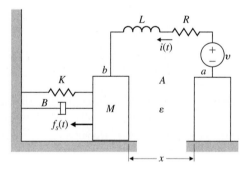

2.20 A very typical problem of electromechanical position control is an electric motor driving a load that has one dominant vibration mode. The problem arises in computer-disk-head control, reel-to-reel tape drives, and many other applications. A schematic diagram is sketched in Fig. 2.51. The motor has an electrical constant K_e, a torque constant K_t, an armature inductance L_a, and a resistance R_a. The rotor has an inertia J_1 and a viscous friction B. The load has an inertia J_2. The two inertias are connected by a shaft with a spring constant k and an equivalent viscous damping b. Write the equations of motion.

Figure 2.51

Motor with a flexible load

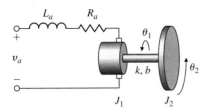

△ *Problems for Section 2.4: Heat and Fluid-Flow Models*

2.21 A precision table-leveling scheme shown in Fig. 2.52 relies on thermal expansion of actuators under two corners to level the table by raising or lowering their respective corners. The parameters are as follows:

$$T_{act} = \text{actuator temperature,}$$
$$T_{amb} = \text{ambient air temperature,}$$
$$R_f = \text{heat-flow coefficient between the actuator and the air,}$$
$$C = \text{thermal capacity of the actuator,}$$
$$R = \text{resistance of the heater.}$$

Assume that (1) the actuator acts as a pure electric resistance, (2) the heat flow into the actuator is proportional to the electric power input, and (3) the motion d is proportional to the difference between T_{act} and T_{amb} due to thermal expansion. Find the differential equations relating the height of the actuator d versus the applied voltage v_i.

Figure 2.52

(a) Precision table kept level by actuators; (b) side view of one actuator

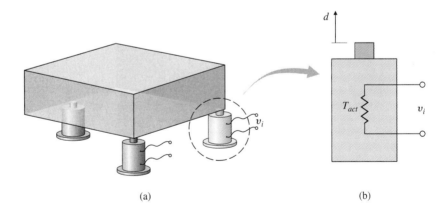

(a) (b)

2.22 An air conditioner supplies cold air at the same temperature to each room on the fourth floor of the high-rise building shown in Fig. 2.53(a). The floor plan is shown in Fig. 2.53(b). The cold airflow produces an equal amount of heat flow q out of each room. Write a set of differential equations governing the temperature in each room, where

$$T_o = \text{temperature outside the building,}$$
$$R_o = \text{resistance to heat flow through the outer walls,}$$
$$R_i = \text{resistance to heat flow through the inner walls.}$$

Figure 2.53

Building air
conditioning:
(a) high-rise building;
(b) floor plan of the
fourth floor

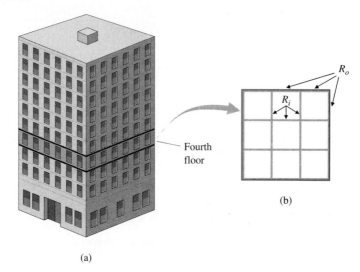

(a)

Assume that (1) all rooms are perfect squares, (2) there is no heat flow through the floors
or ceilings, and (3) the temperature in each room is uniform throughout the room. Take
advantage of symmetry to reduce the number of differential equations to three.

2.23 For the two-tank fluid-flow system shown in Fig. 2.54, find the differential equations
relating the flow into the first tank to the flow out of the second tank.

Figure 2.54

Two-tank fluid-flow
system for Problem 2.23

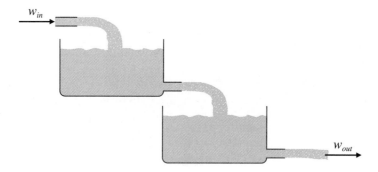

2.24 A laboratory experiment in the flow of water through two tanks is sketched in Fig. 2.55.
Assume that Eq. (2.74) describes flow through the equal-sized holes at points A, B, or C.

    (a) With holes at A and C, but none at B, write the equations of motion for this system in
terms of h_1 and h_2. Assume that $h_3 = 20$ cm, $h_1 > 20$ cm, and $h_2 < 20$ cm. When
$h_2 = 10$ cm, the outflow is 200 g/min.

    (b) At $h_1 = 30$ cm and $h_2 = 10$ cm, compute a linearized model and the transfer function
from pump flow (in cubic centimeters per minute) to h_2.

    (c) Repeat parts (a) and (b) assuming hole A is closed and hole B is open.

Figure 2.55

Two-tank fluid-flow
system for Problem 2.24

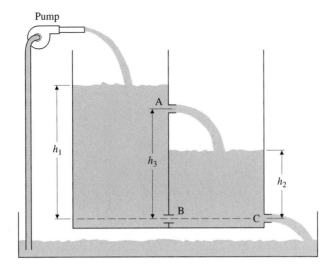

Pump

A

h_1

h_3

h_2

B

C

2.25 The equations for heating a house are given by Eqs. (2.62) and (2.63), and in a particular case can be written with time in *hours* as

$$C\frac{dT_h}{dt} = Ku - \frac{T_h - T_o}{R},$$

where

(a) C is the thermal capacity of the house, BTU/°F,

(b) T_h is the temperature in the house, °F,

(c) T_o is the temperature outside the house, °F,

(d) K is the heat rating of the furnace, $= 90,000$ BTU/hour,

(e) R is the thermal resistance, °F per BTU/hour,

(f) u is the furnace switch, $= 1$ if the furnace is on and $= 0$ if the furnace is off.

It is measured that, with the outside temperature at 32°F and the house at 60°F, the furnace raises the temperature 2°F in 6 minutes (0.1 hour). With the furnace off, the house temperature falls 2°F in 40 minutes. What are the values of C and R for the house?

3

Dynamic Response

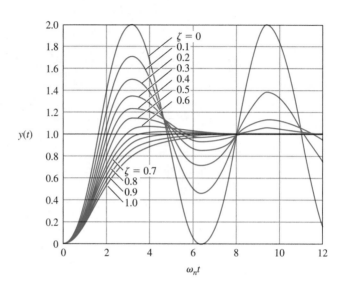

A Perspective on System Response

We saw in Chapter 2 how to obtain the dynamic model of a system. In designing a control system, it is important to see how well a trial design matches the desired performance. We do this by solving the equations of the system model.

There are two ways to approach solving the dynamic equations. For a quick, *approximate* analysis we use linear analysis techniques. The resulting approximations of system response provide insight into why the solution has certain features and how the system might be changed to modify the response in a desired direction. In contrast, a *precise* picture of the system response typically calls for numerical simulation of nonlinear equations of motion using computer aids. This chapter focuses on linear analysis and computer tools that can be used to solve for the time response of linear systems.

There are three domains within which to study dynamic response: the **s-plane**, the **frequency response**, and the **state space** (analysis using the state-variable description). The well-prepared control engineer needs to be fluent in all of them, so they will be treated in depth in Chapters 5, 6, and 7, respectively. The purpose of this chapter is to discuss some of the fundamental mathematical tools needed before studying analysis in the *s*-plane, frequency response, and state space.

Chapter Overview

The Laplace transform, reviewed in Section 3.1 (and Appendix A), is the mathematical tool for transforming differential equations into an easier-to-manipulate algebraic form. In addition to the mathematical tools at our disposal, there are graphical tools that can help us to visualize the model of a system and evaluate the pertinent mathematical relationships between elements of the system. One approach is the block diagram, which was introduced in Chapter 1. Block diagram manipulation is discussed in Section 3.2 and allows the determination of transfer functions.

Once the transfer function has been determined, we can identify its poles and zeros, which tell us a great deal about system characteristics, including its frequency response introduced in Section 3.1. Sections 3.3 to 3.5 focus on poles and zeros and some of the ways for manipulating them to steer system characteristics in a desired way. When feedback is introduced, the possibility that the system may become *unstable* is introduced. To study this effect, in Section 3.6 we consider the definition of stability and Routh's test, which can determine stability by examining the coefficients of the system's characteristic equation. Development of a model based on experimental time-response data is discussed in Section 3.7. Section 3.8 discusses amplitude and time scaling. Finally, Section 3.9 provides the historical perspective for the material in this chapter. An alternative representation of a system in graphical form is the signal-flow graph and flow graphs that allow the determination of complicated transfer functions are discussed in Appendix W3 on the web.

3.1   Review of Laplace Transforms

Two attributes of linear time-invariant systems (LTIs) form the basis for almost all analytical techniques applied to these systems:

1. A linear system response obeys the principle of superposition.
2. The response of an LTI system can be expressed as the convolution of the input with the unit impulse response of the system.

The concepts of superposition, convolution, and impulse response will be defined shortly.

From the second property (as we will show), it follows immediately that the response of an LTI system to an exponential input is also exponential. This result is the principal reason for the usefulness of Fourier and Laplace transforms in the study of LTI systems.

3.1.1   Response by Convolution

Superposition

The **principle of superposition** states that if the system has an input that can be expressed as a sum of signals, then the response of the system can be expressed as the sum of the individual responses to the respective signals. We can express superposition mathematically. Consider the system to have input u and output y. Suppose further that, with the system at rest, we apply the input $u_1(t)$ and observe the output $y_1(t)$. After restoring the system to rest, we apply a second input $u_2(t)$ and again observe the output, which we call $y_2(t)$. Then, we form the composite input $u(t) = \alpha_1 u_1(t) + \alpha_2 u_2(t)$. Finally, if superposition applies, then the response will be $y(t) = \alpha_1 y_1(t) + \alpha_2 y_2(t)$. Superposition will apply if and only if the system is linear.

EXAMPLE 3.1 *Superposition*

Show that superposition holds for the system modeled by the first-order linear differential equation

$$\dot{y} + ky = u.$$

Solution. We let $u = \alpha_1 u_1 + \alpha_2 u_2$ and assume that $y = \alpha_1 y_1 + \alpha_2 y_2$. Then $\dot{y} = \alpha_1 \dot{y}_1 + \alpha_2 \dot{y}_2$. If we substitute these expressions into the system equation, we get

$$\alpha_1 \dot{y}_1 + \alpha_2 \dot{y}_2 + k(\alpha_1 y_1 + \alpha_2 y_2) = \alpha_1 u_1 + \alpha_2 u_2.$$

From this it follows that

$$\alpha_1 (\dot{y}_1 + ky_1 - u_1) + \alpha_2 (\dot{y}_2 + ky_2 - u_2) = 0. \tag{3.1}$$

If y_1 is the solution with input u_1 and y_2 is the solution with input u_2, then Eq. (3.1) is satisfied, the response is the sum of the individual responses, and superposition holds.

Notice that the superposition result of Eq. (3.1) would also hold if k were a function of time. If it were constant, we call the system *time invariant*. In that case, it follows that if the input is delayed or shifted in time, then the output is unchanged except also being shifted by exactly the same amount. Mathematically, this is expressed by saying that, if $y_1(t)$ is the output caused by $u_1(t)$ then $y_1(t - \tau)$ will be the response to $u_1(t - \tau)$.

EXAMPLE 3.2 *Time Invariance*

Consider

$$\dot{y}_1(t) + k(t)y_1(t) = u_1(t) \tag{3.2}$$

and

$$\dot{y}_2(t) + k(t)y_2(t) = u_1(t - \tau),$$

where τ is a constant shift. Assume that $y_2(t) = y_1(t - \tau)$; then

$$\frac{dy_1(t - \tau)}{dt} + k(t)y_1(t - \tau) = u_1(t - \tau).$$

Let us make the change of variable $t - \tau = \eta$, then

$$\frac{dy_1(\eta)}{d\eta} + k(\eta + \tau)y_1(\eta) = u_1(\eta).$$

If $k(\eta + \tau) = k = $ constant, then

$$\frac{dy_1(\eta)}{d\eta} + ky_1(\eta) = u(\eta),$$

which is Eq. (3.1). Therefore, we conclude that if the system is time invariant $y(t - \tau)$ will be the response to $u(t - \tau)$; that is if the input is delayed by τ sec, then the output is also delayed by τ sec.

We are able to solve for the response of a linear system to a general signal simply by decomposing the given signal into a sum of the elementary components and, by superposition, concluding that the response to the general signal is the sum of the responses to the elementary signals. In order for this process to work, the elementary signals need to be sufficiently "rich" that any reasonable signal can be expressed as a sum of them, and their responses have to be easy to find. The most common candidates for elementary signals for use in linear systems are the impulse and the exponential.

Suppose the input signal to an LTI system is $u_1(t) = p(t)$, and the corresponding output signal is $y_1(t) = h(t)$ as shown in Fig. 3.1(a). Now if the input is scaled to $u_1(t) = u(0)p(t)$, then by the scaling property of superposition, the output response will be $y_1(t) = u(0)h(t)$. We showed that an LTI system obeys time invariance. If we delay the short pulse signal in time by τ, then the input is of the form $u_2(t) = p(t - \tau)$ and the output response will also be delayed by the same amount $y_2(t) = h(t - \tau)$ as shown in Fig. 3.1(b). Now by superposition, the response to the two short pulses will be the sum of the two individual outputs as shown in Fig. 3.1(c). If we have four pulses as the input, then the output will be the sum of the four individual responses as shown in Fig. 3.1(d). Any arbitrary input signal $u(t)$ may be approximated by a series of pulses as shown in Fig. 3.2. We define a short pulse $p_\Delta(t)$ as a rectangular pulse having *unit area* such that

Short pulse

$$p_\Delta(t) = \begin{cases} \frac{1}{\Delta}, & 0 \le t \le \Delta \\ 0, & \text{elsewhere} \end{cases} \tag{3.3}$$

as shown in Fig. 3.1(a). Suppose the response of the system to $p_\Delta(t)$ is defined as $h_\Delta(t)$. The response at time $n\Delta$ to $\Delta u(k\Delta)p_\Delta(k\Delta)$ is

$$\Delta u(k\Delta)h_\Delta(\Delta n - \Delta k).$$

By superposition, the total response to the series of the short pulses at time t is given by

$$y(t) = \sum_{k=0}^{k=\infty} \Delta u(k\Delta)h_\Delta(t - \Delta k). \tag{3.4}$$

If we take the limit as $\Delta \to 0$, the basic pulse gets more and more narrow and taller and taller while holding a *constant area*. We then have the concept of an **impulse signal**, $\delta(t)$, and that will allow us to treat continuous signals. In that case we have,

$$\lim_{\Delta \to 0} p_\Delta(t) = \delta(t), \tag{3.5}$$

$$\lim_{\Delta \to 0} h_\Delta(t) = h(t) = \text{the impulse response.} \tag{3.6}$$

Moreover, in the limit as $\Delta \to 0$, the summation in Eq. (3.4) is replaced by the integral

$$y(t) = \int_0^\infty u(\tau)h(t - \tau)\,d\tau, \tag{3.7}$$

that is the convolution integral.

Figure 3.1

Illustration of convolution as the response of a system to a series of short pulse (impulse) input signals

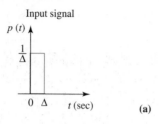

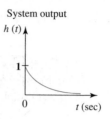

(a)

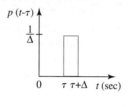

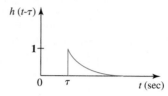

(b)

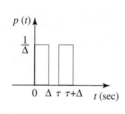

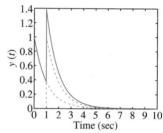

(c)

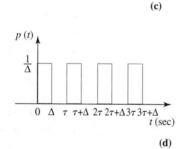

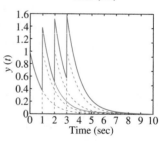

(d)

The idea for the impulse comes from dynamics. Suppose we wish to study the motion of a baseball hit by a bat. The details of the collision between the bat and ball can be very complex as the ball deforms and the bat bends; however, for purposes of computing the path of the ball, we can summarize the effect of the collision as the net velocity change of the ball over a very short time period. We assume that the ball is subjected to an **impulse**, a very intense force for a very short time. The physicist Paul Dirac suggested that such forces could be represented by the mathematical concept of an impulse $\delta(t)$, which has the property that

Impulse response

Definition of impulse

$$\delta(t) = 0 \quad t \neq 0, \tag{3.8}$$

$$\int_{-\infty}^{\infty} \delta(t)dt = 1. \tag{3.9}$$

Sifting property of impulse

If $f(t)$ is continuous at $t = \tau$, then it has the "sifting property."

Figure 3.2

Illustration of the representation of a general input signal as the sum of short pulses

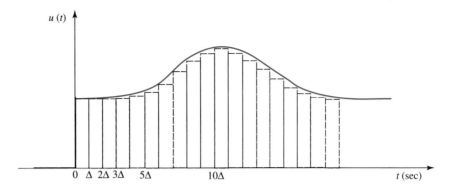

$$\int_{-\infty}^{\infty} f(\tau)\delta(t-\tau)\,d\tau = f(t). \tag{3.10}$$

In other words, the impulse is so short and so intense that no value of f matters except over the short range where the δ occurs. Since integration is a limit of a summation process, Eq. (3.10) can be viewed as representing the function f as a sum of impulses. If we replace f by u, then Eq. (3.10) represents an input $u(t)$ as a sum of impulses of intensity $u(t-\tau)$. To find the response to an arbitrary input, the principle of superposition tells us that we need only find the response to a unit impulse.

If the system is not only linear but also time invariant (LTI), then the impulse response is given by $h(t-\tau)$ because the response at t to an input applied at τ depends only on the difference between the time the impulse is applied and the time we are observing the response, i.e. the *elapsed time*. Time invariant systems are called shift invariant for this reason. For time invariant systems, the output for a general input is given by the integral

$$y(t) = \int_{-\infty}^{\infty} u(\tau)h(t-\tau)\,d\tau, \tag{3.11}$$

or by changing of variables as $\tau_1 = t - \tau$

$$y(t) = \int_{\infty}^{-\infty} u(t-\tau_1)h(\tau_1)\,(-d\tau_1) = \int_{-\infty}^{\infty} h(\tau)u(t-\tau)\,d\tau. \tag{3.12}$$

The convolution integral

This is the **convolution integral**.

EXAMPLE 3.3

Convolution

We can illustrate convolution with a simple system. Determine the impulse response for the system described by the differential equation

$$\dot{y} + ky = u = \delta(t),$$

with an initial condition of $y(0) = 0$ before the impulse.

Solution. Because $\delta(t)$ has an effect only near $t = 0$, we can integrate this equation from just before zero to just after zero with the result that

$$\int_{0^-}^{0^+} \dot{y}\, dt + k \int_{0^-}^{0^+} y\, dt = \int_{0^-}^{0^+} \delta(t)\, dt.$$

The integral of $\dot{y}$ is simply y, the integral of y over so small a range is zero, and the integral of the impulse over the same range is unity. Therefore,

$$y(0^+) - y(0^-) = 1.$$

Because the system was at rest before application of the impulse, $y(0^-) = 0$. Thus the effect of the impulse is that $y(0^+) = 1$. For positive time we have the differential equation

$$\dot{y} + ky = 0, \qquad y(0^+) = 1.$$

If we assume a solution $y = Ae^{st}$, then $\dot{y} = Ase^{st}$. The preceding equation then becomes

$$Ase^{st} + kAe^{st} = 0,$$

$$s + k = 0,$$

$$s = -k.$$

Because $y(0^+) = 1$, it is necessary that $A = 1$. Thus the solution for the impulse response is $y(t) = h(t) = e^{-kt}$ for $t > 0$. To take care of the fact that $h(t) = 0$ for negative time, we define the **unit-step function**

Unit step

$$1(t) = \begin{cases} 0, & t < 0, \\ 1, & t \geq 0. \end{cases}$$

With this definition, the impulse response of the first-order system becomes

$$h(t) = e^{-kt} 1(t).$$

The response of this system to a general input is given by the convolution of this impulse response with the input:

$$y(t) = \int_{-\infty}^{\infty} h(\tau) u(t - \tau)\, d\tau$$

$$= \int_{-\infty}^{\infty} e^{-k\tau} 1(\tau) u(t - \tau)\, d\tau$$

$$= \int_{0}^{\infty} e^{-k\tau} u(t - \tau)\, d\tau.$$

3.1.2   Transfer Functions and Frequency Response

An immediate consequence of convolution is that an input of the form e^{st} results in an output $H(s)e^{st}$. Note that both input and output are exponential time functions, and that the output differs from the input only in the amplitude $H(s)$. $H(s)$ is the transfer function of the system. The constant s may be complex, expressed as $s = \sigma + j\omega$.

Thus, both the input and the output may be complex. If we let $u(t) = e^{st}$ in Eq. (3.12), then

$$y(t) = \int_{-\infty}^{\infty} h(\tau)u(t - \tau)\, d\tau$$

$$= \int_{-\infty}^{\infty} h(\tau)e^{s(t-\tau)}\, d\tau$$

$$= \int_{-\infty}^{\infty} h(\tau)e^{st}e^{-s\tau}\, d\tau$$

$$= \int_{-\infty}^{\infty} h(\tau)e^{-s\tau}\, d\tau\, e^{st}$$

$$= H(s)e^{st}, \tag{3.13}$$

where[1]

$$H(s) = \int_{-\infty}^{\infty} h(\tau)e^{-s\tau}\, d\tau. \tag{3.14}$$

The integral in Eq. (3.14) does not need to be computed to find the transfer function of a system. Instead, one can assume a solution of the form of Eq. (3.13), substitute that into the differential equation of the system, then solve for the transfer function $H(s)$.

The transfer function can be formally defined as follows: The function $H(s)$, which is the transfer gain from $U(s)$ to $Y(s)$—input to output—is called the **transfer function** of the system. It is the ratio of the Laplace transform of the output to the Laplace transform of the input,

$$\frac{Y(s)}{U(s)} = H(s), \tag{3.15}$$

with the *key assumption* that all of the initial conditions on the system are zero. If the input $u(t)$ is the unit impulse $\delta(t)$, then $y(t)$ is the unit impulse response. The Laplace transform of $u(t)$ is 1 and the transform of $y(t)$ is $H(s)$ because

$$Y(s) = H(s). \tag{3.16}$$

In words, this is to say

Transfer function

The transfer function $H(s)$ is the Laplace transform of the unit impulse response $h(t)$.

Thus if one wishes to characterize an LTI system, one applies a unit impulse and the resulting response is a description (the inverse Laplace transform) of the transfer function.

EXAMPLE 3.4   *Transfer Function*

Compute the transfer function for the system of Example 3.1, and find the output y for the input $u = e^{st}$.

---

[1] Notice that this input is exponential for all time and Eq. (3.14) represents the response for all time. If the system is causal, then $h(t) = 0$ for $t < 0$, and the integral reduces to $H(s) = \int_0^{\infty} h(\tau)e^{-s\tau}\, d\tau$.

Solution. The system equation from Example 3.3 is

$$\dot{y}(t) + ky(t) = u(t) = e^{st}. \tag{3.17}$$

We assume that we can express $y(t)$ as $H(s)e^{st}$. With this form, we have $\dot{y} = sH(s)e^{st}$, and Eq. (3.17) reduces to

$$sH(s)e^{st} + kH(s)e^{st} = e^{st}. \tag{3.18}$$

Solving for the transfer function $H(s)$, we get

$$H(s) = \frac{1}{s+k}.$$

Substituting this back into Eq. (3.13) yields the output

$$y = \frac{e^{st}}{s+k}.$$

Frequency response

A very common way to use the exponential response of LTIs is in finding the **frequency response**, or response to a sinusoid. First we express the sinusoid as a sum of two exponential expressions (Euler's relation):

$$A\cos(\omega t) = \frac{A}{2}(e^{j\omega t} + e^{-j\omega t}).$$

If we let $s = j\omega$ in the basic response formula Eq. (3.13), then the response to $u(t) = e^{j\omega t}$ is $y(t) = H(j\omega)e^{j\omega t}$; similarly, the response to $u(t) = e^{-j\omega t}$ is $H(-j\omega)e^{-j\omega t}$. By superposition, the response to the sum of these two exponentials, which make up the cosine signal, is the sum of the responses:

$$y(t) = \frac{A}{2}[H(j\omega)e^{j\omega t} + H(-j\omega)e^{-j\omega t}]. \tag{3.19}$$

The transfer function $H(j\omega)$ is a complex number that can be represented in polar form or in magnitude-and-phase form as $H(j\omega) = M(\omega)e^{j\varphi(\omega)}$, or simply $H = Me^{j\varphi}$. With this substitution, Eq. (3.19) becomes

$$y(t) = \frac{A}{2}M\left(e^{j(\omega t + \varphi)} + e^{-j(\omega t + \varphi)}\right)$$

$$= AM\cos(\omega t + \varphi), \tag{3.20}$$

where

$$M = |H(j\omega)|, \quad \varphi = \angle H(j\omega).$$

This means that if a system represented by the transfer function $H(s)$ has a sinusoidal input with magnitude A, the output will be sinusoidal at the same frequency with magnitude AM and will be shifted in phase by the angle φ.

EXAMPLE 3.5   *Frequency Response*

For the system in Example 3.1, find the response to the sinusoidal input $u = A\cos(\omega t)$. That is,

a. find the frequency response and plot the response for $k = 1$,
b. determine the complete response due to the sinusoidal input $u(t) = \sin(10t)$ again with $k = 1$.

Solution. In Example 3.4 we found the transfer function. To find the frequency response, we let $s = j\omega$ so that

$$H(s) = \frac{1}{s+k} \implies H(j\omega) = \frac{1}{j\omega + k}.$$

From this we get

$$M = \frac{1}{\sqrt{\omega^2 + k^2}} \quad \text{and} \quad \varphi = -\tan^{-1}\left(\frac{\omega}{k}\right).$$

Therefore, the response of this system to a sinusoid will be

$$y(t) = AM\cos(\omega t + \varphi). \tag{3.21}$$

M is usually referred to as the **amplitude ratio** and φ is referred to as the **phase** and they are both functions of the input frequency, ω. The MATLAB® program that follows is used to compute the amplitude ratio and phase for $k = 1$, as shown in Fig. 3.3. The logspace command is used to set the frequency range (on a logarithmic scale) and the bode command is used to compute the frequency response in MATLAB. Presenting frequency response in this manner (i.e., on a log–log scale) was originated by H. W. Bode; thus, these plots are referred to as "Bode plots."[2] (See Chapter 6, Section 6.1.)

```
k = 1;
numH = 1;                    % form numerator
denH = [1 k];                % form denominator
```

Figure 3.3

Frequency response for $k = 1$

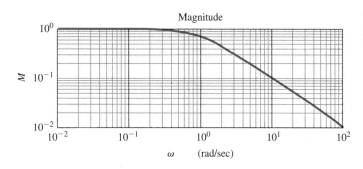

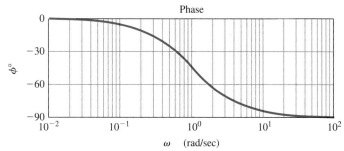

---

[2]Note that % is used in MATLAB to denote comments.

```
sysH = tf(numH,denH);                % define system by its numerator
                                        and denominator
w = logspace(−2,2);                  % set frequency w to 50 values from
                                        10⁻² to 10⁺²
[mag,phase] = bode(sysH,w);          % compute frequency response
loglog(w,squeeze(mag));              % log–log plot of magnitude
semilogx(w,squeeze(phase));          % semi log plot of phase
```

To determine the response to an input that begins at $t = 0$ as $u(t) = \sin(10t)1(t)$, notice that from Laplace transform tables (Appendix A, Table A.2), we have

$$\mathcal{L}\{u(t)\} = \mathcal{L}\{\sin(10t)\} = \frac{10}{s^2 + 100},$$

where $\mathcal{L}$ denotes the Laplace transform, and the output of the system using partial fraction expansion (see Section 3.1.5) is given by

$$Y(s) = H(s)U(s)$$

$$= \frac{1}{s+1}\frac{10}{s^2+100}$$

$$= \frac{\alpha_1}{s+1} + \frac{\alpha_0}{s+j10} + \frac{\alpha_0^*}{s-j10}$$

$$= \frac{\frac{10}{101}}{s+1} + \frac{\frac{j}{2(1-j10)}}{s+j10} + \frac{\frac{-j}{2(1+j10)}}{s-j10}.$$

The inverse Laplace transform of the output is given by (see Appendix A)

$$y(t) = \frac{10}{101}e^{-t} + \frac{1}{\sqrt{101}}\sin(10t + \varphi)$$

$$= y_1(t) + y_2(t),$$

where

$$\varphi = \tan^{-1}(-10) = -84.2°.$$

The component $y_1(t)$ is called the *transient* response as it decays to zero as time goes on and the component $y_2(t)$ is called the *steady state* and equals the response given by Eq. (3.21). Figure 3.4(a) is a plot of the time history of the output showing the different components (y_1, y_2) and the composite (y) output response. The output frequency is 10 rad/sec and the steady-state phase difference measured from Fig. 3.4(b) is approximately $10*\delta t = 1.47$ rad $= 84.2°$[3]. Figure 3.4(b) shows the output lags the input by 84.2°. Figure 3.4(b) shows that the steady-state amplitude of the output is the amplitude ratio $\frac{1}{\sqrt{101}} = 0.0995$ (i.e., the amplitude of the input signal times the magnitude of the transfer function evaluated at $\omega = 10$ rad/sec).

This example illustrates that the response of an LTI system to a sinusoid of frequency ω is a sinusoid with the *same* frequency and with an amplitude ratio equal

---

[3]The phase difference may also be determined by a Lissajous pattern.

Figure 3.4

(a) Complete transient response; (b) phase lag between output and input

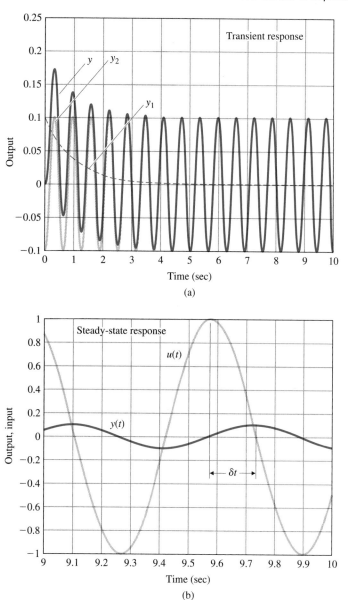

(a)

(b)

to the magnitude of the transfer function evaluated at the input frequency. Furthermore, the phase difference between input and output signals is given by the phase of the transfer function evaluated at the input frequency. The magnitude ratio and phase difference can be computed from the transfer function as just discussed; they can also be measured experimentally quite easily in the laboratory by driving the system with a known sinusoidal input and measuring the steady-state amplitude and phase of the system's output. The input frequency is set to sufficiently many values so that curves such as the one in Fig. 3.3 are obtained.

We can generalize the frequency response by defining the **Laplace transform** of a signal $f(t)$ as

$$F(s) = \int_{-\infty}^{\infty} f(t)e^{-st}\,dt. \tag{3.22}$$

If we apply this definition to both $u(t)$ and $y(t)$ and use the convolution integral Eq. (3.12), we find that

$$Y(s) = H(s)U(s), \tag{3.23}$$

where $Y(s)$ and $U(s)$ are the Laplace transforms of $y(t)$ and $u(t)$, respectively. We prove this result in Appendix A.

Laplace transforms such as Eq. (3.22) can be used to study the complete response characteristics of feedback systems, including the **transient response**—that is, the time response to an initial condition or suddenly applied signal. This is in contrast to the use of Fourier transforms, where the steady-state response is the main concern. A standard problem in control is to find the response $y(t)$ of a system given the input $u(t)$ and a model of the system. With Eq. (3.22) we have a means for computing the response of LTI systems to quite general inputs. Given any input into a system, we compute the transform of the input and the transfer function for the system. The transform of the output is then given by Eq. (3.23) as the product of these two. If we wanted the time function of the output, we would need to "invert" $Y(s)$ to get what is called the **inverse transform**; this step is typically not carried out explicitly. Nevertheless, understanding the process necessary for deriving $y(t)$ from $Y(s)$ is important because it leads to insight into the behavior of linear systems. Hence, given a general linear system with transfer function $H(s)$ and an input signal $u(t)$, the procedure for determining $y(t)$ using the Laplace transform is given by the following steps:

STEP 1. Determine the transfer function: $H(s) = \mathcal{L}\{\text{impulse response of the system}\}$. Compute $H(s)$ by the following steps:

(a) Take the Laplace transform of the equations of motion. A table of transform properties is frequently useful in this process.

(b) Solve the resulting algebraic equations. Often this step is greatly helped by drawing the corresponding block diagram and solving the equations by graphical manipulation of the blocks or using MATLAB.

STEP 2. Determine the Laplace transform of the input signal: $U(s) = \mathcal{L}\{u(t)\}$.

STEP 3. Determine the Laplace transform of the output: $Y(s) = H(s)U(s)$.

STEP 4. Break up $Y(s)$ by partial-fraction expansion.

STEP 5. Find the output of the system by computing the inverse Laplace transform of $Y(s)$ in Step 4, $y(t) = \mathcal{L}^{-1}\{Y(s)\}$ [i.e., invert $Y(s)$ to get $y(t)$]:

(a) Look up the components of $y(t)$ in a table of transform–time function pairs.

(b) Combine the components to give the total solution in the desired form.

As already mentioned, Steps 4 and 5 are almost never carried out in practice, and a modified solution for a *qualitative* rather than a quantitative solution is often adequate and *almost always used for control design purposes*. The process begins with the first three steps as before. However, rather than inverting $Y(s)$, one can use prior knowledge and intuition about the effects of pole and zero locations in $Y(s)$ on the response $y(t)$ to estimate key parameters of $y(t)$. That is, we get information about $y(t)$ from the pole–zero constellation of $Y(s)$ without actually inverting it, as discussed in the rest of this chapter.

While it is possible to determine the transient response properties of the system using Eq. (3.22), it is generally more useful to use a simpler version of the Laplace transform based on the input beginning at time zero.

3.1.3 The $\mathcal{L}_-$ Laplace Transform

Definition of Laplace transform

In many applications it is useful to define a **one-sided** (or **unilateral**) **Laplace transform**, which uses 0^- (that is, a value just before $t = 0$) as the lower limit of integration in Eq. (3.22). The $\mathcal{L}_-$ Laplace transform of $f(t)$, denoted by $\mathcal{L}_-\{f(t)\} = F(s)$, is a function of the complex variable $s = \sigma + j\omega$, where

$$F(s) \triangleq \int_{0^-}^{\infty} f(t)e^{-st}\,dt. \tag{3.24}$$

The decaying exponential term in the integrand in effect provides a built-in convergence factor if $\sigma > 0$. This means that even if $f(t)$ does not vanish as $t \to \infty$, the integrand will vanish for sufficiently large values of σ if f does not grow at a faster-than-exponential rate. The fact that the lower limit of integration is at 0^- allows the use of an impulse function at $t = 0$, as illustrated in Example 3.3; however, this distinction between $t = 0^-$ and $t = 0$ does not usually come up in practice. We will therefore for the most part drop the minus superscript on $t = 0$; however, we will return to using the notation $t = 0^-$ when an impulse at $t = 0$ is involved and the distinction is of practical value.

If Eq. (3.24) is a one-sided transform, then by extension, Eq. (3.22) is a **two-sided Laplace transform**.[4] We will use the $\mathcal{L}$ symbol from here on to mean $\mathcal{L}_-$.

On the basis of the formal definition in Eq. (3.24), we can ascertain the properties of Laplace transforms and compute the transforms of common time functions. The analysis of linear systems by means of Laplace transforms usually involves using tables of common properties and time functions, so we have provided this information in Appendix A. The tables of time functions and their Laplace transforms, together with the table of properties, permit us to find transforms of complex signals from simpler ones. For a thorough study of Laplace transforms and extensive tables, see Churchill (1972) and Campbell and Foster (1948). For more study of the two-sided transform, see Van der Pol and Bremmer (1955). These authors show that the time function can be obtained from the Laplace transform by the inverse relation

$$f(t) = \frac{1}{2\pi j} \int_{\sigma_c - j\infty}^{\sigma_c + j\infty} F(s)e^{st}\,ds, \tag{3.25}$$

where σ_c is a selected value to the right of all the singularities of $F(s)$ in the s-plane. In practice, this relation is seldom used. Instead, complex Laplace transforms are broken down into simpler ones that are listed in the tables along with their corresponding time responses.

---

[4]The other possible one-sided transform is, of course, $\mathcal{L}_+$, in which the lower limit of the integral is 0^+. It is sometimes used in other applications.

Let us compute a few Laplace transforms of some typical time functions.

EXAMPLE 3.6　　　*Step and Ramp Transforms*

Find the Laplace transform of the step $a1(t)$ and ramp $bt1(t)$ functions.

Solution. For a step of size a, $f(t) = a1(t)$, and from Eq. (3.24) we have

$$F(s) = \int_0^\infty ae^{-st}\, dt = \left.\frac{-ae^{-st}}{s}\right|_0^\infty = 0 - \frac{-a}{s} = \frac{a}{s}, \qquad \text{Re}(s) > 0.$$

For the ramp signal $f(t) = bt1(t)$, again from Eq. (3.24) we have

$$F(s) = \int_0^\infty bte^{-st}\, dt = \left[-\frac{bte^{-st}}{s} - \frac{be^{-st}}{s^2} \right]_0^\infty = \frac{b}{s^2}, \qquad \text{Re}(s) > 0,$$

where we employed the technique of integration by parts,

$$\int u\, dv = uv - \int v\, du,$$

with $u = bt$ and $dv = e^{-st}\, dt$. We can then extend the domain of the validity of $F(s)$ to the entire s-plane except at the pole location namely the origin (see Appendix A).

A more subtle example is that of the impulse function.

EXAMPLE 3.7　　　*Impulse Function Transform*

Find the Laplace transform of the unit-impulse function.

Solution. From Eq. (3.24) we get

$$F(s) = \int_{0^-}^\infty \delta(t)e^{-st}\, dt = \int_{0^-}^{0^+} \delta(t)\, dt = 1. \tag{3.26}$$

It is the transform of the unit-impulse function that led us to choose the $\mathcal{L}_-$ transform rather than the $\mathcal{L}_+$ transform.

EXAMPLE 3.8　　　*Sinusoid Transform*

Find the Laplace transform of the sinusoid function.

Solution. Again, we use Eq. (3.24) to get

$$\mathcal{L}\{\sin \omega t\} = \int_0^\infty (\sin \omega t)e^{-st}\, dt. \tag{3.27}$$

If we substitute the relation from Eq. (D.34) in Appendix WD,

$$\sin \omega t = \frac{e^{j\omega t} - e^{-j\omega t}}{2j}$$

into Eq. (3.27), we find that

$$\mathcal{L}\{\sin \omega t\} = \int_0^\infty \left(\frac{e^{j\omega t} - e^{-j\omega t}}{2j} \right) e^{-st} \, dt$$

$$= \frac{1}{2j} \int_0^\infty \left(e^{(j\omega - s)t} - e^{-(j\omega + s)t} \right) dt$$

$$= \frac{1}{2j} \left[\frac{1}{j\omega - s} e^{(j\omega - s)t} - \frac{1}{j\omega + s} e^{-(j\omega + s)t} \right] \Big|_0^\infty$$

$$= \frac{\omega}{s^2 + \omega^2}, \qquad \text{Re}(s) > 0.$$

We can then extend the domain of the validity of computed Laplace transform to the entire s-plane except at the pole locations $s = \pm j\omega$ (see Appendix A).

Table A.2 in Appendix A lists Laplace transforms for elementary time functions. Each entry in the table follows from direct application of the transform definition of Eq. (3.24), as demonstrated by Examples 3.6 to 3.8.

3.1.4 Properties of Laplace Transforms

In this section we will address each of the significant properties of the Laplace transform listed in Table A.1. For the proofs of these properties and related examples as well as the Initial Value Theorem, the reader is referred to Appendix A.

1. Superposition

One of the more important properties of the Laplace transform is that it is linear:

$$\mathcal{L}\{\alpha f_1(t) + \beta f_2(t)\} = \alpha F_1(s) + \beta F_2(s). \tag{3.28}$$

The amplitude scaling property is a special case of this; that is,

$$\mathcal{L}\{\alpha f(t)\} = \alpha F(s). \tag{3.29}$$

2. Time Delay

Suppose a function $f(t)$ is delayed by $\lambda > 0$ units of time. Its Laplace transform is

$$F_1(s) = \int_0^\infty f(t - \lambda)e^{-st} \, dt = e^{-s\lambda} F(s). \tag{3.30}$$

From this result we see that a time delay of λ corresponds to multiplication of the transform by $e^{-s\lambda}$.

3. Time Scaling

It is sometimes useful to time-scale equations of motion. For example, in the control system of a disk drive, it is meaningful to measure time in milliseconds (see also Chapter 10). If the time t is scaled by a factor a, then the Laplace transform of the time-scaled signal is

$$F_1(s) = \int_0^\infty f(at)e^{-st} \, dt = \frac{1}{|a|} F\left(\frac{s}{a} \right). \tag{3.31}$$

4. Shift in Frequency

Multiplication (modulation) of $f(t)$ by an exponential expression in the time domain corresponds to a shift in frequency:

$$F_1(s) = \int_0^\infty e^{-at} f(t) e^{-st}\, dt = F(s+a). \tag{3.32}$$

5. Differentiation

The transform of the derivative of a signal is related to its Laplace transform and its initial condition as follows:

$$\mathcal{L}\left\{\frac{df}{dt}\right\} = \int_{0^-}^\infty \left(\frac{df}{dt}\right) e^{-st}\, dt = -f(0^-) + sF(s). \tag{3.33}$$

Another application of Eq. (3.33) leads to

$$\mathcal{L}\{\ddot{f}\} = s^2 F(s) - sf(0^-) - \dot{f}(0^-). \tag{3.34}$$

Repeated application of Eq. (3.33) leads to

$$\mathcal{L}\{f^m(t)\} = s^m F(s) - s^{m-1} f(0^-) - s^{m-2}\dot{f}(0^-) - \cdots - f^{(m-1)}(0^-), \tag{3.35}$$

where $f^m(t)$ denotes the mth derivative of $f(t)$ with respect to time.

6. Integration

Let us assume that we wish to determine the Laplace transform of the integral of a time function $f(t)$; that is,

$$F_1(s) = \mathcal{L}\left\{\int_0^t f(\xi)\, d\xi\right\} = \frac{1}{s} F(s), \tag{3.36}$$

which means that we simply multiply the function's Laplace transform by $\frac{1}{s}$.

7. Convolution

We have seen previously that the response of a system is determined by convolving the input with the impulse response of the system, or by forming the product of the transfer function and the Laplace transform of the input. The discussion that follows extends this concept to various time functions.

Convolution in the time domain corresponds to multiplication in the frequency domain. Assume that $\mathcal{L}\{f_1(t)\} = F_1(s)$ and $\mathcal{L}\{f_2(t)\} = F_2(s)$. Then

$$\mathcal{L}\{f_1(t) * f_2(t)\} = \int_0^\infty f_1(t) * f_2(t) e^{-st}\, dt = F_1(s)F_2(s). \tag{3.37}$$

This implies that

$$\mathcal{L}^{-1}\{F_1(s)F_2(s)\} = f_1(t) * f_2(t). \tag{3.38}$$

A similar, or dual, of this result is discussed next.

8. Time Product

Multiplication in the time domain corresponds to convolution in the frequency domain:

$$\mathcal{L}\{f_1(t)f_2(t)\} = \frac{1}{2\pi j} F_1(s) * F_2(s). \tag{3.39}$$

9. Multiplication by Time

Multiplication by time corresponds to differentiation in the frequency domain:

$$\mathcal{L}\{tf(t)\} = -\frac{d}{ds}F(s). \tag{3.40}$$

3.1.5 Inverse Laplace Transform by Partial-Fraction Expansion

The easiest way to find $f(t)$ from its Laplace transform $F(s)$, if $F(s)$ is rational, is to expand $F(s)$ as a sum of simpler terms that can be found in the tables. The basic tool for performing this operation is called **partial-fraction expansion**. Consider the general form for the rational function $F(s)$ consisting of the ratio of two polynomials:

$$F(s) = \frac{b_1 s^m + b_2 s^{m-1} + \cdots + b_{m+1}}{s^n + a_1 s^{n-1} + \cdots + a_n}. \tag{3.41}$$

By factoring the polynomials, this same function could also be expressed in terms of the product of factors as

$$F(s) = K\frac{\Pi_{i=1}^{m}(s - z_i)}{\Pi_{i=1}^{n}(s - p_i)}. \tag{3.42}$$

We will discuss the simple case of distinct poles here. For a transform $F(s)$ representing the response of any physical system, $m \leq n$. When $s = z_i$, s is referred to as a **zero** of the function, and when $s = p_i$, s is referred to as a **pole** of the function. Assuming for now that the poles $\{p_i\}$ are real or complex but distinct, we rewrite $F(s)$ as the partial fraction

Zeros and poles

$$F(s) = \frac{C_1}{s - p_1} + \frac{C_2}{s - p_2} + \cdots + \frac{C_n}{s - p_n}. \tag{3.43}$$

Next, we determine the set of constants $\{C_i\}$. We multiply both sides of Eq. (3.43) by the factor $s - p_1$ to get

$$(s - p_1)F(s) = C_1 + \frac{s - p_1}{s - p_2}C_2 + \cdots + \frac{(s - p_1)C_n}{s - p_n}. \tag{3.44}$$

If we let $s = p_1$ on both sides of Eq. (3.44), then all the C_i terms will equal zero except for the first one. For this term,

$$C_1 = (s - p_1)F(s)|_{s=p_1}. \tag{3.45}$$

The other coefficients can be expressed in a similar form:

$$C_i = (s - p_i)F(s)|_{s=p_i}.$$

The cover-up method of determining coefficients

This process is called the **cover-up method** because, in the factored form of $F(s)$ [Eq. (3.42)], we can cover up the individual denominator terms, evaluate the rest of the expression with $s = p_i$, and determine the coefficients C_i. Once this has been completed, the time function becomes

$$f(t) = \sum_{i=1}^{n} C_i e^{p_i t} 1(t)$$

because, as entry 7 in Table A.2 shows, if

$$F(s) = \frac{1}{s - p_i},$$

then

$$f(t) = e^{p_i t} 1(t).$$

For the cases of quadratic factors or repeated roots in the denominator, see Appendix A.

EXAMPLE 3.9   *Partial-Fraction Expansion: Distinct Real Roots*

Suppose you have computed $Y(s)$ and found that

$$Y(s) = \frac{(s + 2)(s + 4)}{s(s + 1)(s + 3)}.$$

Find $y(t)$.

Solution.   We may write $Y(s)$ in terms of its partial-fraction expansion:

$$Y(s) = \frac{C_1}{s} + \frac{C_2}{s + 1} + \frac{C_3}{s + 3}.$$

Using the cover-up method, we get

$$C_1 = \frac{(s + 2)(s + 4)}{(s + 1)(s + 3)}\bigg|_{s=0} = \frac{8}{3}.$$

In a similar fashion,

$$C_2 = \frac{(s + 2)(s + 4)}{s(s + 3)}\bigg|_{s=-1} = -\frac{3}{2}$$

and

$$C_3 = \frac{(s + 2)(s + 4)}{s(s + 1)}\bigg|_{s=-3} = -\frac{1}{6}.$$

We can check the correctness of the result by adding the components again to verify that the original function has been recovered. With the partial fraction the solution can be looked up in the tables at once to be

$$y(t) = \frac{8}{3}1(t) - \frac{3}{2}e^{-t}1(t) - \frac{1}{6}e^{-3t}1(t).$$

The partial fraction expansion may be computed using the residue function in MATLAB:

```
num = conv([1 2],[1 4]);        % form numerator polynomial
den = conv([1 1 0],[1 3]);      % form denominator polynomial
[r,p,k] = residue(num,den);     % compute the residues
```

which yields the result

r = [−0.1667 −1.5000 2.6667]';   p = [−3 −1 0]';   k = [ ];

and agrees with the hand calculations. Note that the conv function in MATLAB is used to multiply two polynomials. (The arguments of the function are the polynomial coefficients.)

3.1.6 The Final Value Theorem

An especially useful property of the Laplace transform in control known as the **Final Value Theorem** allows us to compute the constant steady-state value of a time function given its Laplace transform. The theorem follows from the development of partial-fraction expansion. Suppose we have a transform $Y(s)$ of a signal $y(t)$ and wish to know the final value $y(t)$ from $Y(s)$. There are three possibilities for the limit. It can be constant, undefined, or unbounded. If $Y(s)$ has any poles (i.e., denominator roots, as described in Section 3.1.5) in the right half of the s-plane—that is, if the real part of any $p_i > 0$—then $y(t)$ will grow and the limit will be unbounded. If $Y(s)$ has a pair of poles on the imaginary axis of the s-plane (i.e., $p_i = \pm j\omega$), then $y(t)$ will contain a sinusoid that persists forever and the final value will not be defined. Only one case can provide a nonzero constant final value: If all poles of $Y(s)$ are in the left half of the s-plane, except for one at $s = 0$, then all terms of $y(t)$ will decay to zero except the term corresponding to the pole at $s = 0$, and that term corresponds to a constant in time. Thus, the final value is given by the coefficient associated with the pole at $s = 0$. Therefore, the Final Value Theorem is as follows:

The Final Value Theorem

If all poles of $sY(s)$ are in the left half of the s-plane, then

$$\lim_{t \to \infty} y(t) = \lim_{s \to 0} sY(s). \qquad (3.46)$$

This relationship is proved in Appendix A.

EXAMPLE 3.10   *Final Value Theorem*

Find the final value of the system corresponding to

$$Y(s) = \frac{3(s+2)}{s(s^2 + 2s + 10)}.$$

Solution.   Applying the Final Value Theorem, we obtain

$$y(\infty) = sY(s)|_{s=0} = \frac{3 \cdot 2}{10} = 0.6.$$

Thus, after the transients have decayed to zero, $y(t)$ will settle to a constant value of 0.6.

Use the Final Value Theorem on stable systems only

Care must be taken to apply the Final Value Theorem only to stable systems (see Section 3.6). While one could use Eq. (3.46) on any $Y(s)$, doing so could result in erroneous results, as shown in the next example.

EXAMPLE 3.11   *Incorrect Use of the Final Value Theorem*

Find the final value of the signal corresponding to

$$Y(s) = \frac{3}{s(s-2)}.$$

Solution. If we blindly apply Eq. (3.46), we obtain

$$y(\infty) = sY(s)|_{s=0} = -\frac{3}{2}.$$

However,

$$y(t) = \left(-\frac{3}{2} + \frac{3}{2}e^{2t}\right)1(t),$$

and Eq. (3.46) yields the constant term only. Of course, the true final value is unbounded.

Calculating DC gain by the Final Value Theorem

The theorem can also be used to find the DC gain of a system. The **DC gain** is the ratio of the output of a system to its input (presumed constant) after all transients have decayed. To find the DC gain, we assume that there is a unit-step input [$U(s) = 1/s$] and we use the Final Value Theorem to compute the steady-state value of the output. Therefore, for a system transfer function $G(s)$,

$$\text{DC gain} = \lim_{s \to 0} sG(s)\frac{1}{s} = \lim_{s \to 0} G(s). \tag{3.47}$$

EXAMPLE 3.12   *DC Gain*

Find the DC gain of the system whose transfer function is

$$G(s) = \frac{3(s+2)}{(s^2 + 2s + 10)}.$$

Solution. Applying Eq. (3.47), we get

$$\text{DC gain} = G(s)|_{s=0} = \frac{3 \cdot 2}{10} = 0.6.$$

3.1.7 Using Laplace Transforms to Solve Problems

Laplace transforms can be used to solve differential equations using the properties described in Appendix A. First, we find the Laplace transform of the differential equation using the differentiation properties in Eqs. (A.12) and (A.13) in Appendix A. Then we find the Laplace transform of the output; using partial-fraction expansion and Table A.2, this can be converted to a time response function. We will illustrate this with three examples.

EXAMPLE 3.13   *Homogeneous Differential Equation Solution*

Find the solution to the differential equation

$$\ddot{y}(t) + y(t) = 0, \qquad \text{where}\quad y(0) = \alpha,\ \dot{y}(0) = \beta.$$

Solution. Using Eq. (3.34), the Laplace transform of the differential equation is

$$s^2Y(s) - \alpha s - \beta + Y(s) = 0,$$

$$(s^2 + 1)Y(s) = \alpha s + \beta,$$

$$Y(s) = \frac{\alpha s}{s^2 + 1} + \frac{\beta}{s^2 + 1}.$$

After looking up in the transform tables (Table A.2, Appendix A) the two terms on the right side of the preceding equation, we get

$$y(t) = [\alpha \cos t + \beta \sin t]1(t),$$

where $1(t)$ denotes a unit step function. We can verify that this solution is correct by substituting it back into the differential equation.

Another example will illustrate the solution when the equations are not homogeneous—that is, when the system is forced.

EXAMPLE 3.14

Forced Differential Equation Solution

Find the solution to the differential equation $\ddot{y}(t) + 5\dot{y}(t) + 4y(t) = 3$, where $y(0) = \alpha$, $\dot{y}(0) = \beta$.

Solution. Taking the Laplace transform of both sides using Eqs. (3.33) and (3.34), we get

$$s^2 Y(s) - s\alpha - \beta + 5[sY(s) - \alpha] + 4Y(s) = \frac{3}{s}.$$

Solving for $Y(s)$ yields

$$Y(s) = \frac{s(s\alpha + \beta + 5\alpha) + 3}{s(s+1)(s+4)}.$$

The partial-fraction expansion using the cover-up method is

$$Y(s) = \frac{\frac{3}{4}}{s} - \frac{\frac{3-\beta-4\alpha}{3}}{s+1} + \frac{\frac{3-4\alpha-4\beta}{12}}{s+4}.$$

Therefore, the time function is given by

$$y(t) = \left(\frac{3}{4} + \frac{-3+\beta+4\alpha}{3}e^{-t} + \frac{3-4\alpha-4\beta}{12}e^{-4t} \right) 1(t).$$

By differentiating this solution twice and substituting the result in the original differential equation, we can verify that this solution satisfies the differential equation.

The solution is especially simple if the initial conditions are all zero.

EXAMPLE 3.15

Forced Equation Solution with Zero Initial Conditions

Find the solution to $\ddot{y}(t) + 5\dot{y}(t) + 4y(t) = u(t)$, $y(0) = 0$, $\dot{y}(0) = 0$, $u(t) = 2e^{-2t}1(t)$,

1. using partial-fraction expansion and
2. using MATLAB.

Solution

1. Taking the Laplace transform of both sides, we get

$$s^2 Y(s) + 5sY(s) + 4Y(s) = \frac{2}{s+2}.$$

Solving for $Y(s)$ yields

$$Y(s) = \frac{2}{(s+2)(s+1)(s+4)}.$$

The partial-fraction expansion using the cover-up method is

$$Y(s) = -\frac{1}{s+2} + \frac{\frac{2}{3}}{s+1} + \frac{\frac{1}{3}}{s+4}.$$

Therefore, the time function is given by

$$y(t) = \left(-1e^{-2t} + \frac{2}{3}e^{-t} + \frac{1}{3}e^{-4t} \right) 1(t).$$

2. The partial-fraction expansion may also be computed using the MATLAB residue function,

```
num = 2;                    % form numerator
den = poly([−2;−1;−4]);     % form denominator polynomial from its roots
[r,p,k] = residue(num,den); % compute the residues
```

which results in the desired answer

```
r = [0.3333 −1 0.6667]';      p = [−4 −2 −1]';      k = [ ];
```

and agrees with the hand calculations.

Poles indicate response character

The primary value of using the Laplace transform method of solving differential equations is that it provides information concerning the qualitative characteristic behavior of the response. Once we know the values of the poles of $Y(s)$, we know what kind of characteristic terms will appear in the response. In the last example the pole at $s = -1$ produced a decaying $y = Ce^{-t}$ term in the response. The pole at $s = -4$ produced a $y = Ce^{-4t}$ term in the response, which decays faster. If there had been a pole at $s = +1$, there would have been a growing $y = Ce^{+t}$ term in the response. Using the pole locations to understand in essence how the system will respond is a powerful tool and will be developed further in Section 3.3. Control systems designers often manipulate design parameters so that the poles have values that would give acceptable responses, and they skip the steps associated with converting those poles to actual time responses until the final stages of the design. They use trial-and-error design methods (as described in Chapter 5) that graphically present how changes in design parameters affect the pole locations. Once a design has been obtained, with pole locations predicted to give acceptable responses, the control designer determines a time response to verify that the design is satisfactory. This is typically done by computer, which solves the differential equations directly by using numerical computer methods.

3.1.8 Poles and Zeros

A rational transfer function can be described either as a ratio of two polynomials in s,

$$H(s) = \frac{b_1 s^m + b_2 s^{m-1} + \cdots + b_{m+1}}{s^n + a_1 s^{n-1} + \cdots + a_n} = \frac{N(s)}{D(s)}, \tag{3.48}$$

or as a ratio in factored zero pole form

$$H(s) = K \frac{\prod_{i=1}^{m}(s - z_i)}{\prod_{i=1}^{n}(s - p_i)}. \tag{3.49}$$

K is called the transfer function gain. The roots of the numerator $z_1, z_2,..., z_m$ are called the finite **zeros** of the system. The zeros are locations in the s-plane where the transfer function is zero. If $s = z_i$, then

Zeros

$$H(s)|_{s=z_i} = 0.$$

The zeros also correspond to the signal transmission-blocking properties of the system and are also called the transmission zeros of the system. The system has the inherent capability to block frequencies coinciding with its zero locations. If we excite the system with the nonzero input, $u = u_0 e^{s_0 t}$, where s_0 is not a pole of the system, then the output is identically zero,[5] $y \equiv 0$, for frequencies where $s_0 = z_i$. The zeros also have a significant effect on the transient properties of the system (see Section 3.5).

Poles

The roots of the denominator, $p_1, p_2, \ldots, p_n$ are called the **poles**[6] of the system. The poles are locations in the s-plane where the magnitude of the transfer function becomes infinite. If $s = p_i$, then

$$|H(s)|_{s=p_i} = \infty.$$

The poles of the system determine its stability properties, as we shall see in Section 3.6. The poles of the system also determine the natural or unforced behavior of the system, referred to as the **modes** of the system. The zeros and poles may be complex quantities, and we may display their locations in a complex plane, which we refer to as the s-plane. The locations of the poles and zeros lie at the heart of feedback control design and have significant practical implications for control system design. The system is said to have $n - m$ zeros at infinity if $m < n$ because the transfer function approaches zero as s approaches infinity. If the zeros at infinity are also counted, the system will have the same number of poles and zeros. No physical system can have $n < m$; otherwise, it would have an infinite response at $\omega = \infty$. If $z_i = p_j$, then there are *cancellations* in the transfer function, which may lead to undesirable system properties as discussed in Chapter 7.

3.1.9 Linear System Analysis Using MATLAB

The first step in analyzing a system is to write down (or generate) the set of time-domain differential equations representing the dynamic behavior of the physical system. These equations are generated from the physical laws governing the system behavior—for example, rigid body dynamics, thermo-fluid mechanics, and electromechanics, as described in Chapter 2. The next step in system analysis is to determine and designate inputs and outputs of the system and then to compute the

---

[5]Identically zero means that the output and all of its derivatives are zero for $t > 0$.

[6]The meaning of the pole can also be appreciated by visualizing a 3-D plot of the transfer function, where the real and imaginary parts of s are plotted on the x and y axes, and the magnitude of the transfer function is plotted on the vertical z axis. For a single pole, the resulting 3-D plot will look like a tent with the "tent-pole" being located at the pole of the transfer function!

transfer function characterizing the input–output behavior of the dynamic system. Earlier in this chapter we saw that a linear dynamic system may also be represented by the Laplace transform of its differential equation—that is, its transfer function. The transfer function may be expressed as a ratio of two polynomials as in Eq. (3.48) or in factored zero–pole form as in Eq. (3.49). By analyzing the transfer function, we can determine the dynamic properties of the system, both in a qualitative and quantitative manner. One way of extracting useful system information is simply to determine the pole–zero locations and deduce the essential characteristics of the dynamic properties of the system. Another way is to determine the time-domain properties of the system by determining the response of the system to typical excitation signals such as impulses, steps, ramps, and sinusoids. Yet another way is to determine the time response analytically by computing the inverse Laplace transform using partial-fraction expansions and Tables A.1 and A.2. Of course, it is also possible to determine the system response to an arbitrary input.

We will now illustrate this type of analysis by carrying out the preceding calculations for some of the physical systems addressed in the examples in Chapter 2 in order of increasing degree of difficulty. We will go back and forth between the different representations of the system, transfer function, and pole–zero, etc., using MATLAB as our computational engine. MATLAB typically accepts the specification of a system in several forms, including transfer function and zero–pole, and refers to these two descriptions as tf and zp, respectively. Furthermore, it can transform the system description from any one form to another.

EXAMPLE 3.16

Cruise Control Transfer Function Using MATLAB

Find the transfer function between the input u and the position of the car x in the cruise control system in Example 2.1.

Solution. From Example 2.1 we find that the transfer function of the system is

$$H(s) = \frac{0s^2 + 0s + 0.001}{s^2 + 0.05s + 0} = \frac{0.001}{s(s + 0.05)}.$$

In MATLAB, the coefficients of the numerator polynomial are displayed as the row vector num and the denominator coefficients are displayed as den. The results for this example are

$$\text{num} = [0 \quad 0 \quad 0.001] \quad \text{and} \quad \text{den} = [1 \quad 0.05 \quad 0].$$

MATLAB printsys

They can be returned by MATLAB in this form using the printsys(num,den) command. The pole–zero description is computed using the MATLAB command

$$[z, p, k] = \text{tf2zp(num, den)}$$

and would result in the transfer function in factored form, where $z = [\ ]$, $p = [0 \quad -0.05]'$, and $k = 0.001$.

EXAMPLE 3.17

DC Motor Transfer Function Using MATLAB

In Example 2.13, assume that $J_m = 0.01$ kg·m², $b = 0.001$ N·m·sec, $K_t = K_e = 1$, $R_a = 10$ Ω, and $L_a = 1$ H. Find the transfer function between the input v_a and

1. the output θ_m,
2. the output $\omega = \dot{\theta}_m$.

Solution

1. Substituting the preceding parameters into Example 2.13, we find that the transfer function of the system is

$$H(s) = \frac{100}{s^3 + 10.1s^2 + 101s}.$$

In MATLAB we display the coefficients of the numerator polynomial as the row vector numa and the denominator as dena. The results for this example are

$$\text{numa} = [\ 0 \quad 0 \quad 0 \quad 100\ ] \quad \text{and} \quad \text{dena} = [\ 1 \quad 10.1 \quad 101 \quad 0\ ].$$

The pole–zero description is computed using the MATLAB command

$$[z, p, k] = \text{tf2zp}(\text{numa}, \text{dena})$$

which results in

$$z = [\ ], \quad p = \begin{bmatrix} 0 & -5.0500 & +8.6889j & -5.0500 & -8.6889j \end{bmatrix}', \quad k = 100,$$

and yields the transfer function in factored form:

$$H(s) = \frac{100}{s(s + 5.05 + j8.6889)(s + 5.05 - j8.6889)}.$$

2. If we consider the velocity $\dot{\theta}_m$ as the output, then we find numb=[0 0 100], denb=[1 10.1 101], which tells us that the transfer function is

$$G(s) = \frac{100s}{s^3 + 10.1s^2 + 101s} = \frac{100}{s^2 + 10.1s + 101}.$$

This is as expected, because $\dot{\theta}_m$ is simply the derivative of θ_m; thus $\mathcal{L}\{\dot{\theta}_m\} = s\mathcal{L}\{\theta_m\}$. For a unit step command in v_a, we can compute the step response in MATLAB (recall Example 2.1):

```
numb=[0 0 100];          % form numerator
denb=[1 10.1 101];       % form denominator
sysb=tf(numb,denb);      % define system by its numerator and denominator
t=0:0.01:5;              % form time vector
y=step(sysb,t)           % compute step response;
plot(t,y)                % plot step response
```

The system yields a steady-state constant angular velocity as shown in Fig. 3.5. Note that there is a slight offset, since the system does not have unity DC gain.

When a dynamic system is represented by a single differential equation of any order, finding the polynomial form of the transfer function from that differential equation is usually easy. Therefore, you will find it best in these cases to specify a system directly in terms of its transfer function.

Figure 3.5

Transient response for DC motor

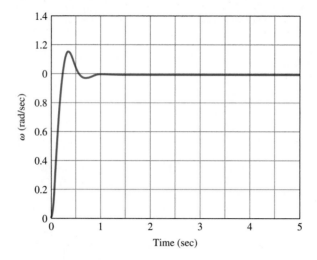

EXAMPLE 3.18   *Transformations Using MATLAB*

Find the transfer function of the system whose differential equation is
$$\ddot{y} + 6\dot{y} + 25y = 9u + 3\dot{u}.$$

Solution. Using the differentiation rules given by Eqs. (3.33) and (3.34), we see by inspection that
$$\frac{Y(s)}{U(s)} = \frac{3s + 9}{s^2 + 6s + 25}.$$
The MATLAB statements are

```
numG = [3 9];       % form numerator
denG = [1 6 25];    % form denominator
```

If the transfer function was desired in factored form, it could be obtained by transforming the tf description. Therefore, the MATLAB statement

```
% convert from numerator-denominator polynomials to pole–zero form
[z,p,k] = tf2zp(numG,denG)
```

would result in $z = -3$, $p = [-3 + 4j \quad -3 - 4j]'$, $k = 3$. This means that the transfer function could also be written as
$$\frac{Y(s)}{U(s)} = \frac{3(s + 3)}{(s + 3 - 4j)(s + 3 + 4j)}.$$
We may also convert from zero–pole representation to the transfer function representation using the MATLAB zp2tf command

```
% convert from pole–zero form to numerator-denominator polynomials
[numG,denG]=zp2tf(z,p,k)
```

For this example, z=[−3], p=[−3+i*4;−3−i*4], k=[3] will yield the numerator and denominator polynomials.

EXAMPLE 3.19   *Satellite Transfer Function Using MATLAB*

1. Find the transfer function between the input F_c and the satellite attitude θ in Example 2.3 and
2. Determine the response of the system to a 25-N pulse of 0.1 sec duration, starting at $t = 5$ sec. Let $d = 1$ m and $I = 5000$ kg-m^2.

Solution

1. From Example 2.3, $\frac{d}{I} = \frac{1}{5000} = 0.0002 \left[\frac{\text{m}}{\text{kg-m}^2} \right]$ and this means that the transfer function of the system is

$$H(s) = \frac{0.0002}{s^2},$$

which can also be determined by inspection for this particular case. We may display the coefficients of the numerator polynomial as the row vector num and the denominator as the row vector den. The results for this example are

$$\text{numG} = [0 \quad 0 \quad 0.0002] \quad \text{and} \quad \text{denG} = [1 \quad 0 \quad 0].$$

2. The following MATLAB statements compute the response of the system to a 25-N, 0.1-sec duration thrust pulse input:

```
numG=[0 0 0.0002];              % form the transfer function
denG=[1 0 0];
sysG=tf(numG,denG);             % define system by its transfer
                                  function

t=0:0.01:10;                    % set up time vector with dt = 0.01 sec
% pulse of 25N, at 5 sec, for 0.1 sec
duration
u1=[zeros(1,500) 25*ones(1,10)
zeros(1,491)];                  % pulse input
[y1]=lsim(sysG,u1,t);           % linear simulation
ff=180/pi;                      % conversion factor from radians
                                  to degrees
y1=ff*y1;                       % output in degrees
plot(t,u1);                     % plot input signal
plot(t,y1);                     % plot output response
```

The system is excited with a short pulse (an impulsive input) that has the effect of imparting a nonzero angle θ_0 at time $t = 5$ sec on the system. Because the system is undamped, in the absence of any control it drifts with constant angular velocity with a value imparted by the impulse at $t = 5$ sec. The time response of the input is shown in Fig. 3.6(a) along with the drift in angle θ in Fig. 3.6(b).

We now excite the system with the same positive-magnitude thrust pulse at time $t = 5$ sec but follow that with a negative pulse with the same magnitude and duration at time $t = 6.1$ sec. [See Figure 3.7(a) for the input thrust.] Then the attitude response of the system is as shown in Figure 3.7(b). This is actually how the satellite attitude angle is controlled in practice. The additional relevant MATLAB statements are

Figure 3.6

Transient response for satellite: (a) thrust input; (b) satellite attitude

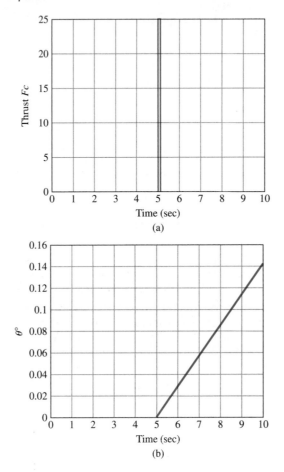

(a)

(b)

```
% double pulse input
u2=[zeros(1,500) 25*ones(1,10) zeros(1,100) −25*ones(1,10) zeros(1,381)];
[y2]=lsim(sysG,u2,t);      % linear simulation
plot(t,u2);                % plot input signal
ff=180/pi;                 % conversion factor from radians to degrees
y2=ff*y2;                  % output in degrees
plot(t,y2);                % plot output response
```

3.2 System Modeling Diagrams

3.2.1 The Block Diagram

To obtain the transfer function, we need to find the Laplace transform of the equations of motion and solve the resulting algebraic equations for the relationship between the input and the output. In many control systems the system equations can be written so that their components do not interact except by having the input of one part be the

Figure 3.7

Transient response for satellite (double-pulse): (a) thrust input; (b) satellite attitude

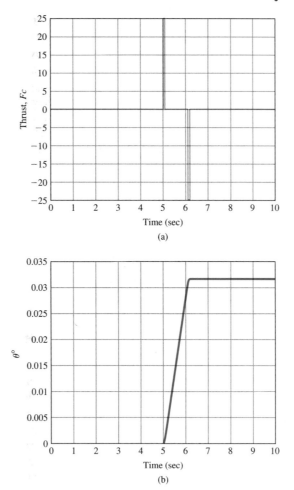

output of another part. In these cases, it is easy to draw a block diagram that represents the mathematical relationships in a manner similar to that used for the component block diagram in Fig. 1.2, Chapter 1. The transfer function of each component is placed in a box, and the input–output relationships between components are indicated by lines and arrows. We can then solve the equations by graphical simplification, which is often easier and more informative than algebraic manipulation, even though the methods are in every way equivalent. Drawings of three elementary block diagrams are seen in Fig. 3.8. It is convenient to think of each block as representing an electronic amplifier with the transfer function printed inside. The interconnections of blocks include summing points, where any number of signals may be added together. These are represented by a circle with the symbol Σ inside. In Fig. 3.8(a) the block with transfer function $G_1(s)$ is in series with the block with transfer function $G_2(s)$, and the overall transfer function is given by the product $G_2 G_1$. In Fig. 3.8(b) two systems are in parallel with their outputs added, and the overall transfer function is given by the sum $G_1 + G_2$. These diagrams derive simply from the equations that describe them.

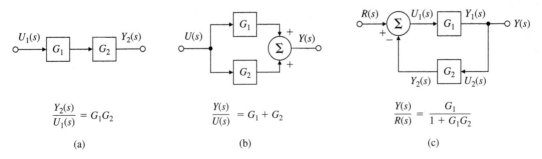

$$\frac{Y_2(s)}{U_1(s)} = G_1 G_2$$

(a)

$$\frac{Y(s)}{U(s)} = G_1 + G_2$$

(b)

$$\frac{Y(s)}{R(s)} = \frac{G_1}{1 + G_1 G_2}$$

(c)

Figure 3.8

Three examples of elementary block diagrams

Negative feedback

Figure 3.8(c) shows a more complicated case. Here the two blocks are connected in a feedback arrangement so that each feeds into the other. When the feedback $Y_2(s)$ is *subtracted*, as shown in the figure, we call it **negative feedback**. As you will see, negative feedback is usually required for system stability. For now we will simply solve the equations and then relate them back to the diagram. The equations are

$$U_1(s) = R(s) - Y_2(s),$$

$$Y_2(s) = G_2(s)G_1(s)U_1(s),$$

$$Y_1(s) = G_1(s)U_1(s),$$

and their solution is

$$Y_1(s) = \frac{G_1(s)}{1 + G_1(s)G_2(s)}R(s). \tag{3.50}$$

We can express the solution by the following rule:

> The gain of a single-loop negative feedback system is given by the forward gain divided by the sum of 1 plus the loop gain.

Positive feedback

When the feedback is added instead of subtracted, we call it **positive feedback**. In this case, the gain is given by the forward gain divided by the sum of 1 minus the loop gain.

The three elementary cases given in Fig. 3.8 can be used in combination to solve, by repeated reduction, any transfer function defined by a block diagram. However, the manipulations can be tedious and subject to error when the topology of the diagram is complicated. Figure 3.9 shows examples of block-diagram algebra that complement those shown in Fig. 3.8. Figures 3.9(a) and (b) show how the interconnections of a block diagram can be manipulated without affecting the mathematical relationships. Figure 3.9(c) shows how the manipulations can be used to convert a general system (on the left) to a system without a component in the feedback path, usually referred to as a **unity feedback system.**

Unity feedback system

In all cases the basic principle is to simplify the topology while maintaining exactly the same relationships among the remaining variables of the block diagram. In relation to the algebra of the underlying linear equations, block-diagram reduction is a pictorial way to solve equations by eliminating variables.

Figure 3.9
Examples of
block-diagram algebra

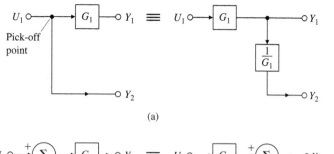

(a)

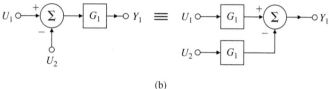

(b)

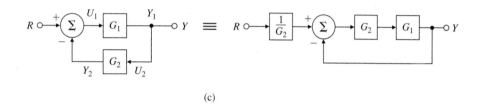

(c)

Figure 3.10
Block diagram of a
second-order system

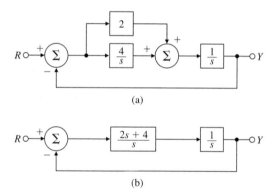

(a)

(b)

EXAMPLE 3.20   *Transfer Function from a Simple Block Diagram*

Find the transfer function of the system shown in Fig. 3.10(a).

Solution. First we simplify the block diagram by reducing the parallel combination of the controller path. This results in the diagram of Fig. 3.10(b), and we use the feedback rule to obtain the closed-loop transfer function:

$$T(s) = \frac{Y(s)}{R(s)} = \frac{\frac{2s+4}{s^2}}{1 + \frac{2s+4}{s^2}} = \frac{2s + 4}{s^2 + 2s + 4}.$$

EXAMPLE 3.21   *Transfer Function from the Block Diagram*

Find the transfer function of the system shown in Fig. 3.11(a).

Solution. First we simplify the block diagram. Using the principles of Eq. (3.50), we replace the feedback loop involving G_1 and G_3 by its equivalent transfer function, noting that it is a positive feedback loop. The result is Fig. 3.11(b). The next step is to move the pick-off point preceding G_2 to its output [see Fig. 3.11(a)], as shown in Fig. 3.11(c). The negative feedback loop on the left is in series with the subsystem on the right, which is composed of the two parallel blocks G_5 and G_6/G_2. The overall transfer function can be written using all three rules for reduction given by Fig. 3.8:

$$T(s) = \frac{Y(s)}{R(s)} = \frac{\frac{G_1 G_2}{1 - G_1 G_3}}{1 + \frac{G_1 G_2 G_4}{1 - G_1 G_3}} \left(G_5 + \frac{G_6}{G_2} \right)$$

$$= \frac{G_1 G_2 G_5 + G_1 G_6}{1 - G_1 G_3 + G_1 G_2 G_4}.$$

Figure 3.11

Example for block-diagram simplification

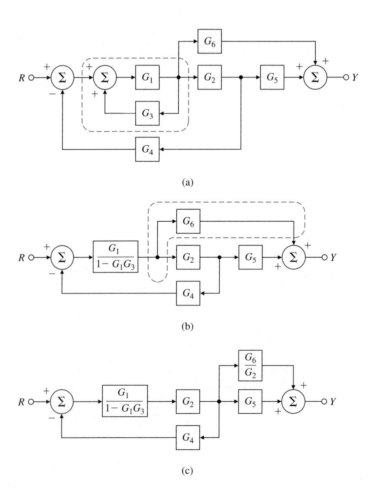

(a)

(b)

(c)

As we have seen, a system of algebraic equations may be represented by a block diagram that represents individual transfer functions by blocks and has interconnections that correspond to the system equations. A block diagram is a convenient tool to visualize the system as a collection of interrelated subsystems that emphasize the relationships among the system variables.

3.2.2   Block Diagram Reduction Using MATLAB

If the individual transfer functions are available for components in a control system, it is possible to use MATLAB commands to compute the transfer functions of interconnected systems. The three commands series, parallel, and feedback can be used for this purpose. They compute the transfer functions of two component block transfer functions in series, parallel, and feedback configurations, respectively. The next simple example illustrates their use.

EXAMPLE 3.22   *Transfer Function of a Simple System Using MATLAB*

Repeat the computation of the transfer function for the block diagram in Fig. 3.10(a) using MATLAB.

Solution.   We label the transfer function of the separate blocks shown in Fig. 3.10(a) as illustrated in Fig. 3.12. Then we combine the two parallel blocks G_1 and G_2 by

```
num1=[2];                        % form G1
den1=[1];
sysG1=tf(num1,den1);             % define subsystem G1
num2=[4];                        % form G2
den2=[1 0];
sysG2=tf(num2,den2);             % define subsystem G2
% parallel combination of G1 and G2 to form subsystem G3
sysG3=parallel(sysG1,sysG2);
```

then we combine the result G3, with the G4 in series by

```
num4=[1];                        % form G4
den4=[1 0];
sysG4=tf(num4,den4);             % define subsystem G4
sysG5=series(sysG3,sysG4);       % series combination of G3 and G4
```

and complete the reduction of the feedback system by

```
num6=[1];                              % form G6
den6=[1];
```

Figure 3.12

Example for block-diagram simplification

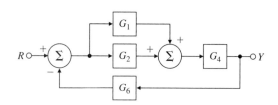

```
sysG6=tf(num6,den6)              % define subsystem G6
[sysCL]=feedback(sysG5,sysG6,-1)  % feedback combination of G5 and G6
```

The MATLAB results are sysCL of the form

$$\frac{Y(s)}{R(s)} = \frac{2s + 4}{s^2 + 2s + 4}$$

and this is the same result as the one obtained by block diagram reduction.

3.3   Effect of Pole Locations

Once the transfer function has been determined by any of the available methods, we can start to analyze the response of the system it represents. When the system equations are simultaneous ordinary differential equations (ODEs), the transfer function that results will be a ratio of polynomials; that is,

$$H(s) = b(s)/a(s).$$

Poles

Zeros

If we assume that b and a have no common factors (as is usually the case), then values of s such that $a(s) = 0$ will represent points where $H(s)$ is infinity. As we saw in Section 3.1.5, these s-values are called poles of $H(s)$. Values of s such that $b(s) = 0$ are points where $H(s) = 0$ and the corresponding s-locations are called zeros. The effect of zeros on the transient response will be discussed in Section 3.5. These poles and zeros completely describe $H(s)$ except for a constant multiplier. Because the impulse response is given by the time function corresponding to the transfer function, we call the impulse response the **natural response** of the system. We can use the poles and zeros to compute the corresponding time response and thus identify time histories with pole locations in the s-plane. For example, the poles identify the classes of signals contained in the impulse response, as may be seen by a partial-fraction expansion of $H(s)$. For a first-order pole,

The impulse response is the natural response.

$$H(s) = \frac{1}{s + \sigma}.$$

First-order system impulse response

Table A.2, entry 7, indicates that the impulse response will be an exponential function; that is,

$$h(t) = e^{-\sigma t} 1(t).$$

Stability

When $\sigma > 0$, the pole is located at $s < 0$, the exponential expression decays, and we say the impulse response is **stable**. If $\sigma < 0$, the pole is to the right of the origin. Because the exponential expression here grows with time, the impulse response is referred to as **unstable** (Section 3.6). Figure 3.13(a) shows a typical stable response and defines the **time constant**

Time constant τ

$$\tau = 1/\sigma \qquad (3.51)$$

as the time when the response is $1/e$ times the initial value. Hence, it is a measure of the rate of decay. The straight line is tangent to the exponential curve at $t = 0$ and terminates at $t = \tau$. This characteristic of an exponential expression is useful in sketching a time plot or checking computer results.

Figure 3.13(b) shows the impulse and step response for a first-order system computed using MATLAB.

Figure 3.13

First-order system response: (a) impulse response; (b) impulse response and step response using MATLAB®

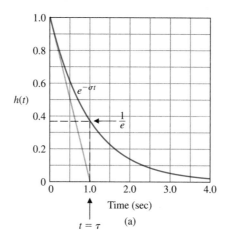

(a)

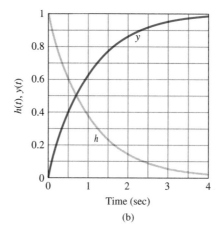

(b)

EXAMPLE 3.23   *Response versus Pole Locations, Real Roots*

Compare the time response with the pole locations for the system with a transfer function between input and output given by

$$H(s) = \frac{2s + 1}{s^2 + 3s + 2}.$$ (3.52)

Solution. The numerator is

$$b(s) = 2\left(s + \frac{1}{2}\right),$$

and the denominator is

$$a(s) = s^2 + 3s + 2 = (s + 1)(s + 2).$$

The poles of $H(s)$ are therefore at $s = -1$ and $s = -2$ and the one (finite) zero is at $s = -\frac{1}{2}$. A complete description of this transfer function is shown by the plot of the locations of the poles and the zeros in the s-plane using the MATLAB pzmap(num,den) function with

Figure 3.14

Sketch of s-plane
showing poles as
crosses and zeros as
circles

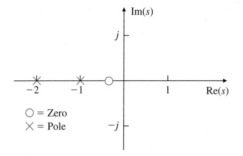

num=[2 1];
den=[1 3 2];

(see Fig. 3.14). A partial-fraction expansion of $H(s)$ results in

$$H(s) = -\frac{1}{s+1} + \frac{3}{s+2}.$$

From Table A.2 we can look up the inverse of each term in $H(s)$, which will give us the time function $h(t)$ that would result if the system input were an impulse. In this case,

$$h(t) = \begin{cases} -e^{-t} + 3e^{-2t} & t \geq 0, \\ 0 & t < 0. \end{cases} \tag{3.53}$$

We see that the shape of the component parts of $h(t)$, which are e^{-t} and e^{-2t}, are determined by the poles at $s = -1$ and -2. This is true of more complicated cases as well: In general, the shapes of the components of the natural response are determined by the locations of the poles of the transfer function.

"Fast poles" and "slow poles" refer to relative rate of signal decay.

A sketch of these pole locations and corresponding natural responses is given in Fig. 3.15, along with other pole locations including complex ones, which will be discussed shortly.

The role of the numerator in the process of partial-fraction expansion is to influence the size of the coefficient that multiplies each component. Because e^{-2t} decays faster than e^{-t}, the signal corresponding to the pole at -2 decays faster than the signal corresponding to the pole at -1. For brevity we simply say that the pole at -2 is faster than the pole at -1. In general, poles farther to the left in the s-plane are associated with natural signals that decay faster than those associated with poles closer to the imaginary axis. If the poles had been located with positive values of s (in the right half of the s-plane), the response would have been a growing exponential function and thus unstable. Figure 3.16 shows that the fast $3e^{-2t}$ term dominates the early part of the time history and that the $-e^{-t}$ term is the primary contributor later on.

Impulse response using MATLAB

The purpose of this example is to illustrate the relationship between the poles and the character of the response, which can be done exactly only by finding the inverse Laplace transform and examining each term as before. However, if we simply wanted to plot the impulse response for this example, the expedient way would be to use the MATLAB sequence

numH = [2 1];   % form numerator
denH = [1 3 2];   % form denominator

Figure 3.15

Time functions associated with points in the s-plane (LHP, left half-plane; RHP, right half-plane)

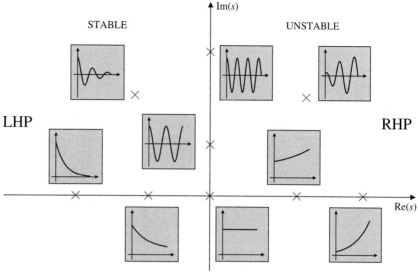

Figure 3.16

Impulse response of Example 3.23 [Eq. (3.52)]

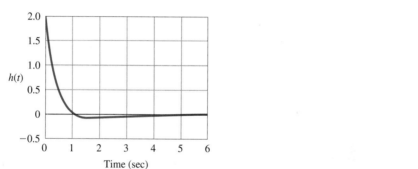

```
sysH=tf(numH,denH);    % define system from its numerator and
                         denominator
impulse(sysH);         % compute impulse response
```

The result is shown in Fig. 3.16.

Complex poles can be defined in terms of their real and imaginary parts, traditionally referred to as

$$s = -\sigma \pm j\omega_d.$$

This means that a pole has a negative real part if σ is positive. Since complex poles always come in complex conjugate pairs, the denominator corresponding to a complex pair will be

$$a(s) = (s + \sigma - j\omega_d)(s + \sigma + j\omega_d) = (s + \sigma)^2 + \omega_d^2. \qquad (3.54)$$

When finding the transfer function from differential equations, we typically write the result in the polynomial form

$$H(s) = \frac{\omega_n^2}{s^2 + 2\zeta\omega_n s + \omega_n^2}. \qquad (3.55)$$

By multiplying out the form given by Eq. (3.54) and comparing it with the coefficients of the denominator of $H(s)$ in Eq. (3.55), we find the correspondence between the parameters to be

$$\sigma = \zeta \omega_n \quad \text{and} \quad \omega_d = \omega_n \sqrt{1 - \zeta^2}, \tag{3.56}$$

Damping ratio; damped and undamped natural frequency

where the parameter ζ is the **damping ratio**[7] and ω_n is the **undamped natural frequency**. The poles of this transfer function are located at a radius ω_n in the s-plane and at an angle $\theta = \sin^{-1} \zeta$, as shown in Fig. 3.17. Therefore, the damping ratio reflects the level of damping as a fraction of the critical damping value where the poles become real. In rectangular coordinates the poles are at $s = -\sigma \pm j\omega_d$. When $\zeta = 0$, we have no damping, $\theta = 0$, and the damped natural frequency $\omega_d = \omega_n$, the undamped natural frequency.

For purposes of finding the time response from Table A.2 corresponding to a complex transfer function, it is easiest to manipulate the $H(s)$ so that the complex poles fit the form of Eq. (3.54), because then the time response can be found directly from the table. Equation (3.55) can be rewritten as

$$H(s) = \frac{\omega_n^2}{(s + \zeta \omega_n)^2 + \omega_n^2(1 - \zeta^2)}. \tag{3.57}$$

Standard second-order system impulse response

Therefore, from entry number 20 in Table A.2 and the definitions in Eq. (3.56), we see that the impulse response is

$$h(t) = \frac{\omega_n}{\sqrt{1 - \zeta^2}} e^{-\sigma t} (\sin \omega_d t) 1(t). \tag{3.58}$$

Figure 3.17

s-plane plot for a pair of complex poles

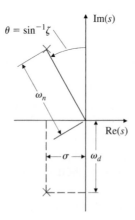

---

[7] In communications and filter engineering, the standard second-order transfer function is written as $H = 1/[1 + Q(s/\omega_n + \omega_n/s)]$. Here, ω_n is called the **band center** and Q is the **quality factor**. Comparison with Eq. (3.55) shows that $Q = 1/2\zeta$.

Figure 3.18(a) plots $h(t)$ for several values of ζ such that time has been normalized to the undamped natural frequency ω_n. Note that the actual frequency ω_d decreases slightly as the damping ratio increases. Note also that for very low damping the response is oscillatory, while for large damping (ζ near 1) the response shows no oscillation. A few of these responses are sketched in Fig. 3.15 to show qualitatively how changing pole locations in the s-plane affect impulse responses. You will find it useful as a control designer to commit the image of Fig. 3.15 to memory so that you can understand instantly how changes in pole locations influence the time response.

Figure 3.18

Responses of second-order systems versus ζ: (a) impulse responses; (b) step responses

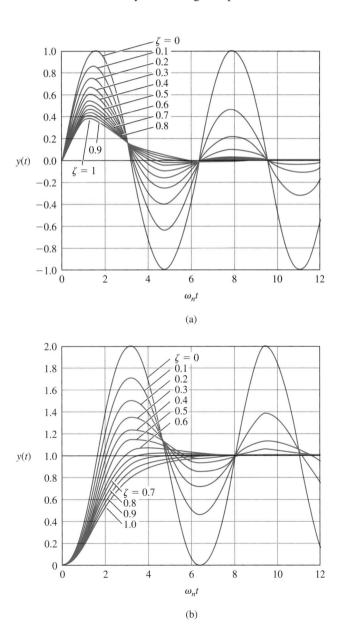

Stability depends on whether natural response grows or decays.

Three pole locations are shown in Fig. 3.19 for comparison with the corresponding impulse responses in Fig. 3.18(a). The negative real part of the pole, σ, determines the decay rate of an exponential envelope that multiplies the sinusoid, as shown in Fig. 3.20. Note that if $\sigma < 0$ (and the pole is in the RHP), then the natural response will grow with time, so, as defined earlier, the system is said to be unstable. If $\sigma = 0$, the natural response neither grows nor decays, so stability is open to debate. If $\sigma > 0$, the natural response decays, so the system is stable.

Step response

It is also interesting to examine the step response of $H(s)$—that is, the response of the system $H(s)$ to a unit step input $u = 1(t)$, where $U(s) = {}^1\!/_s$. The step-response transform is given by $Y(s) = H(s)U(s)$, which is found in Table A.2, entry 21. Figure 3.18(b), which plots $y(t)$ for several values of ζ, shows that the basic transient response characteristics from the impulse response carry over quite well to the step response; the difference between the two responses is that the step response's final value is the commanded unit step.

Figure 3.19

Pole locations corresponding to three values of ζ

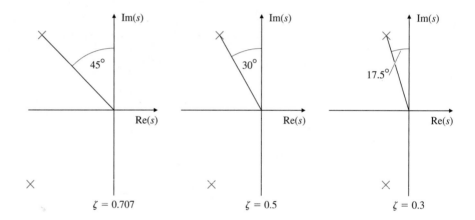

$\zeta = 0.707$ $\zeta = 0.5$ $\zeta = 0.3$

Figure 3.20

Second-order system response with an exponential envelope

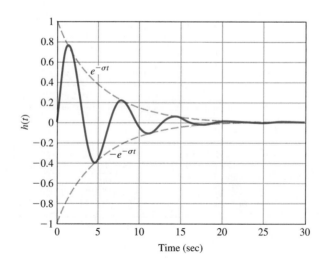

EXAMPLE 3.24   *Oscillatory Time Response*

Discuss the correlation between the poles of

$$H(s) = \frac{2s + 1}{s^2 + 2s + 5} \qquad (3.59)$$

and the impulse response of the system and find the exact impulse response.

Solution. From the form of $H(s)$ given by Eq. (3.55), we see that

$$\omega_n^2 = 5 \Rightarrow \omega_n = \sqrt{5} = 2.24 \text{ rad/sec}$$

and

$$2\zeta\omega_n = 2 \Rightarrow \zeta = \frac{1}{\sqrt{5}} = 0.447.$$

This indicates that we should expect a frequency of around 2 rad/sec with very little oscillatory motion. In order to obtain the exact response, we manipulate $H(s)$ until the denominator is in the form of Eq. (3.54):

$$H(s) = \frac{2s + 1}{s^2 + 2s + 5} = \frac{2s + 1}{(s + 1)^2 + 2^2}.$$

From this equation we see that the poles of the transfer function are complex, with real part -1 and imaginary parts $\pm 2j$. Table A.2 has two entries, numbers 19 and 20, that match the denominator. The right side of the preceding equation needs to be broken into two parts so that they match the numerators of the entries in the table:

$$H(s) = \frac{2s + 1}{(s + 1)^2 + 2^2} = 2\frac{s + 1}{(s + 1)^2 + 2^2} - \frac{1}{2}\frac{2}{(s + 1)^2 + 2^2}.$$

Thus, the impulse response is

$$h(t) = \left(2e^{-t}\cos 2t - \frac{1}{2}e^{-t}\sin 2t\right)1(t).$$

Figure 3.21 is a plot of the response and shows how the envelope attenuates the sinusoid, the domination of the $2\cos 2t$ term, and the small phase shift caused by the $-\frac{1}{2}\sin 2t$ term.

As in the previous example, the expedient way of determining the impulse response would be to use the MATLAB sequence

Impulse response by MATLAB

```
numH = [2 1];        % form numerator
denH = [1 2 5];      % form denominator
sysH=tf(numH,denH);  % define system by its numerator and denominator
t=0:0.1:6;           % form time vector
y=impulse(sysH,t);   % compute impulse response
plot(t,y);           % plot impulse response
```

as shown in Fig. 3.21.

Figure 3.21

System response for Example 3.24

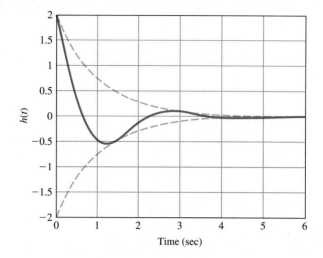

3.4 Time-Domain Specifications

Definitions of rise time, settling time, overshoot, and peak time

Specifications for a control system design often involve certain requirements associated with the time response of the system. The requirements for a step response are expressed in terms of the standard quantities illustrated in Fig. 3.22:

1. The **rise time** t_r is the time it takes the system to reach the vicinity of its new set point.
2. The **settling time** t_s is the time it takes the system transients to decay.
3. The **overshoot** M_p is the maximum amount the system overshoots its final value divided by its final value (and is often expressed as a percentage).
4. The **peak time** t_p is the time it takes the system to reach the maximum overshoot point.

3.4.1 Rise Time

For a second-order system, the time responses shown in Fig. 3.18(b) yield information about the specifications that is too complex to be remembered unless converted to a simpler form. By examining these curves in light of the definitions given in Fig. 3.22, we can relate the curves to the pole-location parameters ζ and ω_n. For example, all the curves rise in roughly the same time. If we consider the curve for $\zeta = 0.5$ to be an average, the rise time from $y = 0.1$ to $y = 0.9$ is approximately $\omega_n t_r = 1.8$. Thus we can say that

Rise time t_r

$$t_r \cong \frac{1.8}{\omega_n}. \tag{3.60}$$

Although this relationship could be embellished by including the effect of the damping ratio, it is important to keep in mind how Eq. (3.60) is typically used. It is accurate only for a second-order system with no zeros; for all other systems it is a rough approximation to the relationship between t_r and ω_n. Most systems being analyzed for control systems design are more complicated than the pure second-order system, so designers use Eq. (3.60) with the knowledge that it is a rough approximation only.

Figure 3.22

Definition of rise time t_r, settling time t_s, and overshoot M_p

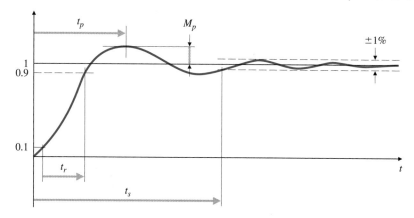

3.4.2 Overshoot and Peak Time

For the overshoot M_p we can be more analytical. This value occurs when the derivative is zero, which can be found from calculus. The time history of the curves in Fig. 3.18(b), found from the inverse Laplace transform of $H(s)/s$, is

$$y(t) = 1 - e^{-\sigma t}\left(\cos \omega_d t + \frac{\sigma}{\omega_d}\sin \omega_d t\right), \tag{3.61}$$

where $\omega_d = \omega_n\sqrt{1 - \zeta^2}$ and $\sigma = \zeta\omega_n$. We may rewrite the preceding equation using the trigonometric identity

$$A\sin(\alpha) + B\cos(\alpha) = C\cos(\alpha - \beta)$$

or

$$C = \sqrt{A^2 + B^2} = \frac{1}{\sqrt{1 - \zeta^2}},$$

$$\beta = \tan^{-1}\left(\frac{A}{B}\right) = \tan^{-1}\left(\frac{\zeta}{\sqrt{1 - \zeta^2}}\right),$$

Standard second-order system step response

with $A = \frac{\sigma}{\omega_d}$, $B = 1$, and $\alpha = \omega_d t$, in a more compact form as

$$y(t) = 1 - \frac{e^{-\sigma t}}{\sqrt{1 - \zeta^2}}\cos(\omega_d t - \beta). \tag{3.62}$$

When $y(t)$ reaches its maximum value, its derivative will be zero:

$$\dot{y}(t) = \sigma e^{-\sigma t}\left(\cos \omega_d t + \frac{\sigma}{\omega_d}\sin \omega_d t\right) - e^{-\sigma t}(-\omega_d \sin \omega_d t + \sigma \cos \omega_d t) = 0$$

$$= e^{-\sigma t}\left(\frac{\sigma^2}{\omega_d} + \omega_d\right)\sin \omega_d t = 0.$$

This occurs when $\sin \omega_d t = 0$, so

$$\omega_d t_p = \pi$$

and thus

$$t_p = \frac{\pi}{\omega_d}. \tag{3.63}$$

Peak time t_p

Substituting Eq. (3.63) into the expression for $y(t)$, we compute

$$y(t_p) \overset{\Delta}{=} 1 + M_p = 1 - e^{-\sigma\pi/\omega_d}\left(\cos\pi + \frac{\sigma}{\omega_d}\sin\pi\right)$$

$$= 1 + e^{-\sigma\pi/\omega_d}.$$

Overshoot M_p

Thus we have the formula

$$M_p = e^{-\pi\zeta/\sqrt{1-\zeta^2}}, \quad 0 \le \zeta < 1, \tag{3.64}$$

which is plotted in Fig. 3.23. Two frequently used values from this curve are $M_p = 0.16$ for $\zeta = 0.5$ and $M_p = 0.05$ for $\zeta = 0.7$.

3.4.3 Settling Time

The final parameter of interest from the transient response is the settling time t_s. This is the time required for the transient to decay to a small value so that $y(t)$ is almost in the steady state. Various measures of smallness are possible. For illustration we will use 1% as a reasonable measure; in other cases 2% or 5% are used. As an analytic computation, we notice that the deviation of y from 1 is the product of the decaying exponential $e^{-\sigma t}$ and the circular functions sine and cosine. The duration of this error

Settling time t_s

is essentially decided by the transient exponential, so we can define the settling time as that value of t_s when the decaying exponential reaches 1%:

$$e^{-\zeta\omega_n t_s} = 0.01.$$

Therefore,

$$\zeta\omega_n t_s = 4.6,$$

or

$$t_s = \frac{4.6}{\zeta\omega_n} = \frac{4.6}{\sigma}, \tag{3.65}$$

where σ is the negative real part of the pole, as may be seen from Fig. 3.17.

Figure 3.23

Overshoot M_p versus damping ratio ζ for the second-order system

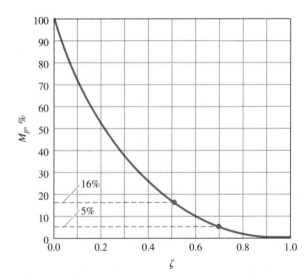

Equations (3.60), (3.64), and (3.65) characterize the transient response of a system having no finite zeros and two complex poles and with undamped natural frequency ω_n, damping ratio ζ, and negative real part σ. In analysis and design, they are used to estimate rise time, overshoot, and settling time, respectively, for just about any system. In design synthesis we wish to specify t_r, M_p, and t_s and to ask where the poles need to be so that the actual responses are less than or equal to these specifications. For specified values of t_r, M_p, and t_s, the synthesis form of the equation is then

Design synthesis

$$\omega_n \geq \frac{1.8}{t_r}, \tag{3.66}$$

$$\zeta \geq \zeta(M_p) \quad \text{(from Fig. 3.23)}, \tag{3.67}$$

$$\sigma \geq \frac{4.6}{t_s}. \tag{3.68}$$

These equations, which can be graphed in the s-plane as shown in Fig. 3.24(a–c), will be used in later chapters to guide the selection of pole and zero locations to meet control system specifications for dynamic response.

It is important to keep in mind that Eqs. (3.66)–(3.68) are qualitative guides and not precise design formulas. They are meant to provide only a starting point for the design iteration. After the control design is complete, the time response should always be checked by an exact calculation, usually by numerical simulation, to verify whether the time specifications have actually been met. If not, another iteration of the design is required.

First-order system step response

For a first-order system,

$$H(s) = \frac{\sigma}{s + \sigma},$$

and the transform of the step response is

$$Y(s) = \frac{\sigma}{s(s + \sigma)}.$$

We see from entry 11 in Table A.2 that $Y(s)$ corresponds to

$$y(t) = (1 - e^{-\sigma t})1(t). \tag{3.69}$$

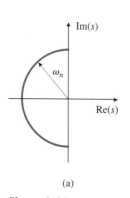

(a)

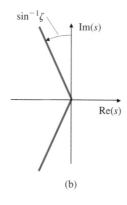

(b)

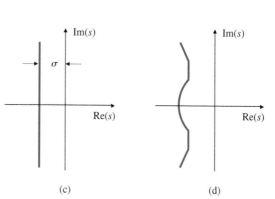

(c)   (d)

Figure 3.24

Graphs of regions in the s-plane delineated by certain transient requirements: (a) rise time; (b) overshoot; (c) settling time; (d) composite of all three requirements

Comparison with the development for Eq. (3.65) shows that the value of t_s for a first-order system is the same:

$$t_s = \frac{4.6}{\sigma}.$$

No overshoot is possible, so $M_p = 0$. The rise time from $y = 0.1$ to $y = 0.9$ can be seen from Fig. 3.13 to be

$$t_r = \frac{\ln 0.9 - \ln 0.1}{\sigma} = \frac{2.2}{\sigma}.$$

Time constant τ

However, it is more typical to describe a first-order system in terms of its time constant, which was defined in Fig. 3.13 to be $\tau = 1/\sigma$.

EXAMPLE 3.25 *Transformation of the Specifications to the s-Plane*

Find the allowable regions in the s-plane for the poles of a transfer function of a system if the system response requirements are $t_r \leq 0.6$ sec, $M_p \leq 10\%$, and $t_s \leq 3$ sec.

Solution. Without knowing whether or not the system is second order with no zeros, it is impossible to find the allowable region accurately. Regardless of the system, we can obtain a first approximation using the relationships for a second-order system. Equation (3.66) indicates that

$$\omega_n \geq \frac{1.8}{t_r} = 3.0 \text{ rad/sec,}$$

Eq. (3.67) and Fig. 3.23 indicate that

$$\zeta \geq 0.6,$$

and Eq. (3.68) indicates that

$$\sigma \geq \frac{4.6}{3} = 1.5 \text{ sec.}$$

The allowable region is anywhere to the left of the solid line in Fig. 3.25. Note that any pole meeting the ζ and ω_n restrictions will automatically meet the σ restriction.

3.5 Effects of Zeros and Additional Poles

Relationships such as those shown in Fig. 3.24 are correct for the simple second-order system; for more complicated systems they can be used only as guidelines. If a certain design has an inadequate rise time (is too slow), we must raise the natural frequency; if the transient has too much overshoot, then the damping needs to be increased; if the transient persists too long, the poles need to be moved to the left in the s-plane.

Effect of zeros

The effect of zeros near poles

Thus far only the poles of $H(s)$ have entered into the discussion. There may also be zeros of $H(s)$.[8] At the level of transient analysis, the zeros exert their influence by modifying the coefficients of the exponential terms whose shape is decided by the poles, as seen in Example 3.23. To illustrate this further, consider the following two transfer functions, which have the same poles but different zeros:

[8]We assume that $b(s)$ and $a(s)$ have no common factors. If this is not so, it is possible for $b(s)$ and $a(s)$ to be zero at the same location and for $H(s)$ to not equal zero there. The implications of this case will be discussed in Chapter 7, when we have a state-space description.

Figure 3.25

Allowable region in
s-plane for
Example 3.25

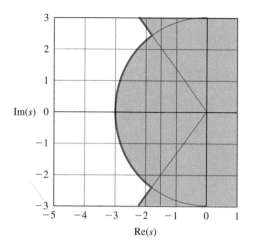

$$H_1(s) = \frac{2}{(s+1)(s+2)}$$

$$= \frac{2}{s+1} - \frac{2}{s+2}, \tag{3.70}$$

$$H_2(s) = \frac{2(s+1.1)}{1.1(s+1)(s+2)}$$

$$= \frac{2}{1.1}\left(\frac{0.1}{s+1} + \frac{0.9}{s+2}\right)$$

$$= \frac{0.18}{s+1} + \frac{1.64}{s+2}. \tag{3.71}$$

They are normalized to have the same DC gain (i.e., gain at $s = 0$). Notice that the
coefficient of the $(s+1)$ term has been modified from 2 in $H_1(s)$ to 0.18 in $H_2(s)$.
This dramatic reduction is brought about by the zero at $s = -1.1$ in $H_2(s)$, which
almost cancels the pole at $s = -1$. If we put the zero exactly at $s = -1$, this term will
vanish completely. In general, a zero near a pole reduces the amount of that term in the
total response. From the equation for the coefficients in a partial-fraction expansion,
Eq. (3.43),

$$C_1 = (s - p_1)F(s)|_{s=p_1},$$

we can see that if $F(s)$ has a zero near the pole at $s = p_1$, the value of $F(s)$ will be
small because the value of s is near the zero. Therefore, the coefficient C_1, which
reflects how much of that term appears in the response, will be small.

In order to take into account how zeros affect the transient response when designing a control system, we consider transfer functions with two complex poles and one
zero. To expedite the plotting for a wide range of cases, we write the transform in a
form with normalized time and zero locations:

$$H(s) = \frac{(s/\alpha\zeta\omega_n) + 1}{(s/\omega_n)^2 + 2\zeta(s/\omega_n) + 1}. \tag{3.72}$$

The zero is located at $s = -\alpha \zeta \omega_n = -\alpha \sigma$. If α is large, the zero will be far removed from the poles and the zero will have little effect on the response. If $\alpha \cong 1$, the value of the zero will be close to that of the real part of the poles and can be expected to have a substantial influence on the response. The step-response curves for $\zeta = 0.5$ and for several values of α are plotted in Fig. 3.26. We see that the major effect of the zero is to increase the overshoot M_p, whereas it has very little influence on the settling time. A plot of M_p versus α is given in Fig. 3.27. The plot shows that the zero has very little effect on M_p if $\alpha > 3$, but as α decreases below 3, it has an increasing effect, especially when $\alpha = 1$ or less.

Figure 3.26

Plots of the step response of a second-order system with a zero ($\zeta = 0.5$)

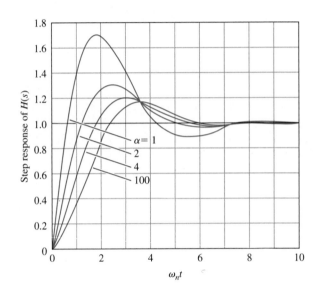

Figure 3.27

Plot of overshoot M_p as a function of normalized zero location α. At $\alpha = 1$, the real part of the zero equals the real part of the poles

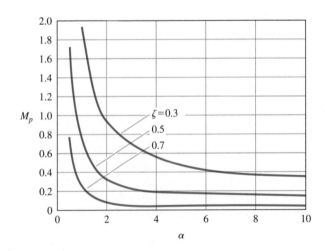

Figure 3.28
Second-order step
responses $y(t)$ of the
transfer functions $H(s)$,
$H_0(s)$, and $H_d(s)$

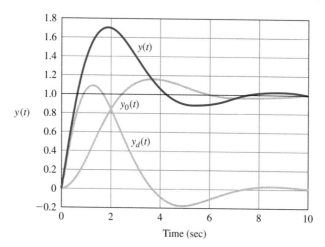

Figure 3.26 can be explained in terms of Laplace-transform analysis. First we
replace s/ω_n with s:

$$H(s) = \frac{s/\alpha\zeta + 1}{s^2 + 2\zeta s + 1}.$$

This has the effect of normalizing frequency in the transfer function and normalizing
time in the corresponding step responses; thus $\tau = \omega_n t$. We then rewrite the transfer
function as the sum of two terms:

$$H(s) = \frac{1}{s^2 + 2\zeta s + 1} + \frac{1}{\alpha\zeta}\frac{s}{s^2 + 2\zeta s + 1}. \tag{3.73}$$

The first term, which we shall call $H_0(s)$, is the original term (having no finite zero),
and the second term $H_d(s)$, which is introduced by the zero, is a product of a constant
$(1/\alpha\zeta)$ times s times the original term. The Laplace transform of df/dt is $sF(s)$,
so $H_d(s)$ corresponds to a product of a constant times the *derivative* of the original
term, i.e.,

$$y(t) = y_0(t) + y_d(t) = y_0(t) + \frac{1}{\alpha\zeta}\dot{y}_0(t).$$

The step responses of $H_0(s)$ denoted by $y_0(t)$ and $H_d(s)$ denoted by $y_d(t)$ are plotted
in Fig. 3.28. Looking at these curves, we can see why the zero increased the overshoot: The derivative has a large hump in the early part of the curve, and adding this
to the $H_0(s)$ response lifts up the total response of $H(s)$ to produce the overshoot.
This analysis is also very informative for the case when $\alpha < 0$ and the zero is in the
RHP where $s > 0$. (This is typically called an **RHP zero** and is sometimes referred
to as a **nonminimum-phase zero**, a topic to be discussed in more detail in Section
6.1.1.) In this case the derivative term is subtracted rather than added. A typical case
is sketched in Fig. 3.29.

RHP or
nonminimum-phase zero

EXAMPLE 3.26   *Effect of the Proximity of the Zero to the Pole Locations
on the Transient Response*

Consider the second-order system with a finite zero and unity DC gain,

$$H(s) = \frac{24}{z}\frac{(s + z)}{(s + 4)(s + 6)}.$$

Figure 3.29

Step responses $y(t)$ of a second-order system with a zero in the RHP: a nonminimum-phase system

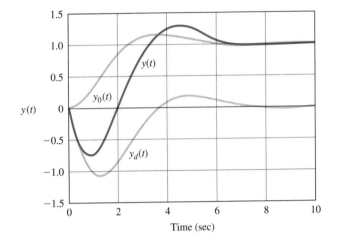

Determine the effect of the zero location ($s = -z$) on the unit step response when $z = \{1, 2, 3, 4, 5, 6\}$.

Solution. The step response is the inverse Laplace transform of

$$H_1(s) = H(s)\frac{1}{s} = \frac{24}{z}\frac{(s+z)}{s(s+4)(s+6)} = \frac{24}{z}\frac{s}{s(s+4)(s+6)} + \frac{24}{s(s+4)(s+6)}$$

and is the sum of the two parts,

$$y(t) = y_1(t) + y_2(t),$$

where

$$y_1(t) = \frac{12}{z}e^{-4t} - \frac{12}{z}e^{-6t},$$

$$y_2(t) = z\int_0^t y_1(\tau)d\tau = -3e^{-4t} + 2e^{-6t} + 1,$$

and

$$y(t) = 1 + \left(\frac{12}{z} - 3\right)e^{-4t} + \left(2 - \frac{12}{z}\right)e^{-6t}.$$

It is seen that if $z = 4$ or $z = 6$, one of the modes of the system is absent from the output, and the response is first order due to the pole–zero cancellations. The step responses of the system is shown in Fig. 3.30 ($z = 4$, dashed, $z = 6$ dot dashed). It is seen that the effect of the zero is most pronounced in terms of the additional overshoot for $z = 1$ (zero location closest to the origin). The system also has overshoot for $z = 2, 3$. For $z = 4$ or $z = 6$ the responses are first order as expected. It is interesting that for $z = 5$, where the zero is located between the two poles, there is no overshoot.

Figure 3.30

Effect of zero on
transient response

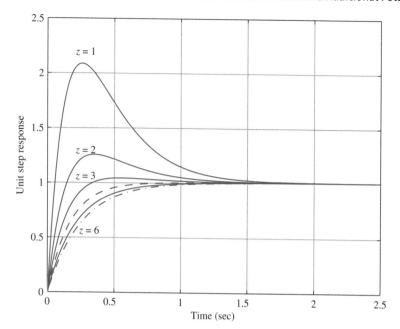

EXAMPLE 3.27

*Effect of the Proximity of the Complex Zeros to the
Lightly Damped Poles*

Consider the third-order feedback system with a pair of lightly damped poles and a
pair of complex zeros with the transfer function,

$$H(s) = \frac{(s+\alpha)^2 + \beta^2}{(s+1)\left[(s+0.1)^2 + 1\right]}.$$

Determine the effect of the complex zero locations ($s = -\alpha \pm j\beta$) on the
unit step response of the system for the three different zero locations $(\alpha, \beta) = (0.1, 1.0), (\alpha, \beta) = (0.25, 1.0)$, and $(\alpha, \beta) = (0.5, 1.0)$ as shown in Fig. 3.31.

Solution. We plot the three unit step responses using MATLAB as shown in Fig. 3.32.
The effect of the lightly damped modes are clearly seen as oscillations in the step
responses for the cases where $(\alpha, \beta) = (0.25, 1.0)$ or $(\alpha, \beta) = (0.5, 1.0)$, that is,
when the complex zeros are not close to the locations of the lightly damped poles
as shown in Fig. 3.31. On the other hand, if the complex zeros cancel the lightly
damped poles exactly as is the case for $(\alpha, \beta) = (0.1, 1.0)$, the oscillations are
completely eliminated in the step response. In practice, the locations of the lightly
damped poles are not known precisely and exact cancellation is not really possible.
However, placing the complex zeros near the locations of the lightly damped poles
may provide sufficient improvement in step response performance. We will come back
to this technique later in Chapters 5, 7, and 10 in the context of dynamic compensator
design.

Figure 3.31

Locations of complex zeros

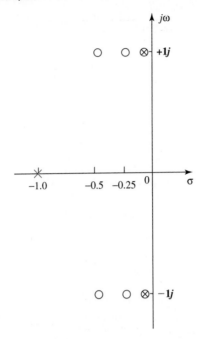

Figure 3.32

Effect of complex zeros on transient response

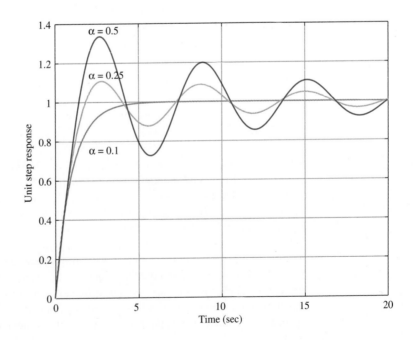

EXAMPLE 3.28   *Aircraft Response Using MATLAB*

The transfer function between the elevator and altitude of the Boeing 747 aircraft described in Section 10.3.2 can be approximated as

$$\frac{h(s)}{\delta_e(s)} = \frac{30(s - 6)}{s(s^2 + 4s + 13)}.$$

1. Use MATLAB to plot the altitude time history for a $1°$ impulsive elevator input. Explain the response, noting the physical reasons for the nonminimum-phase nature of the response.
2. Examine the accuracy of the approximations for t_r, t_s, and M_p [Eqs. (3.60) and (3.65) and Fig. 3.23].

Solution

1. The MATLAB statements to create the impulse response for this case are

```
u = −1;                    % u = delta e
numG = u*30*[1 −6];        % form numerator
denG = [1 4 13 0];         % form denominator
sysG=tf(numG,denG)         % define system by its numerator and denominator
y=impulse(sysG);           % compute impulse response; y = h
plot(y);                   % plot impulse response
```

The result is the plot shown in Fig. 3.33. Notice how the altitude drops initially and then rises to a new final value. The final value is predicted by the Final Value Theorem:

$$h(\infty) = s \left.\frac{30(s - 6)(-1)}{s(s^2 + 4s + 13)}\right|_{s=0} = \frac{30(-6)(-1)}{13} = +13.8.$$

The fact that the response has a finite final value for an impulsive input is due to the *s*-term in the denominator. This represents a pure integration and the integral of an impulse function is a finite value. If the input had been a step, the altitude

Response of a nonminimum-phase system

Figure 3.33

Response of an airplane's altitude to an impulsive elevator input

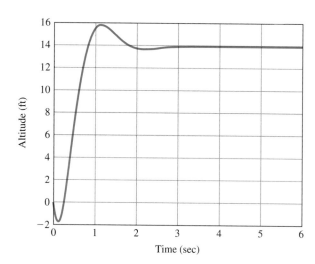

Altitude (ft)

Time (sec)

would have continued to increase with time; in other words the integral of a step function is a ramp function.

The initial drop is predicted by the RHP zero in the transfer function. The negative elevator deflection is defined to be upward by convention (see Fig. 10.30). The upward deflection of the elevators drives the tail down, which rotates the craft nose up and produces the climb. The deflection at the initial instant causes a downward force before the craft has rotated; therefore, the initial altitude response is down. After rotation, the increased lift resulting from the increased angle of attack of the wings causes the airplane to climb.

2. The rise time from Eq. (3.60) is

$$t_r = \frac{1.8}{\omega_n} = \frac{1.8}{\sqrt{13}} = 0.5 \text{ sec.}$$

We find the damping ratio ζ from the relation

$$2\zeta\omega_n = 4 \Rightarrow \zeta = \frac{2}{\sqrt{13}} = 0.55.$$

From Fig. 3.23 we find the overshoot M_p to be 0.14. Because $2\zeta\omega_n = 2\sigma = 4$, [Eq. (3.65)] shows that

$$t_s = \frac{4.6}{\sigma} = \frac{4.6}{2} = 2.3 \text{ sec.}$$

Detailed examination of the time history $h(t)$ from MATLAB output shows that $t_r \cong 0.43$ sec, $M_p \cong 0.14$, and $t_s \cong 2.6$ sec, which are reasonably close to the estimates. The only significant effect of the nonminimum-phase zero was to cause the initial response to go in the "wrong direction" and make the response somewhat sluggish.

Effect of extra pole

In addition to studying the effects of zeros, it is useful to consider the effects of an extra pole on the standard second-order step response. In this case, we take the transfer function to be

$$H(s) = \frac{1}{(s/\alpha\zeta\omega_n + 1)[(s/\omega_n)^2 + 2\zeta(s/\omega_n) + 1]}. \tag{3.74}$$

Plots of the step response for this case are shown in Fig. 3.34 for $\zeta = 0.5$ and several values of α. In this case the major effect is to increase the rise time. A plot of the rise time versus α is shown in Fig. 3.35 for several values of ζ.

From this discussion we can draw several conclusions about the dynamic response of a simple system as revealed by its pole–zero patterns:

Effects of Pole–Zero Patterns on Dynamic Response

1. For a second-order system with no finite zeros, the transient response parameters are approximated as follows:

Rise time: $\qquad t_r \cong \dfrac{1.8}{\omega_n}$,

Overshoot: $\qquad M_p \cong \begin{cases} 5\%, & \zeta = 0.7, \\ 16\%, & \zeta = 0.5 \\ 35\%, & \zeta = 0.3, \end{cases}$ (see Fig. 3.23),

Figure 3.34

Step responses for several third-order systems with $\zeta = 0.5$

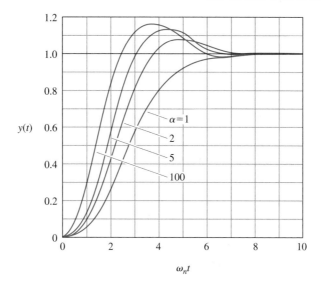

Figure 3.35

Normalized rise time for several locations of an additional pole

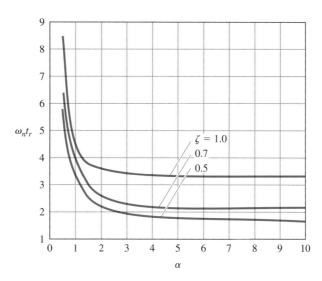

Settling time: $\qquad t_s \cong \dfrac{4.6}{\sigma}.$

2. A zero in the left half-plane (LHP) will increase the overshoot if the zero is within a factor of 4 of the real part of the complex poles. A plot is given in Fig. 3.27.
3. A zero in the RHP will depress the overshoot (and may cause the step response to start out in the wrong direction).
4. An additional pole in the LHP will increase the rise time significantly if the extra pole is within a factor of 4 of the real part of the complex poles. A plot is given in Fig. 3.35.

3.6 Stability

For nonlinear and time-varying systems, the study of stability is a complex and often difficult subject. In this section, we will consider only LTI systems for which we have the following condition for stability:

> An LTI system is said to be stable if all the roots of the transfer function denominator polynomial have negative real parts (i.e., they are *all* in the left hand *s*-plane) and is unstable otherwise.

Stable system

A system is stable if its initial conditions decay to zero and is unstable if they diverge. As just stated, an LTI (constant parameter) system is **stable** if *all* the poles of the system are strictly inside the left half *s*-plane [i.e., all the poles have negative real parts $(s = -\sigma + j\omega, \sigma < 0)$]. If *any* pole of the system is in the right half *s*-plane (i.e., has a positive real part, $s = -\sigma + j\omega, \sigma > 0$), then the system is **unstable**, as shown in Fig. 3.15. With any simple pole on the $j\omega$ axis $(\sigma = 0)$, small initial conditions will persist. For any other pole with $\sigma = 0$, oscillatory motion will persist. Therefore, a system is stable if its transient response decays and unstable if it does not. Figure 3.15 shows the time response of a system due to its pole locations.

In later chapters we will address more advanced notions of stability, such as Nyquist's frequency-response stability test (Chapter 6) and Lyapunov stability (Chapter 9).

3.6.1 Bounded Input–Bounded Output Stability

A system is said to have **bounded input–bounded output (BIBO) stability** if every bounded input results in a bounded output (regardless of what goes on inside the system). A test for this property is readily available when the system response is given by convolution. If the system has input $u(t)$, output $y(t)$, and impulse response $h(t)$, then

$$y(t) = \int_{-\infty}^{\infty} h(\tau)u(t - \tau)d\tau. \tag{3.75}$$

If $u(t)$ is bounded, then there is a constant M such that $|u| \le M < \infty$, and the output is bounded by

$$|y| = \left| \int hu\,d\tau \right|$$

$$\le \int |h||u|\,d\tau$$

$$\le M \int_{-\infty}^{\infty} |h(\tau)|\,d\tau.$$

Thus the output will be bounded if $\int_{-\infty}^{\infty} |h|\,d\tau$ is bounded.

On the other hand, suppose the integral is not bounded and the bounded input $u(t - \tau) = +1$ if $h(\tau) > 0$ and $u(t - \tau) = -1$ if $h(\tau) < 0$. In this case,

$$y(t) = \int_{-\infty}^{\infty} |h(\tau)|d\tau, \tag{3.76}$$

and the output is not bounded. We conclude that

Figure 3.36
Capacitor driven by
current source

Mathematical definition
of BIBO stability

The system with impulse response $h(t)$ is BIBO-stable if and only if the integral

$$\int_{-\infty}^{\infty} |h(\tau)|d\tau < \infty.$$

EXAMPLE 3.29

BIBO Stability for a Capacitor

As an example, determine the capacitor driven by a current source sketched in Fig. 3.36. The capacitor voltage is the output and the current is the input.

Solution. The impulse response of this setup is $h(t) = 1(t)$, the unit step. Now for this response,

$$\int_{-\infty}^{\infty} |h(\tau)|d\tau = \int_{0}^{\infty} d\tau \qquad (3.77)$$

is not bounded. The capacitor is not BIBO-stable. Notice that the transfer function of the system is $1/s$ and has a pole on the imaginary axis. Physically we can see that constant input current will cause the voltage to grow, and thus the system response is neither bounded nor stable. In general, if an LTI system has any pole on the imaginary axis or in the RHP, the response will not be BIBO-stable; if every pole is inside the LHP, then the response will be BIBO-stable. Thus for these systems, pole locations of the transfer function can be used to check for stability.

Determination of BIBO
stability by pole location

An alternative to computing the integral of the impulse response or even to locating the roots of the characteristic equation is given by Routh's stability criterion, which we will discuss in Section 3.6.3.

3.6.2 Stability of LTI Systems

Consider the LTI system whose transfer function denominator polynomial leads to the characteristic equation

$$s^n + a_1 s^{n-1} + a_2 s^{n-2} + \cdots + a_n = 0. \qquad (3.78)$$

Assume that the roots $\{p_i\}$ of the characteristic equation are real or complex, but are distinct. Note that Eq. (3.78) shows up as the denominator in the transfer function for the system as follows *before any cancellation of poles by zeros is made*:

$$T(s) = \frac{Y(s)}{R(s)} = \frac{b_0 s^m + b_1 s^{m-1} + \cdots + b_m}{s^n + a_1 s^{n-1} + \cdots + a_n}$$

$$= \frac{K \prod_{i=1}^{m}(s - z_i)}{\prod_{i=1}^{n}(s - p_i)}, \qquad m \le n. \qquad (3.79)$$

The solution to the differential equation whose characteristic equation is given by Eq. (3.78) may be written using partial-fraction expansion as

$$y(t) = \sum_{i=1}^{n} K_i e^{p_i t}, \tag{3.80}$$

where $\{p_i\}$ are the roots of Eq. (3.78) and $\{K_i\}$ depend on the initial conditions and zero locations. If a zero were to cancel a pole in the RHP for the transfer function, the corresponding K_i would equal zero in the output, but the unstable transient would appear in some internal variable.

The system is stable if and only if (necessary and sufficient condition) every term in Eq. (3.80) goes to zero as $t \to \infty$:

$$e^{p_i t} \to 0 \quad \text{for all } p_i.$$

This will happen if all the poles of the system are strictly in the LHP, where

$$\operatorname{Re}\{p_i\} < 0. \tag{3.81}$$

Internal stability occurs when all poles are strictly in the LHP

The $j\omega$ axis is the stability boundary

If any poles are repeated, the response must be changed from that of Eq. (3.80) by including a polynomial in t in place of K_i, but the conclusion is the same. This is called **internal stability**. Therefore, the stability of a system can be determined by computing the location of the roots of the characteristic equation and determining whether they are all in the LHP. If the system has any poles in the RHP, it is **unstable**. Hence the $j\omega$ axis is the stability boundary between asymptotically stable and unstable response. If the system has nonrepeated $j\omega$ axis poles, then it is said to be **neutrally stable**. For example, a pole at the origin (an integrator) results in a nondecaying transient. A pair of complex $j\omega$ axis poles results in an oscillating response (with constant amplitude). If the system has repeated poles on the $j\omega$ axis, then it is **unstable** [as it results in $te^{\pm j\omega_i t}$ terms in Eq. (3.80)]. For example, a pair of poles at the origin (double integrator) results in an unbounded response. MATLAB software makes the computation of the poles, and therefore determination of the stability of the system, relatively easy.

An alternative to locating the roots of the characteristic equation is given by Routh's stability criterion, which we will discuss next.

3.6.3 Routh's Stability Criterion

There are several methods of obtaining information about the locations of the roots of a polynomial without actually solving for the roots. These methods were developed in the 19th century and were especially useful before the availability of MATLAB software. They are still useful for determining the ranges of coefficients of polynomials for stability, especially when the coefficients are in symbolic (nonnumerical) form. Consider the characteristic equation of an nth-order system: [9]

$$a(s) = s^n + a_1 s^{n-1} + a_2 s^{n-2} + \cdots + a_{n-1} s + a_n. \qquad (3.82)$$

It is possible to make certain statements about the stability of the system without actually solving for the roots of the polynomial. This is a classical problem and several methods exist for the solution.

A necessary condition for Routh stability

A *necessary condition for stability* of the system is that all of the roots of Eq. (3.82) have negative real parts, which in turn requires that all the $\{a_i\}$ be positive.[10]

> A necessary (but not sufficient) condition for stability is that all the coefficients of the characteristic polynomial be positive.

If any of the coefficients are missing (are zero) or are negative, then the system will have poles located outside the LHP. This condition can be checked by inspection. Once the elementary necessary conditions have been satisfied, we need a more powerful test. Equivalent tests were independently proposed by Routh in 1874 and Hurwitz in 1895; we will discuss the former version. Routh's formulation requires the computation of a triangular array that is a function of the $\{a_i\}$. He showed that a *necessary and sufficient condition for stability* is that all of the elements in the first column of this array be positive.

A necessary and sufficient condition for stability

> A system is stable if and only if *all* the elements in the first column of the Routh array are positive.

To determine the Routh array, we first arrange the coefficients of the characteristic polynomial in two rows, beginning with the first and second coefficients and followed by the even-numbered and odd-numbered coefficients:

Routh array

$$
\begin{array}{llllll}
s^n & : & 1 & a_2 & a_4 & \cdots \\
s^{n-1} & : & a_1 & a_3 & a_5 & \cdots
\end{array}
$$

We then add subsequent rows to complete the **Routh array**:

Row	n	s^n:	1	a_2	a_4	$\cdots$
Row	$n-1$	s^{n-1}:	a_1	a_3	a_5	$\cdots$
Row	$n-2$	s^{n-2}:	b_1	b_2	b_3	$\cdots$
Row	$n-3$	s^{n-3}:	c_1	c_2	c_3	$\cdots$
	$\vdots$	$\vdots$	$\vdots$	$\vdots$	$\vdots$	
Row	2	s^2:	*	*		
Row	1	s:	*			
Row	0	s^0:	*			

---

[9] Without loss of generality, we can assume the polynomial to be monic (that is, the coefficient of the highest power of s is 1).

[10] This is easy to see if we construct the polynomial as a product of first- and second-order factors.

We compute the elements from the $(n-2)$th and $(n-3)$th rows as follows:

$$b_1 = -\frac{\det \begin{bmatrix} 1 & a_2 \\ a_1 & a_3 \end{bmatrix}}{a_1} = \frac{a_1 a_2 - a_3}{a_1},$$

$$b_2 = -\frac{\det \begin{bmatrix} 1 & a_4 \\ a_1 & a_5 \end{bmatrix}}{a_1} = \frac{a_1 a_4 - a_5}{a_1},$$

$$b_3 = -\frac{\det \begin{bmatrix} 1 & a_6 \\ a_1 & a_7 \end{bmatrix}}{a_1} = \frac{a_1 a_6 - a_7}{a_1},$$

$$c_1 = -\frac{\det \begin{bmatrix} a_1 & a_3 \\ b_1 & b_2 \end{bmatrix}}{b_1} = \frac{b_1 a_3 - a_1 b_2}{b_1},$$

$$c_2 = -\frac{\det \begin{bmatrix} a_1 & a_5 \\ b_1 & b_3 \end{bmatrix}}{b_1} = \frac{b_1 a_5 - a_1 b_3}{b_1},$$

$$c_3 = -\frac{\det \begin{bmatrix} a_1 & a_7 \\ b_1 & b_4 \end{bmatrix}}{b_1} = \frac{b_1 a_7 - a_1 b_4}{b_1}.$$

Note that the elements of the $(n-2)$th row and the rows beneath it are formed from the two previous rows using determinants, with the two elements in the first column and other elements from successive columns. Normally there are $n+1$ elements in the first column when the array terminates. If these are all positive, then all the roots of the characteristic polynomial are in the LHP. However, if the elements of the first column are not all positive, then the number of roots in the RHP equals the number of sign changes in the column. A pattern of $+, -, +$ is counted as *two* sign changes: one change from $+$ to $-$ and another from $-$ to $+$. For a simple proof of the Routh test, the reader is referred to Ho et al. (1998).

EXAMPLE 3.30   *Routh's Test*

The polynomial

$$a(s) = s^6 + 4s^5 + 3s^4 + 2s^3 + s^2 + 4s + 4$$

satisfies the necessary condition for stability since all the $\{a_i\}$ are positive and nonzero. Determine whether any of the roots of the polynomial are in the RHP.

Solution. The Routh array for this polynomial is

s^6:	1	3	1	4
s^5:	4	2	4	0
s^4:	$\dfrac{5}{2} = \dfrac{4\cdot 3 - 1\cdot 2}{4}$	$0 = \dfrac{4\cdot 1 - 4\cdot 1}{4}$	$4 = \dfrac{4\cdot 4 - 1\cdot 0}{4}$	

s^3:
$$2 = \frac{\dfrac{5}{2} \cdot 2 - 4 \cdot 0}{\dfrac{5}{2}} \qquad -\frac{12}{5} = \frac{\dfrac{5}{2} \cdot 4 - 4 \cdot 4}{\dfrac{5}{2}} \qquad 0$$

s^2:
$$3 = \frac{2 \cdot 0 - \dfrac{5}{2}\left(-\dfrac{12}{5}\right)}{2} \qquad 4 = \frac{2 \cdot 4 - \left(\dfrac{5}{2} \cdot 0\right)}{2}$$

s:
$$-\frac{76}{15} = \frac{3\left(-\dfrac{12}{5}\right) - 8}{3} \qquad 0$$

s^0:
$$4 = \frac{-\dfrac{76}{15} \cdot 4 - 0}{-\dfrac{76}{15}}.$$

We conclude that the polynomial has RHP roots, since the elements of the first column are not all positive. In fact, there are two poles in the RHP because there are two sign changes.[11]

Note that, in computing the Routh array, we can simplify the rest of the calculations by multiplying or dividing a row by a positive constant. Also note that the last two rows each have one nonzero element.

Routh's method is also useful in determining the range of parameters for which a feedback system remains stable.

EXAMPLE 3.31   *Stability versus Parameter Range*

Consider the system shown in Fig. 3.37. The stability properties of the system are a function of the proportional feedback gain K. Determine the range of K over which the system is stable.

Solution. The characteristic equation for the system is given by

$$1 + K\frac{s+1}{s(s-1)(s+6)} = 0,$$

or

$$s^3 + 5s^2 + (K-6)s + K = 0.$$

Figure 3.37

A feedback system for testing stability

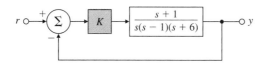

---

[11]The actual roots of the polynomial computed with the MATLAB roots command are -3.2644, $0.7797 \pm 0.7488j$, $-0.6046 \pm 0.9935j$, and -0.8858, which, of course, agree with our conclusion.

The corresponding Routh array is

$$
\begin{array}{ccc}
s^3 : & 1 & K - 6 \\
s^2 : & 5 & K \\
s : & (4K - 30)/5 & \\
s^0 : & K. &
\end{array}
$$

For the system to be stable, it is necessary that

$$
\frac{4K - 30}{5} > 0 \quad \text{and} \quad K > 0,
$$

or

$$
K > 7.5 \quad \text{and} \quad K > 0.
$$

Thus, Routh's method provides an analytical answer to the stability question. Although any gain satisfying this inequality stabilizes the system, the dynamic response could be quite different depending on the specific value of K. Given a specific value of the gain, we may compute the closed-loop poles by finding the roots of the characteristic polynomial. The characteristic polynomial has the coefficients represented by the row vector (in descending powers of s)

denT = [1 5 K–6 K],

Computing roots by
MATLAB

and we may compute the roots using the MATLAB function

roots(denT).

For $K = 7.5$ the roots are at -5 and $\pm 1.22j$, and the system is neutrally stable. Note that Routh's method predicts the presence of poles on the $j\omega$ axis for $K = 7.5$. If we set $K = 13$, the closed-loop poles are at -4.06 and $-0.47 \pm 1.7j$, and for $K = 25$, they are at -1.90 and $-1.54 \pm 3.27j$. In both these cases, the system is stable as predicted by Routh's method. Figure 3.38 shows the transient responses for the three gain values. To obtain these transient responses, we compute the closed-loop transfer function

$$
T(s) = \frac{Y(s)}{R(s)} = \frac{K(s + 1)}{s^3 + 5s^2 + (K - 6)s + K},
$$

Figure 3.38

Transient responses for the system in Fig. 3.37

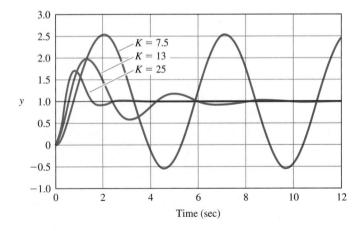

Figure 3.39

System with proportional-integral (PI) control

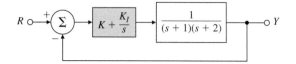

so that the numerator polynomial is expressed as

numT = [K K];   % form numerator

and denT is as before. The MATLAB commands

sysT=tf(numT,denT);   % define system by its numerator and denominator
step(sysT);   % compute step response

produce a plot of the (unit) step response.

EXAMPLE 3.32

Stability Versus Two Parameter Ranges

Find the range of the controller gains (K, K_I) so that the PI (proportional-integral; see Chapter 4) feedback system in Fig. 3.39 is stable.

Solution. The characteristic equation of the closed-loop system is

$$1 + \left(K + \frac{K_I}{s} \right) \frac{1}{(s+1)(s+2)} = 0,$$

which we may rewrite as

$$s^3 + 3s^2 + (2+K)s + K_I = 0.$$

The corresponding Routh array is

$$
\begin{array}{ccc}
s^3 : & 1 & 2+K \\
s^2 : & 3 & K_I \\
s \ : & (6+3K-K_I)/3 & \\
s^0 : & K_I. &
\end{array}
$$

For internal stability we must have

$$K_I > 0 \quad \text{and} \quad K > \frac{1}{3}K_I - 2.$$

The allowable region can be plotted in MATLAB using the ensuing commands

fh=@(ki,k) 6+3*k−ki;
ezplot(fh)
hold on;
f=@(ki,k) ki;
ezplot(f);

and is the shaded area in the (K_I, K) plane shown in Fig. 3.40, which represents an analytical solution to the stability question. This example illustrates the real value of Routh's approach and why it is superior to the numerical approaches. It would

Figure 3.40

Allowable region for stability

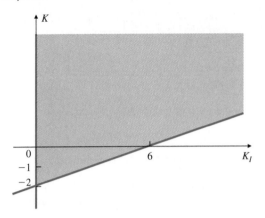

Figure 3.41

Transient response for the system in Fig. 3.39

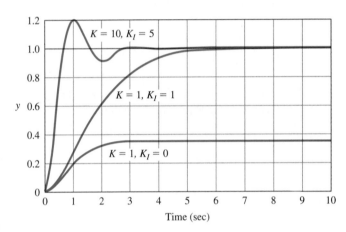

have been more difficult to arrive at these bounds on the gains using numerical search techniques. The closed-loop transfer function is

$$T(s) = \frac{Y(s)}{R(s)} = \frac{Ks + K_I}{s^3 + 3s^2 + (2 + K)s + K_I}.$$

MATLAB roots

As in Example 3.31, we may compute the closed-loop poles for different values of the dynamic compensator gains by using the MATLAB function roots on the denominator polynomial

    denT = [1 3 2+K KI].   % form denominator

Similarly, we may find the zero by finding the root of the numerator polynomial

    numT = [K KI].   % form numerator

The closed-loop zero of the system is at $-K_I/K$. Figure 3.41 shows the transient response for three sets of feedback gains. For $K = 1$ and $K_I = 0$, the closed-loop poles are at 0 and $-1.5 \pm 0.86j$, and there is a zero at the origin. For $K = K_I = 1$, the poles and zeros are all at -1. For $K = 10$ and $K_I = 5$, the closed-loop poles are at -0.46 and $-1.26 \pm 3.3j$ and the zero is at -0.5. The step responses were again obtained using the MATLAB function

```
sysT=tf(numT,denT)     % define system by its numerator and denominator
step(sysT).            % compute step response
```

There is a large steady-state error in this case when $K_I = 0$. (See Chapter 4.)

If the first term in one of the rows is zero or if an entire row is zero, then the standard Routh array cannot be formed, so we have to use one of the special techniques described next.

△ Special Cases

Special case I

If only the first element in one of the rows is zero, then we can replace the zero with a small positive constant $\epsilon > 0$ and proceed as before. We then apply the stability criterion by taking the limit as $\epsilon \to 0$.

EXAMPLE 3.33 *Routh's Test for Special Case I*

Consider the polynomial

$$a(s) = s^5 + 3s^4 + 2s^3 + 6s^2 + 6s + 9.$$

Determine whether any of the roots are in the RHP.

Solution. The Routh array is

$$
\begin{array}{lll}
s^5: & 1 & 2 \quad 6 \\
s^4: & 3 & 6 \quad 9 \\
s^3: & 0 & 3 \quad 0 \\
\text{New } s^3: & \epsilon & 3 \quad 0 \quad \leftarrow \text{Replace zero by } \epsilon \\
s^2: & \frac{2\epsilon-3}{\epsilon} & 3 \quad 0 \\
s: & 3 - \frac{3\epsilon^2}{2\epsilon-3} & 0 \quad 0 \\
s^0: & 3 & 0
\end{array}
$$

There are two sign changes in the first column of the array, which means there are two poles not in the LHP.[12]

Special case II

Another special case occurs when an entire row of the Routh array is zero. This indicates that there are complex conjugate pairs of roots that are mirror images of each other with respect to the imaginary axis. If the ith row is zero, we form an auxiliary equation from the previous (nonzero) row:

$$a_1(s) = \beta_1 s^{i+1} + \beta_2 s^{i-1} + \beta_3 s^{i-3} + \cdots . \tag{3.83}$$

Here $\{\beta_i\}$ are the coefficients of the $(i + 1)$th row in the array. We then replace the ith row by the coefficients of the *derivative* of the auxiliary polynomial and complete

---

[12]The actual roots computed with MATLAB are at $-2.9043, 0.6567 \pm 1.2881j, -0.7046 \pm 0.9929j$.

the array. However, the roots of the auxiliary polynomial in Eq. (3.83) are also roots of the characteristic equation, and these must be tested separately.

EXAMPLE 3.34   *Routh Test for Special Case II*

For the polynomial

$$a(s) = s^5 + 5s^4 + 11s^3 + 23s^2 + 28s + 12,$$

determine whether there are any roots on the $j\omega$ axis or in the RHP.

Solution.   The Routh array is

s^5:	1	11	28	
s^4:	5	23	12	
s^3:	6.4	25.6	0	
s^2:	3	12		
s:	0	0		$\leftarrow a_1(s) = 3s^2 + 12$
New s:	6	0		$\leftarrow \dfrac{da_1(s)}{ds} = 6s$
s^0:	12,			

There are no sign changes in the first column. Hence all the roots have negative real parts except for a pair on the imaginary axis. We may deduce this as follows: When we replace the zero in the first column by $\epsilon > 0$, there are no sign changes. If we let $\epsilon < 0$, then there are two sign changes. Thus, if $\epsilon = 0$, there are two poles on the imaginary axis, which are the roots of

$$a_1(s) = s^2 + 4 = 0,$$

or

$$s = \pm j2.$$

This agrees with the fact that the actual roots are at $-3, \pm 2j, -1$, and -1, as computed using the roots command in MATLAB.

The Routh–Hurwitz result assumes that the characteristic polynomial coefficients are known precisely. It is well–known that the roots of a polynomial can be very sensitive to even slight perturbations in the polynomial coefficients. If the range of variation on each one of the polynomial coefficients is known, then a remarkable result called the Kharitonov Theorem (1978) allows one to test just four so-called Kharitonov polynomials, using the Routh test, to see if the polynomial coefficient variations result in instability.

△ 3.7 Obtaining Models from Experimental Data

There are several reasons for using experimental data to obtain a model of the dynamic system to be controlled. In the first place, the best theoretical model built from equations of motion is still only an approximation of reality. Sometimes, as in the case of a very rigid spacecraft, the theoretical model is extremely good. Other times, as with many chemical processes such as papermaking or metalworking, the theoretical model is very approximate. In every case, before the final control design is done, it is important and prudent to verify the theoretical model with experimental data. Second, in situations for which the theoretical model is especially complicated or the physics of the process is poorly understood, the only reliable information on which to base the control design is the experimental data. Finally, the system is sometimes subject to online changes, which occur when the environment of the system changes. Examples include when an aircraft changes altitude or speed, a paper machine is given a different composition of fiber, or a nonlinear system moves to a new operating point. On these occasions we need to "retune" the controller by changing the control parameters. This requires a model for the new conditions and experimental data are often the most effective, if not the only, information available for the new model.

Four sources of experimental data

There are four kinds of experimental data for generating a model:

1. **transient response**, such as comes from an impulse or a step;
2. **frequency-response data**, which result from exciting the system with sinusoidal inputs at many frequencies;
3. **stochastic steady-state information**, as might come from flying an aircraft through turbulent weather or from some other natural source of randomness;
4. **pseudorandom-noise data**, as may be generated in a digital computer.

Each class of experimental data has its properties, advantages, and disadvantages.

Transient response

Transient response data are quick and relatively easy to obtain. They are also often representative of the natural signals to which the system is subjected. Thus a model derived from such data can be reliable for designing the control system. On the other hand, in order for the signal-to-noise ratio to be sufficiently high, the transient response must be highly noticeable. Consequently, the method is rarely suitable for normal operations, so the data must be collected as part of special tests. A second disadvantage is that the data do not come in a form suitable for standard control systems designs, and some parts of the model, such as poles and zeros, must be computed from the data.[13] This computation can be simple in special cases or complex in the general case.

Frequency response

Frequency-response data (see Chapter 6) are simple to obtain but substantially more time consuming than transient-response information. This is especially so if the time constants of the process are large, as often occurs in chemical processing industries. As with the transient-response data, it is important to have a good signal-to-noise ratio, so obtaining frequency-response data can be very expensive. On the other hand, as we will see in Chapter 6, frequency-response data are exactly in the right form for frequency-response design methods; so once the data have been obtained, the control design can proceed immediately.

Stochastic steady-state

Normal operating records from a natural stochastic environment at first appear to be an attractive basis for modeling systems, since such records are by definition nondisruptive and inexpensive to obtain. Unfortunately, the quality of such data is inconsistent, tending to be worst just when the control is best, because then the upsets are minimal and the signals are smooth. At such times, some or even most of the system

---

[13]Ziegler and Nichols (1943), building on the earlier work of Callender et al. (1936), use the step response directly in designing the controls for certain classes of processes. See Chapter 4 for details.

dynamics are hardly excited. Because they contribute little to the system output, they will not be found in the model constructed to explain the signals. The result is a model that represents only part of the system and is sometimes unsuitable for control. In some instances, as occurs when trying to model the dynamics of the electroencephalogram (brain waves) of a sleeping or anesthetized person to locate the frequency and intensity of alpha waves, normal records are the only possibility. Usually they are the last choice for control purposes.

Pseudorandom noise
(PRBS)

Finally, the pseudorandom signals that can be constructed using digital logic have much appeal. Especially interesting for model making is the pseudorandom binary signal (PRBS). The PRBS takes on the value $+A$ or $-A$ according to the output (1 or 0) of a feedback shift register. The feedback to the register is a binary sum of various states of the register that have been selected to make the output period (which must repeat itself in finite time) as long as possible. For example, with a register of 20 bits, $2^{20} - 1$ (over a million) steps are produced before the pattern repeats. Analysis beyond the scope of this text has revealed that the resulting signal is almost like a broadband random signal. Yet this signal is entirely under the control of the engineer who can set the level (A) and the length (bits in the register) of the signal. The data obtained from tests with a PRBS must be analyzed by computer and both special-purpose hardware and programs for general-purpose computers have been developed to perform this analysis.

3.7.1 Models from Transient-Response Data

To obtain a model from transient data we assume that a step response is available. If the transient is a simple combination of elementary transients, then a reasonable low-order model can be estimated using hand calculations. For example, consider the step response shown in Fig. 3.42. The response is monotonic and smooth. If we assume that it is given by a sum of exponentials, we can write

$$y(t) = y(\infty) + Ae^{-\alpha t} + Be^{-\beta t} + Ce^{-\gamma t} + \cdots . \qquad (3.84)$$

Subtracting off the final value and assuming that $-\alpha$ is the slowest pole, we write

$$y - y(\infty) \cong Ae^{-\alpha t},$$
$$\log_{10}[y - y(\infty)] \cong \log_{10} A - \alpha t \log_{10} e,$$
$$\cong \log_{10} A - 0.4343\alpha t. \qquad (3.85)$$

This is the equation of a line whose slope determines α and intercept determines A. If we fit a line to the plot of $\log_{10}[y - y(\infty)]$ (or $\log_{10}[y(\infty) - y]$ if A is negative), then we can estimate A and α. Once these are estimated, we plot $y - [y(\infty) + Ae^{-\alpha t}]$, which as a curve approximates $Be^{-\beta t}$ and on the log plot is equivalent to $\log_{10} B - 0.4345\beta t$. We repeat the process, each time removing the slowest remaining term, until the data stop being accurate. Then we plot the final model step response and compare it with data so we can assess the quality of the computed model. It is possible to get a good fit to the step response and yet be far off from the true time constants (poles) of the system. However, the method gives a good approximation for control of processes whose step responses look like Fig. 3.42.

Figure 3.42
A step response characteristic of many chemical processes

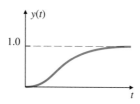

TABLE 3.1

Step Response Data

t	y(t)	t	y(t)
0.1	0.000	1.0	0.510
0.1	0.005	1.5	0.700
0.2	0.034	2.0	0.817
0.3	0.085	2.5	0.890
0.4	0.140	3.0	0.932
0.5	0.215	4.0	0.975
		∞	1.000

Sinha and Kuszta (1983).

EXAMPLE 3.35   *Determining the Model from Time–Response Data*

Find the transfer function that generates the data given in Table 3.1 and which are plotted in Fig. 3.43.

Solution.   Table 3.1 shows and Fig. 3.43 implies that the final value of the data is $y(\infty) = 1$. We know that A is negative because $y(\infty)$ is greater than $y(t)$. Therefore, the first step in the process is to plot $\log_{10}[y(\infty) - y]$, which is shown in Fig. 3.44. From the line (fitted by eye) the values are

$$\log_{10}|A| = 0.125,$$

$$0.4343\alpha = \frac{1.602 - 1.167}{\Delta t} = \frac{0.435}{1} \Rightarrow \alpha \cong 1.$$

Thus

$$A = -1.33,$$

$$\alpha = 1.0.$$

If we now subtract $1 + Ae^{\alpha t}$ from the data and plot the log of the result, we find the plot of Fig. 3.45. Here we estimate

$$\log_{10} B = -0.48,$$

$$0.4343\beta = \frac{-0.48 - (-1.7)}{0.5} = 2.5,$$

$$\beta \cong 5.8,$$

$$B = 0.33.$$

Figure 3.43

Step response data in Table 3.1

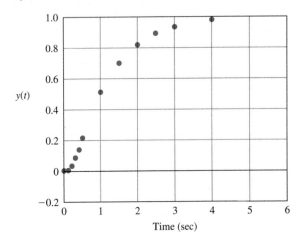

Figure 3.44

$\log_{10}[y(\infty) - y]$ versus t

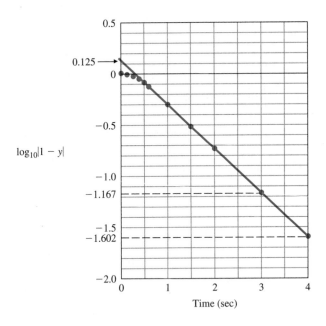

Combining these results, we arrive at the y estimate

$$\hat{y}(t) \cong 1 - 1.33e^{-t} + 0.33e^{-5.8t}. \qquad (3.86)$$

Equation (3.86) is plotted as the colored line in Fig. 3.46 and shows a reasonable fit to the data, although some error is noticeable near $t = 0$.

From $\hat{y}(t)$ we compute

Figure 3.45

$\log_{10}[y - (1 + Ae^{-\alpha t})]$ versus t

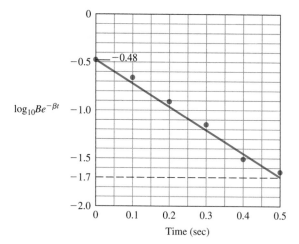

Figure 3.46

Model fits to the experimental data

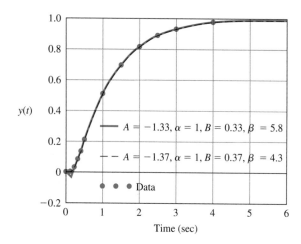

$$\hat{Y}(s) = \frac{1}{s} - \frac{1.33}{s+1} + \frac{0.33}{s+5.8}$$

$$= \frac{(s+1)(s+5.8) - 1.33s(s+5.8) + 0.33s(s+1)}{s(s+1)(s+5.8)}$$

$$= \frac{-0.58s + 5.8}{s(s+1)(s+5.8)}.$$

The resulting transfer function is

$$G(s) = \frac{-0.58(s - 10)}{(s+1)(s+5.8)}.$$

Notice that this method has given us a system with a zero in the RHP, even though the data showed no values of y that were negative. Very small differences in the estimated value for A, all of which approximately fit the data, can cause values of

β to range from 4 to 6. This illustrates the sensitivity of pole locations to the quality of the data and emphasizes the need for a good signal-to-noise ratio.

By using a computer to perform the plotting, we are better able to iterate the four parameters to achieve the best overall fit. The data presentation in Figs. 3.44 and 3.45 can be obtained directly by using a semilog plot. This eliminates having to calculate $\log_{10}$ and the exponential expression to find the values of the parameters. The equations of the lines to be fit to the data are $y(t) = Ae^{\alpha t}$ and $y(t) = Be^{\beta t}$, which are straight lines on a semilog plot. The parameters A and α, or B and β, are iteratively selected so that the straight line comes as close as possible to passing through the data. This process produces the improved fit shown by the dashed black line in Fig. 3.46. The revised parameters, $A = -1.37$, $B = 0.37$, and $\beta = 4.3$ result in the transfer function

$$G(s) = \frac{-0.22s + 4.3}{(s+1)(s+4.3)}.$$

The RHP zero is still present but it is now located at $s \cong +20$ and has no noticeable effect on the time response.

This set of data was fitted quite well by a second-order model. In many cases a higher-order model is required to explain the data and the modes may not be as well separated.

If the transient response has oscillatory modes, then these can sometimes be estimated by comparing them with the standard plots of Fig. 3.18. The period will give the frequency ω_d and the decay from one period to the next will afford an estimate of the damping ratio. If the response has a mixture of modes not well separated in frequency, then more sophisticated methods need to be used. One such is **least-squares system identification**, in which a numerical optimization routine selects the best combination of system parameters so as to minimize the fit error. The fit error is defined to be a scalar **cost function**

$$J = \sum_i (y_{data} - y_{model})^2, i = 1, 2, 3, \cdots, \text{ for each data point,}$$

so that fit errors at all data points are taken into account in determining the best value for the system parameters.

Least-squares system identification

3.7.2   Models from Other Data

As mentioned early in Section 3.1.2, we can also generate a model using frequency-response data, which are obtained by exciting the system with a set of sinusoids and plotting $H(j\omega)$. In Chapter 6 we will show how such plots can be used directly for design. Alternatively, we can use the frequency response to estimate the poles and zeros of a transfer function using straight-line asymptotes on a logarithmic plot.

The construction of dynamic models from normal stochastic operating records or from the response to a PRBS can be based either on the concept of cross-correlation or on the least-squares fit of a discrete equivalent model, both topics in the field of **system identification**. They require substantial presentation and background that are beyond the scope of this text. An introduction to system identification can be found in Chapter 8 of Franklin et al. (1998), and a comprehensive treatment is given in

Ljüng (1999). Based largely on the work of Professor Ljüng, the MATLAB Toolbox on Identification provides substantial software to perform system identification and to verify the quality of the proposed models.

△ 3.8   Amplitude and Time Scaling

The magnitude of the values of the variables in a problem is often very different, sometimes so much so that numerical difficulties arise. This was a serious problem years ago when equations were solved using analog computers and it was routine to *scale* the variables so that all had similar magnitudes. Today's widespread use of digital computers for solving differential equations has largely eliminated the need to scale a problem unless the number of variables is very large, because computers are now capable of accurately handling numbers with wide variations in magnitude. Nevertheless, it is wise to understand the principle of scaling for the few cases in which extreme variations in magnitude exist and scaling is necessary or the computer word size is limited.

3.8.1   Amplitude Scaling

There are two types of scaling that are sometimes carried out: amplitude scaling and time scaling, as we have already seen in Section 3.1.4. **Amplitude scaling** is usually performed unwittingly by simply picking units that make sense for the problem at hand. For the ball levitator, expressing the motion in millimeters and the current in milliamps would keep the numbers within a range that is easy to work with. Equations of motion are sometimes developed in the standard SI units such as meters, kilograms, and amperes, but when computing the motion of a rocket going into orbit, using kilometers makes more sense. The equations of motion are usually solved using computer-aided design software, which is often capable of working in any units. For higher-order systems it becomes important to scale the problem so that system variables have similar numerical variations. A method for accomplishing the best scaling for a complex system is first to estimate the maximum values for each system variable and then to scale the system so that each variable varies between -1 and 1.

In general, we can perform amplitude scaling by defining the scaled variables for each state element: If

$$x' = S_x x, \tag{3.87}$$

then

$$\dot{x}' = S_x \dot{x} \quad \text{and} \quad \ddot{x}' = S_x \ddot{x}. \tag{3.88}$$

We then pick S_x to result in the appropriate scale change, substitute Eqs. (3.87) and (3.88) into the equations of motion, and recompute the coefficients.

EXAMPLE 3.36   *Scaling for the Ball Levitator*

The linearized equation of motion for the ball levitator (see Example 9.2, Chapter 9) is

$$\delta\ddot{x} = 1667\delta x + 47.6\delta i, \tag{3.89}$$

where δx is in units of meters and δi is in units of amperes. Scale the variables for the ball levitator to result in units of millimeters and milliamps instead of meters and amps.

Solution.   Referring to Eq. (3.87), we define

$$\delta x' = S_x \delta x \quad \text{and} \quad \delta i' = S_i \delta i$$

such that both S_x and S_i have a value of 1000 in order to convert δx and δi in meters and amps to $\delta x'$ and $\delta i'$ in millimeters and milliamps. Substituting these relations into Eq. (3.89) and taking note of Eq. (3.88) yields

$$\delta \ddot{x}' = 1667\delta x' + 47.6\frac{S_x}{S_i}\delta i'.$$

In this case $S_x = S_i$, so Eq. (3.89) remains unchanged. Had we scaled the two quantities by different amounts, there would have been a change in the last coefficient in the equation.

3.8.2 Time Scaling

The unit of time when using SI units or English units is seconds. Computer-aided design software is *usually* able to compute results accurately no matter how fast or slow the particular problem at hand. However, if a dynamic system responds in a few microseconds, or if there are characteristic frequencies in the system on the order of several megahertz, the problem may become ill conditioned, so that the numerical routines produce errors. This can be particularly troublesome for high-order systems. The same holds true for an extremely slow system. It is therefore useful to know how to change the units of time should you encounter an ill-conditioned problem.

We define the new scaled time to be

$$\tau = \omega_o t \tag{3.90}$$

such that, if t is measured in seconds and $\omega_o = 1000$, then τ will be measured in milliseconds. The effect of the **time scaling** is to change the differentiation so that

$$\dot{x} = \frac{dx}{dt} = \frac{dx}{d(\tau/\omega_o)} = \omega_o\frac{dx}{d\tau} \tag{3.91}$$

and

$$\ddot{x} = \frac{d^2x}{dt^2} = \omega_o^2\frac{d^2x}{d\tau^2}. \tag{3.92}$$

EXAMPLE 3.37   *Time Scaling an Oscillator*

The equation for an oscillator was derived in Example 2.5. For a case with a very fast natural frequency $\omega_n = 15{,}000$ rad/sec (about 2 kHz), Eq. (2.23) can be rewritten as

$$\ddot{\theta} + 15000^2 \cdot \theta = 10^6 \cdot T_c.$$

Determine the time-scaled equation so that the unit of time is milliseconds.

Solution. The value of ω_o in Eq. (3.90) is 1000. Equation (3.92) shows that

$$\frac{d^2\theta}{d\tau^2} = 10^{-6} \cdot \ddot{\theta},$$

and the time-scaled equation becomes

$$\frac{d^2\theta}{d\tau^2} + 15^2 \cdot \theta = T_c.$$

In practice, we would then solve the equation

$$\ddot{\theta} + 15^2 \cdot \theta = T_c \tag{3.93}$$

and label the plots in milliseconds instead of seconds.

3.9   Historical Perspective

Oliver Heaviside (1850–1925) was an eccentric English electrical engineer, mathematician, and physicist. He was self-taught and left school at the age of 16 to become a telegraph operator. He worked mostly outside the scientific community that was hostile to him. He reformulated Maxwell's equations in the form that is used today. He also laid down the foundations of telecommunication and hypothesized the existence of the ionosphere. He developed the symbolic procedure known as Heaviside's operational calculus for solving differential equations. The Heaviside calculus was widely popular among electrical engineers in the 1920s and 1930s. This was later shown to be equivalent to the more rigorous Laplace transform named after the French mathematician Pierre-Simon Laplace (1749–1827) who had worked on operational calculus earlier.

Laplace was also an astronomer and a mathematician who is sometimes referred to as the "The Newton of France." He studied the origin and dynamical stability of the solar system completing Newton's work in his five volume *Méchanique céleste* (Celestial Mechanics). Laplace invented the general concept of potential as in a gravitational or electric field and described by Laplace's equation. Laplace had a brief political career as Napoleon's Interior Minister. During a famous exchange with Napoleon who asked Laplace why he had not mentioned God in *Méchanique céleste*, Laplace is said to have replied that "Sir, there was no need for that hypothesis." He was an opportunist and changed sides as the political winds shifted. Laplace's operational property transforms a differential equation into an algebraic operation that is much easier to manipulate in engineering applications. It is also applicable to solutions of partial differential equations, the original problem that Laplace was concerned with while developing the transform. Laplace formulated the Laplace's equation with applications to electromagnetic theory, fluid dynamics, and astronomy. Laplace also made fundamental contributions to probability theory.

Laplace and Fourier transforms are intimately related (see Appendix A). The Fourier series and the Fourier transform, developed in that order, provide methods for representing signals in terms of exponential functions. Fourier series are used to represent a periodic signal with discrete spectra in terms of a series. Fourier transforms are used to represent a non-periodic signal with continuous spectra in terms of an integral. The Fourier transform is named after the French mathematician Jean Batiste Joseph

Fourier (1768–1830) who used Fourier series to solve the heat conduction equation expressed in terms of Fourier series. Laplace and Fourier were contemporaries and knew each other very well. In fact, Laplace was one of Fourier's teachers. Fourier accompanied Napoleon on his Egyptian expedition in 1798 as a science advisor and is also credited with the discovery of the greenhouse effect.

Transform methods provide a unifying method in applications to solving many engineering problems. Linear transforms such as the Laplace transform and Fourier transform are useful for studying linear systems. While Fourier transforms are useful to study the steady-state behavior, Laplace transforms are used for studying the transient and closed-loop behavior of dynamic systems. The book by Gardner and Barnes in 1942 was influential in popularizing the Laplace transform in the United States.

SUMMARY

- The Laplace transform is the primary tool used to determine the behavior of linear systems. The Laplace transform of a time function $f(t)$ is given by

$$\mathcal{L}[f(t)] = F(s) = \int_{0^-}^{\infty} f(t)e^{-st}\, dt. \tag{3.94}$$

- This relationship leads to the key property of Laplace transforms, namely,

$$\mathcal{L}[\dot{f}(t)] = sF(s) - f(0^-). \tag{3.95}$$

- This property allows us to find the transfer function of a linear ODE. Given the transfer function $G(s)$ of a system and the input $u(t)$, with transform $U(s)$, the system output transform is $Y(s) = G(s)U(s)$.
- Normally, inverse transforms are found by referring to tables such as Table A.2 in Appendix A or by computer. Properties of Laplace transforms and their inverses are summarized in Table A.1 in Appendix A.
- The Final Value Theorem is useful in finding steady-state errors for stable systems: If all the poles of $s\,Y(s)$ are in the LHP, then

$$\lim_{t \to \infty} y(t) = \lim_{s \to 0} s\,Y(s). \tag{3.96}$$

- Block diagrams are a convenient way to show the relationships between the components of a system. They can usually be simplified using the relations in Fig. 3.9 and Eq. (3.50); that is, the transfer function of the block diagram

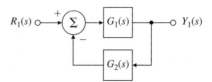

is equivalent to

$$Y_1(s) = \frac{G_1(s)}{1 + G_1(s)G_2(s)} R_1(s). \tag{3.97}$$

- The locations of poles in the s-plane determine the character of the response, as shown in Fig. 3.15.

- The location of a pole in the s-plane is defined by the parameters shown in Fig. 3.22. These parameters are related to the time-domain quantities of rise time t_r, settling time t_s, and overshoot M_p, which are defined in Fig. 3.22. The correspondences between them, for a second-order system with no zeros, are given by

$$t_r \cong \frac{1.8}{\omega_n}, \tag{3.98}$$

$$M_p = e^{-\pi\zeta/\sqrt{1-\zeta^2}}, \tag{3.99}$$

$$t_s = \frac{4.6}{\zeta\omega_n}. \tag{3.100}$$

- When a zero in the LHP is present, the overshoot increases. This effect is summarized in Figs. 3.26 and 3.27.
- When an additional stable pole is present, the system response is more sluggish. This effect is summarized in Figs. 3.34 and 3.35.
- For a stable system, all the closed-loop poles must be in the LHP.
- A system is stable if and only if all the elements in the first column of the Routh array are positive. To determine the Routh array, refer to the formulas in Section 3.6.3.
- Mason's rule is a useful technique to determining transfer functions of complicated interconnected systems.
- Determining a model from experimental data, or verifying an analytically based model by experiment, is an important step in system design.
- **Amplitude and time scaling** (Section 3.8) are methods by which certain complications of dealing with differential equations can be minimized. Scaling of variables results in numerical values that fall within a sufficiently narrow range of magnitude to minimize errors and allow for ease of computation.

REVIEW QUESTIONS

1. What is the definition of "transfer function"?
2. What are the properties of systems whose responses can be described by transfer functions?
3. What is the Laplace transform of $f(t - \lambda)1(t - \lambda)$ if the transform of $f(t)$ is $F(s)$?
4. State the Final Value Theorem.
5. What is the most common use of the Final Value Theorem in control?
6. Given a second-order transfer function with damping ratio ζ and natural frequency ω_n, what is the estimate of the step response rise time? What is the estimate of the percent overshoot in the step response? What is the estimate of the settling time?
7. What is the major effect of a zero in the LHP on the second-order step response?
8. What is the most noticeable effect of a zero in the RHP on the step response of the second-order system?
9. What is the main effect of an extra real pole on the second-order step response?
10. Why is stability an important consideration in control system design?
11. What is the main use of Routh's criterion?

12. Under what conditions might it be important to know how to estimate a transfer function from experimental data?

PROBLEMS

Problems for Section 3.1: Review of Laplace Transforms

3.1 Show that, in a partial-fraction expansion, complex conjugate poles have coefficients that are also complex conjugates. (The result of this relationship is that whenever complex conjugate pairs of poles are present, only one of the coefficients needs to be computed.)

3.2 Find the Laplace transform of the following time functions:

(a) $f(t) = 1 + 2t$

(b) $f(t) = 3 + 7t + t^2 + \delta(t)$

(c) $f(t) = e^{-t} + 2e^{-2t} + te^{-3t}$

(d) $f(t) = (t + 1)^2$

(e) $f(t) = \sinh t$

3.3 Find the Laplace transform of the following time functions:

(a) $f(t) = 3\cos 6t$

(b) $f(t) = \sin 2t + 2\cos 2t + e^{-t}\sin 2t$

(c) $f(t) = t^2 + e^{-2t}\sin 3t$

3.4 Find the Laplace transform of the following time functions:

(a) $f(t) = t\sin t$

(b) $f(t) = t\cos 3t$

(c) $f(t) = te^{-t} + 2t\cos t$

(d) $f(t) = t\sin 3t - 2t\cos t$

(e) $f(t) = 1(t) + 2t\cos 2t$

3.5 Find the Laplace transform of the following time functions (* denotes convolution):

(a) $f(t) = \sin t \sin 3t$

(b) $f(t) = \sin^2 t + 3\cos^2 t$

(c) $f(t) = (\sin t)/t$

(d) $f(t) = \sin t * \sin t$

(e) $f(t) = \int_0^t \cos(t - \tau)\sin \tau \, d\tau$

3.6 Given that the Laplace transform of $f(t)$ is $F(s)$, find the Laplace transform of the following:

(a) $g(t) = f(t)\cos t$

(b) $g(t) = \int_0^t \int_0^{t_1} f(\tau) \, d\tau \, dt_1$

3.7 Find the time function corresponding to each of the following Laplace transforms using partial-fraction expansions:

(a) $F(s) = \dfrac{2}{s(s+2)}$

(b) $F(s) = \dfrac{10}{s(s+1)(s+10)}$

(c) $F(s) = \frac{3s+2}{s^2+4s+20}$

(d) $F(s) = \frac{3s^2+9s+12}{(s+2)(s^2+5s+11)}$

(e) $F(s) = \frac{1}{s^2+4}$

(f) $F(s) = \frac{2(s+2)}{(s+1)(s^2+4)}$

(g) $F(s) = \frac{s+1}{s^2}$

(h) $F(s) = \frac{1}{s^6}$

(i) $F(s) = \frac{4}{s^4+4}$

(j) $F(s) = \frac{e^{-s}}{s^2}$

3.8   Find the time function corresponding to each of the following Laplace transforms:

(a) $F(s) = \frac{1}{s(s+2)^2}$

(b) $F(s) = \frac{2s^2+s+1}{s^3-1}$

(c) $F(s) = \frac{2(s^2+s+1)}{s(s+1)^2}$

(d) $F(s) = \frac{s^3+2s+4}{s^4-16}$

(e) $F(s) = \frac{2(s+2)(s+5)^2}{(s+1)(s^2+4)^2}$

(f) $F(s) = \frac{(s^2-1)}{(s^2+1)^2}$

(g) $F(s) = \tan^{-1}(\frac{1}{s})$

3.9   Solve the following ODEs using Laplace transforms:

(a) $\ddot{y}(t) + \dot{y}(t) + 3y(t) = 0; y(0) = 1, \dot{y}(0) = 2$

(b) $\ddot{y}(t) - 2\dot{y}(t) + 4y(t) = 0; y(0) = 1, \dot{y}(0) = 2$

(c) $\ddot{y}(t) + \dot{y}(t) = \sin t; y(0) = 1, \dot{y}(0) = 2$

(d) $\ddot{y}(t) + 3y(t) = \sin t; y(0) = 1, \dot{y}(0) = 2$

(e) $\ddot{y}(t) + 2\dot{y}(t) = e^t; y(0) = 1, \dot{y}(0) = 2$

(f) $\ddot{y}(t) + y(t) = t; y(0) = 1, \dot{y}(0) = -1$

3.10   Using the convolution integral, find the step response of the system whose impulse response is given below and shown in Fig. 3.47:

Figure 3.47

Impulse response for Problem 3.10

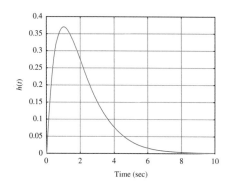

Time (sec)

$$h(t) = \begin{cases} te^{-t} & t \geq 0, \\ 0 & t < 0. \end{cases}$$

3.11 Using the convolution integral, find the step response of the system whose impulse response is given below and shown in Fig. 3.48:

$$h(t) = \begin{cases} 1, & 0 \leq t \leq 2, \\ 0, & t < 0 \text{ and } t > 2. \end{cases}$$

Figure 3.48

Impulse response for Problem 3.11

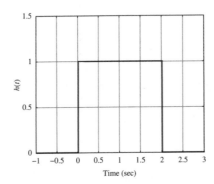

3.12 Consider the standard second-order system

$$G(s) = \frac{\omega_n^2}{s^2 + 2\zeta\omega_n s + \omega_n^2}.$$

(a) Write the Laplace transform of the signal in Fig. 3.49.

(b) What is the transform of the output if this signal is applied to $G(s)$?

(c) Find the output of the system for the input shown in Fig. 3.49.

Figure 3.49

Plot of input for Problem 3.12

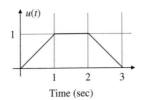

3.13 A rotating load is connected to a field-controlled DC motor with negligible field inductance. A test results in the output load reaching a speed of 1 rad/sec within $1/2$ sec when a constant input of 100 V is applied to the motor terminals. The output steady-state speed from the same test is found to be 2 rad/sec. Determine the transfer function $\frac{\theta(s)}{V_f(s)}$ of the motor.

3.14 A simplified sketch of a computer tape drive is given in Fig. 3.50.

(a) Write the equations of motion in terms of the parameters listed below. K and B represent the spring constant and the damping of tape stretch, respectively, and ω_1 and ω_2 are angular velocities. A positive current applied to the DC motor will provide a torque on the capstan in the clockwise direction as shown by the arrow. Find the value of current that just cancels the force, F, then eliminate the constant

Figure 3.50

Tape drive schematic

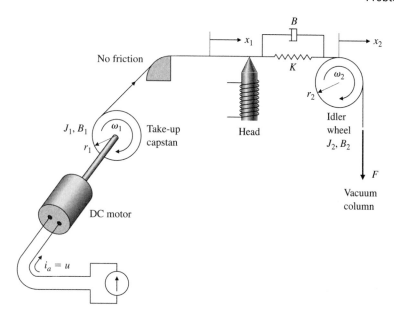

current and its balancing force, F; from your equations. Assume positive angular velocities of the two wheels are in the directions shown by the arrows.

$$J_1 = 5 \times 10^{-5} \text{ kg·m}^2, \text{ motor and capstan inertia}$$

$$B_1 = 1 \times 10^{-2} \text{ N·m·sec, motor damping}$$

$$r_1 = 2 \times 10^{-2} \text{ m}$$

$$K_t = 3 \times 10^{-2} \text{ N·m/A, motor–torque constant}$$

$$K = 2 \times 10^4 \text{ N/m}$$

$$B = 20 \text{ N/m·sec}$$

$$r_2 = 2 \times 10^{-2} \text{ m}$$

$$J_2 = 2 \times 10^{-5} \text{ kg·m}^2$$

$$B_2 = 2 \times 10^{-2} \text{ N·m·sec, viscous damping, idler}$$

$$F = 6 \text{ N, constant force}$$

$$\dot{x}_1 = \text{tape velocity m/sec (variable to be controlled)}$$

(b) Find the transfer function from the motor current to the tape position.

(c) Find the poles and zeros of the transfer function in part (b).

(d) Use MATLAB to find the response of x_1 to a step input in i_a.

3.15 For the system in Fig. 2.51, compute the transfer function from the motor voltage to position θ_2.

3.16 Compute the transfer function for the two-tank system in Fig. 2.55 with holes at A and C.

3.17 For a second-order system with transfer function

$$G(s) = \frac{3}{s^2 + 2s - 3},$$

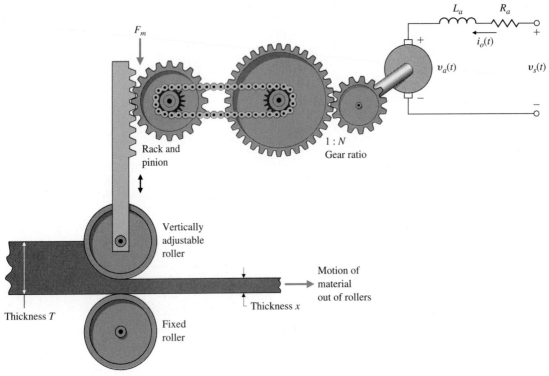

Figure 3.51

Continuous rolling mill

determine the following:

(a) The DC gain;

(b) The final value to a step input.

3.18   Consider the continuous rolling mill depicted in Fig. 3.51. Suppose that the motion of the adjustable roller has a damping coefficient b, and that the force exerted by the rolled material on the adjustable roller is proportional to the material's change in thickness: $F_s = c(T - x)$. Suppose further that the DC motor has a torque constant K_t and a back emf constant K_e, and that the rack-and-pinion has effective radius of R.

(a) What are the inputs to this system? The output?

(b) Without neglecting the effects of gravity on the adjustable roller, draw a block diagram of the system that explicitly shows the following quantities: $V_s(s)$, $I_0(s)$, $F(s)$ (the force the motor exerts on the adjustable roller), and $X(s)$.

(c) Simplify your block diagram as much as possible while still identifying output and each input separately.

Problems for Section 3.2: System Modeling Diagrams

3.19   Consider the block diagram shown in Fig. 3.52. Note that a_i and b_i are constants. Compute the transfer function for this system. This special structure is called the "control canonical form" and will be discussed further in Chapter 7.

Figure 3.52

Block diagram for
Problem 3.19

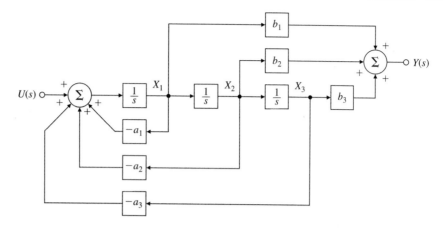

3.20 Find the transfer functions for the block diagrams in Fig. 3.53.

Figure 3.53

Block diagrams for
Problem 3.20

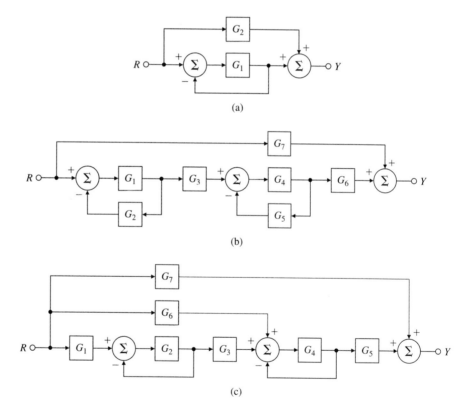

3.21 Find the transfer functions for the block diagrams in Fig. 3.54, using the ideas of block
diagram simplification. The special structure in Fig. 3.54(b) is called the "observer
canonical form" and will be discussed in Chapter 7.

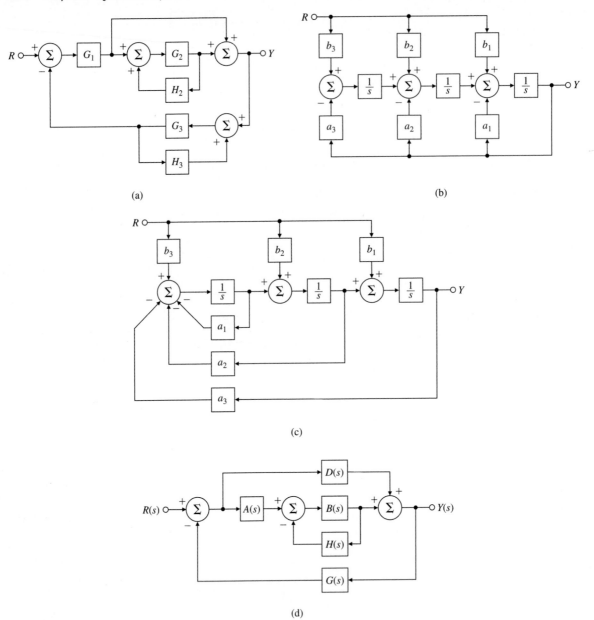

Figure 3.54

Block diagrams for Problem 3.21

3.22 Use block-diagram algebra to determine the transfer function between $R(s)$ and $Y(s)$ in Fig. 3.55.

Figure 3.55

Block diagram for
Problem 3.22

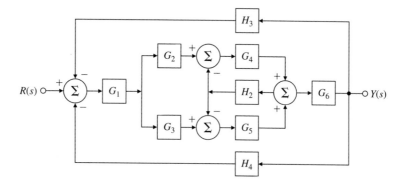

Problems for Section 3.3: Effect of Pole Locations

3.23 For the electric circuit shown in Fig. 3.56, find the following:

  (a) The time-domain equation relating $i(t)$ and $v_1(t)$;

  (b) The time-domain equation relating $i(t)$ and $v_2(t)$;

  (c) Assuming all initial conditions are zero, the transfer function $\frac{V_2(s)}{V_1(s)}$ and the damping ratio ζ and undamped natural frequency ω_n of the system;

  (d) The values of R that will result in $v_2(t)$ having an overshoot of no more than 25%, assuming $v_1(t)$ is a unit step, $L = 10$ mH, and $C = 4 \ \mu$F.

Figure 3.56

Circuit for Problem 3.23

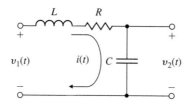

3.24 For the unity feedback system shown in Fig. 3.57, specify the gain K of the proportional controller so that the output $y(t)$ has an overshoot of no more than 10% in response to a unit step.

Figure 3.57

Unity feedback system
for Problem 3.24

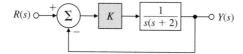

3.25 For the unity feedback system shown in Fig. 3.58, specify the gain and pole location of the compensator so that the overall closed-loop response to a unit-step input has an overshoot of no more than 25%, and a 1% settling time of no more than 0.1 sec. Verify your design using MATLAB.

Figure 3.58

Unity feedback system for Problem 3.25

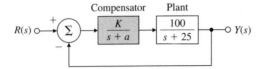

Problems for Section 3.4: Time-Domain Specification

3.26 Suppose you desire the peak time of a given second-order system to be less than t_p'. Draw the region in the s-plane that corresponds to values of the poles that meet the specification $t_p < t_p'$.

3.27 A certain servomechanism system has dynamics dominated by a pair of complex poles and no finite zeros. The time-domain specifications on the rise time (t_r), percent overshoot (M_p), and settling time (t_s) are given by

$$t_r \leq 0.6 \sec,$$

$$M_p \leq 17\%,$$

$$t_s \leq 9.2 \sec.$$

(a) Sketch the region in the s-plane where the poles could be placed so that the system will meet *all* three specifications.

(b) Indicate on your sketch the specific locations (denoted by $\times$) that will have the smallest rise-time and also meet the settling time specification *exactly*.

3.28 Suppose you are to design a unity feedback controller for a first-order plant depicted in Fig. 3.59. (As you will learn in Chapter 4, the configuration shown is referred to as a proportional-integral controller.) You are to design the controller so that the closed-loop poles lie within the shaded regions shown in Fig. 3.60.

(a) What values of ω_n and ζ correspond to the shaded regions in Fig. 3.59? (A simple estimate from the figure is sufficient.)

(b) Let $K_\alpha = \alpha = 2$. Find values for K and K_I so that the poles of the closed-loop system lie within the shaded regions.

Figure 3.59

Unity feedback system for Problem 3.28

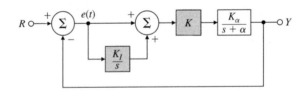

Figure 3.60

Desired closed-loop pole locations for Problem 3.28

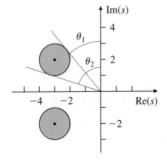

(c) Prove that no matter what the values of K_α and α are, the controller provides enough flexibility to place the poles anywhere in the complex (left-half) plane.

3.29 The open-loop transfer function of a unity feedback system is

$$G(s) = \frac{K}{s(s+2)}.$$

The desired system response to a step input is specified as peak time $t_p = 1$ sec and overshoot $M_p = 5\%$.

(a) Determine whether both specifications can be met simultaneously by selecting the right value of K.

(b) Sketch the associated region in the s-plane where both specifications are met, and indicate what root locations are possible for some likely values of K.

(c) Relax the specifications in part (a) by the same factor and pick a suitable value for K, and use MATLAB to verify that the new specifications are satisfied.

3.30 The equations of motion for the DC motor shown in Fig. 2.32 were given in Eqs. (2.52–2.53) as

$$J_m \ddot{\theta}_m + \left(b + \frac{K_t K_e}{R_a} \right) \dot{\theta}_m = \frac{K_t}{R_a} v_a.$$

Assume that

$$J_m = 0.01 \text{ kg·m}^2,$$

$$b = 0.001 \text{ N·m·sec},$$

$$K_e = 0.02 \text{ V·sec},$$

$$K_t = 0.02 \text{ N·m/A},$$

$$R_a = 10 \text{ }\Omega.$$

(a) Find the transfer function between the applied voltage v_a and the motor speed $\dot{\theta}_m$.

(b) What is the steady-state speed of the motor after a voltage $v_a = 10$ V has been applied?

(c) Find the transfer function between the applied voltage v_a and the shaft angle θ_m.

(d) Suppose feedback is added to the system in part (c) so that it becomes a position servo device such that the applied voltage is given by

$$v_a = K(\theta_r - \theta_m),$$

where K is the feedback gain. Find the transfer function between θ_r and θ_m.

(e) What is the maximum value of K that can be used if an overshoot $M_p < 20\%$ is desired?

(f) What values of K will provide a rise time of less than 4 sec? (Ignore the M_p constraint.)

(g) Use MATLAB to plot the step response of the position servo system for values of the gain $K = 0.5$, 1, and 2. Find the overshoot and rise time for each of the three step responses by examining your plots. Are the plots consistent with your calculations in parts (e) and (f)?

3.31 You wish to control the elevation of the satellite-tracking antenna shown in Figs. 3.61 and 3.62. The antenna and drive parts have a moment of inertia J and a damping B;

Figure 3.61

Satellite-tracking antenna

Source: Courtesy Space Systems/Loral

Figure 3.62

Schematic of antenna for Problem 3.31

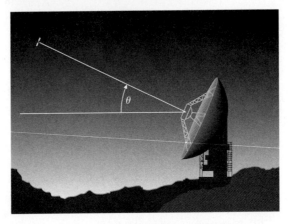

these arise to some extent from bearing and aerodynamic friction, but mostly from the back emf of the DC drive motor. The equations of motion are

$$J\ddot{\theta} + B\dot{\theta} = T_c,$$

where T_c is the torque from the drive motor. Assume that

$$J = 600,000 \text{ kg·m}^2 \quad B = 20,000 \text{ N·m·sec}.$$

(a) Find the transfer function between the applied torque T_c and the antenna angle θ.

(b) Suppose the applied torque is computed so that θ tracks a reference command θ_r according to the feedback law

$$T_c = K(\theta_r - \theta),$$

where K is the feedback gain. Find the transfer function between θ_r and θ.

(c) What is the maximum value of K that can be used if you wish to have an overshoot $M_p < 10\%$?

(d) What values of K will provide a rise time of less than 80 sec? (Ignore the M_p constraint.)

△

(e) Use MATLAB to plot the step response of the antenna system for $K = 200$, 400, 1000, and 2000. Find the overshoot and rise time of the four step responses by examining your plots. Do the plots confirm your calculations in parts (c) and (d)?

3.32   Show that the second-order system

$$\ddot{y} + 2\zeta\omega_n\dot{y} + \omega_n^2 y = 0, \quad y(0) = y_o, \quad \dot{y}(0) = 0,$$

has the response

$$y(t) = y_o \frac{e^{-\sigma t}}{\sqrt{1 - \zeta^2}} \sin(\omega_d t + \cos^{-1}\zeta).$$

Prove that, for the underdamped case ($\zeta < 1$), the response oscillations decay at a predictable rate (see Fig. 3.63) called the **logarithmic decrement**

$$\delta = \ln\frac{y_o}{y_1} = \ln e^{\sigma\tau_d} = \sigma\tau_d = \frac{2\pi\zeta}{\sqrt{1 - \zeta^2}}$$

$$= \ln\frac{\Delta y_1}{y_1} \cong \ln\frac{\Delta y_i}{y_i},$$

where

$$\tau_d = \frac{2\pi}{\omega_d} = \frac{2\pi}{\omega_n\sqrt{1 - \zeta^2}}$$

is the damped natural period of vibration. The damping coefficient in terms of the logarithmic decrement is then

$$\zeta = \frac{\delta}{\sqrt{4\pi^2 + \delta^2}}.$$

Figure 3.63

Definition of logarithmic decrement

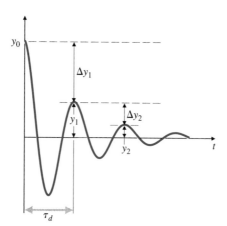

Problems for Section 3.5: Effect of Zeros and Additional Poles

3.33 In aircraft control systems, an ideal pitch response (q_o) versus a pitch command (q_c) is described by the transfer function

$$\frac{Q_o(s)}{Q_c(s)} = \frac{\tau\omega_n^2(s + 1/\tau)}{s^2 + 2\zeta\omega_n s + \omega_n^2}.$$

The actual aircraft response is more complicated than this ideal transfer function; nevertheless, the ideal model is used as a guide for autopilot design. Assume that t_r is the desired rise time and that

$$\omega_n = \frac{1.789}{t_r},$$

$$\frac{1}{\tau} = \frac{1.6}{t_r},$$

$$\zeta = 0.89.$$

Show that this ideal response possesses a fast settling time and minimal overshoot by plotting the step response for $t_r = 0.8, 1.0, 1.2,$ and 1.5 sec.

3.34 Consider the system shown in Fig. 3.64, where

$$G(s) = \frac{1}{s(s + 3)} \quad \text{and} \quad D(s) = \frac{K(s + z)}{s + p}. \tag{3.101}$$

Find K, z, and p so that the closed-loop system has a 10% overshoot to a step input and a settling time of 1.5 sec (1% criterion).

Figure 3.64

Unity feedback system for Problem 3.34

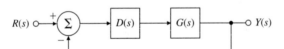

3.35 Sketch the step response of a system with the transfer function

$$G(s) = \frac{s/2 + 1}{(s/40 + 1)[(s/4)^2 + s/4 + 1]}.$$

Justify your answer on the basis of the locations of the poles and zeros. (Do not find the inverse Laplace transform.) Then compare your answer with the step response computed using MATLAB.

3.36 Consider the two nonminimum-phase systems,

$$G_1(s) = -\frac{2(s - 1)}{(s + 1)(s + 2)}, \tag{3.102}$$

$$G_2(s) = \frac{3(s - 1)(s - 2)}{(s + 1)(s + 2)(s + 3)}. \tag{3.103}$$

(a) Sketch the unit step responses for $G_1(s)$ and $G_2(s)$, paying close attention to the transient part of the response.

(b) Explain the difference in the behavior of the two responses as it relates to the zero locations.

(c) Consider a stable, strictly proper system (that is, m zeros and n poles, where $m < n$). Let $y(t)$ denote the step response of the system. The step response is said to have an undershoot if it initially starts off in the "wrong"direction. Prove that a stable, strictly proper system has an undershoot if and only if its transfer function has an *odd* number of *real* RHP zeros.

3.37 Find the relationships for the impulse response and the step response corresponding to Equation (3.57) for the cases where

(a) the roots are repeated.

(b) the roots are both real. Express your answers in terms of hyperbolic functions (sinh, cosh) to best show the properties of the system response.

(c) the value of the damping coefficient, ζ, is negative.

3.38 Consider the following second-order system with an extra pole:

$$H(s) = \frac{\omega_n^2 p}{(s+p)(s^2 + 2\zeta\omega_n s + \omega_n^2)}.$$

Show that the unit-step response is

$$y(t) = 1 + Ae^{-pt} + Be^{-\sigma t}\sin(\omega_d t - \theta),$$

where

$$A = \frac{-\omega_n^2}{\omega_n^2 - 2\zeta\omega_n p + p^2},$$

$$B = \frac{p}{\sqrt{(p^2 - 2\zeta\omega_n p + \omega_n^2)(1 - \zeta^2)}},$$

$$\theta = \tan^{-1}\frac{\sqrt{1 - \zeta^2}}{-\zeta} + \tan^{-1}\frac{\omega_n\sqrt{1 - \zeta^2}}{p - \zeta\omega_n}.$$

(a) Which term dominates $y(t)$ as p gets large?

(b) Give approximate values for A and B for small values of p.

(c) Which term dominates as p gets small? (Small with respect to what?)

(d) Using the preceding explicit expression for $y(t)$ or the step command in MATLAB, and assuming that $\omega_n = 1$ and $\zeta = 0.7$, plot the step response of the preceding system for several values of p ranging from very small to very large. At what point does the extra pole cease to have much effect on the system response?

3.39 Consider the second-order unity DC gain system with an extra zero,

$$H(s) = \frac{\omega_n^2(s+z)}{z(s^2 + 2\zeta\omega_n s + \omega_n^2)}.$$

(a) Show that the unit-step response for the system is given by

$$y(t) = 1 + \frac{\sqrt{1 + \frac{\omega_n^2}{z^2} - \frac{2\zeta\omega_n}{z}}}{\sqrt{1 - \zeta^2}}e^{-\sigma t}\cos(\omega_d t + \beta_1),$$

where

$$\beta_1 = \tan^{-1}\frac{-\zeta + \frac{\omega_n}{z}}{\sqrt{1 - \zeta^2}}.$$

(b) Derive an expression for the step response overshoot, M_p, of this system.

(c) For a given value of overshoot, M_p, how do we solve for ζ and ω_n?

3.40 The block diagram of an autopilot designed to maintain the pitch attitude θ of an aircraft is shown in Fig. 3.65. The transfer function relating the elevator angle δ_e and the pitch attitude θ is

$$\frac{\theta(s)}{\delta_e(s)} = G(s) = \frac{50(s+1)(s+2)}{(s^2 + 5s + 40)(s^2 + 0.03s + 0.06)},$$

where θ is the pitch attitude in degrees and δ_e is the elevator angle in degrees. The autopilot controller uses the pitch attitude error e to adjust the elevator according to the transfer function

$$\frac{\delta_e(s)}{E(s)} = D(s) = \frac{K(s+3)}{s+10}.$$

Using MATLAB, find a value of K that will provide an overshoot of less than 10% and a rise time faster than 0.5 sec for a unit-step change in θ_r. After examining the step response of the system for various values of K, comment on the difficulty associated with making rise time and overshoot measurements for complicated systems.

Figure 3.65

Block diagram of autopilot

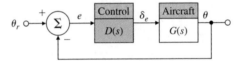

Problems for Section 3.6: Stability

3.41 A measure of the degree of instability in an unstable aircraft response is the amount of time it takes for the *amplitude* of the time response to double (see Fig. 3.66), given some nonzero initial condition.

(a) For a first-order system, show that the **time to double** is

$$\tau_2 = \frac{\ln 2}{p},$$

where p is the pole location in the RHP.

(b) For a second-order system (with two complex poles in the RHP), show that

$$\tau_2 = \frac{\ln 2}{-\zeta\omega_n}.$$

Figure 3.66

Time to double

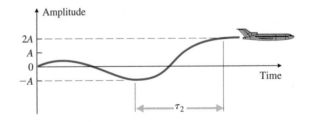

3.42 Suppose that unity feedback is to be applied around the listed open-loop systems. Use Routh's stability criterion to determine whether the resulting closed-loop systems will be stable.

(a) $KG(s) = \dfrac{4(s+2)}{s(s^3+2s^2+3s+4)}$

(b) $KG(s) = \dfrac{2(s+4)}{s^2(s+1)}$

(c) $KG(s) = \dfrac{4(s^3+2s^2+s+1)}{s^2(s^3+2s^2-s-1)}$

3.43 Use Routh's stability criterion to determine how many roots with positive real parts the following equations have:

(a) $s^4 + 8s^3 + 32s^2 + 80s + 100 = 0$.

(b) $s^5 + 10s^4 + 30s^3 + 80s^2 + 344s + 480 = 0$.

(c) $s^4 + 2s^3 + 7s^2 - 2s + 8 = 0$.

(d) $s^3 + s^2 + 20s + 78 = 0$.

(e) $s^4 + 6s^2 + 25 = 0$.

3.44 Find the range of K for which all the roots of the following polynomial are in the LHP:

$$s^5 + 5s^4 + 10s^3 + 10s^2 + 5s + K = 0.$$

Use MATLAB to verify your answer by plotting the roots of the polynomial in the s-plane for various values of K.

3.45 The transfer function of a typical tape-drive system is given by

$$G(s) = \frac{K(s+4)}{s[(s+0.5)(s+1)(s^2+0.4s+4)]},$$

where time is measured in milliseconds. Using Routh's stability criterion, determine the range of K for which this system is stable when the characteristic equation is $1 + G(s) = 0$.

3.46 Consider the closed-loop magnetic levitation system shown in Fig. 3.67. Determine the conditions on the system parameters (a, K, z, p, K_o) to guarantee closed-loop system stability.

Figure 3.67

Magnetic levitation system

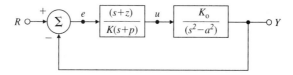

3.47 Consider the system shown in Fig. 3.68.

(a) Compute the closed-loop characteristic equation.

(b) For what values of (T, A) is the system stable? *Hint*: An approximate answer may be found using

$$e^{-Ts} \cong 1 - Ts$$

Figure 3.68

Control system for Problem 3.47

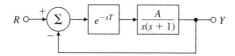

or

$$e^{-Ts} \cong \frac{1 - \frac{T}{2}s}{1 + \frac{T}{2}s}$$

for the pure delay. As an alternative, you could use the computer MATLAB (SIMULINK®) to simulate the system or to find the roots of the system's characteristic equation for various values of T and A.

3.48 Modify the Routh criterion so that it applies to the case in which all the poles are to be to the left of $-\alpha$ when $\alpha > 0$. Apply the modified test to the polynomial

$$s^3 + (6 + K)s^2 + (5 + 6K)s + 5K = 0,$$

finding those values of K for which all poles have a real part less than -1.

3.49 Suppose the characteristic polynomial of a given closed-loop system is computed to be

$$s^4 + (11 + K_2)s^3 + (121 + K_1)s^2 + (K_1 + K_1K_2 + 110K_2 + 210)s + 11K_1 + 100 = 0.$$

Find constraints on the two gains K_1 and K_2 that guarantee a stable closed-loop system, and plot the allowable region(s) in the (K_1, K_2) plane. You may wish to use the computer to help solve this problem.

3.50 Overhead electric power lines sometimes experience a low-frequency, high-amplitude vertical oscillation, or **gallop**, during winter storms when the line conductors become covered with ice. In the presence of wind, this ice can assume aerodynamic lift and drag forces that result in a gallop up to several meters in amplitude. Large-amplitude gallop can cause clashing conductors and structural damage to the line support structures caused by the large dynamic loads. These effects in turn can lead to power outages. Assume that the line conductor is a rigid rod, constrained to vertical motion only, and suspended by springs and dampers as shown in Fig. 3.69. A simple model of this conductor galloping is

$$m\ddot{y} + \frac{D(\alpha)\dot{y} - L(\alpha)v}{(\dot{y}^2 + v^2)^{1/2}} + T\left(\frac{n\pi}{\ell}\right)y = 0,$$

Figure 3.69

Electric power-line conductor

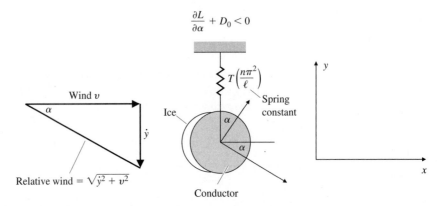

where

$$m = \text{mass of conductor,}$$

$$y = \text{conductor's vertical displacement,}$$

$$D = \text{aerodynamic drag force,}$$

$$L = \text{aerodynamic lift force,}$$

$$v = \text{wind velocity,}$$

$$\alpha = \text{aerodynamic angle of attack} = -\tan^{-1}(\dot{y}/v),$$

$$T = \text{conductor tension,}$$

$$n = \text{number of harmonic frequencies,}$$

$$\ell = \text{length of conductor.}$$

Assume that $L(0) = 0$ and $D(0) = D_0$ (a constant), and linearize the equation around the value $y = \dot{y} = 0$. Use Routh's stability criterion to show that galloping can occur whenever

$$\frac{\partial L}{\partial \alpha} + D_0 < 0.$$

A First Analysis of Feedback

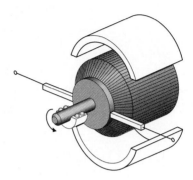

A Perspective on the Analysis of Feedback

In the next three chapters we will introduce three techniques for the design of controllers. Before doing so, it is useful to develop the assumptions to be used and to derive the equations that are common to each of the design approaches we describe. As a general observation, the dynamics of systems to which control is applied are nonlinear and very complex. However, in this initial analysis, we assume that the plant to be controlled as well as the controller can be represented as dynamic systems which are linear and time invariant (LTI). We also assume that they have only single inputs and single outputs, for the most part, and may thus be represented by simple scalar transfer functions. As we mentioned in Chapter 1, our basic concerns for control are **stability**, **tracking**, **regulation**, and **sensitivity**. The goal of the analysis in this chapter is to revisit each of these requirements in a linear dynamic setting and to develop equations that will expose constraints placed on the controller and identify elementary objectives to be suggested for the controllers.

Open-loop and closed-loop control

    The two fundamental structures for realizing controls are the open-loop structure as shown in Fig. 4.1, and the closed-loop structure, also known as feedback control, as shown in Fig. 4.2. The definition of open-loop control is that there is no closed signal path whereby the output influences the control effort. In the structure shown in Fig. 4.1, the controller transfer function modifies the reference input signal before it is applied to the plant. This controller might cancel the unwanted dynamics of the plant and replace them with the more desirable dynamics of the controller. In other cases open-loop control actions are taken on the plant as the environment changes, actions that are calibrated to give a good response but are not dependent on measuring the actual response. An example of this would be an aircraft autopilot whose parameters are changed with altitude or speed but not by feedback of the craft's motion. Feedback control, on the other hand, uses a sensor to measure the output and by feedback indirectly modifies the dynamics of the system. Although it is possible that feedback may cause an otherwise stable system to become unstable (a vicious circle), feedback gives the designer more flexibility and a preferable response to each of our objectives when compared to open-loop control.

Chapter Overview

The chapter begins with consideration of the basic equations of a simple open-loop structure and of an elementary feedback structure. In Section 4.1 the equations for the two structures are presented in general form and compared in turn with respect to **stability, tracking, regulation**, and **sensitivity**. In Section 4.2 the steady-state errors in response to polynomial inputs are analyzed in more detail. As part of the language of steady-state performance, control systems are assigned a **type** number according to the maximum degree of the input polynomial for which the steady-state error is a finite constant. For each type an appropriate error constant is defined, which allows the designer to easily compute the size of this error.

Although Maxwell and Routh developed a mathematical basis for assuring stability of a feedback system, design of controllers from the earliest days was largely trial and error based on experience. From this tradition there emerged an almost universal controller, the **proportional–integral–derivative (PID)** structure considered in Section 4.3. This device has three elements: a **P**roportional term to close the feedback loop, an **I**ntegral term to assure zero error to constant reference and disturbance inputs, and a **D**erivative term to improve (or realize!) stability and good dynamic response. In this section these terms are considered and their respective effects illustrated. As part of the evolution of the PID controller design, a major step was the development of a simple procedure for selecting the three parameters, a process called "tuning the controller." Ziegler and Nichols developed and published a set of experiments to be run, characteristics to be measured, and tuning values to be recommended as a result. These procedures are discussed in this section. Finally, in optional Section 4.4, a brief introduction to the increasingly common digital implementation of controllers is given. Sensitivity of time response to parameter changes is discussed in Appendix W4 on the web.

4.1 The Basic Equations of Control

We begin by collecting a set of equations and transfer functions that will be used throughout the rest of the text. For the open-loop system of Fig. 4.1, if we take the disturbance to be at the input of the plant, the output is given by

Figure 4.1

Open-loop system showing reference, R, control, U, disturbance, W, and output Y

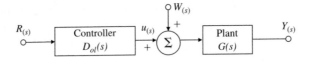

Figure 4.2

Closed-loop system showing the reference, R, control, U, disturbance, W, output, Y, and sensor noise, V

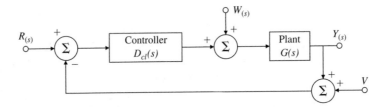

$$Y_{ol} = GD_{ol}R + GW \tag{4.1}$$

and the error, the difference between reference input and system output, is given by

$$E_{ol} = R - Y_{ol} \tag{4.2}$$

$$= R - [GD_{ol}R + GW] \tag{4.3}$$

$$= [1 - GD_{ol}]R - GW. \tag{4.4}$$

The open-loop transfer function in this case is $T_{ol}(s) = G(s)D_{ol}(s)$.

For feedback control, Fig. 4.2 gives the basic unity feedback structure of interest. There are three external inputs: the reference, R, which the output is expected to track, the plant disturbance, W, which the control is expected to counteract so it does not disturb the output, and the sensor noise, V, which the controller is supposed to ignore.

For the feedback block diagram of Fig. 4.2, the equations for the output and the control are given by the superposition of the responses to the three inputs individually, as follows:

$$Y_{cl} = \frac{GD_{cl}}{1 + GD_{cl}}R + \frac{G}{1 + GD_{cl}}W - \frac{GD_{cl}}{1 + GD_{cl}}V. \tag{4.5}$$

$$U = \frac{D_{cl}}{1 + GD_{cl}}R - \frac{GD_{cl}}{1 + GD_{cl}}W - \frac{D_{cl}}{1 + GD_{cl}}V. \tag{4.6}$$

Perhaps more important than these is the equation for the error, $E_{cl} = R - Y_{cl}$.

$$E_{cl} = R - \left[\frac{GD_{cl}}{1 + GD_{cl}}R + \frac{G}{1 + GD_{cl}}W - \frac{GD_{cl}}{1 + GD_{cl}}V\right] \tag{4.7}$$

$$= \frac{1}{1 + GD_{cl}}R - \frac{G}{1 + GD_{cl}}W + \frac{GD_{cl}}{1 + GD_{cl}}V. \tag{4.8}$$

In this case, the closed-loop transfer function is $T_{cl} = \dfrac{GD_{cl}}{1 + GD_{cl}}$.

With these equations we will explore the four basic objectives of **stability**, **tracking**, **regulation**, and **sensitivity** for both the open-loop and the closed-loop cases.

no bias and can be constructed to have very little noise over the entire range of low frequencies of interest. Thus, using this information, we design the controller transfer function to be large at the low frequencies, where it will reduce the effect of w, and we make it small at the higher frequencies, where it will reduce the effects of the high frequency sensor noise. The control engineer must determine in each case the best place on the frequency scale to make the cross over from amplifying to attenuation.

Exercise. Show that if w is a constant bias and if D_{cl} has a pole at $s = 0$ then the error due to this bias will be zero. However, show that if G has a pole at zero, it does not help with a disturbance bias.

4.1.4 Sensitivity

Suppose a plant is designed with gain G at a particular frequency but in operation it changes to be $G + \delta G$. This represents a fractional or percent change of gain of $\delta G/G$. For the purposes of this analysis, we set the frequency at zero and take the open-loop controller gain to be fixed at $D_{ol}(0)$. In the open-loop case the nominal overall gain is thus $T_{ol} = GD_{ol}$, and with the perturbed plant gain, the overall gain would be

$$T_{ol} + \delta T_{ol} = D_{ol}(G + \delta G) = D_{ol}G + D_{ol}\delta G = T_{ol} + D_{ol}\delta G.$$

Therefore, the gain change is $\delta T_{ol} = D_{ol}\delta G$. The sensitivity, S_G^T, of a transfer function, T_{ol}, to a plant gain, G, is *defined* to be the ratio of the fractional change in T_{ol} defined as $\frac{\delta T_{ol}}{T_{ol}}$ to the fractional change in G. In equation form

$$S_G^T = \frac{\dfrac{\delta T_{ol}}{T_{ol}}}{\dfrac{\delta G}{G}} \tag{4.13}$$

$$= \frac{G}{T_{ol}} \frac{\delta T_{ol}}{\delta G}. \tag{4.14}$$

Substituting the values, we find that

$$\frac{\delta T_{ol}}{T_{ol}} = \frac{D_{ol}\delta G}{D_{ol}G} = \frac{\delta G}{G}. \tag{4.15}$$

This means that a 10% error in G would yield a 10% error in T_{ol}. In the open-loop case, therefore, we have computed that $S = 1$.

From Eq. (4.5), the same change in G in the feedback case yields the new steady-state feedback gain as

$$T_{cl} + \delta T_{cl} = \frac{(G + \delta G)D_{cl}}{1 + (G + \delta G)D_{cl}},$$

where T_{cl} is the closed-loop gain. We can compute the sensitivity of this closed-loop gain directly using differential calculus. The closed-loop steady-state gain is

$$T_{cl} = \frac{GD_{cl}}{1 + GD_{cl}}.$$

The first-order variation is proportional to the derivative and is given by

$$\delta T_{cl} = \frac{dT_{cl}}{dG}\delta G.$$

The general expression for sensitivity from Eq. (4.13) is given by

$$S_G^{T_{cl}} \triangleq \text{sensitivity of } T_{cl} \text{ with respect to } G,$$

$$S_G^{T_{cl}} \triangleq \frac{G}{T_{cl}}\frac{dT_{cl}}{dG}, \tag{4.16}$$

so

$$S_G^{T_{cl}} = \frac{G}{GD_{cl}/(1+GD_{cl})}\frac{(1+GD_{cl})D_{cl}-D_{cl}(GD_{cl})}{(1+GD_{cl})^2}$$

$$= \frac{1}{1+GD_{cl}}. \tag{4.17}$$

Advantage of feedback

This result exhibits a major advantage of feedback:[2]

> In feedback control, the error in the overall transfer function gain is less sensitive to variations in the plant gain by a factor of $\mathcal{S} = \frac{1}{1+DG}$ compared to errors in open-loop control gain.

If the gain is such that $1 + DG = 100$, a 10% change in plant gain G will cause only a 0.1% change in the steady-state gain. The open-loop controller is 100 times more sensitive to gain changes than the closed-loop system with loop gain of 100. The example of the unity feedback case is so common that we will refer to the result of Eq. (4.17) simply as the sensitivity, $\mathcal{S}$, without subscripts or superscripts.

The results in this section so far have been computed under the assumption of the steady-state error in the presence of constant inputs, either reference or disturbance. Very similar results can be obtained for the steady-state behavior in the presence of a sinusoidal reference or disturbance signal. This is important because there are times when such signals naturally occur as, for example, with a disturbance of 60 Hertz due to power-line interference in an electronic system. The concept is also important because more complex signals can be described as containing sinusoidal components over a band of frequencies and analyzed using superposition of one frequency at a time. For example, it is well known that human hearing is restricted to signals in the frequency range of about 60 to 15,000 Hertz. A feedback amplifier and loudspeaker system designed for high-fidelity sound must accurately track any sinusoidal (pure tone) signal in this range. If we take the controller in the feedback system shown in Fig. 4.2 to have the transfer function $D(s)$ and we take the process to have the transfer function $G(s)$, then the steady-state open-loop gain at the sinusoidal signal of frequency ω_o will be $|G(j\omega_o)D(j\omega_o)|$ and the error of the feedback system will be

$$|E(j\omega_o)| = |R(j\omega_o)|\left|\frac{1}{1+G(j\omega_o)D(j\omega_o)}\right|. \tag{4.18}$$

---

[2]Bode, who developed the theory of sensitivity as well as many other properties of feedback, defined sensitivity as $S = 1 + GD$, the inverse of our choice.

Thus, to reduce errors to 1% of the input at the frequency ω_o, we must make $|1+DG| \geq$ 100 or, effectively, $|D(j\omega_o)G(j\omega_o)| \gtrsim 100$ and a good audio amplifier must have this loop gain over the range $2\pi 60 \leq \omega \leq 2\pi 15000$. We will revisit this concept in Chapter 6 as part of the design based on frequency response techniques.

The Filtered Case

Thus far the analysis has been based on the simplest open- and closed-loop structures. A more general case includes a dynamic filter on the input and also dynamics in the sensor. The filtered open-loop structure is shown in Fig. 4.3 having the transfer function $T_{ol} = GD_{ol}F$. In this case, the open-loop controller transfer function has been simply replaced by DF and the discussion given for the unfiltered open-loop case is easily applied to this change.

For the filtered feedback case shown in Fig. 4.4, the changes are more siginificant. In that case, the transform of the system output is given by

$$Y = \frac{GD_{cl}F}{1 + GD_{cl}H}R + \frac{G}{1 + GD_{cl}H}W - \frac{HGD_{cl}}{1 + GD_{cl}H}V. \qquad (4.19)$$

As is evident from this equation, the sensor dynamics, H is part of the loop transfer function and enters into the question of stability with $D_{cl}H$ replacing the D_{cl} of the unity feedback case. In fact, if $F = H$ then, with respect to stability, tracking, and regulation, the filtered case is identical to the unity case with $D_{cl}H$ replacing D_{cl}. On the other hand, the filter transfer function F can play the role of the open-loop controller except that here the filter F would be called on to modify the entire loop transfer function, $\frac{GD_{cl}}{1+GD_{cl}H}$, rather than simply GD_{ol}. Therefore the filtered closed-loop structure can realize the best properties of both the open-loop and the unity feedback closed-loop cases. The controller, D_{cl}, can be designed to effectively regulate the system for the disturbance W and the sensor noise, V, while the filter F is designed to improve the tracking accuracy. If the sensor dynamics, H, are accessible to the designer, this term can also be designed to improve the response to the sensor noise. The remaining issue is sensitivity.

Figure 4.3

Filtered open-loop system

Figure 4.4

Filtered closed-loop. R = reference, u = control, Y = output, and V = sensor noise

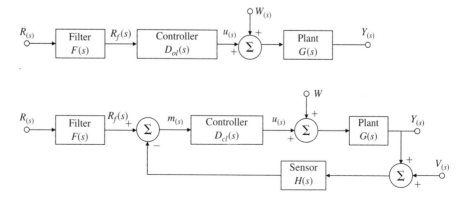

Using the formula given in Eq. (4.13), with changes in the parameter of interest, we can compute

$$S_F^{T_{cl}} = 1.0, \tag{4.20}$$

$$S_G^{T_{cl}} = \frac{1}{1 + GD_{cl}H}, \tag{4.21}$$

$$S_H^{T_{cl}} = \frac{GD_{cl}H}{1 + GD_{cl}H}. \tag{4.22}$$

Of these, the most interesting is the last. Notice that with respect to H, the sensitivity approaches unity as the loop gain grows. Therefore it is particularly important that the transfer function of the sensor be not only low in noise but also very stable in gain. Money spent on the sensor is money well spent.

4.2 Control of Steady-State Error to Polynomial Inputs: System Type

In studying the regulator problem, the reference input is taken to be a constant. It is also the case that the most common plant disturbance is a constant bias. Even in the general tracking problem the reference input is often constant for long periods of time or may be adequately approximated as if it were a polynomial in time, usually one of low degree. For example, when an antenna is tracking the elevation angle to a satellite, the time history as the satellite approaches overhead is an S-shaped curve as sketched in Fig. 4.5. This signal may be approximated by a linear function of time (called a ramp function or velocity input) for a significant time relative to the speed of response of the servomechanism. As another example, the position control of an elevator has a ramp function reference input, which will direct the elevator to move with constant speed until it comes near the next floor. In rare cases, the input can even be approximated over a substantial period as having a constant acceleration. Consideration of these cases leads us to consider steady-state errors in stable systems with polynomial inputs.

As part of the study of steady-state errors to polynomial inputs, a terminology has been developed to express the results. For example, we classify systems as to "**type**" according to the degree of the polynomial that they can reasonably track. For example, a system that can track a polynomial of degree 1 with a constant error is called *Type 1*. Also, to quantify the tracking error, several "**error constants**" are defined. In all of the following analysis, it is assumed that the systems are stable, else the analysis makes no sense at all.

Figure 4.5

Signal for satellite tracking

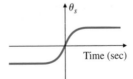

4.2.1 System Type for Tracking

In the unity feedback case shown in Fig. 4.2, the system error is given by Eq. (4.8). If we consider tracking the reference input alone and set $W = V = 0$, then the equation for the error is simply

$$E = \frac{1}{1 + GD_{cl}}R = \mathcal{S}R. \tag{4.23}$$

To consider polynomial inputs, we let $r(t) = {}^{t^k}/{}_{k!}1(t)$ for which the transform is $R = \frac{1}{s^{k+1}}$. We take a mechanical system as the basis for a generic reference nomenclature, calling step inputs for which $k = 0$ "position" inputs, ramp inputs for which $k = 1$ are called "velocity" inputs and if $k = 2$, the inputs are called "acceleration" inputs, regardless of the units of the actual signals. Application of the Final Value Theorem to the error formula gives the result

$$\lim_{t \to \infty} e(t) = e_{ss} = \lim_{s \to 0} sE(s) \tag{4.24}$$

$$= \lim_{s \to 0} s \frac{1}{1 + GD_{cl}}R(s) \tag{4.25}$$

$$= \lim_{s \to 0} s \frac{1}{1 + GD_{cl}} \frac{1}{s^{k+1}}. \tag{4.26}$$

We consider first a system for which GD_{cl} has no pole at the origin and a step input for which $R(s) = {}^1/{}_s$. Thus $r(t)$ is a polynomial of degree 0. In this case, Eq. (4.26) reduces to

$$e_{ss} = \lim_{s \to 0} s \frac{1}{1 + GD_{cl}} \frac{1}{s} \tag{4.27}$$

$$= \frac{1}{1 + GD_{cl}(0)}. \tag{4.28}$$

We define this system to be *Type 0* and we define the constant, $GD_{cl}(0) \triangleq K_p$ as the "*position error constant.*" Notice that if the input should be a polynomial of degree higher than 1, the resulting error would grow without bound. A polynomial of degree 0 is the highest degree a system of *Type 0* can track at all. If $GD_{cl}(s)$ has one pole at the origin, we could continue this line of argument and consider first-degree polynomial inputs but it is quite straightforward to evaluate Eq. (4.26) in a general setting. For this case, it is necessary to describe the behavior of the controller and plant as s approaches 0. For this purpose, we collect all the terms except the pole(s) at the origin into a function $GD_{clo}(s)$, which is finite at $s = 0$ so that we can define the constant $GD_{clo}(0) = K_n$ and write the loop transfer function as

$$GD_{cl}(s) = \frac{GD_{clo}(s)}{s^n}. \tag{4.29}$$

For example, if GD_{cl} has no integrator, then $n = 0$. If the system has one integrator, then $n = 1$, and so forth. Substituting this expression into Eq. (4.26),

$$e_{ss} = \lim_{s \to 0} s \frac{1}{1 + \dfrac{GD_{clo}(s)}{s^n} \dfrac{1}{s^{k+1}}} \tag{4.30}$$

$$= \lim_{s \to 0} \frac{s^n}{s^n + K_n} \frac{1}{s^k}. \tag{4.31}$$

From this equation we can see at once that if $n > k$ then $e = 0$ and if $n < k$ then $e \to \infty$. If $n = k = 0$, then $e_{ss} = \frac{1}{1+K_0}$ and if $n = k \neq 0$, then $e_{ss} = {}^1/K_n$. As we saw above, if $n = k = 0$, the input is a zero-degree polynomial otherwise known as a step or position, the constant K_o is called the "*position constant*" written as K_p, and the system is classified as "*Type 0.*" If $n = k = 1$, the input is a first-degree polynomial otherwise known as a ramp or velocity input and the constant K_1 is called the "*velocity constant*" written as K_v. This system is classified "*Type 1*" (read "type one"). In a similar way, systems of *Type 2* and higher types may be defined. A clear picture of the situation is given by the plot in Fig. 4.6 for a system of Type 1 having a ramp reference input. The error between input and output of size $\frac{1}{K_v}$ is clearly marked.

Using Eq (4.29), these results can be summarized by the equations:

$$K_p = \lim_{s \to 0} GD_{cl}(s), \qquad n = 0, \tag{4.32}$$

$$K_v = \lim_{s \to 0} sGD_{cl}(s), \qquad n = 1, \tag{4.33}$$

$$K_a = \lim_{s \to 0} s^2 GD_{cl}(s), \qquad n = 2. \tag{4.34}$$

The type information can also be usefully gathered in a table of error values as a function of the degree of the input polynomial and the type of the system as shown in Table 4.1.

Figure 4.6

Relationship between ramp response and K_v

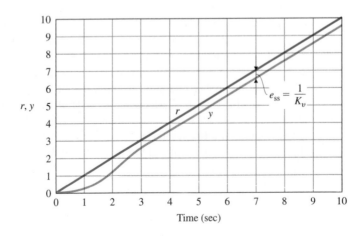

TABLE 4.1

Errors as a Function of System Type			
Type Input	Step (position)	Ramp (velocity)	Parabola (acceleration)
Type 0	$\dfrac{1}{1+K_p}$	∞	∞
Type 1	0	$\dfrac{1}{K_v}$	∞
Type 2	0	0	$\dfrac{1}{K_a}$

EXAMPLE 4.1   *System Type for Speed Control*

Determine the system type and the relevant error constant for speed control with proportional feedback given by $D(s) = k_p$. The plant transfer function is $G = \frac{A}{\tau s+1}$.

Solution. In this case, $GD_{cl} = \frac{k_p A}{\tau s+1}$ and applying Eq. (4.32) we see that $n = 0$ in this case as there is no pole at $s = 0$. Thus the system is Type 0 and the error constant is a position constant given by $K_p = k_p A$.

EXAMPLE 4.2   *System Type Using Integral Control*

Determine the system type and the relevant error constant for the speed control example with proportional plus integral control having controller given by $D_c = k_p + k_I/s$. The plant transfer function is $G = \frac{A}{\tau s+1}$.

Solution. In this case, the loop transfer function is $GD_{cl}(s) = \frac{A(k_p s+k_I)}{s(\tau s+1)}$ and, as a unity feedback system with a single pole at $s = 0$, the system is immediately seen as Type 1. The velocity constant is given by Eq. (4.33) to be $K_v = \lim_{s\to 0} sGD_{cl}(s) = Ak_I$.

The definition of system type helps us to identify quickly the ability of a system to track polynomials. In the unity feedback structure, if the process parameters change without removing the pole at the origin in a Type 1 system, the velocity constant will change but the system will still have zero steady-state error in response to a constant input and will still be Type 1. Similar statements can be made for systems of Type 2 or higher. Thus we can say that system type is a **robust property** with respect to parameter changes in the unity feedback structure. Robustness is a major reason for preferring unity feedback over other kinds of control structure.

Robustness of system type

Another form of the formula for the error constants can be developed directly in terms of the closed-loop transfer function $T(s)$. From Fig. 4.4 the transfer function including a sensor transfer function is

$$\frac{Y(s)}{R(s)} = T(s) = \frac{GD}{1+GDH}, \qquad (4.35)$$

and the system error is

$$E(s) = R(s) - Y(s) = R(s) - T(s)R(s).$$

The reference-to-error transfer function is thus

$$\frac{E(s)}{R(s)} = 1 - T(s),$$

and the system error transform is

$$E(s) = [1 - T(s)]R(s).$$

We assume that the conditions of the Final Value Theorem are satisfied, namely that all poles of $sE(s)$ are in the LHP. In that case the steady-state error is given by applying the Final Value Theorem to get

$$e_{ss} = \lim_{t \to \infty} e(t) = \lim_{s \to 0} sE(s) = \lim_{s \to 0} s[1 - T(s)]R(s). \tag{4.36}$$

If the reference input is a polynomial of degree k, the error transform becomes

$$E(s) = \frac{1}{s^{k+1}}[1 - T(s)]$$

and the steady-state error is given again by the Final Value Theorem:

$$e_{ss} = \lim_{s \to 0} s\frac{1 - T(s)}{s^{k+1}} = \lim_{s \to 0} \frac{1 - T(s)}{s^{k}}. \tag{4.37}$$

As before, the result of evaluating the limit in Eq. (4.37) can be zero, a nonzero constant, or infinite and if the solution to Eq. (4.37) is a nonzero constant, the system is referred to as *Type k*. Notice that a system of Type 1 or higher has a closed-loop DC gain of 1.0, which means that $T(0) = 1$ in these cases.

EXAMPLE 4.3   *System Type for a Servo with Tachometer Feedback*

Consider an electric motor position control problem including a non-unity feedback system caused by having a tachometer fixed to the motor shaft and its voltage (which is proportional to shaft speed) is fed back as part of the control. The parameters are

$$G(s) = \frac{1}{s(\tau s + 1)},$$

$$D(s) = k_p,$$

$$H(s) = 1 + k_t s,$$

$$F(s) = 1.$$

Determine the system type and relevant error constant with respect to reference inputs.

Solution.   The system error is

$$E(s) = R(s) - Y(s)$$

$$= R(s) - T(s)R(s)$$

$$= R(s) - \frac{DG(s)}{1 + HDG(s)}R(s)$$

$$= \frac{1 + (H(s) - 1)DG(s)}{1 + HDG(s)}R(s).$$

The steady-state system error from Eq. (4.37) is

$$e_{ss} = \lim_{s \to 0} sR(s)[1 - T(s)].$$

For a polynomial reference input, $R(s) = 1/s^{k+1}$ and hence

$$e_{ss} = \lim_{s \to 0} \frac{[1 - T(s)]}{s^k} = \lim_{s \to 0} \frac{1}{s^k} \frac{s(\tau s + 1) + (1 + k_t s - 1)k_p}{s(\tau s + 1) + (1 + k_t s)k_p}$$

$$= 0, \qquad k = 0$$

$$= \frac{1 + k_t k_p}{k_p}, \qquad k = 1;$$

therefore the system is Type 1 and the velocity constant is $K_v = \frac{k_p}{1+k_t k_p}$. Notice that if $k_t > 0$, perhaps to improve stability or dynamic response, the velocity constant is smaller than with simply the unity feedback value of k_p. The conclusion is that if tachometer feedback is used to improve dynamic response, the steady-state error is usually increased.

4.2.2 System Type for Regulation and Disturbance Rejection

A system can also be classified with respect to its ability to reject polynomial disturbance inputs in a way analogous to the classification scheme based on reference inputs. The transfer function from the disturbance input $W(s)$ to the error $E(s)$ is

$$\frac{E(s)}{W(s)} = \frac{-Y(s)}{W(s)} = T_w(s) \tag{4.38}$$

because, if the reference is equal to zero, the output is the error. In a similar way as for reference inputs, the system is Type 0 if a step disturbance input results in a nonzero constant steady-state error and is Type 1 if a ramp disturbance input results in a steady-state value of the error that is a non zero constant, etc. In general, following the same approach used in developing Eq. (4.31), we assume that a constant n and a function $T_{o,w}(s)$ can be defined with the properties that $T_{o,w}(0) = 1/K_{n,w}$ and that the disturbance-to-error transfer function can be written as

$$T_w(s) = s^n T_{o,w}(s). \tag{4.39}$$

Then the steady-state error to a disturbance input, which is a polynomial of degree k, is

$$y_{ss} = \lim_{s \to 0} \left[s T_w(s) \frac{1}{s^{k+1}} \right]$$

$$= \lim_{s \to 0} \left[T_{o,w}(s) \frac{s^n}{s^k} \right]. \tag{4.40}$$

From Eq. (4.40), if $n > k$, then the error is zero and if $n < k$, the error is unbounded. If $n = k$, the system is type k and the error is given by $1/K_{n,w}$.

Figure 4.7

DC motor with unity
feedback

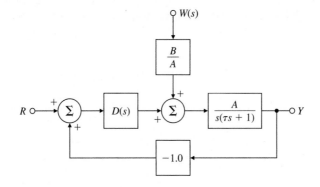

EXAMPLE 4.4   *System Type for a DC Motor Position Control*

Consider the simplified model of a DC motor in unity feedback as shown in
Fig. 4.7, where the disturbance torque is labeled $W(s)$. This case was considered
in Example 2.11.

(a) Use the controller

$$D(s) = k_p, \tag{4.41}$$

and determine the system type and steady-state error properties with respect to
disturbance inputs.

(b) Let the controller transfer function be given by

$$D(s) = k_p + \frac{k_I}{s}, \tag{4.42}$$

and determine the system type and the steady-state error properties for disturbance
inputs.

Solution. (a) The closed-loop transfer function from W to E (where $R = 0$) is

$$T_w(s) = \frac{-B}{s(\tau s + 1) + Ak_p}$$

$$= s^0 T_{o,w},$$

$$n = 0,$$

$$K_{o,w} = \frac{-Ak_p}{B}.$$

Applying Eq. (4.40) we see that the system is Type 0 and the steady-state error to a
unit step torque input is $e_{ss} = {-B}/{Ak_p}$. From the earlier section, this system is seen
to be Type 1 for reference inputs and illustrates that system type can be different for
different inputs to the same system.

(b) For this controller the disturbance error transfer function is

$$T_w(s) = \frac{-Bs}{s^2(\tau s + 1) + (k_p s + k_I)A}, \tag{4.43}$$

$$n = 1, \tag{4.44}$$

$$K_{n,w} = \frac{Ak_I}{-B}. \tag{4.45}$$

and therefore the system is Type 1 and the error to a unit ramp disturbance input will be

$$e_{ss} = \frac{-B}{Ak_I}. \tag{4.46}$$

Truxal's Formula for the Error Constants

Truxal (1955) derived a formula for the velocity constant of a Type 1 system in terms of the closed-loop poles and zeros, a formula that connects the steady-state error to the system's dynamic response. Since control design often requires a trade-off between these two characteristics, Truxal's formula can be useful to know. Its derivation is quite direct. Suppose the closed-loop transfer function $T(s)$ of a Type 1 system is

$$T(s) = K \frac{(s - z_1)(s - z_2) \cdots (s - z_m)}{(s - p_1)(s - p_2) \cdots (s - p_n)}. \tag{4.47}$$

Since the steady-state error in response to a step input in a Type 1 system is zero, the DC gain is unity; thus

$$T(0) = 1. \tag{4.48}$$

The system error is given by

$$E(s) \triangleq R(s) - Y(s) = R(s) \left[1 - \frac{Y(s)}{R(s)} \right] = R(s)[1 - T(s)]. \tag{4.49}$$

The system error due to a unit ramp input is given by

$$E(s) = \frac{1 - T(s)}{s^2}. \tag{4.50}$$

Using the Final Value Theorem, we get

$$e_{ss} = \lim_{s \to 0} \frac{1 - T(s)}{s}. \tag{4.51}$$

Using L'Hôpital's rule we rewrite Eq. (4.51) as

$$e_{ss} = - \lim_{s \to 0} \frac{dT}{ds} \tag{4.52}$$

or

$$e_{ss} = - \lim_{s \to 0} \frac{dT}{ds} = \frac{1}{K_v}. \tag{4.53}$$

Equation (4.53) implies that $1/K_v$ is related to the slope of the transfer function at the origin, a result that will also be shown in Section 6.1.2. Using Eq. (4.48), we can rewrite Eq. (4.53) as

$$e_{ss} = - \lim_{s \to 0} \frac{dT}{ds} \frac{1}{T} \tag{4.54}$$

or

$$e_{ss} = - \lim_{s \to 0} \frac{d}{ds} [\ln T(s)]. \tag{4.55}$$

Substituting Eq. (4.47) into Eq. (4.55), we get

$$e_{ss} = - \lim_{s \to 0} \frac{d}{ds} \left\{ \ln \left[K \frac{\prod_{i=1}^{m}(s - z_i)}{\prod_{i=1}^{n}(s - p_i)} \right] \right\} \tag{4.56}$$

$$= -\lim_{s \to 0} \frac{d}{ds} \left[K + \sum_{i=1}^{m} \ln(s - z_i) - \sum_{i=1}^{m} \ln(s - p_i) \right] \qquad (4.57)$$

or

$$\frac{1}{K_v} = -\frac{d \ln T}{ds}\bigg|_{s=0} = \sum_{i=1}^{n} -\frac{1}{p_i} + \sum_{i=1}^{m} \frac{1}{z_i}. \qquad (4.58)$$

We observe from Eq. (4.58) that K_v increases as the closed-loop poles move away from the origin. Similar relationships exist for other error coefficients, and these are explored in the problems.

EXAMPLE 4.5

Truxal's formula

Truxal's Formula

A third-order Type 1 system has closed-loop poles at $-2 \pm 2j$ and -0.1. The system has only one closed-loop zero. Where should the zero be if a $K_v = 10$ is desired?

Solution. From Truxal's formula we have,

$$\frac{1}{K_v} = -\frac{1}{-2+2j} - \frac{1}{-2-2j} - \frac{1}{-0.1} + \frac{1}{z}$$

or

$$0.1 = 0.5 + 10 + \frac{1}{z},$$

$$\frac{1}{z} = 0.1 - 0.5 - 10,$$

$$= -10.4$$

Therefore, the closed-loop zero should be at $z = {}^1/_{-10.4} = -0.0962$.

4.3 The Three-Term Controller: PID Control

In later chapters we will study three general analytic and graphical design techniques based on the root locus, the frequency response, and the state space formulation of the equations. Here we describe a control method having an older pedigree that was developed through long experience and by trial and error. Starting with simple proportional feedback, engineers early discovered integral control action as a means of eliminating bias offset. Then, finding poor dynamic response in many cases, an "anticipatory" term based on the derivative was added. The result is called the three-term or PID controller and has the transfer function[3]

$$D(s) = k_p + \frac{k_I}{s} + k_D s, \qquad (4.59)$$

where k_p is the proportional term, k_I is the integral term, and k_D is the derivative term. We'll discuss them in turn.

---

[3]The derivative term alone makes this transfer function nonproper and impractical. However adding a high-frequency pole to make the term proper only slightly modifies the performance.

4.3.1 Proportional Control (P)

When the feedback control signal is linearly proportional to the system error, we call the result **proportional feedback**. This was the case for the feedback used in the controller of speed in Section 4.1 for which the controller transfer function is

$$\frac{U(s)}{E(s)} = D_{cl}(s) = k_p. \tag{4.60}$$

If the plant is second order, as, for example, for a motor with nonnegligible inductance, then the plant transfer function can be written as

$$G(s) = \frac{A}{s^2 + a_1 s + a_2}. \tag{4.61}$$

In this case, the characteristic equation with proportional control is

$$1 + k_p G(s) = 0, \tag{4.62}$$

$$s^2 + a_1 s + a_2 + k_p = 0. \tag{4.63}$$

The designer can control the constant term in this equation, which determines the natural frequency, but cannot control the damping of the equation. The system is Type 0 and if k_p is made large to get adequately small steady-state error, the damping may be much too low for satisfactory transient response with proportional control alone.

4.3.2 Proportional Plus Integral Control (PI)

Proportional plus integral control

Adding an integral term to the controller to get the automatic reset effect results in the **proportional plus integral** control equation in the time domain:

$$u(t) = k_p e + k_I \int_{t_0}^{t} e(\tau)\, d\tau, \tag{4.64}$$

for which the $D_{cl}(s)$ in Fig. 4.2 becomes

$$\frac{U(s)}{E(s)} = D_{cl}(s) = k_p + \frac{k_I}{s}. \tag{4.65}$$

Introduction of the integral term raises the type to Type 1 and the system can therefore reject completely constant bias disturbances. For example, consider PI control in a speed control example, where the plant is described by

$$Y = \frac{A}{\tau s + 1}(U + W). \tag{4.66}$$

The transform equation for the controller is

$$U = k_p(R - Y) + k_I \frac{R - Y}{s}, \tag{4.67}$$

and the system transform equation with this controller is

$$(\tau s + 1)Y = A\left(k_p + \frac{k_I}{s}\right)(R - Y) + AW, \tag{4.68}$$

and, if we multiply by s and collect terms,

$$(\tau s^2 + (Ak_p + 1)s + Ak_I)Y = A(k_p s + k_I)R + sAW. \tag{4.69}$$

Because the PI controller includes dynamics, use of this controller will change the dynamic response. This we can understand by considering the characteristic equation given by

$$\tau s^2 + (Ak_p + 1)s + Ak_I = 0. \tag{4.70}$$

The two roots of this equation may be complex and, if so, the natural frequency is $\omega_n = \sqrt{Ak_I/\tau}$ and the damping ratio is $\zeta = \frac{Ak_p+1}{2\tau\omega_n}$. These parameters may both be determined by the controller gains. On the other hand, if the plant is second order, described by

$$G(s) = \frac{A}{s^2 + a_1 s + a_2}, \tag{4.71}$$

then the characteristic equation of the system is

$$1 + \frac{k_p s + k_I}{s} \frac{A}{s^2 + a_1 s + a_2} = 0, \tag{4.72}$$

$$s^3 + a_1 s^2 + a_2 s + Ak_p s + Ak_I = 0. \tag{4.73}$$

In this case, the controller parameters can be used to set two of the coefficients but not the third. For this we need derivative control.

4.3.3 PID Control

The final term in the classical controller is derivative control, **D**. An important effect of this term is that it gives a sharp response to suddenly changing signals. Because of this, the "**D**" term is sometimes introduced into the feedback path as shown in Fig. 4.8(a). This could be either a part of the standard controller or could describe a velocity sensor such as a tachometer on the shaft of a motor. The closed-loop characteristic equation is the same as if the term were in the forward path as given by Eq. (4.59) and drawn in Fig. 4.8(b). It is important to notice that the *zeros* from the reference to the output are

Figure 4.8

Block diagram of the PID controller: (a) with the D-term in the feedback path; and (b) with the D-term in the forward path

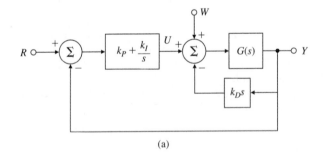

(a)

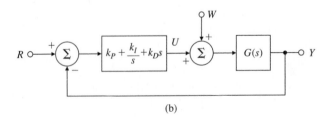

(b)

different in the two cases. With the derivative in the feedback path, the reference is not differentiated, which is how the undesirable response to sudden changes is avoided.

To illustrate the effect of a derivative term on PID control, consider speed control but with the second-order plant. In that case, the characteristic equation is

$$s^2 + a_1 s + a_2 + A\left(k_p + \frac{k_I}{s} + k_D s\right) = 0,$$

$$s^3 + a_1 s^2 + a_2 s + A(k_p s + k_I + k_D s^2) = 0. \tag{4.74}$$

Collecting terms results in

$$s^3 + (a_1 + Ak_D)s^2 + (a_2 + Ak_p)s + Ak_I = 0. \tag{4.75}$$

The point here is that this equation, whose three roots determine the nature of the dynamic response of the system, has three free parameters in k_p, k_I, and k_D and that by selection of these parameters, the roots can be uniquely and, in theory, arbitrarily determined. Without the derivative term, there would be only two free parameters, but with three roots, the choice of roots of the characteristic equation would be restricted. To illustrate the effect more concretely, a numerical example is useful.

EXAMPLE 4.6   *PID Control of Motor Speed*

Consider the DC motor speed control with parameters[4]

$$J_m = 1.13 \times 10^{-2} \qquad b = 0.028 \text{ N·m·sec/rad}, \quad L_a = 10^{-1} \text{henry},$$
$$\text{N·m· sec}^2 \text{/rad},$$
$$R_a = 0.45 \text{ ohms}, \qquad K_t = 0.067 \text{ N·m/amp}, \qquad K_e = 0.067 \text{ volt·sec/rad} \tag{4.76}$$

These parameters were defined in Example 2.11 in Chapter 2. Use the controller parameters

$$k_p = 3, \qquad k_I = 15 \text{ sec}, \qquad k_D = 0.3 \text{ sec}. \tag{4.77}$$

Discuss the effects of P, PI, and PID control on the responses of this system to steps in the disturbance torque and steps in the reference input. Let the unused controller parameters be zero.

Solution.   Figure 4.9(a) illustrates the effects of P, PI, and PID feedback on the step disturbance response of the system. Note that adding the integral term increases the oscillatory behavior but eliminates the steady-state error and that adding the derivative term reduces the oscillation while maintaining zero steady-state error. Figure 4.9(b) illustrates the effects of P, PI, and PID feedback on the step reference response with similar results. The step responses can be computed by forming the numerator and denominator coefficient vectors (in descending powers of s) and using the step function in MATLAB.®

---

[4]These values have been scaled to measure time in milliseconds by multiplying the true L_a and J_m by 1000 each.

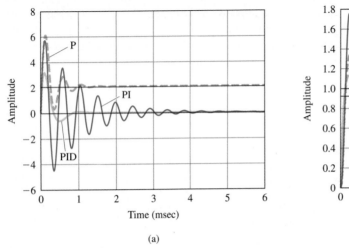
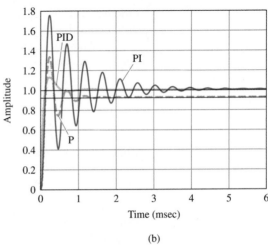

(a)   (b)

Figure 4.9

Responses of P, PI, and PID control to (a) step disturbance input (b) step reference input

EXAMPLE 4.7

PI Control for a DC Motor Position Control

Consider the simplified model of a DC motor in unity feedback as shown in Fig. 4.7 where the disturbance torque is labeled $W(s)$. Let the sensor be $-h$ rather than -1.

(a) Use the proportional controller

$$D(s) = k_p \tag{4.78}$$

and determine the system type and steady-state error properties with respect to disturbance inputs.

(b) Let the control be PI as given by

$$D(s) = k_p + \frac{k_I}{s} \tag{4.79}$$

and determine the system type and the steady-state error properties for disturbance inputs.

Solution.   (a) The closed-loop transfer function from W to E (where $R = 0$) is

$$T_w(s) = \frac{-B}{s(\tau s + 1) + Ak_p h}$$

$$= s^0 T_{o,w},$$

$$n = 0,$$

$$K_{o,w} = \frac{-Ak_p h}{B}.$$

Applying Eq. (4.40) we see that the system is Type 0 and the steady-state error to a unit step torque input is $e_{ss} = {-B}/{Ak_{ph}}$. From the earlier section, this system is seen to be Type 1 for reference inputs and illustrates that system type can be different for different inputs to the same system. However, in this case the system is Type 0 for reference inputs.

(b) If the controller is PI, the disturbance error transfer function is

$$T_w(s) = \frac{-Bs}{s^2(\tau s + 1) + (k_p s + k_I)Ah},$$ (4.80)

$$n = 1,$$ (4.81)

$$K_{n,w} = \frac{Ak_I h}{-B},$$ (4.82)

and therefore the system is Type 1 and the error to a unit ramp disturbance input in this case will be

$$e_{ss} = \frac{-B}{Ak_I h}.$$ (4.83)

EXAMPLE 4.8

Satellite Attitude Control

Consider the model of a satellite attitude control system shown in Fig. 4.10 (a) where

$$J = \text{moment of inertia,}$$

$$W = \text{disturbance torque,}$$

$$K = \text{sensor and reference gain,}$$

$$D(s) = \text{the compensator.}$$

With equal input filter and sensor scale factors, the system with PD control can be redrawn with unity feedback as in Fig. 4.10(b) and with PID control drawn as in Fig. 4.10(c). Assume that the control results in a stable system and determine the system types and error responses to disturbances of the control system for

(a) System Fig. 4.10(b) Proportional plus derivative control where $D(s) = k_P + k_D s$

(b) System Fig. 4.10(c) Proportional plus integral plus derivative control where $D = k_p + k_I/s + k_D s.$[5]

Solution. (a) We see from inspection of Fig. 4.10(b) that with two poles at the origin in the plant, the system is Type 2 with respect to reference inputs. The transfer function from disturbance to error is

$$T_w(s) = \frac{1}{Js^2 + k_D s + k_p}$$ (4.84)

$$= T_{o,w}(s)$$ (4.85)

for which $n = 0$ and $K_{o,w} = k_p$. The system is Type 0 and the error to a unit disturbance step is $1/k_p$.

---

[5] Notice that these controller transfer functions have more zeros than poles and are therefore not practical. In practice, the derivative term would have a high-frequency pole, which has been omitted for simplicity in these examples.

Figure 4.10

Model of a satellite attitude control: (a) basic system; (b) PD control; (c) PID control

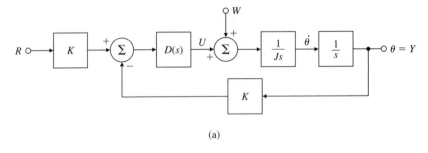

(a)

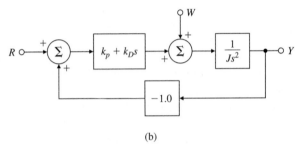

(b)

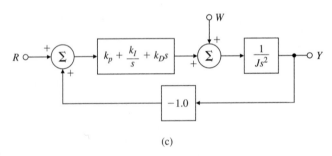

(c)

(b) With PID control, the forward gain has three poles at the origin, so this system is Type 3 for reference inputs but the disturbance transfer function is

$$T_w(s) = \frac{s}{Js^3 + k_D s^2 + k_p s + k_I}, \tag{4.86}$$

$$n = 1, \tag{4.87}$$

$$T_{o,w}(s) = \frac{1}{Js^3 + k_D s^2 + k_p s + k_I} \tag{4.88}$$

from which the system is Type 1 and the error constant is k_I; so the error to a disturbance ramp of unit slope will be $1/k_I$.

4.3.4 Ziegler–Nichols Tuning of the PID Controller

When the PID controller was being developed, selecting values for the several terms (known as "tuning" the controller) was often a hit and miss affair. To bring order to the situation and make life easier for plant operators, control engineers looked for

Figure 4.11

Process reaction curve

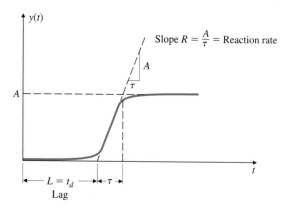

ways to make the tuning more systematic. Callender et al. (1936) proposed a design for PID controllers by specifying satisfactory values for the terms based on estimates of the plant parameters that an operating engineer could make from experiments on the process itself. This approach was extended by J. G. Ziegler and N. B. Nichols (1942, 1943) who recognized that the step responses of a large number of process control systems exhibit a **process reaction curve** like that shown in Fig. 4.11, which can be generated from experimental step response data. The S-shape of the curve is characteristic of many systems and can be approximated by the step response of a plant with transfer function

Transfer function for a high-order system with a characteristic process reaction curve

$$\frac{Y(s)}{U(s)} = \frac{Ae^{-st_d}}{s},\tag{4.89}$$

which is a first-order system with a time delay or "transportation lag" of t_d sec. The constants in Eq. (4.89) can be determined from the unit step response of the process. If a tangent is drawn at the inflection point of the reaction curve, then the slope of the line is $R = {}^A\!/_\tau$, the intersection of the tangent line with the time axis identifies the time delay $L = t_d$ and the final value gives the value of A.[6]

Ziegler and Nichols gave two methods for tuning the PID controller for such a model. In the first method the choice of controller parameters is designed to result in a closed-loop step response transient with a decay ratio of approximately 0.25. This means that the transient decays to a quarter of its value after one period of oscillation, as shown in Fig. 4.12. A quarter decay corresponds to $\zeta = 0.21$ and, while low for many applications, was seen as a reasonable compromise between quick response and adequate stability margins for the process controls being considered. The authors simulated the equations for the system on an analog computer and adjusted the controller parameters until the transients showed the decay of 25% in one period. The regulator parameters suggested by Ziegler and Nichols for the controller terms defined by

Tuning by decay ratio of 0.25

$$D_c(s) = k_p(1 + \frac{1}{T_I s} + T_D s)\tag{4.90}$$

are given in Table 4.2.

---

[6]K. J. Astrom and others have pointed out that a time constant, τ, can also be estimated from the curve and claim that a more effective tuning can be done by including that parameter.

Figure 4.12

Quarter decay ratio

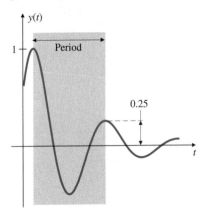

TABLE 4.2

Ziegler–Nichols Tuning for the Regulator
$D(s) = K(1 + 1/T_I s + T_D s)$, **for a Decay Ratio of 0.25**

Type of Controller	Optimum Gain
P	$k_p = 1/RL$
PI	$\begin{cases} k_p = 0.9/RL \\ T_I = L/0.3 \end{cases}$
PID	$\begin{cases} k_p = 1.2/RL \\ T_I = 2L \\ T_D = 0.5L \end{cases}$

Figure 4.13

Determination of ultimate gain and period

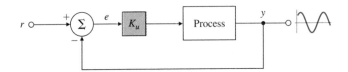

Tuning by evaluation at limit of stability (ultimate sensitivity method)

In the **ultimate sensitivity method** the criteria for adjusting the parameters are based on evaluating the amplitude and frequency of the oscillations of the system at the limit of stability rather than on taking a step response. To use the method, the proportional gain is increased until the system becomes marginally stable and continuous oscillations just begin with amplitude limited by the saturation of the actuator. The corresponding gain is defined as K_u (called the **ultimate gain**) and the period of oscillation is P_u (called the **ultimate period**). These are determined as shown in Figs. 4.13 and 4.14. P_u should be measured when the amplitude of oscillation is as small as possible. Then the tuning parameters are selected as shown in Table 4.3.

Experience has shown that the controller settings according to Ziegler–Nichols rules provide acceptable closed-loop response for many systems. The process operator will often do final tuning of the controller iteratively on the actual process to yield satisfactory control.

Figure 4.14

Neutrally stable system

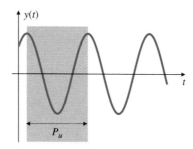

TABLE 4.3

Ziegler–Nichols Tuning for the Regulator
$D_c(s) = k_p(1 + 1/T_I s + T_D s)$, **Based on the Ultimate Sensitivity Method**

Type of Controller	Optimum Gain
P	$k_p = 0.5K_u$
PI	$\begin{cases} k_p = 0.45K_u \\ T_I = \dfrac{P_u}{1.2} \end{cases}$
PID	$\begin{cases} k_p = 1.6K_u \\ T_I = 0.5P_u \\ T_D = 0.125P_u \end{cases}$

Figure 4.15

A measured process reaction curve

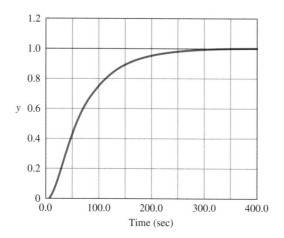

EXAMPLE 4.9

Tuning of a Heat Exchanger: Quarter Decay Ratio

Consider the heat exchanger discussed in Chapter 2. The process reaction curve of this system is shown in Fig. 4.15. Determine proportional and PI regulator gains for the system using the Zeigler–Nichols rules to achieve a quarter decay ratio. Plot the corresponding step responses.

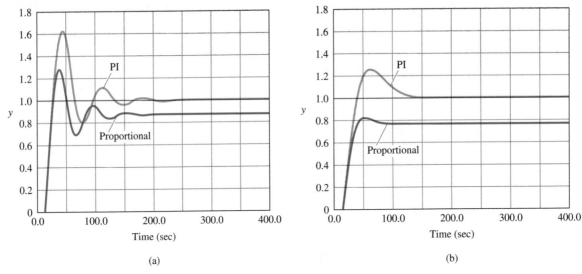

Figure 4.16

Closed-loop step responses

Solution. From the process reaction curve we measure the maximum slope to be $R \cong \frac{1}{90}$ and the time delay to be $L \cong 13$ sec. According to the Zeigler–Nichols rules of Table 4.2 the gains are

$$\text{Proportional}: k_p = \frac{1}{RL} = \frac{90}{13} = 6.92,$$

$$\text{PI}: k_p = \frac{0.9}{RL} = 6.22 \quad \text{and} \quad T_I = \frac{L}{0.3} = \frac{13}{0.3} = 43.3.$$

Figure 4.16(a) shows the step responses of the closed-loop system to these two regulators. Note that the proportional regulator results in a steady-state offset, while the PI regulator tracks the step exactly in the steady state. Both regulators are rather oscillatory and have considerable overshoot. If we arbitrarily reduce the gain k_p by a factor of 2 in each case, the overshoot and oscillatory behaviors are substantially reduced, as shown in Fig. 4.16(b).

EXAMPLE 4.10   *Tuning of a Heat Exchanger: Oscillatory Behavior*

Proportional feedback was applied to the heat exchanger in the previous example until the system showed nondecaying oscillations in response to a short pulse (impulse) input, as shown in Fig. 4.17. The ultimate gain is measured to be $K_u = 15.3$, and the period was measured at $P_u = 42$ *sec*. Determine the proportional and PI regulators according to the Zeigler–Nichols rules based on the ultimate sensitivity method. Plot the corresponding step responses.

Figure 4.17

Ultimate period of heat exchanger

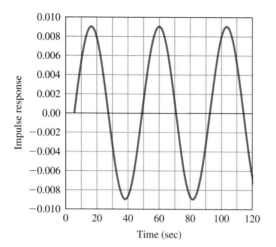

Figure 4.18

Closed-loop step response

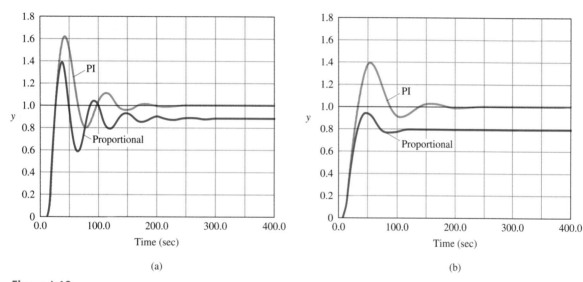

(a)   (b)

Solution. The regulators from Table 4.3 are

$$\text{Proportional}: k_p = 0.5, \quad K_u = 7.65,$$

$$\text{PI}: k_p = 0.45, \quad K_u = 6.885 \quad \text{and} \quad T_I = \frac{1}{1.2}P_u = 35.$$

The step responses of the closed-loop system are shown in Fig. 4.18(a). Note that the responses are similar to those in Example 4.9. If we reduce k_p by 50%, then the overshoot is substantially reduced, as shown in Fig. 4.18(b).

4.4 Introduction to Digital Control

As a result of the revolution in the cost-effectiveness of digital computers, there has been an increasing use of digital logic in embedded applications such as controllers in feedback systems. A digital controller gives the designer much more flexibility to make modifications to the control law after the hardware design is fixed because the formula for calculating the control signal is in the software rather than the hardware. In many instances, this means that the hardware and software designs can proceed almost independently, saving a great deal of time. Also, it is relatively easy to include binary logic and nonlinear operations as part of the function of a digital controller as compared to an analog controller. Special processors designed for real-time signal processing and known as digital signal processors (DSPs) are particularly well suited for use as real-time controllers. Chapter 8 includes a more extensive introduction to the math and concepts associated with the analysis and design of digital controllers and digital control systems. However, in order to be able to compare the analog designs of the next three chapters with reasonable digital equivalents, we give here a brief introduction to the most simple techniques for digital designs.

A digital controller differs from an analog controller in that the signals must be **sampled** and **quantized**.[7] A signal to be used in digital logic needs to be sampled first and then the samples need to be converted by an analog-to-digital converter or A/D[8] into a quantized digital number. Once the digital computer has calculated the proper next control signal value, this value needs to be converted back into a voltage and held constant or otherwise extrapolated by a digital-to-analog converter or D/A[9] in order to be applied to the actuator of the process. The control signal is not changed until the next sampling period. As a result of sampling, there are strict limits on the speed and bandwidth of a digital controller. Discrete design methods that tend to minimize these limitations are described in Chapter 8, which tend to minimize these limitations. A reasonable rule of thumb for selecting the sampling period is that during the rise-time of the response to a step, the input to the discrete controller should be sampled approximately six times. By adjusting the controller for the effects of sampling, the sample period can be as large as two to three times per rise time. This corresponds to a sampling frequency that is 10 to 20 times the system's closed-loop bandwidth. The quantization of the controller signals introduces an equivalent extra noise into the system and to keep this interference at an acceptable level, the A/D converter usually has an accuracy of 10 to 12 bits although inexpensive systems have been designed with only 8 bits. For a first analysis, the effects of the quantization are usually ignored, as they will be in this introduction. A simplified block diagram of a system with a digital controller is shown in Fig. 4.19.

For this introduction to digital control, we will describe a simplified technique for finding a discrete (sampled but not quantized) equivalent to a given continuous controller. The method depends on the sampling period, T_s, being short enough that the reconstructed control signal is close to the signal that the original analog controller

---

[7]A controller that operates on signals that are sampled but *not* quantized is called **discrete** while one that operates on signals that are both sampled and quantized is called **digital**.

[8]Pronounced "A to D."

[9]Often spelled DAC and pronounced as one word to rhyme with quack.

Figure 4.19

Block diagram of a digital controller

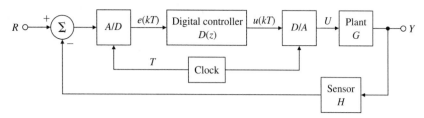

would have produced. We also assume that the numbers used in the digital logic have enough accurate bits so that the quantization implied in the A/D and D/A processes can be ignored. While there are good analysis tools to determine how well these requirements are met, here we will test our results by simulation, following the well-known advice that "The proof of the pudding is in the eating."

Finding a discrete equivalent to a given analog controller is equivalent to finding a recurrence equation for the samples of the control, which will approximate the differential equation of the controller. The assumption is that we have the transfer function of an analog controller and wish to replace it with a discrete controller that will accept samples of the controller input, $e(kT_s)$, from a sampler and, using past values of the control signal, $u(kT_s)$ and present and past samples of the input, $e(kT_s)$ will compute the next control signal to be sent to the actuator. As an example, consider a PID controller with the transfer function

$$U(s) = (k_p + \frac{k_I}{s} + k_D s)E(s), \qquad (4.91)$$

which is equivalent to the three terms of the time-domain expression

$$u(t) = k_p e(t) + k_I \int_0^t e(\tau)d\tau + k_D \dot{e}(t) \qquad (4.92)$$

$$= u_P + u_I + u_D. \qquad (4.93)$$

Based on these terms and the fact that the system is linear, the next control sample can be computed term-by-term. The proportional term is immediate:

$$u_P(kT_s + T_s) = k_p e(kT_s + T_s). \qquad (4.94)$$

The integral term can be computed by breaking the integral into two parts and approximating the second part, which is the integral over one sample period, as follows.

$$u_I(kT_s + T_s) = k_I \int_0^{kT_s+T_s} e(\tau)d\tau \qquad (4.95)$$

$$= k_I \int_0^{kT_s} e(\tau)d\tau + k_I \int_{kT_s}^{kT_s+T_s} e(\tau)d\tau \qquad (4.96)$$

$$= u_I(kT_s) + \{\text{area under } e(\tau) \text{ over one period}\} \qquad (4.97)$$

$$\cong u_I(kT_s) + k_I \frac{T_s}{2}\{e(kT_s + T_s) + e(kT_s)\}. \qquad (4.98)$$

In Eq. (4.98) the area in question has been approximated by that of the trapezoid formed by the base T_s and vertices $e(kT_s + T_s)$ and $e(kT_s)$ as shown by the dashed line in Fig. 4.20.

Figure 4.20

Graphical interpretation of numerical integration

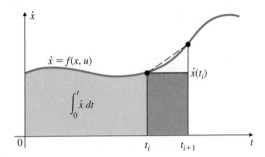

The area can also be approximated by the rectangle of amplitude $e(kT_s)$ and width T_s shown by the solid blue in Fig. 4.20 to give $u_I(kT_s + T_s) = u_I(kT_s) + k_IT_se(kT_s)$. These and other possibilities are considered in Chapter 8.

In the derivative term, the roles of u and e are reversed from integration and the consistent approximation can be written down at once from Eq. (4.98) and Eq. (4.92) as

$$\frac{T_s}{2}\{u_D(kT_s + T_s) + u_D(kT_s)\} = k_D\{e(kT_s + T_s) - e(kT_s)\}. \tag{4.99}$$

As with linear analog transfer functions, these relations are greatly simplified and generalized by the use of transform ideas. At this time, the discrete transform will be introduced simply as a prediction operator z much as if we described the Laplace transform variable, s, as a differential operator. Here we define the operator z as the forward shift operator in the sense that if $U(z)$ is the transform of $u(kT_s)$ then $zU(z)$ will be the transform of $u(kT_s + T_s)$. With this definition, the integral term can be written as

$$zU_I(z) = U_I(z) + k_I\frac{T_s}{2}[zE(z) + E(z)], \tag{4.100}$$

$$U_I(z) = k_I\frac{T_s}{2}\frac{z+1}{z-1}E(z), \tag{4.101}$$

and from Eq. (4.99), the derivative term becomes the inverse as

$$U_D(z) = k_D\frac{2}{T_s}\frac{z-1}{z+1}E(z). \tag{4.102}$$

The complete discrete PID controller is thus described by

$$U(z) = \left(k_p + k_I\frac{T_s}{2}\frac{z+1}{z-1} + k_D\frac{2}{T_s}\frac{z-1}{z+1}\right)E(z). \tag{4.103}$$

Comparing the two discrete equivalents of integration and differentiation with the corresponding analog terms, it is seen that the effect of the discrete approximation in the z domain is as if everywhere in the analog transfer function, the operator s has been replaced by the composite operator $\frac{2}{T_s}\frac{z-1}{z+1}$. This is the trapezoid rule[10] of discrete equivalents:

Trapezoid rule

The discrete equivalent to $D_a(s)$ is

$$D_d(z) = D_a\left(\frac{2}{T_s}\frac{z-1}{z+1}\right) \tag{4.104}$$

[10]The formula is also called Tustin's Method after the English engineer who used the technique to study the responses of nonlinear circuits.

EXAMPLE 4.11   *Discrete Equivalent*

Find the discrete equivalent to the analog controller having transfer function

$$D(s) = \frac{U(s)}{E(s)} = \frac{11s + 1}{3s + 1} \tag{4.105}$$

using the sample period $T_s = 1$.

Solution.   The discrete operator is $\frac{2(z-1)}{z+1}$ and thus the discrete transfer function is

$$D_d(z) = \frac{U(z)}{E(z)} = D(s)\Big|_{s = \frac{2}{T_s}\frac{z-1}{z+1}} \tag{4.106}$$

$$= \frac{11\left[\frac{2(z-1)}{z+1}\right] + 1}{3\left[\frac{2(z-1)}{z+1}\right] + 1}. \tag{4.107}$$

Clearing fractions, the discrete transfer function is

$$D_d(z) = \frac{U(z)}{E(z)} = \frac{23z - 21}{7z - 5}. \tag{4.108}$$

Converting the discrete transfer function to a discrete difference equation using the definition of z as the forward shift operator is done as follows. First we cross-multiply in Eq. (4.108) to obtain

$$(7z - 5)U(z) = (23z - 21)E(z) \tag{4.109}$$

and, interpreting z as a shift operator, this is equivalent to the difference equation[11]

$$7u(k + 1) - 5u(k) = 23e(k + 1) - 21e(k), \tag{4.110}$$

where we have replaced $kT_s + T_s$, with $k + 1$ to simplify the notation. To compute the next control at time $kT_s + T_s$, therefore, we solve the difference equation

$$u(k + 1) = \frac{5}{7}u(k) + \frac{23}{7}e(k + 1) - \frac{21}{7}e(k). \tag{4.111}$$

Now let's apply these results to a control problem. Fortunately MATLAB® provides us with the SIMULINK® capability to simulate both continuous and discrete systems allowing us to compare the responses of the systems with continuous and discrete controllers.

EXAMPLE 4.12   *Equivalent Discrete Controller for Speed Control*

A motor speed control is found to have the plant transfer function

$$\frac{Y}{U} = \frac{45}{(s + 9)(s + 5)}. \tag{4.112}$$

---

[11]The process is entirely similar to that used in Chapter 3 to find the ordinary differential equation to which a rational Laplace transform corresponds.

A PI controller designed for this system has the transfer function

$$D(s) = \frac{U}{E} = 1.4\frac{s+6}{s}. \tag{4.113}$$

The closed-loop system has a rise time of about 0.2 sec and an overshoot of about 20%. Design a discrete equivalent to this controller and compare the step responses and control signals of the two systems. (a) Compare the responses if the sample period is 0.07, which is about three samples per rise time. (b) Compare the responses with a sample period of $T_s = 0.035$, which corresponds to about six samples per rise time.

Solution.   (a) Using the substitution given by Eq. (4.104), the discrete equivalent for $T_s = 0.07$ is given by replacing s by $s \leftarrow \frac{2}{0.07}\frac{z-1}{z+1}$ in $D(s)$ as follows:

$$D_d(z) = 1.4\frac{\dfrac{2}{0.07}\dfrac{z-1}{z+1}+6}{\dfrac{2}{0.07}\dfrac{z-1}{z+1}} \tag{4.114}$$

$$= 1.4\frac{2(z-1)+6*0.07(z+1)}{2(z-1)} \tag{4.115}$$

$$= 1.4\frac{1.21z-0.79}{(z-1)}. \tag{4.116}$$

Based on this expression, the equation for the control is (the sample period is suppressed)

$$u(k+1) = u(k) + 1.4*[1.21e(k+1)-0.79e(k)]. \tag{4.117}$$

(b) For $T_s = 0.035$, the discrete transfer function is

$$D_d = 1.4\frac{1.105z-0.895}{z-1}, \tag{4.118}$$

for which the difference equation is

$$u(k+1) = u(k) + 1.4[1.105\,e(k+1)-0.895\,e(k)].$$

A SIMULINK block diagram for simulating the two systems is given in Fig. 4.21 and plots of the step responses are given in Fig. 4.22(a). The respective control signals are plotted in Fig. 4.22(b). Notice that the discrete controller for $T_s = 0.07$ results in a substantial increase in the overshoot in the step response while with $T_s = 0.035$ the digital controller matches the performance of the analog controller fairly well.

For controllers with many poles and zeros, making the continuous-to-discrete substitution called for in Eq. (4.104) can be very tedious. Fortunately, MATLAB provides a command that does all the work. If one has a continuous transfer function given by $D_c(s) = \frac{numD}{denD}$ represented in MATLAB as sysDa = tf(numD,denD), then the discrete equivalent with sampling period T_s is given by

$$sysDd = c2d(sysDa, T_s, \text{`t'}). \tag{4.119}$$

In this expression, of course, the polynomials are represented in MATLAB form. The last parameter in the c2d function given by 't' calls for the conversion to be done

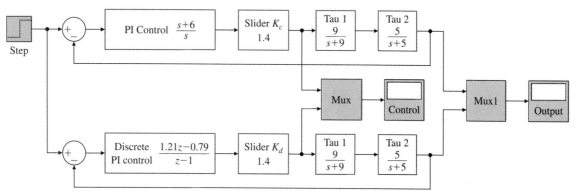

Figure 4.21

SIMULINK® block diagram to compare continuous and discrete controllers.

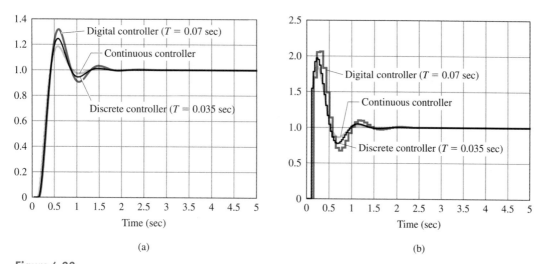

Figure 4.22

Comparison plots of a speed control system with continuous and discrete controllers: (a) output responses. (b) control signals

using the trapezoid method. The alternatives can be found by asking MATLAB for help c2d. For example, to compute the polynomials for $T_s = 0.07$ for the example above, the commands would be

```
numDa = [1 6];
denDa = [1 - 0];
sysDa = tf(numD,denD)
sysDd = c2d( sysDa,0.07,'t')
```

4.5 Historical Perspective

The field of control is characterized by two paths: theory and practice. Control theory is basically the application of mathematics to solve control problems while control practice, as used here, is the practical application of feedback in devices where it is found to be useful. Historically, practical applications have come first with control being introduced by trial and error. Although the applicable mathematics is often known, the theory describing how the control works and pointing the way to improvements has typically been applied later. For example, James Watt's company began manufacturing steam engines using the fly-ball governor in 1788 but it was not until 1840 that G. B. Airy described instability in a similar device and not until 1868 than J. C. Maxwell published "On Governors" with a theoretical description of the problem. Then it was not until 1877, almost 100 years after the steam engine control was introduced, that E. J. Routh published a solution giving the requirements for stability. This situation has been called the "Gap between Theory and Practice" and continues to this day as a source of creative tension that stimulates both theory and practice.

Regulation is central to the process industries, from making beer to making gasoline. In these industries there are a host of variables that need to be kept constant. Typical examples are temperature, pressure, volume, flow rates, composition, and chemical properties such as pH level. However, before one can regulate by feedback, one must be able to measure the variable of interest and before there was control there were sensors. In 1851, George Taylor and David Kendall founded the company that later became the Taylor Instrument Company in Rochester, NY, to make thermometers and barometers for weather forecasting. In 1855 they were making thermometers for several industries, including the brewing industry where they were used for manual control. Other early entries into the instrument field were the Bristol Company, founded in Naugatuck, CT, in 1889 by William Bristol, and the Foxboro Company, founded in Foxboro, MA, in 1908 by William's father and two of his brothers. For example, one of Bristol's instruments was used by Henry Ford to measure (and presumably control) steam pressure while he worked at the Detroit Edison Company. The Bristol Company pioneered in telemetry that permitted instruments to be placed at a distance from the process so a plant manager could monitor several variables at once. As the instruments became more sophisticated, and devices such as motor-driven valves became available, they were used in feedback control often using simple on–off methods as described in Chapter 1 for the home furnace. An important fact was that the several instrument companies agreed upon standards for the variables used so a plant could mix and match instruments and controllers from different suppliers. In 1920 Foxboro introduced a controller based on compressed air that included reset or integral action. Eventually, each of these companies introduced instruments and controllers that could implement full PID action. A major step was taken for tuning PID controllers in 1942 when Ziegler and Nichols, working for Taylor Instruments, published their method for tuning based on experimental data.

The poster child for the tracking problem was that of the anti-aircraft gun, whether on land or at sea. The idea was to use radar to track the target and to have a controller that would predict the path of the aircraft and aim the gun to a position such that the projectile would hit the target when it got there. The Radiation Laboratory was set up at MIT during World War II to develop such radars, one of which was the SCR-584. Interestingly, one of the major contributors to the control methods developed for this project was none other than Nick Nichols who had earlier worked on tuning PID controllers. When the record of the Rad Lab was written, Nichols was selected to be one of the editors of volume 25 on control.

H. S. Black joined Bell Laboratories in 1921 and was assigned to find a design for an electronic amplifier suitable for use as a repeater on the long lines of the telephone company. The basic problem was that the gain of the vacuum tube components he had available drifted over time and he needed a design that, over the audio frequency range, maintained a specific gain with great precision in the face of these drifts. Over the next few years he tried many approaches, including a feed-forward technique designed to cancel the tube distortion. While this worked in the laboratory, it was much too sensitive to be practical in the field. Finally, in August of 1927,[12] while on the ferry boat from Staten Island to Manhattan, he realized that negative feedback might work and he wrote the equations on the only paper available, a page of the New York Times. He applied for a patent in 1928 but it was not issued until December 1937.[13] The theory of sensitivity and many other theories of feedback were worked out by H. W. Bode.

SUMMARY

- The most important measure of the performance of a control system is the system error to all inputs.
- Compared to open-loop control, feedback can be used to stabilize an otherwise unstable system, to reduce errors to plant disturbances, to improve the tracking of reference inputs and to reduce the system's transfer function sensitivity to parameter variations.
- Sensor noise introduces a conflict between efforts to reduce the error caused by plant disturbances and efforts to reduce the errors caused by the sensor noise.
- Classifying a system as Type k indicates the ability of the system to achieve zero steady-state error to polynomials of degree less than but not equal to k. A stable *unity* feedback system is Type k with respect to reference inputs if the loop gain $G(s)D(s)$ has k poles at the origin in which case we can write

$$G(s)D(s) = \frac{A(s+z_1)(s+z_2)\cdots}{s^k(s+p_1)(s+p_2)\cdots}$$

and the error constant is given by

$$K_k = \lim_{s \to 0} s^k G(s)D(s) \qquad (4.120)$$

- A table of steady-state errors for unity feedback systems of Types 0, 1, and 2 to reference inputs is given in Table 4.1.
- Systems can be classified as to type for rejecting disturbances by computing the system error to polynomial disturbance inputs. The system is Type k to disturbances if the error is zero to all disturbance polynomials of degree less than k but nonzero for a polynomial of degree k.

---

[12]Black was 29 years old at the time.

[13]According to the story, many of Black's colleagues at the Bell laboratories did not believe it was possible to feed back a signal 100 times as large as was the input and still keep the system stable. As will be discussed in Chapter 6, this dilemma was solved by H. Nyquist, also at the Labs.

- Increasing the proportional feedback gain reduces steady-state errors but high gain almost always destabilizes the system. Integral control provides robust reduction in steady-state errors, but also may make the system less stable. Derivative control increases damping and improves stability. These three kinds of control combined form the classical PID controller.
- The standard PID controller is described by the equations

$$U(s) = \left(k_p + \frac{k_I}{s} + k_D s \right) E(s) \quad \text{or}$$

$$U(s) = k_p \left(1 + \frac{1}{T_I s} + T_D s \right) E(s) = D(s)E(s).$$

This latter form is ubiquitous in the process-control industry and describes the basic controller in many control systems.
- Useful guidelines for tuning PID controllers were presented in Tables 4.2 and 4.3.
- A difference equation describing a digital controller to be used to replace a given analog controller can be found by replacing s with $\frac{2}{T_s} \frac{z-1}{z+1}$ in the transfer function and using z as a forward shift operator. Thus, if $U(z)$ corresponds to $u(kT_s)$ then $zU(z)$ corresponds to $u(kT_s + T_s)$.
- MATLAB can compute a discrete equivalent with the command c2d.

REVIEW QUESTIONS

1. Give three advantages of feedback in control.
2. Give two disadvantages of feedback in control.
3. A temperature control system is found to have zero error to a constant tracking input and an error of 0.5°C to a tracking input that is linear in time, rising at the rate of 40°C/ sec. What is the system type of this control system and what is the relevant error constant (K_p or K_v or etc.)?
4. What are the units of K_p, K_v, and K_a?
5. What is the definition of system type with respect to reference inputs?
6. What is the definition of system type with respect to disturbance inputs?
7. Why does system type depend on where the external signal enters the system?
8. What is the main objective of introducing integral control?
9. What is the major objective of adding derivative control?
10. Why might a designer wish to put the derivative term in the feedback rather than in the error path?
11. What is the advantage of having a "tuning rule" for PID controllers?
12. Give two reasons to use a digital controller rather than an analog controller.
13. Give two disadvantages to using a digital controller.
14. Give the substitution in the discrete operator z for the Laplace operator s if the approximation to the integral in Eq. (4.98) is taken to be the rectangle of height $e(kT_s)$ and base T_s.

PROBLEMS

Problems for Section 4.1: The Basic Equations of Control

4.1   If S is the sensitivity of the unity feedback system to changes in the plant transfer function and T is the transfer function from reference to output, show that $S + T = 1$.

4.2   We define the sensitivity of a transfer function G to one of its parameters k as the ratio of percent change in G to percent change in k.

$$S_k^G = \frac{dG/G}{dk/k} = \frac{d\ln G}{d\ln k} = \frac{k}{G}\frac{dG}{dk}.$$

The purpose of this problem is to examine the effect of feedback on sensitivity. In particular, we would like to compare the topologies shown in Fig. 4.23 for connecting three amplifier stages with a gain of $-K$ into a single amplifier with a gain of -10.

(a) For each topology in Fig. 4.23, compute β_i so that if $K = 10$, $Y = -10R$.

(b) For each topology, compute S_k^G when $G = {}^Y/_R$. [Use the respective β_i values found in part (a).] Which case is the *least* sensitive?

(c) Compute the sensitivities of the systems in Fig. 4.23(b,c) to β_2 and β_3. Using your results, comment on the relative need for precision in sensors and actuators.

Figure 4.23

Three-amplifier topologies for Problem 4.2

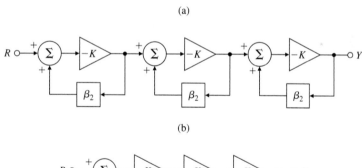

(a)

(b)

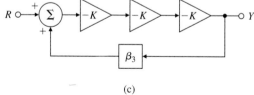

(c)

4.3   Compare the two structures shown in Fig. 4.24 with respect to sensitivity to changes in the overall gain due to changes in the amplifier gain. Use the relation

$$S = \frac{d\ln F}{d\ln K} = \frac{K}{F}\frac{dF}{dK}$$

as the measure. Select H_1 and H_2 so that the nominal system outputs satisfy $F_1 = F_2$, and assume $KH_1 > 0$.

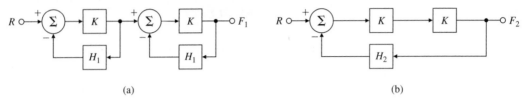

Figure 4.24

Block diagrams for Problem 4.3

4.4 A unity feedback control system has the open-loop transfer function

$$G(s) = \frac{A}{s(s+a)}.$$

(a) Compute the sensitivity of the closed-loop transfer function to changes in the parameter A.

(b) Compute the sensitivity of the closed-loop transfer function to changes in the parameter a.

(c) If the unity gain in the feedback changes to a value of $\beta \neq 1$, compute the sensitivity of the closed-loop transfer function with respect to β.

4.5 Compute the equation for the system error for the filtered feedback system shown in Fig. 4.4.

4.6 If S is the sensitivity of the filtered feedback system to changes in the plant transfer function and T is the transfer function from reference to output, compute the sum of $S + T$. Show that $S + T = 1$ if $F = H$.

(a) Compute the sensitivity of the filtered feedback system shown in Fig. 4.4 with respect to changes in the plant transfer function, G.

(b) Compute the sensitivity of the filtered feedback system shown in Fig. 4.4 with respect to changes in the controller transfer function, D_{cl}.

(c) Compute the sensitivity of the filtered feedback system shown in Fig. 4.4 with respect to changes in the filter transfer function, F.

(d) Compute the sensitivity of the filtered feedback system shown in Fig. 4.4 with respect to changes in the sensor transfer function, H.

Problems for Section 4.2: Control of Steady-State Error

4.7 Consider the DC-motor control system with rate (tachometer) feedback shown in Fig. 4.25(a).

(a) Find values for K' and k_t' so that the system of Fig. 4.25(b) has the same transfer function as the system of Fig. 4.25(a).

(b) Determine the system type with respect to tracking θ_r and compute the system K_v in terms of parameters K' and k_t'.

(c) Does the addition of tachometer feedback with positive k_t increase or decrease K_v?

4.8 Consider the system shown in Fig. 4.26, where

$$D(s) = K\frac{(s+\alpha)^2}{s^2 + \omega_0^2}.$$

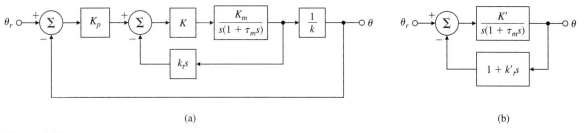

Figure 4.25

Control system for Problem 4.7

Figure 4.26

Control system for Problem 4.8

(a) Prove that if the system is stable, it is capable of tracking a sinusoidal reference input $r = \sin \omega_o t$ with zero steady-state error. (Look at the transfer function from R to E and consider the gain at ω_o.)

(b) Use Routh's criterion to find the range of K such that the closed-loop system remains stable if $\omega_o = 1$ and $\alpha = 0.25$.

4.9 Consider the system shown in Fig. 4.27, which represents control of the angle of a pendulum that has no damping.

(a) What condition must $D(s)$ satisfy so that the system can track a ramp reference input with constant steady-state error?

(b) For a transfer function $D(s)$ that stabilizes the system and satisfies the condition in part (a), find the class of disturbances $w(t)$ that the system can reject with zero steady-state error.

Figure 4.27

Control system for Problem 4.9

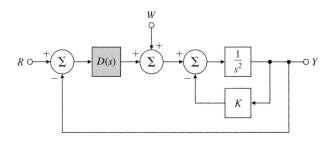

4.10 A unity feedback system has the overall transfer function

$$\frac{Y(s)}{R(s)} = T(s) = \frac{\omega_n^2}{s^2 + 2\zeta\omega_n s + \omega_n^2}.$$

Give the system type and corresponding error constant for tracking polynomial reference inputs in terms of ζ and ω_n.

4.11 Consider the second-order system

$$G(s) = \frac{1}{s^2 + 2\zeta s + 1}.$$

We would like to add a transfer function of the form $D(s) = \frac{K(s+a)}{(s+b)}$ in series with $G(s)$ in a unity feedback structure.

(a) Ignoring stability for the moment, what are the constraints on K, a, and b so that the system is Type 1?

(b) What are the constraints placed on K, a, and b so that the system is both stable and Type 1?

(c) What are the constraints on a and b so that the system is both Type 1 and remains stable for every positive value for K?

4.12 Consider the system shown in Fig. 4.28(a).

(a) What is the system type? Compute the steady-state tracking error due to a ramp input $r(t) = r_0 t 1(t)$.

(b) For the modified system with a feed-forward path shown in Fig. 4.28(b), give the value of H_f so the system is Type 2 for reference inputs and compute the K_a in this case.

(c) Is the resulting Type 2 property of this system robust with respect to changes in H_f? i.e., will the system remain Type 2 if H_f changes slightly?

Figure 4.28

Control system for Problem 4.12

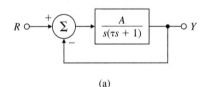

(a)

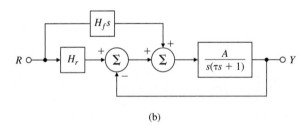

(b)

4.13 A controller for a satellite attitude control with transfer function $G = 1/s^2$ has been designed with a unity feedback structure and has the transfer function $D(s) = \frac{10(s+2)}{s+5}$.

(a) Find the system type for reference tracking and the corresponding error constant for this system.

(b) If a disturbance torque adds to the control so that the input to the process is $u + w$, what is the system type and corresponding error constant with respect to disturbance rejection?

4.14 A compensated motor position control system is shown in Fig. 4.29. Assume that the sensor dynamics are $H(s) = 1$.

Figure 4.29

Control system for
Problem 4.14

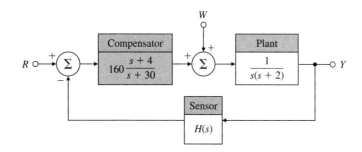

(a) Can the system track a step reference input r with zero steady-state error? If yes, give the value of the velocity constant.

(b) Can the system reject a step disturbance w with zero steady-state error? If yes, give the value of the velocity constant.

(c) Compute the sensitivity of the closed-loop transfer function to changes in the plant pole at -2.

(d) In some instances there are dynamics in the sensor. Repeat parts (a) to (c) for $H(s) = \frac{20}{(s+20)}$ and compare the corresponding velocity constants.

4.15 The general unity feedback system shown in Fig. 4.30 has disturbance inputs w_1, w_2, and w_3 and is asymptotically stable. Also,

$$G_1(s) = \frac{K_1 \prod_{i=1}^{m_1}(s + z_{1i})}{s^{l_1} \prod_{i=1}^{m_1}(s + p_{1i})}, \quad G_2(s) = \frac{K_2 \prod_{i=1}^{m_1}(s + z_{2i})}{s^{l_2} \prod_{i=1}^{m_1}(s + p_{2i})}.$$

(a) Show that the system is of Type 0, Type l_1, and Type $(l_1 + l_2)$ with respect to disturbance inputs w_1, w_2, and w_3 respectively.

Figure 4.30

Single input–single
output unity feedback
system with disturbance
inputs

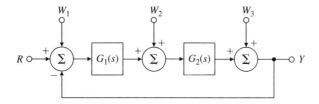

4.16 One possible representation of an automobile speed-control system with integral control is shown in Fig. 4.31.

Figure 4.31

System using integral
control

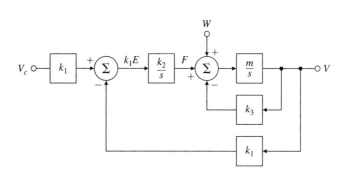

(a) With a zero reference velocity input ($v_c = 0$), find the transfer function relating the output speed v to the wind disturbance w.

(b) What is the steady-state response of v if w is a unit ramp function?

(c) What type is this system in relation to reference inputs? What is the value of the corresponding error constant?

(d) What is the type and corresponding error constant of this system in relation to tracking the disturbance w?

4.17   For the feedback system shown in Fig. 4.32, find the value of α that will make the system Type 1 for $K = 5$. Give the corresponding velocity constant. Show that the system is not robust by using this value of α and computing the tracking error $e = r - y$ to a step reference for $K = 4$ and $K = 6$.

Figure 4.32

Control system for Problem 4.17

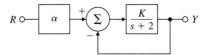

4.18   Suppose you are given the system depicted in Fig. 4.33(a), where the plant parameter a is subject to variations.

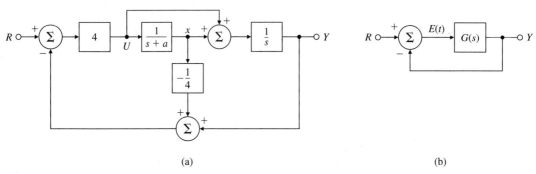

(a)   (b)

Figure 4.33

Control system for Problem 4.18

(a) Find $G(s)$ so that the system shown in Fig. 4.33(b) has the same transfer function from r to y as the system in Fig. 4.33(a).

(b) Assume that $a = 1$ is the nominal value of the plant parameter. What is the system type and the error constant in this case?

(c) Now assume that $a = 1 + \delta a$, where δa is some perturbation to the plant parameter. What is the system type and the error constant for the perturbed system?

4.19   Two feedback systems are shown in Fig. 4.34.

(a) Determine values for K_1, K_2, and K_3 so that

   (i) both systems exhibit zero steady-state error to step inputs (that is, both are Type 1), and

   (ii) their static velocity error constant $K_v = 1$ when $K_0 = 1$.

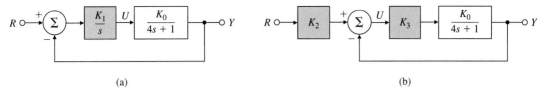

(a)   (b)

Figure 4.34

Two feedback systems for Problem 4.19

(b) Suppose K_0 undergoes a small perturbation: $K_0 \rightarrow K_0 + \delta K_0$. What effect does this have on the system type in each case? Which system has a type which is robust? Which system do you think would be preferred?

4.20 You are given the system shown in Fig. 4.35, where the feedback gain β is subject to variations. You are to design a controller for this system so that the output $y(t)$ accurately tracks the reference input $r(t)$.

Figure 4.35

Control system for Problem 4.20

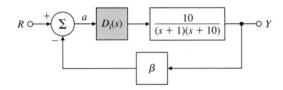

(a) Let $\beta = 1$. You are given the following three options for the controller $D_i(s)$:

$$D_1(s) = k_p, \quad D_2(s) = \frac{k_p s + k_I}{s}, \quad D_3(s) = \frac{k_p s^2 + k_I s + k_2}{s^2}.$$

Choose the controller (including particular values for the controller constants) that will result in a Type 1 system with a steady-state error to a unit reference ramp of less than $\frac{1}{10}$.

(b) Next, suppose that there is some attenuation in the feedback path that is modeled by $\beta = 0.9$. Find the steady-state error due to a ramp input for your choice of $D_i(s)$ in part (a).

(c) If $\beta = 0.9$, what is the system type for part (b)? What are the values of the appropriate error constant?

4.21 Consider the system shown in Fig. 4.36.

(a) Find the transfer function from the reference input to the tracking error.

(b) For this system to respond to inputs of the form $r(t) = t^n 1(t)$ (where $n < q$) with zero steady-state error, what constraint is placed on the open-loop poles $p_1, p_2, \ldots, p_q$?

Figure 4.36

Control system for Problem 4.21

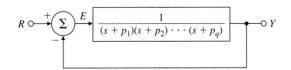

4.22 A linear ODE model of the DC motor with negligible armature inductance ($L_a = 0$) and with a disturbance torque w was given earlier in the chapter; it is restated here, in slightly different form, as

$$\frac{JR_a}{K_t}\ddot{\theta}_m + K_e\dot{\theta}_m = v_a + \frac{R_a}{K_t}w,$$

where θ_m is measured in radians. Dividing through by the coefficient of $\ddot{\theta}_m$, we obtain

$$\ddot{\theta}_m + a_1\dot{\theta}_m = b_0 v_a + c_0 w,$$

where

$$a_1 = \frac{K_t K_e}{JR_a}, \quad b_0 = \frac{K_t}{JR_a}, \quad c_0 = \frac{1}{J}.$$

With rotating potentiometers, it is possible to measure the positioning error between θ and the reference angle θ_r or $e = \theta_{ref} - \theta_m$. With a tachometer we can measure the motor speed $\dot{\theta}_m$. Consider using feedback of the error e and the motor speed $\dot{\theta}_m$ in the form

$$v_a = K(e - T_D\dot{\theta}_m),$$

where K and T_D are controller gains to be determined.

(a) Draw a block diagram of the resulting feedback system showing both θ_m and $\dot{\theta}_m$ as variables in the diagram representing the motor.

(b) Suppose the numbers work out so that $a_1 = 65$, $b_0 = 200$, and $c_0 = 10$. If there is no load torque ($w = 0$), what speed (in rpm) results from $v_a = 100$ V?

(c) Using the parameter values given in part (b), let the control be $D = k_p + k_D s$ and find k_p and k_D so that, using the results of Chapter 3, a step change in θ_{ref} with zero load torque results in a transient that has an approximately 17% overshoot and that settles to within 5% of steady-state in less than 0.05 sec.

(d) Derive an expression for the steady-state error to a reference angle input, and compute its value for your design in part (c) assuming $\theta_{ref} = 1$ rad.

(e) Derive an expression for the steady-state error to a constant disturbance torque when $\theta_{ref} = 0$ and compute its value for your design in part (c) assuming $w = 1.0$.

4.23 We wish to design an automatic speed control for an automobile. Assume that (1) the car has a mass m of 1000 kg, (2) the accelerator is the control U and supplies a force on the automobile of 10 N per degree of accelerator motion, and (3) air drag provides a friction force proportional to velocity of 10 N · sec/m.

(a) Obtain the transfer function from control input U to the velocity of the automobile.

(b) Assume the velocity changes are given by

$$V(s) = \frac{1}{s + 0.02}U(s) + \frac{0.05}{s + 0.02}W(s),$$

where V is given in meters per second, U is in degrees, and W is the percent grade of the road. Design a proportional control law $U = -k_p V$ that will maintain a velocity error of less than 1 m/sec in the presence of a constant 2% grade.

(c) Discuss what advantage (if any) integral control would have for this problem.

(d) Assuming that pure integral control (that is, no proportional term) is advantageous, select the feedback gain so that the roots have critical damping ($\zeta = 1$).

4.24 Consider the automobile speed control system depicted in Fig. 4.37.

Figure 4.37

Automobile
speed-control system

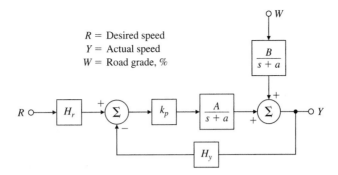

$R =$ Desired speed
$Y =$ Actual speed
$W =$ Road grade, %

(a) Find the transfer functions from $W(s)$ and from $R(s)$ to $Y(s)$.

(b) Assume that the desired speed is a constant reference r, so that $R(s) = r_0/s$. Assume that the road is level, so $w(t) = 0$. Compute values of the gains K, H_r, and H_f to guarantee that

$$\lim_{t \to \infty} y(t) = r_0.$$

Include both the open-loop (assuming $H_y = 0$) and feedback cases ($H_y \neq 0$) in your discussion.

(c) Repeat part (b) assuming that a constant grade disturbance $W(s) = w_0/s$ is present *in addition to* the reference input. In particular, find the variation in speed due to the grade change for both the feed-forward and feedback cases. Use your results to explain (1) why feedback control is necessary and (2) how the gain k_p should be chosen to reduce steady-state error.

(d) Assume that $w(t) = 0$ and that the gain A undergoes the perturbation $A + \delta A$. Determine the error in speed due to the gain change for both the feed-forward and feedback cases. How should the gains be chosen in this case to reduce the effects of δA?

4.25 Consider the multivariable system shown in Fig. 4.38. Assume that the system is stable. Find the transfer functions from each disturbance input to each output and determine the steady-state values of y_1 and y_2 for constant disturbances. We define a multivariable system to be type k with respect to polynomial inputs at w_i if the steady-state value of *every* output is zero for any combination of inputs of degree less than k and at least one input is a non zero constant for an input of degree k. What is the system type with respect to disturbance rejection at w_1? At w_2?

Figure 4.38

Multivariable system

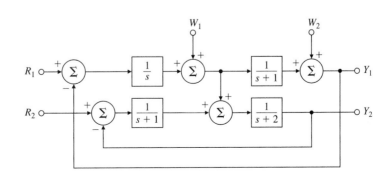

Problems for Section 4.3: The Three-Term Controller. PID Control

4.26 The transfer functions of speed control for a magnetic tape-drive system are shown in Fig. 4.39. The speed sensor is fast enough that its dynamics can be neglected and the diagram shows the equivalent unity feedback system.

(a) Assuming the reference is zero, what is the steady-state error due to a step distur-bance torque of 1 N · m? What must the amplifier gain K be in order to make the steady-state error $e_{ss} \leq 0.01$ rad/sec?

(b) Plot the roots of the closed-loop system in the complex plane, and accurately sketch the time response of the output for a step reference input using the gain K computed in part (a).

(c) Plot the region in the complex plane of acceptable closed-loop poles corresponding to the specifications of a 1% settling time of $t_s \leq 0.1$ sec and an overshoot $M_p \leq 5\%$.

(d) Give values for k_p and k_D for a PD controller, which will meet the specifications.

(e) How would the disturbance-induced steady-state error change with the new control scheme in part (d)? How could the steady-state error to a disturbance torque be eliminated entirely?

Figure 4.39

Speed-control system for a magnetic tape-drive

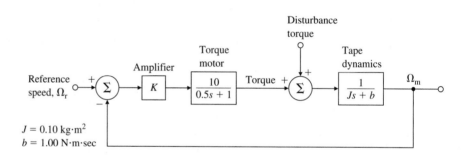

$J = 0.10$ kg·m²
$b = 1.00$ N·m·sec

4.27 Consider the system shown in Fig. 4.40 with PI control.

(a) Determine the transfer function from R to Y.

(b) Determine the transfer function from W to Y.

(c) What is the system type and error constant with respect to reference tracking?

(d) What is the system type and error constant with respect to disturbance rejection?

Figure 4.40

Control system for Problem 4.27

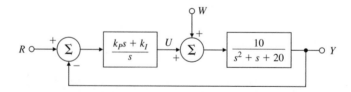

4.28 Consider the second-order plant with transfer function

$$G(s) = \frac{1}{(s+1)(5s+1)}$$

and in a unity feedback structure.

(a) Determine the system type and error constant with respect to tracking polynomial reference inputs of the system for P [$D = k_p$], PD [$D = k_p + k_D s$], and PID [$D = k_p + k_I/s + k_D s$] controllers. Let $k_p = 19$, $k_I = 0.5$, and $k_D = 4/19$.

(b) Determine the system type and error constant of the system with respect to disturbance inputs for each of the three regulators in part (a) with respect to rejecting polynomial disturbances $w(t)$ at the *input* to the plant.

(c) Is this system better at tracking references or rejecting disturbances? Explain your response briefly.

(d) Verify your results for parts (a) and (b) using MATLAB by plotting unit step and ramp responses for both tracking and disturbance rejection.

4.29 The DC-motor speed control shown in Fig. 4.41 is described by the differential equation

$$\dot{y} + 60y = 600v_a - 1500w,$$

where y is the motor speed, v_a is the armature voltage, and w is the load torque. Assume the armature voltage is computed using the PI control law

$$v_a = -\left(k_p e + k_I \int_0^t e\, dt\right),$$

where $e = r - y$.

(a) Compute the transfer function from W to Y as a function of k_p and k_I.

(b) Compute values for k_p and k_I so that the characteristic equation of the closed-loop system will have roots at $-60 \pm 60j$.

Figure 4.41

DC Motor speed-control block diagram for Problems 4.29 and 4.30

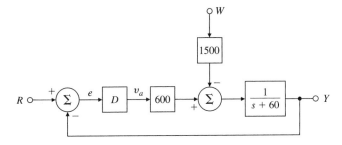

4.30 For the system in Problem 4.29, compute the following steady-state errors:

(a) to a unit-step reference input;

(b) to a unit-ramp reference input;

(c) to a unit-step disturbance input;

(d) for a unit-ramp disturbance input.

(e) Verify your answers to (a) and (d) using MATLAB. Note that a ramp response can be generated as a step response of a system modified by an added integrator at the reference input.

4.31 Consider the satellite-attitude control problem shown in Fig. 4.42 where the normalized parameters are

$$J = 10 \quad \text{spacecraft inertia, N·m·sec}^2/\text{rad}$$

$$\theta_r = \text{reference satellite attitude, rad.}$$

Figure 4.42

Satellite attitude control

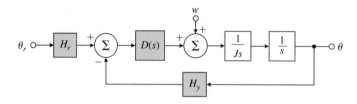

θ = actual satellite attitude, rad.

$H_y = 1$ sensor scale, factor V/rad.

$H_r = 1$ reference sensor scale factor, V/rad.

w = disturbance torque. N·m

(a) Use proportional control, P, with $D(s) = k_p$, and give the range of values for k_p for which the system will be stable.

(b) Use PD control and let $D(s) = (k_p + k_D s)$ and determine the system type and error constant with respect to reference inputs.

(c) Use PD control, let $D(s) = (k_p + k_D s)$ and determine the system type and error constant with respect to disturbance inputs.

(d) Use PI control, let $D(s) = (k_p + k_I/s)$, and determine the system type and error constant with respect to reference inputs.

(e) Use PI control, let $D(s) = (k_p + k_I/s)$, and determine the system type and error constant with respect to disturbance inputs.

(f) Use PID control, let $D(s) = (k_p + k_I/s + k_D s)$ and determine the system type and error constant with respect to reference inputs.

(g) Use PID control, let $D(s) = (k_p + k_I/s + k_D s)$ and determine the system type and error constant with respect to disturbance inputs.

4.32 The unit-step response of a paper machine is shown in Fig. 4.43(a) where the input into the system is stock flow onto the wire and the output is basis weight (thickness). The time delay and slope of the transient response may be determined from the figure.

(a) Find the proportional, PI, and PID-controller parameters using the Zeigler–Nichols transient-response method.

(b) Using proportional feedback control, control designers have obtained a closed-loop system with the unit impulse response shown in Fig. 4.43(b). When the gain $K_u = 8.556$, the system is on the verge of instability. Determine the proportional-, PI-, and PID-controller parameters according to the Zeigler–Nichols ultimate sensitivity method.

4.33 A paper machine has the transfer function

$$G(s) = \frac{e^{-2s}}{3s + 1},$$

where the input is stock flow onto the wire and the output is basis weight or thickness.

(a) Find the PID-controller parameters using the Zeigler–Nichols tuning rules.

(b) The system becomes marginally stable for a proportional gain of $K_u = 3.044$ as shown by the unit impulse response in Fig. 4.44. Find the optimal PID-controller parameters according to the Zeigler–Nichols tuning rules.

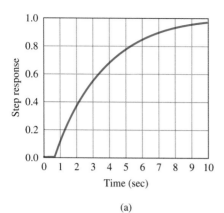
(a)

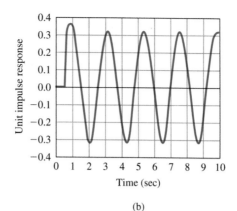
(b)

Figure 4.43

Paper-machine response data for Problem 4.32

Figure 4.44

Unit impulse response for the paper machine in Problem 4.33

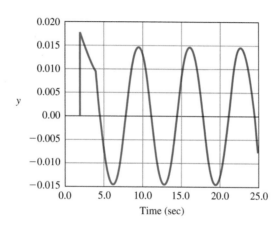

△ *Problems for Section 4.4: Introduction to Digital Control*

4.34 Compute the discrete equivalents for the following possible controllers using the trapezoid rule of Eq. (4.104). Let $T_s = 0.05$ in each case.

(a) $D_1(s) = (s + 2)/2$

(b) $D_2(s) = 2\dfrac{s + 2}{s + 4}$

(c) $D_3(s) = 5\dfrac{(s + 2)}{s + 10}$

(d) $D_4(s) = 5\dfrac{(s + 2)(s + 0.1)}{(s + 10)(s + 0.01)}$

4.35 Give the difference equations corresponding to the discrete controllers found in Problem 4.34 respectively.

(a) part 1

(b) part 2

(c) part 3

(d) part 4

5

The Root-Locus Design Method

A Perspective on the Root-Locus Design Method

In Chapter 3 we related the features of a step response, such as rise time, overshoot, and settling time, to pole locations in the s-plane of the transform of a second-order system characterized by the natural frequency ω_n, the damping ratio ζ, and the real part σ. This relationship is shown graphically in Fig. 3.15. We also examined the changes in these transient-response features when a pole or a zero is added to the transfer function. In Chapter 4 we saw how feedback can improve steady-state errors and can also influence dynamic response by changing the system's pole locations. In this chapter we present a specific technique that shows how changes in one of a system's parameters will modify the *roots of the characteristic equation*, which are the closed-loop poles, and thus change the system's dynamic response. The method was developed by W. R. Evans who gave rules for plotting the paths of the roots, a plot he called the **Root Locus**. With the development of MATLAB® and similar software the rules are no longer needed for detailed plotting, but we feel it is essential for a control designer to understand how proposed dynamic controllers will influence a locus as a guide in the design process. We also feel that it is important to understand the basics of how loci are generated in order to perform sanity checks on the computer results. For these reasons, study of the Evans rules is important.

The root locus is most commonly used to study the effect of loop gain variations; however, the method is general and can be used to plot the roots of any polynomial with respect to any one real parameter that enters the equation linearly. For example, the root-locus method can be used to plot the roots of a characteristic equation as the gain of a velocity sensor feedback changes, or the parameter can be a physical parameter such as motor inertia or armature inductance. Finally, a root locus can be plotted for a characteristic equation that results from the analysis of digital control systems using the z-transform, a topic we introduced in Chapter 4 and will discuss further in Chapter 8.

Chapter Overview

We open in Section 5.1 by illustrating the root locus for some simple feedback systems for which the equations can be solved directly. In Section 5.2 we show how to put an equation into the proper form for developing the rules for the root-locus behavior. In Section 5.3 this approach is applied to determine the locus for a number of typical control problems, which illustrate the factors that influence the final shape. MATLAB is used for detailed plotting of specific loci. When adjustment of the selected parameter alone cannot produce a satisfactory design, designs using other parameters can be studied or dynamic elements such as lead, lag, or notch compensations can be introduced, as described in Section 5.4. In Section 5.5 the uses of the root locus for design are summarized by a comprehensive design for the attitude control of a small airplane. In Section 5.6, the root-locus method is extended to guide the design of systems with a negative parameter, systems with more than one variable parameter, and systems with simple time delay. Finally, Section 5.7 gives historical notes on the origin of root-locus design.

5.1 Root Locus of a Basic Feedback System

We begin with the basic feedback system shown in Fig. 5.1. For this system, the closed-loop transfer function is

$$\frac{Y(s)}{R(s)} = T(s) = \frac{D(s)G(s)}{1 + D(s)G(s)H(s)}, \tag{5.1}$$

and the characteristic equation, whose roots are the poles of this transfer function, is

$$1 + D(s)G(s)H(s) = 0. \tag{5.2}$$

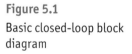

Figure 5.1

Basic closed-loop block diagram

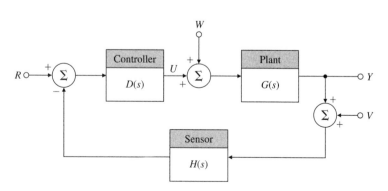

To put the equation in a form suitable for study of the roots as a parameter changes, we first put the equation in polynomial form and select the parameter of interest, which we will call K. We assume that we can define component polynomials $a(s)$ and $b(s)$ so that the characteristic polynomial is in the form $a(s) + Kb(s)$. We then define the transfer function $L(s) = \frac{b(s)}{a(s)}$ so that the characteristic equation can be written as[1]

$$1 + KL(s) = 0. \tag{5.3}$$

If, as is often the case, the parameter is the gain of the controller, then $L(s)$ is simply proportional to $D(s)G(s)H(s)$. Evans suggested that we plot the locus of *all possible* roots of Eq. (5.3) as K varies from zero to infinity and then use the resulting plot to aid us in selecting the best value of K. Furthermore, by studying the effects of additional poles and zeros on this graph, we can determine the consequences of additional dynamics added to $D(s)$ as compensation in the loop. We thus have a tool not only for selecting the specific parameter value but for designing the dynamic compensation as well. The graph of all possible roots of Eq. (5.3) relative to parameter K is called the **root locus**, and the set of rules to construct this graph is called the **root-locus method of Evans**. We begin our discussion of the method with the mechanics of constructing a root locus, using the equation in the form of Eq. (5.3) and K as the variable parameter.

Evans's method

To set the notation for our study, we assume here that the transfer function $L(s)$ is a rational function whose numerator is a monic[2] polynomial $b(s)$ of degree m and whose denominator is a monic polynomial $a(s)$ of degree n such that[3] $n \geq m$. We can factor these polynomials as

$$
\begin{aligned}
b(s) &= s^m + b_1 s^{m-1} + \cdots + b_m \\
&= (s - z_1)(s - z_2) \cdots (s - z_m) \\
&= \prod_{i=1}^{m} (s - z_i), \tag{5.4}
\end{aligned}
$$

$$
\begin{aligned}
a(s) &= s^n + a_1 s^{n-1} + \cdots + a_n \\
&= \prod_{i=1}^{n} (s - p_i).
\end{aligned}
$$

The roots of $b(s) = 0$ are the zeros of $L(s)$ and are labeled z_i, and the roots of $a(s) = 0$ are the poles of $L(s)$ and are labeled p_i. The roots of the characteristic equation itself are r_i from the factored form $(n > m)$,

$$a(s) + Kb(s) = (s - r_1)(s - r_2) \cdots (s - r_n). \tag{5.5}$$

---

[1] In the most common case, $L(s)$ is the loop transfer function of the feedback system and K is the gain of the controller–plant combination. However, the root locus is a general method suitable for the study of any polynomial and any parameter that can be put in the form of Eq. (5.3).

[2] Monic means that the coefficient of the highest power of s is 1.

[3] If $L(s)$ is the transfer function of a physical system, it is necessary that $n \geq m$ or else the system would have an infinite response to a finite input. If the parameter should be chosen so that $n < m$, then we can consider the equivalent equation $1 + K^{-1}L(s)^{-1} = 0$.

We may now state the root-locus problem expressed in Eq. (5.3) in several equivalent but useful ways. Each of the following equations has the same roots:

$$1 + KL(s) = 0, \tag{5.6}$$

$$1 + K\frac{b(s)}{a(s)} = 0, \tag{5.7}$$

$$a(s) + Kb(s) = 0, \tag{5.8}$$

$$L(s) = -\frac{1}{K}. \tag{5.9}$$

Root-locus forms

Equations (5.6)–(5.9) are sometimes referred to as the **root-locus form** or Evans form of a characteristic equation. The root locus is the set of values of s for which Eqs. (5.6)–(5.9) hold for some positive real value[4] of K. Because the solutions to Eqs. (5.6)–(5.9) are the roots of the closed-loop system characteristic equation and are thus closed-loop poles of the system, the root-locus method can be thought of as a method for inferring dynamic properties of the closed-loop system as the parameter K changes.

EXAMPLE 5.1

Root Locus of a Motor Position Control

In Chapter 2 we saw that a normalized transfer function of a DC motor voltage-to-position can be

$$\frac{\Theta_m(s)}{V_a(s)} = \frac{Y(s)}{U(s)} = G(s) = \frac{A}{s(s+c)}.$$

Solve for the root locus of closed-loop poles of the system created by feeding back the output Θ_m as shown in Fig. 5.1 with respect to the parameter A if $D(s) = H(s) = 1$ and also $c = 1$.

Solution. In terms of our notation, the values are

$$L(s) = \frac{1}{s(s+1)}, \qquad b(s) = 1, \qquad m = 0, \qquad z_i = \{empty\}, \tag{5.10}$$

$$K = A, \qquad a(s) = s^2 + s, \qquad n = 2, \qquad p_i = 0, -1.$$

From Eq. (5.6) the root locus is a graph of the roots of the quadratic equation

$$a(s) + Kb(s) = s^2 + s + K = 0. \tag{5.11}$$

Using the quadratic formula, we can immediately express the roots of Eq. (5.11) as

$$r_1, r_2 = -\frac{1}{2} \pm \frac{\sqrt{1 - 4K}}{2}. \tag{5.12}$$

A plot of the corresponding root locus is shown in Fig. 5.2. For $0 \le K \le 1/4$, the roots are real between -1 and 0. At $K = 1/4$ there are two roots at $-1/2$, and for $K > 1/4$ the roots become complex with real parts constant at $-1/2$ and imaginary parts that increase essentially in proportion to the square root of K. The dashed lines in Fig. 5.2 correspond to roots with a damping ratio $\zeta = 0.5$. The poles of $L(s)$ at $s = 0$ and

---

[4]If K is positive, the locus is called the "positive" locus. We will consider later the simple changes if $K < 0$, resulting in a "negative" locus.

Figure 5.2

Root locus for
$L(s) = \frac{1}{s(s+1)}$

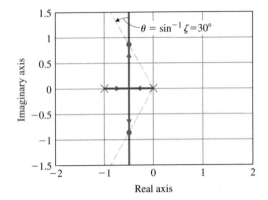

$s = -1$ are marked by the symbol $\times$, and the points where the locus crosses the lines where the damping ratio equals 0.5 are marked with dots ($\bullet$). We can compute K at the point where the locus crosses $\zeta = 0.5$ because we know that, if $\zeta = 0.5$, then $\theta = 30°$ and the magnitude of the imaginary part of the root is $\sqrt{3}$ times the magnitude of the real part. Since the size of the real part is $\frac{1}{2}$, from Eq. (5.12) we have

$$\frac{\sqrt{4K-1}}{2} = \frac{\sqrt{3}}{2},$$

and, therefore, $K = 1$.

We can observe several features of this simple locus by looking at Eqs. (5.11) and (5.12) and Fig. 5.2. First, there are two roots and thus two branches of the root locus. At $K = 0$ these branches begin at the poles of $L(s)$ (which are at 0 and -1), as they should, since for $K = 0$ the system is open loop and the characteristic equation is $a(s) = 0$. As K is increased, the roots move toward each other, coming together at $s = -\frac{1}{2}$, and at that point they break away from the real axis. After the **breakaway point** the roots move off to infinity with equal real parts, so the sum of the two roots is always -1. From the viewpoint of design, we see that by altering the value of the parameter K, we can cause the closed-loop poles to be at any point along the locus in Fig. 5.2. If some points along this locus correspond to a satisfactory transient response, then we can complete the design by choosing the corresponding value of K; otherwise, we are forced to consider a more complex controller. As we pointed out earlier, the root locus technique is not limited to focusing on the system gain ($K = A$ in Example 5.1); the same ideas are applicable for finding the locus with respect to *any* parameter that enters linearly in the characteristic equation.

Breakaway points are where roots move away from the real axis

EXAMPLE 5.2

Root Locus with Respect to a Plant Open-Loop Pole

Consider the characteristic equation as in Example 5.1, except that now let $D(s) = H(s) = 1$ and also let $A = 1$. Select c as the parameter of interest in the equation

$$1 + G(s) = 1 + \frac{1}{s(s+c)}. \qquad (5.13)$$

Find the root locus of the characteristic equation with respect to c.

Solution. The corresponding closed-loop characteristic equation in polynomial form is

$$s^2 + cs + 1 = 0. \tag{5.14}$$

The alternatives of Eq. (5.6) with the associated definitions of poles and zeros will apply if we let

$$L = \frac{s}{s^2 + 1}, \qquad b(s) = s, \qquad m = 1, \qquad z_i = 0,$$

$$K = c, \qquad a(s) = s^2 + 1, \qquad n = 2, \qquad p_i = +j, -j. \tag{5.15}$$

Thus, the root-locus form of the characteristic equation is

$$1 + c\frac{s}{s^2 + 1} = 0.$$

The solutions to Eq. (5.14) are easily computed as

$$r_1, r_2 = -\frac{c}{2} \pm \frac{\sqrt{c^2 - 4}}{2}. \tag{5.16}$$

The locus of solutions is shown in Fig. 5.3, with the poles [roots of $a(s)$] again indicated by $\times$'s and the zero [root of $b(s)$] by the circle (O) symbol. Note that when $c = 0$, the roots are at the $\times$'s on the imaginary axis and the corresponding response would be oscillatory. The damping ratio ζ grows as c increases from 0. At $c = 2$, there are two roots at $s = -1$, and the two locus segments abruptly change direction and move in opposite directions along the real axis; this point of multiple roots where two or more roots come into the real axis is called a **break-in point.**

Break-in point

Of course, computing the root locus for a quadratic equation is easy to do since we can solve the characteristic equation for the roots, as was done in Eqs. (5.12) and (5.16), and directly plot these as a function of the parameter K. To be useful, the method must be suitable for higher-order systems for which explicit solutions are difficult to obtain and rules for the construction of a general root locus were developed by Evans. With the availability of MATLAB, these rules are no longer necessary to plot a specific locus. The command rlocus(sys) will do that. However, in control design we are interested not only in a specific locus but also in how to modify the dynamics in

Figure 5.3

Root locus vs. damping factor c for $1 + G(s) = 1 + \frac{1}{s(s+c)} = 0$

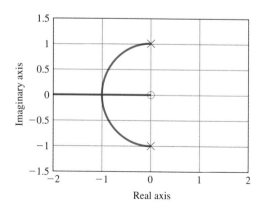

Real axis

such a way as to propose a system that will meet the dynamic response specifications for good control performance. For this purpose, it is very useful to be able to roughly sketch a locus so as to be able to evaluate the consequences of possible compensation alternatives. It is also important to be able to quickly evaluate the correctness of a computer-generated locus to verify that what is plotted by MATLAB is in fact what was meant to be plotted. It is easy to get a constant wrong or to leave out a term and GIGO[5] is the well-known first rule of computation.

5.2   Guidelines for Determining a Root Locus

We begin with a formal definition of a root locus. From the form of Eq. (5.6), we define the root locus this way:

> **Definition I.**   The root locus is the set of values of s for which $1 + KL(s) = 0$ is satisfied as the real parameter K varies from 0 to $+\infty$. Typically, $1 + KL(s) = 0$ is the characteristic equation of the system, and in this case the roots on the locus are the closed-loop poles of that system.

Now suppose we look at Eq. (5.9). If K is to be real and positive, $L(s)$ must be real and negative. In other words, if we arrange $L(s)$ in polar form as magnitude and phase, then the phase of $L(s)$ must be $180°$ in order to satisfy Eq. (5.9). We can thus define the root locus in terms of this **phase condition** as follows.

The basic root-locus rule; the phase of $L(s) = 180°$

> **Definition II.**   The root locus of $L(s)$ is the set of points in the s-plane where the phase of $L(s)$ is $180°$. If we define the angle to the test point from a zero as ψ_i and the angle to the test point from a pole as ϕ_i then Definition II is expressed as those points in the s-plane where, for integer l,

$$\sum \psi_i - \sum \phi_i = 180° + 360°(l - 1). \tag{5.17}$$

The immense merit of Definition II is that, while it is very difficult to solve a high-order polynomial by hand, computing the phase of a transfer function is relatively easy. The usual case is when K is real and positive, and we call this case the **positive or 180° locus**. When K is real and negative, $L(s)$ must be real and positive with a phase of $0°$, and this case is called the **negative or 0° locus**.

From Definition II we can, in principle, determine a positive root locus for a complex transfer function by measuring the phase and marking those places where we find $180°$. This direct approach can be illustrated by considering the example

$$L(s) = \frac{s + 1}{s(s + 5)[(s + 2)^2 + 4]}. \tag{5.18}$$

In Fig. 5.4 the poles of this $L(s)$ are marked $\times$ and the zero is marked $\bigcirc$. Suppose we select the test point $s_0 = -1 + 2j$. We would like to test whether or not s_0 lies on

---

[5] Garbage in, Garbage out.

Figure 5.4

Measuring the phase of
Eq. (5.18)

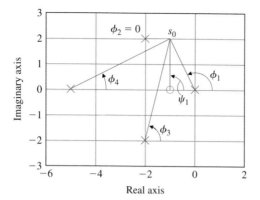

the root locus for some value of K. For this point to be on the locus, we must have $\angle L(s_0) = 180° + 360°(l-1)$ for some integer l, or equivalently, from Eq. (5.18),

$$\angle(s_0+1) - \angle s_0 - \angle(s_0+5) - \angle[(s_0+2)^2+4] = 180° + 360°(l-1). \quad (5.19)$$

The angle from the zero term $s_0 + 1$ can be computed[6] by drawing a line from the location of the zero at -1 to the test point s_0. In this case the line is vertical and has a phase angle marked $\psi_1 = 90°$ in Fig. 5.4. In a similar fashion, the vector from the pole at $s = 0$ to the test point s_0 is shown with angle ϕ_1, and the angles of the two vectors from the complex poles at $-2 \pm 2j$ to s_0 are shown with angles ϕ_2 and ϕ_3. The phase of the vector $s_0 + 5$ is shown with angle ϕ_4. From Eq. (5.19) we find the total phase of $L(s)$ at $s = s_0$ to be the sum of the phases of the numerator term corresponding to the zero minus the phases of the denominator terms corresponding to the poles:

$$\angle L = \psi_1 - \phi_1 - \phi_2 - \phi_3 - \phi_4$$
$$= 90° - 116.6° - 0° - 76° - 26.6°$$
$$= -129.2°.$$

Since the phase of $L(s)$ is not 180°, we conclude that s_0 is *not* on the root locus, so we must select another point and try again. Although measuring phase is not particularly hard, measuring phase at every point in the s-plane is hardly practical. Therefore, to make the method practical, we need some general guidelines for determining where the root locus is. Evans developed a set of rules for the purpose, which we will illustrate by applying them to the root locus for

$$L(s) = \frac{1}{s[(s+4)^2+16]}. \quad (5.20)$$

We begin by considering the positive locus, which is by far the most common case.[7] The first three rules are relatively simple to remember and are essential for any reasonable sketch. The last two are less useful but are used occasionally. As usual, we assume that MATLAB or its equivalent is always available to make an accurate plot of a promising locus.

---

[6]The graphical evaluation of the magnitude and phase of a complex number is reviewed in Appendix WD, Section 3.

[7]The negative locus will be considered in Section 5.6.

5.2.1   Rules for Plotting a Positive (180°) Root Locus

RULE 1. The n branches of the locus start at the poles of $L(s)$ and m of these branches end on the zeros of $L(s)$.

From the equation $a(s) + Kb(s) = 0$, if $K = 0$, the equation reduces to $a(s) = 0$, whose roots are the poles. When K approaches infinity, s must be such that either $b(s) = 0$ or $s \to \infty$. Since there are m zeros where $b(s) = 0$, m branches can end in these places. The case for $s \to \infty$ is considered in Rule 3.

RULE 2. The loci are on the real axis to the left of an odd number of poles and zeros.

If we take a test point on the real axis, such as s_0 in Fig. 5.5, we find that the angles ϕ_1 and ϕ_2 of the two complex poles cancel each other, as would the angles from complex conjugate zeros. Angles from real poles or zeros are 0° if the test point is to the right and 180° if the test point is to the left of a given pole or zero. Therefore, for the total angle to add to $180° + 360°(l - 1)$, the test point must be to the left of an odd number of real-axis poles plus zeros as shown in Fig. 5.5.

RULE 3. For large s and K, $n - m$ of the loci are asymptotic to lines at angles ϕ_l radiating out from the point $s = \alpha$ on the real axis, where

$$\phi_l = \frac{180° + 360°(l - 1)}{n - m}, \qquad l = 1, 2, \ldots, n - m, \qquad (5.21)$$

$$\alpha = \frac{\sum p_i - \sum z_i}{n - m}.$$

As $K \to \infty$, the equation

$$L(s) = -\frac{1}{K}$$

can be satisfied only if $L(s) = 0$. This can occur in two apparently different ways. In the first instance, as discussed in Rule 1, m roots will be found to approach the zeros of $L(s)$. The second manner in which $L(s)$ may go to zero is if $s \to \infty$ since, by assumption, n is larger than m. The asymptotes describe how these $n - m$ roots approach $s \to \infty$. For large s, the equation

$$1 + K\frac{s^m + b_1 s^{m-1} + \cdots + b_m}{s^n + a_1 s^{n-1} + \cdots + a_n} = 0 \qquad (5.22)$$

Figure 5.5

Rule 2. The real-axis parts of the locus are to the left of an odd number of poles and zeros

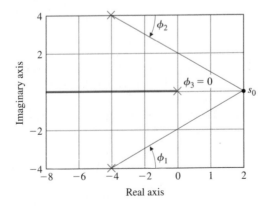

can be approximated[8] by

$$1 + K\frac{1}{(s-\alpha)^{n-m}} = 0. \tag{5.23}$$

This is the equation for a system in which there are $n - m$ poles, all clustered at $s = \alpha$. Another way to visualize this same result is to consider the picture we would see if we could observe the locations of poles and zeros from a vantage point of very large s: They would appear to cluster near the s-plane origin. Thus, m zeros would cancel the effects of m of the poles, and the other $n - m$ poles would appear to be in the same place. We say that the locus of Eq. (5.22) is asymptotic to the locus of Eq. (5.23) for large values of K and s. We need to compute α and to find the locus for the resulting asymptotic system. To find the locus, we choose our search point s_0 such that $s_0 = Re^{j\phi}$ for some large fixed value of R and variable ϕ. Since all poles of this simple system are in the same place, the angle of its transfer function is $180°$ if all $n - m$ angles, each equal to ϕ_l, sum to $180°$. Therefore, ϕ_l is given by

$$(n-m)\phi_l = 180° + 360°(l-1)$$

for some integer l. Thus, the asymptotic root locus consists of radial lines at the $n - m$ distinct angles given by

The angles of the asymptotes

$$\phi_l = \frac{180° + 360°(l-1)}{n-m}, \quad l = 1, 2, \ldots, n-m. \tag{5.24}$$

For the system described by Eq. (5.20), $n - m = 3$ and $\phi_{1,2,3} = 60°$, $180°$, and $300°$ or $\pm 60°$, $180°$.

The lines of the asymptotic locus come from $s_0 = \alpha$ on the real axis. To determine α, we make use of a simple property of polynomials. Suppose we consider the monic polynomial $a(s)$ with coefficients a_i and roots p_i, as in Eq. (5.4), and we equate the polynomial form with the factored form

$$s^n + a_1 s^{n-1} + a_2 s^{n-2} + \cdots + a_n = (s - p_1)(s - p_2) \cdots (s - p_n).$$

If we multiply out the factors on the right side of this equation, we see that the coefficient of s^{n-1} is $-p_1 - p_2 - \cdots - p_n$. On the left side of the equation, we see that this term is a_1. Thus $a_1 = -\sum p_i$; in other words, the coefficient of the *second* highest term in a monic polynomial is the negative sum of its roots—in this case, the poles of $L(s)$. Applying this result to the polynomial $b(s)$, we find the negative sum of the zeros to be b_1. These results can be written as

$$\begin{aligned} -b_1 &= \sum z_i, \\ -a_1 &= \sum p_i. \end{aligned} \tag{5.25}$$

Finally, we apply this result to the closed-loop characteristic polynomial obtained from Eq. (5.22):

$$s^n + a_1 s^{n-1} + \cdots + a_n + K(s^m + b_1 s^{m-1} + \cdots + b_m) \tag{5.26}$$

$$= (s - r_1)(s - r_2) \cdots (s - r_n) = 0.$$

Note that the sum of the roots is the negative of the coefficient of s^{n-1} and **is independent of K if $m < n - 1$**. Therefore, if $L(s)$ has at least two more poles than zeros,

---

[8]This approximation can be obtained by dividing $a(s)$ by $b(s)$ and matching the dominant two terms (highest powers in s) to the expansion of $(s - \alpha)^{n-m}$.

we have $a_1 = -\sum r_i$. We have thus shown that the center point of the roots *does not change with K* if $m < n - 1$ and that the open-loop and closed-loop sum is the same and is equal to $-a_1$, which can be expressed as

$$-\sum r_i = -\sum p_i. \tag{5.27}$$

For large values of K, we have seen that m of the roots r_i approach the zeros z_i and $n - m$ of the roots approach the branches of the asymptotic system $\frac{1}{(s-\alpha)^{n-m}}$ whose poles add up to $(n - m)\alpha$. Combining these results we conclude that the sum of all the roots equals the sum of those roots that go to infinity plus the sum of those roots that go to the zeros of $L(s)$:

$$-\sum r_i = -(n - m)\alpha - \sum z_i = -\sum p_i.$$

The center of the asymptotes

Solving for α, we get

$$\alpha = \frac{\sum p_i - \sum z_i}{n - m}. \tag{5.28}$$

Notice that in the sums $\sum p_i$ and $\sum z_i$ the imaginary parts *always* add to zero, since complex poles and zeros always occur in complex conjugate pairs. Thus Eq. (5.28) requires information about the real parts only. For Eq. (5.20),

$$\alpha = \frac{-4 - 4 + 0}{3 - 0}$$

$$= -\frac{8}{3} = -2.67.$$

The asymptotes at $\pm 60°$ are shown dashed in Fig. 5.6. Notice that they cross the imaginary axis at $\pm (2.67)j\sqrt{3} = \pm 4.62j$. The asymptote at $180°$ was already found on the real axis by Rule 2.

RULE 4. The angle(s) of departure of a branch of the locus from a pole of multiplicity q is given by

$$q\phi_{l,dep} = \sum \psi_i - \sum_{i \neq l} \phi_i - 180° - 360°(l - 1), \tag{5.29}$$

and the angle(s) of arrival of a branch at a zero of multiplicity q is given by

$$q\psi_{l,arr} = \sum \phi_i - \sum_{i \neq l} \psi_i + 180° + 360°(l - 1). \tag{5.30}$$

Figure 5.6

The asymptotes are $n - m$ radial lines from α at equal angles

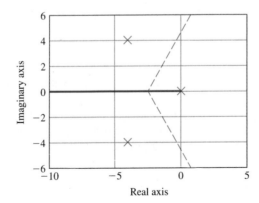

Figure 5.7

The departure and arrival angles are found by looking near a pole or zero

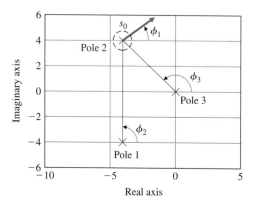

If a system has poles near the imaginary axis it can be important to know if the locus, which starts at such a pole, starts off toward the stable left half-plane (LHP) or heads toward the unstable right half-plane (RHP). To compute the angle by which a branch of the locus departs from one of the poles we take a test point s_o very near the pole in question, define the angle from that pole to the test point as $\phi_{l,\text{dep}}$ and transpose all other terms of Eq.(5.17) to the right-hand side. We can illustrate the process by taking the test point s_o to be near the pole at $-4 + 4j$ of our example and computing the angle of $L(s_0)$. The situation is sketched in Fig. 5.7, and the angle from $-4 + 4j$ to the test point we define as ϕ_1. We select the test point close enough to the pole that the angles ϕ_2 and ϕ_3 to the test point can be considered the same as those angles to the pole. Thus, $\phi_2 = 90°$, $\phi_3 = 135°$, and ϕ_1 can be calculated from the angle condition as *whatever it takes* to make the total be $180°$. The calculation is ($l = 1$)

$$\phi_1 = -90° - 135° - 180° \tag{5.31}$$

$$= -405° \tag{5.32}$$

$$= -45°. \tag{5.33}$$

By the complex conjugate symmetry of the plots, the angle of departure of the locus near the pole at $-4 - 4j$ will be $+45°$.

If there had been zeros in $L(s)$, the angles from the pole to the zeros would have been added to the right side of Eq. (5.31). For the general case, we can see from Eq. (5.31) that the angle of departure from a single pole is

Rule for departure angles

$$\phi_{1,dep} = \sum \psi_i - \sum_{i \neq 1} \phi_i - 180°, \tag{5.34}$$

where $\sum \phi_i$ is the sum of the angles to the remaining poles and $\sum \psi_i$ is the sum of the angles to all the zeros. For a multiple pole of order q, we must count the angle from the pole q times. This alters Eq. (5.34) to

$$q\phi_{l,dep} = \sum \psi_i - \sum_{i \neq l} \phi_i - 180° - 360°(l - 1), \tag{5.35}$$

where l takes on q values because there are q branches of the locus that depart from such a multiple pole.

Rule for arrival angles

The process of calculating a departure angle for small values of K, as shown in Fig. 5.7, is also valid for computing the angle by which a root locus arrives at a zero of $L(s)$ for large values of K. The general formula that results is

$$q\psi_{l,arr} = \sum \phi_i - \sum_{i \neq l} \psi_i + 180° + 360°(l - 1) \tag{5.36}$$

where $\sum \phi_i$ is the sum of the angles to all the poles, $\sum \psi_i$ is the sum of the angles to the remaining zeros, and l is an integer as before.

RULE 5.   The locus can have multiple roots at points on the locus and the branches will approach a point of q roots at angles separated by

$$\frac{180° + 360°(l - 1)}{q} \tag{5.37}$$

and will depart at angles with the same separation. As with any polynomial, it is possible for a characteristic polynomial of a degree greater than 1 to have multiple roots. For example, in the second-order locus of Fig. 5.2, there are two roots at $s = -1/2$ when $K = 1/4$. Here the horizontal branches of the locus come together and the vertical branches break away from the real axis, becoming complex for $K > 1/4$. The locus arrives at $0°$ and $180°$ and departs at $+90°$ and $-90°$.

Continuation locus

In order to compute the angles of arrival and departure from a point of multiple roots, it is useful to use a trick we call the **continuation locus**. We can imagine plotting a root locus for an initial range of K, perhaps for $0 \leq K \leq K_1$. If we let $K = K_1 + K_2$, we can then plot a new locus with parameter K_2, a locus which is the *continuation of the original locus* and whose starting poles are the roots of the original system at $K = K_1$. To see how this works, we return to the second-order root locus of Eq. (5.11) and let K_1 be the value corresponding to the breakaway point $K_1 = 1/4$. If we let $K = 1/4 + K_2$, we have the locus equation $s^2 + s + 1/4 + K_2 = 0$, or

$$\left(s + \frac{1}{2}\right)^2 + K_2 = 0. \tag{5.38}$$

The steps for plotting this locus are, of course, the same as for any other, except that now the initial departure of the locus of Eq. (5.38) corresponds to the breakaway point of the original locus of Eq. (5.11). Applying the rule for departure angles [Eq. (5.35)] from the double pole at $s = -1/2$, we find that

$$2\phi_{dep} = -180° - 360°(l - 1), \tag{5.39}$$

$$\phi_{dep} = -90° - 180°(l - 1), \tag{5.40}$$

$$\phi_{dep} = \pm 90° \text{ (departure angles at breakaway)}. \tag{5.41}$$

In this case, the arrival angles at $s = -1/2$ are, from the original root locus, along the real axis and are clearly $0°$ and $180°$.

The complete locus for our third-order example is drawn in Fig. 5.8. It combines all the results found so far—that is, the real-axis segment, the center of the asymptotes and their angles, and the angles of departure from the poles. It is usually sufficient to draw the locus by using only Rules 1 to 3, which should be memorized. Rule 4 is sometimes useful to understand how locus segments will depart, especially if there is a pole near the $j\omega$ axis. Rule 5 is sometimes useful to help interpret plots that come

Figure 5.8

Root locus for
$L(s) = \frac{1}{s(s^2+8s+32)}$

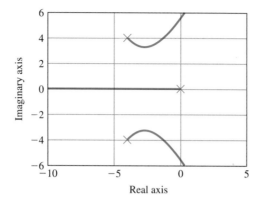

from the computer and, as we shall see in the next section, to explain qualitative changes in some loci as a pole or zero is moved. The actual locus in Fig. 5.8 was drawn using the MATLAB commands

```
numL = [1];
denL = [1 8 32 0];
sysL = tf(numL,denL);
rlocus(sysL)
```

We will next summarize the rules for drawing a root locus.

5.2.2 Summary of the Rules for Determining a Root Locus

RULE 1. The n branches of the locus start at the poles of $L(s)$ and m branches end on the zeros of $L(s)$.

RULE 2. The loci are on the real axis to the left of an odd number of poles and zeros.

RULE 3. For large s and K, $n - m$ of the loci are asymptotic to lines at angles ϕ_l radiating out from the center point $s = \alpha$ on the real axis, where

$$\phi_l = \frac{180° + 360°(l - 1)}{n - m}, \quad l = 1, 2, \ldots, n - m, \tag{5.42}$$

$$\alpha = \frac{\sum p_i - \sum z_i}{n - m}. \tag{5.43}$$

RULE 4. The angle(s) of departure of a branch of the locus from a pole of multiplicity q is given by

$$q\phi_{l,dep} = \sum \psi_i - \sum \phi_i - 180° - 360°(l - 1), \tag{5.44}$$

and the angle(s) of arrival of a branch at a zero of multiplicity q is given by

$$q\psi_{l,arr} = \sum \phi_i - \sum \psi_i + 180° + 360°(l - 1). \tag{5.45}$$

RULE 5. The locus can have multiple roots at points on the locus of multiplicity q. The branches will approach a point of q roots at angles separated by

$$\frac{180° + 360°(l - 1)}{q} \tag{5.46}$$

and will depart at angles with the same separation, forming an array of $2q$ rays equally spaced. If the point is on the real axis, then the orientation of this array is given by the real-axis rule. If the point is in the complex plane, then the angle of departure rule must be applied.

5.2.3 Selecting the Parameter Value

The positive root locus is a plot of *all possible locations* for roots to the equation $1 + KL(s) = 0$ for some real positive value of K. The purpose of design is to select a particular value of K that will meet the specifications for static and dynamic response. We now turn to the issue of selecting K from a particular locus so that the roots are at specific places. Although we shall show how the gain selection can be made by hand calculations from a plot of the locus, this is almost never done by hand because the determination can be accomplished easily by MATLAB. It is useful, however, to be able to perform a rough sanity check on the computer-based results by hand.

Using Definition II of the locus, we developed rules to sketch a root locus from the phase of $L(s)$ alone. If the equation is actually to have a root at a particular place when the phase of $L(s)$ is $180°$, then a **magnitude condition** must also be satisfied. This condition is given by Eq. (5.9), rearranged as

$$K = -\frac{1}{L(s)}.$$

For values of s on the root locus, the phase of $L(s)$ is $180°$, so we can write the magnitude condition as

$$K = \frac{1}{|L|}. \tag{5.47}$$

Equation (5.47) has both an algebraic and a graphical interpretation. To see the latter, consider the locus of $1 + KL(s)$, where

$$L(s) = \frac{1}{s[(s+4)^2 + 16]}. \tag{5.48}$$

For this transfer function, the locus is plotted in Fig. 5.9. In Fig. 5.9, the lines corresponding to a damping ratio of $\zeta = 0.5$ are sketched and the points where the locus crosses these lines are marked with dots (•). Suppose we wish to set the gain so that the roots are located at the dots. This corresponds to selecting the gain so that two of the closed-loop system poles have a damping ratio of $\zeta = 0.5$. (We will find the third pole shortly.) What is the value of K when a root is at the dot? From Eq. (5.47), the value of K is given by 1 over the magnitude of $L(s_0)$, where s_0 is the coordinate of the dot. On the figure we have plotted three vectors marked $s_0 - s_1$, $s_0 - s_2$, and $s_0 - s_3$, which are the vectors from the poles of $L(s)$ to the point s_0. (Since $s_1 = 0$, the first vector equals s_0.) Algebraically, we have

$$L(s_0) = \frac{1}{s_0(s_0 - s_2)(s_0 - s_3)}. \tag{5.49}$$

Using Eq. (5.47), this becomes

$$K = \frac{1}{|L(s_0)|} = |s_0||s_0 - s_2||s_0 - s_3|. \tag{5.50}$$

Figure 5.9

Root locus for
$$L(s) = \frac{1}{s[(s+4)^2+16]}$$
showing calculations of
gain K

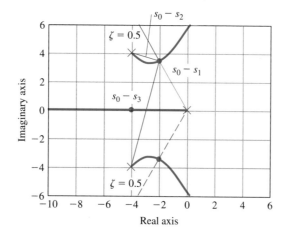

Graphical calculation of
the desired gain

The graphical interpretation of Eq. (5.50) shows that its three magnitudes are the lengths of the corresponding vectors drawn on Fig. 5.9 (see Appendix WD). Hence we can compute the gain to place the roots at the dot ($s = s_0$) by measuring the lengths of these vectors and multiplying the lengths together, *provided that the scales of the imaginary and real axes are identical.* Using the scale of the figure, we estimate that

$$|s_0| \cong 4.0,$$
$$|s_0 - s_2| \cong 2.1,$$
$$|s_0 - s_3| \cong 7.7.$$

Thus the gain is estimated to be

$$K = 4.0(2.1)(7.7) \cong 65.$$

We conclude that if K is set to the value 65, then a root of $1+KL$ will be at s_0, which has the desired damping ratio of 0.5. Another root is at the conjugate of s_0. Where is the third root? The third branch of the locus lies along the negative real axis. If performing the calculations by hand, we would need to take a test point, compute a trial gain, and repeat this process until we have found the point where $K = 65$. However, if performing a check on MATLAB's determination, it is sufficient to merely use the procedure above to verify the gain at the root location indicated by the computer.

To use MATLAB, plot the locus using the command rlocus(sysL), for example, then the command [K,p] = rlocfind(sysL) will produce a crosshair on the plot and, when spotted at the desired location of the root and selected with a mouse click, the value of the gain K is returned as well as the roots corresponding to that K in the variable p. The use of rltool makes this even easier, and will be discussed in more detail in Example 5.7.

Finally, with the gain selected, it is possible to compute the error constant of the control system. A process with the transfer function given by Eq. (5.48) has one integrator and, in a unity feedback configuration, will be a Type 1 control system.

In this case the steady-state error in tracking a ramp input is given by the velocity constant:

$$K_v = \lim_{s \to 0} sKL(s) \tag{5.51}$$

$$= \lim_{s \to 0} s \frac{K}{s[(s+4)^2 + 16]} \tag{5.52}$$

$$= \frac{K}{32}. \tag{5.53}$$

With the gain set for complex roots at a damping $\zeta = 0.5$, the root-locus gain is $K = 65$, so from Eq. (5.53) we get $K_v = 65/32 \cong 2$. If the closed-loop dynamic response, as determined by the root locations, is satisfactory and the steady-state accuracy, as measured by K_v, is good enough, then the design can be completed by gain selection alone. However, if no value of K satisfies all of the constraints, as is typically the case, then additional modifications are necessary to meet the system specifications.

5.3 Selected Illustrative Root Loci

A number of important control problems are characterized by a process with the simple "double integrator" transfer function

$$G(s) = \frac{1}{s^2}. \tag{5.54}$$

The control of attitude of a satellite is described by this equation. Also, the read/write head assembly of a computer hard-disk drive is typically floating on an air bearing so that friction is negligible for all but the smallest motion. The motor is typically driven by a current source so the back emf does not affect the torque. The result is a plant described by Eq. (5.54). If we form a unity feedback system with this plant, and a proportional controller, the root locus with respect to controller gain is

$$1 + k_p \frac{1}{s^2} = 0. \tag{5.55}$$

If we apply the rules to this (trivial) case, the results are as follows:

RULE 1. The locus has two branches that start at $s = 0$.

RULE 2. There are no parts of the locus on the real axis.

RULE 3. The two asymptotes have origin at $s = 0$ and are at the angles of $\pm 90°$.

RULE 4. The loci depart from $s = 0$ at the angles of $\pm 90°$.

Conclusion: The locus consists of the imaginary axis and the transient would be oscillatory for any value of k_p. A more useful design results with the use of proportional plus derivative control.

EXAMPLE 5.3   *Root Locus for Satellite Attitude Control with PD Control*

The characteristic equation with PD control is

$$1 + [k_p + k_D s] \frac{1}{s^2} = 0. \tag{5.56}$$

Figure 5.10

Root locus for
$L(s) = G(s) = \frac{(s+1)}{s^2}$

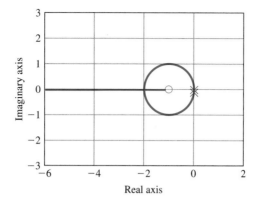

To put the equation in root-locus form, we define $K = k_D$, and for the moment arbitrarily select the gain ratio[9] as $k_p/k_D = 1$, which results in the root-locus form

$$1 + K\frac{s+1}{s^2} = 0. \tag{5.57}$$

Solution. Again we compute the results of the rules:

RULE 1. There are two branches that start at $s = 0$, one of which terminates on the zero at $s = -1$ and the other of which approaches infinity.

RULE 2. The real axis to the left of $s = -1$ is on the locus.

RULE 3. Since $n - m = 1$, there is one asymptote along the negative real axis.

RULE 4. The angles of departure from the double pole at $s = 0$ are $\pm 90°$.

RULE 5. From Rules 1–4, it should be clear that the locus will curl around the zero, rejoin the real axis to the left of the zero, and terminate as indicated by Rule 1. It turns out that the locus segments rejoin the real axis at $s = -2$, which creates a point of multiple roots. Evaluation of the angle of arrival at this point will show that the segments arrive at $\pm 90°$, from which on the locus from Rule 2: it is a point of multiple roots, in this case a point of break in. We conclude that two branches of the locus leave the origin going north and south and that they curve around[10] without passing into the RHP and break into the real axis at $s = -2$, from which point one branch goes west toward infinity and the other goes east to rendezvous with the zero at $s = -1$. The locus is plotted in Fig. 5.10 with the commands

```
numS = [1 1];
denS = [1 0 0];
sysS = tf(numS,denS);
rlocus( sysS)
```

[9]Given a specific physical system, this number would be selected with consideration of the specified rise time of the design or the maximum control signal (control authority) of the actuator.

[10]You can prove that the path is a circle by assuming that $s + 1 = e^{j\theta}$ and showing that the equation has a solution for a range of positive K and real θ under this assumption. (See Problem 5.18.)

Comparing this case with that for the simple $1/s^2$, we see that

The addition of the zero has pulled the locus into the LHP, a point of general importance in constructing a compensation.

In the previous case, we considered pure PD control. However, as we have mentioned earlier, the physical operation of differentiation is not practical and in practice PD control is approximated by

$$D(s) = k_p + \frac{k_D s}{s/p + 1}, \tag{5.58}$$

which can be put in root-locus form by defining $K = k_p + p k_D$ and $z = p k_p/K$ so that [11]

$$D(s) = K \frac{s + z}{s + p}. \tag{5.59}$$

For reasons we will see when we consider design by frequency response, this controller transfer function is called a "lead compensator" or, referring to the frequent implementation by electrical components, a "lead network." The characteristic equation for the $1/s^2$ plant with this controller is

$$1 + D(s)G(s) = 1 + KL(s) = 0,$$

$$1 + K \frac{s + z}{s^2(s + p)} = 0.$$

EXAMPLE 5.4   *Root Locus of the Satellite Control with Modified PD or Lead Compensation*

To evaluate the effect of the added pole, we will again set $z = 1$ and consider three different values for p. We begin with a somewhat large value, $p = 12$, and consider the root locus for

$$1 + K \frac{s + 1}{s^2(s + 12)}. \tag{5.60}$$

Solution.   Again, we apply the rules for plotting a root locus:

RULE 1. There are now three branches to the locus, two starting at $s = 0$ and one starting at $s = -12$.

RULE 2. The real axis segment $-12 \le s \le -1$ is part of the locus.

RULE 3. There are $n - m = 3 - 1 = 2$ asymptotes centered at $\alpha = \frac{-12 - (-1)}{2} = -11/2$ and at the angles $\pm 90°$.

RULE 4. The angles of departure of the branches at $s = 0$ are again $\pm 90°$. The angle of departure from the pole at $s = -12$ is at $0°$.

There are several possibilities on how the locus segments behave while still adhering to the guidance above. MATLAB is the expedient way to discover the paths. The MATLAB commands

---

[11] The use of z here for zero is not to be confused with the use of the operator z used in defining the discrete transfer function needed to describe digital controllers.

Figure 5.11

Root locus for
$L(s) = \frac{(s+1)}{s^2(s+12)}$

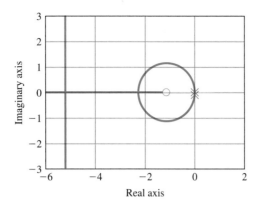

```
numL = [1 1];
denL = [1 12 0 0];
sysL = tf(numL,denL);
rlocus(sysL)
```
show that two branches of locus break vertically from the poles at $s = 0$, curve around to the left without passing into the RHP, and break in at $s = -2.3$, where one branch goes right to meet the zero at $s = -1$ and the other goes left, where it is met by the root that left the pole at $s = -12$. These two form a multiple root at $s = -5.2$ and break away there and approach the vertical asymptotes located at $s = -5.5$. The locus is plotted in Fig. 5.11.

Considering this locus, we see that the effect of the added pole has been to distort the simple circle of the PD control but, for points near the origin, the locus is quite similar to the earlier case. The situation changes when the pole is brought closer in.

EXAMPLE 5.5

Root Locus of the Satellite Control with Lead Having a Relatively Small Value for the Pole

Now consider $p = 4$ and draw the root locus for

$$1 + K\frac{s+1}{s^2(s+4)} = 0. \tag{5.61}$$

Solution. Again, by the rules, we have the following:

RULE 1. There are again three branches to the locus, two starting from $s = 0$ and one from $s = -4$.

RULE 2. The segment of the real axis $-4 \le s \le -1$ is part of the locus.

RULE 3. There are two asymptotes centered at $\alpha = -3/2$ and at the angles $\pm 90°$.

RULE 4. The branches again depart from the poles at $s = 0$ at $\pm 90°$.

RULE 5. The MATLAB commands

```
numL=[1 1];
denL=[1 4 0 0];
```

Figure 5.12

Root locus for
$L(s) = \frac{(s+1)}{s^2(s+4)}$

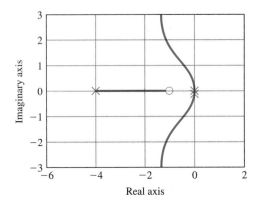

```
sysL=tf(numL,denL)
rlocus(sysL)
```

show that two branches of this locus break away vertically from the poles at $s = 0$, curve slightly to the left and join the asymptotes going north and south. The locus segment from the root at $s = -4$ goes east and terminates at the zero. In this case, the locus differs from the case when $p = -12$ in that there are no break-in or breakaway points on the real axis as part of the locus. The MATLAB plot is given in Fig. 5.12.

In these two cases we have similar systems, but in one case, $p = -12$, there were both a break-in and a breakaway on the real axis, whereas for $p = -4$, these features have disappeared. A logical question might be to ask at what point they went away. As a matter of fact, it happens at $p = 9$, and we'll look at that locus next.

EXAMPLE 5.6   *The Root Locus for the Satellite with a Transition Value for the Pole*

Plot the root locus for

$$1 + K \frac{s+1}{s^2(s+9)} = 0. \tag{5.62}$$

Solution

RULE 1. The locus has three branches, starting from $s = 0$ and $s = -9$.

RULE 2. The real axis segment $-9 \leq s \leq -1$ is part of the locus.

RULE 3. The two asymptotes are centered at $\alpha = -\,{}^8/_2 = -4$.

RULE 4. The departures are, as before, at $\pm 90°$ from $s = 0$.

RULE 5. The MATLAB commands

```
numL=[1 1];
denL=[1 9 0 0]
sysL=tf(numL,denL);
rlocus(sysL)
```

produces the locus in Fig. 5.13. It shows the two branches of this locus break away vertically from the poles at $s = 0$ and curl around and join the real axis again at

Figure 5.13

Root locus for
$L(s) = \frac{(s+1)}{s^2(s+9)}$

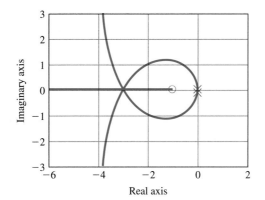

$s = -3$ with an angle of arrival of $\pm 60°$ while the branch from the pole at $s = -9$ heads east and joins the other two poles at $s = -3$ with an angle of arrival of $0°$. These three locus segments continue on by splitting out of $s = -3$ at the departure angles of $0°$ and $\pm 120°$, with one heading into the zero and the other two heading away to the northwest to join the asymptotes. Using Rule 5 would confirm these angles of arrival and departure.[12]

From Figs. 5.11 through 5.13, it is evident that when the third pole is near the zero (p near 1), there is only a modest distortion of the locus that would result for $D(s)G(s) \cong K\frac{1}{s^2}$, which consists of two straight-line locus branches departing at $\pm 90°$ from the two poles at $s = 0$. Then, as we increase p, the locus changes until at $p = 9$ the locus breaks in at -3 in a triple multiple root. As the pole p is moved to the left beyond -9, the locus exhibits distinct break-in and breakaway points, approaching, as p gets very large, the circular locus of one zero and two poles. Figure 5.13, when $p = 9$, is thus a transition locus between the two second-order extremes, which occur at $p = 1$ (when the zero is canceled) and $p \to \infty$ (where the extra pole has no effect).

EXAMPLE 5.7   *An Exercise to Repeat the Prior Examples Using RLTOOL*

Repeat Examples 5.3 through 5.6 using MATLAB's RLTOOL feature.

Solution. RLTOOL is an interactive root-locus design tool in MATLAB that provides a graphical user interface (GUI) for performing root-locus analysis and design. RLTOOL provides an easy way to design feedback controllers because it allows rapid iterations and quickly shows their effect on the resulting root locus. To illustrate the use of the tool, the MATLAB commands

```
numL=[1  1];
denL=[1  0  0];
sysL=tf(numL,denL)
rltool(sysL)
```

---

[12]The shape of this special root locus is a trisectrix of Maclaurin, a plane curve that can be used to trisect an angle.

Figure 5.14

RLTOOL graphical user interface

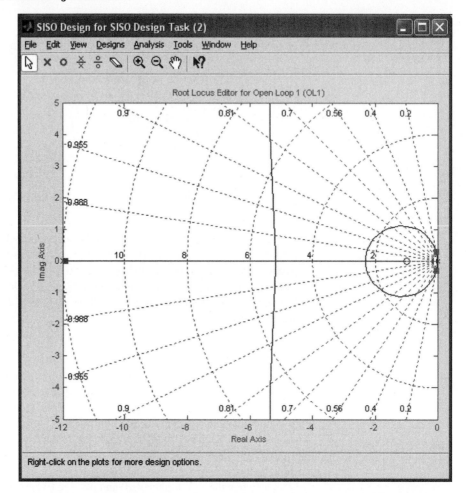

will initiate the GUI and produce the root locus shown in Fig. 5.10, which is similar to Examples 5.4 through 5.6, but without the pole on the negative real axis that was moved around for illustration purposes in the three prior examples. By clicking on "Compensator Editor" in the "Control and Estimation Tools Manager" window, right clicking on the "Dynamics" dialog window and selecting "add pole/zero", you can add a pole at the location $s = -12$. This will produce the locus that is shown in Fig. 5.11 and Fig. 5.14. Now put your mouse on the pole at $s = -12$, hold down the mouse button, and slide it from $s = -12$ to $s = -4$ slowly, so you can examine the locus shapes at all intermediate points. Be especially careful (and slow) as you pass through $s = -9$ because the locus shape changes very quickly with the pole in this region. Note that you can also put your mouse on one of the closed-loop poles (squares) and slide that along the locus. It will show you the location of the other roots that correspond to that value of the gain, K, and the frequency and damping of the closed-loop roots will be shown for when the roots are complex pairs. More detail can be found in the RLTOOL Tutorial in Appendix WR.

A useful conclusion drawn from this example is the following:

> An additional pole moving in from the far left tends to push the locus branches to the right as it approaches a given locus.

The double integrator is the simplest model of the examples, assuming a rigid body with no friction. A more realistic case would include the effects of flexibility in the satellite attitude control, where at least the solar panels would be flexible. In the case of the disk drive read/write mechanism, the head and supporting arm assembly always has flexibility and usually a very complex behavior with a number of lightly damped modes, which can often be usefully approximated by a single dominant mode. In Section 2.1 it was shown that flexibility in the disk drive added a set of complex poles to the $1/s^2$ model. Generally there are two possibilities, depending on whether the sensor is on the same rigid body as the actuator, which is called the collocated case,[13] or is on another body, in which case we have the noncollocated case.[14] We begin with consideration of the collocated case similar to that given by Eq. (2.20). As we saw in Chapter 2, the transfer function in the collocated case has not only a pair of complex poles but also a pair of nearby complex zeros located at a lower natural frequency than the poles. The numbers in the examples that follow are chosen more to illustrate the root-locus properties than to represent particular physical models.

EXAMPLE 5.8   *Root Locus of the Satellite Control with a Collocated Flexibility*

Plot the root locus of the characteristic equation $1 + G(s)D(s) = 0$, where

$$G(s) = \frac{(s + 0.1)^2 + 6^2}{s^2[(s + 0.1)^2 + 6.6^2]} \tag{5.63}$$

is in a unity feedback structure with the controller transfer function

$$D(s) = K\frac{s + 1}{s + 12}. \tag{5.64}$$

Solution.   In this case

$$L(s) = \frac{s + 1}{s + 12}\frac{(s + 0.1)^2 + 6^2}{s^2[(s + 0.1)^2 + 6.6^2]}$$

has both poles and zeros near the imaginary axis and we should expect to find the departure angles of particular importance.

RULE 1. There are five branches to the locus, three of which approach finite zeros and two of which approach asymptotes.

RULE 2. The real-axis segment $-12 \leq s \leq -1$ is part of the locus.

---

[13]Typical of the satellite attitude control, where the flexibility arises from solar panels and both actuator and sensor act on the main body of the satellite.

[14]Typical of the satellite, where the flexibility arises from a scientific package whose attitude is to be controlled from a command body coupled to the package by a flexible strut. This case is also typical of computer hard-disk read/write head control, where the motor is on one end of the arm and the head is on the other.

Figure 5.15

Figure for computing a departure angle for

$$L(s) = \frac{s+1}{s+12} \frac{(s+0.1)^2+6^2}{s^2(s+0.1)^2+6.6^2}$$

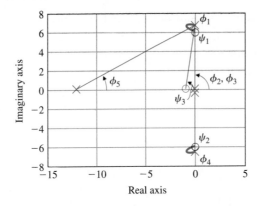

RULE 3. The center of the two asymptotes is at

$$\alpha = \frac{-12 - 0.1 - 0.1 - (-0.1 - 0.1 - 1)}{5 - 3} = -\frac{11}{2}.$$

The angle of the asymptotes is $\pm 90°$.

RULE 4. We compute the departure angle from the pole at $s = -0.1 + j6.6$. The angle at this pole we will define to be ϕ_1. The other angles are marked on Fig. 5.15. The root-locus condition is

$$\phi_1 = \psi_1 + \psi_2 + \psi_3 - (\phi_2 + \phi_3 + \phi_4 + \phi_5) - 180°,$$

$$\phi_1 = 90° + 90° + \tan^{-1}(6.6) - [90° + 90° + 90°$$

$$+ \tan^{-1}\left(\frac{6.6}{12}\right)] - 180°, \qquad (5.65)$$

$$\phi_1 = 81.4° - 90° - 28.8° - 180°,$$

$$= -217.4° = 142.6°,$$

so the root leaves this pole up and to the left, into the stable region of the plane. An interesting exercise would be to compute the arrival angle at the zero located at $s = -0.1 + j6$.

Using MATLAB, the locus is plotted in Fig. 5.16. Note that all the attributes that were determined using the simple rules were exhibited by the plot, thus verifying in part that the data were entered correctly.

The previous example showed that

> In the collocated case, the presence of a single flexible mode introduces a lightly damped root to the characteristic equation but does not cause the system to be unstable.

The departure angle calculation showed that the root departs from the pole introduced by the flexible mode toward the LHP. Next, let's consider the noncollocated

Figure 5.16

Root locus for
$L(s) = \frac{s+1}{s+12}$
$\frac{(s+0.1)^2+6^2}{s^2(s+0.1)^2+6.6^2}$

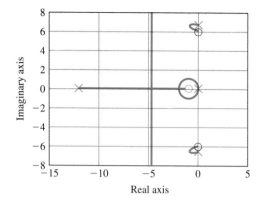

case, for which we take the plant transfer function to be

$$G(s) = \frac{1}{s^2[(s+0.1)^2 + 6.6^2]},\tag{5.66}$$

compensated again by the lead

$$D(s) = K\frac{s+1}{s+12}.\tag{5.67}$$

As these equations show, the noncollocated transfer function has the complex poles but does not have the associated complex zeros as occurred in the previous example and that we also saw for the collated case of Chapter 2 in Eq. (2.20). This will have a substantial effect, as illustrated by Example 5.9.

EXAMPLE 5.9   *Root Locus for the Noncollocated Case*

Apply the rules and draw the root locus for

$$KL(s) = DG = K\frac{s+1}{s+12}\frac{1}{s^2[(s+0.1)^2 + 6.6^2]},\tag{5.68}$$

paying special attention to the departure angles from the complex poles.

RULE 1. There are five branches to the root locus, of which one approaches the zero and four approach the asymptotes.

RULE 2. The real-axis segment defined by $-12 \leq s \leq -1$ is part of the locus.

RULE 3. The center of the asymptotes is located at

$$\alpha = \frac{-12 - 0.2 - (-1)}{5 - 1} = \frac{-11.2}{4},$$

and the angles for the four asymptotic branches are at $\pm 45°, \pm 135°$.

RULE 4. We again compute the departure angle from the pole at $s = -0.1 + j6.6$. The angle at this pole we will define to be ϕ_1. The other angles are marked on Fig. 5.17.

Figure 5.17

Figure to compute a departure angle for $L(s) = \frac{s+1}{s+12} \frac{1}{s^2(s+0.1)^2+6.6^2}$

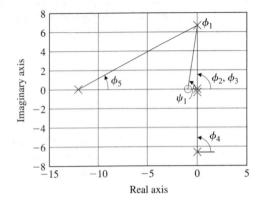

Figure 5.18

Root locus for $L(s) = \frac{s+1}{s+12} \frac{1}{s^2(s+0.1)^2+6.6^2}$

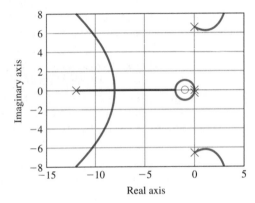

The root locus condition is

$$\phi_1 = \psi_1 - (\phi_2 + \phi_3 + \phi_4 + \phi_5) - 180°,$$

$$\phi_1 = \tan^{-1}(6.6) - \left[90° + 90° + 90° + \tan^{-1}\left(\frac{6.6}{12}\right)\right] - 180°,$$

$$\phi_1 = 81.4° - 90° - 90° - 90° - 28.8° - 180°, \tag{5.69}$$

$$\phi_1 = 81.4° - 90° - 28.8° - 360°,$$

$$\phi_1 = -37.4°.$$

In this case, the root leaves the pole down and to the *right*, toward the unstable region. We would expect the system to soon become unstable as gain is increased.

RULE 5. The locus is plotted in Fig. 5.18 with the commands

```
numG = 1;
denG = [1.0  0.20  43.57  0  0];
sysG = tf(numG,denG);
numD = [1  1];
denD = [1  12];
sysD = tf(numD,denD);
sysL = sysD*sysG;
rlocfind(sysL)
```

and is seen to agree with the calculations above. By using RLTOOL, we see that the locus from the complex poles enter into the RHP almost immediately as the gain is increased. Furthermore, by selecting those roots so that they are just to the left of the imaginary axis, it can be seen that the dominant slow roots down near the origin have extremely low damping. Therefore, this system will have a very lightly damped response with very oscillatory flexible modes. It would not be considered acceptable with the lead compensator as chosen for this example.

A Locus with Complex Multiple Roots

We have seen loci with break-in and breakaway points on the real axis. Of course, an equation of fourth or higher order can have multiple roots that are complex. Although such a feature of a root locus is a rare event, it is an interesting curiosity that is illustrated by the next example.

EXAMPLE 5.10   *Root Locus Having Complex Multiple Roots*

Sketch the root locus of $1 + KL(s) = 0$, where

$$L(s) = \frac{1}{s(s+2)[(s+1)^2 + 4]}.$$

Solution

RULE 1. There are four branches of the locus, all of which approach asymptotes.

RULE 2. The real-axis segment $-2 \leq s \leq 0$ is on the locus.

RULE 3. The center of the asymptotes is at

$$\alpha = \frac{-2 - 1 - 1 - 0 + 0}{4 - 0} = -1$$

and the angles are $\phi_l = 45°, 135°, -45°, -135°$.

RULE 4. The departure angle ϕ_{dep} from the pole at $= -1 + 2j$, based on Fig. 5.19, is

Figure 5.19

Figure to compute departure angle for $L(s) = \frac{1}{s(s+2)(s+1)^2+4}$

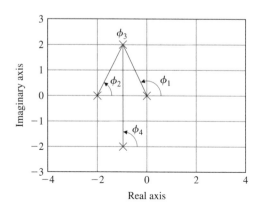

Figure 5.20

Root locus for

$L(s) = \dfrac{1}{s(s+2)(s+1)^2+4}$

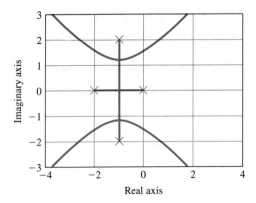

$$\phi_{dep} = \phi_3 = -\phi_1 - \phi_2 - \phi_4 + 180°$$

$$= -\tan^{-1}\left(\frac{2}{-1}\right) - \tan^{-1}\left(\frac{2}{1}\right) - 90° + 180°$$

$$- 116.6° - 63.4° - 90° + 180°$$

$$= -90°.$$

We can observe at once that, along the line $s = -1 + j\omega$, ϕ_2 and ϕ_1 are angles of an isosceles triangle and always add to $180°$. Hence, the entire line from one complex pole to the other is on the locus *in this special case*.

RULE 5. Using MATLAB, we see that there are multiple roots at $s = -1 \pm 1.22j$, and branches of the locus come together at $-1 \pm 1.22j$. Using Rule 5, we can verify that the locus segments break away at $0°$ and $180°$ as shown by MATLAB.

The locus in this example is a transition between two types of loci: one where the complex poles are to the left of the example case and approach the asymptotes at $\pm135°$ and another where the complex poles are to the right of their positions in the example and approach the asymptotes at $\pm45°$.

5.4   Design Using Dynamic Compensation

Consideration of control design begins with the design of the process itself. The importance of early consideration of potential control problems in the design of the process and selection of the actuator and sensor cannot be overemphasized. It is not uncommon for a first study of the control to suggest that the process itself can be changed by, for example, adding damping or stiffness to a structure to make a flexibility easier to control. Once these factors have been taken into account, the design of the controller begins. If the process dynamics are of such a nature that a satisfactory design cannot be obtained by adjustment of the proportional gain alone, then some modification or compensation of the dynamics is indicated. While the variety of possible compensation schemes is great, three categories have been found to

Lead and lag
compensations

be particularly simple and effective. These are lead, lag, and notch compensations.[15] **Lead compensation** approximates the function of PD control and acts mainly to speed up a response by lowering rise time and decreasing the transient overshoot. **Lag compensation** approximates the function of PI control and is usually used to improve the steady-state accuracy of the system. **Notch compensation** will be used to achieve stability for systems with lightly damped flexible modes, as we saw with the satellite attitude control having noncollocated actuator and sensor. In this section we will examine techniques to select the parameters of these three schemes. Lead, lag, and notch compensations have historically been implemented using analog electronics and hence were often referred to as networks. Today, however, most new control system designs use digital computer technology, in which the compensation is implemented in the software. In this case, one needs to compute discrete equivalents to the analog transfer functions, as described in Chapter 4 and discussed further in Chapter 8 and in Franklin et al. (1998).

Compensation with a transfer function of the form

$$D(s) = K \frac{s+z}{s+p} \tag{5.70}$$

is called lead compensation if $z < p$ and lag compensation if $z > p$. Compensation is typically placed in series with the plant in the feed-forward path, as shown in Fig. 5.21. It can also be placed in the feedback path and in that location has the same effect on the overall system poles but results in different transient responses from reference inputs. The characteristic equation of the system in Fig. 5.21 is

$$1 + D(s)G(s) = 0,$$

$$1 + KL(s) = 0,$$

where K and $L(s)$ are selected to put the equation in root-locus form as before.

5.4.1 Design Using Lead Compensation

To explain the basic stabilizing effect of lead compensation on a system, we first consider proportional control for which $D(s) = K$. If we apply this compensation to a second-order position control system with normalized transfer function

$$G(s) = \frac{1}{s(s+1)},$$

Figure 5.21

Feedback system with compensation

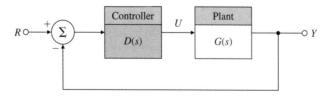

---

[15]The names of these compensation schemes derive from their frequency (sinusoidal) responses, wherein the output leads the input in one case (a positive phase shift) and lags the input in another (a negative phase shift). The frequency response of the third looks as if a notch had been cut in an otherwise flat frequency response. See Chapter 6.

Figure 5.22

Root loci for
$1 + D(s)G(s) = 0$,
$G(s) = \frac{1}{s(s+1)}$: with
compensation $D(s) = K$
(solid lines) and with
$D(s) = K(s + 2)$
(dashed lines)

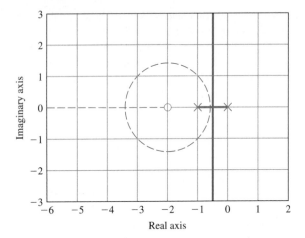

the root locus with respect to K is shown as the solid-line portion of the locus in Fig. 5.22. Also shown in Fig. 5.22 is the locus produced by proportional plus derivative control, where $D(s) = K(s+2)$. The modified locus is the circle sketched with dashed lines. As we saw in the examples, the effect of the zero is to move the locus to the left, toward the more stable part of the s-plane. If, now, our speed-of-response specification calls for $\omega_n \cong 2$, then proportional control alone ($D = K$) can produce only a very low value of damping ratio ζ when the roots are put at the required value of ω_n. Hence, at the required gain, the transient overshoot will be substantial. However, by adding the zero of PD control we can move the locus to a position having closed-loop roots at $\omega_n = 2$ and damping ratio $\zeta \geq 0.5$. We have "compensated" the given dynamics by using $D(s) = K(s + 2)$.

As we observed earlier, pure derivative control is not normally practical because of the amplification of sensor noise implied by the differentiation and must be approximated. If the pole of the lead compensation is placed well outside the range of the design ω_n, then we would not expect it to upset the dynamic response of the design in a serious way. For example, consider the lead compensation

$$D(s) = K\frac{s+2}{s+p}.$$

The root loci for two cases with $p = 10$ and $p = 20$ are shown in Fig. 5.23, along with the locus for PD control. The important fact about these loci is that for small gains, before the real root departing from $-p$ approaches -2, the loci with lead compensation are almost identical to the locus for which $D(s) = K(s+2)$. Note that the effect of the pole is to lower the damping, but for the early part of the locus, the effect of the pole is not great if $p > 10$.

Selection of the zero and pole of a lead

Selecting exact values of z and p in Eq. (5.70) for particular cases is usually done by trial and error, which can be minimized with experience. In general, the zero is placed in the neighborhood of the closed-loop ω_n, as determined by rise-time or settling-time requirements, and the pole is located at a distance 5 to 20 times the value of the zero location. The choice of the exact pole location is a compromise between the conflicting effects of noise suppression, for which one wants a small value for p, and compensation effectiveness for which one wants a large p. In general, if the

Figure 5.23

Root loci for three cases
with $G(s) = \frac{1}{s(s+1)}$:

(a) $D(s) = \frac{(s+2)}{(s+20)}$;

(b) $D(s) = \frac{(s+2)}{(s+10)}$;

(c) $D(s) = s + 2$ (solid
lines)

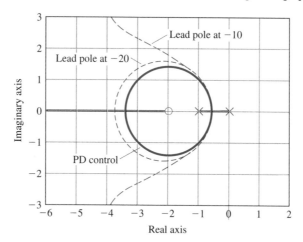

pole is too close to the zero, then, as seen in Fig. 5.23, the root locus moves back too far toward its uncompensated shape and the zero is not successful in doing its job. On the other hand, for reasons that are perhaps easier to understand from the frequency response, when the pole is too far to the left, the magnification of sensor noise appearing at the output of $D(s)$ is too great and the motor or other actuator of the process can be overheated by noise energy in the control signal, $u(t)$. With a large value of p, the lead compensation approaches pure PD control. A simple example will illustrate the approach.

EXAMPLE 5.11   *Design Using Lead Compensation*

Find a compensation for $G(s) = 1/[s(s + 1)]$ that will provide overshoot of no more than 20% and rise time of no more than 0.3 sec.

Solution.   From Chapter 3, we estimate that a damping ratio of $\zeta \geq 0.5$ and a natural frequency of $\omega_n \cong \frac{1.8}{0.3} \cong 6$ should satisfy the requirements. To provide some margin, we will shoot for $\zeta \geq 0.5$ and $\omega_n \geq 7$ rad/sec. Considering the root loci plotted in Fig. 5.23, we will first try

$$D(s) = K\frac{s+2}{s+10}.$$

Figure 5.24 shows that $K = 70$ will yield $\zeta = 0.56$ and $\omega_n = 7.7$ rad/sec, which satisfies the goals based the initial estimates. The third pole will be at $s = -2.4$ with $K = 70$. Because this third pole is so near the lead zero at -2, the overshoot should not be increased very much from the second-order case. However, Fig. 5.25 shows the step response of the system exceeds the overshoot specification a small amount. Typically, lead compensation in the feed-forward path will increase the step-response overshoot because the zero of the compensation has a differentiating effect, as discussed in Chapter 3. The rise-time specification has been met because the time for the amplitude to go from 0.1 to 0.9 is less than 0.3 sec.

   We want to tune the compensator to achieve better damping in order to reduce the overshoot in the transient response. The expedient way to do this is to use RLTOOL,

Figure 5.24

Root locus for lead design

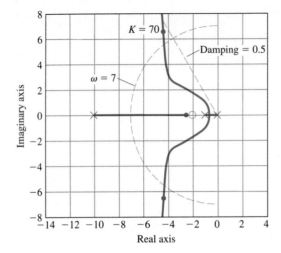

Figure 5.25

Step response for Example 5.11

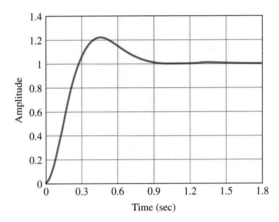

```
sysG=tf(1,[1 1 0]);
sysD=tf([1 2],[1 10]);
rltool(sysG,sysD)
```

By moving the pole of the lead compensator more to the left in order to pull the locus in that direction, and selecting $K = 91$, we obtain

$$D(s) = 91\frac{(s+2)}{(s+13)},$$

which will provide more damping than the previous design iteration. Figure 5.26 shows the root locus with the s-plane regions superimposed on the same plot from RLTOOL. The transient response from RLTOOL is shown in Fig. 5.27 and demonstrates that the overshoot specification is now met (in fact exceeded) with $M_p = 17\%$ and the rise time has degraded some from the previous iteration, but still satisfies the 0.3-sec specification.

Figure 5.26

Illustration of the tuning of the dynamic lead compensator using RLTOOL

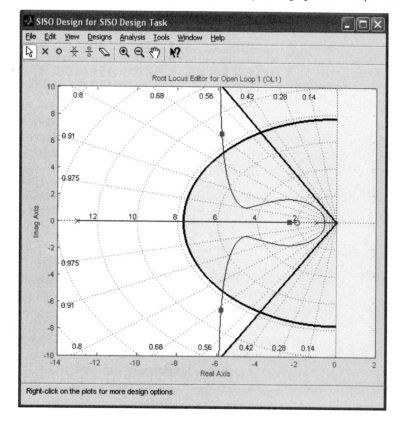

Figure 5.27

Step response for $K = 91$ and $L(s) = \frac{(s+2)}{(s+13)} \frac{1}{s(s+1)}$

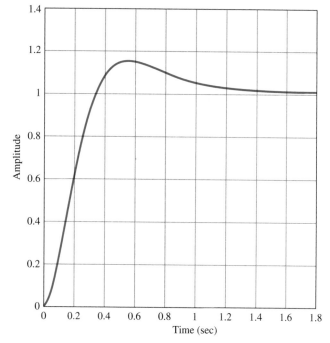

As stated earlier, the name *lead compensation* is a reflection of the fact that to sinusoidal signals, these transfer functions impart phase lead. For example, the phase of Eq. (5.70) at $s = j\omega$ is given by

$$\phi = \tan^{-1}\left(\frac{\omega}{z}\right) - \tan^{-1}\left(\frac{\omega}{p}\right). \tag{5.71}$$

If $z < p$, then ϕ is positive, which by definition indicates phase lead. The details of design using the phase angle of the lead compensation will be treated in Chapter 6.

5.4.2 Design Using Lag Compensation

Once satisfactory dynamic response has been obtained, perhaps by using one or more lead compensations, we may discover that the low-frequency gain—the value of the relevant steady-state error constant, such as K_v—is still too low. As we saw in Chapter 4, the system type, which determines the degree of the polynomial the system is capable of following, is determined by the order of the pole of the transfer function $D(s)G(s)$ at $s = 0$. If the system is Type 1, the velocity-error constant, which determines the magnitude of the error to a ramp input, is given by $\lim_{s\to 0} sD(s)G(s)$. In order to increase this constant, it is necessary to do so in a way that does not upset the already satisfactory dynamic response. Thus, we want an expression for $D(s)$ that will yield a significant gain at $s = 0$ to raise K_v (or some other steady-state error constant) but is nearly unity (no effect) at the higher frequency ω_n, where dynamic response is determined. The result is

$$D(s) = \frac{s+z}{s+p}, \quad z > p, \tag{5.72}$$

where the values of z and p are small compared with ω_n, yet $D(0) = z/p = 3$ to 10 (the value depending on the extent to which the steady-state gain requires boosting). Because $z > p$, the phase ϕ given by Eq. (5.71) is negative, corresponding to phase lag. Hence a device with this transfer function is called lag compensation.

An example of lag compensation

The effects of lag compensation on dynamic response can be studied by looking at the corresponding root locus. Again, we take $G(s) = \frac{1}{[s(s+1)]}$, include the lead compensation $KD_1(s) = \frac{K(s+2)}{(s+13)}$ that produced the locus in Fig. 5.26. With the gain of $K = 91$ from the previous tuned example, we find that the velocity constant is

$$K_v = \lim_{s\to 0} sKD_1G$$

$$= \lim_{s\to 0} s(91)\frac{s+2}{s+13}\frac{1}{s(s+1)}$$

$$= \frac{91*2}{13} = 14.$$

Suppose we require that $K_v = 70$. To obtain this, we require a lag compensation with $z/p = 5$ in order to increase the velocity constant by a factor of 5. This can be accomplished with a pole at $p = -0.01$ and a zero at $z = -0.05$, which keeps the values of both z and p very small so that $D_2(s)$ would have little effect on the portions of the locus representing the dominant dynamics around $\omega_n = 7$. The result is a lag

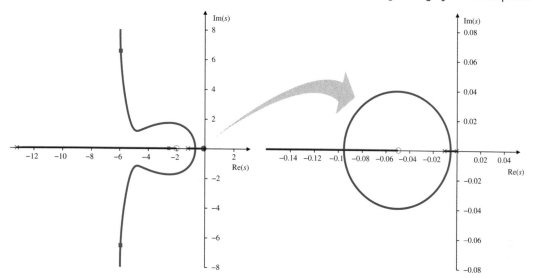

Figure 5.28
Root locus with both lead and lag compensations

compensation with the transfer function of $D_2(s) = \frac{(s+0.05)}{(s+0.01)}$. The root locus with both lead and lag compensation is plotted in Fig. 5.28 and we see that, for the large scale on the left, the locus is not noticeably different from that in Fig. 5.26. This was the result of selecting very small values for the pole and zero. With $K = 91$, the dominant roots are at $-5.8 \pm j6.5$. The effect of the lag compensation can be seen by expanding the region of the locus around the origin as shown on the right side of Fig. 5.28. Here we can see the circular locus that is a result of the small pole and zero. A closed-loop root remains very near the lag-compensation zero at $-0.05 + 0j$; therefore, the transient response corresponding to this root will be a very slowly decaying term, which will have a small magnitude because the zero will almost cancel the pole in the transfer function. Still, the decay is so slow that this term may seriously influence the settling time. Furthermore, the zero will *not* be present in the step response to a disturbance torque and the slow transient will be much more evident there. Because of this effect, it is important to place the lag pole–zero combination at as high a frequency as possible without causing major shifts in the dominant root locations.

5.4.3 Design Using Notch Compensation

Suppose the design has been completed with lead and lag compensation given by

$$KD(s) = 91\frac{s+2}{s+13}\frac{s+0.05}{s+0.01}, \tag{5.73}$$

but is found to have a substantial oscillation at about 50 rad/sec when tested, because there was an unsuspected flexibility of the noncollocated type at a natural frequency of $\omega_n = 50$. On reexamination, the plant transfer function, including the effect of the flexibility, is estimated to be

$$G(s) = \frac{2500}{s(s+1)(s^2+s+2500)}. \tag{5.74}$$

A mechanical engineer claims that some of the "control energy" has spilled over into the lightly damped flexible mode and caused it to be excited. In other words, as we saw from the similar system whose root locus is shown in Fig. 5.18, the very lightly damped roots at 50 rad/sec have been made even less damped or perhaps unstable by the feedback. The best method to fix this situation is to modify the structure so that there is a mechanical increase in damping. Unfortunately, this is often not possible because it is found too late in the design cycle. If it isn't possible, how else can this oscillation be corrected? There are at least two possibilities. An additional lag compensation might lower the loop gain far enough that there is greatly reduced spillover and the oscillation is eliminated. Reducing the gain at the high frequency

is called **gain stabilization**. If the response time resulting from gain stabilization is too long, a second alternative is to add a zero near the resonance so as to shift the departure angles from the resonant poles so as to cause the closed-loop root to move into the LHP, thus causing the associated transient to die out. This approach is called **phase stabilization**, and its action is similar to that of flexibility in the collocated motion control discussed earlier. Gain and phase stabilization are explained more precisely by their effect on the frequency response (Chapter 6) where these methods of stabilization will be discussed further. For phase stabilization, the result is called a notch compensation, and an example has a transfer function

$$D_{notch}(s) = \frac{s^2 + 2\zeta\omega_o s + \omega_o^2}{(s+\omega_o)^2}. \tag{5.75}$$

A necessary design decision is whether to place the notch frequency above or below that of the natural resonance of the flexibility in order to get the necessary phase. A check of the angle of departure shows that with the plant as compensated by Eq. (5.73) and the notch as given, it is necessary to place the frequency of the notch *above* that of the resonance to get the departure angle to point toward the LHP. Thus the compensation is added with the transfer function

$$D_{notch}(s) = \frac{s^2 + 0.8s + 3600}{(s+60)^2}. \tag{5.76}$$

The gain of the notch at $s = 0$ has been kept at 1 so as not to change the K_v. The new root locus is shown in Fig. 5.29 and the step response is shown in Fig 5.30. Note from the step response that the oscillations are well damped, the rise-time specification is still met, but the overshoot has degraded. To rectify the increased overshoot and strictly meet all the specifications, further iteration should be carried out in order to provide more damping of the fast roots in the vicinity of $\omega_n = 7$ rad/sec.

When considering notch or phase stabilization, it is important to understand that its success depends on maintaining the correct phase at the frequency of the resonance. If that frequency is subject to significant change, which is common in many cases, then the notch needs to be removed far enough from the nominal frequency in order to work for all cases. The result may be interference of the notch with the rest of the

Figure 5.29

Root locus with lead, lag, and notch compensations

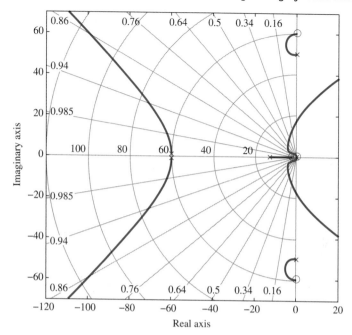

Figure 5.30

Step response with lead, lag, and notch compensations

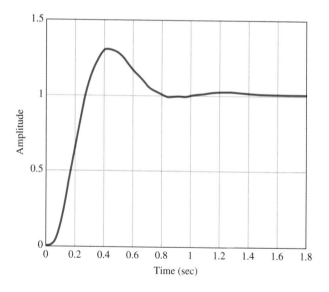

dynamics and poor performance. As a general rule, gain stabilization is substantially more robust to plant changes than is phase stabilization.

5.4.4 Analog and Digital Implementations

Lead compensation can be implemented using analog electronics, but digital computers are preferred.

Lead compensation can be physically realized in many ways. In analog electronics a common method is to use an operational amplifier, an example of which is shown

Figure 5.31

Possible circuit of a lead compensation

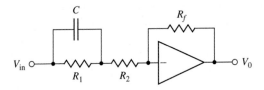

in Fig. 5.31. The transfer function of the circuit in Fig. 5.31 is readily found by the methods of Chapter 2 to be

$$D_{lead}(s) = -a\frac{s+z}{s+p}, \tag{5.77}$$

where

$$a = \frac{p}{z}, \quad \text{if} \quad R_f = R_1 + R_2,$$

$$z = \frac{1}{R_1C},$$

$$p = \frac{R_1 + R_2}{R_2} \cdot \frac{1}{R_1C}.$$

If a design for $D(s)$ is complete and a digital implementation is desired, then the technique of Chapter 4 can be used by first selecting a sampling period T_s and then making substitution of $\frac{2}{T_s}\frac{z-1}{z+1}$ for s. For example, consider the lead compensation $D(s) = \frac{s+2}{s+13}$. Then, since the rise time is about 0.3, a sampling period of six samples per rise time results in the selection of $T_s = 0.05$ sec. With the substitution of $\frac{2}{0.05}\frac{z-1}{z+1}$ for s into this transfer function, the discrete transfer function is

$$\frac{U(z)}{E(z)} = \frac{40\dfrac{z-1}{z+1} + 2}{40\dfrac{z-1}{z+1} + 13}$$

$$= \frac{1.55z - 1.4}{1.96z - 1}. \tag{5.78}$$

Clearing fractions and using the fact that operating on the time functions $zu(kT_s) = u(kT_s + T_s)$, we see that Eq. (5.78) is equivalent to the formula for the controller given by

$$u(kT_s + T_s) = \frac{1}{1.96}u(kT_s) + \frac{1.55}{1.96}e(kT_s + T_s) - \frac{1.4}{1.96}e(kT_s). \tag{5.79}$$

The MATLAB commands to generate the discrete equivalent controller are

```
sysC=tf([1 2],[1 13]);
sysD=c2D(sysC,0.05)
```

Fig. 5.32 shows the SIMULINK diagram for implementing the digital controller. The result of the simulation is contained in Fig. 5.33, which shows the comparison of analog and digital control outputs, and Fig. 5.34, which shows the analog and digital control outputs.

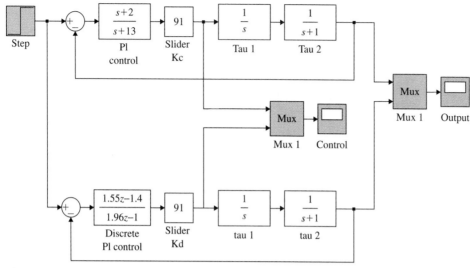

Figure 5.32
SIMULINK® diagram for comparison of analog and digital control

Figure 5.33
Comparison of analog
and digital
control output
responses

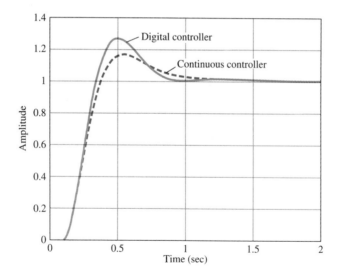

As with lead compensation, lag or notch compensation can be implemented using a digital computer and following the same procedure. However, they, too, can be implemented using analog electronics, and a circuit diagram of a lag network is given in Fig. 5.35. The transfer function of this circuit can be shown to be

$$D(s) = -a\frac{s+z}{s+p},$$

Figure 5.34

Comparison of analog and digital control time histories

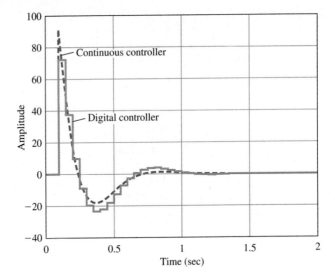

Figure 5.35

Possible circuit of lag compensation

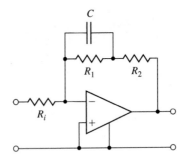

where

$$a = \frac{R_2}{R_i},$$

$$z = \frac{R_1 + R_2}{R_1 R_2 C},$$

$$p = \frac{1}{R_1 C}.$$

Usually $R_i = R_2$, so the high-frequency gain is unity, or $a = 1$, and the low-frequency increase in gain to enhance K_v or other error constant is set by $k = a\frac{z}{p} = \frac{R_1 + R_2}{R_2}$.

5.5 A Design Example Using the Root Locus

EXAMPLE 5.12

Control of a Small Airplane

For the Piper Dakota shown in Fig. 5.36, the transfer function between the elevator input and the pitch attitude is

Figure 5.36

Autopilot design in the Piper Dakota, showing elevator and trim tab

Source: Photo courtesy of Denise Freeman

(a)

Trim tab δ_t →

Elevator δ_e →

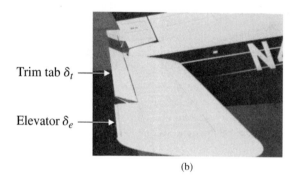

(b)

$$G(s) = \frac{\theta(s)}{\delta_e(s)} = \frac{160(s + 2.5)(s + 0.7)}{(s^2 + 5s + 40)(s^2 + 0.03s + 0.06)}, \qquad (5.80)$$

where

$$\theta = \text{pitch attitude, degrees (see Fig. 10.30)},$$

$$\delta_e = \text{elevator angle, degrees.}$$

(For a more detailed discussion of longitudinal aircraft motion, refer to Section 10.3.)

1. Design an autopilot so that the response to a step elevator input has a rise time of 1 sec or less and an overshoot less than 10%.

2. When there is a constant disturbing moment acting on the aircraft so that the pilot must supply a constant force on the controls for steady flight, it is said to be out of trim. The transfer function between the disturbing moment and the attitude is the same as that due to the elevator; that is,

$$\frac{\theta(s)}{M_d(s)} = \frac{160(s + 2.5)(s + 0.7)}{(s^2 + 5s + 40)(s^2 + 0.03s + 0.06)}, \qquad (5.81)$$

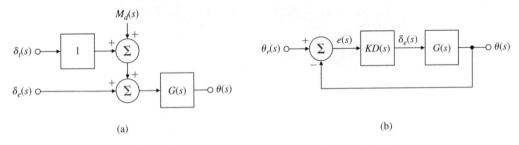

Figure 5.37

Block diagrams for autopilot design: (a) open loop; (b) feedback scheme excluding trim control

where M_d is the moment acting on the aircraft. There is a separate aerodynamic surface for trimming, δ_t, that can be actuated and will change the moment on the aircraft. It is shown in the close-up of the tail in Fig. 5.36. Its influence is depicted in the block diagram shown in Fig. 5.37(a). For both manual and autopilot flight, it is desirable to adjust the trim so that there is no steady-state control effort required from the elevator (that is, so $\delta_e = 0$). In manual flight, this means that no force is required by the pilot to keep the aircraft at a constant altitude, whereas in autopilot control it means reducing the amount of electrical power required and saving wear and tear on the servomotor that drives the elevator. Design an autopilot that will command the trim δ_t so as to drive the steady-state value of δ_e to zero for an arbitrary constant moment M_d as well as meet the specifications in part (a).

Solution

1. To satisfy the requirement that the rise time $t_r \leq 1$ sec, Eq. (3.60) indicates that, for the ideal second-order case, ω_n must be greater than 1.8 rad/sec. And to provide an overshoot of less than 10%, Fig. (3.23) indicates that ζ should be greater than 0.6, again, for the ideal second-order case. In the design process, we can examine a root locus for a candidate for feedback compensation and then look at the resulting time response when the roots appear to satisfy the design guidelines. However, since this is a fourth-order system, the design guidelines might not be sufficient, or they might be overly restrictive.

   To initiate the design process, it is often instructive to look at the system characteristics with proportional feedback, that is, where $D(s) = 1$ in Fig. 5.37(b). The statements in MATLAB to create a root locus with respect to K and a time response for the proportional feedback case with $K = 0.3$ are as follows:

```
numG = 160*conv ([1  2.5],[1  0.7]);
denG = conv([1  5  40],[1  0.03  0.06]);
sysG = tf(numG,denG);
rlocus(sysG)
K = 0.3
sysL = K*sysG
sysH = tf(1,1);
```

[sysT] = feedback (sysL,sysH)
step(sysT)

The resulting root locus and time response are shown with dashed lines in Figs. 5.38 and 5.39. Notice from Fig. 5.38 that the two faster roots will always have a damping ratio ζ that is less than 0.4; therefore, proportional feedback will not be acceptable. Also, the slower roots have some effect on the time response

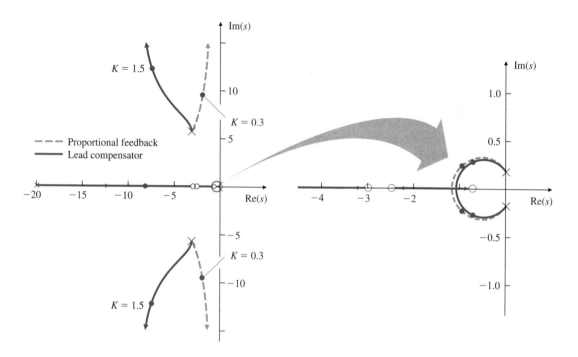

Figure 5.38
Root loci for autopilot design

Figure 5.39
Time–response plots for autopilot design

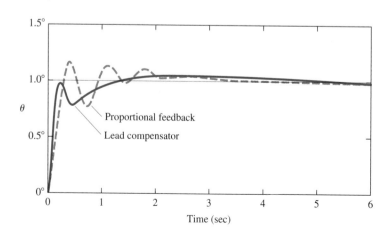

shown in Fig. 5.39 (dashed curve) with $K = 0.3$ in that they cause a long-term settling. However, the dominating characteristic of the response that determines whether or not the compensation meets the specifications is the behavior in the first few seconds, which is dictated by the fast roots. The low damping of the fast roots causes the time response to be oscillatory, which leads to excess overshoot and a longer settling time than desired.

We saw in Section 5.4.1 that lead compensation causes the locus to shift to the left, a change needed here to increase the damping. Some trial and error will be required to arrive at a suitable pole and zero location. Values of $z = 3$ and $p = 20$ in Eq. (5.70) have a substantial effect in moving the fast branches of the locus to the left; thus

$$D(s) = \frac{s + 3}{s + 20}.$$

Lead compensation via MATLAB

Trial and error is also required to arrive at a value of K that meets the specifications. The statements in MATLAB to add this compensation are

```
numD = [1  3];
denD = [1  20];
sysD = tf(numD,denD);
sysDG = sysD*sysG
rlocus(sysDG)
K = 1.5;
sysKDG = K*sysDG;
sysH = tf(1,1)
sysT = feedback(sysKDG,sysH)
step(sysT)
```

The root locus for this case and the corresponding time response are also shown in Figs. 5.38 and 5.39 by the solid lines. Note that the damping of the fast roots that corresponds to $K = 1.5$ is $\zeta = 0.52$, which is slightly lower than we would like; also, the natural frequency is $\omega_n = 15$ rad/sec, much faster than we need. However, these values are close enough to meeting the guidelines to suggest a look at the time response. In fact, the time response shows that $t_r \cong 0.9$ sec and $M_p \cong 8\%$, both within the specifications, although by a very slim margin.

In sum, the primary design path consisted of adjusting the compensation to influence the fast roots, examining their effect on the time response, and continuing the design iteration until the time specifications were satisfied.

2. The purpose of the trim is to provide a moment that will eliminate a steady-state nonzero value of the elevator. Therefore, if we integrate the elevator command δ_e and feed this integral to the trim device, the trim should eventually provide the moment required to hold an arbitrary altitude, thus eliminating the need for a steady-state δ_e. This idea is shown in Fig. 5.40(a). If the gain on the integral term K_I is small enough, the destabilizing effect of adding the integral should be small and the system should behave approximately as before, since that feedback loop has been left intact. The block diagram in Fig. 5.40(a) can be reduced to that in

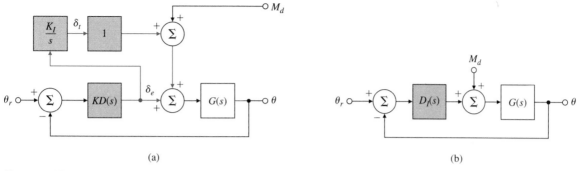

Figure 5.40

Block diagram showing the trim-command loop

Fig. 5.40(b) for analysis purposes by defining the compensation to include the PI form

$$D_I(s) = KD(s)\left(1 + \frac{K_I}{s}\right).$$

However, it is important to keep in mind that, physically, there will be two outputs from the compensation: δ_e (used by the elevator servomotor) and δ_t (used by the trim servomotor).

The characteristic equation of the system with the integral term is

$$1 + KDG + \frac{K_I}{s}KDG = 0.$$

To aid in the design process, it is desirable to find the locus of roots with respect to K_I, but the characteristic equation is not in any of the root-locus forms given by Eqs. (5.6)–(5.9). Therefore, dividing by $1 + KDG$ yields

$$1 + \frac{(K_I/s)KDG}{1 + KDG} = 0.$$

To put this system in root locus form, we define

$$L(s) = \frac{1}{s}\frac{KDG}{1 + KDG}. \tag{5.82}$$

In MATLAB, with $\frac{KDG}{1+KGD}$ already computed as sysT, we construct the integrator as sysIn = tf(1,[1 0]), the loop gain of the system with respect to K_I as sysL = sysIn*sysT, and the root locus with respect to K_I is found with rltool(sysL).

It can be seen from the locus in Fig. 5.41 that the damping of the fast roots decreases as K_I increases, as is typically the case when integral control is added. This shows the necessity for keeping the value of K_I as low as possible. After some trial and error, we select $K_I = 0.15$. This value has little effect on the roots—note the roots are virtually on top of the previous roots obtained without the integral term—and little effect on the short-term behavior of the step response, as shown in Fig. 5.42(a), so the specifications are still met. $K_I = 0.15$ does cause the longer-term attitude behavior to approach the commanded value with no error, as we would expect with integral control. It also causes δ_e to

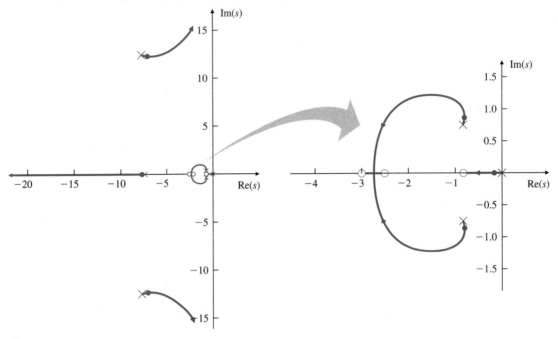

Figure 5.41
Root locus versus K_I: assumes an added integral term and lead compensation with a gain $K = 1.5$; roots for $K_I = 0.15$ marked with •

approach zero [Fig. 5.42(b) shows it settling in approximately 30 sec], which is good because this is the reason for choosing integral control in the first place. The time for the integral to reach the correct value is predicted by the new, slow real root that is added by the integral term at $s = -0.14$. The time constant associated with this root is $\tau = 1/0.14 \cong 7$ sec. The settling time to 1% for a root with $\sigma = 0.14$ is shown by Eq. (3.65) to be $t_s = 33$ sec, which agrees with the behavior in Fig. 5.42(b).

5.6 Extensions of the Root-Locus Method

As we have seen in this chapter, the root-locus technique is a graphical scheme to show locations of possible roots of an algebraic equation as a single real parameter varies. The method can be extended to consider negative values of the parameter, a sequential consideration of more than one parameter, and systems with time delay. In this section we examine these possibilities. Another interesting extension to nonlinear systems is in Chapter 9.

5.6.1 Rules for Plotting a Negative (0°) Root Locus

We now consider modifying the root-locus procedure to permit analysis of negative values of the parameter. In a number of important cases, the transfer function of the

Figure 5.42

Step response for the case with an integral term and 5° command

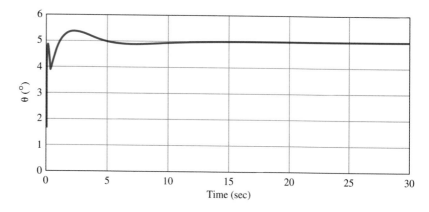

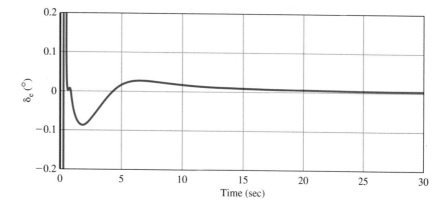

plant has a zero in the RHP and is said to be nonminimum phase. The result is often a locus of the form $1 + A(z_i - s)G'(s) = 1 + (-A)(s - z_i)G'(s) = 0$, and in the standard form the parameter $K = -A$ must be negative. Another important issue calling for understanding the negative locus arises in building a control system. In any physical implementation of a control system there are inevitably a number of amplifiers and components whose gain sign must be selected. By Murphy's Law,[16] when the loop is first closed, the sign will be wrong and the behavior will be unexpected unless the engineer understands how the response will go if the gain which should be positive is instead negative. So what are the rules for a negative locus (a root locus relative to a negative parameter)? First of all, Eqs. (5.6)–(5.9) must be satisfied for negative values of K, which implies that $L(s)$ is real and *positive*. In other words, for the negative locus, the phase condition is

Definition of a Negative Root Locus

The angle of $L(s)$ is $0° + 360°(l - 1)$ for s on the negative locus.

The steps for plotting a negative locus are essentially the same as for the positive locus, except that we search for places where the angle of $L(s)$ is $0° + 360°(l - 1)$

---

[16]Anything that *can* go wrong, *will* go wrong.

instead of $180° + 360°(l - 1)$. For this reason, a negative locus is also referred to as a 0° root locus. This time we find that the locus is to the left of an *even* number of real poles plus zeros (the number zero being even). Computation of the asymptotes for large values of s is, as before, given by

$$\alpha = \frac{\sum p_i - \sum z_i}{n - m}, \tag{5.83}$$

but we modify the angles to be

$$\phi_l = \frac{360°(l - 1)}{n - m}, l = 1, 2, 3, \dots, n - m$$

(shifted by $\frac{180°}{(n-m)}$ from the 180° locus). Following are the guidelines for plotting a 0° locus:

RULE 1. (As before) The n branches of the locus leave the poles and m approach the zeros and $n - m$ approach asymptotes to infinity.

RULE 2. The locus is on the real axis to the left of an *even* number of real poles plus zeros.

RULE 3. The asymptotes are described by

$$\alpha = \frac{\sum p_i - \sum z_i}{n - m} = \frac{-a_1 + b_1}{n - m},$$

$$\phi_l = \frac{360°(l - 1)}{n - m}, \quad l = 1, 2, 3, \dots, n - m.$$

Notice that the angle condition here is measured from 0° rather than from 180° as it was in the positive locus.

RULE 4. Departure angles from poles and arrival angles to zeros are found by searching in the near neighborhood of the pole or zero where the phase of $L(s)$ is 0°, so that

$$q\phi_{\text{dep}} = \sum \psi_i - \sum \phi_i - 360°(l - 1),$$

$$q\psi_{\text{arr}} = \sum \phi_i - \sum \psi_i + 360°(l - 1),$$

where q is the order of the pole or zero and l takes on q integer values such that the angles are between $\pm 180°$.

RULE 5. The locus can have multiple roots at points on the locus and the branches will approach a point of q roots at angle separated by

$$\frac{180° + 360°(l - 1)}{q}$$

and will depart at angles with the same separation.

The result of extending the guidelines for constructing root loci to include negative parameters is that we can visualize the root locus as a set of continuous curves showing the location of possible solutions to the equation $1 + KL(s) = 0$ for *all real values of K*, both positive and negative. One branch of the locus departs from every pole in one direction for positive values of K, and another branch departs from the same pole in another direction for negative K. Likewise, all zeros will have two branches arriving, one with positive and the other with negative values of K. For the $n - m$ excess poles, there will be $2(n - m)$ branches of the locus asymptotically approaching infinity as K approaches positive and negative infinity, respectively. For a single pole or zero, the angles of departure or arrival for the two locus branches will be 180° apart. For a double pole or zero, the two positive branches will be 180° apart and the two negative branches will be at 90° to the positive branches.

The negative locus is often required when studying a nonminimum phase transfer function. A well-known example is that of the control of liquid level in the boiler of a steam power plant. If the level is too low, the actuator valve adds (relatively) cold water to the boiling water in the vessel. The initial effect of the addition is to slow down the rate of boiling, which reduces the number and size of the bubbles and causes the level to fall momentarily before the added volume and heat cause it to rise again to the new increased level. This initial underflow is typical of nonminimum phase systems. Another typical nonminimum phase transfer function is that of the altitude control of an airplane. To make the plane climb, the upward deflection of the elevators initially causes the plane to drop before it rotates and climbs. A Boeing 747 in this mode can be described by the scaled and normalized transfer function

$$G(s) = \frac{6 - s}{s(s^2 + 4s + 13)}. \tag{5.84}$$

To put $1 + KG(s)$ in root-locus form, we need to multiply by -1 to get

$$G(s) = -\frac{s - 6}{s(s^2 + 4s + 13)}. \tag{5.85}$$

EXAMPLE 5.13   *Negative Root Locus for an Airplane*

Sketch the negative root locus for the equation

$$1 + K\frac{s - 6}{s(s^2 + 4s + 13)} = 0. \tag{5.86}$$

Solution

RULE 1. There are three branches and two asymptotes.

RULE 2. A real-axis segment is to the right of $s = 6$ and a segment is to the left of $s = 0$.

RULE 3. The angles of the asymptotes are $\phi_l = \frac{(l-1)360°}{2} = 0°$, 180°, and the center of the asymptotes is at $\alpha = \frac{-2-2-(6)}{3-1} = -5$.

RULE 4. The branch departs the pole at $s = -2 + j3$ at the angle

$$\phi = \tan^{-1}\left(\frac{3}{-6}\right) - \tan^{-1}\left(\frac{3}{-2}\right) - 90° + 360°(l - 1),$$

$$\phi = 159.4 - 123.7 - 90 + 360°(l - 1),$$

$$\phi = -54.3°.$$

The locus is plotted in Fig. 5.43 by MATLAB, which is seen to be consistent with these values.

Figure 5.43

Negative root locus corresponding to $L(s) = (s - 6)/s(s^2 + 4s + 13)$

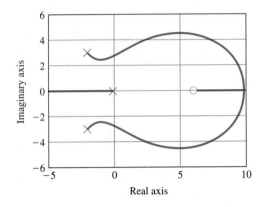

Real axis

△ **5.6.2 Consideration of Two Parameters**

An important technique for practical control is to consider a structure with two loops, an inner loop around an actuator or part of the process dynamics and an outer loop around the entire plant-plus-inner-controller. The process is called **successive loop closure**. A controller is selected for the inner loop to be robust and give good response alone, and then the outer loop can be designed to be simpler and more effective than if the entire control was done without the aid of the inner loop. The use of the root locus to study such a system with two parameters can be illustrated by a simple example.

Successive loop closure

EXAMPLE 5.14

Root Locus Using Two Parameters in Succession

A block diagram of a relatively common servomechanism structure is shown in Fig. 5.44. Here a speed-measuring device (a tachometer) is available and the problem is to use the root locus to guide the selection of the tachometer gain K_T as well as the amplifier gain K_A. The characteristic equation of the system in Fig. 5.44 is

$$1 + \frac{K_A}{s(s+1)} + \frac{K_T}{s+1} = 0,$$

which is not in the standard $1 + KL(s)$ form. After clearing fractions, the characteristic equation becomes

$$s^2 + s + K_A + K_T s = 0, \tag{5.87}$$

which is a function of two parameters, whereas the root locus technique can consider only one parameter at a time. In this case, we set the gain K_A to a nominal value of 4

Figure 5.44

Block diagram of a servomechanism structure, including tachometer feedback

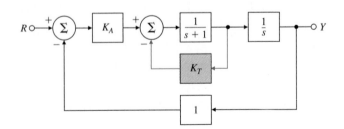

and consider first the locus with respect to K_T. With $K_A = 4$, Eq. (5.87) can be put into root-locus form for a root-locus study with respect to K_T with $L(s) = \frac{s}{s^2+s+4}$, or

$$1 + K_T \frac{s}{s^2 + s + 4} = 0. \tag{5.88}$$

For this root locus, the zero is at $s = 0$ and the poles are at the roots of $s^2 + s + 4 = 0$, or $s = -\frac{1}{2} \pm 1.94j$. A sketch of the locus using the rules as before is shown in Fig. 5.45.

From this locus, we can select K_T so the complex roots have a specific damping ratio or take any other value of K_T that would result in satisfactory roots for the characteristic equation. Consider $K_T = 1$. Having selected a trial value of K_T, we can now re-form the equation to consider the effects of changing from $K_A = 4$ by taking the new parameter to be K_1 so that $K_A = 4 + K_1$. The locus with respect to K_1 is governed by Eq. (5.50), now with $L(s) = \frac{1}{s^2+2s+4}$, so that the locus is for the equation

$$1 + K_1 \frac{1}{s^2 + 2s + 4} = 0. \tag{5.89}$$

Note that the *poles* of the new locus corresponding to Eq. (5.89) are the *roots* of the previous locus, which was drawn versus K_T, and the roots were taken at $K_T = 1$. The locus is sketched in Fig. 5.46, with the previous locus versus K_T left dashed. We

Figure 5.45

Root locus of closed-loop poles of the system in Fig. 5.44 versus K_T

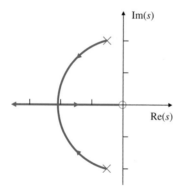

Figure 5.46

Root locus versus $K_1 = K_A + 4$ after choosing $K_T = 1$

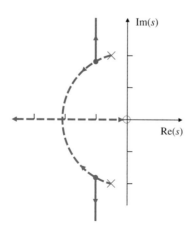

could draw a locus with respect to K_1 for a while, stop, resolve the equation, and continue the locus with respect to K_T, in a sort of see-saw between the parameters K_A and K_T, and thus use the root locus to study the effects of two parameters on the roots of a characteristic equation. Notice, of course, that we can also plot the root locus for negative values of K_1 and thus consider values of K_A less than 4.

△ 5.6.3 Time Delay

Time delays often arise in control systems, both from delays in the process itself and from delays in the processing of sensed signals. Chemical plants often have processes with a time delay representing the time material takes to be transported via pipes or other conveyer. In measuring the attitude of a spacecraft en route to Mars, there is a significant time delay for the sensed quantity to arrive back on Earth due to the speed of light. There is also a small time delay in any digital control system due to the cycle time of the computer and the fact that data is processed at discrete intervals.

Time delays always reduce the stability of a system.

Time delay *always* reduces the stability of a system; therefore, it is important to be able to analyze its effect. In this section we discuss how to use the root locus for such analysis. Although an exact method of analyzing time delay is available in the frequency-response methods to be described in Chapter 6, knowing several different ways to analyze a design provides the control designer with more flexibility and an ability to check the candidate solutions.

An example of a root locus with time delay

Consider the problem of designing a control system for the temperature of the heat exchanger described in Chapter 2. The transfer function between the control A_s and the measured output temperature T_m is described by two first-order terms plus a time delay T_d of 5 sec. The time delay results because the temperature sensor is physically located downstream from the exchanger, so that there is a delay in its reading. The transfer function is

$$G(s) = \frac{e^{-5s}}{(10s + 1)(60s + 1)}, \tag{5.90}$$

where the e^{-5s} term arises from the time delay.[17]

The corresponding root-locus equations with respect to proportional gain K are

$$1 + KG(s) = 0,$$

$$1 + K\frac{e^{-5s}}{(10s + 1)(60s + 1)} = 0,$$

$$600s^2 + 70s + 1 + Ke^{-5s} = 0. \tag{5.91}$$

How would we plot the root locus corresponding to Eq. (5.91)? Since it is not a polynomial, we cannot proceed with the methods used in previous examples. So we reduce the given problem to one we have previously solved by approximating the nonrational function e^{-5s} with a rational function. Since we are concerned with control systems and hence typically with low frequencies, we want an approximation that will be good

---

[17]Time delay is often referred to as "transportation lag" in the process industries.

for small s.[18] The most common means for finding such an approximation is attributed to H. Padé. It consists of matching the series expansion of the transcendental function e^{-5s} with the series expansion of a rational function whose numerator is a polynomial of degree p and whose denominator is a polynomial of degree q. The result is called a (p,q) **Padé approximant**[19] to e^{-5s}. We will initially compute the approximants to e^{-s}, and in the final result we will substitute $T_d s$ for s to allow for any desired delay.

The resulting $(1,1)$ Padé approximant ($p = q = 1$) is (see Appendix W5 for details)

$$e^{-T_d s} \simeq \frac{1 - (T_d s/2)}{1 + (T_d s/2)}. \qquad (5.92)$$

If we assume $p = q = 2$, we have five parameters and a better match is possible. In this case we have the $(2,2)$ approximant, which has the transfer function

$$e^{-T_d s} \simeq \frac{1 - T_d s/2 + (T_d s)^2/12}{1 + T_d s/2 + (T_d s)^2/12}. \qquad (5.93)$$

The comparison of these approximants can be seen from their pole–zero configurations as plotted in Fig. 5.47. The locations of the poles are in the LHP and the zeros are in the RHP at the reflections of the poles.

In some cases a very crude approximation is acceptable. For small delays the $(0,1)$ approximant can be used, which is simply a first-order lag given by

$$e^{-T_d s} \simeq \frac{1}{1 + T_d s}. \qquad (5.94)$$

To illustrate the effect of a delay and the accuracy of the different approximations, root loci for the heat exchanger are drawn in Fig. 5.48 for four cases. Notice that, for low gains and up to the point where the loci cross the imaginary axis, the approximate curves are very close to the exact. However, the $(2,2)$ Padé curve follows the exact curve much further than does the first-order lag, and its increased accuracy would be useful if the delay were larger. All analyses of the delay show its destabilizing effect and how it limits the achievable response time of the system.

Padé approximant

Contrasting methods of approximating delay

Figure 5.47

Poles and zeros of the Padé approximants to e^{-s}, with superscripts identifying the corresponding approximants; for example, x^1 represents the $(1,1)$ approximant

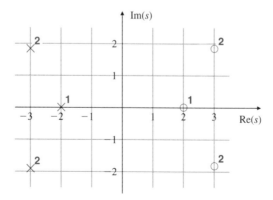

---

[18]The nonrational function e^{-5s} is analytic for all finite values of s and so may be approximated by a rational function. If nonanalytic functions such as $\sqrt{s}$ were involved, great caution would be needed in selecting an approximation valid near $s = 0$.

[19]The (p,p) Padé approximant for a delay of T seconds is most commonly used and is computed by the MATLAB command [num,den] = pade(T, P).

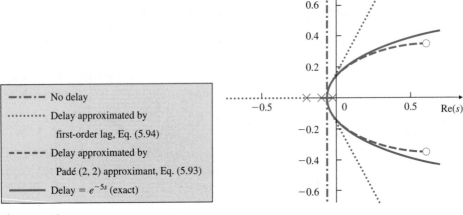

Figure 5.48

Root loci for the heat exchanger with and without time delay

While the Padé approximation leads to a rational transfer function, in theory it is not necessary for plotting a root locus. A direct application of the phase condition can be used to plot portions of an exact locus of a system with time delay. The phase-angle condition does not change if the transfer function of the process is nonrational, so we still must search for values of s for which the phase is $180° + 360°l$. If we write the transfer function as

$$G(s) = e^{-T_d s}\bar{G}(s),$$

the phase of $G(s)$ is the phase of $\bar{G}(s)$ minus $\lambda\omega$ for $s = \sigma + j\omega$. Thus we can formulate a root-locus problem as searching for locations where the phase of $\bar{G}(s)$ is $180° + T_d\omega + 360°(l - 1)$. To plot such a locus, we would fix ω and search along a horizontal line in the s-plane until we found a point on the locus, then raise the value of ω, change the target angle, and repeat. Similarly, the departure angles are modified by $T_d\omega$, where ω is the imaginary part of the pole from which the departure is being computed. MATLAB does not provide a program to plot the root locus of systems with delay, so we must be satisfied here with Padé approximants. Since it is possible to plot the frequency response (or Bode plot) of delay exactly and easily, if the designer feels that the Padé approximant is not satisfactory, the expedient approach is to use the frequency-response design methods described in Chapter 6.

5.7 Historical Perspective

In Chapter 1 we gave an overview of the early development of feedback control analysis and design including frequency response and root-locus design. Root-locus design was introduced in 1948 by Walter R. Evans, who was working in the field of guidance and control of aircraft and missiles at the Autonetics Division of North American Aviation (now a part of The Boeing Co.). Many of his problems involved unstable or neutrally stable dynamics, which made the frequency methods difficult, so he suggested returning to the study of the characteristic equation that had been the basis of the work of Maxwell and Routh nearly 70 years earlier. However, rather than treat the algebraic problem, Evans posed it as a graphical problem in the complex s-plane. Evans was also interested in the character of the dynamic response of the aerospace vehicles being controlled; therefore he wanted to solve for the closed loop roots in order to understand the dynamic behavior. To facilitate this understanding, Evans developed techniques and rules allowing one to follow graphically the paths of the roots of the characteristic equation as a parameter was changed. His method is suitable for design as well as for stability analysis and remains an important technique today. Originally, it enabled the solutions to be carried out by hand since computers were not available to design engineers during the 1940s; however, they remain an important tool today for aiding the design process. As we learned in this chapter, Evans method involves finding a locus of points where the angles to the other poles and zeros add up to a certain value. To aid in this determination, Evans invented the "Spirule" that is shown in Fig. 5.49. The device could be used to measure the angles and to perform the addition or subtraction very quickly. A skilled controls engineer could evaluate whether the angle criterion was met for a fairly complex design problem in a few seconds. In addition, the spiral curve on the rectangular portion of the device allowed the designer to multiply distances in order to determine the gain at a selected spot on the locus in a manner analogous to a slide rule.

Evans was clearly motivated to aid the engineer who had no access to a computer in their design and analysis of control systems. Computers were virtually nonexistent in the 1940s. Large mainframe computers started being used somewhat for large-scale data processing by corporations in the 1950s, but there were no courses in engineering programs that taught the use of computers for analysis and design until about 1960. Engineering usage became commonplace through the 1960s, but the process involved submitting a job to a mainframe computer via a large deck of punched cards and waiting for the results for hours or overnight, a situation that was not conducive to any kind of design iteration. Mainframe computers in that era were just transitioning from vacuum tubes to transistors, random access memory would be in the neighborhood of 32k!, and the long-term data storage was by a magnetic tape drive. Random access drums and disks arrived during that decade, thus greatly speeding up the process of retrieving data. A big step forward in computing for engineers occurred when the batch processing based on punched cards was replaced by time share with many users at remote terminals during the late 1960s and early 1970s. Mechanical calculators were also available through the 1940s, 1950s and 1960s that could add, subtract, multiply, and divide and cost about $2000 in 1960. The very high-end devices could

Figure 5.49

A Spirule: used to sketch a root locus before computers

Source: Photo courtesy of David Powell

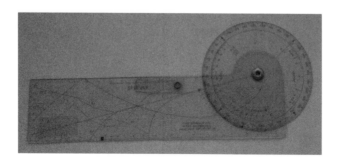

Figure 5.50

The Frieden mechanical calculator

Source: Courtesy of the Computer History Museum

also do square roots. These machines were the basis for the complex computations done at Los Alamos during World War II. They were the size of a typewriter, had a large carriage that went back and forth during the calculations, and would occasionally ring a bell at the end of the carriage stroke (see Fig. 5.50). They were accurate to eight or more decimal places and were often used after the advent of computers to perform spot checks of the results, but a square root could take tens of seconds to complete, the machines were noisy, and the process was tedious. Enterprising engineers learned which particular calculations played certain tunes and it was not unusual to hear favorites, such as Jingle Bells.

The personal computer arrived in the late 1970s, although the ones at that time utilized an audio cassette tape for data storage and had very limited random access memory, usually less than 16k. But as these desktop machines matured over the ensuing decade, the age of the computer for engineering design came into its own. First came the floppy disk for long-term data storage, followed by the hard drive toward the mid- and late-1980s. Initially, the BASIC and APL languages was the primary methods of programming. MATLAB was introduced by Cleve Moler in the 1970s. Two things happened in 1984: Apple introduced the point and click MacIntosh and PC-MATLAB was introduced by The Mathworks, which was specifically founded to commercialize MATLAB on personal computers. Initially, The Mathworks' MATLAB was primarily written for control system analysis, but has branched out into many fields since the initial introduction. At that point in the evolution, the engineer could truly perform design iterations with little or no time between trials. Other similar programs were available for mainframe computers before that time; two being CTRL-C and MATRIXx; however, those programs have not adapted to the personal computer revolution and are fading from general use.

SUMMARY

- A root locus is a graph of the values of s that are solutions to the equation

$$1 + KL(s) = 0$$

with respect to a real parameter K.

1. When $K > 0$, s is on the locus if $\angle L(s) = 180°$, producing a $180°$ or positive K locus.

2. When $K < 0$, s is on the locus if $\angle L(s) = 0°$, producing a $0°$ or negative K locus.

- If $KL(s)$ is the loop transfer function of a system with negative feedback, then the characteristic equation of the closed-loop system is

$$1 + KL(s) = 0,$$

and the root-locus method displays the effect of changing the gain K on the closed-loop system roots.

- A *specific* locus for a system sysL in MATLAB notation can be plotted by rlocus(sysL) and rltool(sysL)

- A working knowledge of how to determine a root locus is useful for verifying computer results and for suggesting design alternatives.

- The key features for the aid in sketching a $180°$ locus are as follows:

    1. The locus is on the real axis to the left of an odd number of poles plus zeros.
    2. Of the n branches, m approach the zeros of $L(s)$ and $n - m$ branches approach asymptotes centered at α and leaving at angles ϕ_l:

    $$n = \text{number of poles},$$

    $$m = \text{number of zeros},$$

    $$n - m = \text{number of asymptotes},$$

    $$\alpha = \frac{\sum p_i - \sum z_i}{n - m},$$

    $$\phi_l = \frac{180° + 360°(l - 1)}{n - m}, \quad l = 1, 2, \ldots, n - m.$$

    3. Branches of the locus depart from the poles of order q and arrive at the zeros of order q with angles

    $$\phi_{l,dep} = \frac{1}{q}\left(\sum \psi_i - \sum_{i \neq l} \phi_i - 180° - 360°(l - 1) \right),$$

    $$\psi_{l,arr} = \frac{1}{q}\left(\sum \phi_i - \sum_{i \neq l} \psi_i + 180° + 360°(l - 1) \right),$$

    where

    $$q = \text{order of the pole or zero},$$

    $$\psi_i = \text{angles from the zeros},$$

    $$\phi_i = \text{angles from the poles}.$$

- The parameter K corresponding to a root at a particular point s_0 on the locus can be found from

$$K = \frac{1}{|L(s_0)|},$$

where $|L(s_0)|$ can be found graphically by measuring the distances from s_0 to each of the poles and zeros.

- For a locus drawn with rlocus(sysL), the parameter and corresponding roots can be found with [K, p] = rlocfind(sysL) or with rltool.
- Lead compensation, given by

$$D(s) = \frac{s+z}{s+p}, \quad z < p,$$

approximates proportional–derivative (PD) control. For a fixed error coefficient, it generally moves the locus to the left and improves the system damping.
- Lag compensation, given by

$$D(s) = \frac{s+z}{s+p}, \quad z > p,$$

approximates proportional–integral (PI) control. It generally improves the steady-state error for fixed speed of response by increasing the low-frequency gain and typically degrades stability.
- △ The root locus can be used to analyze successive loop closures by studying two (or more) parameters in succession.
- △ The root locus can be used to approximate the effect of time delay.

REVIEW QUESTIONS

1. Give two definitions for the root locus.
2. Define the negative root locus.
3. Where are the sections of the (positive) root locus on the real axis?
4. What are the angles of departure from two coincident poles at $s = -a$ on the real axis? There are no poles or zeros to the right of $-a$.
5. What are the angles of departure from *three* coincident poles at $s = -a$ on the real axis? There are no poles or zeros to the right of $-a$.
6. What is the principal effect of a lead compensation on a root locus?
7. What is the principal effect of a lag compensation on a root locus in the vicinity of the dominant closed-loop roots?
8. What is the principal effect of a lag compensation on the steady-state error to a polynomial reference input?
9. Why is the angle of departure from a pole near the imaginary axis especially important?
10. Define a conditionally stable system.
11. Show, with a root-locus argument, that a system having three poles at the origin MUST be conditionally stable.

PROBLEMS

Problems for Section 5.1: Root Locus of a Basic Feedback System

5.1 Set up the listed characteristic equations in the form suited to Evans's root-locus method. Give $L(s)$, $a(s)$, and $b(s)$ and the parameter K in terms of the original parameters in each case. Be sure to select K so that $a(s)$ and $b(s)$ are monic in each case and the degree of $b(s)$ is not greater than that of $a(s)$.

(a) $s + (1/\tau) = 0$ versus parameter τ

(b) $s^2 + cs + c + 1 = 0$ versus parameter c

(c) $(s + c)^3 + A(Ts + 1) = 0$

    (i) versus parameter A,

    (ii) versus parameter T,

    (iii) versus the parameter c, if possible. Say why you can or cannot. Can a plot of the roots be drawn versus c for given constant values of A and T by any means at all?

(d) $1 + \left[k_p + \frac{k_I}{s} + \frac{k_D s}{\tau s + 1} \right] G(s) = 0$. Assume that $G(s) = A \frac{c(s)}{d(s)}$, where $c(s)$ and $d(s)$ are monic polynomials with the degree of $d(s)$ greater than that of $c(s)$.

    (i) versus k_p

    (ii) versus k_I

    (iii) versus k_D

    (iv) versus τ

Problems for Section 5.2: Guidelines for Sketching a Root Locus

5.2 Roughly sketch the root loci for the pole–zero maps as shown in Fig. 5.51 without the aid of a computer. Show your estimates of the center and angles of the asymptotes, a rough evaluation of arrival and departure angles for complex poles and zeros, and the loci for positive values of the parameter K. Each pole–zero map is from a characteristic equation of the form

$$1 + K \frac{b(s)}{a(s)} = 0,$$

where the roots of the numerator $b(s)$ are shown as small circles ○ and the roots of the denominator $a(s)$ are shown as ×'s on the s-plane. Note that in Fig. 5.51(c) there are two poles at the origin.

Figure 5.51
Pole–zero maps

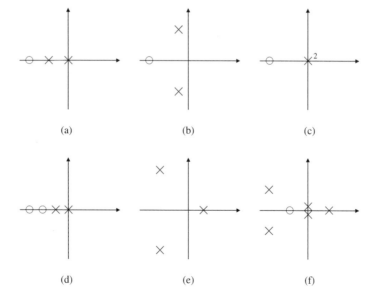

    (a)   (b)   (c)

    (d)   (e)   (f)

5.3   For the characteristic equation

$$1 + \frac{K}{s(s+1)(s+5)} = 0,$$

(a) Draw the real-axis segments of the corresponding root locus.

(b) Sketch the asymptotes of the locus for $K \to \infty$.

(c) Sketch the locus?

(d) Verify your sketch with a MATLAB plot.

5.4   *Real poles and zeros.* Sketch the root locus with respect to K for the equation $1 + KL(s) = 0$ and the listed choices for $L(s)$. Be sure to give the asymptotes, and the arrival and departure angles at any complex zero or pole. After completing each hand sketch, verify your results using MATLAB. Turn in your hand sketches and the MATLAB results on the same scales.

(a) $L(s) = \dfrac{(s+2)}{s(s+1)(s+5)(s+10)}$

(b) $L(s) = \dfrac{1}{s(s+1)(s+5)(s+10)}$

(c) $L(s) = \dfrac{(s+2)(s+6)}{s(s+1)(s+5)(s+10)}$

(d) $L(s) = \dfrac{(s+2)(s+4)}{s(s+1)(s+5)(s+10)}$

5.5   *Complex poles and zeros.* Sketch the root locus with respect to K for the equation $1 + KL(s) = 0$ and the listed choices for $L(s)$. Be sure to give the asymptotes and the arrival and departure angles at any complex zero or pole. After completing each hand sketch, verify your results using MATLAB. Turn in your hand sketches and the MATLAB results on the same scales.

(a) $L(s) = \dfrac{1}{s^2 + 3s + 10}$

(b) $L(s) = \dfrac{1}{s(s^2 + 3s + 10)}$

(c) $L(s) = \dfrac{(s^2 + 2s + 8)}{s(s^2 + 2s + 10)}$

(d) $L(s) = \dfrac{(s^2 + 2s + 12)}{s(s^2 + 2s + 10)}$

(e) $L(s) = \dfrac{(s^2 + 1)}{s(s^2 + 4)}$

(f) $L(s) = \dfrac{(s^2 + 4)}{s(s^2 + 1)}$

5.6   *Multiple poles at the origin.* Sketch the root locus with respect to K for the equation $1 + KL(s) = 0$ and the listed choices for $L(s)$. Be sure to give the asymptotes and the arrival and departure angles at any complex zero or pole. After completing each hand sketch, verify your results using MATLAB. Turn in your hand sketches and the MATLAB results on the same scales.

(a) $L(s) = \dfrac{1}{s^2(s+8)}$

(b) $L(s) = \dfrac{1}{s^3(s+8)}$

(c) $L(s) = \dfrac{1}{s^4(s+8)}$

(d) $L(s) = \dfrac{(s+3)}{s^2(s+8)}$

(e) $L(s) = \dfrac{(s+3)}{s^3(s+4)}$

(f) $L(s) = \dfrac{(s+1)^2}{s^3(s+4)}$

(g) $L(s) = \dfrac{(s+1)^2}{s^3(s+10)^2}$

5.7 *Mixed real and complex poles.* Sketch the root locus with respect to K for the equation $1 + KL(s) = 0$ and the listed choices for $L(s)$. Be sure to give the asymptotes and the arrival and departure angles at any complex zero or pole. After completing each hand sketch, verify your results using MATLAB. Turn in your hand sketches and the MATLAB results on the same scales.

(a) $L(s) = \dfrac{(s+2)}{s(s+10)(s^2+2s+2)}$

(b) $L(s) = \dfrac{(s+2)}{s^2(s+10)(s^2+6s+25)}$

(c) $L(s) = \dfrac{(s+2)^2}{s^2(s+10)(s^2+6s+25)}$

(d) $L(s) = \dfrac{(s+2)(s^2+4s+68)}{s^2(s+10)(s^2+4s+85)}$

(e) $L(s) = \dfrac{[(s+1)^2+1]}{s^2(s+2)(s+3)}$

5.8 *RHP and zeros.* Sketch the root locus with respect to K for the equation $1 + KL(s) = 0$ and the listed choices for $L(s)$. Be sure to give the asymptotes and the arrival and departure angles at any complex zero or pole. After completing each hand sketch, verify your results using MATLAB. Turn in your hand sketches and the MATLAB results on the same scales.

(a) $L(s) = \dfrac{s+2}{s+10}\dfrac{1}{s^2-1}$; the model for a case of magnetic levitation with lead compensation.

(b) $L(s) = \dfrac{s+2}{s(s+10)}\dfrac{1}{(s^2-1)}$; the magnetic levitation system with integral control and lead compensation.

(c) $L(s) = \dfrac{s-1}{s^2}$

(d) $L(s) = \dfrac{s^2+2s+1}{s(s+20)^2(s^2-2s+2)}$. What is the largest value that can be obtained for the damping ratio of the stable complex roots on this locus?

(e) $L(s) = \dfrac{(s+2)}{s(s-1)(s+6)^2}$

(f) $L(s) = \dfrac{1}{(s-1)[(s+2)^2+3]}$

5.9 Put the characteristic equation of the system shown in Fig. 5.52 in root-locus form with respect to the parameter α, and identify the corresponding $L(s)$, $a(s)$, and $b(s)$. Sketch

Figure 5.52

Control system for Problem 5.9

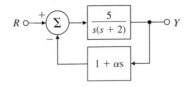

the root locus with respect to the parameter α, estimate the closed-loop pole locations, and sketch the corresponding step responses when $\alpha = 0, 0.5$, and 2. Use MATLAB to check the accuracy of your approximate step responses.

5.10 Use the MATLAB function rltool to study the behavior of the root locus of $1 + KL(s)$ for

$$L(s) = \frac{(s + a)}{s(s + 1)(s^2 + 8s + 52)}$$

as the parameter a is varied from 0 to 10, paying particular attention to the region between 2.5 and 3.5. Verify that a multiple root occurs at a complex value of s for some value of a in this range.

5.11 Use Routh's criterion to find the range of the gain K for which the systems in Fig. 5.53 are stable, and use the root locus to confirm your calculations.

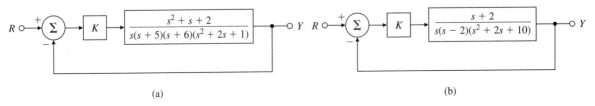

(a)   (b)

Figure 5.53

Feedback systems for Problem 5.11

5.12 Sketch the root locus for the characteristic equation of the system for which

$$L(s) = \frac{(s + 2)}{s(s + 1)(s + 5)},$$

and determine the value of the root-locus gain for which the complex conjugate poles have a damping ratio of 0.5.

5.13 For the system in Fig. 5.54,

(a) Find the locus of closed-loop roots with respect to K.

(b) Is there a value of K that will cause all roots to have a damping ratio greater than 0.5?

(c) Find the values of K that yield closed-loop poles with the damping ratio $\zeta = 0.707$.

(d) Use MATLAB to plot the response of the resulting design to a reference step.

Figure 5.54

Feedback system for Problem 5.13

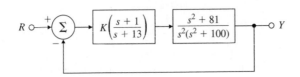

5.14 For the feedback system shown in Fig. 5.55, find the value of the gain K that results in dominant closed-loop poles with a damping ratio $\zeta = 0.5$.

Figure 5.55
Feedback system for Problem 5.14

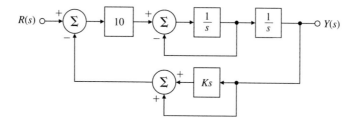

Problems for Section 5.3: Selected Illustrative Root Loci

5.15 A simplified model of the longitudinal motion of a certain helicopter near hover has the transfer function

$$G(s) = \frac{9.8(s^2 - 0.5s + 6.3)}{(s + 0.66)(s^2 - 0.24s + 0.15)}$$

and the characteristic equation $1 + D(s)G(s) = 0$. Let $D(s) = k_p$ at first.

(a) Compute the departure and arrival angles at the complex poles and zeros.

(b) Sketch the root locus for this system for parameter $K = 9.8k_p$. Use axes $-4 \leq x \leq 4;\ -3 \leq y \leq 3$.

(c) Verify your answer using MATLAB. Use the command axis([−4 4 −3 3]) to get the right scales.

(d) Suggest a practical (at least as many poles as zeros) alternative compensation $D(s)$ that will at least result in a stable system.

5.16 (a) For the system given in Fig. 5.56, plot the root locus of the characteristic equation as the parameter K_1 is varied from 0 to ∞ with $\lambda = 2$. Give the corresponding $L(s)$, $a(s)$, and $b(s)$.

(b) Repeat part (a) with $\lambda = 5$. Is there anything special about this value?

(c) Repeat part (a) for fixed $K_1 = 2$, with the parameter $K = \lambda$ varying from 0 to ∞.

Figure 5.56
Control system for Problem 5.16

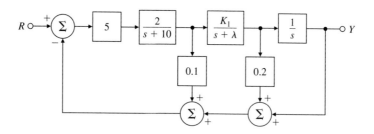

5.17 For the system shown in Fig. 5.57, determine the characteristic equation and sketch the root locus of it with respect to positive values of the parameter c. Give $L(s)$, $a(s)$, and $b(s)$, and be sure to show with arrows the direction in which c increases on the locus.

Figure 5.57

Control system for
Problem 5.17

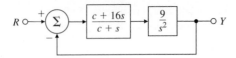

5.18   Suppose you are given a system with the transfer function

$$L(s) = \frac{(s+z)}{(s+p)^2},$$

where z and p are real and $z > p$. Show that the root locus for $1 + KL(s) = 0$ with respect to K is a circle centered at z with radius given by

$$r = (z - p).$$

Hint: Assume $s + z = re^{j\phi}$ and show that $L(s)$ is real and negative for real ϕ under this assumption.

5.19   The loop transmission of a system has two poles at $s = -1$ and a zero at $s = -2$. There is a third real-axis pole p located somewhere to the *left* of the zero. Several different root loci are possible, depending on the exact location of the third pole. The extreme cases occur when the pole is located at infinity or when it is located at $s = -2$. Give values for p and sketch the three distinct types of loci.

5.20   For the feedback configuration of Fig. 5.58, use asymptotes, center of asymptotes, angles of departure and arrival, and the Routh array to sketch root loci for the characteristic equations of the listed feedback control systems versus the parameter K. Use MATLAB to verify your results.

(a) $G(s) = \dfrac{1}{s(s+1+3j)(s+1-3j)}$, $H(s) = \dfrac{s+2}{s+8}$

(b) $G(s) = \dfrac{1}{s^2}$, $H(s) = \dfrac{s+1}{s+3}$

(c) $G(s) = \dfrac{(s+5)}{(s+1)}$, $H(s) = \dfrac{s+7}{s+3}$

(d) $G(s) = \dfrac{(s+3+4j)(s+3-4j)}{s(s+1+2j)(s+1-2j)}$, $H(s) = 1 + 3s$

Figure 5.58

Feedback system for
Problem 5.20

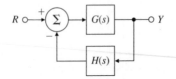

5.21   Consider the system in Fig. 5.59.

(a) Using Routh's stability criterion, determine all values of K for which the system is stable.

(b) Use MATLAB to draw the root locus versus K and find the values of K at the imaginary-axis crossings.

Figure 5.59

Feedback system for Problem 5.21

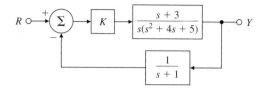

Problems for Section 5.4: Design Using Dynamic Compensation

5.22 Let

$$G(s) = \frac{1}{(s+2)(s+3)} \quad \text{and} \quad D(s) = K\frac{s+a}{s+b}.$$

Using root-locus techniques, find values for the parameters a, b, and K of the compensation $D(s)$ that will produce closed-loop poles at $s = -1 \pm j$ for the system shown in Fig. 5.60.

Figure 5.60

Unity feedback system for Problems 5.22 to 5.28 and 5.33

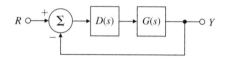

5.23 Suppose that in Fig. 5.60

$$G(s) = \frac{1}{s(s^2 + 2s + 2)} \quad \text{and} \quad D(s) = \frac{K}{s+2}.$$

Sketch the root locus with respect to K of the characteristic equation for the closed-loop system, paying particular attention to points that generate multiple roots if $KL(s) = D(s)G(s)$.

5.24 Suppose the unity feedback system of Fig. 5.60 has an open-loop plant given by $G(s) = 1/s^2$. Design a lead compensation $D(s) = K\frac{s+z}{s+p}$ to be added in series with the plant so that the dominant poles of the closed-loop system are located at $s = -2 \pm 2j$.

5.25 Assume that the unity feedback system of Fig. 5.60 has the open-loop plant

$$G(s) = \frac{1}{s(s+3)(s+6)}.$$

Design a lag compensation to meet the following specifications:

- The step response settling time is to be less than 5 sec.
- The step response overshoot is to be less than 17%.
- The steady-state error to a unit-ramp input must not exceed 10%.

5.26 A numerically controlled machine tool positioning servomechanism has a normalized and scaled transfer function given by

$$G(s) = \frac{1}{s(s+1)}.$$

Performance specifications of the system in the unity feedback configuration of Fig. 5.60 are satisfied if the closed-loop poles are located at $s = -1 \pm j\sqrt{3}$.

(a) Show that this specification cannot be achieved by choosing proportional control alone, $D(s) = k_p$.

(b) Design a lead compensator $D(s) = K\frac{s+z}{s+p}$ that will meet the specification.

5.27 A servomechanism position control has the plant transfer function

$$G(s) = \frac{10}{s(s+1)(s+10)}.$$

You are to design a series compensation transfer function $D(s)$ in the unity feedback configuration to meet the following closed-loop specifications:

- The response to a reference step input is to have no more than 16% overshoot.
- The response to a reference step input is to have a rise time of no more than 0.4 sec.
- The steady-state error to a unit ramp at the reference input must be less than 0.02.

(a) Design a lead compensation that will cause the system to meet the dynamic response specifications.

(b) If $D(s)$ is proportional control, $D(s) = k_p$, what is the velocity constant K_v?

(c) Design a lag compensation to be used in series with the lead you have designed to cause the system to meet the steady-state error specification.

(d) Give the MATLAB plot of the root locus of your final design.

(e) Give the MATLAB response of your final design to a reference step.

5.28 Assume that the closed-loop system of Fig. 5.60 has a feed-forward transfer function

$$G(s) = \frac{1}{s(s+2)}.$$

Design a lag compensation so that the dominant poles of the closed-loop system are located at $s = -1 \pm j$ and the steady-state error to a unit-ramp input is less than 0.2.

5.29 An elementary magnetic suspension scheme is depicted in Fig. 5.61. For small motions near the reference position, the voltage e on the photo detector is related to the ball displacement x (in meters) by $e = 100x$. The upward force (in newtons) on the ball caused by the current i (in amperes) may be approximated by $f = 0.5i + 20x$. The mass of the ball is 20 g and the gravitational force is 9.8 N/kg. The power amplifier is a voltage-to-current device with an output (in amperes) of $i = u + V_0$.

(a) Write the equations of motion for this set-up.

(b) Give the value of the bias V_0 that results in the ball being in equilibrium at $x = 0$.

(c) What is the transfer function from u to e?

(d) Suppose that the control input u is given by $u = -Ke$. Sketch the root locus of the closed-loop system as a function of K.

Figure 5.61
Elementary magnetic suspension

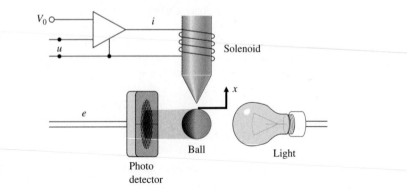

Ball

Photo detector

Solenoid

Light

(e) Assume that a lead compensation is available in the form $\frac{U}{E} = D(s) = K\frac{s+z}{s+p}$. Give values of K, z, and p that yield improved performance over the one proposed in part (d).

5.30 A certain plant with the nonminimum phase transfer function

$$G(s) = \frac{4-2s}{s^2+s+9}$$

is in a unity positive feedback system with the controller transfer function $D(s)$.

(a) Use MATLAB to determine a (negative) value for $D(s) = K$ so that the closed-loop system with negative feedback has a damping ratio $\zeta = 0.707$.

(b) Use MATLAB to plot the system's response to a reference step.

5.31 Consider the rocket-positioning system shown in Fig. 5.62.

(a) Show that if the sensor that measures x has a unity transfer function, the lead compensator

$$H(s) = K\frac{s+2}{s+4}$$

stabilizes the system.

(b) Assume that the sensor transfer function is modeled by a single pole with a 0.1 sec time constant and unity DC gain. Using the root-locus procedure, find a value for the gain K that will provide the maximum damping ratio.

Figure 5.62

Block diagram for rocket-positioning control system

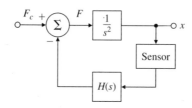

5.32 For the system in Fig. 5.63,

(a) Find the locus of closed-loop roots with respect to K.

(b) Find the maximum value of K for which the system is stable. Assume $K = 2$ for the remaining parts of this problem.

(c) What is the steady-state error ($e = r - y$) for a step change in r?

(d) What is the steady-state error in y for a constant disturbance w_1?

Figure 5.63

Control system for Problem 5.32

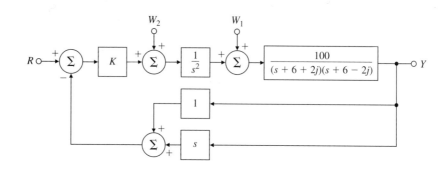

(e) What is the steady-state error in y for a constant disturbance w_2?

(f) If you wished to have more damping, what changes would you make to the system?

5.33 Consider the plant transfer function

$$G(s) = \frac{bs + k}{s^2[mMs^2 + (M + m)bs + (M + m)k]}$$

to be put in the unity feedback loop of Fig. 5.60. This is the transfer function relating the input force $u(t)$ and the position $y(t)$ of mass M in the noncollocated sensor and actuator problem. In this problem we will use root-locus techniques to design a controller $D(s)$ so that the closed-loop step response has a rise time of less than 0.1 sec and an overshoot of less than 10%. You may use MATLAB for any of the following questions:

(a) Approximate $G(s)$ by assuming that $m \cong 0$, and let $M = 1$, $k = 1$, $b = 0.1$, and $D(s) = K$. Can K be chosen to satisfy the performance specifications? Why or why not?

(b) Repeat part (a) assuming that $D(s) = K(s+z)$, and show that K and z can be chosen to meet the specifications.

(c) Repeat part (b), but with a practical controller given by the transfer function

$$D(s) = K\frac{p(s + z)}{s + p}.$$

Pick p so that the values for K and z computed in part (b) remain more or less valid.

(d) Now suppose that the small mass m is not negligible, but is given by $m = M/10$. Check to see if the controller you designed in part (c) still meets the given specifications. If not, adjust the controller parameters so that the specifications are met.

5.34 Consider the Type 1 system drawn in Fig. 5.64. We would like to design the compensation $D(s)$ to meet the following requirements: (1) The steady-state value of y due to a constant unit disturbance w should be less than $\frac{4}{5}$, and (2) the damping ratio $\zeta = 0.7$. Using root-locus techniques,

(a) Show that proportional control alone is not adequate.

(b) Show that proportional–derivative control will work.

(c) Find values of the gains k_p and k_D for $D(s) = k_p + k_D s$ that meet the design specifications.

Figure 5.64

Control system for Problem 5.34

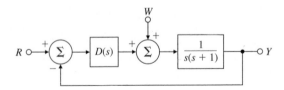

Problems for Section 5.5: A Design Example Using the Root Locus

5.35 Consider the positioning servomechanism system shown in Fig. 5.65, where

$$e_i = K_{pot}\theta_i, \quad e_o = K_{pot}\theta_o, \quad K_{pot} = 10 \text{ V/rad},$$

$$T = \text{motor torque} = K_t i_a,$$

$$k_m = K_t = \text{torque constant} = 0.1 \text{ N·m/A},$$

$$R_a = \text{armature resistance} = 10 \ \Omega,$$

$$\text{Gear ratio} = 1 : 1,$$

$$J_L + J_m = \text{total inertia} = 10^{-3} \text{ kg·m}^2,$$

$$C = 200 \ \mu\text{F},$$

$$v_a = K_A(e_i - e_f).$$

(a) What is the range of the amplifier gain K_A for which the system is stable? Estimate the upper limit graphically using a root-locus plot.

(b) Choose a gain K_A that gives roots at $\zeta = 0.7$. Where are all three closed-loop root locations for this value of K_A?

Figure 5.65

Positioning servomechanism

Source: Reprinted from Clark, 1962, with permission

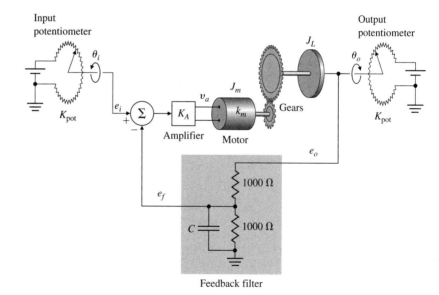

Feedback filter

5.36 We wish to design a velocity control for a tape-drive servomechanism. The transfer function from current $I(s)$ to tape velocity $\Omega(s)$ (in millimeters per millisecond per ampere) is

$$\frac{\Omega(s)}{I(s)} = \frac{15(s^2 + 0.9s + 0.8)}{(s + 1)(s^2 + 1.1s + 1)}.$$

We wish to design a Type 1 feedback system so that the response to a reference step satisfies

$$t_r \leq 4 \text{ msec}, \quad t_s \leq 15 \text{ msec}, \quad M_p \leq 0.05.$$

(a) Use the integral compensator k_I/s to achieve Type 1 behavior, and sketch the root locus with respect to k_I. Show on the same plot the region of acceptable pole locations corresponding to the specifications.

(b) Assume a proportional-integral compensator of the form $k_p(s + \alpha)/s$, and select the best possible values of k_p and α you can find. Sketch the root-locus plot of your design, giving values for k_p and α, and the velocity constant K_v your design achieves. On your plot, indicate the closed-loop poles with a dot ($\bullet$) and include the boundary of the region of acceptable root locations.

5.37   The normalized, scaled equations of a cart as drawn in Fig. 5.66 of mass m_c holding an inverted uniform pendulum of mass m_p and length ℓ with no friction are

$$\ddot{\theta} - \theta = -v, \tag{5.95}$$

$$\ddot{y} + \beta\theta = v,$$

where $\beta = \frac{3m_p}{4(m_c+m_p)}$ is a mass ratio bounded by $0 < \beta < 0.75$. Time is measured in terms of $\tau = \omega_o t$ where $\omega_o^2 = \frac{3g(m_c+m_p)}{\ell(4m_c+m_p)}$. The cart motion y is measured in units of pendulum length as $y = \frac{3x}{4\ell}$ and the input is force normalized by the system weight $v = \frac{u}{g(m_c+m_p)}$. These equations can be used to compute the transfer functions

$$\frac{\Theta}{V} = -\frac{1}{s^2 - 1}, \tag{5.96}$$

$$\frac{Y}{V} = \frac{s^2 - 1 + \beta}{s^2(s^2 - 1)}. \tag{5.97}$$

In this problem you are to design a control for the system by first closing a loop around the pendulum, Eq. (5.96), and then, with this loop closed, closing a second loop around the cart plus pendulum, Eq. (5.97). For this problem, let the mass ratio be $m_c = 5m_p$.

(a) Draw a block diagram for the system with V input and both Y and Θ as outputs.

(b) Design a lead compensation $D(s) = K_p\frac{s+z}{s+p}$ for the Θ loop to cancel the pole at $s = -1$ and place the two remaining poles at $-4 \pm j4$. The new control is $U(s)$, where the force is $V(s) = U(s) + D(s)\Theta(s)$. Draw the root locus of the angle loop.

(c) Compute the transfer function of the new plant from U to Y with $D(s)$ in place.

(d) Design a controller $D_c(s)$ for the cart position with the pendulum loop closed. Draw the root locus with respect to the gain of $D_c(s)$.

(e) Use MATLAB to plot the control, cart position, and pendulum position for a unit step change in cart position.

Figure 5.66

Figure of cart pendulum for Problem 5.37

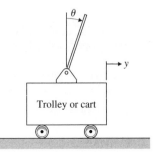

5.38   Consider the 270-ft U.S. Coast Guard cutter *Tampa* (902) shown in Fig. 5.67. Parameter identification based on sea-trials data (Trankle, 1987) was used to estimate the hydrodynamic coefficients in the equations of motion. The result is that the response of the heading angle of the ship ψ to rudder angle δ and wind changes w can be described by the second-order transfer functions

$$G_\delta(s) = \frac{\psi(s)}{\delta(s)} = \frac{-0.0184(s + 0.0068)}{s(s + 0.2647)(s + 0.0063)},$$

$$G_w(s) = \frac{\psi(s)}{w(s)} = \frac{0.0000064}{s(s + 0.2647)(s + 0.0063)},$$

where

ψ = heading angle, rad,

r = reference heading angle, rad,

r = yaw rate, rad/sec,

δ = rudder angle, rad,

w = wind speed, m/sec.

(a) Determine the open-loop settling time of r for a step change in δ.

(b) In order to regulate the heading angle ψ, design a compensator that uses ψ and the measurement provided by a yaw-rate gyroscope (that is, by $\dot{\psi} = r$). The settling time of ψ to a step change in ψ_r is specified to be less than 50 sec, and for a 5° change in heading, the maximum allowable rudder angle deflection is specified to be less than 10°.

(c) Check the response of the closed-loop system you designed in part (b) to a wind gust disturbance of 10 m/sec. (Model the disturbance as a step input.) If the *steady-state* value of the heading due to this wind gust is more than 0.5°, modify your design so that it meets this specification as well.

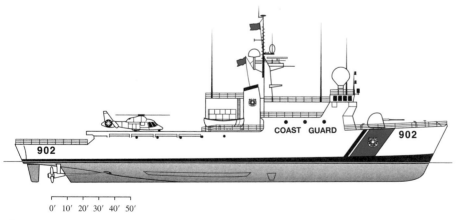

0' 10' 20' 30' 40' 50'

Figure 5.67
USCG cutter *Tampa* (902) for Problem 5.38

5.39 Golden Nugget Airlines has opened a free bar in the tail of their airplanes in an attempt to lure customers. In order to automatically adjust for the sudden weight shift due to passengers rushing to the bar when it first opens, the airline is mechanizing a pitch-attitude autopilot. Figure 5.68 shows the block diagram of the proposed arrangement. We will model the passenger moment as a step disturbance $M_p(s) = M_0/s$, with a maximum expected value for M_0 of 0.6.

(a) What value of K is required to keep the steady-state error in θ to less than 0.02 rad ($\cong 1°$)? (Assume the system is stable.)

(b) Draw a root locus with respect to K.

(c) Based on your root locus, what is the value of K when the system becomes unstable?

(d) Suppose the value of K required for acceptable steady-state behavior is 600. Show that this value yields an unstable system with roots at

$$s = -2.9, -13.5, +1.2 \pm 6.6j.$$

(e) You are given a black box with *rate gyro* written on the side and told that, when installed, it provides a perfect measure of $\dot{\theta}$, with output $K_T\dot{\theta}$. Assume that $K = 600$ as in part (d) and draw a block diagram indicating how you would incorporate the rate gyro into the autopilot. (Include transfer functions in boxes.)

(f) For the rate gyro in part (e), sketch a root locus with respect to K_T.

(g) What is the maximum damping factor of the complex roots obtainable with the configuration in part (e)?

(h) What is the value of K_T for part (g)?

(i) Suppose you are not satisfied with the steady-state errors and damping ratio of the system with a rate gyro in parts (e) through (h). Discuss the advantages and disadvantages of adding an integral term and extra lead networks in the control law. Support your comments using MATLAB or with rough root-locus sketches.

Figure 5.68

Golden Nugget Airlines autopilot

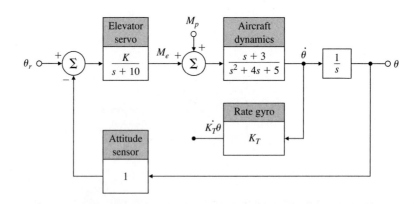

5.40 Consider the instrument servomechanism with the parameters given in Fig. 5.69. For each of the following cases, draw a root locus with respect to the parameter K, and indicate the location of the roots corresponding to your final design:

(a) *Lead network*: Let

$$H(s) = 1, \quad D(s) = K\frac{s+z}{s+p}, \quad \frac{p}{z} = 6.$$

Figure 5.69

Control system for Problem 5.40

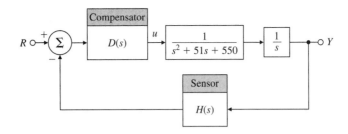

Select z and K so that the roots nearest the origin (the dominant roots) yield

$$\zeta \geq 0.4, \quad -\sigma \leq -7, \quad K_v \geq 16\frac{2}{3}\text{ sec.}$$

(b) *Output-velocity (tachometer) feedback*: Let

$$H(s) = 1 + K_T s \quad \text{and} \quad D(s) = K.$$

Select K_T and K so that the dominant roots are in the same location as those of part (a). Compute K_v. If you can, give a physical reason explaining the reduction in K_v when output derivative feedback is used.

(c) *Lag network*: Let

$$H(s) = 1 \quad \text{and} \quad D(s) = K\frac{s+1}{s+p}.$$

Using proportional control, it is possible to obtain a $K_v = 12$ at $\zeta = 0.4$. Select K and p so that the dominant roots correspond to the proportional-control case but with $K_v = 100$ rather than $K_v = 12$.

Problems for Section 5.6: Extensions of the Root Locus Method

5.41 Plot the loci for the $0°$ locus or negative K for each of the following:

(a) The examples given in Problem 5.3

(b) The examples given in Problem 5.4

(c) The examples given in Problem 5.5

(d) The examples given in Problem 5.6

(e) The examples given in Problem 5.7

(f) The examples given in Problem 5.8

5.42 Suppose you are given the plant

$$L(s) = \frac{1}{s^2 + (1+\alpha)s + (1+\alpha)},$$

where α is a system parameter that is subject to variations. Use both positive and negative root-locus methods to determine what variations in α can be tolerated before instability occurs.

5.43 Consider the system in Fig. 5.70.

(a) Use Routh's criterion to determine the regions in the (K_1, K_2) plane for which the system is stable.

(b) Use RLTOOL to verify your answer to part (a).

Figure 5.70

Feedback system for
Problem 5.43

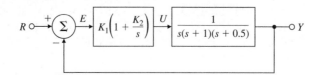

△ **5.44** The block diagram of a positioning servomechanism is shown in Fig. 5.71.

(a) Sketch the root locus with respect to K when no tachometer feedback is present ($K_T = 0$).

(b) Indicate the root locations corresponding to $K = 16$ on the locus of part (a). For these locations, estimate the transient-response parameters t_r, M_p, and t_s. Compare your estimates to measurements obtained using the step command in MATLAB.

(c) For $K = 16$, draw the root locus with respect to K_T.

(d) For $K = 16$ and with K_T set so that $M_p = 0.05$ ($\zeta = 0.707$), estimate t_r and t_s. Compare your estimates to the actual values of t_r and t_s obtained using MATLAB.

(e) For the values of K and K_T in part (d), what is the velocity constant K_v of this system?

Figure 5.71

Control system for
Problem 5.44

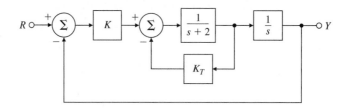

△ **5.45** Consider the mechanical system shown in Fig. 5.72, where g and a_0 are gains. The feedback path containing gs controls the amount of rate feedback. For a fixed value of a_0, adjusting g corresponds to varying the location of a zero in the s-plane.

(a) With $g = 0$ and $\tau = 1$, find a value for a_0 such that the poles are complex.

(b) Fix a_0 at this value, and construct a root locus that demonstrates the effect of varying g.

Figure 5.72

Control system for
Problem 5.45

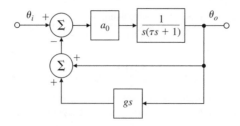

△ **5.46** Sketch the root locus with respect to K for the system in Fig. 5.73 using the Padé(1,1) approximation and the first-order lag approximation. For both approximations, what is the range of values of K for which the system is unstable?

Figure 5.73
Control system for
Problem 5.46

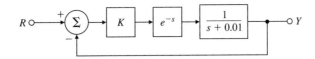

△   **5.47**   Prove that the plant $G(s) = 1/s^3$ *cannot* be made unconditionally stable if pole cancellation is forbidden.

△   **5.48**   For the equation $1 + KG(s)$, where

$$G(s) = \frac{1}{s(s - p)[(s + 1)^2 + 4]},$$

use MATLAB to examine the root locus as a function of K for p in the range from $p = 1$ to $p = 10$, making sure to include the point $p = 2$.

6

The Frequency-Response Design Method

A Perspective on the Frequency-Response Design Method

The design of feedback control systems in industry is probably accomplished using frequency-response methods more often than any other. Frequency-response design is popular primarily because it provides good designs in the face of uncertainty in the plant model. For example, for systems with poorly known or changing high-frequency resonances, we can temper the feedback compensation to alleviate the effects of those uncertainties. Currently, this tempering is carried out more easily using frequency-response design than with any other method.

Another advantage of using frequency response is the ease with which experimental information can be used for design purposes. Raw measurements of the output amplitude and phase of a plant undergoing a sinusoidal input excitation are sufficient to design a suitable feedback control. No intermediate processing of the data (such as finding poles and zeros or determining system matrices) is required to arrive at the system model. The wide availability of computers has rendered this advantage less important now than it was years ago; however, for relatively simple systems, frequency response is often

Photo courtesy of Cirrus Design Corporation

296

still the most cost-effective design method. The method is most effective for systems that are stable in open loop.

Yet another advantage is that it is the easiest method to use for designing compensation. A simple rule can be used to provide reasonable designs with a minimum of trial and error.

Although the underlying theory is somewhat challenging and requires a rather broad knowledge of complex variables, the methodology of frequency-response design is easy, and the insights gained by learning the theory are well worth the struggle.

Chapter Overview

The chapter opens with a discussion of how to obtain the frequency response of a system by analyzing its poles and zeros. An important extension of this discussion is how to use Bode plots to graphically display the frequency response. In Sections 6.2 and 6.3 we discuss stability briefly and then in more depth the use of the Nyquist stability criterion. In Sections 6.4 through 6.6 we introduce the notion of stability margins, discuss Bode's gain–phase relationship, and study the closed-loop frequency response of dynamic systems. The gain–phase relationship suggests a very simple rule for compensation design: Shape the frequency-response magnitude so that it crosses magnitude 1 with a slope of −1. As with our treatment of the root-locus method, we describe how adding dynamic compensation can adjust the frequency response (Section 6.7) and improve system stability and/or error characteristics. We also show how to implement compensation digitally in an example.

In optional Sections 6.7.7 and 6.7.8 we discuss issues of sensitivity that relate to the frequency response, including material on sensitivity functions and stability robustness. The next two sections on analyzing time delays in the system and Nichols charts represents additional, somewhat advanced material that may also be considered optional. The final Section 6.10 is a short history of the Frequency Response design method.

6.1 Frequency Response

The basic concepts of frequency response were discussed in Section 3.1.2. In this section we will review those ideas and extend the concepts for use in control system design.

Frequency response

A linear system's response to sinusoidal inputs—called the system's **frequency response**—can be obtained from knowledge of its pole and zero locations.

To review the ideas, we consider a system described by

$$\frac{Y(s)}{U(s)} = G(s),$$

where the input $u(t)$ is a sine wave with an amplitude A:

$$u(t) = A \sin(\omega_o t) 1(t).$$

This sine wave has a Laplace transform

$$U(s) = \frac{A\omega_o}{s^2 + \omega_o^2}.$$

With zero initial conditions, the Laplace transform of the output is

$$Y(s) = G(s)\frac{A\omega_o}{s^2 + \omega_o^2}. \qquad (6.1)$$

Partial-fraction expansion   A partial-fraction expansion of Eq. (6.1) [assuming that the poles of $G(s)$ are distinct] will result in an equation of the form

$$Y(s) = \frac{\alpha_1}{s - p_1} + \frac{\alpha_2}{s - p_2} + \cdots + \frac{\alpha_n}{s - p_n} + \frac{\alpha_o}{s + j\omega_o} + \frac{\alpha_o^*}{s - j\omega_o}, \qquad (6.2)$$

where $p_1, p_2, \ldots, p_n$ are the poles of $G(s)$, α_o would be found by performing the partial-fraction expansion, and α_o^* is the complex conjugate of α_o. The time response that corresponds to $Y(s)$ is

$$y(t) = \alpha_1 e^{p_1 t} + \alpha_2 e^{p_2 t} + \cdots + \alpha_n e^{p_n t} + 2|\alpha_o| \cos(\omega_o t + \phi), \qquad t \geq 0, \qquad (6.3)$$

where

$$\phi = \tan^{-1}\left[\frac{\text{Im}(\alpha_o)}{\text{Re}(\alpha_o)}\right].$$

If all the poles of the system represent stable behavior (the real parts of $p_1, p_2, \ldots, p_n < 0$), the natural unforced response will die out eventually, and therefore the steady-state response of the system will be due solely to the sinusoidal term in Eq. (6.3), which is caused by the sinusoidal excitation. Example 3.5 determined the response of the system $G(s) = \frac{1}{(s+1)}$ to the input $u = \sin 10t$ and showed that response in Fig. 3.4, which is repeated here as Fig. 6.1. It shows that e^{-t}, the natural part of the response associated with $G(s)$, disappears after several time constants, and the pure sinusoidal response is essentially all that remains. Example 3.5 showed that the remaining sinusoidal term in Eq. (6.3) can be expressed as

$$y(t) = AM\cos(\omega_o t + \phi), \qquad (6.4)$$

where

$$M = |G(j\omega_o)| = |G(s)|_{s=j\omega_o} = \sqrt{\{\text{Re}[G(j\omega_o)]\}^2 + \{\text{Im}[G(j\omega_o)]\}^2}, \qquad (6.5)$$

$$\phi = \tan^{-1}\left[\frac{\text{Im}[G(j\omega_o)]}{\text{Re}[G(j\omega_o)]}\right] = \angle G(j\omega_o). \qquad (6.6)$$

Figure 6.1

Response of $G(s) = \frac{1}{(s+1)}$ to $\sin 10t$

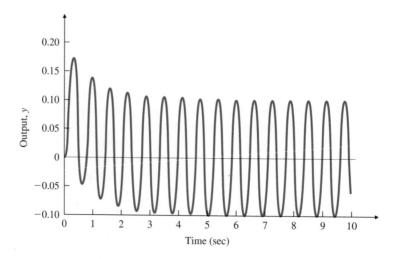

Time (sec)

In polar form,

$$G(j\omega_o) = Me^{j\phi}. \tag{6.7}$$

Equation (6.4) shows that a stable system with transfer function $G(s)$ excited by a sinusoid with unit amplitude and frequency ω_o will, after the response has reached steady state, exhibit a sinusoidal output with a magnitude $M(\omega_o)$ and a phase $\phi(\omega_o)$ at

Frequency-response plot   the frequency ω_o. The facts that the output y is a sinusoid with the *same* frequency as the input u and that the magnitude ratio M and phase ϕ of the output are independent of the amplitude A of the input are a consequence of $G(s)$ being a linear constant system. If the system being excited were a nonlinear or time-varying system, the output might contain frequencies other than the input frequency, and the output–input ratio might be dependent on the input magnitude.

Magnitude and phase   More generally, the **magnitude** M is given by $|G(j\omega)|$, and the **phase** ϕ is given by $\angle[G(j\omega)]$; that is, the magnitude and angle of the complex quantity $G(s)$ are evaluated with s taking on values along the imaginary axis ($s = j\omega$). The frequency response of a system consists of these functions of frequency that tell us how a system will respond to a sinusoidal input of any frequency. We are interested in analyzing the frequency response not only because it will help us understand how a system responds to a sinusoidal input, but also because evaluating $G(s)$ with s taking on values along the $j\omega$ axis will prove to be very useful in determining the stability of a closed-loop system. As we saw in Chapter 3, the $j\omega$ axis is the boundary between stability and instability; we will see in Section 6.4 that evaluating $G(j\omega)$ provides information that allows us to determine closed-loop stability from the open-loop $G(s)$.

EXAMPLE 6.1   *Frequency-Response Characteristics of a Capacitor*

Consider the capacitor described by the equation

$$i = C\frac{dv}{dt},$$

where v is the input and i is the output. Determine the sinusoidal steady-state response of the capacitor.

Solution. The transfer function of this circuit is

$$\frac{I(s)}{V(s)} = G(s) = Cs,$$

so

$$G(j\omega) = Cj\omega.$$

Computing the magnitude and phase, we find that

$$M = |Cj\omega| = C\omega \quad \text{and} \quad \phi = \angle(Cj\omega) = 90°.$$

For a unit-amplitude sinusoidal input v, the output i will be a sinusoid with magnitude $C\omega$, and the phase of the output will lead the input by $90°$. Note that for this example the magnitude is proportional to the input frequency while the phase is independent of frequency.

EXAMPLE 6.2   *Frequency-Response Characteristics of a Lead Compensator*

Recall from Chapter 5 [Eq. (5.70)] the transfer function of the lead compensation, which is equivalent to

$$D(s) = K\frac{Ts + 1}{\alpha Ts + 1}, \quad \alpha < 1. \tag{6.8}$$

1. Analytically determine its frequency-response characteristics and discuss what you would expect from the result.
2. Use MATLAB® to plot $D(j\omega)$ with $K = 1$, $T = 1$, and $\alpha = 0.1$ for $0.1 \le \omega \le 100$, and verify the features predicted from the analysis in 1, above.

Solution

1. **Analytical evaluation:** Substituting $s = j\omega$ into Eq. (6.8), we get

$$D(j\omega) = K\frac{Tj\omega + 1}{\alpha Tj\omega + 1}.$$

From Eqs. (6.5) and (6.6) the amplitude is

$$M = |D| = |K|\frac{\sqrt{1 + (\omega T)^2}}{\sqrt{1 + (\alpha\omega T)^2}},$$

and the phase is given by

$$\phi = \angle(1 + j\omega T) - \angle(1 + j\alpha\omega T)$$
$$= \tan^{-1}(\omega T) - \tan^{-1}(\alpha\omega T).$$

At very low frequencies the amplitude is just $|K|$, and at very high frequencies it is $|K/\alpha|$. Therefore, the amplitude is higher at very high frequency. The phase is zero at very low frequencies and goes back to zero at very high frequencies. At intermediate frequencies, evaluation of the $\tan^{-1}(\cdot)$ functions would reveal that ϕ becomes positive. These are the general characteristics of lead compensation.

2. **Computer evaluation:** A MATLAB script for frequency-response evaluation was shown for Example 3.5. A similar script for the lead compensation:

```
num = [1 1];
den = [0.1 1];
sysD = tf(num,den);
w=logspace(−1,2);          % determines frequencies over range
                             of interest
[mag,phase] = bode(sysD,w);  % computes magnitude and phase over
                             frequency range of interest
loglog(w,squeeze(mag)),grid;
semilogx(w,squeeze(phase)),grid;
```

produces the frequency-response magnitude and phase plots shown in Fig 6.2.

Figure 6.2
(a) Magnitude;
(b) phase for the lead compensation in Example 6.2

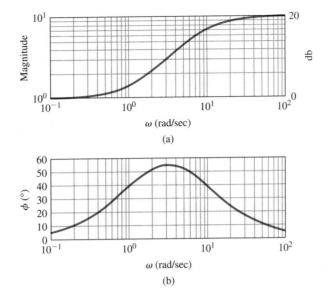

The analysis indicated that the low-frequency magnitude should be K ($=1$) and the high-frequency magnitude should be $K/\alpha(= 10)$, which are both verified by the magnitude plot. The phase plot also verifies that the value approaches zero at high and low frequencies and that the intermediate values are positive.

In the cases for which we do not have a good model of the system and wish to determine the frequency-response magnitude and phase experimentally, we can excite the system with a sinusoid varying in frequency. The magnitude $M(\omega)$ is obtained by measuring the ratio of the output sinusoid to input sinusoid in the steady state at each frequency. The phase $\phi(\omega)$ is the measured difference in phase between input and output signals.[1]

A great deal can be learned about the dynamic response of a system from knowledge of the magnitude $M(\omega)$ and the phase $\phi(\omega)$ of its transfer function. In the obvious case, if the signal is a sinusoid, then M and ϕ completely describe the response. Furthermore, if the input is periodic, then a Fourier series can be constructed to decompose the input into a sum of sinusoids, and again $M(\omega)$ and $\phi(\omega)$ can be used with each component to construct the total response. For transient inputs, our best path to understanding the meaning of M and ϕ is to relate the frequency response $G(j\omega)$ to the transient responses calculated by the Laplace transform. For example, in Fig.3.18(b) we plotted the step response of a system having the transfer function

$$G(s) = \frac{1}{(s/\omega_n)^2 + 2\zeta(s/\omega_n) + 1}, \tag{6.9}$$

---

[1] Agilent Technologies produces instruments called spectral analyzers that automate this experimental procedure and greatly speed up the process.

for various values of ζ. These transient curves were normalized with respect to time as $\omega_n t$. In Fig. 6.3 we plot $M(\omega)$ and $\phi(\omega)$ for these same values of ζ to help us see what features of the frequency response correspond to the transient-response characteristics. Specifically, Figs. 3.18(b) and 6.3 indicate the effect of damping on system time response and the corresponding effect on the frequency response. They show that the damping of the system can be determined from the transient response overshoot or from the peak in the magnitude of the frequency response [(Fig. 6.3 (a)]. Furthermore, from the frequency response, we see that ω_n is approximately equal to the bandwidth—the frequency where the magnitude starts to fall off from its low-frequency value. (We will define bandwidth more formally in the next paragraph.) Therefore, the rise time can be estimated from the bandwidth. We also see that the peak overshoot in frequency is approximately $1/2\zeta$ for $\zeta < 0.5$, so the peak overshoot in the step response can be estimated from the peak overshoot in the frequency response. Thus, we see that essentially the same information is contained in the frequency-response curve as is found in the transient-response curve.

Bandwidth

A natural specification for system performance in terms of frequency response is the **bandwidth**, defined to be the maximum frequency at which the output of a system will track an input sinusoid in a satisfactory manner. By convention, for the system shown in Fig. 6.4 with a sinusoidal input r, the bandwidth is the frequency of r at which the output y is attenuated to a factor of 0.707 times the input.[2] Figure 6.5 depicts the idea graphically for the frequency response of the *closed-loop* transfer function

$$\frac{Y(s)}{R(s)} \triangleq T(s) = \frac{KG(s)}{1 + KG(s)}.$$

The plot is typical of most closed-loop systems in that (1) the output follows the input ($|T| \cong 1$) at the lower excitation frequencies, and (2) the output ceases to follow the input ($|T| < 1$) at the higher excitation frequencies. The maximum value of the frequency-response magnitude is referred to as the **resonant peak** M_r.

Bandwidth is a measure of speed of response and is therefore similar to time-domain measures such as rise time and peak time or the s-plane measure of dominant-root(s) natural frequency. In fact, if the $KG(s)$ in Fig. 6.4 is such that the closed-loop response is given by Fig. 6.3, we can see that the bandwidth will equal the natural frequency of the closed-loop root (that is, $\omega_{BW} = \omega_n$ for a closed-loop damping ratio of $\zeta = 0.7$). For other damping ratios, the bandwidth is approximately equal to the natural frequency of the closed-loop roots, with an error typically less than a factor of 2.

The definition of the bandwidth stated here is meaningful for systems that have a low-pass filter behavior, as is the case for any physical control system. In other applications the bandwidth may be defined differently. Also, if the ideal model of the system does not have a high-frequency roll-off (e.g., if it has an equal number of poles and zeros), the bandwidth is infinite; however, this does not occur in nature as nothing responds well at infinite frequencies.

In many cases, the designer's primary concern is the error in the system due to disturbances rather than the ability to track an input. For error analysis, we are

---

[2]If the output is a voltage across a 1-Ω resistor, the power is v^2 and when $|v| = 0.707$, the power is reduced by a factor of 2. By convention, this is called the half-power point.

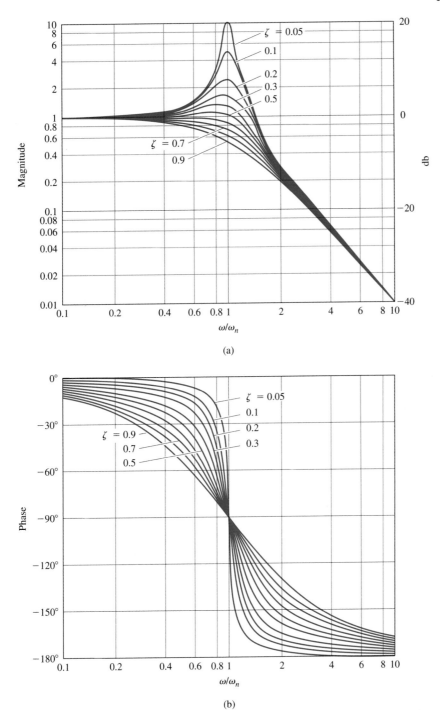

(a)

(b)

Figure 6.3
(a) Magnitude; (b) phase of Eq. (6.9)

Figure 6.4

Simplified system definition

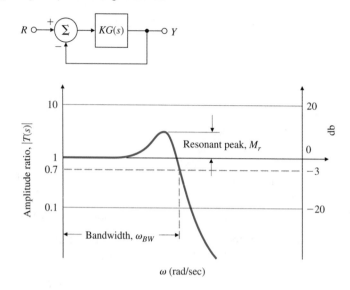

Figure 6.5

Definitions of bandwidth and resonant peak

more interested in one of the sensitivity functions defined in Section 4.1, $S(s)$, rather than $T(s)$. For most open-loop systems with high gain at low frequencies, $S(s)$ for a disturbance input will have very low values at low frequencies and grows as the frequency of the input or disturbance approaches the bandwidth. For analysis of either $T(s)$ or $S(s)$, it is typical to plot their response versus the frequency of the input. Either frequency response for control systems design can be evaluated using the computer, or can be quickly sketched for simple systems using the efficient methods described in the following Section 6.1.1. The methods described next are also useful to expedite the design process as well as to perform sanity checks on the computer output.

6.1.1   Bode Plot Techniques

Display of frequency response is a problem that has been studied for a long time. Before computers, this was accomplished by hand; therefore, it was useful to be able to accomplish this quickly. The most useful technique for hand plotting was developed by H. W. Bode at Bell Laboratories between 1932 and 1942. This technique allows plotting that is quick and yet sufficiently accurate for control systems design. Most control systems designers now have access to computer programs that diminish the need for hand plotting; however, it is still important to develop good intuition so that you can quickly identify erroneous computer results, and for this you need the ability to perform a sanity check and in some cases to determine approximate results by hand. The idea in Bode's method is to plot magnitude curves using a logarithmic scale and phase curves using a linear scale. This strategy allows us to plot a high-order $G(j\omega)$ by simply adding the separate terms graphically, as discussed in Appendix B. This addition is possible because a complex expression with zero and pole factors can be written in polar (or phasor) form as

$$G(j\omega) = \frac{\vec{s}_1 \vec{s}_2}{\vec{s}_3 \vec{s}_4 \vec{s}_5} = \frac{r_1 e^{j\theta_1} r_2 e^{j\theta_2}}{r_3 e^{j\theta_3} r_4 e^{j\theta_4} r_5 e^{j\theta_5}} = \left(\frac{r_1 r_2}{r_3 r_4 r_5}\right) e^{j(\theta_1 + \theta_2 - \theta_3 - \theta_4 - \theta_5)}. \qquad (6.10)$$

Composite plot from
individual terms

(The overhead arrow indicates a phasor.) Note from Eq. (6.10) that the phases of the individual terms are added directly to obtain the phase of the **composite** expression, $G(j\omega)$. Furthermore, because

$$|G(j\omega)| = \frac{r_1 r_2}{r_3 r_4 r_5},$$

it follows that

$$\log_{10}|G(j\omega)| = \log_{10} r_1 + \log_{10} r_2 - \log_{10} r_3 - \log_{10} r_4 - \log_{10} r_5. \quad (6.11)$$

We see that addition of the logarithms of the individual terms provides the logarithm of the magnitude of the composite expression. The frequency response is typically presented as two curves; the logarithm of magnitude versus $\log \omega$ and the phase versus

Bode plot

$\log \omega$. Together these two curves constitute a **Bode plot** of the system. Because

$$\log_{10} Me^{j\phi} = \log_{10} M + j\phi \log_{10} e, \quad (6.12)$$

we see that the Bode plot shows the real and imaginary parts of the logarithm of

Decibel

$G(j\omega)$. In communications it is standard to measure the power gain in decibels (db):[3]

$$|G|_{db} = 10 \log_{10} \frac{P_2}{P_1}. \quad (6.13)$$

Here P_1 and P_2 are the input and output powers. Because power is proportional to the square of the voltage, the power gain is also given by

$$|G|_{db} = 20 \log_{10} \frac{V_2}{V_1}. \quad (6.14)$$

Hence we can present a Bode plot as the magnitude in decibels versus $\log \omega$ and the phase in degrees versus $\log \omega$.[4] In this book we give Bode plots in the form $\log |G|$ versus $\log \omega$; also, we mark an axis in decibels on the right-hand side of the magnitude plot to give you the choice of working with the representation you prefer. However, for frequency-response plots, we are not actually plotting power, and use of Eq. (6.14) can be somewhat misleading. If the magnitude data are derived in terms of $\log |G|$, it is conventional to plot them on a log scale but identify the scale in terms of $|G|$ only (without "log"). If the magnitude data are given in decibels, the vertical scale is linear such that each decade of $|G|$ represents 20 db.

Advantages of Working with Frequency Response in Terms of Bode Plots

Advantages of Bode plots

1. Dynamic compensator design can be based entirely on Bode plots.
2. Bode plots can be determined experimentally.
3. Bode plots of systems in series (or tandem) simply add, which is quite convenient.
4. The use of a log scale permits a much wider range of frequencies to be displayed on a single plot than is possible with linear scales.

It is important for the control systems engineer to understand the Bode plot techniques for several reasons: This knowledge allows the engineer not only to deal

---

[3]Researchers at Bell Laboratories first defined the unit of power gain as a **bel** (named for Alexander Graham Bell, the founder of the company). However, this unit proved to be too large, and hence a **decibel or db** ($1/10$ of a bel) was selected as a more useful unit. The abbreviation dB is also sometimes used; however, Bode used db and we choose to follow his lead.

[4]Henceforth we will drop the base of the logarithm; it is understood to be 10.

with simple problems, but also to perform a sanity check on computer results for more complicated cases. Often approximations can be used to quickly sketch the frequency response and deduce stability, as well as to determine the form of the needed dynamic compensations. Finally, an understanding of the plotting method is useful in interpreting frequency-response data that have been generated experimentally.

In Chapter 5 we wrote the open-loop transfer function in the form

$$KG(s) = K\frac{(s - z_1)(s - z_2) \cdots}{(s - p_1)(s - p_2) \cdots} \tag{6.15}$$

because it was the most convenient form for determining the degree of stability from the root locus with respect to the gain K. In working with frequency response, it is more convenient to replace s with $j\omega$ and to write the transfer functions in the **Bode form**

Bode form of the transfer function

$$KG(j\omega) = K_o\frac{(j\omega\tau_1 + 1)(j\omega\tau_2 + 1) \cdots}{(j\omega\tau_a + 1)(j\omega\tau_b + 1) \cdots} \tag{6.16}$$

because the gain K_o in this form is directly related to the transfer-function magnitude at very low frequencies. In fact, for Type 0 systems, K_o is the gain at $\omega = 0$ in Eq. (6.16) and is also equal to the DC gain of the system. Although a straightforward calculation will convert a transfer function in the form of Eq. (6.15) to an equivalent transfer function in the form of Eq. (6.16), note that K and K_o will not usually have the same value in the two expressions.

Transfer functions can also be rewritten according to Eqs. (6.10) and (6.11). As an example, suppose that

$$KG(j\omega) = K_o\frac{j\omega\tau_1 + 1}{(j\omega)^2(j\omega\tau_a + 1)}. \tag{6.17}$$

Then

$$\angle KG(j\omega) = \angle K_o + \angle(j\omega\tau_1 + 1) - \angle(j\omega)^2 - \angle(j\omega\tau_a + 1) \tag{6.18}$$

and

$$\log|KG(j\omega)| = \log|K_o| + \log|j\omega\tau_1 + 1| - \log|(j\omega)^2| - \log|j\omega\tau_a + 1|. \tag{6.19}$$

In decibels, Eq. (6.19) becomes

$$|KG(j\omega)|_{db} = 20\log|K_o| + 20\log|j\omega\tau_1 + 1| - 20\log|(j\omega)^2|$$
$$- 20\log|j\omega\tau_a + 1|. \tag{6.20}$$

All transfer functions for the kinds of systems we have talked about so far are composed of three classes of terms:

Classes of terms of transfer functions

1. $K_o(j\omega)^n$.
2. $(j\omega\tau + 1)^{\pm 1}$.
3. $\left[\left(\frac{j\omega}{\omega_n}\right)^2 + 2\zeta\frac{j\omega}{\omega_n} + 1\right]^{\pm 1}$.

First we will discuss the plotting of each individual term and how the terms affect the composite plot including all the terms; then we will discuss how to draw the composite curve.

Figure 6.6

Magnitude of $(j\omega)^n$

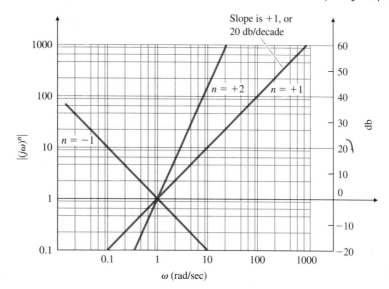

Class 1: singularities at the origin

1. $K_o(j\omega)^n$ Because

$$\log K_o|(j\omega)^n| = \log K_o + n\log|j\omega|,$$

the magnitude plot of this term is a straight line with a slope $n \times$ (20 db per decade). Examples for different values of n are shown in Fig. 6.6. $K_o(j\omega)^n$ is the only class of term that affects the slope at the lowest frequencies, because all other terms are constant in that region. The easiest way to draw the curve is to locate $\omega = 1$ and plot $\log K_o$ at that frequency. Then draw the line with slope n through that point.[5] The phase of $(j\omega)^n$ is $\phi = n \times 90°$; it is independent of frequency and is thus a horizontal line: $-90°$ for $n = -1$, $-180°$ for $n = -2$, $+90°$ for $n = +1$, and so forth.

Class 2: first-order term

2. $j\omega\tau + 1$ The magnitude of this term approaches one asymptote at very low frequencies and another asymptote at very high frequencies:

(a) For $\omega\tau \ll 1$, $j\omega\tau + 1 \cong 1$.
(b) For $\omega\tau \gg 1$, $j\omega\tau + 1 \cong j\omega\tau$.

Break point

If we call $\omega = 1/\tau$ the **break point**, then we see that below the break point the magnitude curve is approximately constant ($= 1$), while above the break point the magnitude curve behaves approximately like the class 1 term $K_o(j\omega)$. The example plotted in Fig. 6.7, $G(s) = 10s+1$, shows how the two asymptotes cross at the break point and how the actual magnitude curve lies above that point by a factor of 1.4 (or $+3$ db). (If the term were in the denominator, it would be below the break point by a factor of 0.707 or -3 db.) Note that this term will have only a small effect on the composite magnitude curve below the break point, because its value is equal to 1 ($= 0$ db) in this region. The slope at high frequencies is $+1$

---

[5] In decibels the slopes are $n \times 20$ db per decade or $n \times 6$ db per octave (an octave is a change in frequency by a factor of 2).

Figure 6.7

Magnitude plot for $j\omega\tau + 1; \tau = 10$

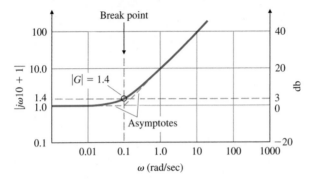

Figure 6.8

Phase plot for $j\omega\tau + 1$; $\tau = 10$

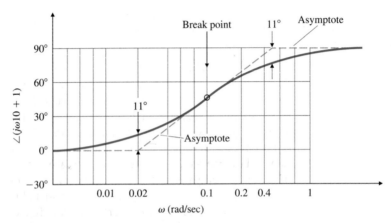

(or +20 db per decade). The phase curve can also be easily drawn by using the following low- and high-frequency asymptotes:

(a) For $\omega\tau \ll 1$, $\angle 1 = 0°$.
(b) For $\omega\tau \gg 1$, $\angle j\omega\tau = 90°$.
(c) For $\omega\tau \cong 1$, $\angle(j\omega\tau + 1) \cong 45°$.

For $\omega\tau \cong 1$, the $\angle(j\omega + 1)$ curve is tangent to an asymptote going from 0° at $\omega\tau = 0.2$ to 90° at $\omega\tau = 5$, as shown in Fig. 6.8. The figure also illustrates the three asymptotes (dashed lines) used for the phase plot and how the actual curve deviates from the asymptotes by 11° at their intersections. Both the composite phase and magnitude curves are unaffected by this class of term at frequencies below the break point by more than a factor of 10 because the term's magnitude is 1 (or 0 db) and its phase is less than 5°.

Class 3: second-order term

3. $[(j\omega/\omega_n)^2 + 2\zeta(j\omega/\omega_n) + 1]^{\pm 1}$ This term behaves in a manner similar to the class 2 term, with differences in detail: The break point is now $\omega = \omega_n$. The magnitude changes slope by a factor of +2 (or +40 db per decade) at the break point (and −2, or −40 db per decade, when the term is in the denominator). The phase changes by ±180°, and the transition through the break-point region varies with the damping ratio ζ. Figure 6.3 shows the magnitude and phase for several different damping ratios when the term is in the denominator. Note that the magnitude asymptote for frequencies above the break point has a slope of −2

Peak amplitude

(or -40 db per decade), and that the transition through the break-point region has a large dependence on the damping ratio. A rough determination of this transition can be made by noting that

$$|G(j\omega)| = \frac{1}{2\zeta} \quad \text{at} \quad \omega = \omega_n \qquad (6.21)$$

for this class of second-order term in the denominator. If the term was in the numerator, the magnitude would be the reciprocal of the curve plotted in Fig. 6.3(a).

No such handy rule as Eq. (6.21) exists for sketching in the transition for the phase curve; therefore, we would have to resort to Fig. 6.3(b) for an accurate plot of the phase. However, a very rough idea of the transition can be gained by noting that it is a step function for $\zeta = 0$, while it obeys the rule for two first-order (class 2) terms when $\zeta = 1$ with simultaneous break-point frequencies. All intermediate values of ζ fall between these two extremes. The phase of a second-order term is always $\pm 90°$ at ω_n.

Composite curve

When the system has several poles and several zeros, plotting the frequency response requires that the components be combined into a composite curve. To plot the composite magnitude curve, it is useful to note that the slope of the asymptotes is equal to the sum of the slopes of the individual curves. Therefore, the composite asymptote curve has integer slope changes at each break-point frequency: $+1$ for a first-order term in the numerator, -1 for a first-order term in the denominator, and ± 2 for second-order terms. Furthermore, the lowest-frequency portion of the asymptote has a slope determined by the value of n in the $(j\omega)^n$ term and is located by plotting the point $K_o \omega^n$ at $\omega = 1$. Therefore, the complete procedure consists of plotting the lowest-frequency portion of the asymptote, then sequentially changing the asymptote's slope at each break point in order of ascending frequency, and finally drawing the actual curve by using the transition rules discussed earlier for classes 2 and 3.

The composite phase curve is the sum of the individual curves. Addition of the individual phase curves graphically is made possible by locating the curves so that the composite phase approaches the individual curve as closely as possible. A quick but crude sketch of the composite phase can be found by starting the phase curve below the lowest break point and setting it equal to $n \times 90°$. The phase is then stepped at each break point in order of ascending frequency. The amount of the phase step is $\pm 90°$ for a first-order term and $\pm 180°$ for a second-order term. Break points in the numerator indicate a positive step in phase, while break points in the denominator indicate a negative phase step.[6] The plotting rules so far have only considered poles and zeros in the left half-plane (LHP). Changes for singularities in the right half-plane (RHP) will be discussed at the end of the section.

Summary of Bode Plot Rules

1. Manipulate the transfer function into the Bode form given by Eq. (6.16).
2. Determine the value of n for the $K_o(j\omega)^n$ term (class 1). Plot the low-frequency magnitude asymptote through the point K_o at $\omega = 1$ with a slope of n (or $n \times 20$ db per decade).

---

[6]This approximate method was pointed out to us by our Parisian colleagues.

3. Complete the composite magnitude asymptotes: Extend the low-frequency asymptote until the first frequency break point. Then step the slope by ± 1 or ± 2, depending on whether the break point is from a first- or second-order term in the numerator or denominator. Continue through all break points in ascending order.

4. The approximate magnitude curve is increased from the asymptote value by a factor of 1.4 (+3 db) at first-order numerator break points, and decreased by a factor of 0.707 (−3 db) at first-order denominator break points. At second-order break points, the resonant peak (or valley) occurs according to Fig. 6.3(a), using the relation $|G(j\omega)| = 1/2\zeta$ at denominator, (or $|G(j\omega)| = 2\zeta$ at numerator) break points.

5. Plot the low-frequency asymptote of the phase curve, $\phi = n \times 90°$.

6. As a guide, the approximate phase curve changes by $\pm 90°$ or $\pm 180°$ at each break point in ascending order. For first-order terms in the numerator, the change of phase is $+90°$; for those in the denominator the change is $-90°$. For second-order terms, the change is $\pm 180°$.

7. Locate the asymptotes for each individual phase curve so that their phase change corresponds to the steps in the phase toward or away from the approximate curve indicated by Step 6. Each individual phase curve occurs as indicated by Fig. 6.8 or Fig. 6.3(b).

8. Graphically add each phase curve. Use grids if an accuracy of about $\pm 5°$ is desired. If less accuracy is acceptable, the composite curve can be done by eye. Keep in mind that the curve will start at the lowest-frequency asymptote and end on the highest-frequency asymptote and will approach the intermediate asymptotes to an extent that is determined by how close the break points are to each other.

EXAMPLE 6.3   *Bode Plot for Real Poles and Zeros*

Plot the Bode magnitude and phase for the system with the transfer function

$$KG(s) = \frac{2000(s + 0.5)}{s(s + 10)(s + 50)}.$$

Solution

1. We convert the function to the Bode form of Eq. (6.16):

$$KG(j\omega) = \frac{2[(j\omega/0.5) + 1]}{j\omega[(j\omega/10) + 1][(j\omega/50) + 1]}.$$

2. We note that the term in $j\omega$ is first order and in the denominator, so $n = -1$. Therefore, the low-frequency asymptote is defined by the first term:

$$KG(j\omega) = \frac{2}{j\omega}.$$

This asymptote is valid for $\omega < 0.1$, because the lowest break point is at $\omega = 0.5$. The magnitude plot of this term has the slope of -1 (or -20 db per decade). We locate the magnitude by passing through the value 2 at $\omega = 1$ even though the composite curve will not go through this point because of the break point at $\omega = 0.5$. This is shown in Fig. 6.9(a).

3. We obtain the remainder of the asymptotes, also shown in Fig. 6.9(a): The first break point is at $\omega = 0.5$ and is a first-order term in the numerator, which thus calls for a change in slope of $+1$. We therefore draw a line with 0 slope that intersects the original -1 slope. Then we draw a -1 slope line that intersects the previous one at $\omega = 10$. Finally, we draw a -2 slope line that intersects the previous -1 slope at $\omega = 50$.

4. The actual curve is approximately tangent to the asymptotes when far away from the break points, a factor of 1.4 ($+3$ db) above the asymptote at the $\omega = 0.5$ break point, and a factor of 0.7 (-3 db) below the asymptote at the $\omega = 10$ and $\omega = 50$ break points.

5. Because the phase of $2/j\omega$ is $-90°$, the phase curve in Fig. 6.9(b) starts at $-90°$ at the lowest frequencies.

6. The result is shown in Fig. 6.9(c).

7. The individual phase curves, shown dashed in Fig. 6.9(b), have the correct phase change for each term and are aligned vertically so that their phase change corresponds to the steps in the phase from the approximate curve in Fig. 6.9(c). Note that the composite curve approaches each individual term.

8. The graphical addition of each dashed curve results in the solid composite curve in Fig. 6.9(b). As can be seen from the figure, the vertical placement of each individual phase curve makes the required graphical addition particularly easy because the composite curve approaches each individual phase curve in turn.

EXAMPLE 6.4   *Bode Plot with Complex Poles*

As a second example, draw the frequency response for the system

$$KG(s) = \frac{10}{s[s^2 + 0.4s + 4]}. \qquad (6.22)$$

Solution. A system like this is more difficult to plot than the one in the previous example because the transition between asymptotes is dependent on the damping ratio; however, the same basic ideas illustrated in Example 6.3 apply.

This system contains a second-order term in the denominator. Proceeding through the steps, we convert Eq. (6.22) to the Bode form of Eq. (6.16):

$$KG(s) = \frac{10}{4} \frac{1}{s(s^2/4 + 2(0.1)s/2 + 1)}.$$

Starting with the low-frequency asymptote, we have $n = -1$ and $|G(j\omega)| \cong 2.5/\omega$. The magnitude plot of this term has a slope of -1 (-20 db per decade) and passes through the value of 2.5 at $\omega = 1$ as shown in Fig. 6.10(a). For the second-order pole, note that $\omega_n = 2$ and $\zeta = 0.1$. At the break-point frequency of the poles, $\omega = 2$, the slope shifts to -3 (-60 db per decade). At the pole break point the magnitude ratio above the asymptote is $1/2\zeta = 1/0.2 = 5$. The phase curve for this case starts at $\phi = -90°$, corresponding to the $1/s$ term, falls to $\phi = -180°$ at $\omega = 2$ due to the pole as shown in Fig. 6.10(b), and then approaches $\phi = -270°$ for higher frequencies. Because the damping is small, the stepwise approximation is a very good one. The true composite phase curve is shown in Fig. 6.10(b).

Figure 6.9

Composite plots:
(a) magnitude;
(b) phase;
(c) approximate phase

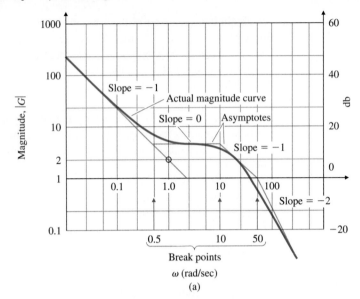

(a)

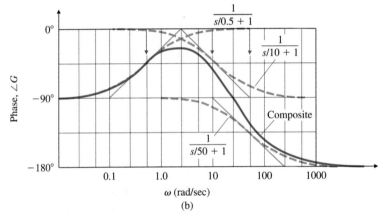

(b)

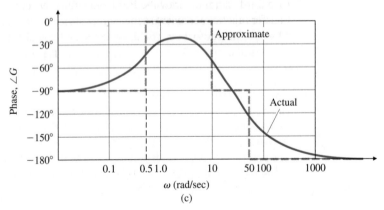

(c)

Figure 6.10

Bode plot for a transfer function with complex poles: (a) magnitude; (b) phase

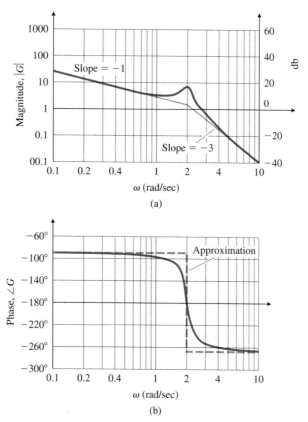

(a)

(b)

EXAMPLE 6.5

Bode Plot for Complex Poles and Zeros: Satellite with Flexible Appendages

As a third example, draw the Bode plots for a system with second-order terms. The transfer function represents a mechanical system with two equal masses coupled with a lightly damped spring. The applied force and position measurement are collocated on the same mass. For the transfer function, the time scale has been chosen so that the resonant frequency of the complex zeros is equal to 1. The transfer function is

$$KG(s) = \frac{0.01(s^2 + 0.01s + 1)}{s^2[(s^2/4) + 0.02(s/2) + 1]}.$$

Solution. Proceeding through the steps, we start with the low-frequency asymptote, $0.01/\omega^2$. It has a slope of -2 (-40 db per decade) and passes through magnitude $= 0.01$ at $\omega = 1$, as shown in Fig. 6.11(a). At the break-point frequency of the zero, $\omega = 1$, the slope shifts to zero until the break point of the pole, which is located at $\omega = 2$, when the slope returns to a slope of -2. To interpolate the true curve, we plot

Figure 6.11

Bode plot for a transfer function with complex poles and zeros:
(a) magnitude;
(b) phase

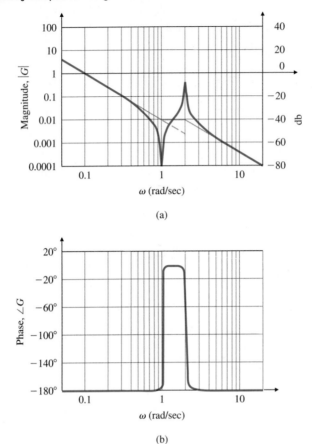

(a)

(b)

the point at the zero break point, $\omega = 1$, with a magnitude ratio below the asymptote of $2\zeta = 0.01$. At the pole break point, the magnitude ratio above the asymptote is $1/2\zeta = 1/0.02 = 50$. The magnitude curve is a "doublet" of a negative pulse followed by a positive pulse. Figure 6.11(b) shows that the phase curve for this system starts at $-180°$ (corresponding to the $1/s^2$ term), jumps $180°$ to $\phi = 0$ at $\omega = 1$, due to the zeros, and then falls $180°$ back to $\phi = -180°$ at $\omega = 2$, due to the pole. With such small damping ratios the stepwise approximation is quite good. (We haven't drawn this on Fig. 6.11(b), because it would not be easily distinguishable from the true phase curve.) Thus, the true composite phase curve is a nearly square pulse between $\omega = 1$ and $\omega = 2$.

In actual designs, Bode plots are made with a computer. However, acquiring the ability to determine how Bode plots should behave is a useful skill, because it gives the designer insight into how changes in the compensation parameters will affect the frequency response. This allows the designer to iterate to the best designs more quickly.

EXAMPLE 6.6   *Computer-Aided Bode Plot for Complex Poles and Zeros*

Repeat Example 6.5 using MATLAB.

Solution.   To obtain Bode plots using MATLAB, we call the function bode as follows:

```
numG = 0.01*[1  0.01  1];
denG = [0.25  0.01  1  0  0];
sysG = tf(numG,denG);
[mag, phase, w] = bode(sysG);
loglog(w,squeeze(mag))
semilogx(w,squeeze(phase))
```

These commands will result in a Bode plot that matches that in Fig. 6.11 very closely. To obtain the magnitude plot in decibels, the last three lines can be replaced with

```
bode(sysG)
```

Nonminimum-Phase Systems

A system with a zero in the RHP undergoes a net change in phase when evaluated for frequency inputs between zero and infinity, which, for an associated magnitude plot, is greater than if all poles and zeros were in the LHP. Such a system is called **nonminimum phase**. As can be seen from the construction in Fig. WD.3 in Appendix WD, if the zero is in the RHP, then the phase *decreases* at the zero break point instead of exhibiting the usual phase increase that occurs for an LHP zero. Consider the transfer functions

$$G_1(s) = 10\frac{s+1}{s+10},$$

$$G_2(s) = 10\frac{s-1}{s+10}.$$

Both transfer functions have the same magnitude for all frequencies; that is,

$$|G_1(j\omega)| = |G_2(j\omega)|,$$

as shown in Fig. 6.12(a). But the phases of the two transfer functions are drastically different [Fig. 6.12(b)]. A minimum-phase system (all zeros in the LHP) with a given magnitude curve will produce the smallest net change in the associated phase, as shown in G_1, compared with what the nonminimum-phase system will produce, as shown by the phase of G_2. Hence, G_2 is nonminimum phase. The discrepancy between G_1 and G_2 with regard to the phase change would be greater if two or more zeros of the plant were in the RHP.

6.1.2   Steady-State Errors

We saw in Section 4.2 that the steady-state error of a feedback system decreases as the gain of the open-loop transfer function increases. In plotting a composite magnitude

Figure 6.12

Bode plot minimum- and nonminimum-phase systems: for (a) magnitude; (b) phase

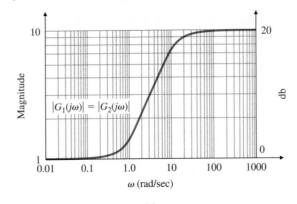

(a)

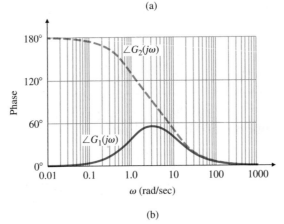

(b)

curve, we saw in Section 6.1.1 that the open-loop transfer function, at very low frequencies, is approximated by

$$KG(j\omega) \cong K_o(j\omega)^n. \tag{6.23}$$

Therefore, we can conclude that the larger the value of the magnitude on the low-frequency asymptote, the lower the steady-state errors will be for the closed-loop system. This relationship is very useful in the design of compensation: Often we want to evaluate several alternate ways to improve stability and to do so we want to be able to see quickly how changes in the compensation will affect the steady-state errors.

Position error constant

For a system of the form given by Eq. (6.16)—that is, where $n = 0$ in Eq. (6.23) (a Type 0 system)—the low-frequency asymptote is a constant and the gain K_o of the open-loop system is equal to the position-error constant K_p. For a unity feedback system with a unit-step input, the Final Value Theorem (Section 3.1.6) was used in Section 4.2.1 to show that the steady-state error is given by

$$e_{ss} = \frac{1}{1 + K_p}.$$

Velocity error coefficient

For a unity-feedback system in which $n = -1$ in Eq. (6.23), defined to be a Type 1 system in Section 4.2.1, the low-frequency asymptote has a slope of -1. The magnitude of the low-frequency asymptote is related to the gain according to Eq. (6.23);

Figure 6.13

Determination of K_v from the Bode plot for the system $KG(s) = \frac{10}{s(s+1)}$

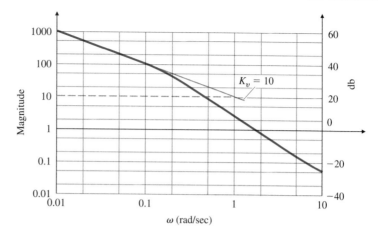

ω (rad/sec)

therefore, we can again read the gain, K_o/ω, directly from the Bode magnitude plot. Equation (4.33) tells us that the velocity-error constant

$$K_v = K_o,$$

where, for a unity-feedback system with a unit-ramp input, the steady-state error is

$$e_{ss} = \frac{1}{K_v}.$$

The easiest way of determining the value of K_v in a type 1 system is to read the magnitude of the low-frequency asymptote at $\omega = 1$ rad/sec, because this asymptote is $A(\omega) = K_v/\omega$. In some cases the lowest-frequency break point will be below $\omega = 1$ rad/sec; therefore, the asymptote needs to extend to $\omega = 1$ rad/sec in order to read K_v directly. Alternately, we could read the magnitude at any frequency on the low-frequency asymptote and compute it from $K_v = \omega A(\omega)$.

EXAMPLE 6.7   *Computation of K_v*

As an example of the determination of steady-state errors, a Bode magnitude plot of an open-loop system is shown in Fig. 6.13. Assuming that there is unity feedback as in Fig. 6.4, find the velocity-error constant, K_v.

Solution. Because the slope at the low frequencies is -1, we know that the system is Type 1. The extension of the low-frequency asymptote crosses $\omega = 1$ rad/sec at a magnitude of 10. Therefore, $K_v = 10$ and the steady-state error to a unit ramp for a unity-feedback system would be 0.1. Alternatively, at $\omega = 0.01$ we have $|A(\omega)| = 1000$; therefore, from Eq. (6.23) we have

$$K_o = K_v \cong \omega|A(\omega)| = 0.01(1000) = 10.$$

6.2 Neutral Stability

In the early days of electronic communications, most instruments were judged in terms of their frequency response. It is therefore natural that when the feedback amplifier was introduced, techniques to determine stability in the presence of feedback were based on this response.

Suppose the closed-loop transfer function of a system is known. We can determine the stability of a system by simply inspecting the denominator in factored form (because the factors give the system roots directly) to observe whether the real parts are positive or negative. However, the closed-loop transfer function is usually not known; in fact, the whole purpose behind understanding the root-locus technique is to be able to find the factors of the denominator in the closed-loop transfer function, given only the open-loop transfer function. Another way to determine closed-loop stability is to evaluate the frequency response of the *open-loop* transfer function $KG(j\omega)$ and then perform a test on that response. Note that this method also does not require factoring the denominator of the closed-loop transfer function. In this section we will explain the principles of this method.

Suppose we have a system defined by Fig. 6.14(a) and whose root locus behaves as shown in Fig. 6.14(b); that is, instability results if K is larger than 2. The neutrally stable points lie on the imaginary axis—that is, where $K = 2$ and $s = j1.0$. Furthermore, we saw in Section 5.1 that all points on the locus have the property that

$$|KG(s)| = 1 \quad \text{and} \quad \angle G(s) = 180°.$$

At the point of neutral stability we see that these root-locus conditions hold for $s = j\omega$, so

$$|KG(j\omega)| = 1 \quad \text{and} \quad \angle G(j\omega) = 180°. \tag{6.24}$$

Thus a Bode plot of a system that is neutrally stable (that is, with K defined such that a closed-loop root falls on the imaginary axis) will satisfy the conditions of Eq. (6.24). Figure 6.15 shows the frequency response for the system whose root locus is plotted in Fig. 6.14 for various values of K. The magnitude response corresponding to $K = 2$ passes through 1 at the same frequency ($\omega = 1$ rad/sec) at which the phase passes through 180°, as predicted by Eq. (6.24).

Having determined the point of neutral stability, we turn to a key question: Does increasing the gain increase or decrease the system's stability? We can see from the root locus in Fig. 6.14(b) that any value of K less than the value at the neutrally stable point will result in a stable system. At the frequency ω where the phase $\angle G(j\omega) = -180°$ ($\omega = 1$ rad/sec), the magnitude $|KG(j\omega)| < 1.0$ for stable

Figure 6.14

Stability example:
(a) system definition;
(b) root locus

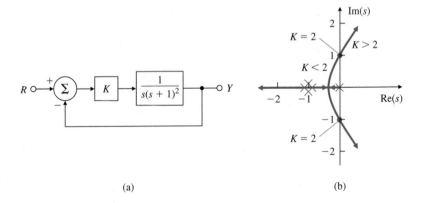

(a)

(b)

Figure 6.15

Frequency-response magnitude and phase for the system in Fig. 6.14

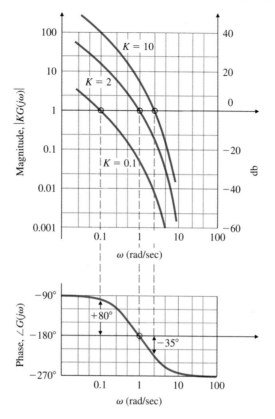

Stability condition

values of K and > 1 for unstable values of K. Therefore, we have the following trial stability condition, based on the character of the open-loop frequency response:

$$|KG(j\omega)| < 1 \quad \text{at} \quad \angle G(j\omega) = -180°. \tag{6.25}$$

This stability criterion holds for all systems for which increasing gain leads to instability and $|KG(j\omega)|$ crosses the magnitude ($=1$) once, the most common situation. However, there are systems for which an increasing gain can lead from instability to stability; in this case, the stability condition is

$$|KG(j\omega)| > 1 \quad \text{at} \quad \angle G(j\omega) = -180°. \tag{6.26}$$

There are also cases when $|KG(j\omega)|$ crosses magnitude ($=1$) more than once. One way to resolve the ambiguity that is usually sufficient is to perform a rough sketch of the root locus. Another, more rigorous, way to resolve the ambiguity is to use the Nyquist stability criterion, the subject of the next section. However, because the Nyquist criterion is fairly complex, it is important while studying it to bear in mind the theme of this section, namely, that for most systems a simple relationship exists between closed-loop stability and the open-loop frequency response.

6.3 The Nyquist Stability Criterion

For most systems, as we saw in the previous section, an increasing gain eventually causes instability. In the very early days of feedback control design, this relationship between gain and stability margins was assumed to be universal. However, designers found occasionally that in the laboratory the relationship reversed itself; that is, the amplifier would become unstable when the gain was decreased. The confusion caused by these conflicting observations motivated Harry Nyquist of the Bell Telephone Laboratories to study the problem in 1932. His study explained the occasional reversals and resulted in a more sophisticated analysis with no loopholes. Not surprisingly, his test has come to be called the **Nyquist stability criterion**. It is based on a result from complex variable theory known as the **argument principle**,[7] as we briefly explain in this section and in more detail in Appendix WD.

The Nyquist stability criterion relates the open-loop frequency response to the number of closed-loop poles of the system in the RHP. Study of the Nyquist criterion will allow you to determine stability from the frequency response of a complex system, perhaps with one or more resonances, where the magnitude curve crosses 1 several times and/or the phase crosses 180° several times. It is also very useful in dealing with open-loop, unstable systems, nonminimum-phase systems, and systems with pure delays (transportation lags).

6.3.1 The Argument Principle

Consider the transfer function $H_1(s)$ whose poles and zeros are indicated in the s-plane in Fig. 6.16(a). We wish to evaluate H_1 for values of s on the clockwise contour C_1. (Hence this is called a **contour evaluation**.) We choose the test point s_o for evaluation. The resulting complex quantity has the form $H_1(s_o) = \vec{v} = |\vec{v}|e^{j\alpha}$. The value of the argument of $H_1(s_o)$ is

$$\alpha = \theta_1 + \theta_2 - (\phi_1 + \phi_2).$$

As s traverses C_1 in the clockwise direction starting at s_o, the angle α of $H_1(s)$ in Fig. 6.16(b) will change (decrease or increase), but it will not undergo a net change of 360° as long as there are no poles or zeros within C_1. This is because none of the angles that make up α go through a net revolution. The angles θ_1, θ_2, ϕ_1, and ϕ_2 increase or decrease as s traverses around C_1, but they return to their original values as s returns to s_o without rotating through 360°. This means that the plot of $H_1(s)$ [Fig. 6.16(b)] will not encircle the origin. This conclusion follows from the fact that α is the sum of the angles indicated in Fig. 6.16(a), so the only way that α can be changed by 360° after s executes one full traverse of C_1 is for C_1 to contain a pole or zero.

Now consider the function $H_2(s)$, whose pole–zero pattern is shown in Fig. 6.16(c). Note that it has a singularity (pole) within C_1. Again, we start at the test point s_o. As s traverses in the clockwise direction around C_1, the contributions from the angles θ_1, θ_2, and ϕ_1 change, but they return to their original values as soon as s returns to s_o. In contrast, ϕ_2, the angle from the pole within C_1, undergoes a net change of $-360°$ after one full traverse of C_1. Therefore, the argument of $H_2(s)$ undergoes the same change, causing H_2 to encircle the origin in the counterclockwise direction, as shown in Fig. 6.16(d). The behavior would be similar if the contour C_1 had enclosed a zero instead of a pole. The mapping of C_1 would again enclose the origin once in the $H_2(s)$-plane, except it would do so in the clockwise direction.

---

[7]Sometimes referred to as "Cauchy's Principle of the Argument."

Figure 6.16

Contour evaluations:
(a) s-plane plot of poles
and zeros of $H_1(s)$ and
the contour C_1;
(b) $H_1(s)$ for s on C_1;
(c) s-plane plot of poles
and zeros of $H_2(s)$ and
the contour C_1;
(d) $H_2(s)$ for s on C_1

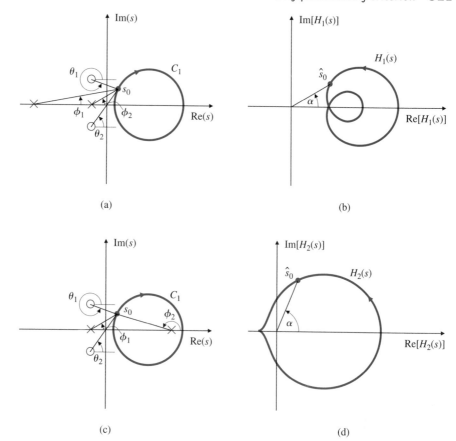

(a)

(b)

(c)

(d)

Thus we have the essence of the argument principle:

Argument principle

A contour map of a complex function will encircle the origin $Z - P$ times, where Z is the number of zeros and P is the number of poles of the function inside the contour.

For example, if the number of poles and zeros within C_1 is the same, the net angles cancel and there will be no net encirclement of the origin.

6.3.2 Application to Control Design

To apply the principle to control design, we let the C_1 contour in the s-plane encircle the entire RHP, the region in the s-plane where a pole would cause an unstable system (Fig. 6.17). The resulting evaluation of $H(s)$ will encircle the origin only if $H(s)$ has an RHP pole or zero.

As stated earlier, what makes all this contour behavior useful is that a contour evaluation of an *open-loop KG(s)* can be used to determine stability of the *closed-loop*

Figure 6.17

An s-plane plot of a contour C_1 that encircles the entire RHP

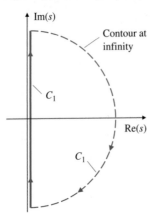

Figure 6.18

Block diagram for $Y(s)/R(s) = KG(s)/[1 + KG(s)]$

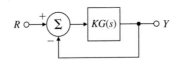

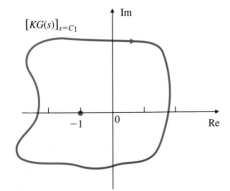

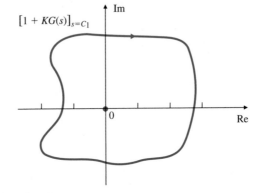

Figure 6.19

Evaluations of $KG(s)$ and $1 + KG(s)$: Nyquist plots

system. Specifically, for the system in Fig. 6.18, the closed-loop transfer function is

$$\frac{Y(s)}{R(s)} = T(s) = \frac{KG(s)}{1 + KG(s)}.$$

Therefore, the closed-loop roots are the solutions of

$$1 + KG(s) = 0,$$

and we apply the principle of the argument to the function $1 + KG(s)$. If the evaluation contour of this function of s enclosing the entire RHP contains a zero or pole of $1 + KG(s)$, then the evaluated contour of $1 + KG(s)$ will encircle the origin. Notice that $1 + KG(s)$ is simply $KG(s)$ shifted to the right 1 unit, as shown in Fig. 6.19. Therefore, if the plot of $1 + KG(s)$ encircles the origin, the plot of $KG(s)$ will encircle

−1 on the real axis. Therefore, we can plot the contour evaluation of the open-loop $KG(s)$, examine its encirclements of −1, and draw conclusions about the origin encirclements of the closed-loop function $1 + KG(s)$. Presentation of the evaluation of $KG(s)$ in this manner is often referred to as a **Nyquist plot**, or **polar plot**, because we plot the magnitude of $KG(s)$ versus the angle of $KG(s)$.

To determine whether an encirclement is due to a pole or zero, we write $1+KG(s)$ in terms of poles and zeros of $KG(s)$:

$$1 + KG(s) = 1 + K\frac{b(s)}{a(s)} = \frac{a(s) + Kb(s)}{a(s)}. \tag{6.27}$$

Equation (6.27) shows that the poles of $1 + KG(s)$ are also the poles of $G(s)$. Because it is safe to assume that the poles of $G(s)$ [or factors of $a(s)$] are known, the (rare) existence of any of these poles in the RHP can be accounted for. Assuming for now that there are no poles of $G(s)$ in the RHP, an encirclement of −1 by $KG(s)$ indicates a zero of $1 + KG(s)$ in the RHP, and thus an unstable root of the closed-loop system.

We can generalize this basic idea by noting that a clockwise contour C_1 enclosing a zero of $1+KG(s)$—that is, a closed-loop system root—will result in $KG(s)$ encircling the −1 point in a clockwise direction. Likewise, if C_1 encloses a pole of $1 + KG(s)$—that is, if there is an unstable open-loop pole—there will be a counterclockwise $KG(s)$ encirclement of −1. Furthermore, if two poles or two zeros are in the RHP, $KG(s)$ will encircle −1 twice, and so on. The net number of clockwise encirclements, N, equals the number of zeros (closed-loop system roots) in the RHP, Z, minus the number of open-loop poles in the RHP, P:

$$N = Z - P.$$

This is the key concept of the Nyquist stability criterion.

A simplification in the plotting of $KG(s)$ results from the fact that any $KG(s)$ that represents a physical system will have zero response at infinite frequency (i.e., has more poles than zeros). This means that the big arc of C_1 corresponding to s at infinity (Fig. 6.17) results in $KG(s)$ being a point of infinitesimally small value near the origin for that portion of C_1. Therefore, we accomplish a complete evaluation of a physical system $KG(s)$ by letting s traverse the imaginary axis from $-j\infty$ to $+j\infty$ (actually, from $-j\omega_h$ to $+j\omega_h$, where ω_h is large enough that $|KG(j\omega)|$ is much less than 1 for all $\omega > \omega_h$). The evaluation of $KG(s)$ from $s = 0$ to $s = j\infty$ has already been discussed in Section 6.1 under the context of finding the frequency response of $KG(s)$. Because $G(-j\omega)$ is the complex conjugate of $G(j\omega)$, we can easily obtain the entire plot of $KG(s)$ by reflecting the $0 \le s \le +j\infty$ portion about the real axis, to get the $(-j\infty \le s < 0)$ portion. Hence we see that closed-loop stability can be determined in all cases by examination of the frequency response of the open-loop transfer function on a polar plot. In some applications, models of physical systems are simplified so as to eliminate some high-frequency dynamics. The resulting reduced-order transfer function might have an equal number of poles and zeros. In that case the big arc of C_1 at infinity needs to be considered.

In practice, many systems behave like those discussed in Section 6.2, so you need not carry out a complete evaluation of $KG(s)$ with subsequent inspection of the −1 encirclements; a simple look at the frequency response may suffice to determine stability. However, in the case of a complex system for which the simplistic rules given in Section 6.2 become ambiguous, you will want to perform the complete analysis, summarized as follows:

Procedure for Determining Nyquist Stability

1. Plot $KG(s)$ for $-j\infty \leq s \leq +j\infty$. Do this by first evaluating $KG(j\omega)$ for $\omega = 0$ to ω_h, where ω_h is so large that the magnitude of $KG(j\omega)$ is negligibly small for $\omega > \omega_h$, then reflecting the image about the real axis and adding it to the preceding image. The magnitude of $KG(j\omega)$ will be small at high frequencies for any physical system. The Nyquist plot will always be symmetric with respect to the real axis. The plot is normally created by the NYQUIST MATLAB m-file.
2. Evaluate the number of clockwise encirclements of -1, and call that number N. Do this by drawing a straight line in any direction from -1 to ∞. Then count the net number of left-to-right crossings of the straight line by $KG(s)$. If encirclements are in the counterclockwise direction, N is negative.
3. Determine the number of unstable (RHP) poles of $G(s)$, and call that number P.
4. Calculate the number of unstable closed-loop roots Z:

$$Z = N + P. \tag{6.28}$$

For stability we wish to have $Z = 0$; that is, no characteristic equation roots in the RHP.

Let us now examine a rigorous application of the procedure for determining stability using Nyquist plots for some examples.

EXAMPLE 6.8   *Nyquist Plot for a Second-Order System*

Determine the stability properties of the system defined in Fig. 6.20.

Solution.   The root locus of the system in Fig. 6.20 is shown in Fig. 6.21. It shows that the system is stable for all values of K. The magnitude of the frequency response of $KG(s)$ is plotted in Fig. 6.22(a) for $K = 1$, and the phase is plotted in Fig. 6.22(b); this is the typical Bode method of presenting frequency response and represents the evaluation of $G(s)$ over the interesting range of frequencies. The same information is replotted in Fig. 6.23[8] in the Nyquist (polar) plot form. Note how the points A, B, C,

Figure 6.20

Control system for Example 6.8

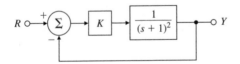

---

[8]The shape of this Nyquist plot is a cardioid, meaning "heart-shaped," plane curve. The name was first used by de Castillon in the *Philosophical Transactions of the Royal Society* in 1741. The cardioid is also used in optics.

Figure 6.21

Root locus of $G(s) = \frac{1}{(s+1)^2}$ with respect to K

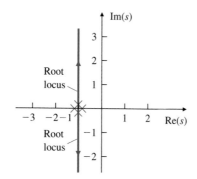

D, and E are mapped from the Bode plot to the Nyquist plot in Fig. 6.23. The arc from $G(s) = +1$ ($\omega = 0$) to $G(s) = 0$ ($\omega = \infty$) that lies below the real axis is derived from Fig. 6.22. The portion of the C_1 arc at infinity from Fig. 6.17 transforms into $G(s) = 0$ in Fig. 6.23; therefore, a continuous evaluation of $G(s)$ with s traversing

Figure 6.22

Open-loop Bode plot for $G(s) = \frac{1}{(s+1)^2}$

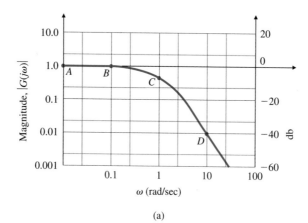

(a)

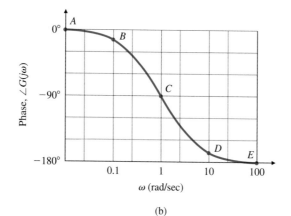

(b)

Figure 6.23

Nyquist plot of the evaluation of $KG(s)$ for $s = C_1$ and $K = 1$

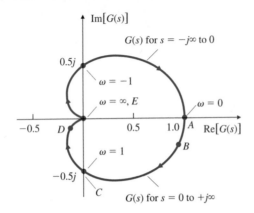

C_1 is completed by simply reflecting the lower arc about the real axis. This creates the portion of the contour above the real axis and completes the Nyquist (polar) plot. Because the plot does not encircle -1, $N = 0$. Also, there are no poles of $G(s)$ in the RHP, so $P = 0$. From Eq. (6.28), we conclude that $Z = 0$, which indicates there are no unstable roots of the closed-loop system for $K = 1$. Furthermore, different values of K would simply change the magnitude of the polar plot, but no positive value of K would cause the plot to encircle -1, because the polar plot will always cross the negative real axis when $KG(s) = 0$. Thus the Nyquist stability criterion confirms what the root locus indicated: the closed-loop system is stable for all $K > 0$.

The MATLAB statements that will produce this Nyquist plot are

```
numG = 1;
denG = [1  2  1];
sysG = tf(numG,denG);
w=logspace(-2,2);
nyquist(sysG,w);
```

Often the control systems engineer is more interested in determining a range of gains K for which the system is stable than in testing for stability at a specific value of K. To accommodate this requirement, but to avoid drawing multiple Nyquist plots for various values of the gain, the test can be slightly modified. To do so, we scale $KG(s)$ by K and examine $G(s)$ to determine stability for a range of gains K. This is possible because an encirclement of -1 by $KG(s)$ is equivalent to an encirclement of $-1/K$ by $G(s)$. Therefore, instead of having to deal with $KG(s)$, we need only consider $G(s)$, and count the number of the encirclements of the $-1/K$ point.

Applying this idea to Example 6.8, we see that the Nyquist plot cannot encircle the $-1/K$ point. For positive K, the $-1/K$ point will move along the negative real axis, so there will not be an encirclement of $G(s)$ for any value of $K > 0$.

(There are also values of $K < 0$ for which the Nyquist plot shows the system to be stable; specifically, $-1 < K < 0$. This result may be verified by drawing the $0°$ locus.)

EXAMPLE 6.9   *Nyquist Plot for a Third-Order System*

As a second example, consider the system $G(s) = 1/s(s + 1)^2$ for which the closed-loop system is defined in Fig. 6.24. Determine its stability properties using the Nyquist criterion.

Figure 6.24

Control system for Example 6.9

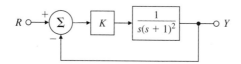

Solution. This is the same system discussed in Section 6.2. The root locus in Fig. 6.14(b) shows that this system is stable for small values of K but unstable for large values of K. The magnitude and phase of $G(s)$ in Fig. 6.25 are transformed into the Nyquist plot shown in Fig. 6.26. Note how the points A, B, C, D, and E on the Bode plot of Fig. 6.25 map into those on the Nyquist plot of Fig. 6.26. Also note the large arc at infinity that arises from the open-loop pole at $s = 0$. This pole creates an infinite magnitude of $G(s)$ at $\omega = 0$; in fact, a pole anywhere on the imaginary axis will create an arc at infinity. To correctly determine the number of $-1/K$ point encirclements, we must draw this arc in the proper half-plane: Should it cross the positive real axis, as shown in Fig. 6.26, or the negative one? It is also necessary to assess whether the arc should sweep out 180° (as in Fig. 6.26), 360°, or 540°.

A simple artifice suffices to answer these questions. We modify the C_1 contour to take a small detour around the pole either to the right (Fig. 6.27) or to the left. It makes

Figure 6.25

Bode plot for $G(s) = 1/s(s + 1)^2$

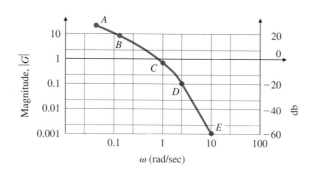

(a)

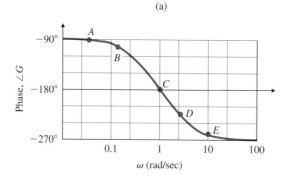

(b)

Figure 6.26

Nyquist plot[9] for
$G(s) = 1/s(s+1)^2$

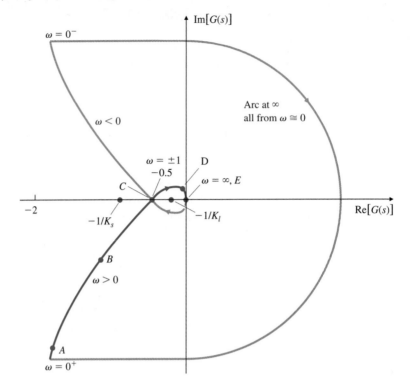

no difference to the final stability question which way, but it is more convenient to go to the right because then no poles are introduced within the C_1 contour, keeping the value of P equal to 0. Because the phase of $G(s)$ is the negative of the sum of the angles from all of the poles, we see that the evaluation results in a Nyquist plot moving from $+90°$ for s just below the pole at $s = 0$, across the positive real axis to $-90°$ for s just above the pole. Had there been two poles at $s = 0$, the Nyquist plot at infinity would have executed a full $360°$ arc, and so on for three or more poles. Furthermore, for a pole elsewhere on the imaginary axis, a $180°$ clockwise arc would also result but would be oriented differently than the example shown in Fig. 6.26.

The Nyquist plot crosses the real axis at $\omega = 1$ with $|G| = 0.5$, as indicated by the Bode plot. For $K > 0$, there are two possibilities for the location of $-1/K$: inside the two loops of the Nyquist plot, or outside the Nyquist contour completely. For large values of K (K_l in Fig. 6.26), $-0.5 < -1/K_l < 0$ will lie inside the two loops; hence $N = 2$, and therefore, $Z = 2$, indicating that there are two unstable roots. This happens for $K > 2$. For small values of K (K_s in Fig. 6.26), $-1/K$ lies outside the loops; thus $N = 0$, and all roots are stable. All this information is in agreement with the root locus in Fig. 6.14(b). (When $K < 0$, $-1/K$ lies on the positive real axis, then $N = 1$, which means $Z = 1$ and the system has one unstable root. The $0°$ root locus will verify this result.)

---

[9]The shape of this Nyquist plot is a translated strophoid plane curve, meaning "a belt with a twist." The curve was first studied by Barrow in 1670.

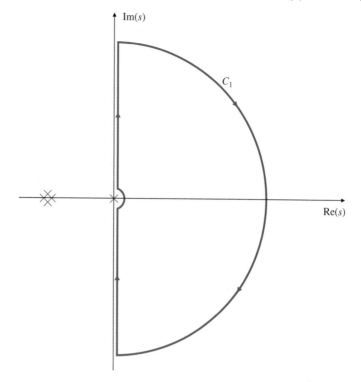

For this and many similar systems, we can see that the encirclement criterion
reduces to a very simple test for stability based on the open-loop frequency response:
The system is stable if $|KG(j\omega)| < 1$ when the phase of $G(j\omega)$ is $180°$. Note that this
relation is identical to the stability criterion given in Eq. (6.25); however, by using the
Nyquist criterion, we don't require the root locus to determine whether $|KG(j\omega)| < 1$
or $|KG(j\omega)| > 1$.

We draw the Nyquist plot using MATLAB, with

```
numG = 1 ;
denG = [1  2  1  0];
sysG = tf(numG,denG);
nyquist(sysG)
axis([−3 3 −3 3]);
```

The axis command scaled the plot so that only points between $+3$ and -3 on
the real and imaginary axes were included. Without manual scaling, the plot would
be scaled based on the maximum values computed by MATLAB and the essential
features in the vicinity of the -1 region would be lost.

For systems that are open-loop unstable, care must be taken because now $P \neq 0$ in
Eq. (6.28). We shall see that the simple rules from Section 6.2 will need to be revised
in this case.

EXAMPLE 6.10

Nyquist Plot for an Open-Loop Unstable System

The third example is defined in Fig. 6.28. Determine its stability properties using the Nyquist criterion.

Figure 6.28

Control system for Example 6.10

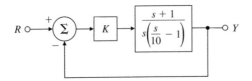

Solution. The root locus for this system is sketched in Fig. 6.29. The open-loop system is unstable because it has a pole in the RHP. The open-loop Bode plot is shown in Fig. 6.30. Note in the Bode that $|KG(j\omega)|$ behaves exactly the same as if the pole had been in the LHP. However, $\angle G(j\omega)$ increases by 90° instead of the usual decrease at a pole. Any system with a pole in the RHP is unstable; hence it would be impossible to determine its frequency response experimentally because the system would never reach a steady-state sinusoidal response for a sinusoidal input. It is, however, possible to compute the magnitude and phase of the transfer function according to the rules in Section 6.1. The pole in the RHP affects the Nyquist encirclement criterion, because the value of P in Eq. (6.28) is $+1$.

We convert the frequency-response information of Fig. 6.30 into the Nyquist plot (Fig. 6.31) as in the previous examples. As before, the C_1 detour around the pole at $s = 0$ in Fig. 6.32 creates a large arc at infinity in Fig. 6.31. This arc crosses the *negative* real-axis because of the 180° phase contribution of the pole in the RHP.

The real-axis crossing occurs at $|G(s)| = 1$ because in the Bode plot $|G(s)| = 1$ when $\angle G(s) = 180°$, which happens to be at $\omega \cong 3$ rad/sec.

The contour shows two different behaviors, depending on the values of K (> 0). For large values of K (K_1 in Fig. 6.31), there is one counterclockwise encirclement; hence $N = -1$. However, because $P = 1$ from the RHP pole, $Z = N + P = 0$, so there are no unstable system roots and the system is stable for $K > 1$. For small values of K (K_s in Fig. 6.31), $N = +1$ because of the clockwise encirclement and $Z = 2$, indicating two unstable roots. This happens if $K > 1$. These results can be verified

Figure 6.29

Root locus for $G(s) = \dfrac{(s+1)}{s(s/10-1)}$

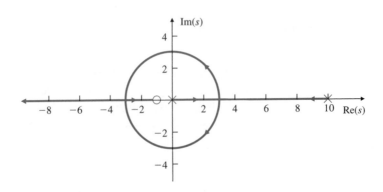

Figure 6.30

Bode plot for
$G(s) = \frac{(s+1)}{s(s/10-1)}$

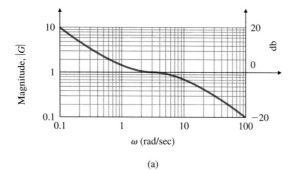

(a)

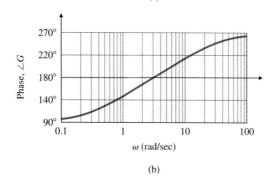

(b)

Figure 6.31

Nyquist plot[10] for
$G(s) = \frac{(s+1)}{s(s/10-1)}$

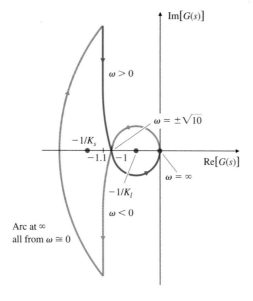

---

[10]The shape of this Nyquist plot is a strophoid.

Figure 6.32

C_1 contour for Example 6.10

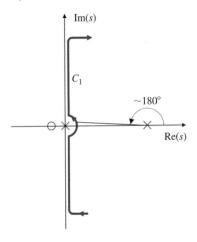

qualitatively by the root locus in Fig. 6.29. (If $K < 0$, $-1/K$ is on the positive real-axis so that $N = 0$ and $Z = 1$, indicating the system will have one unstable closed-loop pole, which can be verified by a $0°$ root locus.)

As with all systems, the stability boundary occurs at $|KG(j\omega)| = 1$ for the phase of $\angle G(j\omega) = 180°$. However, in this case, $|KG(j\omega)|$ must be greater than 1 to yield the correct number of -1 point encirclements to achieve stability.

To draw the Nyquist plot using MATLAB, use the following commands:

```
numG = [1  1];
denG = [0.1 −1  0];
sysG = tf(numG,denG);
nyquist(sysG)
axis([−3  3  −3  3])
```

The existence of the RHP pole in Example 6.10 affected the Bode plotting rules of the phase curve and affected the relationship between encirclements and unstable closed-loop roots because $P = 1$ in Eq. (6.28). But we apply the Nyquist stability criterion without any modifications. The same is true for systems with a RHP zero; that is, a nonminimum-phase zero has no effect on the Nyquist stability criterion, but the Bode plotting rules are affected.

EXAMPLE 6.11   *Nyquist Plot Characteristics*

Find the Nyquist plot for the third-order system

$$G(s) = K \frac{s^2 + 3}{(s + 1)^2}$$

and reconcile the plot with the characteristics of $G(s)$. If the $G(s)$ is to be included in a feedback system as shown in Fig. 6.18, then determine whether the system is stable for all positive values of K.

Solution.   To draw the Nyquist plot using MATLAB, use the following commands:

Figure 6.33

Nyquist plot for
Example 6.11

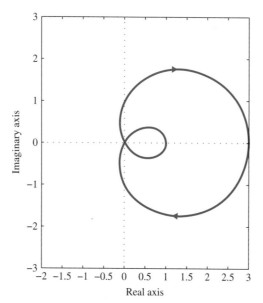

```
numG = [1  0  3];
denG = [1  2  1];
sysG = tf(numG,denG);
nyquist(sysG)
axis([−2 3 −3 3])
```

The result is shown in Fig. 6.33.[11] Note that there are no arcs at infinity for this case due to the lack of any poles at the origin or on the $j\omega$ axis. Also note that the Nyquist curve associated with the Bode plot ($s = +j\omega$) starts at $(3, 0)$, ends at $(1, 0)$, and, therefore, starts and ends with a phase angle of $0°$. This is as it should be since the numerator and denominator of $G(s)$ are equal order and there are no singularities at the origin. So the Bode plot should start and end with a zero phase. Also note that the Nyquist plot goes through $(0, 0)$ as s goes through $s = +j\sqrt{3}$, as it should since the magnitude equals zero when s is at a zero. Furthermore, note that the phase goes from $-120°$ as s approaches $(0, 0)$ to $+60°$ as s departs from $(0, 0)$. This behavior follows since a Bode plot phase will jump by $+180°$ instantaneously as s passes through a zero on the $j\omega$ axis. The phase initially decreases as the plot leaves the starting point at $(3, 0)$ because the lowest frequency singularity is the pole at $s = -1$.

Changing the gain, K, will increase or decrease the magnitude of the Nyquist plot but it can never cross the negative real axis. Therefore, the closed-loop system will always be stable for positive K. Exercise: Verify this result by making a rough root locus sketch by hand.

---

[11] The shape of this Nyquist plot is a limaçon, a fact pointed out by the third author's 17-year-old son, who had recently learned about them in his 10th grade trigonometry class. Limaçon means "snail" in French from the Latin "limax," and was first investigated by Dürer in 1525.

6.4 Stability Margins

A large fraction of control system designs behave in a pattern roughly similar to that of the system in Section 6.2 and Example 6.9 in Section 6.3; that is, the system is stable for all small gain values and becomes unstable if the gain increases past a certain critical point. Two commonly used quantities that measure the stability margin for such systems are directly related to the stability criterion of Eq. (6.25): gain margin and phase margin. In this section we will define and use these two concepts to study system design. Another measure of stability, originally defined by O. J. M. Smith (1958), combines these two margins into one and gives a better indication of stability for complicated cases.

Gain margin

The **gain margin (GM)** is the factor by which the gain can be raised before instability results. For the typical case, it can be read directly from the Bode plot (for example, see Fig. 6.15) by measuring the vertical distance between the $|KG(j\omega)|$ curve and the $|KG(j\omega)| = 1$ line at the frequency where $\angle G(j\omega) = 180°$. We see from the figure that when $K = 0.1$, the system is stable and GM $= 20$ (or 26 db). When $K = 2$, the system is neutrally stable with GM $= 1$ (0 db), while $K = 10$ results in an unstable system with GM $= 0.2$ (-14 db). Note that GM is the *factor* by which the gain K can be raised before instability results; therefore, $|GM| < 1$ (or $|GM| < 0$ db) indicates an unstable system. The GM can also be determined from a root locus with respect to K by noting two values of K: (1) at the point where the locus crosses the $j\omega$-axis, and (2) at the nominal closed-loop poles. The GM is the ratio of these two values.

Phase margin

Another measure that is used to indicate the stability margin in a system is the **phase margin (PM)**. It is the amount by which the phase of $G(j\omega)$ exceeds $-180°$ when $|KG(j\omega)| = 1$, which is an alternative way of measuring the degree to which the stability conditions of Eq. (6.25) are met. For the case in Fig. 6.15, we see that PM $\cong 80°$ for $K = 0.1$, PM $= 0°$ for $K = 2$, and PM $= -35°$ for $K = 10$. A positive PM is required for stability.

Note that the two stability measures, PM and GM, together determine how far the complex quantity $G(j\omega)$ passes from the -1 point, which is another way of stating the neutral-stability point specified by Eq. (6.24).

The stability margins may also be defined in terms of the Nyquist plot. Figure 6.34 shows that GM and PM are measures of how close the Nyquist plot comes to encircling the -1 point. Again we can see that the GM indicates how much the gain can be raised before instability results in a system like the one in Example 6.9. The PM is

Figure 6.34

Nyquist plot for defining GM and PM

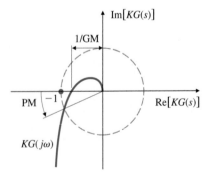

the difference between the phase of $G(j\omega)$ and 180° when $KG(j\omega)$ crosses the circle $|KG(s)| = 1$; the positive value of PM is assigned to the stable case (i.e., with no Nyquist encirclements).

It is easier to determine these margins directly from the Bode plot than from the Nyquist plot. The term **crossover frequency**, ω_c, is often used to refer to the frequency at which the gain is unity, or 0 db. Figure 6.35 shows the same data plotted in Fig. 6.25, but for the case with $K = 1$. The same values of PM (= 22°) and GM (= 2) may be obtained from the Nyquist plot shown in Fig. 6.26. The real-axis crossing at −0.5 corresponds to a GM of $1/0.5$ or 2 and the PM could be computed graphically by measuring the angle of $G(j\omega)$ as it crosses $|G(j\omega)| = 1$ circle.

One of the useful aspects of frequency-response design is the ease with which we can evaluate the effects of gain changes. In fact, we can determine the PM from Fig. 6.35 for any value of K without redrawing the magnitude or phase information. We need only indicate on the figure where $|KG(j\omega)| = 1$ for selected trial values of K, as has been done with dashed lines in Fig. 6.36. Now we can see that $K = 5$ yields an unstable PM of −22°, while a gain of $K = 0.5$ yields a PM of +45°. Furthermore, if we wish a certain PM (say 70°), we simply read the value of $|G(j\omega)|$ corresponding

Figure 6.35

GM and PM from the magnitude and phase plots

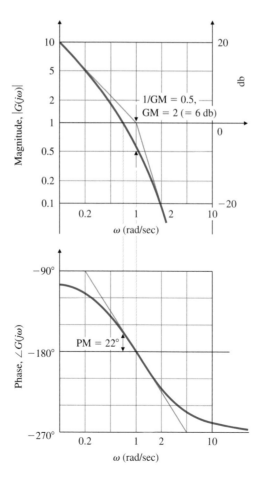

Figure 6.36

PM versus K from the frequency-response data

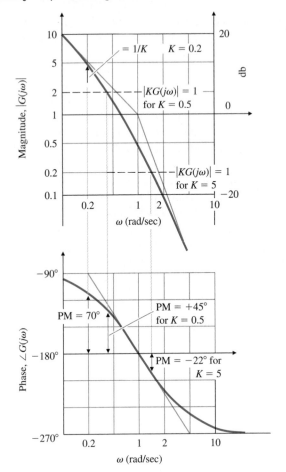

to the frequency that would create the desired PM (here $\omega = 0.2$ rad/sec yields 70°, where $|G(j\omega)| = 5$), and note that the magnitude at this frequency is $1/K$. Therefore, a PM of 70° will be achieved with $K = 0.2$.

The PM is more commonly used to specify control system performance because it is most closely related to the damping ratio of the system. This can be seen for the open-loop second-order system

$$G(s) = \frac{\omega_n^2}{s(s + 2\zeta\omega_n)}, \tag{6.29}$$

which, with unity feedback, produces the closed-loop system

$$T(s) = \frac{\omega_n^2}{s^2 + 2\zeta\omega_n s + \omega_n^2}. \tag{6.30}$$

It can be shown that the relationship between the PM and ζ in this system is

$$PM = \tan^{-1}\left[\frac{2\zeta}{\sqrt{\sqrt{1 + 4\zeta^4} - 2\zeta^2}}\right], \tag{6.31}$$

and this function is plotted in Fig. 6.37. Note that the function is approximately a straight line up to about PM $= 60°$. The dashed line shows a straight-line approximation to the function, where

$$\zeta \cong \frac{\text{PM}}{100}. \tag{6.32}$$

It is clear that the approximation holds only for PM below about 70°. Furthermore, Eq. (6.31) is only accurate for the second-order system of Eq. (6.30). In spite of these limitations, Eq. (6.32) is often used as a rule of thumb for relating the closed-loop damping ratio to PM. It is useful as a starting point; however, it is important always to check the actual damping of a design, as well as other aspects of the performance, before calling the design complete.

The gain margin for the second-order system [given by Eq. (6.29)] is infinite (GM $= \infty$), because the phase curve does not cross $-180°$ as the frequency increases. This would also be true for any first- or second-order system.

Additional data to aid in evaluating a control system based on its PM can be derived from the relationship between the resonant peak M_r and ζ seen in Fig. 6.3. Note that this figure was derived for the same system [Eq. (6.9)] as Eq. (6.30). We can convert the information in Fig. 6.37 into a form relating M_r to the PM. This is depicted in Fig. 6.38, along with the step-response overshoot M_p. Therefore, we see that, given the PM, one can infer information about what the overshoot of the closed-loop step response would be.

Figure 6.37

Damping ratio versus PM

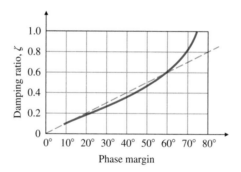

Figure 6.38

Transient-response overshoot (M_p) and frequency-response resonant peak (M_r) versus PM for
$$T(s) = \frac{\omega_n^2}{(s^2 + 2\zeta\omega_n s + \omega_n^2)}$$

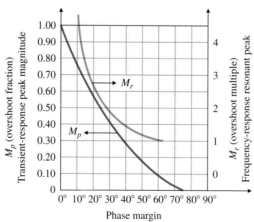

Many engineers think directly in terms of the PM when judging whether a control system is adequately stabilized. In these terms, a PM = 30° is often judged to be the lowest adequate value. In addition to testing the stability of a system design using the PM, a designer would typically also be concerned with meeting a speed-of-response specification such as bandwidth, as discussed in Section 6.1. In terms of the frequency-response parameters discussed so far, the crossover frequency would best describe a system's speed of response. This idea will be discussed further in Sections 6.6 and 6.7.

In some cases the PM and GM are not helpful indicators of stability. For first- and second-order systems, the phase never crosses the 180° line; hence, the GM is always ∞ and not a useful design parameter. For higher-order systems it is possible to have more than one frequency where $|KG(j\omega)| = 1$ or where $\angle KG(j\omega) = 180°$, and the margins as previously defined need clarification. An example of this can be seen in Fig. 10.12, where the magnitude crosses 1 three times. A decision was made to define PM by the first crossing, because the PM at this crossing was the smallest of the three values and thus the most conservative assessment of stability. A Nyquist plot based on the data in Fig. 10.12 would show that the portion of the Nyquist curve closest to the −1 point was the critical indicator of stability, and therefore use of the crossover frequency yielding the minimum value of PM was the logical choice. At best, a designer needs to be judicious when applying the margin definitions described in Fig. 6.34. In fact, the actual stability margin of a system can be rigorously assessed only by examining the Nyquist plot to determine its closest approach to the −1 point.

Vector margin

To aid in this analysis, O. J. M. Smith (1958) introduced the **vector margin**, which he defined to be the distance to the −1 point from the closest approach of the Nyquist plot.[12] Figure 6.39 illustrates the idea graphically. Because the vector margin is a single margin parameter, it removes all the ambiguities in assessing stability that come with using GM and PM in combination. In the past it has not been used extensively due to difficulties in computing it. However, with the widespread availability of computer aids, the idea of using the vector margin to describe the degree of stability is much more feasible.

Conditionally stable systems

There are certain practical examples in which an increase in the gain can make the system stable. As we saw in Chapter 5, these systems are called **conditionally**

Figure 6.39

Definition of the vector margin on the Nyquist plot

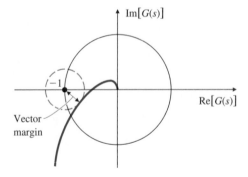

---

[12]This value is closely related to the use of the sensitivity function for design and the concept of stability robustness, to be discussed in optional Section 6.7.7.

stable. A representative root-locus plot for such systems is shown in Fig. 6.40. For a point on the root locus, such as A, an increase in the gain would make the system stable by bringing the unstable roots into the LHP. For point B, either a gain increase or decrease could make the system become unstable. Therefore, several gain margins exist that correspond to either gain reduction or gain increase, and the definition of the GM in Fig. 6.34 is not valid.

Figure 6.40

Root locus for a conditionally stable system

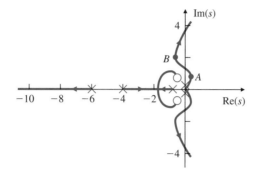

EXAMPLE 6.12

Stability Properties for a Conditionally Stable System

Determine the stability properties as a function of the gain K for the system with the open-loop transfer function

$$KG(s) = \frac{K(s+10)^2}{s^3}.$$

Solution.   This is a system for which increasing gain causes a transition from instability to stability. The root locus in Fig. 6.41(a)[13] shows that the system is unstable

Figure 6.41

System in which increasing gain leads from instability to stability: (a) root locus; (b) Nyquist plot

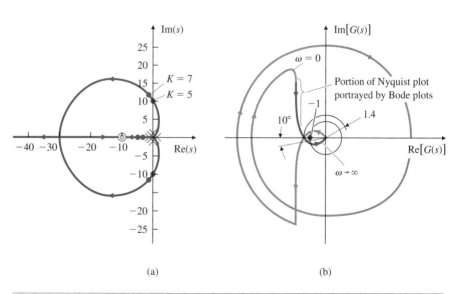

(a)   (b)

---

[13]The shape of this root locus is the plane curve limaçon.

for $K < 5$ and stable for $K > 5$. The Nyquist plot in Fig. 6.41(b) was drawn for the stable value $K = 7$. Determination of the margins according to Fig. 6.34 yields PM $= +10°$ (stable) and GM $= 0.7$ (unstable). According to the rules for stability discussed earlier, these two margins yield conflicting signals on the system's stability.

We resolve the conflict by counting the Nyquist encirclements in Fig. 6.41(b). There is one clockwise encirclement and one counterclockwise encirclement of the -1 point. Hence there are no net encirclements, which confirms that the system is stable for $K = 7$. For systems like this it is best to resort to the root locus and/or Nyquist plot (rather than the Bode plot) to determine stability.

EXAMPLE 6.13   *Nyquist Plot for a System with Multiple Crossover Frequencies*

Draw the Nyquist plot for the system

$$G(s) = \frac{85(s+1)(s^2+2s+43.25)}{s^2(s^2+2s+82)(s^2+2s+101)}$$

$$= \frac{85(s+1)(s+1\pm6.5j)}{s^2(s+1\pm9j)(s+1\pm10j)},$$

and determine the stability margins.

Solution.   The Nyquist plot (Fig. 6.42) shows that there are three crossover frequencies ($\omega = 0.75$, 9.0, and 10.1 rad/sec) with three corresponding PM values of $37°$, $80°$, and $40°$, respectively. However, the key indicator of stability is the proximity of the Nyquist plot as it approaches the -1 point while crossing the real-axis. In this case, only the GM indicates the poor stability margins of this system. The Bode plot for this system (Fig. 6.43) shows the same three crossings of magnitude $= 1$ at 0.75, 9.0, and 10.1 rad/sec. The GM value of 1.26 from the Bode plot corresponding to $\omega = 10.4$ rad/sec qualitatively agrees with the GM from the Nyquist plot and would be the most useful and unambiguous margin for this example.

In summary, many systems behave roughly like Example 6.9, and for them, the GM and PM are well defined and useful. There are also frequent instances of

Figure 6.42

Nyquist plot of the complex system in Example 6.13

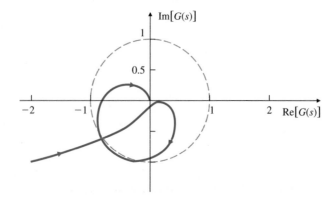

Figure 6.43

Bode plot of the system in Example 6.13

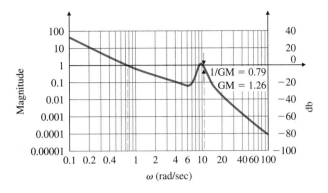

(a)

(b)

more complicated systems with multiple magnitude 1 crossovers or unstable open-loop systems for which the stability criteria defined by Fig. 6.34 are ambiguous or incorrect; therefore, we need to verify the GM and PM as previously defined, and/or modify them by reverting back to the Nyquist stability criterion.

6.5 Bode's Gain–Phase Relationship

One of Bode's important contributions is the following theorem:

Bode's theorem

For any stable minimum-phase system (i.e., one with no RHP zeros or poles), the phase of $G(j\omega)$ is uniquely related to the magnitude of $G(j\omega)$.

When the slope of $|G(j\omega)|$ versus ω on a log–log scale persists at a constant value for approximately a decade of frequency, the relationship is particularly simple and is given by

$$\angle G(j\omega) \cong n \times 90°, \tag{6.33}$$

where n is the slope of $|G(j\omega)|$ in units of decade of amplitude per decade of frequency. For example, in considering the magnitude curve alone in Fig. 6.44, we see that Eq. (6.33) can be applied to the two frequencies $\omega_1 = 0.1$ (where $n = -2$) and

Figure 6.44

An approximate gain–phase relationship demonstration

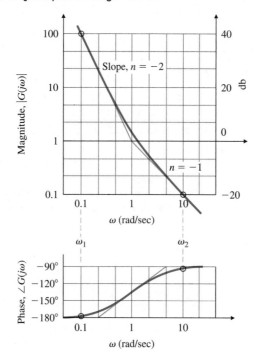

$\omega_2 = 10$ (where $n = -1$), which are a decade removed from the change in slope, to yield the approximate values of phase, $-180°$ and $-90°$. The exact phase curve shown in the figure verifies that indeed the approximation is quite good. It also shows that the approximation will degrade if the evaluation is performed at frequencies closer to the change in slope.

An exact statement of the Bode gain–phase theorem is

$$\angle G(j\omega_o) = \frac{1}{\pi} \int_{-\infty}^{+\infty} \left(\frac{dM}{du} \right) W(u)\, du \quad \text{(in radians)}, \qquad (6.34)$$

where

$$M = \log \text{ magnitude} = \ln |G(j\omega)|,$$

$$u = \text{normalized frequency} = \ln(\omega/\omega_o),$$

$$dM/du \cong \text{slope } n, \text{ as defined in Eq. (6.33)},$$

$$W(u) = \text{weighting function} = \ln(\coth|u|/2).$$

Figure 6.45 is a plot of the weighting function $W(u)$ and shows how the phase is most dependent on the slope at ω_o; it is also dependent, though to a lesser degree, on slopes at neighboring frequencies. The figure also suggests that the weighting could be approximated by an impulse function centered at ω_o. We may approximate the weighting function as

$$W(u) \cong \frac{\pi^2}{2} \delta(u),$$

which is precisely the approximation made to arrive at Eq. (6.33) using the "sifting" property of the impulse function (and conversion from radians to degrees).

Figure 6.45

Weighting function in Bode's gain–phase theorem

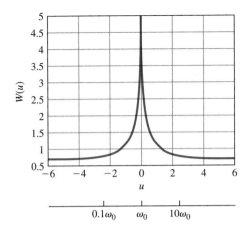

In practice, Eq. (6.34) is never used, but Eq. (6.33) *is* used as a guide to infer stability from $|G(\omega)|$ alone. When $|KG(j\omega)| = 1$,

$$\angle G(j\omega) \cong -90° \quad \text{if } n = -1,$$

$$\angle G(j\omega) \cong -180° \quad \text{if } n = -2.$$

For stability we want $\angle G(j\omega) > -180°$ for PM > 0. Therefore, we adjust the $|KG(j\omega)|$ curve so that it has a slope of -1 at the "crossover" frequency, ω_c, (i.e., where $|KG(j\omega)| = 1$). If the slope is -1 for a decade above and below the crossover frequency, then PM $\cong 90°$; however, to ensure a reasonable PM, it is usually necessary only to insist that a -1 slope (-20 db per decade) persist for a decade in frequency that is centered at the crossover frequency. We therefore see that there is a very simple design criterion:

Crossover frequency

Crossover at -1 slope

> Adjust the slope of the magnitude curve $|KG(j\omega)|$ so that it crosses over magnitude 1 with a slope of -1 for a decade around ω_c.

This criterion will usually be sufficient to provide an acceptable PM, and hence provide adequate system damping. To achieve the desired speed of response, the system gain is adjusted so that the crossover point is at a frequency that will yield the desired bandwidth or speed of response as determined by Eq. (3.60). Recall that the natural frequency ω_n, bandwidth, and crossover frequency are all approximately equal, as will be discussed further in Section 6.6.

EXAMPLE 6.14

Use of Simple Design Criterion for Spacecraft Attitude Control

For the spacecraft attitude-control problem defined in Fig. 6.46, find a suitable expression for $KD(s)$ that will provide good damping and a bandwidth of approximately 0.2 rad/sec. Also determine the value of the sensitivity function, $\mathcal{S}$, at $\omega = 0.05$ rad/sec in order to evaluate the magnitude of the tracking error for a reference input at that frequency.

Figure 6.46

Spacecraft
attitude-control system

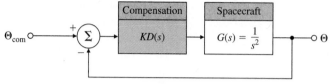

Figure 6.47

Magnitude of the
spacecraft's frequency
response

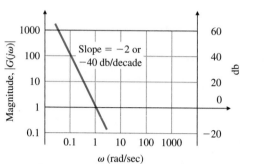

Solution. The magnitude of the frequency response of the spacecraft (Fig. 6.47) clearly requires some reshaping, because it has a slope of -2 (or -40 db per decade) everywhere. The simplest compensation to do the job consists of using proportional and derivative terms (a PD compensator), which produces the relation

$$KD(s) = K(T_D s + 1). \tag{6.35}$$

We will adjust the gain K to produce the desired bandwidth, and adjust break point $\omega_1 = 1/T_D$ to provide the -1 slope at the crossover frequency. The actual design process to achieve the desired specifications is now very simple: We pick a value of K to provide a crossover at 0.2 rad/sec and choose a value of ω_1 that is about four times lower than the crossover frequency, so that the slope will be -1 in the vicinity of the crossover. Figure 6.48 shows the steps we take to arrive at the final compensation:

1. Plot $|G(j\omega)|$.
2. Modify the plot to include $|D(j\omega)|$, with $\omega_1 = 0.05$ rad/sec ($T_D = 20$), so that the slope will be $\cong -1$ at $\omega = 0.2$ rad/sec.
3. Determine that $|DG| = 100$, where the $|DG|$ curve crosses the line $\omega = 0.2$ rad/sec, which is where we want magnitude 1 crossover to be.
4. In order for crossover to be at $\omega = 0.2$ rad/sec, compute

$$K = \frac{1}{[|DG|]_{\omega=0.2}} = \frac{1}{100} = 0.01.$$

Therefore,

$$KD(s) = 0.01(20s + 1)$$

will meet the specifications, thus completing the design.

If we were to draw the phase curve of KDG, we would find that PM $= 75°$, which is certainly quite adequate. A plot of the closed-loop frequency-response magnitude (Fig. 6.49) shows that, indeed, the crossover frequency and the bandwidth are almost identical in this case. The sensitivity function was defined by Eq. (4.17) and for this problem is

Figure 6.48

Compensated open-loop transfer function

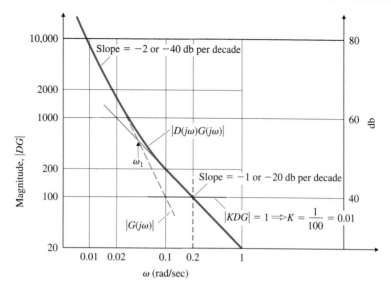

Figure 6.49

Closed-loop frequency response of $T(s)$ and $S(s)$

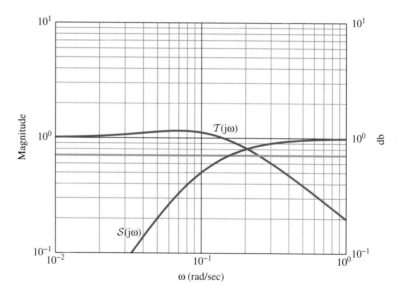

$$S = \frac{1}{1 + KDG}.$$

It is shown on the graph along with $T(s)$, the output response versus in input command. The frequency response of T confirms that the design achieved the desired bandwidth of 0.2 rad/sec, and it can also be seen that S has the value of 0.2 at $\omega = 0.05$ rad/sec. The step response of the closed-loop system is shown in Fig. 6.50 and its 14% overshoot confirms the adequate damping.

Figure 6.50

Step response for PD compensation

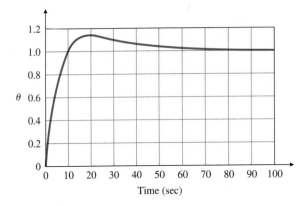

6.6 Closed-Loop Frequency Response

The closed-loop bandwidth was defined in Section 6.1 and in Fig. 6.5. Figure 6.3 showed that the natural frequency is always within a factor of two of the bandwidth for a second-order system. In Example 6.14, we designed the compensation so that the crossover frequency was at the desired bandwidth and verified by computation that the bandwidth was identical to the crossover frequency. Generally, the match between the crossover frequency and the bandwidth is not as good as in Example 6.14. We can help establish a more exact correspondence by making a few observations. Consider a system in which $|KG(j\omega)|$ shows the typical behavior

$$|KG(j\omega)| \gg 1 \quad \text{for} \quad \omega \ll \omega_c,$$

$$|KG(j\omega)| \ll 1 \quad \text{for} \quad \omega \gg \omega_c,$$

where ω_c is the crossover frequency. The closed-loop frequency-response magnitude is approximated by

$$|T(j\omega)| = \left| \frac{KG(j\omega)}{1 + KG(j\omega)} \right| \cong \begin{cases} 1, & \omega \ll \omega_c, \\ |KG|, & \omega \gg \omega_c. \end{cases} \tag{6.36}$$

In the vicinity of crossover, where $|KG(j\omega)| = 1$, $|T(j\omega)|$ depends heavily on the PM. A PM of 90° means that $\angle G(j\omega_c) = -90°$, and therefore $|T(j\omega_c)| = 0.707$. On the other hand, PM $= 45°$ yields $|T(j\omega_c)| = 1.31$.

The exact evaluation of Eq. (6.36) was used to generate the curves of $|T(j\omega)|$ in Fig. 6.51. It shows that the bandwidth for smaller values of PM is typically somewhat greater than ω_c, though usually it is less than $2\omega_c$; thus

$$\omega_c \le \omega_{BW} \le 2\omega_c.$$

Another specification related to the closed-loop frequency response is the resonant-peak magnitude M_r, defined in Fig. 6.5. Figures 6.3 and 6.38 show that, for linear systems, M_r is generally related to the damping of the system. In practice, M_r is rarely used; most designers prefer to use the PM to specify the damping of a system, because the imperfections that make systems nonlinear or cause delays usually erode the phase more significantly than the magnitude.

As demonstrated in the last example, it is also important in the design to achieve certain error characteristics and these are often evaluated as a function of the input

Figure 6.51

Closed-loop bandwidth with respect to PM

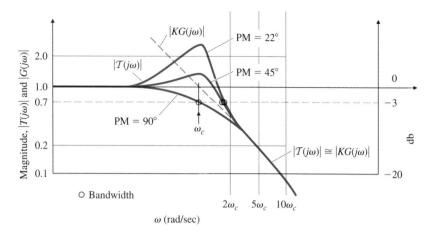

or disturbance frequency. In some cases, the primary function of the control system is to regulate the output to a certain constant input in the presence of disturbances. For these situations, the key item of interest for the design would be the closed-loop frequency response of the error with respect to disturbance inputs.

6.7 Compensation

As we discussed in Chapters 4 and 5, dynamic elements (or compensation) are typically added to feedback controllers to improve the system's stability and error characteristics because the process itself cannot be made to have acceptable characteristics with proportional feedback alone.

Section 4.3 discussed the basic types of feedback: proportional, derivative, and integral. Section 5.4 discussed three kinds of dynamic compensation: lead compensation, which approximates proportional-derivative (PD) feedback, lag compensation, which approximates proportional-integral (PI) control, and notch compensation, which has special characteristics for dealing with resonances. In this section we discuss these and other kinds of compensation in terms of their frequency-response characteristics. In most cases, the compensation will be implemented in a microprocessor. Techniques for converting the continuous compensation $D(s)$ into a form that can be coded in the computer was briefly discussed in Section 4.4. It will be illustrated further in this section and will be discussed in more detail in Chapter 8.

The frequency-response stability analysis to this point has usually considered the closed-loop system to have the characteristic equation $1 + KG(s) = 0$. With the introduction of compensation, the closed-loop characteristic equation becomes $1 + KD(s)G(s) = 0$, and all the previous discussion in this chapter pertaining to the frequency response of $KG(s)$ applies directly to the compensated case if we apply it to the frequency response of $KD(s)G(s)$. We call this quantity $L(s)$, the "loop gain," or open-loop transfer function of the system, where $L(s) = KD(s)G(s)$.

Figure 6.52

Frequency response of PD control

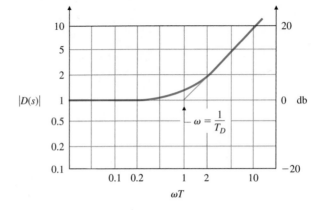

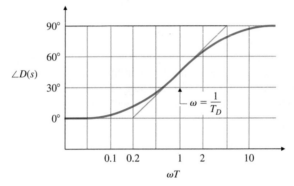

6.7.1 PD Compensation

PD compensation

We will start the discussion of compensation design by using the frequency response with PD control. The compensator transfer function, given by

$$D(s) = (T_D s + 1), \qquad (6.37)$$

was shown in Fig. 5.22 to have a stabilizing effect on the root locus of a second-order system. The frequency-response characteristics of Eq. (6.37) are shown in Fig. 6.52. A stabilizing influence is apparent by the increase in phase and the corresponding +1 slope at frequencies above the break point $1/T_D$. We use this compensation by locating $1/T_D$ so that the increased phase occurs in the vicinity of crossover (that is, where $|KD(s)G(s)| = 1$), thus increasing the PM.

Note that the magnitude of the compensation continues to grow with increasing frequency. This feature is undesirable because it amplifies the high-frequency noise that is typically present in any real system and, as a continuous transfer function, cannot be realized with physical elements. It is also the reason we stated in Section 5.4 that pure derivative compensation gives trouble.

6.7.2 Lead Compensation

In order to alleviate the high-frequency amplification of the PD compensation, a first-order pole is added in the denominator at frequencies substantially higher than the

Lead compensation

break point of the PD compensator. Thus the phase increase (or lead) still occurs, but the amplification at high frequencies is limited. The resulting **lead compensation** has a transfer function of

$$D(s) = \frac{Ts + 1}{\alpha Ts + 1}, \qquad \alpha < 1, \tag{6.38}$$

where $1/\alpha$ is the ratio between the pole/zero break-point frequencies. Figure 6.53 shows the frequency response of this lead compensation. Note that a significant amount of phase lead is still provided, but with much less amplification at high frequencies. A lead compensator is generally used whenever a substantial improvement in damping of the system is required.

The phase contributed by the lead compensation in Eq. (6.38) is given by

$$\phi = \tan^{-1}(T\omega) - \tan^{-1}(\alpha T\omega).$$

It can be shown (see Problem 6.44) that the frequency at which the phase is maximum is given by

$$\omega_{max} = \frac{1}{T\sqrt{\alpha}}. \tag{6.39}$$

The maximum phase contribution—that is, the peak of the $\angle D(s)$ curve in Fig. 6.53—corresponds to

$$\sin \phi_{max} = \frac{1 - \alpha}{1 + \alpha}, \tag{6.40}$$

or

$$\alpha = \frac{1 - \sin \phi_{max}}{1 + \sin \phi_{max}}.$$

Figure 6.53

Lead-compensation frequency response with $1/\alpha = 10$

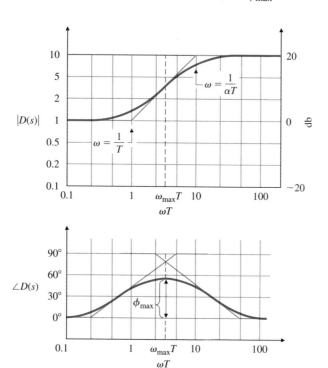

Another way to look at this is the following: The maximum phase occurs at a frequency that lies midway between the two break-point frequencies (sometimes called corner frequencies) on a logarithmic scale,

$$\log \omega_{max} = \log \frac{1/\sqrt{T}}{\sqrt{\alpha T}}$$

$$= \log \frac{1}{\sqrt{T}} + \log \frac{1}{\sqrt{\alpha T}}$$

$$= \frac{1}{2} \left[\log \left(\frac{1}{T} \right) + \log \left(\frac{1}{\alpha T} \right) \right], \quad (6.41)$$

as shown in Fig. 6.53. Alternatively, we may state these results in terms of the pole–zero locations. Rewriting $D(s)$ in the form used for root-locus analysis, we have

$$D(s) = \frac{s + z}{s + p}. \quad (6.42)$$

Problem 6.44 shows that

$$\omega_{max} = \sqrt{|z| \, |p|} \quad (6.43)$$

and

$$\log \omega_{max} = \frac{1}{2} (\log |z| + \log |p|). \quad (6.44)$$

These results agree with the previous ones if we let $z = -1/T$ and $p = -1/\alpha T$ in Eqs. (6.39) and (6.41).

For example, a lead compensator with a zero at $s = -2$ ($T = 0.5$) and a pole at $s = -10$ ($\alpha T = 0.1$) (and thus $\alpha = \frac{1}{5}$) would yield the maximum phase lead at

$$\omega_{max} = \sqrt{2 \cdot 10} = 4.47 \text{ rad/sec}.$$

Lead ratio $= \frac{1}{\alpha}$

The amount of phase lead at the midpoint depends only on α in Eq. (6.40) and is plotted in Fig. 6.54. For $\alpha = 1/5$, Fig. 6.54 shows that $\phi_{max} = 40°$. Note from the figure that we could increase the phase lead up to 90° using higher values of the **lead ratio**, $1/\alpha$; however, Fig. 6.53 shows that increasing values of $1/\alpha$ also produces higher amplifications at higher frequencies. Thus our task is to select a value of $1/\alpha$ that is a good compromise between an acceptable PM and an acceptable noise sensitivity at high frequencies. Usually the compromise suggests that a lead compensation should

Figure 6.54

Maximum phase increase for lead compensation

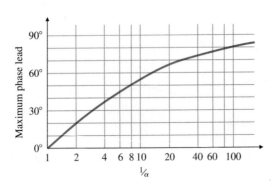

contribute a maximum of 70° to the phase. If a greater phase lead is needed, then a double-lead compensation would be suggested, where

$$D(s) = \left(\frac{Ts + 1}{\alpha Ts + 1} \right)^2.$$

Even if a system had negligible amounts of noise present and the pure derivative compensation of Eq. (6.37) were acceptable, a continuous compensation would look more like Eq. (6.38) than Eq. (6.37) because of the impossibility of building a pure differentiator. No physical system—mechanical or electrical—responds with infinite amplitude at infinite frequencies, so there will be a limit in the frequency range (or bandwidth) for which derivative information (or phase lead) can be provided. This is also true with a digital implementation. Here, the sample rate limits the high-frequency amplification and essentially places a pole in the compensation transfer function.

EXAMPLE 6.15   *Lead Compensation for a DC Motor*

As an example of designing a lead compensator, let us repeat the design of compensation for the DC motor with the transfer function

$$G(s) = \frac{1}{s(s + 1)}$$

that was carried out in Section 5.4.1. This also represents the model of a satellite tracking antenna (see Fig. 3.61). This time we wish to obtain a steady-state error of less than 0.1 for a unit-ramp input. Furthermore, we desire an overshoot $M_p < 25\%$.

1. Determine the lead compensation satisfying the specifications.
2. Determine the digital version of the compensation with $T_s = 0.05$ sec.
3. Compare the step and ramp responses of both implementations.

Solution

1. The steady-state error is given by

$$e_{ss} = \lim_{s \to 0} s \left[\frac{1}{1 + KD(s)G(s)} \right] R(s), \tag{6.45}$$

where $R(s) = 1/s^2$ for a unit ramp, so Eq. (6.45) reduces to

$$e_{ss} = \lim_{s \to 0} \left\{ \frac{1}{s + KD(s)[1/(s + 1)]} \right\} = \frac{1}{KD(0)}.$$

Therefore, we find that $KD(0)$, the steady-state gain of the compensation, cannot be less than 10 ($K_v \geq 10$) if it is to meet the error criterion, so we pick $K = 10$. To relate the overshoot requirement to PM, Fig. 6.38 shows that a PM of 45° should suffice. The frequency response of $KG(s)$ in Fig. 6.55 shows that the PM = 20° if no phase lead is added by compensation. If it were possible to simply add phase without affecting the magnitude, we would need an additional phase of only 25° at the $KG(s)$ crossover frequency of $\omega = 3$ rad/sec. However, maintaining the same low-frequency gain and adding a compensator zero would increase the crossover frequency; hence more than a 25° phase contribution will

Figure 6.55

Frequency response for lead-compensation design

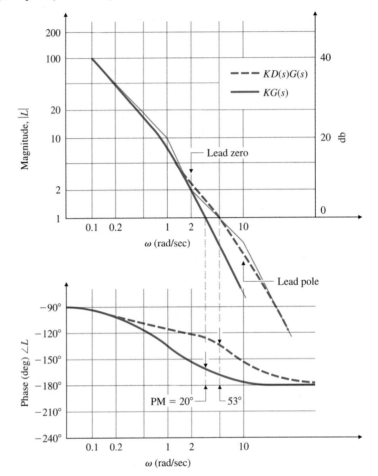

be required from the lead compensation. To be safe, we will design the lead compensator so that it supplies a maximum phase lead of 40°. Fig. 6.54 shows that $1/\alpha = 5$ will accomplish that goal. We will derive the greatest benefit from the compensation if the maximum phase lead from the compensator occurs at the crossover frequency. With some trial and error, we determine that placing the zero at $\omega = 2$ rad/sec and the pole at $\omega = 10$ rad/sec causes the maximum phase lead to be at the crossover frequency. The compensation, therefore, is

$$KD(s) = 10\frac{s/2 + 1}{s/10 + 1}.$$

The frequency-response characteristics of $L(s) = KD(s)G(s)$ in Fig. 6.55 can be seen to yield a PM of 53°, which satisfies the design goals.

The root locus for this design, originally given as Fig. 5.24, is repeated here as Fig. 6.56, with the root locations marked for $K = 10$. The locus is not needed for the frequency-response design procedure; it is presented here only for comparison with the root locus design method presented in Chapter 5. The entire process can be expedited by the use of MATLAB's SISOTOOL routine, which

Figure 6.56

Root locus for lead
compensation design

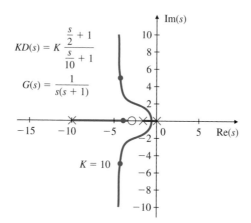

$$KD(s) = K \frac{\frac{s}{2} + 1}{\frac{s}{10} + 1}$$

$$G(s) = \frac{1}{s(s+1)}$$

simultaneously provides the root locus and the Bode plot through an interactive GUI interface. For this example, the MATLAB statements

```
G=tf(1,[1 1 0]);
D=tf(10*[1/2 1],[1/10 1]);
sisotool(G,D)
```

will provide the plots as shown in Fig. 6.57. It can also be used to generate the Nyquist and time-response plots if desired.

2. To find the discrete equivalent of $D(s)$, we use the trapezoidal rule given by Eq. (4.104). That is,

$$D_d(z) = \frac{\frac{2}{T_s} \frac{z-1}{z+1}/2 + 1}{\frac{2}{T_s} \frac{z-1}{z+1}/10 + 1}, \tag{6.46}$$

which, with $T_s = 0.05$ sec, reduces to

$$D_d(z) = \frac{4.2z - 3.8}{z - 0.6}. \tag{6.47}$$

This same result can be obtained by the MATLAB statement

```
sysD = tf([0.5 1],[0.1 1]);
sysDd = c2d(sysD, 0.05, 'tustin').
```

Because

$$\frac{U(z)}{E(z)} = KD_d(z), \tag{6.48}$$

the discrete control equation that results is

$$u(k + 1) = 0.6u(k) + 10(4.2e(k + 1) - 3.8e(k)). \tag{6.49}$$

3. The SIMULINK® block diagram of the continuous and discrete versions of $D(s)$ controlling the DC motor is shown in Fig. 6.58. The step responses of the two controllers are plotted together in Fig. 6.59(a) and are reasonably close to one another; however, the discrete controller does exhibit slightly increased overshoot, as is often the case. Both overshoots are less than 25%, and thus

meet the specifications. The ramp responses of the two controllers, shown in Fig. 6.59(b), are essentially identical, and both meet the 0.1 specified error.

Figure 6.57
SISOTOOL graphical user interface for Example 6.15

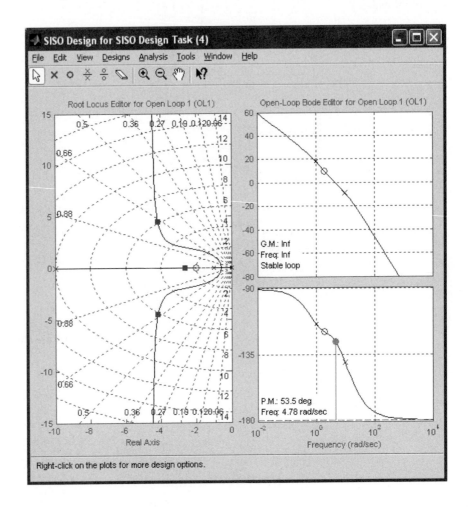

Figure 6.58
SIMULINK® block diagram for transient response of lead-compensation design

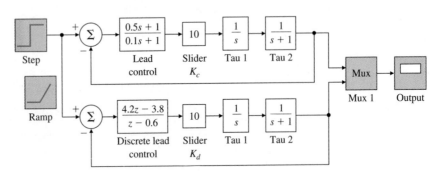

Figure 6.59

Lead-compensation design: (a) step response; (b) ramp response

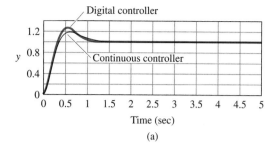

(a)

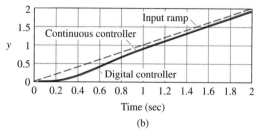

(b)

The design procedure used in Example 6.15 can be summarized as follows:

1. Determine the low-frequency gain so that the steady-state errors are within specification.
2. Select the combination of lead ratio $1/\alpha$ and zero values ($1/T$) that achieves an acceptable PM at crossover.
3. The pole location is then at ($1/\alpha T$).

This design procedure will apply to many cases; however, keep in mind that the specific procedure followed in any particular design may need to be tailored to its particular set of specifications.

In Example 6.15 there were two specifications: peak overshoot and steady-state error. We transformed the overshoot specification into a PM, but the steady-state error specification we used directly. No speed-of-response type of specification was given; however, it would have impacted the design in the same way that the steady-state error specification did. The speed of response or bandwidth of a system is directly related to the crossover frequency, as we pointed out earlier in Section 6.6. Figure 6.55 shows that the crossover frequency was ~ 5 rad/sec. We could have increased it by raising the gain K and increasing the frequency of the lead compensator pole and zero in order to keep the slope of -1 at the crossover frequency. Raising the gain would also have decreased the steady-state error to be better than the specified limit. The gain margin was never introduced into the problem because the stability was adequately specified by the PM alone. Furthermore, the gain margin would not have been useful for this system because the phase never crossed the 180° line and the GM was always infinite.

Design parameters for lead networks

In lead-compensation designs there are three primary design parameters:

1. The crossover frequency ω_c, which determines bandwidth ω_{BW}, rise time t_r, and settling time t_s;
2. The PM, which determines the damping coefficient ζ and the overshoot M_p;
3. The low-frequency gain, which determines the steady-state error characteristics.

The design problem is to find the best values for the parameters, given the requirements. In essence, lead compensation increases the value of $\omega_c/L(0)$ ($=\omega_c/K_v$ for a Type 1 system). That means that, if the low-frequency gain is kept the same, the crossover frequency will increase. Or if the crossover frequency is kept the same, the low-frequency gain will decrease. Keeping this interaction in mind, the designer can assume a fixed value of one of these three design parameters and then adjust the other two iteratively until the specifications are met. One approach is to set the low-frequency gain to meet the error specifications and add a lead compensator to increase PM at the crossover frequency. An alternative is to pick the crossover frequency to meet a time response specification, then adjust the gain and lead characteristics so that the PM specification is met. A step-by-step procedure is outlined next for these two cases. They apply to a sizable class of problems for which a single lead is sufficient. As with all such design procedures, it provides only a starting point; the designer will typically find it necessary to go through several design iterations in order to meet all the specifications.

Design Procedure for Lead Compensation

1. Determine open-loop gain K to satisfy error or bandwidth requirements:

   (a) to meet error requirement, pick K to satisfy error constants (K_p, K_v, or K_a) so that e_{ss} error specification is met, or alternatively,

   (b) to meet bandwidth requirement, pick K so that the open-loop crossover frequency is a factor of two below the desired closed-loop bandwidth.

2. Evaluate the PM of the uncompensated system using the value of K obtained from Step 1.

3. Allow for extra margin (about $10°$), and determine the needed phase lead ϕ_{max}.

4. Determine α from Eq. (6.40) or Fig. 6.54.

5. Pick ω_{max} to be at the crossover frequency; thus the zero is at $1/T = \omega_{max}\sqrt{\alpha}$ and the pole is at $1/\alpha T = \omega_{max}/\sqrt{\alpha}$.

6. Draw the compensated frequency response and check the PM.

7. Iterate on the design. Adjust compensator parameters (poles, zeros, and gain) until all specifications are met. Add an additional lead compensator (that is, a double-lead compensation) if necessary.

While these guidelines will not apply to all the systems you will encounter in practice, they do suggest a systematic trial-and-error process to search for a satisfactory compensator that will usually be successful.

EXAMPLE 6.16   *Lead Compensator for a Temperature Control System*

The third-order system

$$KG(s) = \frac{K}{(s/0.5 + 1)(s + 1)(s/2 + 1)}$$

is representative of a typical temperature control system. Design a lead compensator such that $K_p = 9$ and the PM is at least $25°$.

Solution.   Let us follow the design procedure:

1. Given the specification for K_p, we solve for K:

$$K_p = \lim_{s \to 0} KG(s) = K = 9.$$

2. The Bode plot of the uncompensated system, $KG(s)$, with $K = 9$ can be created by the MATLAB statements below and is shown in Fig. 6.60 along with the two compensated cases.

```
numG = 9;
den2 = conv([2 1],[1 1]);
denG = conv(den2,[0.5 1]);
sysG = tf(numG,denG);
w=logspace(-1,1);
[mag,phas] = bode(sysG,w);
loglog(w,squeeze(mag),grid;
semilogx(w,squeeze(phase),grid;
```

It is difficult to read the PM and crossover frequencies accurately from the Bode plots; therefore, the MATLAB command

```
[GM,PM,Wcg,Wcp] = margin(mag,phas,w)
```

can be invoked. The quantity PM is the phase margin and Wcp is the frequency at which the gain crosses magnitude 1. (GM and Wcg are the open-loop gain margin and the frequency at which the phase crosses 180.) For this example, the output is

GM =1.25, PM = 7.12, Wcg = 1.87, Wcp = 1.68,

which says that the PM of the uncompensated system is 7° and that this occurs at a crossover frequency of 1.7 rad/sec.

3. Allowing for 10° of extra margin, we want the lead compensator to contribute $25° + 10° - 7° = 28°$ at the crossover frequency. The extra margin is typically

Figure 6.60

Bode plot for the lead-compensation design in Example 6.16

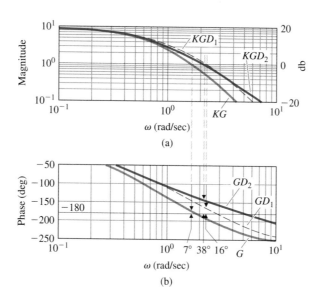

required because the lead will increase the crossover frequency from the open-loop case, at which point more phase increase will be required.

4. From Fig. 6.54 we see that $\alpha = 1/3$ will produce approximately 30° phase increase midway between the zero and pole.

5. As a first cut, let's place the zero at 1 rad/sec ($T = 1$) and the pole at 3 rad/sec ($\alpha T = 1/3$), thus bracketing the open-loop crossover frequency and preserving the factor of 3 between pole and zero, as indicated by $\alpha = 1/3$. The lead compensator is

$$D_1(s) = \frac{s+1}{s/3+1} = \frac{1}{0.333}\left(\frac{s+1}{s+3}\right).$$

6. The Bode plot of the system with $D_1(s)$ (Fig. 6.60, middle curve) has a PM of 16°. We did not achieve the desired PM of 30°, because the lead shifted the crossover frequency from 1.7 rad/sec to 2.3 rad/sec, thus increasing the required phase increase from the lead. The step response of the system with $D_1(s)$ (Fig. 6.61) shows a very oscillatory response, as we might expect from the low PM of 16°.

7. We repeat the design with extra phase increase and move the zero location slightly to the right so that the crossover frequency won't be shifted so much. We choose $\alpha = 1/10$ with the zero at $s = -1.5$, so

$$D_2(s) = \frac{s/1.5+1}{s/15+1} = \frac{1}{0.1}\left(\frac{s+1.5}{s+15}\right).$$

This compensation produces a PM = 38°, and the crossover frequency lowered slightly to 2.2 rad/sec. Figure 6.60 (upper curve) shows the frequency response of the revised design. Figure 6.61 shows a substantial reduction in the oscillations, which you should expect from the higher PM value.

Figure 6.61

Step response for lead-compensation design

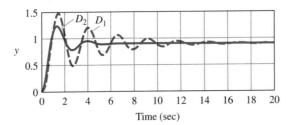

Time (sec)

EXAMPLE 6.17   *Lead-Compensator Design for a Type 1 Servomechanism System*

Consider the third-order system

$$KG(s) = K\frac{10}{s(s/2.5+1)(s/6+1)}.$$

This type of system would result for a DC motor with a lag in the shaft position sensor. Design a lead compensator so that the PM = 45° and $K_v = 10$.

Solution.   Again, we follow the design procedure given earlier:

1. As given, $KG(s)$ will yield $K_v = 10$ if $K = 1$. Therefore, the K_v requirement is met by $K = 1$ and the low-frequency gain of the compensation should be 1.

2. The Bode plot of the system is shown in Fig. 6.62. The PM of the uncompensated system (lower curve) is approximately $-4°$, and the crossover frequency is at $\omega_c \cong 4$ rad/sec.

3. Allowing for $5°$ of extra PM, we need PM $= 45° + 5° - (-4°) = 54°$ to be contributed by the lead compensator.

4. From Fig. 6.54 we find that α must be 0.1 to achieve a maximum phase lead of $54°$.

5. The new gain crossover frequency will be higher than the open-loop value of $\omega_c = 4$ rad/sec, so let's select the pole and zero of the lead compensation to be at 20 and 2 rad/sec, respectively. So the candidate compensator is

$$D_1(s) = \frac{s/2 + 1}{s/20 + 1} = \frac{1}{0.1} \frac{s + 2}{s + 20}.$$

6. The Bode plot of the compensated system (Fig. 6.62, middle curve) shows a PM of $23°$. Further iteration will show that a single-lead compensator cannot meet the specification because of the high-frequency slope of -3.

7. We need a double-lead compensator in this system. If we try a compensator of the form

$$D_2(s) = \frac{1}{(0.1)^2} \frac{(s + 2)(s + 4)}{(s + 20)(s + 40)} = \frac{(s/2 + 1)(s/4 + 1)}{(s/20 + 1)(s/40 + 1)},$$

we obtain PM $= 46°$. The Bode plot for this case is shown as the upper curve in Fig. 6.62.

Figure 6.62

Bode plot for the lead-compensation design in Example 6.17

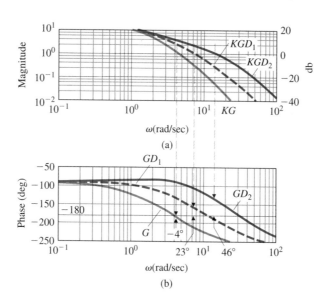

Both Examples 6.16 and 6.17 are third order. Example 6.17 was more difficult to design compensation for, because the error requirement, K_v, forced the crossover frequency, ω_c, to be so high that a single lead could not provide enough PM.

Figure 6.63

Frequency response
of PI control

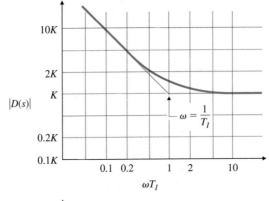

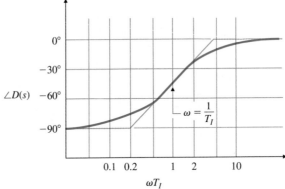

6.7.3 PI Compensation

PI compensation

In many problems it is important to keep the bandwidth low and also to reduce the steady-state error. For this purpose, a proportional-integral (PI) or lag compensator is useful. From Eq. (4.65), we see that PI control has the transfer function

$$D(s) = \frac{K}{s}\left(s + \frac{1}{T_I}\right),\tag{6.50}$$

which results in the frequency-response characteristics shown in Fig. 6.63. The desirable aspect of this compensation is the infinite gain at zero frequency, which reduces the steady-state errors. This is accomplished, however, at the cost of a phase decrease at frequencies lower than the break point at $\omega = 1/T_I$. Therefore, $1/T_I$ is usually located at a frequency substantially less than the crossover frequency so that the system's PM is not affected significantly.

6.7.4 Lag Compensation

Lag compensation

As we discussed in Section 5.4, **lag compensation** approximates PI control. Its transfer function was given by Eq. (5.72) for root-locus design, but for frequency-response design, it is more convenient to write the transfer function of the lag compensation *alone* in the Bode form

$$D(s) = \alpha\frac{Ts + 1}{\alpha Ts + 1}, \qquad \alpha > 1,\tag{6.51}$$

Figure 6.64

Frequency response of lag compensation with $\alpha = 10$

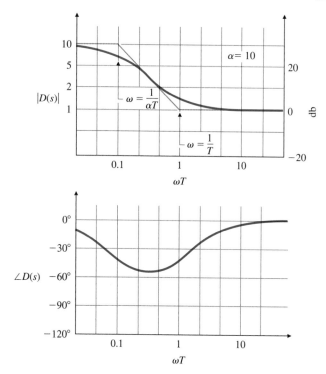

where α is the ratio between the zero/pole break-point frequencies. The complete controller will almost always include an overall gain K and perhaps other dynamics in addition to the lag compensation. Although Eq. (6.51) looks very similar to the lead compensation in Eq. (6.38), the fact is that $\alpha > 1$ causes the pole to have a lower break-point frequency than the zero. This relationship produces the low-frequency increase in amplitude and phase decrease (lag) apparent in the frequency-response plot in Fig. 6.64 and gives the compensation the essential feature of integral control—an increased low-frequency gain. The typical objective of lag-compensation design is to provide additional gain of α in the low-frequency range and to leave the system sufficient PM. Of course, phase lag is not a useful effect, and the pole and zero of the lag compensator are selected to be at much lower frequencies than the uncompensated system crossover frequency in order to keep the effect on the PM to a minimum. Thus, the lag compensator increases the open-loop DC gain, thereby improving the steady-state response characteristics, without changing the transient response characteristics significantly. If the pole and zero are relatively close together and near the origin (that is, if the value of T is large), we can increase the low-frequency gain (and thus K_p, K_v, or K_a) by a factor α without moving the closed-loop poles appreciably. Hence, the transient response remains approximately the same while the steady-state response is improved.

We now summarize a step-by-step procedure for lag-compensator design.

Design Procedure for Lag Compensation

1. Determine the open-loop gain K that will meet the PM requirement without compensation.

2. Draw the Bode plot of the uncompensated system with crossover frequency from Step 1, and evaluate the low-frequency gain.
3. Determine α to meet the low-frequency gain error requirement.
4. Choose the corner frequency $\omega = 1/T$ (the zero of the lag compensator) to be one octave to one decade below the new crossover frequency ω_c.
5. The other corner frequency (the pole location of the lag compensator) is then $\omega = 1/\alpha T$.
6. Iterate on the design. Adjust compensator parameters (poles, zeros, and gain) to meet all the specifications.

EXAMPLE 6.18

Lag-Compensator Design for Temperature Control System

Again consider the third-order system of Example 6.16:

$$KG(s) = \frac{K}{\left(\frac{1}{0.5}s + 1\right)(s + 1)\left(\frac{1}{2}s + 1\right)}.$$

Design a lag compensator so the PM is at least $40°$ and $K_p = 9$.

Solution. We follow the design procedure previously enumerated.

1. From the open-loop plot of $KG(s)$, shown for $K = 9$ in Fig. 6.60, it can be seen that a PM $> 40°$ will be achieved if the crossover frequency $\omega_c \lesssim 1$ rad/sec. This will be the case if $K = 3$. So we pick $K = 3$ in order to meet the PM specification.
2. The Bode plot of $KG(s)$ in Fig. 6.65 with $K = 3$ shows that the PM is $\approx 50°$ and the low-frequency gain is now 3. Exact calculation of the PM using MATLAB's margin shows that PM $= 53°$.

Figure 6.65

Frequency response of lag-compensation design in Example 6.18

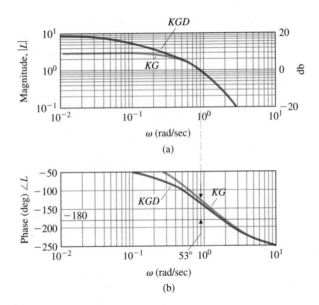

(a)

(b)

Figure 6.66

Step response of
lag-compensation
design in Example 6.18

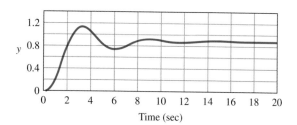

3. The low frequency gain should be raised by a factor of 3, which means the lag compensation needs to have $\alpha = 3$.
4. We choose the corner frequency for the zero to be approximately a factor of 5 slower than the expected crossover frequency—that is, at 0.2 rad/sec. So, $1/T = 0.2$, or $T = 5$.
5. We then have the value for the other corner frequency: $\omega = 1/\alpha T = \frac{1}{(3)(5)} = 1/15$ rad/sec. The compensator is thus

$$D(s) = 3\frac{5s+1}{15s+1}.$$

The compensated frequency response is also shown in Fig. 6.65. The low-frequency gain of $KD(0)G(0) = 3K = 9$, thus $K_p = 9$ and the PM lowers slightly to 44°, which satisfies the specifications. The step response of the system, shown in Fig. 6.66, illustrates the reasonable damping that we would expect from PM $= 44°$.
6. No iteration is required in this case.

Note that Examples 6.16 and 6.18 are both for the same plant, and both had the same steady-state error requirement. One was compensated with lead and one was compensated with lag. The result is that the bandwidth of the lead-compensated design is higher than that for the lag-compensated design by approximately a factor of 3. This result can be seen by comparing the crossover frequencies of the two designs.

A beneficial effect of lag compensation, an increase in the low-frequency gain for better error characteristics, was just demonstrated in Example 6.18. However, in essence, lag compensation reduces the value of $\omega_c/L(0)$ $(=\omega_c/K_v$ for a Type 1 system). That means that, if the crossover frequency is kept the same, the low-frequency gain will increase. Likewise, if the low-frequency gain is kept the same, the crossover frequency will decrease. Therefore, lag compensation could also be interpreted to reduce the crossover frequency and thus obtain a better PM. The procedure for design in this case is partially modified. First, pick the low-frequency gain to meet error requirements, then locate the lag compensation pole and zero in order to provide a crossover frequency with adequate PM. The next example illustrates this design procedure. The end result of the design will be the same no matter what procedure is followed.

EXAMPLE 6.19   *Lag Compensation of the DC Motor*

Repeat the design of the DC motor control in Example 6.15, this time using lag compensation. Fix the low-frequency gain in order to meet the error requirement of $K_v = 10$; then use the lag compensation to meet the PM requirement of 45°.

Figure 6.67

Frequency response of
lag-compensation
design in Example 6.19

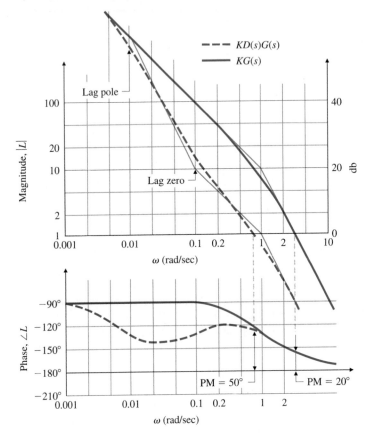

Solution.   The frequency response of the system $KG(s)$, with the required gain of $K = 10$, is shown in Fig. 6.67. The uncompensated system has a crossover frequency at approximately 3 rad/sec where the PM $= 20°$. The designer's task is to select the lag compensation break points so that the crossover frequency is lowered and more favorable PM results. To prevent detrimental effects from the compensation phase lag, the pole and zero position values of the compensation need to be substantially lower than the new crossover frequency. One possible choice is shown in Fig. 6.67: The lag zero is at 0.1 rad/sec, and the lag pole is at 0.01 rad/sec. This selection of parameters produces a PM of 50°, thus satisfying the specifications. Here the stabilization is achieved by keeping the crossover frequency to a region where $G(s)$ has favorable phase characteristics. The criterion for selecting the pole and zero locations $1/T$ is to make them low enough to minimize the effects of the phase lag from the compensation at the crossover frequency. Generally, however, the pole and zero are located no lower than necessary, because the additional system root (compare with the root locus of a similar system design in Fig. 5.28) introduced by the lag will be in the same frequency range as the compensation zero and will have some effect on the output response, especially the response to disturbance inputs.

The response of the system to a step reference input is shown in Fig. 6.68. It shows no steady-state error to a step input, because this is a Type 1 system. However,

Figure 6.68

Step response of
lag-compensation
design in Example 6.19

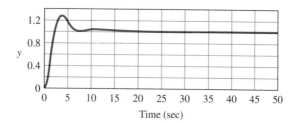

the introduction of the slow root from the lag compensation has caused the response to require about 25 sec to settle down to the zero steady-state value. The overshoot M_p is somewhat larger than you would expect from the guidelines, based on a second-order system shown in Fig. 6.38 for a PM $= 50°$; however, the performance is adequate.

As we saw previously for a similar situation, Examples 6.15 and 6.19 meet an identical set of specifications for the same plant in very different ways. In the first case the specifications are met with a lead compensation, and a crossover frequency $\omega_c = 5$ rad/sec ($\omega_{BW} \cong 6$ rad/sec) results. In the second case the same specifications are met with a lag compensation, and $\omega_c \cong 0.8$ rad/sec ($\omega_{BW} \cong 1$ rad/sec) results. Clearly, had there been specifications for rise time or bandwidth, they would have influenced the choice of compensation (lead or lag). Likewise, if the slow settling to the steady-state value was a problem, it might have suggested the use of lead compensation instead of lag.

In more realistic systems, dynamic elements usually represent the actuator and sensor as well as the process itself, so it is typically impossible to raise the crossover frequency much beyond the value representing the speed of response of the components being used. Although linear analysis seems to suggest that almost any system can be compensated, in fact, if we attempt to drive a set of components much faster than their natural frequencies, the system will saturate, the linearity assumptions will no longer be valid, and the linear design will represent little more than wishful thinking. With this behavior in mind, we see that simply increasing the gain of a system and adding lead compensators to achieve an adequate PM may not always be possible. It may be preferable to satisfy error requirements by adding a lag network so that the closed-loop bandwidth is kept at a more reasonable frequency.

6.7.5 PID Compensation

PID compensation

For problems that need PM improvement at ω_c *and* low-frequency gain improvement, it is effective to use both derivative and integral control. By combining Eqs. (6.37) and (6.50), we obtain PID control. Its transfer function is

$$D(s) = \frac{K}{s}\left[(T_D s + 1)\left(s + \frac{1}{T_I}\right)\right], \tag{6.52}$$

and its frequency-response characteristics are shown in Fig. 6.69. This form is slightly different from that given by Eq. (4.59); however, the effect of the difference is inconsequential. This compensation is roughly equivalent to combining lead and

Figure 6.69
Frequency response of
PID compensation with
$T_I/T_D = 20$

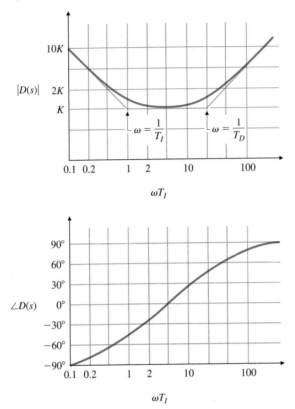

lag compensators in the same design, and so is sometimes referred to as a **lead–lag compensator**. Hence, it can provide simultaneous improvement in transient and steady-state responses.

EXAMPLE 6.20

PID Compensation Design for Spacecraft Attitude Control

A simplified design for spacecraft attitude control was presented in Section 6.5; however, here we have a more realistic situation that includes a sensor lag and a disturbing torque. Figure 6.70 defines the system. Design a PID controller to have zero steady-state error to a constant-disturbance torque, a PM of 65°, and as high a bandwidth as is reasonably possible. Also evaluate the pointing errors versus frequency and compare them to the errors that would result if the system is open loop. For a torque disturbance from solar pressure that acts as a sinusoid at the orbital rate ($\omega = 0.001$ rad/sec or $\approx$100-minute period), determine the fractional improvement by the feedback system.

Solution. First, let us take care of the steady-state error. For the spacecraft to be at a steady final value, the total input torque, $T_d + T_c$, must equal zero. Therefore, if $T_d \neq 0$, then $T_c = -T_d$. The only way this can be true with no error ($e = 0$) is for $D(s)$ to contain an integral term. Hence, including integral control in the compensation will meet the steady-state requirement. This could also be verified mathematically by use of the Final Value Theorem (see Problem 6.47).

Figure 6.70

Block diagram of spacecraft control using PID design, Example 6.20

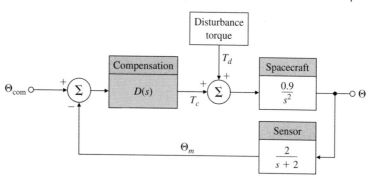

The frequency response of the spacecraft and sensor,

$$G(s) = \frac{0.9}{s^2}\left(\frac{2}{s+2}\right),\tag{6.53}$$

is shown in Fig. 6.71. The slopes of -2 (that is, -40 db per decade) and -3 (-60 db per decade) show that the system would be unstable for any value of K if no derivative feedback were used. This is clear because of Bode's gain–phase relationship, which shows that the phase would be $-180°$ for the -2 slope and $-270°$ for the -3 slope and which would correspond to a PM of $0°$ or $-90°$. Therefore, derivative control is required to bring the slope to -1 at the crossover frequency that was shown in Section 6.5 to be a requirement for stability. The problem now is to pick values for the three parameters in Eq. (6.52)—K, T_D, and T_I—that will satisfy the specifications.

The easiest approach is to work first on the phase so that PM $= 65°$ is achieved at a reasonably high frequency. This can be accomplished primarily by adjusting T_D, noting that T_I has a minor effect if sufficiently larger than T_D. Once the phase is adjusted, we establish the crossover frequency; then we can easily determine the gain K.

We examine the phase of the PID controller in Fig. 6.69 to determine what would happen to the compensated spacecraft system, $D(s)G(s)$, as T_D is varied. If $1/T_D \geq 2$ rad/sec, the phase lead from the PID control would simply cancel the sensor phase lag, and the composite phase would never exceed $-180°$, an unacceptable situation. If $1/T_D \leq 0.01$, the composite phase would approach $-90°$ for some range of frequencies and would exceed $-115°$ for an even wider range of frequencies; the latter threshold would provide a PM of $65°$. In the compensated phase curve shown in Fig. 6.71, $1/T_D = 0.1$, which is the largest value of $1/T_D$ that could provide the required PM of $65°$. The phase would never cross the $-115°$ ($65°$ PM) line for any $1/T_D > 0.1$. For $1/T_D = 0.1$, the crossover frequency ω_c that produces the $65°$ PM is 0.5 rad/sec. For a value of $1/T_D \ll 0.05$, the phase essentially follows the dotted curve in Fig. 6.71, which indicates that the maximum possible ω_c is approximately 1 rad/sec and is provided by $1/T_D = 0.05$. Therefore, $0.05 < 1/T_D < 0.1$ is the only sensible range for $1/T_D$; anything less than 0.05 would provide no significant increase in bandwidth, while anything more than 0.1 could not meet the PM specification. Although the final choice is somewhat arbitrary, we have chosen $1/T_D = 0.1$ for our final design.

Our choice for $1/T_I$ is a factor of 20 lower than $1/T_D$; that is, $1/T_I = 0.005$. A factor less than 20 would negatively impact the phase at crossover, thus lowering the PM.

Figure 6.71

Compensation for PID
design in Example 6.20

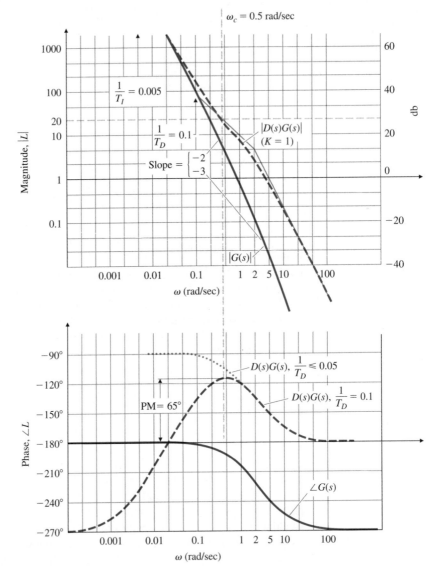

Furthermore, it is generally desirable to keep the compensated magnitude as large
as possible at frequencies below ω_c in order to have a faster transient response and
smaller errors; maintaining $1/T_D$ and $1/T_I$ at the highest possible frequencies will bring
this about.

The only remaining task is to determine the proportional part of the PID controller,
or K. Unlike the system in Example 6.18, where we selected K in order to meet a
steady-state error specification, here we select a value of K that will yield a crossover
frequency at the point corresponding to the required PM of 65°. The basic procedure
for finding K, discussed in Section 6.6, consists of plotting the compensated system
amplitude with $K = 1$, finding the amplitude value at crossover, then setting $1/K$ equal

to that value. Figure 6.71 shows that when $K = 1$, $|D(s)G(s)| = 20$ at the desired crossover frequency $\omega_c = 0.5$ rad/sec. Therefore,

$$\frac{1}{K} = 20, \quad \text{so} \quad K = \frac{1}{20} = 0.05.$$

The compensation equation that satisfies all of the specifications is now complete:

$$D(s) = \frac{0.05}{s}[(10s + 1)(s + 0.005)].$$

It is interesting to note that this system would become unstable if the gain were lowered so that $\omega_c \leq 0.02$ rad/sec, the region in Fig. 6.71 where the phase of the compensated system is less than $-180°$. As mentioned in Section 6.4, this situation is referred to as a conditionally stable system. A root locus with respect to K for this and any conditionally stable system would show the portion of the locus corresponding to very low gains in the RHP. The response of the system for a unit step θ_{com} is shown in Fig. 6.72(a) and exhibits well damped behavior, as should be expected with a 65° PM.

The response of the system for a step disturbance torque $T_d = 0.1$ N is shown in Fig. 6.72(b). Note that the integral control term does eventually drive the error to zero; however, it is slow due to the presence of a closed-loop pole in the vicinity of the zero at $s = -0.005$. Recall from the design process that this zero was located in order that the integral term not impact the PM unduly. So if the slow disturbance response is not acceptable, speeding up this pole will decrease the PM and damping of the system. Compromise is often a necessity in control system design!

The frequency response of the error characteristics is shown in Fig. 6.73. The top curve is the open-loop error characteristics and the bottom curve is the closed-loop response. The error is attenuated by almost a factor of 10^6 by the feedback for a disturbance at the orbital rate; there is decreasing error attenuation as the disturbance frequency increases, and there is almost no error attenuation at the system bandwidth of ≈ 0.5 rad/sec, as you would expect. Note from the design process that the bandwidth was limited by the response characteristics of the sensor, which had a bandwidth of 2 rad/sec. Therefore, the only way to improve the error characteristics would be to

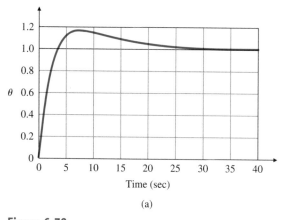

(a)

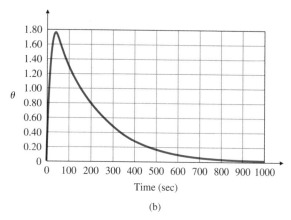

(b)

Figure 6.72

Transient response for PID example: (a) step response; (b) step-disturbance response

Figure 6.73
Frequency response of
the error due to a
disturbance input,
open-loop and
closed-loop

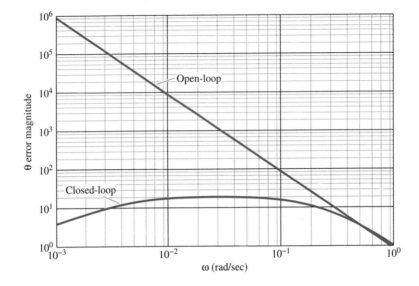

increase the bandwidth of the sensor. On the other hand, increasing the bandwidth
of the sensor may introduce jitter from the high-frequency sensor noise. Thus we
see the classic trade-off dilemma: the designer has to make a judgment as to which
feature (low errors due to disturbances or low errors due to sensor noise) is the more
important to the overall system performance.

**Summary of
Compensation
Characteristics**

1. *PD control* adds phase lead at all frequencies above the break point. If there
   is no change in gain on the low-frequency asymptote, PD compensation will
   increase the crossover frequency and the speed of response. The increase in
   magnitude of the frequency response at the higher frequencies will increase
   the system's sensitivity to noise.
2. *Lead compensation* adds phase lead at a frequency band between the two
   break points, which are usually selected to bracket the crossover frequency.
   If there is no change in gain on the low-frequency asymptote, lead compensation will increase both the crossover frequency and the speed of response
   over the uncompensated system.
3. *PI control* increases the frequency-response magnitude at frequencies below
   the break point, thereby decreasing steady-state errors. It also contributes
   phase lag below the break point, which must be kept at a low enough
   frequency to avoid degrading the stability excessively.
4. *Lag compensation* increases the frequency-response magnitude at frequencies below the two break points, thereby decreasing steady-state errors.
   Alternatively, with suitable adjustments in K, lag compensation can be used
   to decrease the frequency-response magnitude at frequencies above the two
   break points, so that ω_c yields an acceptable PM. Lag compensation also
   contributes phase lag between the two break points, which must be kept at

frequencies low enough to keep the phase decrease from degrading the PM excessively. This compensation will typically provide a slower response than using lead compensation.

6.7.6 Design Considerations

We have seen in the preceding designs that characteristics of the open-loop Bode plot of the loop gain, $L(s)$ $(= KDG)$, determine performance with respect to steady-state errors, low-frequency errors, and dynamic response. Other properties of feedback, developed in Chapter 4, include reducing the effects of sensor noise and parameter changes on the performance of the system.

The consideration of steady-state errors or low-frequency errors due to command inputs and disturbances has been an important design component in the different design methods presented. Design for acceptable errors due to command inputs and disturbances can be thought of as placing a lower bound on the low-frequency gain of the open loop system. Another aspect of the sensitivity issue concerns the high-frequency portion of the system. So far, Chapter 4 and Sections 5.4 and 6.7 have briefly discussed the idea that, to alleviate the effects of sensor noise, the gain of the system at high frequencies must be kept low. In fact, in the development of lead compensation, we added a pole to pure derivative control specifically to reduce the effects of sensor noise at the higher frequencies. It is not unusual for designers to place an extra pole in the compensation, that is, to use the relation

$$D(s) = \frac{Ts + 1}{(\alpha Ts + 1)^2},$$

in order to introduce even more attenuation for noise reduction.

A second consideration affecting high-frequency gains is that many systems have high-frequency dynamic phenomena, such as mechanical resonances, that could have an impact on the stability of a system. In very-high-performance designs, these high-frequency dynamics are included in the plant model, and a compensator is designed with a specific knowledge of those dynamics. A standard approach to designing for unknown high-frequency dynamics is to keep the high-frequency gain low, just as we did for sensor-noise reduction. The reason for this can be seen from the gain–frequency relationship of a typical system, shown in Fig. 6.74. The only way instability can result from high-frequency dynamics is if an unknown high-frequency resonance causes the magnitude to rise above 1. Conversely, if all unknown high-frequency phenomena are guaranteed to remain below a magnitude of 1, stability can be guaranteed. The likelihood of an unknown resonance in the plant G rising above 1 can be reduced if the nominal high-frequency loop gain (L) is lowered by the addition of extra poles in $D(s)$. When the stability of a system with resonances is assured by tailoring the high-frequency magnitude never to exceed 1, we refer to this process as **amplitude** or **gain stabilization**. Of course, if the resonance characteristics are known exactly, a specially tailored compensation, such as one with a notch at the resonant frequency, can be used to change the phase at a specific frequency to avoid encirclements of -1, thus stabilizing the system even though the amplitude does exceed magnitude 1. This method of stabilization is referred to as **phase stabilization**. A drawback to phase

Gain stabilization

Phase stabilization

Figure 6.74

Effect of high-frequency plant uncertainty

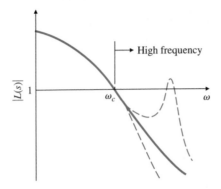

stabilization is that the resonance information is often not available with adequate precision or varies with time; therefore, the method is more susceptible to errors in the plant model used in the design. Thus, we see that sensitivity to plant uncertainty and sensor noise are both reduced by sufficiently low loop gain at high-frequency.

These two aspects of sensitivity—high- and low-frequency behavior—can be depicted graphically, as shown in Fig. 6.75. There is a minimum low-frequency gain allowable for acceptable steady-state and low-frequency error performance and a maximum high-frequency gain allowable for acceptable noise performance and for low probability of instabilities caused by plant-modeling errors. We define the low-frequency lower bound on the frequency response as W_1 and the upper bound as W_2^{-1}, as shown in the figure. Between these two bounds the control engineer must achieve a gain crossover near the required bandwidth; as we have seen, the crossover must occur at a slope of -1 or slightly steeper for good PM and hence damping.

For example, if a control system was required to follow a sinusoidal reference input with frequencies from 0 to ω_1 with errors no greater than 1%, the function W_1 would be 100 from $\omega = 0$ to ω_1. Similar ideas enter into defining possible values for the W_2^{-1} function which would constrain the open-loop gain to be below W_2^{-1} for frequencies above ω_2. These ideas will be discussed further in the following subsections.

Figure 6.75

Design criteria for low sensitivity

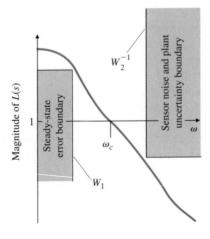

△ **6.7.7 Specifications in Terms of the Sensitivity Function**

We have seen how the gain and phase margins give useful information about the relative stability of nominal systems and can be used to guide the design of lead and
lag compensations. However, the GM and PM are only two numbers and have limitations as guides to the design of realistic control problems. We can express more
complete design specifications in the frequency domain if we first give frequency
descriptions for the external signals, such as the reference and disturbance, and consider the sensitivity function defined in Section 4.1. For example, we have so far
described dynamic performance by the transient response to simple steps and ramps.
A more realistic description of the actual complex input signals is to represent them
as random processes with corresponding frequency power density spectra. A less
sophisticated description, which is adequate for our purposes, is to assume that the
signals can be represented as a sum of sinusoids with frequencies in a specified range.
For example, we can usually describe the frequency content of the reference input as
a sum of sinusoids with relative amplitudes given by a magnitude function $|R|$ such
as that plotted in Fig. 6.76, which represents a signal with sinusoidal components
having about the same amplitudes up to some value ω_1 and very small amplitudes for
frequencies above that. With this assumption, the response tracking specification can
be expressed by a statement such as "the magnitude of the system error is to be less
than the bound e_b (a value such as 0.01) for any sinusoid of frequency ω_o in the range
$0 \leq \omega_o \leq \omega_1$ and of amplitude given by $|R(j\omega_o)|$." To express such a performance
requirement in terms that can be used in design, we consider again the unity-feedback
system drawn in Fig. 6.77. For this system, the error is given by

$$E(j\omega) = \frac{1}{1+DG}R \triangleq \mathcal{S}(j\omega)R, \tag{6.54}$$

Sensitivity function

where we have used the **sensitivity function**

Figure 6.76

Plot of typical reference
spectrum

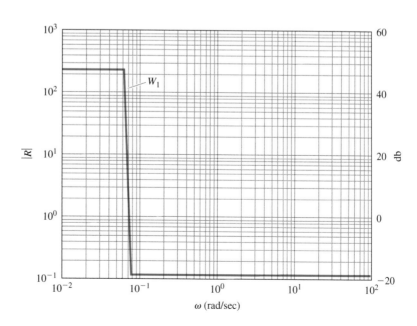

Figure 6.77

Closed-loop block diagram

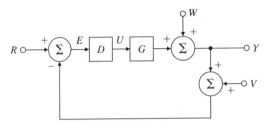

$$S \triangleq \frac{1}{1 + DG}. \tag{6.55}$$

In addition to being the factor multiplying the system error, the sensitivity function is also the reciprocal of the distance of the Nyquist curve, DG, from the critical point -1. A large value for S indicates a Nyquist plot that comes close to the point of instability. The frequency-based error specification based on Eq. (6.54) can be expressed as $|E| = |S| \, |R| \le e_b$. In order to normalize the problem without needing to define both the spectrum R and the error bound each time, we define the real function of frequency $W_1(\omega) = |R| / e_b$ and the requirement can be written as

$$\boxed{|S| W_1 \le 1.} \tag{6.56}$$

EXAMPLE 6.21   *Performance Bound Function*

A unity-feedback system is to have an error less than 0.005 for all unity amplitude sinusoids below frequency 100 Hertz. Draw the performance frequency function $W_1(\omega)$ for this design.

Solution.   The spectrum, from the problem description, is unity for $0 \le \omega \le 200\pi$ rad/sec. Because $e_b = 0.005$, the required function is given by a rectangle of amplitude $1/0.005 = 200$ over the given range. The function is plotted in Fig. 6.78.

The expression in Eq. (6.56) can be translated to the more familiar Bode plot coordinates and given as a requirement on loop gain by observing that over the frequency range when errors are small the loop gain is large. In that case $|S| \approx 1/|DG|$, and the requirement is approximately

$$\frac{W_1}{|DG|} \le 1,$$

$$\boxed{|DG| \ge W_1.} \tag{6.57}$$

This requirement can be seen as an extension of the steady-state error requirement from just $\omega = 0$ to the range $0 \le \omega_o \le \omega_1$.

In addition to the requirement on dynamic performance, the designer is usually required to design for **stability robustness**. By this we mean that, while the design is done for a nominal plant transfer function, the actual system is expected to be stable for an entire class of transfer functions that represents the range of changes that are

Figure 6.78

Plot of example performance function, W_1

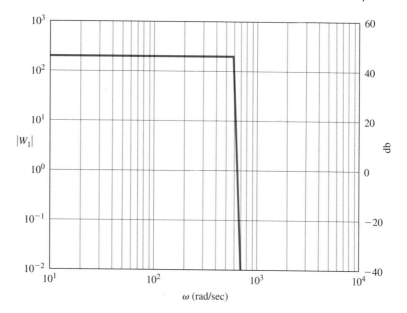

expected to be faced as temperature, age, and other operational and environmental factors vary the plant dynamics from the nominal case. A realistic way to express this uncertainty is to describe the plant transfer function as having a multiplicative uncertainty:

$$G(j\omega) = G_o(j\omega)[1 + W_2(\omega)\Delta(j\omega)]. \qquad (6.58)$$

In Eq. (6.58), the real function W_2 is a magnitude function that expresses the size of changes as a function of frequency that the transfer function is expected to experience. In terms of G and G_o, the expression is

$$W_2 = \left| \frac{G - G_o}{G_o} \right|. \qquad (6.59)$$

The shape of W_2 is almost always very small for low frequencies (we know the model very well there) and increases substantially as we go to higher frequencies, where parasitic parameters come into play and unmodeled structural flexibility is common. A typical shape is sketched in Fig. 6.79. The complex function, $\Delta(j\omega)$, represents the uncertainty in phase and is restricted only by the constraint

$$0 \le |\Delta| \le 1. \qquad (6.60)$$

We assume that the nominal design has been done and is stable, so that the Nyquist plot of DG_o satisfies the Nyquist stability criterion. In this case, the nominal characteristic equation $1+DG_o = 0$ is never satisfied for any real frequency. If the system is to have stability robustness, the characteristic equation using the uncertain plant as described by Eq. (6.58) must not go to zero for any real frequency for any value of Δ. The requirement can be written as

$$1 + DG \ne 0, \qquad (6.61)$$

$$1 + DG_o[1 + W_2\Delta] \ne 0,$$

$$(1 + DG_o)(1 + TW_2\Delta) \ne 0,$$

Figure 6.79

Plot of typical plant uncertainty, W_2

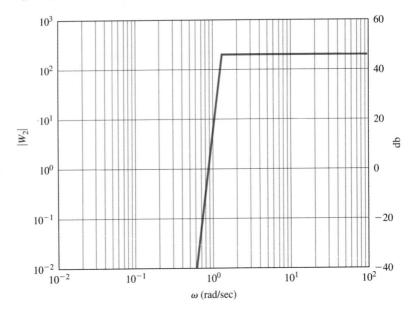

where we have defined the **complementary sensitivity function** as

$$T(j\omega) \triangleq DG_o/(1+DG_o) = 1 - \mathcal{S}. \tag{6.62}$$

Because the nominal system is stable, the first term in Eq. (6.61), $(1+DG_o)$, is never zero. Thus, if Eq. (6.61) is not to be zero for any frequency and any $\triangle$, then it is necessary and sufficient that

$$|TW_2\triangle| < 1,$$

which reduces to

$$|T|\,W_2 < 1, \tag{6.63}$$

making use of Eq. (6.60). As with the performance specification, for single-input–single-output unity-feedback systems this requirement can be approximated by a more convenient form. Over the range of high frequencies where W_2 is non-negligible because there is significant model uncertainty, DG_o is small. Therefore we can approximate $T \approx DG_o$, and the constraint reduces to

$$|DG_o|\,W_2 < 1,$$

$$\boxed{|DG_o| < \frac{1}{W_2}.} \tag{6.64}$$

The robustness issue is important to design and can affect the high-frequency open-loop frequency response, as discussed above. However, as discussed earlier, it is also important to limit the high-frequency magnitude in order to attenuate noise effects.

Figure 6.80

Plot of constraint on
$|DG_o|$ $(= |W_2^{-1}|)$

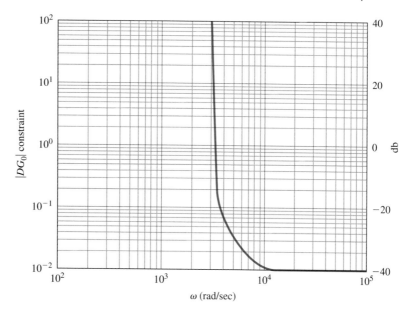

ω (rad/sec)

EXAMPLE 6.22

Typical Plant Uncertainty

The uncertainty in a plant model is described by a function W_2 that is zero until
$\omega = 3000$, increases linearly from there to a value of 100 at $\omega = 10,000$, and remains
at 100 for higher frequencies. Plot the constraint on DG_o to meet this requirement.

Solution. Where $W_2 = 0$, there is no constraint on the magnitude of loop gain;
above $\omega = 3000$, $1/W_2 = DG_o$ is a hyperbola from ∞ to 0.01 at $\omega = 10,000$ and
remains at 0.01 for $\omega > 10,000$. The bound is sketched in Fig. 6.80.

In practice, the magnitude of the loop gain is plotted on log–log (Bode) coordinates, and the constraints of Eqs. (6.57) and (6.64) are included on the same plot. A
typical sketch is drawn in Fig. 6.75. The designer is expected to construct a loop gain
that will stay above W_1 for frequencies below ω_1, cross over the magnitude-1 line
$(|DG| = 0)$ in the range $\omega_1 \leq \omega \leq \omega_2$, and stay below $1/W_2$ for frequencies above ω_2.

$\triangle$ **6.7.8 Limitations on Design in Terms of the Sensitivity Function**

One of the major contributions of Bode was to derive important limitations on transfer
functions that set limits on achievable design specifications. For example, one would
like to have the system error kept small for the widest possible range of frequencies
and yet have a system that is robustly stable for a very uncertain plant. In terms of the
plot in Fig. 6.81, we want W_1 and W_2 to be very large in their respective frequency
ranges and for ω_1 to be pushed up close to ω_2. Thus the loop gain is expected to
plunge with a large negative slope from being greater than W_1 to being less than
$1/W_2$ in a very short span, while maintaining a good PM to assure stability and good

Figure 6.81

Tracking and stability robustness constraints on the Bode plot; an example of impossible constraints

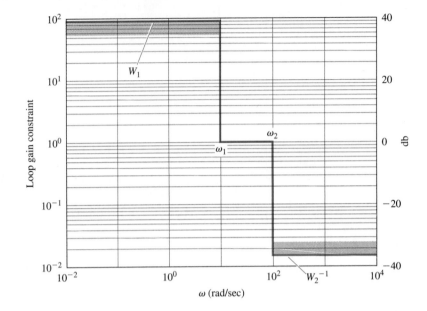

dynamic performance. The Bode gain–phase formula given earlier shows that this is *impossible* with a linear controller, by showing that the minimum possible phase is determined by an integral depending on the slope of the magnitude curve. If the slope is constant for a substantial range around ω_o, then Eq. (6.34) can be approximated by

$$\phi(\omega_o) \approx \frac{\pi}{2} \frac{dM}{du}\bigg|_{u=0}, \tag{6.65}$$

where M is the log magnitude and $u = \log \omega/\omega_o$. If, for example, the phase is to be kept above $-150°$ to maintain a $30°$ PM, then the magnitude slope near ω_o is estimated to be

$$\frac{dM}{du} \approx \frac{2}{\pi}\left(-150\frac{\pi}{180}\right)$$

$$\approx -1.667.$$

If we try to make the average slope steeper (more negative) than this, we will lose the PM. From this condition, there developed the design rule that the asymptotes of the Bode plot magnitude, which are restricted to be integral values for rational functions, should be made to cross over the zero-db line at a slope of -1 over a frequency range of about one decade around the crossover frequency, as already discussed in Section 6.5. Modifications to this rule need to be made in particular cases, of course, but the limitation implied by Eq. (6.65) is a hard limit that cannot be avoided. Thus, it is clear that it would be impossible to stabilize the system of Fig. 6.81.

EXAMPLE 6.23   *Robustness Constraints*

If $W_1 = W_2 = 100$, and we want PM $= 30°$, what is the minimum ratio of ω_2/ω_1?

Solution. The slope is

$$\frac{\log W_1 - \log \frac{1}{W_2}}{\log \omega_1 - \log \omega_2} = \frac{2+2}{\log \frac{\omega_1}{\omega_2}} = -1.667.$$

Thus, the log of the ratio is $\log \omega_1/\omega_2 = -2.40$ and $\omega_2 = 251\omega_1$.

An alternative to the standard Bode plot as a design guide can be based on a plot of the sensitivity function as a function of frequency. In this format, Eq. (6.56) requires that $|\mathcal{S}| < \frac{1}{W_1}$ over the range $0 \le \omega \le \omega_1$ for performance, and Eq. (6.64) requires that $|\mathcal{S}| \approx 1$ over the range $\omega_2 \le \omega$ for stability robustness. It should come as no surprise that Bode found a limitation on the possibilities in this case, too. The constraint, extended by Freudenberg and Looze, shows that an integral of the sensitivity function is determined by the presence of poles in the RHP. Suppose the loop gain DG_o has n_p poles, p_i, in the RHP and "rolls off" at high frequencies at a slope faster than -1. For rational functions, this means that there is an excess of at least two more finite poles than zeros. Then it can be shown that

$$\int_0^\infty \ln(|\mathcal{S}|)\, d\omega = \pi \sum_{i=1}^{n_p} \mathrm{Re}\{p_i\}. \tag{6.66}$$

If there are no RHP poles, then the integral is zero. This means that if we make the log of the sensitivity function very negative over some frequency band to reduce errors in that band, then, *of necessity*, $\ln|\mathcal{S}|$ will be positive over another part of the band, and errors will be amplified there. If there are unstable poles, the situation is worse, because the positive area where sensitivity magnifies the error must *exceed* the negative area where the error is reduced by the feedback. If the system is minimum phase, then it is, in principle, possible to keep the magnitude of the sensitivity small by spreading the sensitivity increase over all positive frequencies to infinity, but such a design requires an excessive bandwidth and is rarely practical. If a specific bandwidth is imposed, then the sensitivity function is constrained to take on a finite, possibly large, positive value at some point below the bandwidth. As implied by the definition of the vector margin (VM) in Section 6.4 (Fig. 6.39), a large $\mathcal{S}_{\max}$ corresponds to a Nyquist plot that comes close to the -1 critical point and a system having a small vector margin, because

Vector margin

$$\mathrm{VM} = \frac{\mathcal{S}_{\max}}{\mathcal{S}_{\max} - 1}. \tag{6.67}$$

If the system is not minimum-phase, the situation is worse. An alternative to Eq. (6.66) is true if there is a nonminimum-phase zero of DG_o, a zero in the RHP. Suppose that the zero is located at $z_o = \sigma_o + j\omega_o$, where $\sigma_o > 0$. Again, we assume there are n_p RHP poles at locations p_i with conjugate values $\overline{p}_i$. Now the condition can be expressed as a two-sided weighted integral

$$\int_{-\infty}^{\infty} \ln(|\mathcal{S}|) \frac{\sigma_o}{\sigma_o^2 + (\omega - \omega_o)^2}\, d\omega = \pi \sum_{i=1}^{n_p} \ln \left| \frac{\overline{p}_i + z_o}{p_i - z_o} \right|. \tag{6.68}$$

Figure 6.82

Sensitivity function for Example 6.24

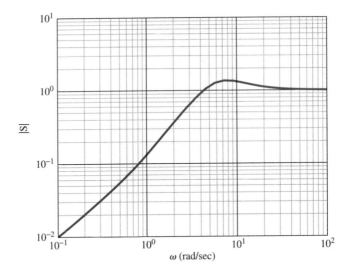

In this case, we do not have the "roll-off" restriction, and there is no possibility of spreading the positive area over high frequencies, because the weighting function goes to zero with frequency. The important point about this integral is that if the nonminimum-phase zero is close to a RHP pole, the right side of the integral can be very large, and the excess of positive area is required to be correspondingly large. Based on this result, one expects especially great difficulty meeting both tracking and robustness specifications on sensitivity with a system having RHP poles and zeros close together.

EXAMPLE 6.24   *Sensitivity Function for Antenna*

Compute and plot the sensitivity function for the design of the antenna for which $G(s) = 1/s(s + 1)$ and $D(s) = 10(0.5s + 1)/(0.1s + 1)$.

Solution.   The sensitivity function for this case is

$$S = \frac{s(s + 1)(s + 10)}{s^3 + 11s^2 + 60s + 100},$$ (6.69)

and the plot shown in Fig. 6.82 is given by the MATLAB commands

```
numS = [1 11 10 0];
denS = [1 11 60 100];
sysS = tf(numS,denS);
[mag,ph,w] = bode(sysS);
loglog(w,squeeze(mag)),grid
```

The largest value of S is given by $M = \max(\text{mag})$ and is 1.366, from which the vector margin is $VM = 3.73$.

△ 6.8 Time Delay

The Laplace transform of a pure time delay is $G_D(s) = e^{-sT_d}$ and was approximated by a rational function (Padé approximate) in our earlier discussion of root-locus analysis in Chapter 5. Although this same approximation could be used with frequency-response methods, an exact analysis of the delay is possible with the Nyquist criterion and Bode plots.

The frequency response of the delay is given by the magnitude and phase of $e^{-sT_d}\vert_{s=j\omega}$. The magnitude is

Time-delay magnitude

$$|G_D(j\omega)| = |e^{-j\omega T_d}| = 1, \quad \text{for all } \omega. \tag{6.70}$$

Time-delay phase

This result is expected, because a time delay merely shifts the signal in time and has no effect on its magnitude. The phase is

$$\angle G_D(j\omega) = -\omega T_d \tag{6.71}$$

in radians, and it grows increasingly negative in proportion to the frequency. This, too, is expected, because a fixed time delay T_d becomes a larger fraction or multiple of a sine wave as the period drops, due to increasing frequency. A plot of $\angle G_D(j\omega)$ is drawn in Fig. 6.83. Note that the phase lag is greater than 270° for values of ωT_d greater than about 5 rad. This trend implies that it would be virtually impossible to stabilize a system (or to achieve a positive PM) with a crossover frequency greater than $\omega = 5/T_d$, and it would be difficult for frequencies greater than $\omega \cong 3/T_d$. These characteristics essentially place a constraint on the achievable bandwidth of any system with a time delay. (See Problem 6.69 for an illustration of this constraint.)

The frequency domain concepts such as the Nyquist criterion apply directly to systems with pure time delay. This means that no approximations (Padé type or otherwise) are needed and the exact effect of time delay can be applied to a Bode plot, as shown in the following example.

EXAMPLE 6.25   *Effect of Sampling on Stability*

Determine the additional phase lag due to the digital sampling in Example 6.15 and reconcile that difference with the observed performance of the continuous and digital

Figure 6.83
Phase lag due to pure time delay

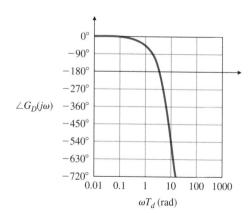

implementations shown in the example. How slowly could you sample if it was necessary to limit the decrease in the PM to less than 20°?

Solution. The sample rate in Example 6.15 was selected to be $T_s = 0.05$ sec. We can see from Fig. 4.22 that the effect of the sampling is to hold the application of the control over one sample period, thus the actual delay varies between zero and one full sample period. Therefore, on the average, the effect of the sampling is to inject a time delay of $T_s/2 = {}^{0.05}/_2 = 0.025 = T_d$ sec. From Eq. (6.71), we see that the phase lag due to this sampling at the crossover frequency of 5 rad/sec, where we measure the PM, is $\angle G_D = -\omega T_d = -(5)(0.025) = -0.125$ rad $= -7°$. Therefore, the PM will decrease from 45° for the continuous implementation to 38° for the digital implementation. Fig. 6.59(a) shows that the overshoot, M_p, degraded from 1.2 for the continuous case to ≈ 1.27 for the digital case, which is predicted by Eq. (6.32) and Fig. 6.38.

In order to limit the phase lag to 20° at $\omega = 5$ rad/sec, we see from Eq. (6.71) that the maximum tolerable $T_d = 20/(5 * 57.3) = 0.07$ sec, so that the slowest sampling acceptable would be $T_s = 0.14$ sec. Note, however, that this large decrease in the PM would result in the overshoot increasing from $\approx 20\%$ to $\approx 40\%$.

The example illustrates that a time delay, whether introduced by digital sampling or by any other source, has a very severe effect on the achievable bandwidth. Evaluation of the effect using Eq. (6.71) or Fig. 6.83 is simple and straightforward, thus giving a quick analysis of the limitations imposed by any delay in the system. One can also evaluate the effect of a delay using a Nyquist Diagram, and this is shown in Appendix W6.

△ 6.9 Alternative Presentation of Data

Before computers were widely available, other ways to present frequency-response data were developed to aid both in understanding design and in easing the designer's work load. The widespread availability of computers has virtually eliminated the need for these methods. One technique used was the Nichols chart, which we examine in this section because of its place in history. For those interested, we also present the inverse Nyquist method in Appendix W6.

6.9.1 Nichols Chart

A rectangular plot of $\log |G(j\omega)|$ versus $\angle G(j\omega)$ can be drawn by simply transferring the information directly from the separate magnitude and phase portions in a Bode plot; one point on the new curve thus results from a given value of the frequency ω. This means that the new curve is parameterized as a function of frequency. As with the Bode plots, the magnitude information is plotted on a logarithmic scale, while the phase information is plotted on a linear scale. This template was suggested by N. Nichols and is usually referred to as a **Nichols chart**. The idea of plotting the magnitude of $G(j\omega)$ versus its phase is similar to the concept of plotting the real and imaginary parts of $G(j\omega)$, which formed the basis for the Nyquist plots shown in Sections 6.3 and 6.4. However, it is difficult to capture all the pertinent characteristics of $G(j\omega)$ on the linear scale of the Nyquist plot. The log scale for magnitude in the Nichols chart alleviates this difficulty, allowing this kind of presentation to be useful for design.

For any value of the complex transfer function $G(j\omega)$, Section 6.6 showed that there is a unique mapping to the unity-feedback closed-loop transfer function

$$T(j\omega) = \frac{G(j\omega)}{1 + G(j\omega)}, \tag{6.72}$$

or in polar form,

$$T(j\omega) = M(\omega)e^{j\alpha(\omega)}, \tag{6.73}$$

where $M(\omega)$ is the magnitude of the closed-loop transfer function and $\alpha(\omega)$ is the phase of the closed-loop transfer function. Specifically,

$$M = \left| \frac{G}{1 + G} \right|, \tag{6.74}$$

$$\alpha = \tan^{-1}(N) = \angle\frac{G}{1 + G}. \tag{6.75}$$

It can be proven that the contours of constant closed-loop magnitude and phase are circles when $G(j\omega)$ is presented in the *linear* Nyquist plot. These circles are

M and N circles   referred to as the **M and N circles**, respectively.

The Nichols chart also contains contours of constant *closed-loop* magnitude and phase based on these relationships, as shown in Fig. 6.84; however, they are no longer circles, because the Nichols charts are semilog plots of magnitude versus linear phase. A designer can therefore graphically determine the bandwidth of a closed-loop system from the plot of the open-loop data on a Nichols chart by noting where the open-loop curve crosses the 0.70 contour of the closed-loop magnitude and determining the frequency of the corresponding data point. Likewise, a designer can determine the resonant peak amplitude M_r by noting the value of the magnitude of the highest closed-loop contour tangent to the curve. The frequency associated with the magnitude and

Resonant frequency   phase at the point of tangency is sometimes referred to as the **resonant frequency** ω_r. Similarly, a designer can determine the GM by observing the value of the gain where the Nichols plot crosses the $-180°$ line, and the PM by observing the phase where the plot crosses the amplitude 1 line.[14] MATLAB provides for easy drawing of a Nichols chart via the nichols m-file.

EXAMPLE 6.26   *Nichols Chart for PID Example*

Determine the bandwidth and resonant peak magnitude of the compensated system whose frequency response is shown in Fig. 6.71.

Solution. The magnitude and phase information of the compensated design example seen in Fig. 6.71 is shown on a Nichols chart in Fig. 6.85. When comparing the two figures, it is important to divide the magnitudes in Fig. 6.71 by a factor of 20 in order to obtain $|D(s)G(s)|$ rather than the normalized values used in Fig. 6.71. Because the curve crosses the closed-loop magnitude 0.70 contour at $\omega = 0.8$ rad/sec, we see that

---

[14] James, H. M., N. B. Nichols, and R. S. Phillips (1947).

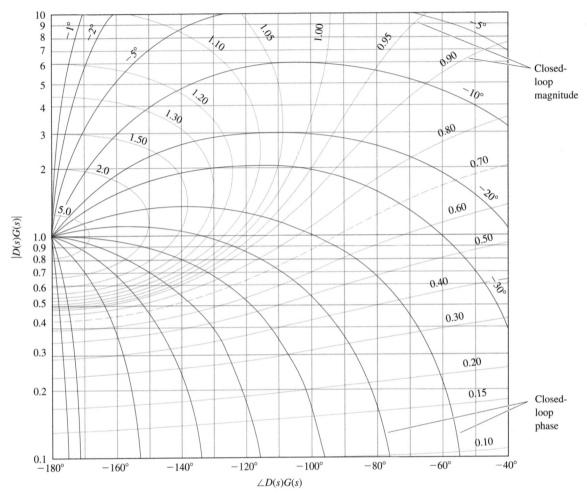

Figure 6.84
Nichols chart

the bandwidth of this system is 0.8 rad/sec. Because the largest-magnitude contour touched by the curve is 1.20, we also see that $M_r = 1.2$.

This presentation of data was particularly valuable when a designer had to generate plots and perform calculations by hand. A change in gain, for example, could be evaluated by sliding the curve vertically on transparent paper over a standard Nichols chart as shown in Fig. 6.84. The GM, PM, and bandwidth were then easy to read off the chart, thus allowing evaluations of several values of gain with a minimal amount of effort. With access to computer-aided methods, however, we can now calculate the bandwidth and perform many repetitive evaluations of the gain or any other parameter with a few key strokes. Today the Nichols chart is used primarily as an alternative

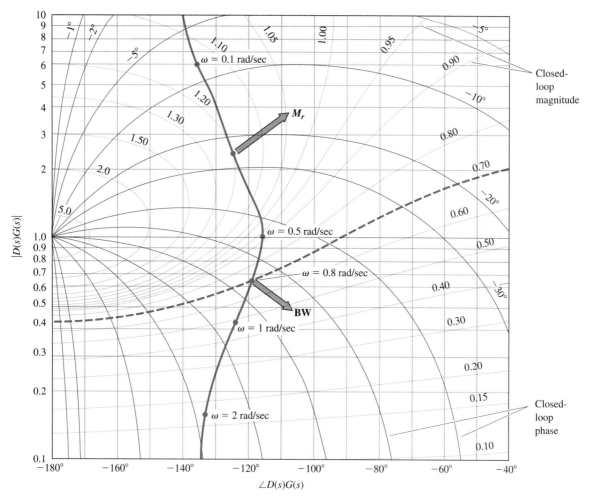

Figure 6.85

Example plot on the Nichols chart for determining bandwidth and M_r

way to present the information in a Nyquist plot. For complex systems for which the -1 encirclements need to be evaluated, the magnitude log scale of the Nichols chart enables us to examine a wider range of frequencies than a Nyquist plot does, as well as allowing us to read the gain and phase margins directly. Although MATLAB will directly compute PM and GM, the algorithm may lead to suspicious results for very complex cases and the analyst may want to verify the result using the MATLAB nichols m-file so the actual encirclements can be examined and the bases for the PM and GM better understood. An example of the use of a Nichols chart for a complex case is shown in Appendix W6.

Another presentation of data is the Inverse Nyquist Diagram, which simplifies the determination of the GM. This is described in more detail in Appendix W6 as well.

6.10 Historical Perspective

As discussed in Chapter 5, engineers before 1960s did not have access to computers to help in their analyses. Therefore, any method that allowed the determination of stability or response characteristics that did not require factoring the characteristic equation was highly useful. The invention of the electronic feedback amplifier by H. S. Black in 1927 at Bell Telephone Laboratories provided extra incentive to develop methods and the development of the frequency response method was the first that enabled design iteration for feedback control design.

The development of the feedback amplifier is briefly described in an interesting article based on a talk by Hendrik W. Bode (1960) reproduced in Bellman and Kalaba (1964). With the introduction of electronic amplifiers, long-distance telephoning became possible in the decades following World War I. However, as distances increased, so did the loss of electrical energy; in spite of using larger-diameter wire, increasing numbers of amplifiers were needed to replace the lost energy. Unfortunately, large numbers of amplifiers resulted in much distortion since the small nonlinearity of the vacuum tubes then used in electronic amplifiers was multiplied many times. To solve the problem of reducing distortion, Black proposed the feedback amplifier. As discussed earlier in Chapter 4, the more we wish to reduce errors (or distortion), the higher the feedback needs to be. The loop gain from actuator to plant to sensor to actuator must be made very large. But the designers found that too high a gain produced a squeal and the feedback loop became unstable. In this technology the dynamics were so complex (with differential equations of order 50 being common) that Routh's criterion, the only way of solving for stability at the time, was not very helpful. So the communications engineers at Bell Telephone Laboratories, familiar with the concept of frequency response and the mathematics of complex variables, turned to complex analysis. In 1932 H. Nyquist published a paper describing how to determine stability from a graphical plot of the open-loop frequency response. Bode then developed his plotting methods in 1938 that made them easy to create without extensive calculations or help from a computer. From the plotting methods and Nyquist's stability theory there developed an extensive methodology of feedback amplifier design described by Bode (1945) and extensively used still in the design of feedback controls. The reasons for using the method today are primarily to allow for a good design no matter what the unmodeled dynamics are and to expedite the design process, even when carried out with a computer that is fully capable of solving the characteristic equation. After developing the frequency-response design methods prior to World War II, Bode went on to help in electronic fire control devices during the war. The methods that he had developed for feedback amplifiers proved highly applicable to servomechanisms for the effort. Bode characterized this crossover of control system design methods as being a "sort of shotgun marriage."

SUMMARY

- The frequency-response **Bode plot** is a graph of the transfer function magnitude in logarithmic scale and the phase in linear scale versus frequency in logarithmic scale. For a transfer function $G(s)$,

$$A = |G(j\omega)| = |G(s)|_{s=j\omega}$$

$$= \sqrt{\{\text{Re}[G(j\omega)]\}^2 + \{\text{Im}[G(j\omega)]\}^2}$$

$$\phi = \tan^{-1}\left[\frac{\text{Im}[G(j\omega)]}{\text{Re}[G(j\omega)]}\right] = \angle G(j\omega).$$

- For a transfer function in Bode form,

$$KG(\omega) = K_0 \frac{(j\omega\tau_1 + 1)(j\omega\tau_2 + 1)\cdots}{(j\omega\tau_a + 1)(j\omega\tau_b + 1)\cdots},$$

  the Bode frequency response can be easily plotted by hand using the rules described in Section 6.1.1.
- Bode plots can be obtained using computer algorithms (bode in MATLAB), but hand-plotting skills are still extremely helpful.
- For a second-order system, the peak magnitude of the Bode plot is related to the damping by

$$|G(j\omega)| = \frac{1}{2\zeta} \quad \text{at } \omega = \omega_n.$$

- A method of determining the stability of a closed-loop system based on the frequency response of the system's open-loop transfer function is the **Nyquist stability criterion**. Rules for plotting the **Nyquist plot** are described in Section 6.3. The number of RHP closed-loop roots is given by

$$Z = N + P,$$

  where

$$N = \text{number of clockwise encirclements of the } -1 \text{ point,}$$

$$P = \text{number of open-loop poles in the RHP.}$$

- The Nyquist plot may be obtained using computer algorithms (nyquist in MATLAB).
- The **gain margin** (GM) and **phase margin** (PM) can be determined directly by inspecting the open-loop Bode plot or the Nyquist plot. Also, use of MATLAB's margin function determines the values directly.
- For a standard second-order system, the PM is related to the closed-loop damping by Eq. (6.32),

$$\zeta \cong \frac{\text{PM}}{100}.$$

- The **bandwidth** of the system is a measure of speed of response. For control systems, it is defined as the frequency corresponding to 0.707 (-3 db) in the closed-loop magnitude Bode plot and is approximately given by the crossover frequency ω_c, which is the frequency at which the open-loop gain curve crosses magnitude 1.
- The **vector margin** is a single-parameter stability margin based on the closest point of the Nyquist plot to the critical point $-1/K$.
- For a stable minimum-phase system, Bode's gain–phase relationship uniquely relates the phase to the gain of the system and is approximated by Eq. (6.33),

$$\angle G(j\omega) \cong n \times 90°,$$

Figure 6.86

Typical system

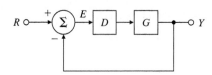

where n is the slope of $|G(j\omega)|$ in units of decade of amplitude per decade of frequency. The relationship shows that, in most cases, stability is ensured if the gain plot crosses the magnitude 1 line with a slope of -1.

- Experimental frequency-response data of the open-loop system can be used directly for analysis and design of a closed-loop control system with no analytical model.

- For the system shown in Fig. 6.86, the open-loop Bode plot is the frequency response of GD, and the closed-loop frequency response is obtained from $T(s) = GD/(1 + GD)$.

- The frequency-response characteristics of several types of compensation have been described, and examples of design using these characteristics have been discussed. Design procedures were given for lead and lag compensators in Section 6.7. The examples in that section show the ease of selecting specific values of design variables, a result of using frequency-response methods. A summary was provided at the end of Section 6.7.5.

- **Lead compensation**, given by Eq. (6.38),

$$D(s) = \frac{Ts + 1}{\alpha Ts + 1}, \quad \alpha < 1,$$

  is a high-pass filter and approximates PD control. It is used whenever substantial improvement in damping of the system is required. It tends to increase the speed of response of a system for a fixed low-frequency gain.

- **Lag compensation**, given by Eq. (6.51),

$$D(s) = \alpha \frac{Ts + 1}{\alpha Ts + 1}, \quad \alpha > 1, \tag{6.76}$$

  is a low-pass filter and approximates PI control. It is usually used to increase the low-frequency gain of the system so as to improve steady-state response for fixed bandwidth. For a fixed low-frequency gain, it will decrease the speed of response of a system.

- Tracking-error reduction and disturbance rejection can be specified in terms of the low-frequency gain of the Bode plot. Sensor-noise rejection can be specified in terms of high-frequency attenuation of the Bode plot (see Fig. 6.75).

△ • The **Nichols plot** is an alternate representation of the frequency response as a plot of gain versus phase and is parameterized as a function of frequency.

△ • Time delay can be analyzed exactly in a Bode plot or a Nyquist plot.

REVIEW QUESTIONS

1. Why did Bode suggest plotting the magnitude of a frequency response on log–log coordinates?

2. Define a decibel.

3.   What is the transfer function magnitude if the gain is listed as 14 db?

4.   Define gain crossover.

5.   Define phase crossover.

6.   Define phase margin, PM.

7.   Define gain margin, GM.

8.   What Bode plot characteristic is the best indicator of the closed-loop step response overshoot?

9.   What Bode plot characteristic is the best indicator of the closed-loop step response rise time?

10.   What is the principal effect of a lead compensation on Bode plot performance measures?

11.   What is the principal effect of a lag compensation on Bode plot performance measures?

12.   How do you find the K_v of a Type 1 system from its Bode plot?

13.   Why do we need to know beforehand the number of open-loop unstable poles in order to tell stability from the Nyquist plot?

14.   What is the main advantage in control design of counting the encirclements of $-1/K$ of $D(j\omega)G(j\omega)$ rather than encirclements of -1 of $KD(j\omega)G(j\omega)$?

15.   Define a conditionally stable feedback system. How can you identify one on a Bode plot?

△   16.   A certain control system is required to follow sinusoids, which may be any frequency in the range $0 \leq \omega_\ell \leq 450$ rad/sec and have amplitudes up to 5 units, with (sinusoidal) steady-state error to be never more than 0.01. Sketch (or describe) the corresponding performance function $W_1(\omega)$.

PROBLEMS

Problems for Section 6.1: Frequency Response

6.1   (a) Show that α_0 in Eq. (6.2), with $A = U_o$ and $\omega_o = \omega$, is

$$\alpha_0 = \left[G(s) \frac{U_0\omega}{s - j\omega} \right]_{s=-j\omega} = -U_0 G(-j\omega) \frac{1}{2j},$$

and

$$\alpha_0^* = \left[G(s) \frac{U_0\omega}{s + j\omega} \right]_{s=+j\omega} = U_0 G(j\omega) \frac{1}{2j}.$$

(b) By assuming the output can be written as

$$y(t) = \alpha_0 e^{-j\omega t} + \alpha_0^* e^{j\omega t},$$

derive Eqs. (6.4)–(6.6).

6.2   (a) Calculate the magnitude and phase of

$$G(s) = \frac{1}{s + 10}$$

by hand for $\omega = 1, 2, 5, 10, 20, 50,$ and 100 rad/sec.

(b) Sketch the asymptotes for $G(s)$ according to the Bode plot rules, and compare these with your computed results from part (a).

6.3 Sketch the asymptotes of the Bode plot magnitude and phase for each of the following open-loop transfer functions. After completing the hand sketches, verify your result using MATLAB. Turn in your hand sketches and the MATLAB results on the same scales.

(a) $L(s) = \dfrac{2000}{s(s + 200)}$

(b) $L(s) = \dfrac{100}{s(0.1s + 1)(0.5s + 1)}$

(c) $L(s) = \dfrac{1}{s(s + 1)(0.02s + 1)}$

(d) $L(s) = \dfrac{1}{(s + 1)^2(s^2 + 2s + 4)}$

(e) $L(s) = \dfrac{10(s + 4)}{s(s + 1)(s^2 + 2s + 5)}$

(f) $L(s) = \dfrac{1000(s + 0.1)}{s(s + 1)(s^2 + 8s + 64)}$

(g) $L(s) = \dfrac{(s + 5)(s + 3)}{s(s + 1)(s^2 + s + 4)}$

(h) $L(s) = \dfrac{4s(s + 10)}{(s + 100)(4s^2 + 5s + 4)}$

(i) $L(s) = \dfrac{s}{(s + 1)(s + 10)(s^2 + 2s + 2500)}$

6.4 *Real poles and zeros.* Sketch the asymptotes of the Bode plot magnitude and phase for each of the listed open-loop transfer functions. After completing the hand sketches, verify your result using MATLAB. Turn in your hand sketches and the MATLAB results on the same scales.

(a) $L(s) = \dfrac{1}{s(s + 1)(s + 5)(s + 10)}$

(b) $L(s) = \dfrac{(s + 2)}{s(s + 1)(s + 5)(s + 10)}$

(c) $L(s) = \dfrac{(s + 2)(s + 6)}{s(s + 1)(s + 5)(s + 10)}$

(d) $L(s) = \dfrac{(s + 2)(s + 4)}{s(s + 1)(s + 5)(s + 10)}$

6.5 *Complex poles and zeros.* Sketch the asymptotes of the Bode plot magnitude and phase for each of the listed open-loop transfer functions, and approximate the transition at the second-order break point, based on the value of the damping ratio. After completing the hand sketches, verify your result using MATLAB. Turn in your hand sketches and the MATLAB results on the same scales.

(a) $L(s) = \dfrac{1}{s^2 + 3s + 10}$

(b) $L(s) = \dfrac{1}{s(s^2 + 3s + 10)}$

(c) $L(s) = \dfrac{(s^2 + 2s + 8)}{s(s^2 + 2s + 10)}$

(d) $L(s) = \dfrac{(s^2 + 2s + 12)}{s(s^2 + 2s + 10)}$

(e) $L(s) = \dfrac{(s^2 + 1)}{s(s^2 + 4)}$

(f) $L(s) = \dfrac{(s^2 + 4)}{s(s^2 + 1)}$

6.6 *Multiple poles at the origin.* Sketch the asymptotes of the Bode plot magnitude and phase for each of the listed open-loop transfer functions. After completing the hand sketches, verify your result with MATLAB. Turn in your hand sketches and the MATLAB results on the same scales.

(a) $L(s) = \dfrac{1}{s^2(s + 8)}$

(b) $L(s) = \dfrac{1}{s^3(s + 8)}$

(c) $L(s) = \dfrac{1}{s^4(s + 8)}$

(d) $L(s) = \dfrac{(s + 3)}{s^2(s + 8)}$

(e) $L(s) = \dfrac{(s + 3)}{s^3(s + 4)}$

(f) $L(s) = \dfrac{(s + 1)^2}{s^3(s + 4)}$

(g) $L(s) = \dfrac{(s + 1)^2}{s^3(s + 10)^2}$

6.7 *Mixed real and complex poles.* Sketch the asymptotes of the Bode plot magnitude and phase for each of the listed open-loop transfer functions. Embellish the asymptote plots with a rough estimate of the transitions for each break point. After completing the hand sketches, verify your result with MATLAB. Turn in your hand sketches and the MATLAB results on the same scales.

(a) $L(s) = \dfrac{(s + 2)}{s(s + 10)(s^2 + 2s + 2)}$

(b) $L(s) = \dfrac{(s + 2)}{s^2(s + 10)(s^2 + 6s + 25)}$

(c) $L(s) = \dfrac{(s + 2)^2}{s^2(s + 10)(s^2 + 6s + 25)}$

(d) $L(s) = \dfrac{(s + 2)(s^2 + 4s + 68)}{s^2(s + 10)(s^2 + 4s + 85)}$

(e) $L(s) = \dfrac{[(s + 1)^2 + 1]}{s^2(s + 2)(s + 3)}$

6.8 *Right half-plane poles and zeros.* Sketch the asymptotes of the Bode plot magnitude and phase for each of the listed open-loop transfer functions. Make sure that the phase asymptotes properly take the RHP singularity into account by sketching the complex plane to see how the $\angle L(s)$ changes as s goes from 0 to $+j\infty$. After completing the hand sketches, verify your result with MATLAB. Turn in your hand sketches and the MATLAB results on the same scales.

(a) $L(s) = \dfrac{s+2}{s+10}\dfrac{1}{s^2-1}$ (the model for a case of magnetic levitation with lead compensation)

(b) $L(s) = \dfrac{s+2}{s(s+10)}\dfrac{1}{(s^2-1)}$ (The magnetic levitation system with integral control and lead compensation)

(c) $L(s) = \dfrac{s-1}{s^2}$

(d) $L(s) = \dfrac{s^2+2s+1}{s(s+20)^2(s^2-2s+2)}$

(e) $L(s) = \dfrac{(s+2)}{s(s-1)(s+6)^2}$

(f) $L(s) = \dfrac{1}{(s-1)[(s+2)^2+3]}$

6.9 A certain system is represented by the asymptotic Bode diagram shown in Fig. 6.87. Find and sketch the response of this system to a unit-step input (assuming zero initial conditions).

Figure 6.87

Magnitude portion of Bode plot for Problem 6.9

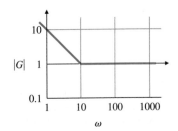

6.10 Prove that a magnitude slope of -1 in a Bode plot corresponds to -20 db per decade or -6 db per octave.

6.11 A normalized second-order system with a damping ratio $\zeta = 0.5$ and an additional zero is given by

$$G(s) = \frac{s/a+1}{s^2+s+1}.$$

Use MATLAB to compare the M_p from the step response of the system for $a = 0.01$, 0.1, 1, 10, and 100 with the M_r from the frequency response of each case. Is there a correlation between M_r and M_p?

6.12 A normalized second-order system with $\zeta = 0.5$ and an additional pole is given by

$$G(s) = \frac{1}{[(s/p)+1](s^2+s+1)}.$$

Draw Bode plots with $p = 0.01$, 0.1, 1, 10, and 100. What conclusions can you draw about the effect of an extra pole on the bandwidth compared with the bandwidth for the second-order system with no extra pole?

6.13 For the closed-loop transfer function

$$T(s) = \frac{\omega_n^2}{s^2 + 2\zeta\omega_n s + \omega_n^2},$$

derive the following expression for the bandwidth ω_{BW} of $T(s)$ in terms of ω_n and ζ:

$$\omega_{BW} = \omega_n\sqrt{1 - 2\zeta^2 + \sqrt{2 + 4\zeta^4 - 4\zeta^2}}.$$

Assuming that $\omega_n = 1$, plot ω_{BW} for $0 \le \zeta \le 1$.

6.14 Consider the system whose transfer function is

$$G(s) = \frac{A_0\omega_0 s}{Qs^2 + \omega_0 s + \omega_0^2 Q}.$$

This is a model of a tuned circuit with *quality factor Q*.

(a) Compute the magnitude and phase of the transfer function analytically, and plot them for $Q = 0.5$, 1, 2, and 5 as a function of the normalized frequency ω/ω_0.

(b) Define the bandwidth as the distance between the frequencies on either side of ω_0 where the magnitude drops to 3 db below its value at ω_0, and show that the bandwidth is given by

$$BW = \frac{1}{2\pi}\left(\frac{\omega_0}{Q}\right).$$

(c) What is the relation between Q and ζ?

6.15 A DC voltmeter schematic is shown in Fig. 6.88. The pointer is damped so that its maximum overshoot to a step input is 10%.

(a) What is the undamped natural frequency of the system?

(b) What is the damped natural frequency of the system?

(c) Plot the frequency response using MATLAB to determine what input frequency will produce the largest magnitude output?

(d) Suppose this meter is now used to measure a 1-V AC input with a frequency of 2 rad/sec. What amplitude will the meter indicate after initial transients have died out? What is the phase lag of the output with respect to the input? Use a Bode plot analysis to answer these questions. Use the lsim command in MATLAB to verify your answer in part (d).

Figure 6.88
Voltmeter schematic

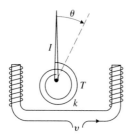

$I = 40 \times 10^{-6}\text{ kg}\cdot\text{m}^2$
$k = 4 \times 10^{-6}\text{ kg}\cdot\text{m}^2/\text{sec}^2$
$T = \text{input torque} = K_m v$
$v = \text{input voltage}$
$K_m = 1\text{ N}\cdot\text{m/V}$

Problems for Section 6.2: Neutral Stability

6.16 Determine the range of K for which the closed-loop systems (see Fig. 6.18) are stable for each of the cases below by making a Bode plot for $K = 1$ and imagining the magnitude plot sliding up or down until instability results. Verify your answers by using a very rough sketch of a root-locus plot.

(a) $KG(s) = \dfrac{K(s+2)}{s+20}$

(b) $KG(s) = \dfrac{K}{(s+10)(s+1)^2}$

(c) $KG(s) = \dfrac{K(s+10)(s+1)}{(s+100)(s+5)^3}$

6.17 Determine the range of K for which each of the listed systems is stable by making a Bode plot for $K = 1$ and imagining the magnitude plot sliding up or down until instability results. Verify your answers by using a very rough sketch of a root-locus plot.

(a) $KG(s) = \dfrac{K(s+1)}{s(s+5)}$

(b) $KG(s) = \dfrac{K(s+1)}{s^2(s+10)}$

(c) $KG(s) = \dfrac{K}{(s+2)(s^2+9)}$

(d) $KG(s) = \dfrac{K(s+1)^2}{s^3(s+10)}$

Problems for Section 6.3: The Nyquist Stability Criterion

6.18 (a) Sketch the Nyquist plot for an open-loop system with transfer function $1/s^2$; that is, sketch

$$\left. \frac{1}{s^2} \right|_{s=C_1},$$

where C_1 is a contour enclosing the entire RHP, as shown in Fig. 6.17. (*Hint:* Assume C_1 takes a small detour around the poles at $s = 0$, as shown in Fig. 6.27.)

(b) Repeat part (a) for an open-loop system whose transfer function is $G(s) = 1/(s^2 + \omega_0^2)$.

6.19 Sketch the Nyquist plot based on the Bode plots for each of the following systems, and then compare your result with that obtained by using the MATLAB command nyquist:

(a) $KG(s) = \dfrac{K(s+2)}{s+10}$

(b) $KG(s) = \dfrac{K}{(s+10)(s+2)^2}$

(c) $KG(s) = \dfrac{K(s+10)(s+1)}{(s+100)(s+2)^3}$

(d) Using your plots, estimate the range of K for which each system is stable, and qualitatively verify your result by using a rough sketch of a root-locus plot.

6.20 Draw a Nyquist plot for

$$KG(s) = \frac{K(s+1)}{s(s+3)}, \tag{6.77}$$

choosing the contour to be to the right of the singularity on the $j\omega$-axis. Next, using the Nyquist criterion, determine the range of K for which the system is stable. Then redo the Nyquist plot, this time choosing the contour to be to the left of the singularity on the imaginary axis. Again, using the Nyquist criterion, check the range of K for which the system is stable. Are the answers the same? Should they be?

6.21 Draw the Nyquist plot for the system in Fig. 6.89. Using the Nyquist stability criterion, determine the range of K for which the system is stable. Consider both positive and negative values of K.

Figure 6.89

Control system for Problem 6.21

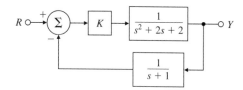

6.22 (a) For $\omega = 0.1$ to 100 rad/sec, sketch the phase of the minimum-phase system

$$G(s) = \frac{s+1}{s+10}\bigg|_{s=j\omega}$$

and the nonminimum-phase system

$$G(s) = -\frac{s-1}{s+10}\bigg|_{s=j\omega},$$

noting that $\angle(j\omega - 1)$ decreases with ω rather than increasing.

(b) Does an RHP zero affect the relationship between the -1 encirclements on a polar plot and the number of unstable closed-loop roots in Eq. (6.28)?

(c) Sketch the phase of the following unstable system for $\omega = 0.1$ to 100 rad/sec:

$$G(s) = \frac{s+1}{s-10}\bigg|_{s=j\omega}.$$

(d) Check the stability of the systems in (a) and (c) using the Nyquist criterion on $KG(s)$. Determine the range of K for which the closed-loop system is stable, and check your results qualitatively by using a rough root-locus sketch.

6.23 *Nyquist plots and the classical plane curves:* Determine the Nyquist plot, using MAT-LAB, for the systems given below, with $K = 1$, and verify that the beginning point and end point for the $j\omega > 0$ portion have the correct magnitude and phase:

(a) The classical curve called Cayley's Sextic, discovered by Maclaurin in 1718:

$$KG(s) = K\frac{1}{(s+1)^3}.$$

(b) The classical curve called the Cissoid, meaning ivy-shaped:

$$KG(s) = K\frac{1}{s(s+1)}.$$

(c) The classical curve called the Folium of Kepler, studied by Kepler in 1609:

$$KG(s) = K\frac{1}{(s-1)(s+1)^2}.$$

(d) The classical curve called the Folium (not Kepler's):

$$KG(s) = K\frac{1}{(s-1)(s+2)}.$$

(e) The classical curve called the Nephroid, meaning kidney-shaped:

$$KG(s) = K\frac{2(s+1)(s^2 - 4s + 1)}{(s-1)^3}.$$

(f) The classical curve called Nephroid of Freeth, named after the English mathematician T. J. Freeth:

$$KG(s) = K\frac{(s+1)(s^2 + 3)}{4(s-1)^3}.$$

(g) A shifted Nephroid of Freeth:

$$KG(s) = K\frac{(s^2 + 1)}{(s-1)^3}.$$

Problems for Section 6.4: Stability Margins

6.24 The Nyquist plot for some actual control systems resembles the one shown in Fig. 6.90. What are the gain and phase margin(s) for the system of Fig. 6.90, given that $\alpha = 0.4$, $\beta = 1.3$, and $\phi = 40°$. Describe what happens to the stability of the system as the gain goes from zero to a very large value. Sketch what the corresponding root locus must look like for such a system. Also, sketch what the corresponding Bode plots would look like for the system.

Figure 6.90

Nyquist plot for Problem 6.24

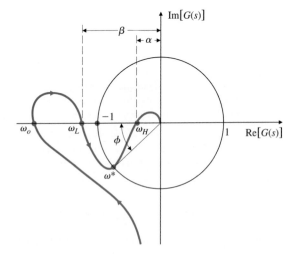

6.25 The Bode plot for

$$G(s) = \frac{100[(s/10) + 1]}{s[(s/1) - 1][(s/100) + 1]}$$

is shown in Fig. 6.91.

(a) Why does the phase start at $-270°$ at the low frequencies?

(b) Sketch the Nyquist plot for $G(s)$.

(c) Is the closed-loop system shown in Fig. 6.91 stable?

(d) Will the system be stable if the gain is lowered by a factor of 100? Make a rough sketch of a root locus for the system, and qualitatively confirm your answer.

Figure 6.91
Bode plot for
Problem 6.25

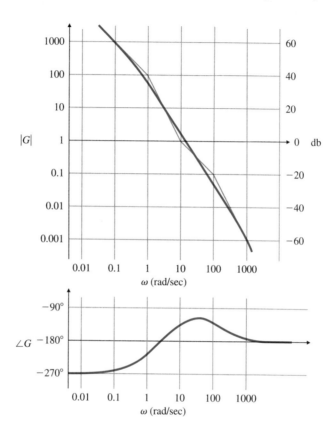

6.26 Suppose that in Fig. 6.92,

$$G(s) = \frac{25(s+1)}{s(s+2)(s^2+2s+16)}.$$

Use MATLAB's margin to calculate the PM and GM for $G(s)$ and, on the basis of the Bode plots, conclude which margin would provide more useful information to the control designer for this system.

Figure 6.92
Control system for
Problem 6.26

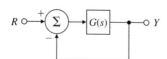

6.27 Consider the system given in Fig. 6.93.

(a) Use MATLAB to obtain Bode plots for $K = 1$, and use the plots to estimate the range of K for which the system will be stable.

(b) Verify the stable range of K by using margin to determine PM for selected values of K.

(c) Use rlocus to determine the values of K at the stability boundaries.

(d) Sketch the Nyquist plot of the system, and use it to verify the number of unstable roots for the unstable ranges of K.

(e) Using Routh's criterion, determine the ranges of K for closed-loop stability of this system.

Figure 6.93

Control system for Problem 6.27

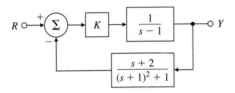

6.28 Suppose that in Fig. 6.92,

$$G(s) = \frac{3.2(s+1)}{s(s+2)(s^2+0.2s+16)}.$$

Use MATLAB's margin to calculate the PM and GM for $G(s)$, and comment on whether you think this system will have well-damped closed-loop roots.

6.29 For a given system, show that the ultimate period P_u and the corresponding ultimate gain K_u for the Zeigler–Nichols method can be found by using the following:

(a) Nyquist diagram

(b) Bode plot

(c) Root locus

6.30 If a system has the open-loop transfer function

$$G(s) = \frac{\omega_n^2}{s(s+2\zeta\omega_n)}$$

with unity feedback, then the closed-loop transfer function is given by

$$T(s) = \frac{\omega_n^2}{s^2+2\zeta\omega_n s+\omega_n^2}.$$

Verify the values of the PM shown in Fig. 6.37 for $\zeta = 0.1, 0.4$, and 0.7.

6.31 Consider the unity-feedback system with the open-loop transfer function

$$G(s) = \frac{K}{s(s+1)[(s^2/25)+0.4(s/5)+1]}.$$

(a) Use MATLAB to draw the Bode plots for $G(j\omega)$, assuming that $K = 1$.

(b) What gain K is required for a PM of $45°$? What is the GM for this value of K?

(c) What is K_v when the gain K is set for PM $= 45°$?

(d) Create a root locus with respect to K, and indicate the roots for a PM of $45°$.

6.32 For the system depicted in Fig. 6.94(a), the transfer-function blocks are defined by

$$G(s) = \frac{1}{(s+2)^2(s+4)} \quad \text{and} \quad H(s) = \frac{1}{s+1}.$$

(a) Using rlocus and rlocfind, determine the value of K at the stability boundary.

(b) Using rlocus and rlocfind, determine the value of K that will produce roots with damping corresponding to $\zeta = 0.707$.

(c) What is the gain margin of the system if the gain is set to the value determined in part (b)? Answer this question *without* using any frequency-response methods.

(d) Create the Bode plots for the system, and determine the gain margin that results for PM $= 65°$. What damping ratio would you expect for this PM?

(e) Sketch a root locus for the system shown in Fig. 6.94(b). How does it differ from the one in part (a)?

(f) For the systems in Figs. 6.94(a) and (b), how does the transfer function $Y_2(s)/R(s)$ differ from $Y_1(s)/R(s)$? Would you expect the step response to $r(t)$ to be different for the two cases?

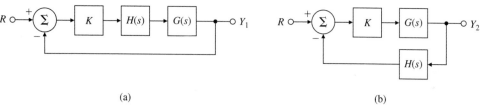

(a)   (b)

Figure 6.94
Block diagram for Problem 6.32: (a) unity feedback; (b) $H(s)$ in feedback

6.33 For the system shown in Fig. 6.95, use Bode and root-locus plots to determine the gain and frequency at which instability occurs. What gain (or gains) gives a PM of $20°$? What is the gain margin when PM $= 20°$?

Figure 6.95
Control system for
Problem 6.33

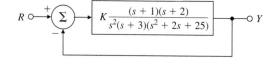

6.34 A magnetic tape-drive speed-control system is shown in Fig. 6.96. The speed sensor is slow enough that its dynamics must be included. The speed-measurement time constant is $\tau_m = 0.5$ sec; the reel time constant is $\tau_r = J/b = 4$ sec, where $b =$ the output shaft damping constant $= 1$ N·m·sec; and the motor time constant is $\tau_1 = 1$ sec.

(a) Determine the gain K required to keep the steady-state speed error to less than 7% of the reference-speed setting.

(b) Determine the gain and phase margins of the system. Is this a good system design?

6.35 For the system in Fig. 6.97, determine the Nyquist plot and apply the Nyquist criterion

(a) to determine the range of values of K (positive and negative) for which the system will be stable, and

Figure 6.96

Magnetic tape-drive
speed control

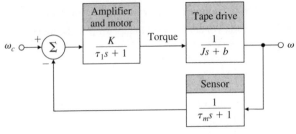

Figure 6.97

Control system for
Problems 6.35, 6.62,
and 6.63

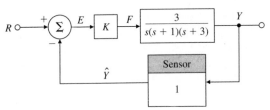

(b) to determine the number of roots in the RHP for those values of K for which the system is unstable. Check your answer by using a rough root-locus sketch.

6.36 For the system shown in Fig. 6.98, determine the Nyquist plot and apply the Nyquist criterion

(a) to determine the range of values of K (positive and negative) for which the system will be stable, and

(b) to determine the number of roots in the RHP for those values of K for which the system is unstable. Check your answer by using a rough root-locus sketch.

Figure 6.98

Control system for
Problem 6.36

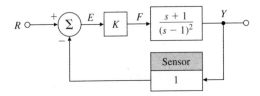

6.37 For the system shown in Fig. 6.99, determine the Nyquist plot and apply the Nyquist criterion

(a) to determine the range of values of K (positive and negative) for which the system will be stable, and

(b) to determine the number of roots in the RHP for those values of K for which the system is unstable. Check your answer by using a rough root-locus sketch.

Figure 6.99

Control system for
Problem 6.37

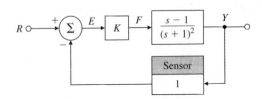

6.38 The Nyquist diagrams for two stable, open-loop systems are sketched in Fig. 6.100. The proposed operating gain is indicated as K_0, and arrows indicate increasing frequency. In each case give a rough estimate of the following quantities for the closed-loop (unity feedback) system:

(a) Phase margin

(b) Damping ratio

(c) Range of gain for stability (if any)

(d) System type (0, 1, or 2)

Figure 6.100

Nyquist plots for Problem 6.38

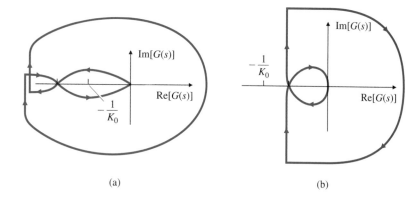

(a)

(b)

6.39 The steering dynamics of a ship are represented by the transfer function

$$\frac{V(s)}{\delta_r(s)} = G(s) = \frac{K[-(s/0.142) + 1]}{s(s/0.325 + 1)(s/0.0362 + 1)},$$

where V is the ship's lateral velocity in meters per second, and δ_r is the rudder angle in radians.

(a) Use the MATLAB command bode to plot the log magnitude and phase of $G(j\omega)$ for $K = 0.2$.

(b) On your plot, indicate the crossover frequency, PM, and GM.

(c) Is the ship steering system stable with $K = 0.2$?

(d) What value of K would yield a PM of $30°$, and what would the crossover frequency be?

6.40 For the open-loop system

$$KG(s) = \frac{K(s + 1)}{s^2(s + 10)^2},$$

determine the value for K at the stability boundary and the values of K at the points where PM $= 30°$.

Problems for Section 6.5: Bode's Gain–Phase Relationship

6.41 The frequency response of a plant in a unity feedback configuration is sketched in Fig. 6.101. Assume that the plant is open-loop stable and minimum-phase.

(a) What is the velocity constant K_v for the system as drawn?

(b) What is the damping ratio of the complex poles at $\omega = 100$?

Figure 6.101

Magnitude frequency response for Problem 6.41

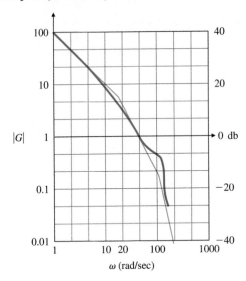

(c) Approximately what is the system error in tracking (following) a sinusoidal input of $\omega = 3$ rad/sec?

(d) What is the PM of the system as drawn? (Estimate to within $\pm 10°$.)

6.42   For the system

$$G(s) = \frac{100(s/a + 1)}{s(s + 1)(s/b + 1)},$$

where $b = 10a$, find the approximate value of a that will yield the best PM by sketching only candidate values of the frequency-response magnitude.

Problem for Section 6.6: Closed-Loop Frequency Response

6.43   For the open-loop system

$$KG(s) = \frac{K(s + 1)}{s^2(s + 10)^2},$$

determine the value for K that will yield PM $\geq 30°$ and the maximum possible closed-loop bandwidth. Use MATLAB to find the bandwidth.

Problems for Section 6.7: Compensation Design

6.44   For the lead compensator

$$D(s) = \frac{Ts + 1}{\alpha Ts + 1},$$

where $\alpha < 1$,

(a) Show that the phase of the lead compensator is given by

$$\phi = \tan^{-1}(T\omega) - \tan^{-1}(\alpha T\omega).$$

(b) Show that the frequency where the phase is maximum is given by

$$\omega_{max} = \frac{1}{T\sqrt{\alpha}}$$

and that the maximum phase corresponds to

$$\sin \phi_{max} = \frac{1-\alpha}{1+\alpha}.$$

(c) Rewrite your expression for ω_{max} to show that the maximum-phase frequency occurs at the geometric mean of the two corner frequencies on a logarithmic scale:

$$\log \omega_{max} = \frac{1}{2}\left(\log \frac{1}{T} + \log \frac{1}{\alpha T}\right).$$

(d) To derive the same results in terms of the pole–zero locations, rewrite $D(s)$ as

$$D(s) = \frac{s+z}{s+p},$$

and then show that the phase is given by

$$\phi = \tan^{-1}\left(\frac{\omega}{|z|}\right) - \tan^{-1}\left(\frac{\omega}{|p|}\right),$$

such that

$$\omega_{max} = \sqrt{|z||p|}.$$

Hence the frequency at which the phase is maximum is the square root of the product of the pole and zero locations.

6.45 For the third-order servo system

$$G(s) = \frac{50{,}000}{s(s+10)(s+50)},$$

use Bode plot sketches to design a lead compensator so that PM $\geq 50°$ and $\omega_{BW} \geq 20$ rad/sec. Then verify and refine your design by using MATLAB.

6.46 For the system shown in Fig. 6.102, suppose that

$$G(s) = \frac{5}{s(s+1)(s/5+1)}.$$

Use Bode plot sketches to design a lead compensation $D(s)$ with unity DC gain so that PM $\geq 40°$. Then verify and refine your design by using MATLAB. What is the approximate bandwidth of the system?

Figure 6.102
Control system for
Problem 6.46

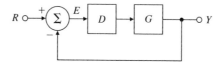

6.47 Derive the transfer function from T_d to θ for the system in Fig. 6.70. Then apply the Final Value Theorem (assuming $T_d = $ constant) to determine whether $\theta(\infty)$ is nonzero for the following two cases:

(a) When $D(s)$ has no integral term: $\lim_{s \to 0} D(s) = $ constant;

(b) When $D(s)$ has an integral term:

$$D(s) = \frac{D'(s)}{s}.$$

In this case, $\lim_{s \to 0} D'(s) = $ constant.

6.48 The inverted pendulum has a transfer function given by Eq. (2.31), which is similar to

$$G(s) = \frac{1}{s^2 - 1}.$$

(a) Use Bode plot sketches to design a lead compensator to achieve a PM of 30°. Then verify and refine your design by using MATLAB.

(b) Sketch a root locus and correlate it with the Bode plot of the system.

(c) Could you obtain the frequency response of this system experimentally?

6.49 The open-loop transfer function of a unity feedback system is

$$G(s) = \frac{K}{s(s/5 + 1)(s/50 + 1)}.$$

(a) Use Bode plot sketches to design a lag compensator for $G(s)$ so that the closed-loop system satisfies the following specifications:

(i) The steady-state error to a unit-ramp reference input is less than 0.01.

(ii) PM $\geq 40°$.

(b) Verify and refine your design by using MATLAB.

6.50 The open-loop transfer function of a unity-feedback system is

$$G(s) = \frac{K}{s(s/5 + 1)(s/200 + 1)}.$$

(a) Use Bode plot sketches to design a lead compensator for $G(s)$ so that the closed-loop system satisfies the following specifications:

(i) The steady-state error to a unit-ramp reference input is less than 0.01.

(ii) For the dominant closed-loop poles, the damping ratio $\zeta \geq 0.4$.

(b) Verify and refine your design using MATLAB, including a direct computation of the damping of the dominant closed-loop poles.

6.51 A DC motor with negligible armature inductance is to be used in a position control system. Its open-loop transfer function is given by

$$G(s) = \frac{50}{s(s/5 + 1)}.$$

(a) Use Bode plot sketches to design a compensator for the motor so that the closed-loop system satisfies the following specifications:

(i) The steady-state error to a unit-ramp input is less than $1/200$.

(ii) The unit-step response has an overshoot of less than 20%.

(iii) The bandwidth of the compensated system is no less than that of the uncompensated system.

(b) Verify and/or refine your design using MATLAB, including a direct computation of the step-response overshoot.

6.52 The open-loop transfer function of a unity-feedback system is

$$G(s) = \frac{K}{s(1 + s/5)(1 + s/20)}.$$

(a) Sketch the system block diagram, including input reference commands and sensor noise.

(b) Use Bode plot sketches to design a compensator for $G(s)$ so that the closed-loop system satisfies the following specifications:

(i) The steady-state error to a unit-ramp input is less than 0.01.

(ii) PM $\geq 45°$.

(iii) The steady-state error for sinusoidal inputs with $\omega < 0.2$ rad/sec is less than $^1/_{250}$.

(iv) Noise components introduced with the sensor signal at frequencies greater than 200 rad/sec are to be attenuated at the output by at least a factor of 100.

(c) Verify and/or refine your design using MATLAB, including a computation of the closed-loop frequency response to verify (iv).

6.53 Consider a Type 1 unity-feedback system with

$$G(s) = \frac{K}{s(s+1)}.$$

Use Bode plot sketches to design a lead compensator so that $K_v = 20$ sec^{-1} and PM $> 40°$. Use MATLAB to verify and/or refine your design so that it meets the specifications.

6.54 Consider a satellite attitude-control system with the transfer function

$$G(s) = \frac{0.05(s+25)}{s^2(s^2+0.1s+4)}.$$

Amplitude-stabilize the system using lead compensation so that GM ≥ 2 (6 db), and PM $\geq 45°$, keeping the bandwidth as high as possible with a single lead.

6.55 In one mode of operation, the autopilot of a jet transport is used to control altitude. For the purpose of designing the altitude portion of the autopilot loop, only the long-period airplane dynamics are important. The linearized relationship between altitude and elevator angle for the long-period dynamics is

$$G(s) = \frac{h(s)}{\delta(s)} = \frac{20(s+0.01)}{s(s^2+0.01s+0.0025)} \frac{\text{ft/ sec}}{\text{deg}}.$$

The autopilot receives from the altimeter an electrical signal proportional to altitude. This signal is compared with a command signal (proportional to the altitude selected by the pilot), and the difference provides an error signal. The error signal is processed through compensation, and the result is used to command the elevator actuators. A block diagram of this system is shown in Fig. 6.103. You have been given the task of designing the compensation. Begin by considering a proportional control law $D(s) = K$.

(a) Use MATLAB to draw a Bode plot of the open-loop system for $D(s) = K = 1$.

(b) What value of K would provide a crossover frequency (i.e., where $|G| = 1$) of 0.16 rad/sec?

(c) For this value of K, would the system be stable if the loop were closed?

(d) What is the PM for this value of K?

Figure 6.103
Control system for
Problem 6.55

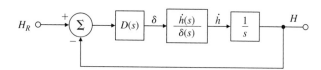

(e) Sketch the Nyquist plot of the system, and locate carefully any points where the phase angle is 180° or the magnitude is unity.

(f) Use MATLAB to plot the root locus with respect to K, and locate the roots for your value of K from part (b).

(g) What steady-state error would result if the command was a step change in altitude of 1000 ft?

For parts (h) and (i), assume a compensator of the form

$$D(s) = K\frac{Ts+1}{\alpha Ts+1}.$$

(h) Choose the parameters K, T, and α so that the crossover frequency is 0.16 rad/sec and the PM is greater than 50°. Verify your design by superimposing a Bode plot of $D(s)G(s)/K$ on top of the Bode plot you obtained for part (a), and measure the PM directly.

(i) Use MATLAB to plot the root locus with respect to K for the system, including the compensator you designed in part (h). Locate the roots for your value of K from part (h).

(j) Altitude autopilots also have a mode in which the rate of climb is sensed directly and commanded by the pilot.

 (i)   Sketch the block diagram for this mode.

 (ii) Define the pertinent $G(s)$.

 (iii) Design $D(s)$ so that the system has the same crossover frequency as the altitude hold mode and the PM is greater than 50°.

6.56 For a system with open-loop transfer function

$$G(s) = \frac{10}{s[(s/1.4)+1][(s/3)+1]},$$

design a lag compensator with unity DC gain so that PM $\geq 40°$. What is the approximate bandwidth of this system?

6.57 For the ship-steering system in Problem 6.39,

(a) Design a compensator that meets the following specifications:

 (i)   Velocity constant $K_v = 2$,

 (ii) PM $\geq 50°$,

 (iii) Unconditional stability (PM > 0 for all $\omega \leq \omega_c$, the crossover frequency).

(b) For your final design, draw a root locus with respect to K, and indicate the location of the closed-loop poles.

6.58 Consider a unity-feedback system with

$$G(s) = \frac{1}{s\,(s/20+1)\,(s^2/100^2+0.5s/100+1)}. \qquad (6.78)$$

(a) A lead compensator is introduced with $\alpha = 1/5$ and a zero at $1/T = 20$. How must the gain be changed to obtain crossover at $\omega_c = 31.6$ rad/sec, and what is the resulting value of K_v?

(b) With the lead compensator in place, what is the required value of K for a lag compensator that will readjust the gain to a K_v value of 100?

(c) Place the pole of the lag compensator at 3.16 rad/sec, and determine the zero location that will maintain the crossover frequency at $\omega_c = 31.6$ rad/sec. Plot the compensated frequency response on the same graph.

(d) Determine the PM of the compensated design.

6.59 Golden Nugget Airlines had great success with their free bar near the tail of the airplane. (See Problem 5.39.) However, when they purchased a much larger airplane to handle the passenger demand, they discovered that there was some flexibility in the fuselage that caused a lot of unpleasant yawing motion at the rear of the airplane when in turbulence, which caused the revelers to spill their drinks. The approximate transfer function for the Dutch roll mode (Section 10.3.1) is

$$\frac{r(s)}{\delta_r(s)} = \frac{8.75(4s^2 + 0.4s + 1)}{(s/0.01 + 1)(s^2 + 0.24s + 1)},$$

where r is the airplane's yaw rate and δ_r is the rudder angle. In performing a finite element analysis (FEA) of the fuselage structure and adding those dynamics to the Dutch roll motion, they found that the transfer function needed additional terms which reflected the fuselage lateral bending that occurred due to excitation from the rudder and turbulence. The revised transfer function is

$$\frac{r(s)}{\delta_r(s)} = \frac{8.75(4s^2 + 0.4s + 1)}{(s/0.01 + 1)(s^2 + 0.24s + 1)} \cdot \frac{1}{(s^2/\omega_b^2 + 2\zeta s/\omega_b + 1)},$$

where ω_b is the frequency of the bending mode ($= 10$ rad/sec) and ζ is the bending mode damping ratio ($=0.02$). Most swept-wing airplanes have a "yaw damper," which essentially feeds back yaw rate measured by a rate gyro to the rudder with a simple proportional control law. For the new Golden Nugget airplane, the proportional feedback gain $K = 1$, where

$$\delta_r(s) = -Kr(s). \tag{6.79}$$

(a) Make a Bode plot of the open-loop system, determine the PM and GM for the nominal design, and plot the step response and Bode magnitude of the closed-loop system. What is the frequency of the lightly damped mode that is causing the difficulty?

(b) Investigate remedies to quiet down the oscillations, but maintain the same low-frequency gain in order not to affect the quality of the Dutch roll damping provided by the yaw rate feedback. Specifically, investigate each of the following, one at a time:

(i) Increasing the damping of the bending mode from $\zeta = 0.02$ to $\zeta = 0.04$ (would require adding energy-absorbing material in the fuselage structure).

(ii) Increasing the frequency of the bending mode from $\omega_b = 10$ rad/sec to $\omega_b = 20$ rad/sec (would require stronger and heavier structural elements).

(iii) Adding a low-pass filter in the feedback—that is, replacing K in Eq. (6.79) with $KD(s)$, where

$$D(s) = \frac{1}{s/\tau_p + 1}. \tag{6.80}$$

Pick τ_p so that the objectionable features of the bending mode are reduced while maintaining the PM $\geq 60°$.

(iv) Adding a notch filter as described in Section 5.4.3. Pick the frequency of the notch zero to be at ω_b, with a damping of $\zeta = 0.04$, and pick the denominator poles to be $(s/100 + 1)^2$, keeping the DC gain of the filter $= 1$.

(c) Investigate the sensitivity of the preceding two compensated designs (iii and iv) by determining the effect of a reduction in the bending mode frequency of -10%. Specifically, reexamine the two designs by tabulating the GM, PM, closed-loop bending mode damping ratio, and resonant-peak amplitude, and qualitatively describe the differences in the step response.

(d) What do you recommend to Golden Nugget to help their customers quit spilling their drinks? (Telling them to get back in their seats is not an acceptable answer for this problem! Make the recommendation in terms of improvements to the yaw damper.)

△ **6.60** Consider a system with the open-loop transfer function (loop gain)

$$G(s) = \frac{1}{s(s+1)(s/10+1)}.$$

(a) Create the Bode plot for the system, and find GM and PM.

(b) Compute the sensitivity function and plot its magnitude frequency response.

(c) Compute the vector margin (VM).

△ **6.61** Prove that the sensitivity function (s) has magnitude greater than 1 inside a circle with a radius of 1 centered at the -1 point. What does this imply about the shape of the Nyquist plot if closed-loop control is to outperform open-loop control at all frequencies?

△ **6.62** Consider the system in Fig. 6.102 with the plant transfer function

$$G(s) = \frac{10}{s(s/10+1)}.$$

(a) We wish to design a compensator $D(s)$ that satisfies the following design specifications:

(i) $K_v = 100$

(ii) $PM \geq 45°$

(iii) Sinusoidal inputs of up to 1 rad/sec to be reproduced with $\leq 2\%$ error

(iv) Sinusoidal inputs with a frequency of greater than 100 rad/sec to be attenuated at the output to $\leq 5\%$ of their input value

(b) Create the Bode plot of $G(s)$, choosing the open-loop gain so that $K_v = 100$.

(c) Show that a *sufficient* condition for meeting the specification on sinusoidal inputs is that the magnitude plot lies outside the shaded regions in Fig. 6.104. Recall that

Figure 6.104

Control system constraints for Problem 6.62

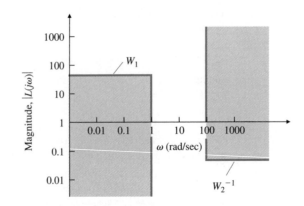

$$\frac{Y}{R} = \frac{KG}{1 + KG} \quad \text{and} \quad \frac{E}{R} = \frac{1}{1 + KG}.$$

(d) Explain why introducing a lead network alone cannot meet the design specifications.

(e) Explain why a lag network alone cannot meet the design specifications.

(f) Develop a full design using a lead–lag compensator that meets all the design specifications without altering the previously chosen low-frequency open-loop gain.

△ *Problems for Section 6.8: Time Delay*

6.63 Assume that the system

$$G(s) = \frac{e^{-T_d s}}{s + 10}$$

has a 0.2-sec time delay ($T_d = 0.2$ sec). While maintaining a phase margin $\geq 40°$, find the maximum possible bandwidth by using the following:

(a) One lead-compensator section

$$D(s) = K\frac{s + a}{s + b},$$

where $b/a = 100$;

(b) Two lead-compensator sections

$$D(s) = K\left(\frac{s + a}{s + b}\right)^2,$$

where $b/a = 10$.

(c) Comment on the statement in the text about the limitations on the bandwidth imposed by a delay.

6.64 Determine the range of K for which the following systems are stable:

(a) $G(s) = K\dfrac{e^{-4s}}{s}$

(b) $G(s) = K\dfrac{e^{-s}}{s(s + 2)}$

6.65 In Chapter 5, we used various approximations for the time delay, one of which is the first order Padé:

$$e^{-T_d s} \cong H_1(s) = \frac{1 - T_d s/2}{1 + T_d s/2}.$$

Using frequency response methods, the exact time delay

$$H_2(s) = e^{-T_d s}$$

can be obtained. Plot the phase of $H_1(s)$ and $H_2(s)$, and discuss the implications.

6.66 Consider the heat exchanger of Example 2.15 with the open-loop transfer function

$$G(s) = \frac{e^{-5s}}{(10s + 1)(60s + 1)}.$$

(a) Design a lead compensator that yields PM $\geq 45°$ and the maximum possible closed-loop bandwidth.

(b) Design a PI compensator that yields PM $\geq 45°$ and the maximum possible closed-loop bandwidth.

△   *Problems for Section 6.9: Alternative Presentations of Data*

6.67 A feedback control system is shown in Fig. 6.105. The closed-loop system is specified to have an overshoot of less than 30% to a step input.

(a) Determine the corresponding PM specification in the frequency domain and the corresponding closed-loop resonant-peak value M_r. (See Fig. 6.38.)

(b) From Bode plots of the system, determine the maximum value of K that satisfies the PM specification.

(c) Plot the data from the Bode plots [adjusted by the K obtained in part (b)] on a copy of the Nichols chart in Fig. 6.84, and determine the resonant peak magnitude M_r. Compare that with the approximate value obtained in part (a).

(d) Use the Nichols chart to determine the resonant-peak frequency ω_r and the closed-loop bandwidth.

Figure 6.105
Control system for
Problem 6.67

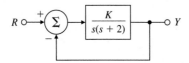

6.68 The Nichols plots of an uncompensated and a compensated system are shown in Fig. 6.106.

(a) What are the resonance peaks of each system?

(b) What are the PM and GM of each system?

(c) What are the bandwidths of each system?

(d) What type of compensation is used?

6.69 Consider the system shown in Fig. 6.97.

(a) Construct an inverse Nyquist plot of $[Y(j\omega)/E(j\omega)]^{-1}$. (See Appendix W6.)

(b) Show how the value of K for neutral stability can be read directly from the inverse Nyquist plot.

(c) For $K = 4, 2$, and 1, determine the gain and phase margins.

(d) Construct a root-locus plot for the system, and identify corresponding points in the two plots. To what damping ratios ζ do the GM and PM of part (c) correspond?

6.70 An unstable plant has the transfer function

$$\frac{Y(s)}{F(s)} = \frac{s+1}{(s-1)^2}.$$

A simple control loop is to be closed around it, in the same manner as in the block diagram in Fig. 6.97.

(a) Construct an inverse Nyquist plot of Y/F. (See Appendix W6.)

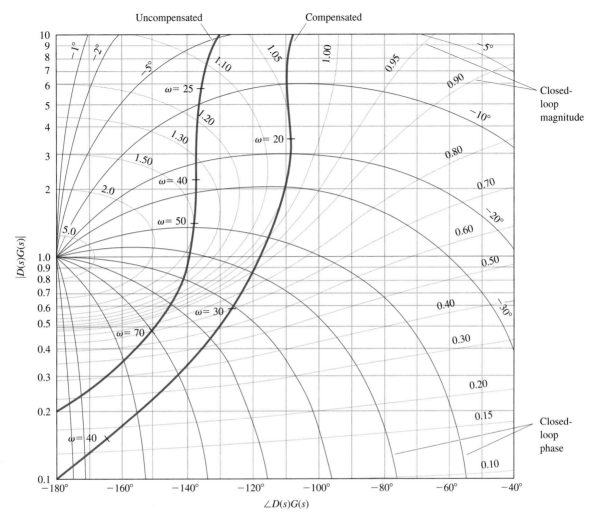

Uncompensated   Compensated

Closed-
loop
magnitude

Closed-
loop
phase

$\angle D(s)G(s)$

Figure 6.106
Nichols plots for Problem 6.68

(b) Choose a value of K to provide a PM of 45°. What is the corresponding GM?

(c) What can you infer from your plot about the stability of the system when $K < 0$?

(d) Construct a root-locus plot for the system, and identify corresponding points in the two plots. In this case, to what value of ζ does PM = 45° correspond?

6.71 Consider the system shown in Fig. 6.107(a).

(a) Construct a Bode plot for the system.

(b) Use your Bode plot to sketch an inverse Nyquist plot. (See Appendix W6.)

(c) Consider closing a control loop around $G(s)$, as shown in Fig. 6.107(b). Using the inverse Nyquist plot as a guide, read from your Bode plot the values of GM and PM when $K = 0.7$, 1.0, 1.4, and 2. What value of K yields PM $= 30°$?

(d) Construct a root-locus plot, and label the same values of K on the locus. To what value of ζ does each pair of PM/GM values correspond? Compare ζ versus PM with the rough approximation in Fig. 6.59.

Figure 6.107

Control system for Problem 6.71

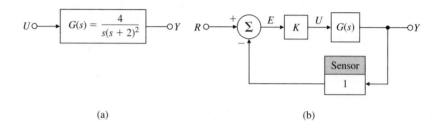

(a)   (b)

7

State-Space Design

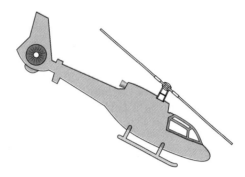

A Perspective on State-Space Design

In addition to the transform techniques of root locus and frequency response, there is a third major method of designing feedback control systems: the state-space method. We will introduce the state-variable method of describing differential equations. In state-space design, the control engineer designs a dynamic compensation by working directly with the state-variable description of the system. Like the transform techniques, the aim of the state-space method is to find a compensation $D(s)$, such as that shown in Fig. 7.1, that satisfies the design specifications. Because the state-space method of describing the plant and computing the compensation is so different from the transform techniques, it may seem at first to be solving an entirely different problem. We selected the examples and analysis given toward the end of this chapter to help convince you that, indeed, state-space design results in a compensator with a transfer function $D(s)$ that is equivalent to those $D(s)$ compensators obtained with the other two methods.

Because it is particularly well suited to the use of computer techniques, state-space design is increasingly studied and used today by control engineers.

Chapter Overview

This chapter begins by considering the purposes and advantages of using state-space design. We discuss selection of state-variables and state-space models for various dynamic systems through several examples in Section 7.2.

413

Figure 7.1

A control system design definition

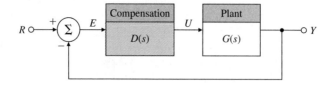

Models in state-variable form enhance our ability to apply the computational efficiency of computer-aided design tools such as MATLAB®. In Section 7.3 we show that it is beneficial to look at the state-variable form in terms of an analog computer simulation model. In Section 7.4 we review the development of state-variable equations from block diagrams. We then solve for the dynamic response, using state equations for both hand and computer analysis. Having covered these preliminary fundamentals, we next proceed to the major task of control system design via state-space. The steps of the design method are as follows:

1. Select closed-loop pole (root as referred to in previous chapters) locations and develop the control law for the closed-loop system that corresponds to satisfactory dynamic response (Sections 7.5 and 7.6).
2. Design an estimator (Section 7.7).
3. Combine the control law and the estimator (Section 7.8).
4. Introduce the reference input (Sections 7.5.2 and 7.9).

After working through the central design steps, we briefly explore the use of integral control in state-space (Section 7.10). The next three sections of this chapter consider briefly some additional concepts pertaining to the state-space method; because they are relatively advanced, they may be considered optional to some courses or readers. Finally Section 7.14 provides some historical perspective for the material in this chapter.

7.1 Advantages of State-Space

The idea of **state-space** comes from the state-variable method of describing differential equations. In this method the differential equations describing a dynamic system are organized as a set of first-order differential equations in the vector-valued state of the system, and the solution is visualized as a trajectory of this state vector in space. **State-space control design** is the technique in which the control engineer designs a dynamic compensation by working directly with the state-variable description of the system. Thus far, we have seen that the ordinary differential equations (ODEs) of physical dynamic systems can be manipulated into state-variable form. In the field of mathematics, where ODEs are studied, the state-variable form is called the **normal form** for the equations. There are several good reasons for studying equations in this form, three of which are listed here:

Normal form

- *To study more general models:* The ODEs do not have to be linear or stationary. Thus, by studying the equations themselves, we can develop methods that are very general. Having them in state-variable form gives us a compact, standard form for study. Furthermore, the techniques of state-space analysis and design easily extend to systems with multiple inputs and/or multiple outputs. Of course, in this text we study mainly linear time-invariant models with single input and output (for the reasons given earlier).

Phase plane

- *To introduce the ideas of geometry into differential equations:* In physics the plane of position versus velocity of a particle or rigid body is called the **phase plane**, and the trajectory of the motion can be plotted as a curve in this plane. The state is a generalization of that idea to include more than two dimensions. While we cannot plot more than three dimensions, the concepts of distance, of orthogonal and parallel lines, and other concepts from geometry can be useful in visualizing the solution of an ODE as a path in state-space.

- *To connect internal and external descriptions:* The state of a dynamic system often directly describes the distribution of internal energy in the system. For example, it is common to select the following as state-variables: position (potential energy), velocity (kinetic energy), capacitor voltage (electric energy), and inductor current (magnetic energy). The internal energy can always be computed from the state-variables. By a system of analysis to be described shortly, we can relate the state to the system inputs and outputs and thus connect the internal variables to the external inputs and to the sensed outputs. In contrast, the transfer function relates only the input to the output and does not show the internal behavior. The state form keeps the latter information, which is sometimes important.

Use of the state-space approach has often been referred to as **modern control design**, and use of transfer-function-based methods, such as root locus and frequency response, referred to as **classical control design**. However, because the state-space method of description for ODEs has been in use for over 100 years and was introduced to control design in the late 1950s, it seems somewhat misleading to refer to it as modern. We prefer to refer to the two approaches to design as state-space methods and transform methods.

Advantages of state-space design are especially apparent when the system to be controlled has more than one control input or more than one sensed output. However, in this book we shall examine the ideas of state-space design using the simpler single-input-single output (SISO) systems. The design approach used for systems described in state form is "divide and conquer." First, we design the control as if all of the state were measured and available for use in the control law. This provides the possibility of assigning arbitrary dynamics for the system. Having a satisfactory control law based on full-state feedback, we introduce the concept of an observer and construct estimates of the state based on the sensed output. We then show that these estimates can be used in place of the actual state-variables. Finally, we introduce the external reference-command inputs, and the structure is complete. Only at this point can we recognize that the resulting compensation has the same essential structure as that developed with transform methods.

Before we can begin the design using state descriptions, it is necessary to develop some analytical results and tools from matrix linear algebra for use throughout the chapter. We assume that you are familiar with such elementary matrix concepts as the identity matrix, triangular and diagonal matrices, and the transpose of a matrix.

We also assume that you have some familiarity with the mechanics of matrix algebra, including adding, multiplying, and inverting matrices. More advanced results will be developed in Section 7.4 in the context of the dynamic response of a linear system. All of the linear algebra results used in this chapter are repeated in Appendix WE for your reference and review.

7.2 System Description in State-Space

The motion of any finite dynamic system can be expressed as a set of first-order ODEs. This is often referred to as the state-variable representation. For example, the use of Newton's law and the free-body diagram in Section 2.1 typically lead to second-order differential equations—that is, equations that contain the second derivative, such as $\ddot{x}$ in Eq. (2.3) or $\ddot{\theta}$ in Eq. (2.15). The latter equation can be expressed as

$$\dot{x}_1 = x_2, \tag{7.1}$$

$$\dot{x}_2 = \frac{u}{I}, \tag{7.2}$$

where

$$u = F_c d + M_D,$$

$$x_1 = \theta,$$

$$x_2 = \dot{\theta},$$

$$\dot{x}_2 = \ddot{\theta}.$$

The output of this system is θ, the satellite attitude.

Standard form of linear differential equations

These same equations can be represented in the **state-variable form** as the vector equation

$$\dot{\mathbf{x}} = \mathbf{F}\mathbf{x} + \mathbf{G}u, \tag{7.3}$$

where the input is u and the output is

$$y = \mathbf{H}\mathbf{x} + Ju. \tag{7.4}$$

The column vector $\mathbf{x}$ is called the **state of the system** and contains n elements for an nth-order system. For mechanical systems, the state vector elements usually consist of the positions and velocities of the separate bodies, as is the case for the example in Eqs. (7.1) and (7.2). The quantity $\mathbf{F}$ is an $n \times n$ **system matrix**, $\mathbf{G}$ is an $n \times 1$ **input matrix**, $\mathbf{H}$ is a $1 \times n$ row matrix referred to as the **output matrix**, and J is a scalar called the **direct transmission term**.[1] To save space, we will sometimes refer to a state vector by its **transpose**,

$$\mathbf{x} = [\ x_1 \quad x_2 \ldots\ ]^T,$$

which is equivalent to

$$\mathbf{x} = \begin{bmatrix} x_1 \\ x_2 \\ \vdots \end{bmatrix}.$$

---

[1] It is also common to use the notation **A, B, C,** and D in place of **F, G, H,** and J. We will typically use **F, G** to represent plant dynamics and **A, B** to represent a general linear system.

The differential equation models of more complex systems, such as those developed in Chapter 2 on mechanical, electrical, and electromecheanical systems, can be described by state-variables through selection of positions, velocities, capacitor voltages, and inductor currents as suitable state-variables.

In this chapter we will consider control systems design using the state-variable form. For the case in which the relationships are nonlinear [such as the case in Eqs. (2.22), (2.75), and (2.79)], the linear form cannot be used directly. One must linearize the equations as we did in Chapter 2 to fit the form (see also Chapter 9).

The state-variable method of specifying differential equations is used by computer-aided control systems design software packages (e.g., MATLAB). Therefore, in order to specify linear differential equations to the computer, you need to know the values of the matrices $\mathbf{F}$, $\mathbf{G}$, $\mathbf{H}$, and the constant J.

EXAMPLE 7.1   *Satellite Attitude Control Model in State-Variable Form*

Determine the $\mathbf{F}$, $\mathbf{G}$, $\mathbf{H}$, J matrices in the state-variable form for the satellite attitude control model in Example 2.3 with $M_D = 0$.

Solution. Define the attitude and the angular velocity of the satellite as the state-variables so that $\mathbf{x} \triangleq [\theta \ \omega]^T$.[2] The single second-order equation (2.15) can then be written in an equivalent way as two first-order equations:

$$\dot{\theta} = \omega,$$

$$\dot{\omega} = \frac{d}{I} F_c.$$

These equations are expressed, using Eq. (7.3), $\dot{\mathbf{x}} = \mathbf{F}\mathbf{x} + \mathbf{G}u$, as

$$\begin{bmatrix} \dot{\theta} \\ \dot{\omega} \end{bmatrix} = \begin{bmatrix} 0 & 1 \\ 0 & 0 \end{bmatrix} \begin{bmatrix} \theta \\ \omega \end{bmatrix} + \begin{bmatrix} 0 \\ d/I \end{bmatrix} F_c.$$

The output of the system is the satellite attitude, $y = \theta$. Using Eq. (7.4), $y = \mathbf{H}\mathbf{x} + Ju$, this relation is expressed as

$$y = [1 \quad 0] \begin{bmatrix} \theta \\ \omega \end{bmatrix}.$$

Therefore, the matrices for the state-variable form are

$$\mathbf{F} = \begin{bmatrix} 0 & 1 \\ 0 & 0 \end{bmatrix}, \quad \mathbf{G} = \begin{bmatrix} 0 \\ d/I \end{bmatrix}, \quad \mathbf{H} = [1 \quad 0], \quad J = 0,$$

and the input $u \triangleq F_c$.

For this very simple example, the state-variable form is a more cumbersome way of writing the differential equation than the second-order version in Eq. (2.15). However, the method is not more cumbersome for most systems, and the advantages of having a standard form for use in computer-aided design have led to widespread use of the state-variable form.

---

[2]The symbol $\triangleq$ means "is to be defined."

The next example has more complexity and shows how to use MATLAB to find the solution of linear differential equations.

EXAMPLE 7.2   *Cruise Control Step Response*

(a) Rewrite the equation of motion from Example 2.1 in state-variable form, where the output is the car position x.

(b) Use MATLAB to find the response of the velocity of the car for the case in which the input jumps from being $u = 0$ at time $t = 0$ to a constant $u = 500$ N thereafter. Assume that the car mass m is 1000 kg and $b = 50$ N·sec/m.

Solution

(a) **Equations of motion:** First we need to express the differential equation describing the plant, Eq. (2.3), as a set of simultaneous first-order equations. To do so, we define the position and the velocity of the car as the state-variables x and v, so that $\mathbf{x} = [x \ v]^T$. The single second-order equation, Eq. (2.3), can then be rewritten as a set of two first-order equations:

$$\dot{x} = v,$$

$$\dot{v} = -\frac{b}{m}v + \frac{1}{m}u.$$

Next, we use the standard form of Eq. (7.3), $\dot{\mathbf{x}} = \mathbf{Fx} + \mathbf{Gu}$, to express these equations:

$$\begin{bmatrix} \dot{x} \\ \dot{v} \end{bmatrix} = \begin{bmatrix} 0 & 1 \\ 0 & -b/m \end{bmatrix} \begin{bmatrix} x \\ v \end{bmatrix} + \begin{bmatrix} 0 \\ 1/m \end{bmatrix} u. \qquad (7.5)$$

The output of the system is the car position $y = x_1 = x$, which is expressed in matrix form as

$$y = \begin{bmatrix} 1 & 0 \end{bmatrix} \begin{bmatrix} x \\ v \end{bmatrix},$$

or

$$y = \mathbf{Hx}.$$

So the state-variable-form matrices defining this example are

$$\mathbf{F} = \begin{bmatrix} 0 & 1 \\ 0 & -b/m \end{bmatrix}, \quad \mathbf{G} = \begin{bmatrix} 0 \\ 1/m \end{bmatrix}, \quad \mathbf{H} = \begin{bmatrix} 1 & 0 \end{bmatrix}, \quad J = 0.$$

(b) **Time response:** The equations of motion are those given in part (a), except that now the output is $v = x_2$. Therefore, the output matrix is

$$\mathbf{H} = \begin{bmatrix} 0 & 1 \end{bmatrix}.$$

The coefficients required are $b/m = 0.05$ and $1/m = 0.001$. The numerical values of the matrices defining the system are thus

$$\mathbf{F} = \begin{bmatrix} 0 & 1 \\ 0 & -0.05 \end{bmatrix}, \quad \mathbf{G} = \begin{bmatrix} 0 \\ 0.001 \end{bmatrix}, \quad \mathbf{H} = \begin{bmatrix} 0 & 1 \end{bmatrix}, \quad J = 0.$$

The step function in MATLAB calculates the time response of a linear system to a unit-step input. Because the system is linear, the output for this case can be multiplied by the magnitude of the input step to derive a step response of any amplitude. Equivalently, the **G** matrix can be multiplied by the magnitude of the input step.

Figure 7.2

Response of the car velocity to a step in u

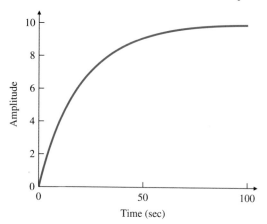

Step response with MATLAB

The statements

```
F = [0  1;0  −0.05];
G = [0;0.001];
H = [0  1];
J = 0;
sys = ss(F, 500*G,H,J); % step gives unit step response, so 500*G
    gives u = 500 N.
step(sys); % plots the step response
```

calculate and plot the time response for an input step with a 500-N magnitude. The step response is shown in Fig. 7.2.

EXAMPLE 7.3

Bridged Tee Circuit in State-Variable Form

Determine the state-space equations for the circuit shown in Fig. 2.25.

Solution. In order to write the equations in the state-variable form (i.e., a set of simultaneous first-order differential equations), we select the capacitor voltages v_1 and v_2 as the state elements (i.e., $\mathbf{x} = [v_1 v_2]^T$) and v_i as the input (i.e., $u = v_i$). Here $v_1 = v_2$, $v_2 = v_1 - v_3$, and still $v_1 = v_i$. Thus $v_1 = v_i$, $v_2 = v_1$, and $v_3 = v_i - v_2$. In terms of v_1 and v_2, Eq. (2.34) is

$$\frac{v_1 - v_i}{R_1} + \frac{v_1 - (v_i - v_2)}{R_2} + C_1\frac{dv_1}{dt} = 0.$$

Rearranging this equation into standard form, we get

$$\frac{dv_1}{dt} = -\frac{1}{C_1}\left(\frac{1}{R_1} + \frac{1}{R_2}\right)v_1 - \frac{1}{C_1}\left(\frac{1}{R_2}\right)v_2 + \frac{1}{C_1}\left(\frac{1}{R_1} + \frac{1}{R_2}\right)v_i. \quad (7.6)$$

In terms of v_1 and v_2, Eq. (2.35) is

$$\frac{v_i - v_2 - v_1}{R_2} + C_2\frac{d}{dt}(v_i - v_2 - v_i) = 0.$$

In standard form, the equation is

$$\frac{dv_2}{dt} = -\frac{v_1}{C_2 R_2} - \frac{v_2}{C_2 R_2} + \frac{v_i}{C_2 R_2}. \tag{7.7}$$

Equations (2.34)–(2.35) are entirely equivalent to the state-variable form, Eqs. (7.6) and (7.7), in describing the circuit. The standard matrix definitions are

$$\mathbf{F} = \begin{bmatrix} -\frac{1}{C_1}\left(\frac{1}{R_1} + \frac{1}{R_2}\right) & -\frac{1}{C_1}\left(\frac{1}{R_2}\right) \\ -\frac{1}{C_2 R_2} & -\frac{1}{C_2 R_2} \end{bmatrix},$$

$$\mathbf{G} = \begin{bmatrix} \frac{1}{C_1}\left(\frac{1}{R_1} + \frac{1}{R_2}\right) \\ \frac{1}{C_2 R_2} \end{bmatrix},$$

$$\mathbf{H} = \begin{bmatrix} 0 & -1 \end{bmatrix}, J = 1.$$

EXAMPLE 7.4   *Loudspeaker with Circuit in State-Variable Form*

For the loudspeaker in Fig. 2.29 and the circuit driving it in Fig. 2.30 find the state-space equations relating the input voltage v_a to the output cone displacement x. Assume that the effective circuit resistance is R and the inductance is L.

Solution.   Recall the two coupled equations, (2.44) and (2.48), that constitute the dynamic model for the loudspeaker:

$$M\ddot{x} + b\dot{x} = 0.63i,$$

$$L\frac{di}{dt} + Ri = v_a - 0.63\dot{x}.$$

A logical state vector for this third-order system would be $\mathbf{x} \triangleq [x \ \dot{x} \ i]^T$, which leads to the standard matrices

$$\mathbf{F} = \begin{bmatrix} 0 & 1 & 0 \\ 0 & -b/M & 0.63/M \\ 0 & -0.63/L & -R/L \end{bmatrix}, \quad \mathbf{G} = \begin{bmatrix} 0 \\ 0 \\ 1/L \end{bmatrix}, \quad \mathbf{H} = \begin{bmatrix} 1 & 0 & 0 \end{bmatrix}, \quad J = 0,$$

where now the input $u \triangleq v_a$.

EXAMPLE 7.5   *Modeling a DC Motor in State-Variable Form*

Find the state-space equations for the DC motor with the equivalent electric circuit shown in Fig. 2.32(a).

Solution.   Recall the equations of motion [Eqs. (2.52) and (2.53)] from Chapter 2:

$$J_m\ddot{\theta}_m + b\dot{\theta}_m = K_t i_a,$$

$$L_a\frac{di_a}{dt} + R_a i_a = v_a - K_e\dot{\theta}_m.$$

A state vector for this third-order system is $\mathbf{x} \triangleq \begin{bmatrix} \theta_m & \dot{\theta}_m & i_a \end{bmatrix}^T$, which leads to the standard matrices

$$
\mathbf{F} = \begin{bmatrix} 0 & 1 & 0 \\ 0 & -\dfrac{b}{J_m} & \dfrac{K_t}{J_m} \\ 0 & -\dfrac{K_e}{L_a} & -\dfrac{R_a}{L_a} \end{bmatrix} \quad \mathbf{G} = \begin{bmatrix} 0 \\ 0 \\ \dfrac{1}{L_a} \end{bmatrix}, \quad \mathbf{H} = \begin{bmatrix} 1 & 0 & 0 \end{bmatrix}, \quad J = 0,
$$

where the input $u \triangleq v_a$.

The state-variable form can be applied to a system of any order. Example 7.6 illustrates the method for a fourth-order system.

EXAMPLE 7.6   *Flexible Disk Drive in State-Variable Form*

Find the state-variable form of the differential equations for Example 2.4, where the output is θ_2.

Solution. Define the state vector to be

$$
\mathbf{x} = \begin{bmatrix} \theta_1 & \dot{\theta}_1 & \theta_2 & \dot{\theta}_2 \end{bmatrix}^T.
$$

Then solve Eqs. (2.17) and (2.18) for $\ddot{\theta}_1$ and $\ddot{\theta}_2$ so that the state-variable form is more apparent. The resulting matrices are

$$
\mathbf{F} = \begin{bmatrix} 0 & 1 & 0 & 0 \\ -\dfrac{k}{I_1} & -\dfrac{b}{I_1} & \dfrac{k}{I_1} & \dfrac{b}{I_1} \\ 0 & 0 & 0 & 1 \\ \dfrac{k}{I_2} & \dfrac{b}{I_2} & -\dfrac{k}{I_2} & -\dfrac{b}{I_2} \end{bmatrix}, \quad \mathbf{G} = \begin{bmatrix} 0 \\ \dfrac{1}{I_1} \\ 0 \\ 0 \end{bmatrix},
$$

$$
\mathbf{H} = \begin{bmatrix} 0 & 0 & 1 & 0 \end{bmatrix}, \quad J = 0.
$$

Difficulty arises if the differential equation contains derivatives of the input u. Techniques to handle this situation will be discussed in Section 7.4.

7.3 Block Diagrams and State-Space

Perhaps the most effective way of understanding the state-variable equations is via an analog computer block-diagram representation. The structure of the representation uses integrators as the central element, which are quite suitable for first-order, state-variable representation of equations of motion for a system. Even though the analog computers are almost extinct, analog computer implementation is still a useful concept for state-variable design, and in the circuit design of analog compensation.[3]

The analog computer was a device composed of electric components designed to simulate ODEs. The basic dynamic component of the analog computer is an **integrator**, constructed from an operational amplifier with a capacitor feedback and a resistor feed-forward as shown in Fig. 2.28. Because an integrator is a device whose

---

[3]As well as due to its historical significance.

Figure 7.3

An integrator

Figure 7.4

Components of an
analog computer

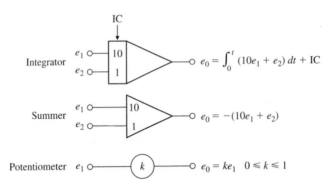

input is the derivative of its output, as shown in Fig. 7.3, if, in an analog-computer simulation, we identify the outputs of the integrators as the state, we will then automatically have the equations in state-variable form. Conversely, if a system is described by state-variables, we can construct an analog-computer simulation of that system by taking one integrator for each state-variable and connecting its input according to the given equation for that state-variable as expressed in the state-variable equations. The analog-computer diagram is a picture of the state equations.

The components of a typical analog computer used to accomplish these functions are shown in Fig. 7.4. Notice that the operational amplifier has a sign change that gives it a negative gain.

EXAMPLE 7.7   *Analog-Computer Implementation*

Find a state-variable description and the transfer function of the third-order system shown in Fig. 7.5 whose differential equation is

$$\dddot{y} + 6\ddot{y} + 11\dot{y} + 6y = 6u.$$

Solution. We solve for the highest derivative term in the ODE to obtain

$$\dddot{y} = -6\ddot{y} - 11\dot{y} - 6y + 6u. \tag{7.8}$$

Now we assume that we have this highest derivative and note that the lower order terms can be obtained by integration as shown in Fig. 7.6(a). Finally, we apply Eq. (7.8) to complete the realization shown in Fig. 7.6(b). To obtain the state description, we simply define the state-variables as the output of the integrators $x_1 = \ddot{y}$, $x_2 = \dot{y}$, $x_3 = y$, to obtain

$$\dot{x}_1 = -6x_1 - 11x_2 - 6x_3 + 6u,$$

$$\dot{x}_2 = x_1,$$

$$\dot{x}_3 = x_2,$$

Figure 7.5

Third-order system

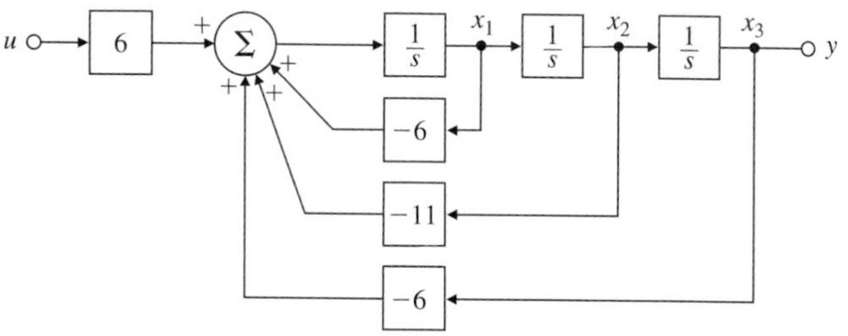

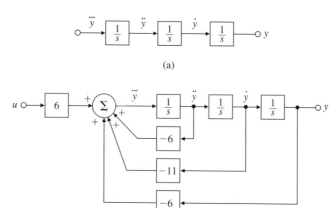

Figure 7.6

Block diagram of a system to solve $\dddot{y} + 6\ddot{y} + 11\dot{y} + 6y = 6u$, using only integrators as dynamic elements: (a) intermediate diagram; (b) final diagram

which provides the state-variable description

$$\mathbf{F} = \begin{bmatrix} -6 & -11 & -6 \\ 1 & 0 & 0 \\ 0 & 1 & 0 \end{bmatrix}, \quad \mathbf{G} = \begin{bmatrix} 6 \\ 0 \\ 0 \end{bmatrix}, \quad \mathbf{H} = [\, 0 \quad 0 \quad 1 \,], \quad J = 0.$$

The MATLAB statement

```
[num,den] = ss2tf(F,G,H,J);
```

will yield the transfer function

$$\frac{Y(s)}{U(s)} = \frac{6}{s^3 + 6s^2 + 11s + 6}.$$

If the transfer function were desired in factored form, it could be obtained by transforming either the ss or tf description. Therefore, either of the MATLAB statements

```
% convert state-variable realization to pole–zero form
[z,p,k] = ss2zp(F,G,H,J)
```

and

```
% convert numerator-denominator to pole–zero form
[z,p,k] = tf2zp(num,den)
```

would result in

$$z = [\,], \quad p = [\, -3 \quad -2 \quad -1 \,]', \quad k = 6.$$

This means that the transfer function could also be written in factored form as

$$\frac{Y(s)}{U(s)} = \frac{6}{(s+1)(s+2)(s+3)}.$$

7.3.1 Time and Amplitude Scaling in State-Space

We have already discussed time and amplitude scaling in Chapter 3 . We now extend the ideas to the state-variable form. Time scaling with $\tau = \omega_o t$ replaces Eq. (7.3) with

$$\frac{d\mathbf{x}}{d\tau} = \frac{1}{\omega_o}\mathbf{F}\mathbf{x} + \frac{1}{\omega_o}\mathbf{G}u = \mathbf{F'}\mathbf{x} + \mathbf{G'}u. \tag{7.9}$$

Amplitude scaling of the state corresponds to replacing $\mathbf{x}$ with $\mathbf{z} = \mathbf{D}_x^{-1}\mathbf{x}$, where $\mathbf{D}_x$ is a diagonal matrix of scale factors. Input scaling corresponds to replacing u with $v = \mathbf{D}_u^{-1}u$. With these substitutions,

$$\mathbf{D}_x\dot{\mathbf{z}} = \frac{1}{\omega_o}\mathbf{F}\mathbf{D}_x\mathbf{z} + \frac{1}{\omega_o}\mathbf{G}\mathbf{D}_u v. \tag{7.10}$$

Then

$$\dot{\mathbf{z}} = \frac{1}{\omega_o}\mathbf{D}_x^{-1}\mathbf{F}\mathbf{D}_x\mathbf{z} + \frac{1}{\omega_o}\mathbf{D}_x^{-1}\mathbf{G}\mathbf{D}_u v = \mathbf{F'}\mathbf{z} + \mathbf{G'}v. \tag{7.11}$$

Equation (7.11) compactly expresses the time- and amplitude-scaling operations. Regrettably, it does not relieve the engineer of the responsibility of actually thinking of good scale factors so that scaled equations are in good shape.

EXAMPLE 7.8   *Time Scaling an Oscillator*

The equation for an oscillator was derived in Example 2.5. For a case with a very fast natural frequency $\omega_n = 15,000$ rad/sec (about 2 kHz), Eq. (2.23) can be rewritten as

$$\ddot{\theta} + 15000^2 \cdot \theta = 10^6 \cdot T_c.$$

Determine the time-scaled equation so that the unit of time is milliseconds.

Solution.   In state-variable form with a state vector $\mathbf{x} = [\theta \ \dot{\theta}]^T$, the unscaled matrices are

$$\mathbf{F} = \begin{bmatrix} 0 & 1 \\ -15000^2 & 0 \end{bmatrix} \quad \text{and} \quad \mathbf{G} = \begin{bmatrix} 0 \\ 10^6 \end{bmatrix}.$$

Applying Eq. (7.9) results in

$$\mathbf{F'} = \begin{bmatrix} 0 & \frac{1}{1000} \\ -\frac{15000^2}{1000} & 0 \end{bmatrix} \quad \text{and} \quad \mathbf{G'} = \begin{bmatrix} 0 \\ 10^3 \end{bmatrix},$$

which yields state-variable equations that are scaled.

7.4 Analysis of the State Equations

In the previous section we introduced and illustrated the process of selecting a state and organizing the equations in state form. In this section we review that process and describe how to analyze the dynamic response using the state description. In Section 7.4.1 we begin by relating the state description to block diagrams and the Laplace transform description and to consider the fact that for a given system the choice of state is not unique. We show how to use this nonuniqueness to select among several canonical forms for the one that will help solve the particular problem at hand; a control canonical form makes feedback gains of the state easy to design. After studying the structure of state equations in Section 7.4.2, we consider the dynamic response and show how transfer-function poles and zeros are related to the matrices of the state descriptions. To illustrate the results with hand calculations, we offer a simple example that represents the model of a thermal system. For more realistic examples, a computer-aided control systems design software package such as MATLAB is especially helpful; relevant MATLAB commands will be described from time to time.

7.4.1 Block Diagrams and Canonical Forms

We begin with a thermal system that has a simple transfer function

$$G = \frac{b(s)}{a(s)} = \frac{s+2}{s^2 + 7s + 12} = \frac{2}{s+4} + \frac{-1}{s+3}. \tag{7.12}$$

The roots of the numerator polynomial $b(s)$ are the zeros of the transfer function, and the roots of the denominator polynomial $a(s)$ are the poles. Notice that we have represented the transfer function in two forms, as a ratio of polynomials and as the result of a partial-fraction expansion. In order to develop a state description of this system (and this is a generally useful technique), we construct a block diagram that corresponds to the transfer function (and the differential equations) *using only isolated integrators as the dynamic elements*. One such block diagram, structured in **control canonical form**, is drawn in Fig. 7.7. The central feature of this structure is that each state-variable is connected by the feedback to the control input.

Once we have drawn the block diagram in this form, we can identify the state description matrices simply by inspection; this is possible because when the output of an integrator is a state-variable, the input of that integrator is the derivative of that

Figure 7.7

A block diagram representing Eq. (7.12) in control form

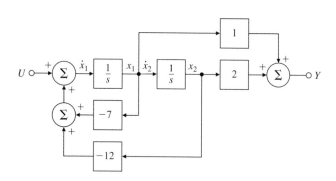

variable. For example, in Fig. 7.7, the equation for the first state-variable is

$$\dot{x}_1 = -7x_1 - 12x_2 + u.$$

Continuing in this fashion, we get

$$\dot{x}_2 = x_1,$$

$$y = x_1 + 2x_2.$$

These three equations can then be rewritten in the matrix form

$$\dot{\mathbf{x}} = \mathbf{A}_c\mathbf{x} + \mathbf{B}_c u, \tag{7.13}$$

$$y = \mathbf{C}_c\mathbf{x}, \tag{7.14}$$

where

$$\mathbf{A}_c = \begin{bmatrix} -7 & -12 \\ 1 & 0 \end{bmatrix}, \qquad \mathbf{B}_c = \begin{bmatrix} 1 \\ 0 \end{bmatrix}, \tag{7.15a}$$

$$\mathbf{C}_c = \begin{bmatrix} 1 & 2 \end{bmatrix}, \qquad D_c = 0, \tag{7.15b}$$

and where the subscript c refers to control canonical form.

Two significant facts about this form are that the coefficients 1 and 2 of the numerator polynomial $b(s)$ appear in the $\mathbf{C}_c$ matrix, and (except for the leading term) the coefficients 7 and 12 of the denominator polynomial $a(s)$ appear (with opposite signs) as the first row of the $\mathbf{A}_c$ matrix. Armed with this knowledge, we can thus write down *by inspection* the state matrices in control canonical form for any system whose transfer function is known as a ratio of numerator and denominator polynomials. If $b(s) = b_1 s^{n-1} + b_2 s^{n-2} + \cdots + b_n$ and $a(s) = s^n + a_1 s^{n-1} + a_2 s^{n-2} + \cdots + a_n$, then the MATLAB steps are

MATLAB tf2ss

```
num = b = [b₁  b₂  ···  bₙ]
den = a = [1  a₁  a₂  ···  aₙ]
[Ac,  Bc,  Cc,  Dc] = tf2ss(num,den).
```

Control canonical form

We read tf2ss as "transfer function to state-space." The result will be

$$\mathbf{A}_c = \begin{bmatrix} -a_1 & -a_2 & \cdots & \cdots & -a_n \\ 1 & 0 & \cdots & \cdots & 0 \\ 0 & 1 & 0 & \cdots & 0 \\ \vdots & & \ddots & 0 & \vdots \\ 0 & 0\cdots & \cdots & 1 & 0 \end{bmatrix}, \qquad \mathbf{B}_c = \begin{bmatrix} 1 \\ 0 \\ 0 \\ \vdots \\ 0 \end{bmatrix}, \tag{7.16a}$$

$$\mathbf{C}_c = \begin{bmatrix} b_1 & b_2 & \cdots & \cdots & b_n \end{bmatrix}, \qquad D_c = 0. \tag{7.16b}$$

The block diagram of Fig. 7.7 and the corresponding matrices of Eq. (7.15) are not the only way to represent the transfer function $G(s)$. A block diagram corresponding to the partial-fraction expansion of $G(s)$ is given in Fig. 7.8. Using the same technique as before, with the state-variables marked as shown in the figure, we can determine the matrices directly from the block diagram as being

$$\dot{\mathbf{z}} = \mathbf{A}_m\mathbf{z} + \mathbf{B}_m u,$$

$$y = \mathbf{C}_m\mathbf{z} + D_m u,$$

Figure 7.8

Block diagram for
Eq. (7.12) in modal
canonical form

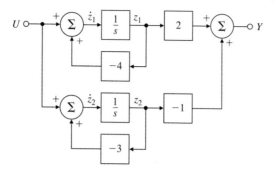

where

$$\mathbf{A}_m = \begin{bmatrix} -4 & 0 \\ 0 & -3 \end{bmatrix}, \qquad \mathbf{B}_m = \begin{bmatrix} 1 \\ 1 \end{bmatrix}, \qquad (7.17a)$$

$$\mathbf{C}_m = [2 \quad -1], \qquad D_m = 0, \qquad (7.17b)$$

Modal form

and the subscript m refers to **modal canonical form**. The name for this form derives from the fact that the poles of the system transfer function are sometimes called the **normal modes** of the system. The important fact about the matrices in this form is that the system poles (here -4 and -3) appear as the elements along the diagonal of the $\mathbf{A}_m$ matrix, and the residues, the numerator terms in the partial-fraction expansion (here 2 and -1), appear in the $\mathbf{C}_m$ matrix.

Expressing a system in modal canonical form can be complicated by two factors: (1) the elements of the matrices will be complex when the poles of the system are complex, and (2) the system matrix *cannot* be diagonal when the partial-fraction expansion has repeated poles. To solve the first problem, we express the complex poles of the partial-fraction expansion as conjugate pairs in second-order terms so that all the elements remain real. The corresponding $\mathbf{A}_m$ matrix will then have 2×2 blocks along the main diagonal representing the local coupling between the variables of the complex-pole set. To handle the second difficulty, we also couple the corresponding state-variables, so that the poles appear along the diagonal with off-diagonal terms indicating the coupling. A simple example of this latter case is the satellite system from Example 7.1, whose transfer function is $G(s) = 1/s^2$. The system matrices for this transfer function in a modal form are

$$\mathbf{F} = \begin{bmatrix} 0 & 1 \\ 0 & 0 \end{bmatrix}, \qquad \mathbf{G} = \begin{bmatrix} 0 \\ 1 \end{bmatrix}, \qquad \mathbf{H} = [\ 1 \quad 0\ ], \qquad J = 0. \qquad (7.18)$$

EXAMPLE 7.9

State Equations in Modal Canonical Form

A "quarter car model" [see Eq. (2.12)] with one resonant mode has a transfer function given by

$$G(s) = \frac{2s + 4}{s^2(s^2 + 2s + 4)} = \frac{1}{s^2} - \frac{1}{s^2 + 2s + 4}. \qquad (7.19)$$

Find state matrices in modal form describing this system.

Figure 7.9

Block diagram for a
fourth-order system in
modal canonical form
with shading indicating
portion in control
canonical form

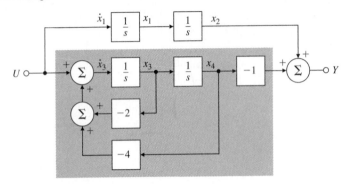

Solution. The transfer function has been given in real partial-fraction form. To get state-description matrices, we draw a corresponding block diagram with integrators only, assign the state, and write down the corresponding matrices. This process is not unique, so there are several acceptable solutions to the problem as stated, but they will differ in only trivial ways. A block diagram with a satisfactory assignment of variables is given in Fig. 7.9.

Notice that the second-order term to represent the complex poles has been realized in control canonical form. There are a number of other possibilities that can be used as alternatives for this part. This particular form allows us to write down the system matrices by inspection:

$$\mathbf{F} = \begin{bmatrix} 0 & 0 & 0 & 0 \\ 1 & 0 & 0 & 0 \\ 0 & 0 & -2 & -4 \\ 0 & 0 & 1 & 0 \end{bmatrix}, \quad \mathbf{G} = \begin{bmatrix} 1 \\ 0 \\ 1 \\ 0 \end{bmatrix},$$

$$\mathbf{H} = [\, 0 \quad 1 \quad 0 \quad -1 \,], \quad J = 0. \tag{7.20}$$

Thus far, we have seen that we can obtain the state description from a transfer function in either control or modal form. Because these matrices represent the same dynamic system, we might ask as to, what is the relationship between the matrices in the two forms (and their corresponding state-variables)? More generally, suppose we have a set of state equations that describe some physical system in no particular form, and we are given a problem for which the control canonical form would be helpful. (We will see such a problem in Section 7.5.) Is it possible to calculate the desired canonical form without obtaining the transfer function first? To answer these questions requires a look at the topic of state transformations.

State description and
output equation

Consider a system described by the state equations

$$\dot{\mathbf{x}} = \mathbf{F}\mathbf{x} + \mathbf{G}u, \tag{7.21a}$$

$$y = \mathbf{H}\mathbf{x} + Ju. \tag{7.21b}$$

As we have seen, this is not a unique description of the dynamic system. We consider a change of state from $\mathbf{x}$ to a new state $\mathbf{z}$ that is a linear transformation of $\mathbf{x}$. For a nonsingular matrix $\mathbf{T}$, we let

$$\mathbf{x} = \mathbf{T}\mathbf{z}. \tag{7.22}$$

By substituting Eq. (7.22) into Eq. (7.21a), we have the equations of motion in terms of the new state $\mathbf{z}$:

$$\dot{\mathbf{x}} = \mathbf{T}\dot{\mathbf{z}} = \mathbf{FTz} + \mathbf{G}u, \tag{7.23a}$$

$$\dot{\mathbf{z}} = \mathbf{T}^{-1}\mathbf{FTz} + \mathbf{T}^{-1}\mathbf{G}u, \tag{7.23b}$$

$$\dot{\mathbf{z}} = \mathbf{Az} + \mathbf{B}u. \tag{7.23c}$$

Transformation of state

In Eq. (7.23c),

$$\mathbf{A} = \mathbf{T}^{-1}\mathbf{FT}, \tag{7.24a}$$

$$\mathbf{B} = \mathbf{T}^{-1}\mathbf{G}. \tag{7.24b}$$

Then we substitute Eq. (7.22) into Eq. (7.21b) to get the output in terms of the new state $\mathbf{z}$:

$$y = \mathbf{HTz} + Ju$$

$$= \mathbf{Cz} + Du.$$

Here

$$\mathbf{C} = \mathbf{HT}, \quad D = J. \tag{7.25}$$

Given the general matrices $\mathbf{F}$, $\mathbf{G}$, and $\mathbf{H}$ and scalar J, we would like to find the transformation matrix $\mathbf{T}$ such that $\mathbf{A}$, $\mathbf{B}$, $\mathbf{C}$, and D are in a particular form, for example, control canonical form. To find such a $\mathbf{T}$, we assume that $\mathbf{A}$, $\mathbf{B}$, $\mathbf{C}$, and D are already in the required form, further assume that the transformation $\mathbf{T}$ has a general form, and match terms. Here we will work out the third-order case; how to extend the analysis to the general case should be clear from the development. It goes like this.

First we rewrite Eq. (7.24a) as

$$\mathbf{AT}^{-1} = \mathbf{T}^{-1}\mathbf{F}.$$

If $\mathbf{A}$ is in control canonical form, and we describe $\mathbf{T}^{-1}$ as a matrix with rows $\mathbf{t}_1$, $\mathbf{t}_2$, and $\mathbf{t}_3$, then

$$\begin{bmatrix} -a_1 & -a_2 & -a_3 \\ 1 & 0 & 0 \\ 0 & 1 & 0 \end{bmatrix} \begin{bmatrix} \mathbf{t}_1 \\ \mathbf{t}_2 \\ \mathbf{t}_3 \end{bmatrix} = \begin{bmatrix} \mathbf{t}_1\mathbf{F} \\ \mathbf{t}_2\mathbf{F} \\ \mathbf{t}_3\mathbf{F} \end{bmatrix}. \tag{7.26}$$

Working out the third and second rows gives the matrix equations

$$\mathbf{t}_2 = \mathbf{t}_3\mathbf{F}, \tag{7.27a}$$

$$\mathbf{t}_1 = \mathbf{t}_2\mathbf{F} = \mathbf{t}_3\mathbf{F}^2. \tag{7.27b}$$

From Eq. (7.24b), assuming that $\mathbf{B}$ is also in control canonical form, we have the relation

$$\mathbf{T}^{-1}\mathbf{G} = \mathbf{B},$$

or

$$\begin{bmatrix} \mathbf{t}_1\mathbf{G} \\ \mathbf{t}_2\mathbf{G} \\ \mathbf{t}_3\mathbf{G} \end{bmatrix} = \begin{bmatrix} 1 \\ 0 \\ 0 \end{bmatrix}. \tag{7.28}$$

Combining Eqs. (7.27) and (7.28), we get

$$\mathbf{t}_3\mathbf{G} = 0,$$

$$\mathbf{t}_2\mathbf{G} = \mathbf{t}_3\mathbf{FG} = 0,$$

$$\mathbf{t}_1\mathbf{G} = \mathbf{t}_3\mathbf{F}^2\mathbf{G} = 1.$$

These equations can in turn be written in matrix form as

$$\mathbf{t}_3[\ \mathbf{G}\quad \mathbf{FG}\quad \mathbf{F}^2\mathbf{G}\ ] = [\ 0\quad 0\quad 1\ ]$$

or

$$\mathbf{t}_3 = [\ 0\quad 0\quad 1\ ]\mathcal{C}^{-1}, \tag{7.29}$$

Controllability matrix
transformation to control
canonical form

where the **controllability matrix** $\mathcal{C} = [\ \mathbf{G}\quad \mathbf{FG}\quad \mathbf{F}^2\mathbf{G}\ ]$. Having $\mathbf{t}_3$, we can now go back to Eq. (7.27) and construct all the rows of $\mathbf{T}^{-1}$.

To sum up, the recipe for converting a general state description of dimension n to control canonical form is as follows:

- From $\mathbf{F}$ and $\mathbf{G}$, form the controllability matrix

$$\mathcal{C} = [\ \mathbf{G}\quad \mathbf{FG}\quad \cdots\quad \mathbf{F}^{n-1}\mathbf{G}\ ]. \tag{7.30}$$

- Compute the last row of the inverse of the transformation matrix as

$$\mathbf{t}_n = [\ 0\quad 0\quad \cdots\quad 1\ ]\mathcal{C}^{-1}. \tag{7.31}$$

- Construct the entire transformation matrix as

$$\mathbf{T}^{-1} = \begin{bmatrix} \mathbf{t}_n\mathbf{F}^{n-1} \\ \mathbf{t}_n\mathbf{F}^{n-2} \\ \vdots \\ \mathbf{t}_n \end{bmatrix}. \tag{7.32}$$

- Compute the new matrices from $\mathbf{T}^{-1}$, using Eqs. (7.24a), (7.24b), and (7.25).

When the controllability matrix $\mathcal{C}$ is nonsingular, the corresponding $\mathbf{F}$ and $\mathbf{G}$ matrices are said to be **controllable**. This is a technical property that usually holds for physical systems and will be important when we consider feedback of the state in Section 7.5. We will also consider a few physical illustrations of loss of controllability at that time.

Controllable systems

Because computing the transformation given by Eq. (7.32) is numerically difficult to do accurately, it is almost never done. The reason for developing this transformation in some detail is to show how such changes of state could be done in theory and to make the following important observation:

> One can *always* transform a given state description to control canonical form if (and only if) the controllability matrix $\mathcal{C}$ is nonsingular.

If we need to test for controllability in a real case with numbers, we use a numerically stable method that depends on converting the system matrices to "staircase" form rather than on trying to compute the controllability matrix. Problem 7.29 at the end of the chapter calls for consideration of this method.

An important question regarding controllability follows directly from our discussion so far: What is the effect of a state transformation on controllability? We can show the result by using Eqs. (7.30), (7.24a), and (7.24b). The controllability matrix of the system $(\mathbf{F}, \mathbf{G})$ is

$$\mathcal{C}_{\mathbf{x}} = [\ \mathbf{G}\quad \mathbf{FG}\quad \cdots\quad \mathbf{F}^{n-1}\mathbf{G}\ ]. \tag{7.33}$$

After the state transformation, the new description matrices are given by Eqs. (7.24a) and (7.24b), and the controllability matrix changes to

$$C_z = [\ B \quad AB \quad \cdots \quad A^{n-1}B\ ] \tag{7.34a}$$

$$= [\ T^{-1}G \quad T^{-1}FTT^{-1}G \quad \cdots \quad T^{-1}F^{n-1}TT^{-1}G\ ] \tag{7.34b}$$

$$= T^{-1}C_x. \tag{7.34c}$$

Thus we see that C_z is nonsingular if and only if C_x is nonsingular, yielding the following observation:

> A change of state by a nonsingular linear transformation does *not* change controllability.

Observer canonical form

We return once again to the transfer function of Eq. (7.12), this time to represent it with the block diagram having the structure known as **observer canonical form** (Fig. 7.10). The corresponding matrices for this form are

$$A_o = \begin{bmatrix} -7 & 1 \\ -12 & 0 \end{bmatrix}, \quad B_o = \begin{bmatrix} 1 \\ 2 \end{bmatrix}, \tag{7.35a}$$

$$C_o = [\ 1 \quad 0\ ], \quad D_o = 0. \tag{7.35b}$$

The significant fact about this canonical form is that all the feedback is from the output to the state-variables.

Let us now consider what happens to the controllability of this system as the zero at -2 is varied. For this purpose, we replace the second element 2 of B_o with the variable zero location $-z_o$ and form the controllability matrix:

$$C_x = [\ B_o \quad A_oB_o\ ] \tag{7.36a}$$

$$= \begin{bmatrix} 1 & -7-z_o \\ -z_o & -12 \end{bmatrix}. \tag{7.36b}$$

The determinant of this matrix is a function of z_o:

$$\det(C_x) = -12 + (z_o)(-7-z_o)$$

$$= -(z_o^2 + 7z_o + 12).$$

This polynomial is zero for $z_o = -3$ or -4, implying that controllability is lost for these values. What does this mean? In terms of the parameter z_o, the transfer function is

Figure 7.10

Observer canonical form

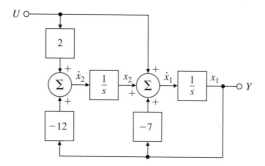

$$G(s) = \frac{s - z_o}{(s+3)(s+4)}.$$

If $z_o = -3$ or -4, there is a pole–zero cancellation and the transfer function reduces from a second-order system to a first-order one. When $z_o = -3$, for example, the mode at -3 is decoupled from the input and control of this mode is lost.

Notice that we have taken the transfer function given by Eq. (7.12) and given it two realizations, one in control canonical form and one in observer canonical form. The control form is always controllable for any value of the zero, while the observer form loses controllability if the zero cancels either of the poles. Thus, these two forms may represent the same transfer function, but it may not be possible to transform the state of one to the state of the other (in this case, from observer to control canonical form). While a transformation of state cannot affect controllability, the particular state selected from a transfer function can:

> Controllability is a function of the *state* of the system and cannot be decided from a transfer function.

To discuss controllability more at this point would take us too far afield. The closely related property of observability and the observer canonical form will be taken up in Section 7.7.1. A more detailed discussion of these properties of dynamic systems is given in the Appendix WF, for those who would like to learn more.

We return now to the modal form for the equations, given by Eqs. (7.17a) and (7.17b) for the example transfer function. As mentioned before, it is not always possible to find a modal form for transfer functions that have repeated poles, so we assume our system has only distinct poles. Furthermore, we assume that the general state equations given by Eqs. (7.21a) and (7.21b) apply. We want to find a transformation matrix $\mathbf{T}$ defined by Eq. (7.22) such that the transformed Eqs. (7.24a) and (7.25) will be in modal form. In this case, we assume that the $\mathbf{A}$ matrix is diagonal and that $\mathbf{T}$ is composed of the *columns* $\mathbf{t}_1$, $\mathbf{t}_2$, and $\mathbf{t}_3$. With this assumption, the state transformation Eq. (7.24a) becomes

$$\mathbf{TA = FT}$$

$$\begin{bmatrix} \mathbf{t}_1 & \mathbf{t}_2 & \mathbf{t}_3 \end{bmatrix} \begin{bmatrix} p_1 & 0 & 0 \\ 0 & p_2 & 0 \\ 0 & 0 & p_3 \end{bmatrix} = \mathbf{F} \begin{bmatrix} \mathbf{t}_1 & \mathbf{t}_2 & \mathbf{t}_3 \end{bmatrix}. \tag{7.37}$$

Transformation to modal form

Equation (7.37) is equivalent to the three vector-matrix equations

$$p_i \mathbf{t}_i = \mathbf{F} \mathbf{t}_i, \quad i = 1, 2, 3. \tag{7.38}$$

Eigenvectors

Eigenvalues

In matrix algebra Eq. (7.38) is a famous equation, whose solution is known as the **eigenvector/eigenvalue problem**. Recall that $\mathbf{t}_i$ is a vector, $\mathbf{F}$ is a matrix, and p_i is a scalar. The vector $\mathbf{t}_i$ is called an **eigenvector** of $\mathbf{F}$, and p_i is called the corresponding **eigenvalue**. Because we saw earlier that the modal form is equivalent to a partial-fraction-expansion representation with the system poles along the diagonal of the state matrix, it should be clear that these "eigenvalues" are precisely the poles of our system. The transformation matrix that will convert the state description matrices to modal form has as its columns the eigenvectors of $\mathbf{F}$, as shown in Eq. (7.37) for

MATLAB eig

the third-order case. As it happens, there are robust, reliable computer algorithms to compute eigenvalues and the eigenvectors of quite large systems using the QR algorithm.[4] In MATLAB, the command p = eig(F) is the way to compute the poles if the system equations are in state form.

Notice also that Eq. (7.38) is homogeneous in that, if $\mathbf{t}_i$ is an eigenvector, so is $\alpha \mathbf{t}_i$, for any scalar α. In most cases the scale factor is selected so that the length (square root of the sum of squares of the magnitudes of the elements) is unity. MATLAB will perform this operation. Another option is to select the scale factors so that the input matrix **B** is composed of all 1's. The latter choice is suggested by a partial-fraction expansion with each part realized in control canonical form. If the system is real, then each element of **F** is real, and if $p = \sigma + j\omega$ is a pole, so is the conjugate, $p^* = \sigma - j\omega$. For these eigenvalues, the eigenvectors are also complex and conjugate. It is possible to compose the transformation matrix using the real and complex parts of the eigenvectors separately, so the modal form is real but has 2×2 blocks for each pair of complex poles. Later, we will see the result of the MATLAB function that does this, but first let us look at the simple real-poles case.

EXAMPLE 7.10   *Transformation of Thermal System from Control to Modal Form*

Find the matrix to transform the control form matrices in Eq. (7.15) into the modal form of Eq. (7.17).

Solution.   According to Eqs. (7.37) and (7.38), we need first to find the eigenvectors and eigenvalues of the $\mathbf{A}_c$ matrix. We take the eigenvectors to be

$$\begin{bmatrix} t_{11} \\ t_{21} \end{bmatrix} \quad \text{and} \quad \begin{bmatrix} t_{12} \\ t_{22} \end{bmatrix}.$$

The equations using the eigenvector on the left are

$$\begin{bmatrix} -7 & -12 \\ 1 & 0 \end{bmatrix} \begin{bmatrix} t_{11} \\ t_{21} \end{bmatrix} = p \begin{bmatrix} t_{11} \\ t_{21} \end{bmatrix}, \tag{7.39a}$$

$$-7t_{11} - 12t_{21} = pt_{11}, \tag{7.39b}$$

$$t_{11} = pt_{21}. \tag{7.39c}$$

Substituting Eq. (7.39c) into Eq. (7.39b) results in

$$-7pt_{21} - 12t_{21} = p^2 t_{21}, \tag{7.40a}$$

$$p^2 t_{21} + 7pt_{21} + 12t_{21} = 0, \tag{7.40b}$$

$$p^2 + 7p + 12 = 0, \tag{7.40c}$$

$$p = -3, -4. \tag{7.40d}$$

We have found (again!) that the eigenvalues (poles) are -3 and -4; furthermore, Eq. (7.39c) tells us that the two eigenvectors are

$$\begin{bmatrix} -4t_{21} \\ t_{21} \end{bmatrix} \quad \text{and} \quad \begin{bmatrix} -3t_{22} \\ t_{22} \end{bmatrix},$$

---

[4]This algorithm is part of MATLAB and all other well-known computer-aided design packages. It is carefully documented in the software package LAPACK (Anderson et al., 1999). See also Strang (1988).

where t_{21} and t_{22} are arbitrary nonzero scale factors. We want to select the two scale factors such that both elements of $\mathbf{B}_m$ in Eq. (7.17a) are unity. The equation for $\mathbf{B}_m$ in terms of $\mathbf{B}_c$ is $\mathbf{TB}_m = \mathbf{B}_c$, and its solution is $t_{21} = -1$ and $t_{22} = 1$. Therefore, the transformation matrix and its inverse[5] are

$$\mathbf{T} = \begin{bmatrix} 4 & -3 \\ -1 & 1 \end{bmatrix}, \quad \mathbf{T}^{-1} = \begin{bmatrix} 1 & 3 \\ 1 & 4 \end{bmatrix}. \tag{7.41}$$

Elementary matrix multiplication shows that, using $\mathbf{T}$ as defined by Eq. (7.41), the matrices of Eqs. (7.15) and (7.17) are related as follows:

$$\mathbf{A}_m = \mathbf{T}^{-1}\mathbf{A}_c\mathbf{T}, \qquad \mathbf{B}_m = \mathbf{T}^{-1}\mathbf{B}_c,$$

$$\mathbf{C}_m = \mathbf{C}_c\mathbf{T}, \qquad \mathbf{D}_m = \mathbf{D}_c. \tag{7.42}$$

These computations can be carried out by using the following MATLAB statements

```
T = [4 −3; −1 1];
Am = inv(T)*Ac*T;
Bm = inv(T)*Bc;
Cm = Cc*T;
Dm = Dc;
```

The next example has five state-variables and, in state-variable form, is too complicated for hand calculations. However, it is a good example for illustrating the use of computer software designed for the purpose. The model we will use is based on a physical state after amplitude and time scaling have been done.

EXAMPLE 7.11   *Using MATLAB to Find Poles and Zeros of Tape-Drive System*

Find the eigenvalues of the system matrix described below for the tape-drive control (see Fig. 3.50). Also, compute the transformation of the equations of the tape drive in their given form to modal canonical form. The system matrices are

$$\mathbf{F} = \begin{bmatrix} 0 & 2 & 0 & 0 & 0 \\ -0.1 & -0.35 & 0.1 & 0.1 & 0.75 \\ 0 & 0 & 0 & 2 & 0 \\ 0.4 & 0.4 & -0.4 & -1.4 & 0 \\ 0 & -0.03 & 0 & 0 & -1 \end{bmatrix}, \quad \mathbf{G} = \begin{bmatrix} 0 \\ 0 \\ 0 \\ 0 \\ 1 \end{bmatrix}, \tag{7.43}$$

$\mathbf{H}_2 = \begin{bmatrix} 0.0 & 0.0 & 1.0 & 0.0 & 0.0 \end{bmatrix}$ Servomotor position output,

$\mathbf{H}_3 = \begin{bmatrix} 0.5 & 0.0 & 0.5 & 0.0 & 0.0 \end{bmatrix}$ Position at read/write head as output,

$\mathbf{H}_T = \begin{bmatrix} -0.2 & -0.2 & 0.2 & 0.2 & 0.0 \end{bmatrix}$ Tension output,

$J = 0.0.$

---

[5]To find the inverse of a 2×2 matrix, you need only interchange the elements subscripted "11" and "22," change the signs of the "12" and the "21" elements, and divide by the determinant [$= 1$ in Eq. (7.41)].

The state vector is defined as

$$\mathbf{x} = \begin{bmatrix} x_1 \text{ (tape position at capstan)} \\ \omega_1 \text{ (speed of the drive wheel)} \\ x_3 \text{ (position of the tape at the head)} \\ \omega_2 \text{ (output speed)} \\ i \quad \text{(current into capstan motor)} \end{bmatrix}.$$

The matrix $\mathbf{H}_3$ corresponds to making x_3 (the position of the tape over the read/write head) the output, and the matrix $\mathbf{H}_T$ corresponds to making tension the output.

Solution. To compute the eigenvalues by using MATLAB, we write

$$P=\text{eig}(F),$$

which results in

$$P = \begin{bmatrix} -0.6371 + 0.6669i \\ -0.6371 - 0.6669i \\ 0.0000 \\ -0.5075 \\ -0.9683 \end{bmatrix}.$$

Notice that the system has all poles in the left half-plane (LHP) except for one pole at the origin. This means that a step input will result in a ramp output, so we conclude the system has Type 1 behavior.

To transform to modal form, we use the MATLAB function canon:

MATLAB canon

```
sysG = ss(F,G,H3,J)
[sysGm, TI] =canon(sysG, 'modal')
[Am,Bm,Cm,Dm]=ssdata(sysGm)
```

The result of this calculation is

$$\text{Am} = \mathbf{A}_m = \begin{bmatrix} -0.6371 & 0.6669 & 0.0000 & 0.0000 & 0.0000 \\ -0.6669 & -0.6371 & 0.0000 & 0.0000 & 0.0000 \\ 0.0000 & 0.0000 & 0.0000 & 0.0000 & 0.0000 \\ 0.0000 & 0.0000 & 0.0000 & -0.5075 & 0.0000 \\ 0.0000 & 0.0000 & 0.0000 & 0.0000 & -0.9683 \end{bmatrix}.$$

Notice that the complex poles appear in the 2×2 block in the upper left corner of $\mathbf{A}_m$, and the real poles fall on the main diagonal of this matrix. The rest of the calculations from canon are

$$\text{Bm} = \mathbf{B}_m = \begin{bmatrix} 0.4785 \\ -0.6274 \\ -1.0150 \\ -3.5980 \\ 4.9133 \end{bmatrix},$$

$$\text{Cm} = \mathbf{C}_m = [\ 1.2569 \quad -1.0817 \quad -2.8284 \quad 1.8233 \quad 0.4903\ ],$$

$$\text{Dm} = D_m = 0,$$

$$\mathsf{TI} = \mathbf{T}^{-1} = \begin{bmatrix} -0.3439 & -0.3264 & 0.3439 & 0.7741 & 0.4785 \\ 0.1847 & -0.7291 & -0.1847 & 0.0969 & -0.6247 \\ -0.1844 & -1.3533 & -0.1692 & -0.3383 & -1.0150 \\ 0.3353 & -2.3627 & -0.3353 & -1.0161 & -3.5980 \\ -0.0017 & 0.2077 & 0.0017 & 0.0561 & 4.9133 \end{bmatrix}.$$

It happens that canon was written to compute the *inverse* of the transformation we are working with (as you can see from TI in the previous equation), so we need to invert our MATLAB results. The inverse is calculated from

MATLAB inv

$$\mathsf{T} = \mathrm{inv}(\mathsf{T\,I})$$

and results in

$$\mathsf{T} = \mathbf{T} = \begin{bmatrix} 0.3805 & 0.8697 & -2.8284 & 1.3406 & 0.4714 \\ -0.4112 & -0.1502 & 0.0000 & -0.3402 & -0.2282 \\ 2.1334 & -3.0330 & -2.8284 & 2.3060 & 0.5093 \\ 0.3317 & 1.6776 & 0.0000 & -0.5851 & -0.2466 \\ 0.0130 & -0.0114 & -0.0000 & 0.0207 & 0.2160 \end{bmatrix}.$$

The eigenvectors computed with [V,P]=eig(F) are

$$V = \mathbf{V}$$
$$= \begin{bmatrix} -0.1168 + 0.1925i & -0.1168 - 0.1925i & -0.7071 & 0.4871 & 0.5887 \\ -0.0270 - 0.1003i & -0.0270 + 0.1003i & -0.0000 & -0.1236 & -0.2850 \\ 0.8797 & 0.8797 & -0.7071 & 0.8379 & 0.6360 \\ -0.2802 + 0.2933i & -0.2802 - 0.2933i & -0.0000 & -0.2126 & -0.3079 \\ 0.0040 + 0.0010i & 0.0040 - 0.0010i & 0.0000 & 0.0075 & 0.2697 \end{bmatrix}.$$

Notice that the first two columns of the real transformation $\mathbf{T}$ are composed of the real and the imaginary parts of the first eigenvector in the first column of $\mathbf{V}$. It is this step that causes the complex roots to appear in the 2×2 block in the upper left of the $\mathbf{A}_m$ matrix. The vectors in $\mathbf{V}$ are normalized to unit length, which results in nonnormalized values in $\mathbf{B}_m$ and $\mathbf{C}_m$. If we found it desirable to do so, we could readily find further transformations to make each element of $\mathbf{B}_m$ equal 1 or to interchange the order in which the poles appear.

7.4.2   Dynamic Response from the State Equations

Having considered the structure of the state-variable equations, we now turn to finding the dynamic response from the state description and to the relationships between the state description and our earlier discussion in Chapter 6 of the frequency response and poles and zeros. Let us begin with the general equations of state given by Eqs. (7.21a) and (7.21b), and consider the problem in the frequency domain. Taking the Laplace transform of

$$\dot{\mathbf{x}} = \mathbf{F}\mathbf{x} + \mathbf{G}u, \tag{7.44}$$

we obtain

$$s\mathbf{X}(s) - \mathbf{x}(0) = \mathbf{F}\mathbf{X}(s) + \mathbf{G}U(s), \tag{7.45}$$

which is now an algebraic equation. If we collect the terms involving $\mathbf{X}(s)$ on the left side of Eq. (7.45), keeping in mind that in matrix multiplication order is very important, we find that[6]

$$(s\mathbf{I} - \mathbf{F})\mathbf{X}(s) = \mathbf{G}U(s) + \mathbf{x}(0).$$

If we premultiply both sides by the inverse of $(s\mathbf{I} - \mathbf{F})$, then

$$\mathbf{X}(s) = (s\mathbf{I} - \mathbf{F})^{-1}\mathbf{G}U(s) + (s\mathbf{I} - \mathbf{F})^{-1}\mathbf{x}(0). \tag{7.46}$$

The output of the system is

$$Y(s) = \mathbf{H}\mathbf{X}(s) + JU(s), \tag{7.47a}$$

$$= \mathbf{H}(s\mathbf{I} - \mathbf{F})^{-1}\mathbf{G}U(s) + \mathbf{H}(s\mathbf{I} - \mathbf{F})^{-1}\mathbf{x}(0) + JU(s). \tag{7.47b}$$

Transfer function from state equations

This equation expresses the output response to both an initial condition and an external forcing input. The coefficient of the external input is the transfer function of the system, which in this case is given by

$$G(s) = \frac{Y(s)}{U(s)} = \mathbf{H}(s\mathbf{I} - \mathbf{F})^{-1}\mathbf{G} + J. \tag{7.48}$$

EXAMPLE 7.12

Thermal System Transfer Function from the State Description

Use Eq. (7.48) to find the transfer function of the thermal system described by Eqs. (7.15a) and (7.15b).

Solution. The state-variable description matrices of the system are

$$\mathbf{F} = \begin{bmatrix} -7 & -12 \\ 1 & 0 \end{bmatrix},$$

$$\mathbf{G} = \begin{bmatrix} 1 \\ 0 \end{bmatrix},$$

$$\mathbf{H} = [\ 1 \quad 2\ ], \quad J = 0.$$

To compute the transfer function according to Eq. (7.48), we form

$$s\mathbf{I} - \mathbf{F} = \begin{bmatrix} s+7 & 12 \\ -1 & s \end{bmatrix}$$

and compute

$$(s\mathbf{I} - \mathbf{F})^{-1} = \frac{\begin{bmatrix} s & -12 \\ 1 & s+7 \end{bmatrix}}{s(s+7) + 12}. \tag{7.49}$$

---

[6]The identity matrix $\mathbf{I}$ is a matrix of ones on the main diagonal and zeros everywhere else; therefore, $\mathbf{Ix} = \mathbf{x}$.

We then substitute Eq. (7.49) into Eq. (7.48) to get

$$G(s) = \frac{[\ 1 \quad 2\ ]\begin{bmatrix} s & -12 \\ 1 & s+7 \end{bmatrix}\begin{bmatrix} 1 \\ 0 \end{bmatrix}}{s(s+7)+12} \tag{7.50}$$

$$= \frac{[\ 1 \quad 2\ ]\begin{bmatrix} s \\ 1 \end{bmatrix}}{s(s+7)+12} \tag{7.51}$$

$$= \frac{(s+2)}{(s+3)(s+4)}. \tag{7.52}$$

The results can also be found using the MATLAB statements,

[num,den] = ss2tf(F,G,H,J)

and yield num = [0 1 2] and den = [1 7 12], which agrees with hand calculations.

Because Eq. (7.48) expresses the transfer function in terms of the general state-space descriptor matrices **F**, **G**, **H**, and J, we are able to express poles and zeros in terms of these matrices. We saw earlier that by transforming the state matrices to diagonal form, the poles appear as the eigenvalues on the main diagonal of the **F** matrix. We now take a systems theory point of view to look at the poles and zeros as they are involved in the transient response of a system.

As we saw in Chapter 3, a pole of the transfer function $G(s)$ is a value of generalized frequency s such that, if $s = p_i$, then the system can respond to an initial condition as $K_i e^{p_i t}$, *with no forcing function*. In this context, p_i is called a **natural frequency** or **natural mode** of the system. If we take the state-space equations (7.21a and 7.21b) and set the forcing function u to zero, we have

$$\dot{\mathbf{x}} = \mathbf{F}\mathbf{x}. \tag{7.53}$$

If we assume some (as yet unknown) initial condition

$$\mathbf{x}(0) = \mathbf{x}_0 \tag{7.54}$$

and that the entire state motion behaves according to the same natural frequency, then the state can be written as $\mathbf{x}(t) = e^{p_i t}\mathbf{x}_0$. It follows from Eq. (7.53) that

$$\dot{\mathbf{x}}(t) = p_i e^{p_i t}\mathbf{x}_0 = \mathbf{F}\mathbf{x} = \mathbf{F}e^{p_i t}\mathbf{x}_0, \tag{7.55}$$

or

$$\mathbf{F}\mathbf{x}_0 = p_i \mathbf{x}_0. \tag{7.56}$$

We can rewrite Eq. (7.56) as

$$(p_i\mathbf{I} - \mathbf{F})\mathbf{x}_0 = 0. \tag{7.57}$$

Equations (7.56) and (7.57) constitute the eigenvector/eigenvalue problem we saw in Eq. (7.38) with eigenvalues p_i and, in this case, eigenvectors $\mathbf{x}_0$ of the matrix **F**. If we are just interested in the eigenvalues, we can use the fact that for a nonzero $\mathbf{x}_0$, Eq. (7.57) has a solution if and only if

Transfer function poles
from state equations

$$\det(p_i\mathbf{I} - \mathbf{F}) = 0. \tag{7.58}$$

These equations show again that the *poles* of the transfer function are the eigenvalues of the system matrix $\mathbf{F}$. The determinant equation (7.58) is a polynomial in the eigenvalues p_i known as the **characteristic equation**. In Example 7.10 we computed the eigenvalues and eigenvectors of a particular matrix in control canonical form. As an alternative computation for the poles of that system, we could solve the characteristic equation (7.58). For the system described by Eqs. (7.15a) and (7.15b), we can find the poles from Eq. (7.58) by solving

$$\det(s\mathbf{I} - \mathbf{F}) = 0, \tag{7.59a}$$

$$\det \begin{bmatrix} s+7 & 12 \\ -1 & s \end{bmatrix} = 0, \tag{7.59b}$$

$$s(s+7) + 12 = (s+3)(s+4) = 0. \tag{7.59c}$$

This confirms again that the poles of the system are the eigenvalues of $\mathbf{F}$.

We can also determine the zeros of a system from the state-variable description matrices $\mathbf{F}, \mathbf{G}, \mathbf{H}$, and J using a systems theory point of view. From this perspective, a zero is a value of generalized frequency s such that the system can have a nonzero input and state and yet have an output of zero. If the input is exponential at the zero frequency z_i, given by

$$u(t) = u_0 e^{z_i t}, \tag{7.60}$$

then the output is identically zero:

$$y(t) \equiv 0. \tag{7.61}$$

The state-space description of Eqs. (7.60) and (7.61) would be

$$u = u_0 e^{z_i t}, \quad \mathbf{x}(t) = \mathbf{x}_0 e^{z_i t}, \quad y(t) \equiv 0. \tag{7.62}$$

Thus

$$\dot{\mathbf{x}} = z_i e^{z_i t} \mathbf{x}_0 = \mathbf{F} e^{z_i t} \mathbf{x}_0 + \mathbf{G} u_0 e^{z_i t}, \tag{7.63}$$

or

$$[z_i \mathbf{I} - \mathbf{F} \quad -\mathbf{G}] \begin{bmatrix} \mathbf{x}_0 \\ u_0 \end{bmatrix} = \mathbf{0} \tag{7.64}$$

and

$$y = \mathbf{H}\mathbf{x} + Ju = \mathbf{H}e^{z_i t}\mathbf{x}_0 + Ju_0 e^{z_i t} \equiv 0 \tag{7.65}$$

Combining Eqs. (7.64) and (7.65), we get

$$\begin{bmatrix} z_i \mathbf{I} - \mathbf{F} & -\mathbf{G} \\ \mathbf{H} & J \end{bmatrix} \begin{bmatrix} \mathbf{x}_0 \\ u_0 \end{bmatrix} = \begin{bmatrix} \mathbf{0} \\ 0 \end{bmatrix}. \tag{7.66}$$

Transfer function zeros from state equations

From Eq. (7.66) we can conclude that a zero of the state-space system is a value of z_i where Eq. (7.66) has a nontrivial solution. With one input and one output, the matrix is square, and a solution to Eq. (7.66) is equivalent to a solution to

$$\det \begin{bmatrix} z_i \mathbf{I} - \mathbf{F} & -\mathbf{G} \\ \mathbf{H} & J \end{bmatrix} = 0. \tag{7.67}$$

EXAMPLE 7.13   *Zeros for the Thermal System from a State Description*

Compute the zero(s) of the thermal system described by Eq. (7.15).

Solution. We use Eq. (7.67) to compute the zeros:

$$\det \begin{bmatrix} s+7 & 12 & -1 \\ -1 & s & 0 \\ 1 & 2 & 0 \end{bmatrix} = 0,$$

$$-2 - s = 0,$$

$$s = -2.$$

Note that this result agrees with the zero of the transfer function given by Eq. (7.12). The result can also be found using the following MATLAB statements:

sysG = ss(Ac,Bc,Cc,Dc);
z = tzero(sysG)

and yields z = −2.0.

Equation (7.58) for the characteristic equation and Eq. (7.67) for the zeros polynomial can be combined to express the transfer function in a compact form from state-description matrices as

$$G(s) = \frac{\det \begin{bmatrix} s\mathbf{I} - \mathbf{F} & -\mathbf{G} \\ \mathbf{H} & J \end{bmatrix}}{\det(s\mathbf{I} - \mathbf{F})}. \tag{7.68}$$

(See Appendix WE for more details.) While Eq. (7.68) is a compact formula for theoretical studies, it is very sensitive to numerical errors. A numerically stable algorithm for computing the transfer function is described in Emami-Naeini and Van Dooren (1982). Given the transfer function, we can compute the frequency response as $G(j\omega)$, and as discussed earlier, we can use Eqs. (7.57) and (7.66) to find the poles and zeros, upon which the transient response depends, as we saw in Chapter 3.

EXAMPLE 7.14   *Analysis of the State Equations of a Tape Drive*

Compute the poles, zeros, and transfer function for the equations of the tape-drive servomechanism given in Example 7.11.

Solution. There are two different ways to compute the answer to this problem. The most direct is to use the MATLAB function ss2tf (state-space to transfer function), which will give the numerator and denominator polynomials directly. This function permits multiple inputs and outputs; the fifth argument of the function tells which input is to be used. We have only one input here but must still provide the argument. The computation of the transfer function from motor-current input to the servomotor position output is

MATLAB ss2tf

[N2, D2] = ss2tf(F, G, H2, J, 1)

which results in

$$N2 = [\ 0 \quad 0 \quad 0.0000 \quad 0 \quad 0.6000 \quad 1.2000\ ],$$

$$D2 = [\ 1.0000 \quad 2.7500 \quad 3.2225 \quad 1.8815 \quad 0.4180 \quad -0.0000\ ].$$

Similarly, for the position at the read/write head, the transfer function polynomials are computed by

[N3, D3] = ss2tf(F, G, H3, J, 1),

which results in

$$N3 = [\ 0\quad -0.0000\quad -0.0000\quad 0.7500\quad 1.3500\quad 1.2000\ ],$$
$$D3 = [\ 1.0000\quad 2.7500\quad 3.2225\quad 1.8815\quad 0.4180\quad -0.0000\ ].$$

Finally, the transfer function to tension is

[NT, DT] = ss2tf(F, G, HT, J, 1)

producing

$$NT = [\ 0\quad -0.0000\quad -0.1500\quad -0.4500\quad -0.3000\quad 0.0000\ ],$$
$$DT = [\ 1.0000\quad 2.7500\quad 3.2225\quad 1.8815\quad 0.4180\quad -0.0000\ ].$$

It is interesting to check to see whether the poles and zeros determined this way agree with those found by other means. For a polynomial, we use the function roots:

MATLAB roots

$$\text{roots(D3)} = \begin{bmatrix} -0.6371 + 0.6669i \\ -0.6371 - 0.6669i \\ -0.9683 \\ -0.5075 \\ 0.0000 \end{bmatrix}.$$

Checking with Example 7.11, we confirm that they agree.

How about the zeros? We can find these by finding the roots of the numerator polynomial. We compute the roots of the polynomial N3:

$$\text{roots(N3)} = \begin{bmatrix} -1.6777 \times 10^7 \\ 1.6777 \times 10^7 \\ -0.9000 + 0.8888i \\ -0.9000 - 0.8888i \end{bmatrix}.$$

Here we notice that roots are given with a magnitude of 10^7, which seems inconsistent with the values given for the polynomial. The problem is that MATLAB has used the very small leading terms in the polynomial as real values and thereby introduced extraneous roots that are for all practical purposes at infinity. The true zeros are found by truncating the polynomial to the significant values using the statement

$$N3R = N3(4:6)$$

to get

$$N3R = [\ 0.7499\quad 1.3499\quad 1.200\ ],$$
$$\text{roots(N3R)} = \begin{bmatrix} -0.9000 + 0.8888i \\ -0.9000 - 0.8888i \end{bmatrix}.$$

The other approach is to compute the poles and zeros separately and, if desired, combine these into a transfer function. The poles were computed with eig in Example 7.11 and are

$$P = \begin{bmatrix} -0.6371 + 0.6669i \\ -0.6371 - 0.6669i \\ 0.0000 \\ -0.5075 \\ -0.9683 \end{bmatrix}.$$

MATLAB tzero

The zeros can be computed by the equivalent of Eq. (7.66) with the function tzero (transmission zeros). The zeros depend on which output is being used, of course, and are respectively given below. For the position of the tape at the servomotor as the output, the statement

$$\text{sysG2} = \text{ss}(F, G, H2, J)$$

$$\text{ZER2} = \text{tzero}(\text{sysG2})$$

yields

$$\text{ZER2} = -2.0000.$$

For the position of the tape over the read/write head as the output, we use the statement

$$\text{sysG3} = \text{ss}(F, G, H3, J)$$

$$\text{ZER3} = \text{tzero}(\text{sysG3})$$

$$\text{ZER3} = \begin{bmatrix} -0.9000 + 0.8888i \\ -0.9000 - 0.8888i \end{bmatrix}.$$

We note that these results agree with the values previously computed from the numerator polynomial N3. Finally, for the tension as output, we use

$$\text{sysGT} = \text{ss}(F, G, HT, J)$$

$$\text{ZERT} = \text{tzero}(\text{sysGT})$$

to get

$$\text{ZERT} = \begin{bmatrix} 0 \\ -1.9999 \\ -1.0000 \end{bmatrix}.$$

From these results we can write down, for example, the transfer function to position x_3 as

$$G(s) = \frac{X_3(s)}{E_1(s)}$$

$$= \frac{0.75s^2 + 1.35s + 1.2}{s^5 + 2.75s^4 + 3.22s^3 + 1.88s^2 + 0.418s}$$

$$= \frac{0.75(s + 0.9 \pm 0.8888j)}{s(s + 0.507)(s + 0.968)(s + 0.637 \pm 0.667j)}. \tag{7.69}$$

7.5 Control-Law Design for Full-State Feedback

One of the attractive features of the state-space design method is that it consists of a sequence of independent steps, as mentioned in the chapter overview. The first step, discussed in Section 7.5.1, is to determine the control. The purpose of the control law is to allow us to assign a set of pole locations for the closed-loop system that will correspond to satisfactory dynamic response in terms of rise time and other measures of transient response. In Section 7.5.2 we will show how to introduce the reference input with full-state feedback, and in Section 7.6 we will describe the process of finding the poles for good design.

Estimator/observer

The second step—necessary if the full state is not available—is to design an **estimator** (sometimes called an **observer**), which computes an estimate of the entire state vector when provided with the measurements of the system indicated by Eq. (7.21b). We will examine estimator design in Section 7.7.

The third step consists of combining the control law and the estimator. Figure 7.11 shows how the control law and the estimator fit together and how the combination

The control law and the estimator together form the compensation

takes the place of what we have been previously referring to as **compensation**. At this stage, the control-law calculations are based on the estimated state rather than the actual state. In Section 7.8 we will show that this substitution is reasonable, and also that using the combined control law and estimator results in closed-loop pole locations that are the same as those determined when designing the control and estimator separately.

The fourth and final step of state-space design is to introduce the reference input in such a way that the plant output will track external commands with acceptable rise-time, overshoot, and settling-time values. At this point in the design, all the closed-loop poles have been selected, and the designer is concerned with the zeros of the overall transfer function. Figure 7.11 shows the command input r introduced in the same relative position as was done with the transform design methods; however, in Section 7.9 we will show how to introduce the reference at another location, resulting in different zeros and (usually) superior control.

7.5.1   Finding the Control Law

Control law

The first step in the state-space design method, as mentioned earlier, is to find the control law as feedback of a linear combination of the state-variables—that is,

Figure 7.11

Schematic diagram of state-space design elements

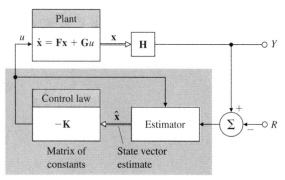

Compensation

$$u = -\mathbf{K}\mathbf{x} = -[\ K_1 \quad K_2 \quad \cdots \quad K_n\ ] \begin{bmatrix} x_1 \\ x_2 \\ \vdots \\ x_n \end{bmatrix}. \tag{7.70}$$

We assume for feedback purposes that all the elements of the state vector are at our disposal. In practice, of course, this would usually be a ridiculous assumption; moreover, a well-trained control designer knows that other design methods do not require so many sensors. The assumption that all state-variables are available merely allows us to proceed with this first step.

Equation (7.70) tells us that the system has a constant matrix in the state-vector feedback path, as shown in Fig. 7.12. For an nth-order system, there will be n feedback gains, $K_1, \ldots, K_n$, and because there are n roots of the system, it is possible that there are enough degrees of freedom to select *arbitrarily* any desired root location by choosing the proper values of K_i. This freedom contrasts sharply with root-locus design, in which we have only one parameter and the closed-loop poles are restricted to the locus.

Substituting the feedback law given by Eq. (7.70) into the system described by Eq. (7.21a) yields

$$\dot{\mathbf{x}} = \mathbf{F}\mathbf{x} - \mathbf{G}\mathbf{K}\mathbf{x}. \tag{7.71}$$

Control characteristic equation

The characteristic equation of this closed-loop system is

$$\det[s\mathbf{I} - (\mathbf{F} - \mathbf{G}\mathbf{K})] = 0. \tag{7.72}$$

When evaluated, this yields an nth-order polynomial in s containing the gains $K_1, \ldots, K_n$. The control-law design then consists of picking the gains $\mathbf{K}$ so that the roots of Eq. (7.72) are in desirable locations. Selecting desirable root locations is an inexact science that may require some iteration by the designer. Issues in their selection are considered in Examples 7.15 to 7.17 as well as in Section 7.6. For now, we assume that the desired locations are known, say,

$$s = s_1, s_2, \ldots, s_n.$$

Then the corresponding desired (control) characteristic equation is

$$\alpha_c(s) = (s - s_1)(s - s_2) \ldots (s - s_n) = 0. \tag{7.73}$$

Hence the required elements of $\mathbf{K}$ are obtained by matching coefficients in Eqs. (7.72) and (7.73). This forces the system's characteristic equation to be identical to the desired characteristic equation and the closed-loop poles to be placed at the desired locations.

Figure 7.12

Assumed system for control-law design

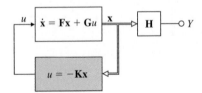

EXAMPLE 7.15   *Control Law for a Pendulum*

Suppose you have a pendulum with frequency ω_0 and a state-space description given by

$$\begin{bmatrix} \dot{x}_1 \\ \dot{x}_2 \end{bmatrix} = \begin{bmatrix} 0 & 1 \\ -\omega_0^2 & 0 \end{bmatrix} \begin{bmatrix} x_1 \\ x_2 \end{bmatrix} + \begin{bmatrix} 0 \\ 1 \end{bmatrix} u. \tag{7.74}$$

Find the control law that places the closed-loop poles of the system so that they are both at $-2\omega_0$. In other words, you wish to double the natural frequency and increase the damping ratio ζ from 0 to 1.

Solution.   From Eq. (7.73) we find that

$$\alpha_c(s) = (s + 2\omega_0)^2 \tag{7.75a}$$

$$= s^2 + 4\omega_0 s + 4\omega_0^2. \tag{7.75b}$$

Equation (7.72) tells us that

$$\det[s\mathbf{I} - (\mathbf{F} - \mathbf{GK})] = \det\left\{ \begin{bmatrix} s & 0 \\ 0 & s \end{bmatrix} - \left(\begin{bmatrix} 0 & 1 \\ -\omega_0^2 & 0 \end{bmatrix} - \begin{bmatrix} 0 \\ 1 \end{bmatrix} \begin{bmatrix} K_1 & K_2 \end{bmatrix} \right) \right\},$$

or

$$s^2 + K_2 s + \omega_0^2 + K_1 = 0. \tag{7.76}$$

Equating the coefficients with like powers of s in Eqs. (7.75b) and (7.76) yields the system of equations

$$K_2 = 4\omega_0,$$

$$\omega_0^2 + K_1 = 4\omega_0^2,$$

and therefore,

$$K_1 = 3\omega_0^2,$$

$$K_2 = 4\omega_0.$$

More concisely, the control law is

$$\mathbf{K} = \begin{bmatrix} K_1 & K_2 \end{bmatrix} = \begin{bmatrix} 3\omega_0^2 & 4\omega_0 \end{bmatrix}.$$

Figure 7.13 shows the response of the closed-loop system to the initial conditions $x_1 = 1.0$, $x_2 = 0.0$, and $\omega_0 = 1$. It shows a very well damped response, as would be expected from having two roots at $s = -2$. The MATLAB command impulse was used to generate the plot.

Calculating the gains by using the technique illustrated in Example 7.15 becomes rather tedious when the order of the system is higher than 3. There are, however, special "canonical" forms of the state-variable equations for which the algebra for finding the gains is especially simple. One such canonical form that is useful in control law design is the *control canonical form*. Consider the third-order system[7]

$$\dddot{y} + a_1\ddot{y} + a_2\dot{y} + a_3 y = b_1\ddot{u} + b_2\dot{u} + b_3 u, \tag{7.77}$$

---

[7]This development is exactly the same for higher-order systems.

Figure 7.13

Impulse response of the undamped oscillator with full-state feedback ($\omega_0 = 1$)

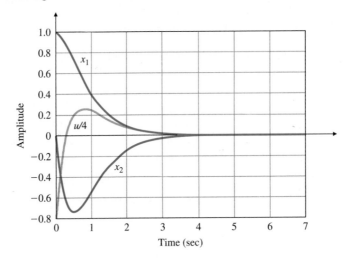

which corresponds to the transfer function

$$G(s) = \frac{Y(s)}{U(s)} = \frac{b_1 s^2 + b_2 s + b_3}{s^3 + a_1 s^2 + a_2 s + a_3} = \frac{b(s)}{a(s)}. \qquad (7.78)$$

Suppose we introduce an auxiliary variable (referred to as the *partial state*) ξ, which relates $a(s)$ and $b(s)$ as shown in Fig. 7.14(a). The transfer function from U to ξ is

$$\frac{\xi(s)}{U(s)} = \frac{1}{a(s)}, \qquad (7.79)$$

or

$$\dddot{\xi} + a_1\ddot{\xi} + a_2\dot{\xi} + a_3\xi = u. \qquad (7.80)$$

It is easy to draw a block diagram corresponding to Eq. (7.80) if we rearrange the equation as follows:

$$\dddot{\xi} = -a_1\ddot{\xi} - a_2\dot{\xi} - a_3\xi + u. \qquad (7.81)$$

The summation is indicated in Fig. 7.14(b), where each ξ on the right-hand side is obtained by sequential integration of $\dddot{\xi}$. To form the output, we go back to Fig. 7.14(a) and note that

$$Y(s) = b(s)\xi(s), \qquad (7.82)$$

which means that

$$y = b_1\ddot{\xi} + b_2\dot{\xi} + b_3\xi. \qquad (7.83)$$

We again pick off the outputs of the integrators, multiply them by $\{b_i\}$'s, and form the right-hand side of Eq. (7.77) by using a summer to yield the output as shown in Fig. 7.14(c). In this case, all the feedback loops return to the point of the application of the input, or "control" variable, and hence the form is referred to as the *control canonical form*. Reduction of the structure by Mason's rule or by elementary block diagram operations verifies that this structure has the transfer function given by $G(s)$.

Taking the state as the outputs of the three integrators numbered, by convention, from the left, namely

$$x_1 = \ddot{\xi}_1, x_2 = \dot{\xi}, x_3 = \xi, \qquad (7.84)$$

Figure 7.14

Derivation of control canonical form

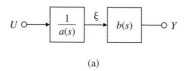

(a)

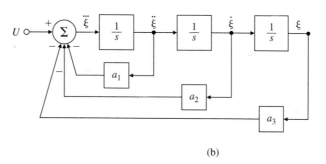

(b)

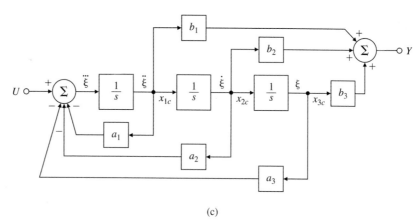

(c)

we obtain

$$\dot{x}_1 = \dddot{\xi} = -a_1 x_1 - a_2 x_2 - a_3 x_3 + u,$$

$$\dot{x}_2 = x_1,$$

$$\dot{x}_3 = x_2. \tag{7.85}$$

We may now write the matrices describing the control canonical form in general:

$$\mathbf{F}_c = \begin{bmatrix} -a_1 & -a_2 & \cdots & \cdots & -a_n \\ 1 & 0 & \cdots & \cdots & 0 \\ 0 & 1 & 0 & \cdots & 0 \\ \vdots & & \ddots & & 0 \\ 0 & 0 & \cdots & 1 & 0 \end{bmatrix}, \qquad \mathbf{G}_c = \begin{bmatrix} 1 \\ 0 \\ 0 \\ \vdots \\ 0 \end{bmatrix}, \tag{7.86a}$$

$$\mathbf{H}_c = \begin{bmatrix} b_1 & b_2 & \cdots & \cdots & b_n \end{bmatrix}, \qquad J_c = 0. \tag{7.86b}$$

Companion form matrix

The special structure of this system matrix is referred to as the **upper companion form** because the characteristic equation is $a(s) = s^n + a_1 s^{n-1} + a_2 s^{n-2} \cdots + a_n$ and the coefficients of this monic "companion" polynomial are the elements in the first row of $\mathbf{F}_c$. If we now form the closed-loop system matrix $\mathbf{F}_c - \mathbf{G}_c \mathbf{K}_c$, we find that

$$\mathbf{F}_c - \mathbf{G}_c \mathbf{K}_c = \begin{bmatrix} -a_1 - K_1 & -a_2 - K_2 & \cdots & \cdots & -a_n - K_n \\ 1 & 0 & \cdots & \cdots & 0 \\ 0 & 1 & 0 & \cdots & 0 \\ \vdots & & \ddots & & \vdots \\ 0 & 0 & \cdots & 1 & 0 \end{bmatrix}. \tag{7.87}$$

By visually comparing Eqs. (7.86a) and (7.87), we see that the closed-loop characteristic equation is

$$s^n + (a_1 + K_1)s^{n-1} + (a_2 + K_2)s^{n-2} + \cdots + (a_n + K_n) = 0. \tag{7.88}$$

Therefore, if the desired pole locations result in the characteristic equation given by

$$\alpha_c(s) = s^n + \alpha_1 s^{n-1} + \alpha_2 s^{n-2} + \cdots + \alpha_n = 0, \tag{7.89}$$

then the necessary feedback gains can be found by equating the coefficients in Eqs. (7.88) and (7.89):

$$K_1 = -a_1 + \alpha_1, \ K_2 = -a_2 + \alpha_2, \ldots, \ K_n = -a_n + \alpha_n. \tag{7.90}$$

We now have the basis for a design procedure. Given a system of order n described by an arbitrary $(\mathbf{F}, \mathbf{G})$ and given a desired nth-order monic characteristic polynomial $\alpha_c(s)$, we (1) transform $(\mathbf{F}, \mathbf{G})$ to control canonical form $(\mathbf{F}_c, \mathbf{G}_c)$ by changing the state $\mathbf{x} = \mathbf{T}\mathbf{z}$, and we (2) solve for the control gains by inspection using Eq. (7.90) to give the control law $u = -\mathbf{K}_c \mathbf{z}$. Because this gain is for the state in the *control* form, we must (3) transform the gain back to the original state to get $\mathbf{K} = \mathbf{K}_c \mathbf{T}^{-1}$.

Ackermann's formula for pole placement

An alternative to this transformation method is given by **Ackermann's formula** (1972), which organizes the three-step process of converting to $(\mathbf{F}_c, \mathbf{G}_c)$, solving for the gains, and converting back again into the very compact form

$$\mathbf{K} = [\ 0 \quad \cdots \quad 0 \quad 1\ ]\mathcal{C}^{-1}\alpha_c(\mathbf{F}), \tag{7.91}$$

such that

$$\mathcal{C} = [\ \mathbf{G} \quad \mathbf{F}\mathbf{G} \quad \mathbf{F}^2\mathbf{G} \quad \cdots \quad \mathbf{F}^{n-1}\mathbf{G}\ ], \tag{7.92}$$

where $\mathcal{C}$ is the controllability matrix we saw in Section 7.4, n gives the order of the system and the number of state-variables, and $\alpha_c(\mathbf{F})$ is a matrix defined as

$$\alpha_c(\mathbf{F}) = \mathbf{F}^n + \alpha_1 \mathbf{F}^{n-1} + \alpha_2 \mathbf{F}^{n-2} + \cdots + \alpha_n \mathbf{I}, \tag{7.93}$$

where the α_i are the coefficients of the desired characteristic polynomial Eq. (7.89). Note that Eq. (7.93) is a matrix equation. Refer to Appendix WG for the derivation of Ackermann's formula.

EXAMPLE 7.16

Ackermann's Formula for Undamped Oscillator

(a) Use Ackermann's formula to solve for the gains for the undamped oscillator of Example 7.15. (b) Verify the calculations with MATLAB for $\omega_0 = 1$.

Solution

(a) The desired characteristic equation is $\alpha_c(s) = (s+2\omega_0)^2$. Therefore, the desired characteristic polynomial coefficients,

$$\alpha_1 = 4\omega_0, \qquad \alpha_2 = 4\omega_0^2,$$

are substituted into Eq. (7.93) and the result is

$$\alpha_c(\mathbf{F}) = \begin{bmatrix} -\omega_0^2 & 0 \\ 0 & -\omega_0^2 \end{bmatrix} + 4\omega_0 \begin{bmatrix} 0 & 1 \\ -\omega_0^2 & 0 \end{bmatrix} + 4\omega_0^2 \begin{bmatrix} 1 & 0 \\ 0 & 1 \end{bmatrix}, \quad (7.94a)$$

$$= \begin{bmatrix} 3\omega_0^2 & 4\omega_0 \\ -4\omega_0^3 & 3\omega_0^2 \end{bmatrix}. \quad (7.94b)$$

The controllability matrix is

$$\mathcal{C} = [\ \mathbf{G} \quad \mathbf{FG}\ ] = \begin{bmatrix} 0 & 1 \\ 1 & 0 \end{bmatrix},$$

which yields

$$\mathcal{C}^{-1} = \begin{bmatrix} 0 & 1 \\ 1 & 0 \end{bmatrix}. \quad (7.95)$$

Finally, we substitute Eqs. (7.95) and (7.94a) into Eq. (7.91) to get

$$\mathbf{K} = [\ K_1 \quad K_2\ ]$$

$$= [\ 0 \quad 1\ ] \begin{bmatrix} 0 & 1 \\ 1 & 0 \end{bmatrix} \begin{bmatrix} 3\omega_0^2 & 4\omega_0 \\ -4\omega_0^3 & 3\omega_0^2 \end{bmatrix}.$$

Therefore

$$\mathbf{K} = [\ 3\omega_0^2 \quad 4\omega_0\ ],$$

which is the same result we obtained previously.

(b) The MATLAB statements

```
wo = 1;
F = [0 1;-wo*wo 0];
G = [0;1];
pc = [-2*wo;-2*wo];
K = acker(F,G,pc)
```

yield $\mathbf{K} = [\ 3 \quad 4\ ]$, which agrees with hand calculations.

MATLAB acker, place

As was mentioned earlier, computation of the controllability matrix has very poor numerical accuracy, and this carries over to Ackermann's formula. Equation (7.91), implemented in MATLAB with the function acker, can be used for the design of single-input-single-output (SISO) systems with a small (≤ 10) number of state-variables. For more complex cases a more reliable formula is available, implemented in MATLAB with place. A modest limitation on place is that, because it is based on assigning closed-loop eigenvectors, none of the desired closed-loop poles may be repeated; that is, the poles must be distinct,[8] a requirement that does not apply to acker.

---

[8] One may get around this restriction by moving the repeated poles by very small amounts to make them distinct.

The fact that we can shift the poles of a system by state feedback to any desired location is a rather remarkable result. The development in this section reveals that this shift is possible if we can transform $(\mathbf{F}, \mathbf{G})$ to the control form $(\mathbf{F}_c, \mathbf{G}_c)$, which in turn is possible if the system is controllable. In rare instances the system may be uncontrollable, in which case no possible control will yield arbitrary pole locations. **Uncontrollable systems** have certain modes, or subsystems, that are unaffected by the control. This usually means that parts of the system are physically disconnected from the input. For example, in modal canonical form for a system with distinct poles, one of the modal state-variables is not connected to the input if there is a zero entry in the $\mathbf{B}_m$ matrix. A good physical understanding of the system being controlled would prevent any attempt to design a controller for an uncontrollable system. As we saw earlier, there are algebraic tests for controllability; however, no mathematical test can replace the control engineer's understanding of the physical system. Often the physical situation is such that every mode is controllable to some degree, and, while the mathematical tests indicate the system is controllable, certain modes are so weakly controllable that designs to control them are virtually useless.

An example of weak controllability

Airplane control is a good example of weak controllability of certain modes. Pitch plane motion $\mathbf{x}_p$ is primarily affected by the elevator δ_e and weakly affected by rolling motion $\mathbf{x}_r$. Rolling motion is essentially affected only by the ailerons δ_a. The state-space description of these relationships is

$$\begin{bmatrix} \dot{\mathbf{x}}_p \\ \dot{\mathbf{x}}_r \end{bmatrix} = \begin{bmatrix} \mathbf{F}_p & \varepsilon \\ 0 & \mathbf{F}_r \end{bmatrix} \begin{bmatrix} \mathbf{x}_p \\ \mathbf{x}_r \end{bmatrix} + \begin{bmatrix} \mathbf{G}_p & 0 \\ 0 & \mathbf{G}_r \end{bmatrix} \begin{bmatrix} \delta_e \\ \delta_a \end{bmatrix}, \qquad (7.96)$$

where the matrix of small numbers ε represents the weak coupling from rolling motion to pitching motion. A mathematical test of controllability for this system would conclude that pitch plane motion (and therefore altitude) is controllable by the ailerons as well as by the elevator! However, it is impractical to attempt to control an airplane's altitude by rolling the aircraft with the ailerons.

Another example will illustrate some of the properties of pole placement by state feedback and the effects of loss of controllability on the process.

EXAMPLE 7.17   *How Zero Location Can Affect the Control Law*

A specific thermal system is described by Eq. (7.35a) in observer canonical form with a zero at $s = z_0$. (a) Find the state feedback gains necessary for placing the poles of this system at the roots of $s^2 + 2\zeta\omega_n s + \omega_n^2$ (i.e., at $-\zeta\omega_n \pm j\omega_n\sqrt{1 - \zeta^2}$). (b) Repeat the computation with MATLAB, using the parameter values $z_0 = 2$, $\zeta = 0.5$, and $\omega_n = 2$ rad/sec.

Solution

(a) The state description matrices are

$$\mathbf{A}_o = \begin{bmatrix} -7 & 1 \\ -12 & 0 \end{bmatrix}, \quad \mathbf{B}_o = \begin{bmatrix} 1 \\ -z_0 \end{bmatrix},$$

$$\mathbf{C}_o = [1 \quad 0], \qquad D_o = 0.$$

First we substitute these matrices into Eq. (7.72) to get the closed-loop characteristic equation in terms of the unknown gains and the zero position:

$$s^2 + (7 + K_1 - z_0 K_2)s + 12 - K_2(7z_0 + 12) - K_1 z_0 = 0.$$

Next we equate this equation to the desired characteristic equation to get the equations

$$K_1 - z_0 K_2 = 2\zeta \omega_n - 7,$$

$$-z_0 K_1 - (7z_0 + 12)K_2 = \omega_n^2 - 12.$$

The solutions to these equations are

$$K_1 = \frac{z_0(14\zeta \omega_n - 37 - \omega_n^2) + 12(2\zeta \omega_n - 7)}{(z_0 + 3)(z_0 + 4)},$$

$$K_2 = \frac{z_0(7 - 2\zeta \omega_n) + 12 - \omega_n^2}{(z_0 + 3)(z_0 + 4)}.$$

(b) The following MATLAB statements can be used to find the solution:

```
Ao = [−7 1;−12 0];
zo = 2;
Bo = [1;−zo];
pc = roots([1 2 4]);
K = place(Ao,Bo,pc)
```

These statements yield K=[−3.80 0.60], which agrees with the hand calculations. If the zero were close to one of the open-loop poles, say $z_0 = -2.99$, then we find K=[2052.5 −688.1].

Two important observations should be made from this example. The first is that the gains grow as the zero z_0 approaches either -3 or -4, the values where this system loses controllability. In other words, as controllability is almost lost, the control gains become very large.

> The system has to work harder and harder to achieve control as controllability slips away.

The second important observation illustrated by the example is that both K_1 and K_2 grow as the desired closed-loop bandwidth given by ω_n is increased. From this, we conclude that

> To move the poles a long way requires large gains.

These observations lead us to a discussion of how we might go about selecting desired pole locations in general. Before we begin that topic, we will complete the design with full-state feedback by showing how the reference input might be applied to such a system and what the resulting response characteristics are.

7.5.2 Introducing the Reference Input with Full-State Feedback

Thus far, the control has been given by Eq. (7.70), or $u = -\mathbf{Kx}$. In order to study the transient response of the pole-placement designs to input commands, it is necessary to introduce the reference input into the system. An obvious way to do this is to change the control to $u = -\mathbf{K}\mathbf{x} + r$. However, the system will now almost surely have a nonzero steady-state error to a step input. The way to correct this problem is to compute the steady-state values of the state and the control input that will result in zero output error and then force them to take these values. If the desired final values of the state and the control input are $\mathbf{x}_{ss}$ and u_{ss} respectively, then the new control formula should be

$$u = u_{ss} - \mathbf{K}(\mathbf{x} - \mathbf{x}_{ss}), \tag{7.97}$$

so that when $\mathbf{x} = \mathbf{x}_{ss}$ (no error), $u = u_{ss}$. To pick the correct final values, we must solve the equations so that the system will have zero steady-state error to *any* constant input. The system differential equations are the standard ones:

$$\dot{\mathbf{x}} = \mathbf{F}\mathbf{x} + \mathbf{G}u, \tag{7.98a}$$

$$y = \mathbf{H}\mathbf{x} + Ju. \tag{7.98b}$$

In the constant steady state, Eqs. (7.98a) and (7.98b) reduce to the pair

$$\mathbf{0} = \mathbf{F}\mathbf{x}_{ss} + \mathbf{G}u_{ss}, \tag{7.99a}$$

$$y_{ss} = \mathbf{H}\mathbf{x}_{ss} + Ju_{ss}. \tag{7.99b}$$

Gain calculation for reference input

We want to solve for the values for which $y_{ss} = r_{ss}$ for any value of r_{ss}. To do this, we make $\mathbf{x}_{ss} = \mathbf{N}_\mathbf{x} r_{ss}$ and $u_{ss} = N_u r_{ss}$. With these substitutions we can write Eqs. (7.99) as a matrix equation; the common factor of r_{ss} cancels out to give the equation for the gains:

$$\begin{bmatrix} \mathbf{F} & \mathbf{G} \\ \mathbf{H} & J \end{bmatrix} \begin{bmatrix} \mathbf{N}_\mathbf{x} \\ N_u \end{bmatrix} = \begin{bmatrix} \mathbf{0} \\ 1 \end{bmatrix}. \tag{7.100}$$

This equation can be solved for $\mathbf{N}_\mathbf{x}$ and N_u to get

$$\begin{bmatrix} \mathbf{N}_\mathbf{x} \\ N_u \end{bmatrix} = \begin{bmatrix} \mathbf{F} & \mathbf{G} \\ \mathbf{H} & J \end{bmatrix}^{-1} \begin{bmatrix} \mathbf{0} \\ 1 \end{bmatrix}.$$

Control equation with reference input

With these values, we finally have the basis for introducing the reference input so as to get zero steady-state error to a step input:

$$u = N_u r - \mathbf{K}(\mathbf{x} - \mathbf{N}_\mathbf{x} r) \tag{7.101a}$$

$$= -\mathbf{K}\mathbf{x} + (N_u + \mathbf{K}\mathbf{N}_\mathbf{x})r. \tag{7.101b}$$

The coefficient of r in parentheses is a constant that can be computed beforehand. We give it the symbol $\bar{N}$, so

$$u = -\mathbf{K}\mathbf{x} + \bar{N}r. \tag{7.102}$$

The block diagram of the system is shown in Fig. 7.15.

EXAMPLE 7.18   *Introducing the Reference Input*

Compute the necessary gains for zero steady-state error to a step command at x_1, and plot the resulting unit step response for the oscillator in Example 7.15 with $\omega_0 = 1$.

Solution. We substitute the matrices of Eq. (7.74) (with $\omega_0 = 1$ and $\mathbf{H} = [\ 1 \quad 0\ ]$ because $y = x_1$) into Eq. (7.100) to get

$$\begin{bmatrix} 0 & 1 & 0 \\ -1 & 0 & 1 \\ 1 & 0 & 0 \end{bmatrix} \begin{bmatrix} \mathbf{N}_\mathbf{x} \\ N_u \end{bmatrix} = \begin{bmatrix} 0 \\ 0 \\ 1 \end{bmatrix}. \tag{7.103}$$

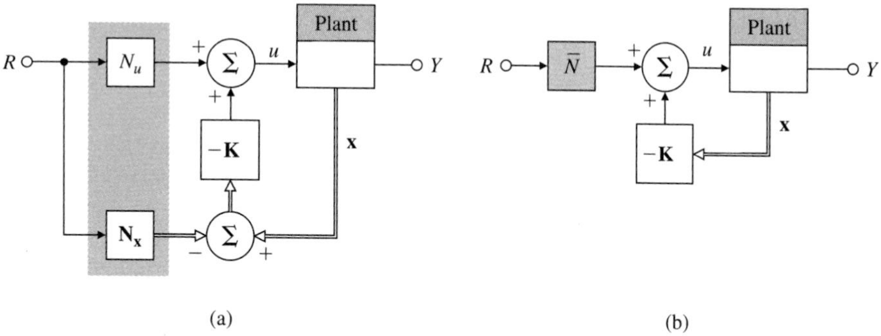

Figure 7.15

Block diagram for introducing the reference input with full-state feedback: (a) with state and control gains; (b) with a single composite gain

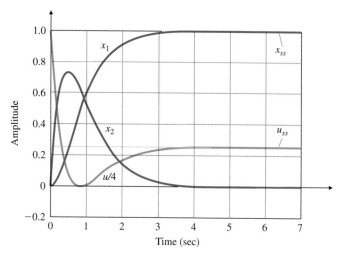

(a)   (b)

Figure 7.16

Step response of oscillator to a reference input

The solution is (x=a\b in MATLAB where a and b are the left- and right-hand side matrices respectively),

$$\mathbf{N_x} = \begin{bmatrix} 1 \\ 0 \end{bmatrix},$$

$$N_u = 1,$$

and, for the given control law, $\mathbf{K} = [3\omega_0^2 \quad 4\omega_0] = [\ 3 \quad 4\ ]$,

$$\bar{N} = N_u + \mathbf{K}\mathbf{N_x} = 4. \tag{7.104}$$

The corresponding step response (using the MATLAB step command) is plotted in Fig. 7.16.

Note that there are two equations for the control—Eqs. (7.101b) and (7.102). While these expressions are equivalent in theory, they differ in practical implementation in that Eq. (7.101b) is usually more robust to parameter errors than Eq. (7.102), particularly when the plant includes a pole at the origin and Type 1 behavior is possible. The difference is most clearly illustrated by the next example.

EXAMPLE 7.19   *Reference Input to a Type 1 System: DC Motor*

DC Motor Eq. (2.52)

Compute the input gains necessary to introduce a reference input with zero steady-state error to a step for the DC motor of Example 5.1, which in state-variable form is described by the matrices:

$$\mathbf{F} = \begin{bmatrix} 0 & 1 \\ 0 & -1 \end{bmatrix}, \quad \mathbf{G} = \begin{bmatrix} 0 \\ 1 \end{bmatrix},$$

$$\mathbf{H} = \begin{bmatrix} 1 & 0 \end{bmatrix}, \quad \mathbf{J} = 0.$$

Assume that the state feedback gain is $\begin{bmatrix} K_1 & K_2 \end{bmatrix}$.

Solution. If we substitute the system matrices of this example into the equation for the input gains, Eq. (7.100), we find that the solution is

$$\mathbf{N}_x = \begin{bmatrix} 1 \\ 0 \end{bmatrix},$$

$$N_u = 0,$$

$$\bar{N} = K_1.$$

With these values, the expression for the control using $\mathbf{N}_x$ and N_u [Eq. (7.101b)] reduces to

$$u = -K_1(x_1 - r) - K_2 x_2,$$

while the one using $\bar{N}$ [Eq. (7.102)] becomes

$$u = -K_1 x_1 - K_2 x_2 + K_1 r.$$

The block diagrams for the systems using each of the control equations are given in Fig. 7.17. When using Eq. (7.102), as shown in Fig. 7.17(b), it is necessary to multiply the input by a gain $K_1 (= \bar{N})$ exactly equal to that used in the feedback. If these two gains do not match exactly, there will be a steady-state error. On the other hand, if we use Eq. (7.101b), as shown in Fig. 7.17(a), there is only one gain to be used on the difference between the reference input and the first state, and zero steady-state error will result even if this gain is slightly in error. The system of Fig. 7.17(a) is more robust than the system of Fig. 7.17(b).

With the reference input in place, the closed-loop system has input r and output y. From the state description we know that the system poles are at the eigenvalues of the closed-loop system matrix, $\mathbf{F} - \mathbf{GK}$. In order to compute the closed-loop transient response, it is necessary to know where the closed-loop zeros of the transfer function from r to y are. They are to be found by applying Eq. (7.67) to the closed-loop description, which we assume has no direct path from input u to output y, so that $J = 0$. The zeros are values of s such that

$$\det \begin{bmatrix} s\mathbf{I} - (\mathbf{F} - \mathbf{GK}) & -\bar{N}\mathbf{G} \\ \mathbf{H} & 0 \end{bmatrix} = 0. \tag{7.105}$$

We can use two elementary facts about determinants to simplify Eq. (7.105). In the first place, if we divide the last column by $\bar{N}$, which is a scalar, then the point where the determinant is zero remains unchanged. The determinant is also not changed if we

Figure 7.17

Alternative structures for introducing the reference input:
(a) Eq. (7.101b);
(b) Eq. (7.102)

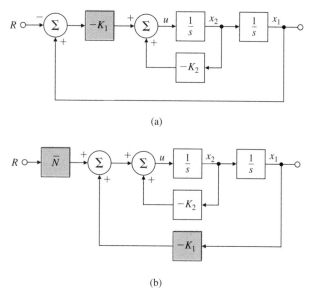

(a)

(b)

multiply the last column by $\mathbf{K}$ and add it to the first (block) column, with the result that the $\mathbf{GK}$ term is cancelled out. Thus the matrix equation for the zeros reduces to

$$\det \begin{bmatrix} s\mathbf{I} - \mathbf{F} & -\mathbf{G} \\ \mathbf{H} & 0 \end{bmatrix} = 0. \qquad (7.106)$$

Equation (7.106) is the same as Eq. (7.67) for the zeros of the plant *before* the feedback was applied. The important conclusion is that

> When full-state feedback is used as in Eq. (7.101b) or (7.102), the zeros remain unchanged by the feedback.

7.6 Selection of Pole Locations for Good Design

The first step in the pole-placement design approach is to decide on the closed-loop pole locations. When selecting pole locations, it is always useful to keep in mind that the control effort required is related to how far the open-loop poles are moved by the feedback. Furthermore, when a zero is near a pole, the system may be nearly uncontrollable, and as we saw in Section 7.5, moving such poles requires large control gains and thus a large control effort; however, the designer is able to temper the choices to take control effort into account. Therefore, a pole-placement philosophy that aims to fix only the undesirable aspects of the open-loop response and avoids either large increases in bandwidth or efforts to move poles that are near zeros will typically allow smaller gains, and thus smaller control actuators, than a philosophy that arbitrarily picks all the poles without regard to the original open-loop pole and zero locations.

In this section we discuss two techniques to aid in the pole-selection process. The first approach—dominant second-order poles—deals with pole selection without explicit regard for their effect on control effort; however, the designer is able to temper

Two methods of pole selection

the choices to take control effort into account. The second method (called optimal control, or symmetric root locus) does specifically address the issue of achieving a balance between good system response and control effort.

7.6.1 Dominant Second-Order Poles

The step response corresponding to the second-order transfer function with complex poles at radius ω_n and damping ratio ζ was discussed in Chapter 3. The rise time, overshoot, and settling time can be deduced directly from the pole locations. We can choose the closed-loop poles for a higher-order system as a desired pair of dominant second-order poles, and select the rest of the poles to have real parts corresponding to sufficiently damped modes, so that the system will mimic a second-order response with reasonable control effort. We also must make sure that the zeros are far enough into the LHP to avoid having any appreciable effect on the second-order behavior. A system with several lightly damped high-frequency vibration modes plus two rigid-body low-frequency modes lends itself to this philosophy. Here we can pick the low-frequency modes to achieve desired values of ω_n and ζ and select the rest of the poles to increase the damping of the high-frequency modes, while holding their frequency constant in order to minimize control effort. To illustrate this design method, we obviously need a system of higher than second-order; we will use the tape drive servomotor described in Example 7.11.

EXAMPLE 7.20 *Pole Placement as a Dominant Second-Order System*

Design the tape servomotor by the dominant second-order poles method to have no more than 5% overshoot and a rise time of no more than 4 sec. Keep the peak tension as low as possible.

Solution. From the plots of the second-order transients in Fig. 3.18, a damping ratio $\zeta = 0.7$ will meet the overshoot requirement and, for this damping ratio, a rise time of 4 sec suggests a natural frequency of about $1/1.5$. There are five poles in all, so the other three need to be placed far to the left of the dominant pair. For our purposes, "far" means the transients due to the fast poles should be over well before the transients due to the dominant poles, and we assume a factor of 4 in the respective undamped natural frequencies to be adequate. From these considerations, the desired poles are given by

$$\text{pc} = \left[-0.707 + 0.707 * j; -0.707 - 0.707 * j; -4; -4; -4 \right]/1.5; \quad (7.107)$$

With these desired poles, we can use the function acker with $\mathbf{F}$ and $\mathbf{G}$ from Example 7.11, Eq. (7.70), to find the control gains

$$\mathbf{K}_2 = [\ 8.5123 \quad 20.3457 \quad -1.4911 \quad -7.8821 \quad 6.1927\ ]. \quad (7.108)$$

MATLAB acker

These are found with the following MATLAB statements:

```
F = [0 2 0 0 0;−.1 −.35 .1 .1.75;0 0 0 2 0;.4 .4 −.4 -1.4 0;0 −.03 0 0 -1];
G = [0;0;0;0;1];
pc = [−.707+.707*j;−.707−.707*j;−4;−4;−4]/1.5;
K2 = acker(F,G,pc)
```

The step response and the corresponding tension plots for this and another design (to be discussed in Section 7.6.2) are given in Fig. 7.18 and Fig. 7.19. Notice that the rise time is approximately 4 sec and the overshoot is about 5%, as specified.

Figure 7.18

Step responses of the tape servomotor designs

Figure 7.18

Step responses of the tape servomotor designs

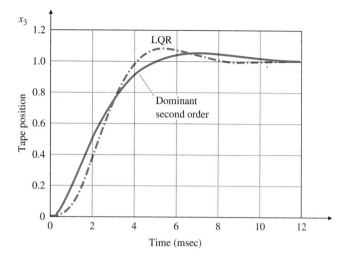

Figure 7.19

Tension plots for tape servomotor step responses

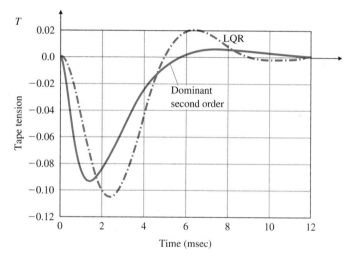

Because the design process is iterative, the poles we selected should be seen as only a first step, to be followed by further modifications to meet the specifications as accurately as necessary.

For this example we happened to select adequate pole locations on the first try.

7.6.2 Symmetric Root Locus (SRL)

LQR design

A most effective and widely used technique of linear control systems design is the optimal **linear quadratic regulator (LQR)**. The simplified version of the LQR problem is to find the control such that the performance index

$$\mathcal{J} = \int_0^\infty [\rho z^2(t) + u^2(t)] \, dt \tag{7.109}$$

Chapter 7 State-Space Design

is minimized for the system

$$\dot{\mathbf{x}} = \mathbf{F}\mathbf{x} + \mathbf{G}u, \tag{7.110a}$$

$$z = \mathbf{H}_1\mathbf{x}, \tag{7.110b}$$

where ρ in Eq. (7.109) is a weighting factor of the designer's choice. A remarkable fact is that the control law that minimizes $\mathcal{J}$ is given by linear-state feedback

$$u = -\mathbf{K}\mathbf{x}. \tag{7.111}$$

Here the optimal value of $\mathbf{K}$ is that which places the closed-loop poles at the stable roots (those in the LHP) of the symmetric root-locus (SRL) equation (Kailath, 1980)

$$1 + \rho G_0(-s)G_0(s) = 0, \tag{7.112}$$

where G_0 is the open-loop transfer function from u to z:

$$G_0(s) = \frac{Z(s)}{U(s)} = \mathbf{H}_1(s\mathbf{I} - \mathbf{F})^{-1}\mathbf{G} = \frac{N(s)}{D(s)}. \tag{7.113}$$

Note that this is a root-locus problem as discussed in Chapter 5 with respect to the parameter ρ, which weights the relative cost of (tracking error) z^2 with respect to the control effort u^2 in the performance index Eq. (7.109). Note also that s and $-s$ affect Eq. (7.112) in an identical manner; therefore, for any root s_0 of Eq. (7.112), there will also be a root at $-s_0$. We call the resulting root locus a **SRL**, since the locus in the LHP will have a mirror image in the right half-plane (RHP); that is, there is symmetry with respect to the imaginary axis. We may thus choose the optimal closed-loop poles by first selecting the matrix $\mathbf{H}_1$, which defines the tracking error and which the designer wishes to keep small, and then choosing ρ, which balances the importance of this tracking error against the control effort. Notice that the output we select as tracking error does *not* need to be the plant sensor output. That is why we call the output in Eq. (7.110) z rather than y.

Selecting a set of stable poles from the solution of Eq. (7.112) results in desired closed-loop poles, which we can then use in a pole-placement calculation such as Ackermann's formula (Eq. 7.91) to obtain $\mathbf{K}$. As with all root loci for real transfer functions G_0, the locus is also symmetric with respect to the real axis; thus there is symmetry with respect to both the real and imaginary axes. We can write the SRL equation in the standard root-locus form

$$1 + \rho \frac{N(-s)N(s)}{D(-s)D(s)} = 0, \tag{7.114}$$

obtain the locus poles and zeros by reflecting the open-loop poles and zeros of the transfer function from U to Z across the imaginary axis (which doubles the number of poles and zeros), and then sketch the locus. Note that the locus could be either a $0°$ or $180°$ locus, depending on the sign of $G_0(-s)G_0(s)$ in Eq. (7.112). A quick way to determine which type of locus to use ($0°$ or $180°$) is to pick the one that *has no part on the imaginary axis*. The real-axis rule of root locus plotting will reveal this right away. For the controllability assumptions we have made here, plus the assumption that all the system modes are present in the chosen output z, the optimal closed-loop system is guaranteed to be stable; thus no part of the locus can be on the imaginary axis.

Figure 7.20
SRL for a first-order
system

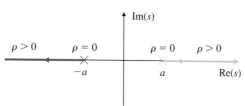

EXAMPLE 7.21

SRL for Servo Speed Control

Plot the SRL for the following servo speed control system with $z = y$:

$$\dot{y} = -ay + u, \tag{7.115a}$$

$$G_0(s) = \frac{1}{s+a}. \tag{7.115b}$$

Solution. The SRL equation [Eq. (7.112)] for this example is

$$1 + \rho \frac{1}{(-s+a)(s+a)} = 0. \tag{7.116}$$

The SRL, shown in Fig. 7.20, is a $0°$ locus. The optimal (stable) pole can be determined explicitly in this case as

$$s = -\sqrt{a^2 + \rho}. \tag{7.117}$$

Thus, the closed-loop root location that minimizes the performance index of Eq. (7.109) lies on the real axis at the distance given by Eq. (7.117) and is always to the left of the open-loop root.

EXAMPLE 7.22

SRL Design for Satellite Attitude Control

Draw the SRL for the satellite system with $z = y$.
Solution. The equations of motion are

$$\dot{\mathbf{x}} = \begin{bmatrix} 0 & 1 \\ 0 & 0 \end{bmatrix} \mathbf{x} + \begin{bmatrix} 0 \\ 1 \end{bmatrix} u, \tag{7.118}$$

$$y = [\ 1 \quad 0\ ]\ \mathbf{x}. \tag{7.119}$$

We then calculate from Eqs. (7.118) and (7.119) so that

$$G_0(s) = \frac{1}{s^2}. \tag{7.120}$$

The symmetric $180°$ loci are shown in Fig. 7.21. The MATLAB statements to generate the SRL are

```
numGG = [1];
denGG = conv([1 0 0],[1 0 0]);
sysGG = tf(numGG,denGG);
rlocus(sysGG);
```

Figure 7.21

SRL for the satellite

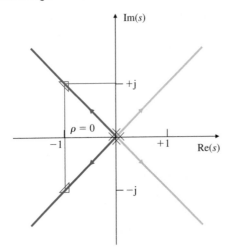

It is interesting to note that the (stable) closed-loop poles have damping of $\zeta =$ 0.707. We would choose two stable roots for a given value of ρ, for example $s = -1 \pm j1$ for $\rho = 4.07$, on the SRL and use them for pole-placement and control-law design.

Choosing different values of ρ can provide us with pole locations that achieve varying balances between a fast response (small values of $\int z^2 \, dt$) and a low control effort (small values of $\int u^2 \, dt$). Figure 7.22 shows the design trade-off curve for the satellite (double-integrator) plant [Eq. (7.18)] for various values of ρ ranging from 0.01 to 100. The curve has two asymptotes (dashed lines) corresponding to low (large ρ) and high (small ρ) penalty on the control usage. In practice, usually a value of ρ corresponding to a point close to the knee of the trade-off curve is chosen. This is because it provides a reasonable compromise between the use of control and the speed of response. For the satellite plant, the value of $\rho = 1$ corresponds to the knee of the curve. In this case the closed-loop poles have a damping ratio of $\zeta = 0.707$! Figure 7.23 shows the associated Nyquist plot, which has a phase margin PM = 65° and infinite gain margin. These excellent stability properties are a general feature of LQR designs.

It is also possible to locate optimal pole locations for the design of an open-loop unstable system using the SRL and LQR method.

EXAMPLE 7.23   *SRL Design for an Inverted Pendulum*

Draw the SRL for the linearized equations of the simple inverted pendulum with $\omega_o = 1$. Take the output, z, to be the sum of twice the position plus the velocity (so as to weight or penalize *both* position and velocity).

Solution.   The equations of motion are

$$\dot{\mathbf{x}} = \begin{bmatrix} 0 & 1 \\ \omega_0^2 & 0 \end{bmatrix} \mathbf{x} + \begin{bmatrix} 0 \\ -1 \end{bmatrix} u. \tag{7.121}$$

Figure 7.22

Design trade-off curve for satellite plant

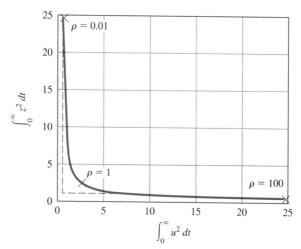

Figure 7.23

Nyquist plot for LQR design

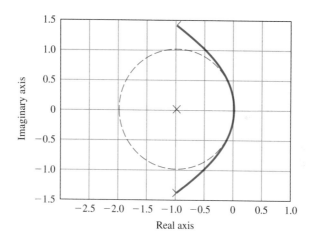

For the specified output of 2 × position + velocity, we let the tracking error be

$$z = [\; 2 \quad 1 \;]\mathbf{x}. \tag{7.122}$$

We then calculate from Eqs. (7.121) and (7.122) so that

$$G_0(s) = -\frac{s+2}{s^2 - \omega_0^2}. \tag{7.123}$$

The symmetric $0°$ loci are shown in Fig. 7.24. The MATLAB statements to generate the SRL are (for $\omega_o = 1$),

```
numGG=conv(-[1 2],-[-1 2]);
denGG=conv([1 0 -1],[1 0 -1]);
sysGG=tf(numGG,denGG);
rlocus(sysGG);
```

Figure 7.24

SRL for the inverted pendulum

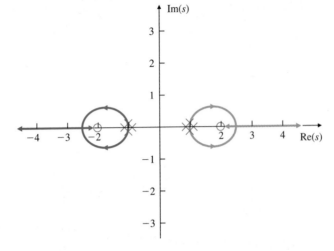

Figure 7.25

Step response for the inverted pendulum

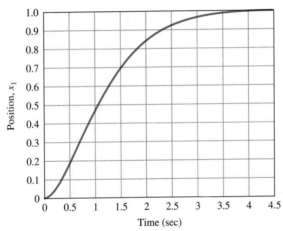

For $\rho = 1$, we find that the closed-loop poles are at $-1.36 \pm j0.606$, corresponding to $\mathbf{K} = [\ -2.23\quad -2.73\ ]$. If we substitute the system matrices of this example into the equation for the input gains, Eq. (7.100), we find that the solution is

$$\mathbf{N}_x = \begin{bmatrix} 1 \\ 0 \end{bmatrix},$$

$$N_u = 1,$$

$$\bar{N} = -1.23.$$

With these values, the expression for the control using $\mathbf{N}_x$ and N_u (Eq. 7.101b) the controller reduces to

$$u = -\mathbf{Kx} + \bar{N}r.$$

The corresponding step response for position is shown in Fig. 7.25.

As a final example in this section, we consider again the tape servomotor and introduce LQR design using the computer directly to solve for the optimal control law. From Eqs. (7.109) and (7.111), we know that the information needed to find the optimal control is given by the system matrices **F** and **G** and the output matrix **H**$_1$. Most computer-aided software packages, including MATLAB, use a more general form of Eq. (7.109):

$$\mathcal{J} = \int_0^\infty (\mathbf{x}^T \mathbf{Q}\mathbf{x} + \mathbf{u}^T \mathbf{R}\mathbf{u})\, dt. \tag{7.124}$$

MATLAB lqr

Equation (7.124) reduces to the simpler form of Eq. (7.109) if we take $\mathbf{Q} = \rho \mathbf{H}_1^T \mathbf{H}_1$ and $\mathbf{R} = 1$. The direct solution for the optimal control gain is the MATLAB statement

$$K = \text{lqr}(F, G, Q, R). \tag{7.125}$$

Bryson's rule

One reasonable method to start the LQR design iteration is suggested by Bryson's rule (Bryson and Ho, 1969). In practice, an appropriate choice to obtain acceptable values of **x** and **u** is to initially choose diagonal matrices **Q** and **R** such that

$$Q_{ii} = 1/\text{maximum acceptable value of } [x_i^2],$$
$$R_{ii} = 1/\text{maximum acceptable value of } [u_i^2].$$

The weighting matrices are then modified during subsequent iterations to achieve an acceptable trade-off between performance and control effort.

EXAMPLE 7.24   *LQR Design for a Tape Drive*

(a) Find the optimal control for the tape drive of Example 7.11, using the position x_3 as the output for the performance index. Let $\rho = 1$. Compare the results with that of dominant second order obtained before.

(b) Compare the LQR designs for $\rho = 0.1, 1, 10$.

Solution

(a) All we need to do here is to substitute the matrices into Eq. (7.125), form the feedback system, and plot the response. The performance index matrix is the scalar $R = 1$; the most difficult part of the problem is finding the state-cost matrix **Q**. With the output-cost variable $z = x_3$, the output matrix from Example 7.11 is

$$\mathbf{H}_3 = [\ 0.5 \quad 0 \quad 0.5 \quad 0 \quad 0\ ],$$

and with $\rho = 1$, the required matrix is

$$\mathbf{Q} = \mathbf{H}_3^T \mathbf{H}_3$$
$$= \begin{bmatrix} 0.25 & 0 & 0.25 & 0 & 0 \\ 0 & 0 & 0 & 0 & 0 \\ 0.25 & 0 & 0.25 & 0 & 0 \\ 0 & 0 & 0 & 0 & 0 \end{bmatrix}.$$

The gain is given by MATLAB, using the following statements:

```
F=[0 2 0 0 0; −.1 −.35 .1 .1.75;0 0 0 2 0;.4 .4 −.4 −1.4;0 −.03 0 0 −1];
G=[0; 0; 0; 0; 1];
H3=[.5 0 .5 0 0];
R=1;
rho=1;
Q=rho*H3'*H3;
K=lqr(F,G,Q,R)
```

The MATLAB calculated gain is

$$\mathbf{K} = [\; 0.6526 \quad 2.1667 \quad 0.3474 \quad 0.5976 \quad 1.0616 \;]. \qquad (7.126)$$

The results of a position step and the corresponding tension are plotted in Figs. 7.18 and 7.19 (using step) with the earlier responses for comparison. Obviously, there is a vast range of choice for the elements of $\mathbf{Q}$ and $\mathbf{R}$, so substantial experience is needed in order to use the LQR method effectively.

(b) The LQR designs may be repeated as in part (a) with the same $\mathbf{Q}$ and $\mathbf{R}$, but with $\rho = 0.1, 10$. Figure 7.26 shows a comparison of position step and the corresponding tension for the three designs. As seen from the results, the smaller values of ρ correspond to higher cost on the control and slower response, whereas the larger values of ρ correspond to lower cost on the control and relatively fast response.

Limiting Behavior of LQR Regulator Poles

It is interesting to consider the limiting behavior of the optimal closed-loop poles as a function of the root-locus parameter (i.e., ρ) although, in practice, neither case would be used.

"*Expensive control*" case ($\rho \to 0$): Equation (7.109) primarily penalizes the use of control energy. If the control is expensive, then the optimal control does not move any of the open-loop poles except for those that are in the RHP. The poles in the RHP are simply moved to their mirror images in the LHP. The optimal control does this to stabilize the system using minimum control effort and makes no attempt to move any of the poles of the system in the LHP. The closed-loop pole locations are simply the starting points on the SRL in the LHP. The optimal control does not speed up the response of the system in this case. For the satellite plant, the vertical dashed line in Fig. 7.22 corresponds to the "expensive control" case and illustrates that the very low control usage results in a very large error in z.

"*Cheap control*" case ($\rho \to \infty$): In this case control energy is no object and arbitrary control effort may be used by the optimal control law. The control law then moves some of the closed-loop pole locations right on top of the zeros in the LHP. The rest are moved to infinity along the SRL asymptotes. If the system is nonminimum phase, some of the closed-loop poles are moved to mirror images of these zeros in the LHP, as shown in Example 7.23. The rest of the poles go to infinity and do so along a Butterworth filter pole pattern, as shown in Example 7.22. The optimal control law provides the fastest possible response time consistent with the LQR cost function. The feedback gain matrix $\mathbf{K}$ becomes unbounded in this case. For the double-integrator

Figure 7.26

(a) Step response of the tape servomotor for LQR designs;
(b) corresponding tension for tape servomotor step responses

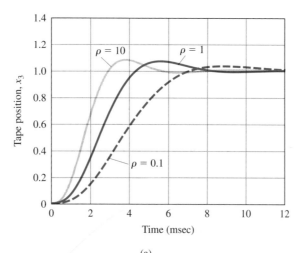

(a)

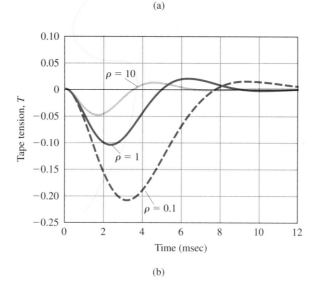

(b)

plant, the horizontal dashed line in Fig. 7.22 corresponds to the "cheap control" case.

Robustness Properties of LQR Regulators

It has been proved (Anderson and Moore, 1990) that the Nyquist plot for LQR design avoids a circle of unity radius centered at the -1 point as shown in Fig. 7.23. This leads to extraordinary phase and gain margin properties. It can be shown (Problem 7.32) that the return difference must satisfy

$$|1 + \mathbf{K}(j\omega\mathbf{I} - \mathbf{F})^{-1}\mathbf{G}| \geq 1. \tag{7.127}$$

Let us rewrite the loop gain as the sum of its real and imaginary parts:

$$L(j\omega) = \mathbf{K}(j\omega\mathbf{I} - \mathbf{F})^{-1}\mathbf{G} = \text{Re}(L(j\omega)) + j\text{Im}(L(j\omega)). \tag{7.128}$$

Equation (7.127) implies that

$$([\text{Re}(L(j\omega))]+1)^2 + [\text{Im}(L(j\omega))]^2 \geq 1, \tag{7.129}$$

which means that the Nyquist plot must indeed avoid a circle centered at -1 with unit radius. This implies that $\frac{1}{2} < \text{GM} < \infty$, which means that the "upward" gain margin is $\text{GM} = \infty$ and the "downward" gain margin is $\text{GM} = \frac{1}{2}$ (see also Problem 6.24 of Chapter 6). Hence the LQR gain matrix, $\mathbf{K}$, can be multiplied by a large scalar or reduced by half with guaranteed closed-loop system stability. The phase margin, PM, is at least $\pm 60°$. These margins are remarkable, and it is not realistic to assume that they can be achieved in practice, because of the presence of modeling errors and lack of sensors!

LQR gain and phase margins

7.6.3 Comments on the Methods

The two methods of pole selection described in Sections 7.6.1 and 7.6.2 are alternatives the designer can use for an initial design by pole placement. Note that the first method (dominant second order) suggests selecting closed-loop poles without regard to the effect on the control effort required to achieve that response. In some cases, therefore, the resulting control effort may be ridiculously high. The second method (SRL), on the other hand, selects poles that result in some balance between system errors and control effort. The designer can easily examine the relationship between shifts in that balance (by changing ρ) and system root locations, time response, and feedback gains. Whatever initial pole-selection method we use, some modification is almost always necessary to achieve the desired balance of bandwidth, overshoot, sensitivity, control effort, and other practical design requirements. Further insight into pole selection will be gained from the examples that illustrate compensation in Section 7.8 and from the case studies in Chapter 10.

7.7 Estimator Design

The control law designed in Section 7.5 assumed that all the state-variables are available for feedback. In most cases, not all the state-variables are measured. The cost of the required sensors may be prohibitive, or it may be physically impossible to measure all of the state-variables, as in, for example, a nuclear power plant. In this section we demonstrate how to reconstruct all of the state-variables of a system from a few measurements. If the estimate of the state is denoted by $\hat{\mathbf{x}}$, it would be convenient if we could replace the true state in the control law given by Eq. (7.102) with the estimates, so that the control becomes $u = -\mathbf{K}\hat{\mathbf{x}} + \bar{N}r$. This is indeed possible, as we shall see in Section 7.8, so construction of a state estimate is a key part of state-space control design.

7.7.1 Full-Order Estimators

One method of estimating the state is to construct a full-order model of the plant dynamics,

$$\dot{\hat{\mathbf{x}}} = \mathbf{F}\hat{\mathbf{x}} + \mathbf{G}u, \tag{7.130}$$

where $\hat{\mathbf{x}}$ is the estimate of the actual state $\mathbf{x}$. We know $\mathbf{F}$, $\mathbf{G}$, and $u(t)$. Hence this estimator will be satisfactory if we can obtain the correct initial condition $\mathbf{x}(0)$ and set

Figure 7.27

Open-loop estimator

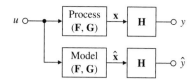

$\hat{\mathbf{x}}(0)$ equal to it. Figure 7.27 depicts this open-loop estimator. However, it is precisely the lack of information about $\mathbf{x}(0)$ that requires the construction of an estimator. Otherwise, the estimated state would track the true state exactly. Thus, if we made a poor estimate for the initial condition, the estimated state would have a continually growing error or an error that goes to zero too slowly to be of use. Furthermore, small errors in our knowledge of the system $(\mathbf{F}, \mathbf{G})$ would also cause the estimate to diverge from the true state.

To study the dynamics of this estimator, we define the error in the estimate to be

$$\tilde{\mathbf{x}} \triangleq \mathbf{x} - \hat{\mathbf{x}}. \tag{7.131}$$

Then the dynamics of this error system are given by

$$\dot{\tilde{\mathbf{x}}} = \mathbf{F}\tilde{\mathbf{x}}, \quad \tilde{\mathbf{x}}(0) = \mathbf{x}(0) - \hat{\mathbf{x}}(0). \tag{7.132}$$

The error converges to zero for a stable system ($\mathbf{F}$ stable), but we have no ability to influence the rate at which the state estimate converges to the true state. Furthermore, the error is converging to zero at the same rate as the natural dynamics of $\mathbf{F}$. If this convergence rate were satisfactory, no control or estimation would be required.

We now invoke the golden rule: When in trouble, use feedback. Consider feeding back the difference between the measured and estimated outputs and correcting the model continuously with this error signal. The equation for this scheme, shown in Fig. 7.28, is

Feed back the output error to correct the state estimate equation.

$$\dot{\hat{\mathbf{x}}} = \mathbf{F}\hat{\mathbf{x}} + \mathbf{G}u + \mathbf{L}(y - \mathbf{H}\hat{\mathbf{x}}). \tag{7.133}$$

Here $\mathbf{L}$ is a proportional gain defined as

$$\mathbf{L} = [l_1, l_2, \dots, l_n]^T \tag{7.134}$$

and is chosen to achieve satisfactory error characteristics. The dynamics of the error can be obtained by subtracting the estimate [Eq. (7.133)] from the state [Eq. (7.44)], to get the error equation

$$\dot{\tilde{\mathbf{x}}} = (\mathbf{F} - \mathbf{L}\mathbf{H})\tilde{\mathbf{x}}. \tag{7.135}$$

Estimate-error characteristic equation

The characteristic equation of the error is now given by

$$\det[s\mathbf{I} - (\mathbf{F} - \mathbf{L}\mathbf{H})] = 0. \tag{7.136}$$

If we can choose $\mathbf{L}$ so that $\mathbf{F} - \mathbf{L}\mathbf{H}$ has stable and reasonably fast eigenvalues, then $\tilde{\mathbf{x}}$ will decay to zero and remain there—independent of the known forcing function $u(t)$ and its effect on the state $\mathbf{x}(t)$ and irrespective of the initial condition $\tilde{\mathbf{x}}(0)$. This means that $\hat{\mathbf{x}}(t)$ will converge to $\mathbf{x}(t)$, regardless of the value of $\hat{\mathbf{x}}(0)$; furthermore, we can choose the dynamics of the error to be stable as well as much faster than the open-loop dynamics determined by $\mathbf{F}$.

Note that in obtaining Eq. (7.135), we have assumed that $\mathbf{F}, \mathbf{G}$, and $\mathbf{H}$ are identical in the physical plant and in the computer implementation of the estimator. If we do

Figure 7.28

Closed-loop estimator

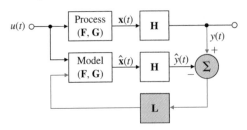

not have an accurate model of the plant $(\mathbf{F}, \mathbf{G}, \mathbf{H})$, the dynamics of the error are no longer governed by Eq. (7.135). However, we can typically choose $\mathbf{L}$ so that the error system is still at least stable and the error remains acceptably small, even with (small) modeling errors and disturbing inputs. It is important to emphasize that the nature of the plant and the estimator are quite different. The plant is a physical system such as a chemical process or servomechanism, whereas the estimator is usually a digital processor computing the estimated state according to Eq. (7.133).

The selection of $\mathbf{L}$ can be approached in exactly the same fashion as $\mathbf{K}$ is selected in the control-law design. If we specify the desired location of the estimator error poles as

$$s_i = \beta_1, \beta_2, \dots, \beta_n,$$

then the desired estimator characteristic equation is

$$\alpha_e(s) \overset{\Delta}{=} (s - \beta_1)(s - \beta_2) \cdots (s - \beta_n). \tag{7.137}$$

We can then solve for $\mathbf{L}$ by comparing coefficients in Eqs. (7.136) and (7.137).

EXAMPLE 7.25   *An Estimator Design for a Simple Pendulum*

Design an estimator for the simple pendulum. Compute the estimator gain matrix that will place both the estimator error poles at $-10\omega_0$ (five times as fast as the controller poles selected in Example 7.15). Verify the result using MATLAB for $\omega_0 = 1$. Evaluate the performance of the estimator.

Solution. The equations of motion are

$$\dot{\mathbf{x}} = \begin{bmatrix} 0 & 1 \\ -\omega_0^2 & 0 \end{bmatrix} \mathbf{x} + \begin{bmatrix} 0 \\ 1 \end{bmatrix} u, \tag{7.138a}$$

$$y = [1 \quad 0]\mathbf{x}. \tag{7.138b}$$

We are asked to place the two estimator error poles at $-10\omega_0$. The corresponding characteristic equation is

$$\alpha_e(s) = (s + 10\omega_0)^2 = s^2 + 20\omega_0 s + 100\omega_0^2. \tag{7.139}$$

From Eq. (7.136), we get

$$\det[s\mathbf{I} - (\mathbf{F} - \mathbf{LH})] = s^2 + l_1 s + l_2 + \omega_0^2. \tag{7.140}$$

Comparing the coefficients in Eqs. (7.139) and (7.140), we find that

$$\mathbf{L} = \begin{bmatrix} l_1 \\ l_2 \end{bmatrix} = \begin{bmatrix} 20\omega_0 \\ 99\omega_0^2 \end{bmatrix}. \tag{7.141}$$

The result can also be found from MATLAB for $\omega_0 = 1$, using the following MATLAB statements:

```
wo=1;
F=[0 1;−wo*wo 0];
H=[1 0];
pe=[−10*wo;−10*wo];
Lt=acker(F',H',pe);
L=Lt'
```

This yields $\mathbf{L} = [20\ 99]^T$ and agrees with the preceding hand calculations.

Performance of the estimator can be tested by adding the actual state feedback to the plant and plotting the estimate errors. Note that this is not the way the system will ultimately be built, but this approach provides a means of validating the estimator performance. Combining Eq. (7.71) of the plant with state feedback with Eq. (7.133) of the estimator with output feedback results in the following overall system equations:

$$\begin{bmatrix} \dot{\mathbf{x}} \\ \dot{\hat{\mathbf{x}}} \end{bmatrix} = \begin{bmatrix} \mathbf{F - GK} & \mathbf{0} \\ \mathbf{LH - GK} & \mathbf{F - LH} \end{bmatrix} \begin{bmatrix} \mathbf{x} \\ \hat{\mathbf{x}} \end{bmatrix}, \tag{7.142}$$

$$y = [\ \mathbf{H} \quad \mathbf{0}\ ] \begin{bmatrix} \mathbf{x} \\ \hat{\mathbf{x}} \end{bmatrix}, \tag{7.143}$$

$$\tilde{y} = [\ \mathbf{H} \quad -\mathbf{H}\ ] \begin{bmatrix} \mathbf{x} \\ \hat{\mathbf{x}} \end{bmatrix}. \tag{7.144}$$

A block diagram of the setup is drawn in Fig. 7.29.

The response of this closed-loop system with $\omega_0 = 1$ to an initial condition $\mathbf{x}_0 = [1.0, 0.0]^T$ and $\hat{\mathbf{x}}_0 = [0, 0]^T$ is shown in Fig. 7.30, where $\mathbf{K}$ is obtained from Example 7.15 and $\mathbf{L}$ comes from Eq. (7.141). The response may be obtained using impulse or initial in MATLAB. Note that the state estimates converge to the actual state-variables after an initial transient even though the initial value of $\hat{\mathbf{x}}$ had a large error. Also note that the estimate error decays approximately five times faster than the decay of the state itself, as we designed it to do.

MATLAB impulse, initial

Observer Canonical Form

As was the case for control-law design, there is a canonical form for which the estimator gain design equations are particularly simple and the existence of a solution is obvious. We introduced this form in Section 7.4.1. The equations are in the observer

Figure 7.29

Estimator connected to the plant

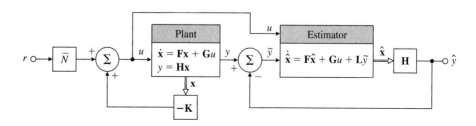

Figure 7.30

Initial-condition response of oscillator showing **x** and **x̂**

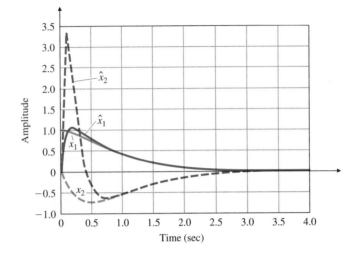

Figure 7.31

Block diagram for observer canonical form of a third-order system

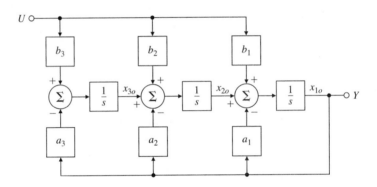

canonical form and have the structure:

$$\dot{\mathbf{x}}_o = \mathbf{F}_o\mathbf{x}_o + \mathbf{G}_o u, \tag{7.145a}$$

$$y = \mathbf{H}_o\mathbf{x}_o, \tag{7.145b}$$

where

$$\mathbf{F}_o = \begin{bmatrix} -a_1 & 1 & 0 & 0 & \cdots & 0 \\ -a_2 & 0 & 1 & 0 & \cdots & \vdots \\ \vdots & \vdots & & \ddots & & 1 \\ -a_n & 0 & & 0 & & 0 \end{bmatrix}, \quad \mathbf{G}_o = \begin{bmatrix} b_1 \\ b_2 \\ \vdots \\ b_n \end{bmatrix},$$

$$\mathbf{H}_o = [\, 1 \quad 0 \quad 0 \quad \cdots \quad 0 \,].$$

Observer canonical form

A block diagram for the third-order case is shown in Fig. 7.31. In observer canonical form, all the feedback loops come from the output, or observed signal. Like control canonical form, observer canonical form is a "direct" form because the values of the significant elements in the matrices are obtained directly from the coefficients of the numerator and denominator polynomials of the corresponding transfer function

$G(s)$. The matrix $\mathbf{F}_o$ is called a **left companion matrix** to the characteristic equation because the coefficients of the equation appear on the left side of the matrix.

One of the advantages of the observer canonical form is that the estimator gains can be obtained from it by inspection. The estimator error closed-loop matrix for the third-order case is

$$\mathbf{F}_o - \mathbf{LH}_o = \begin{bmatrix} -a_1 - l_1 & 1 & 0 \\ -a_2 - l_2 & 0 & 1 \\ -a_3 - l_3 & 0 & 0 \end{bmatrix}, \qquad (7.146)$$

which has the characteristic equation

$$s^3 + (a_1 + l_1)s^2 + (a_2 + l_2)s + (a_3 + l_3) = 0, \qquad (7.147)$$

and the estimator gain can be found by comparing the coefficients of Eq. (7.147) with $\alpha_e(s)$ from Eq. (7.137).

In a development exactly parallel with the control-law case, we can find a transformation to take a given system to observer canonical form if and only if the system has a

Observability structural property that in this case we call **observability**. Roughly speaking, observability refers to our ability to deduce information about all the modes of the system by monitoring only the sensed outputs. Unobservability results when some mode or subsystem is disconnected physically from the output and therefore no longer appears in the measurements. For example, if only derivatives of certain state-variables are measured, and these state-variables do not affect the dynamics, a constant of integration is obscured. This situation occurs with a plant having the transfer function $1/s^2$ if only velocity is measured, for then it is impossible to deduce the initial value of the position. On the other hand, for an oscillator, a velocity measurement is sufficient to estimate position because the acceleration, and consequently the velocity observed, are affected by position. The mathematical test for determining observability is that

Observability matrix the **observability matrix**

$$\mathcal{O} = \begin{bmatrix} \mathbf{H} \\ \mathbf{HF} \\ \vdots \\ \mathbf{HF}^{n-1} \end{bmatrix} \qquad (7.148)$$

must have independent columns. In the one output case we will study, $\mathcal{O}$ is square, so the requirement is that $\mathcal{O}$ be nonsingular or have a nonzero determinant. In general, we can find a transformation to observer canonical form if and only if the observability matrix is nonsingular. Note that this is analogous to our earlier conclusions for transforming system matrices to control canonical form.

As with control-law design, we could find the transformation to observer form, compute the gains from the equivalent of Eq. (7.147), and transform back. An alter-

Ackermann's estimator formula native method of computing $\mathbf{L}$ is to use Ackermann's formula in estimator form, which is

$$\mathbf{L} = \alpha_e(\mathbf{F})\mathcal{O}^{-1} \begin{bmatrix} 0 \\ 0 \\ \vdots \\ 1 \end{bmatrix}, \qquad (7.149)$$

where $\mathcal{O}$ is the observability matrix given in Eq. (7.148).

TABLE 7.1	**Duality**	
	Control	Estimation
	F	F^T
	G	H^T
	H	G^T

Duality

Duality of estimation and control

You may already have noticed from this discussion the considerable resemblance between estimation and control problems. In fact, the two problems are mathematically equivalent. This property is called **duality**. Table 7.1 shows the duality relationships between the estimation and control problems. For example, Ackermann's control formula [Eq. (7.91)] becomes the estimator formula Eq. (7.149) if we make the substitutions given in Table 7.1. We can demonstrate this directly using matrix algebra. The control problem is to select the row matrix $\mathbf{K}$ for satisfactory placement of the poles of the system matrix $\mathbf{F} - \mathbf{GK}$; the estimator problem is to select the column matrix $\mathbf{L}$ for satisfactory placement of the poles of $\mathbf{F} - \mathbf{LH}$. However, the poles of $\mathbf{F} - \mathbf{LH}$ equal those of $(\mathbf{F} - \mathbf{LH})^T = \mathbf{F}^T - \mathbf{H}^T\mathbf{L}^T$, and in this form, the algebra of the design for $\mathbf{L}^T$ is identical to that for $\mathbf{K}$. Therefore, where we used Ackermann's formula or the place algorithm in the forms

MATLAB acker, place

$$K=acker(F, G, p_c),$$

$$K=place(F, G, p_c),$$

for the control problem, we use

$$Lt=acker(F', H', p_e),$$

$$Lt=place(F', H', p_e),$$

$$L=Lt',$$

where p_e is a vector containing the desired estimator error poles for the estimator problem.

Thus duality allows us to use the same design tools for estimator problems as for control problems with proper substitutions. The two canonical forms are also dual, as we can see by comparing the triples $(\mathbf{F}_c, \mathbf{G}_c, \mathbf{H}_c)$ and $(\mathbf{F}_o, \mathbf{G}_o, \mathbf{H}_o)$.

7.7.2   Reduced-Order Estimators

The estimator design method described in Section 7.7.1 reconstructs the entire state vector using measurements of some of the state-variables. If the sensors have no noise, then a full-order estimator contains redundancies, and it seems reasonable to question the necessity for estimating state-variables that are measured directly. Can we reduce the complexity of the estimator by using the state-variables that are measured directly and exactly? Yes. However, it is better to implement a full-order estimator if there is significant noise on the measurements because, in addition to estimating unmeasured state-variables, the estimator filters the measurements.

The **reduced-order estimator** reduces the order of the estimator by the number (1 in this text) of sensed outputs. To derive this estimator, we start with the assumption that the output equals the first state as, for example, $y = x_a$. If this is not so, a preliminary step is required. Transforming to observer form is possible but overkill; any nonsingular transformation with $\mathbf{H}$ as the first row will do. We now partition the state vector into two parts: x_a, which is directly measured, and $\mathbf{x}_b$, which represents the remaining state-variables that need to be estimated. If we partition the system matrices accordingly, the complete description of the system is given by

$$\begin{bmatrix} \dot{x}_a \\ \dot{\mathbf{x}}_b \end{bmatrix} = \begin{bmatrix} F_{aa} & \mathbf{F}_{ab} \\ \mathbf{F}_{ba} & \mathbf{F}_{bb} \end{bmatrix} \begin{bmatrix} x_a \\ \mathbf{x}_b \end{bmatrix} + \begin{bmatrix} G_a \\ \mathbf{G}_b \end{bmatrix} u, \tag{7.150a}$$

$$y = [\; 1 \quad \mathbf{0} \;] \begin{bmatrix} x_a \\ \mathbf{x}_b \end{bmatrix}. \tag{7.150b}$$

The dynamics of the unmeasured state-variables are given by

$$\dot{\mathbf{x}}_b = \mathbf{F}_{bb}\mathbf{x}_b + \underbrace{\mathbf{F}_{ba}x_a + \mathbf{G}_b u}_{\text{known input}}, \tag{7.151}$$

where the right-most two terms are known and can be considered as an input into the $\mathbf{x}_b$ dynamics. Because $x_a = y$, the measured dynamics are given by the scalar equation

$$\dot{x}_a = \dot{y} = F_{aa}y + \mathbf{F}_{ab}\mathbf{x}_b + G_a u. \tag{7.152}$$

If we collect the known terms of Eq. (7.152) on one side, yielding

$$\underbrace{\dot{y} - F_{aa}y - G_a u}_{\text{known measurement}} = \mathbf{F}_{ab}\mathbf{x}_b, \tag{7.153}$$

we obtain a relationship between known quantities on the left side, which we consider measurements, and unknown state-variables on the right. Therefore, Eqs. (7.152) and (7.153) have the same relationship to the state $\mathbf{x}_b$ that the original equation [Eq. (7.150b)] had to the entire state $\mathbf{x}$. Following this line of reasoning, we can establish the following substitutions in the original estimator equations to obtain a (reduced-order) estimator of $\mathbf{x}_b$:

$$\mathbf{x} \leftarrow \mathbf{x}_b, \tag{7.154a}$$

$$\mathbf{F} \leftarrow \mathbf{F}_{bb}, \tag{7.154b}$$

$$\mathbf{G}u \leftarrow \mathbf{F}_{ba}y + \mathbf{G}_b u, \tag{7.154c}$$

$$y \leftarrow \dot{y} - F_{aa}y - G_a u, \tag{7.154d}$$

$$\mathbf{H} \leftarrow \mathbf{F}_{ab}. \tag{7.154e}$$

Therefore, the reduced-order estimator equations are obtained by substituting Eqs. (7.154) into the full-order estimator [Eq. (7.133)]:

$$\dot{\hat{\mathbf{x}}}_b = \mathbf{F}_{bb}\hat{\mathbf{x}}_b + \underbrace{\mathbf{F}_{ba}y + \mathbf{G}_b u}_{\text{input}} + \mathbf{L}\,(\underbrace{\dot{y} - F_{aa}y - G_a u}_{\text{measurement}} - \mathbf{F}_{ab}\hat{\mathbf{x}}_b). \tag{7.155}$$

If we define the estimator error to be

$$\tilde{\mathbf{x}}_b \triangleq \mathbf{x}_b - \hat{\mathbf{x}}_b, \tag{7.156}$$

Figure 7.32

Reduced-order
estimator structure

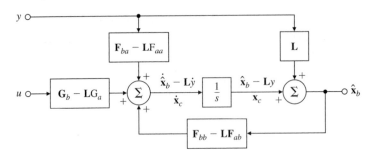

then the dynamics of the error are given by subtracting Eq. (7.151) from Eq. (7.155) to get

$$\dot{\tilde{\mathbf{x}}}_b = (\mathbf{F}_{bb} - \mathbf{L}\mathbf{F}_{ab})\tilde{\mathbf{x}}_b, \qquad (7.157)$$

and its characteristic equation is given by

$$\det[s\mathbf{I} - (\mathbf{F}_{bb} - \mathbf{L}\mathbf{F}_{ab})] = 0. \qquad (7.158)$$

We design the dynamics of this estimator by selecting $\mathbf{L}$ so that Eq. (7.158) matches a reduced order $\alpha_e(s)$. Now Eq. (7.155) can be rewritten as

$$\dot{\hat{\mathbf{x}}}_b = (\mathbf{F}_{bb} - \mathbf{L}\mathbf{F}_{ab})\hat{\mathbf{x}}_b + (\mathbf{F}_{ba} - \mathbf{L}\mathbf{F}_{aa})y + (\mathbf{G}_b - \mathbf{L}\mathbf{G}_a)u + \mathbf{L}\dot{y}. \qquad (7.159)$$

The fact that we must form the derivative of the measurements in Eq. (7.159) appears to present a practical difficulty. It is known that differentiation amplifies noise, so if y is noisy, the use of $\dot{y}$ is unacceptable. To get around this difficulty, we define the new controller state to be

$$\mathbf{x}_c \triangleq \hat{\mathbf{x}}_b - \mathbf{L}y. \qquad (7.160)$$

In terms of this new state, the implementation of the reduced-order estimator is given by

$$\dot{\mathbf{x}}_c = (\mathbf{F}_{bb} - \mathbf{L}\mathbf{F}_{ab})\hat{\mathbf{x}}_b + (\mathbf{F}_{ba} - \mathbf{L}\mathbf{F}_{aa})y + (\mathbf{G}_b - \mathbf{L}\mathbf{G}_a)u, \qquad (7.161)$$

and $\dot{y}$ no longer appears directly. A block-diagram representation of the reduced-order estimator is shown in Fig. 7.32.

EXAMPLE 7.26

Reduced-order estimator

A Reduced-Order Estimator Design for Pendulum

Design a reduced-order estimator for the pendulum that has the error pole at $-10\omega_0$.

Solution. We are given the system equations

$$\begin{bmatrix} \dot{x}_1 \\ \dot{x}_2 \end{bmatrix} = \begin{bmatrix} 0 & 1 \\ -\omega_0^2 & 0 \end{bmatrix} \begin{bmatrix} x_1 \\ x_2 \end{bmatrix} + \begin{bmatrix} 0 \\ 1 \end{bmatrix} u,$$

$$y = \begin{bmatrix} 1 & 0 \end{bmatrix} \begin{bmatrix} x_1 \\ x_2 \end{bmatrix}.$$

The partitioned matrices are

$$\begin{bmatrix} \mathbf{F}_{aa} & \mathbf{F}_{ab} \\ \mathbf{F}_{ba} & \mathbf{F}_{bb} \end{bmatrix} = \begin{bmatrix} 0 & 1 \\ -\omega_0^2 & 0 \end{bmatrix},$$

$$\begin{bmatrix} G_a \\ G_b \end{bmatrix} = \begin{bmatrix} 0 \\ 1 \end{bmatrix}.$$

From Eq. (7.158), we find the characteristic equation in terms of L:

$$s - (0 - L) = 0.$$

We compare it with the desired equation,

$$\alpha_e(s) = s + 10\omega_0 = 0,$$

which yields

$$L = 10\omega_0.$$

The estimator equation, from Eq. (7.161), is

$$\dot{x}_c = -10\omega_0\hat{x}_2 - \omega_0^2 y + u,$$

and the state estimate, from Eq. (7.160), is

$$\hat{x}_2 = x_c + 10\omega_0 y.$$

We use the control law given in the earlier examples. The response of the estimator to a plant initial condition $\mathbf{x}_0 = [1.0, 0.0]^T$ and an estimator initial condition $x_{c0} = 0$ is shown in Fig. 7.33 for $\omega_0 = 1$. The response may be obtained using impulse or initial in MATLAB. Note the similarity of the initial-condition response to that of the full-order estimator plotted in Fig. 7.30.

MATLAB impulse, initial

The reduced-order estimator gain can also be found from MATLAB by using

$$\mathsf{Lt} = \mathsf{acker}(F'_{bb}, F'_{ab}, \mathsf{p}_e),$$

$$\mathsf{Lt} = \mathsf{place}(F'_{bb}, F'_{ab}, \mathsf{p}_e),$$

$$\mathsf{L} = \mathsf{Lt}'.$$

The conditions for the existence of the reduced-order estimator are the same as for the full-order estimator—namely, observability of $(\mathbf{F}, \mathbf{H})$.

Figure 7.33
Initial-condition response of the reduced-order estimator

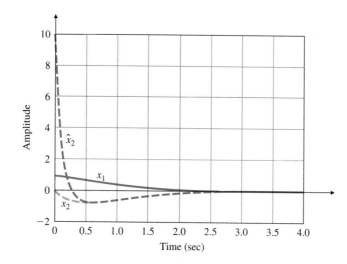

7.7.3   Estimator Pole Selection

We can base our selection of estimator pole locations on the techniques discussed in Section 7.6 for the case of controller poles. As a rule of thumb, the estimator poles can be chosen to be faster than the controller poles by a factor of 2 to 6. This ensures a faster decay of the estimator errors compared with the desired dynamics, thus causing the controller poles to dominate the total response. If sensor noise is large enough to be a major concern, we may choose the estimator poles to be slower than two times the controller poles, which would yield a system with lower bandwidth and more noise smoothing. However, we would expect the total system response in this case to be strongly influenced by the location of the estimator poles. If the estimator poles are slower than the controller poles, we would expect the system response to disturbances to be dominated by the dynamic characteristics of the estimator rather than by those selected by the control law.

In comparison with the selection of controller poles, estimator pole selection requires us to be concerned with a much different relationship than with control effort. As in the controller, there is a feedback term in the estimator that grows in magnitude as the requested speed of response increases. However, this feedback is in the form of an electronic signal or a digital word in a computer, so its growth causes no special difficulty. In the controller, increasing the speed of response increases the control effort; this implies the use of a larger actuator, which in turn increases size, weight, and cost. The important consequence of increasing the speed of response of an estimator is that the bandwidth of the estimator becomes higher, thus causing more sensor noise to pass on to the control actuator. Of course, if $(\mathbf{F}, \mathbf{H})$ are not observable, then no amount of estimator gain can produce a reasonable state estimate. Thus, as with controller design, the best estimator design is a balance between good transient response and low-enough bandwidth that sensor noise does not significantly impair actuator activity. Both dominant second-order and prototype characteristic equation ideas can be used to meet the requirements.

There is also a result for estimator gain design based on the SRL. In optimal estimation theory, the best choice for estimator gain is dependent on the ratio of sensor noise intensity v to process (disturbance) noise intensity [w in Eq. (7.163)]. This is best understood by reexamining the estimator equation

$$\dot{\hat{\mathbf{x}}} = \mathbf{F}\hat{\mathbf{x}} + \mathbf{G}u + \mathbf{L}(y - \mathbf{H}\hat{\mathbf{x}}) \tag{7.162}$$

to see how it interacts with the system when process noise w is present. The plant with process noise is described by

$$\dot{\mathbf{x}} = \mathbf{F}\mathbf{x} + \mathbf{G}u + \mathbf{G}_1 w, \tag{7.163}$$

and the measurement equation with sensor noise v is described by

$$y = \mathbf{H}\mathbf{x} + v. \tag{7.164}$$

The estimator error equation with these additional inputs is found directly by subtracting Eq. (7.162) from Eq. (7.163) and substituting Eq. (7.164) for y:

$$\dot{\tilde{\mathbf{x}}} = (\mathbf{F} - \mathbf{L}\mathbf{H})\tilde{\mathbf{x}} + \mathbf{G}_1 w - \mathbf{L}v. \tag{7.165}$$

In Eq. (7.165) the sensor noise is multiplied by $\mathbf{L}$ and the process noise is not. If $\mathbf{L}$ is very small, then the effect of sensor noise is removed, but the estimator's dynamic response will be "slow," so the error will not reject effects of w very well. The state of a low-gain estimator will not track uncertain plant inputs very well. These results can, with some success, also be applied to model errors in, for example, $\mathbf{F}$ or $\mathbf{G}$. Such model errors will add terms to Eq. (7.165) and act like additional process noise. On the other hand, if $\mathbf{L}$ is large, then the estimator response will be fast and the disturbance or process noise will be rejected, but the sensor noise, multiplied by $\mathbf{L}$, results in large errors. Clearly, a balance between these two effects is required. It turns out that the optimal solution to this balance can be found under very reasonable assumptions by solving an **SRL** equation for the estimator that is very similar to the one for the optimal control formulation [Eq. (7.112)]. The estimator **SRL** equation is

Estimator SRL equation

$$1 + qG_e(-s)G_e(s) = 0, \qquad (7.166)$$

where q is the ratio of input disturbance noise intensity to sensor noise intensity and G_e is the transfer function from the process noise to the sensor output and is given by

$$G_e(s) = \mathbf{H}(s\mathbf{I} - \mathbf{F})^{-1}\mathbf{G}_1. \qquad (7.167)$$

Note from Eqs. (7.112) and (7.166) that $G_e(s)$ is similar to $G_0(s)$. However, a comparison of Eqs. (7.113) and (7.167) shows that $G_e(s)$ has the input matrix $\mathbf{G}_1$ instead of $\mathbf{G}$, and that G_0 is the transfer function from the control input u to *cost* output z and has output matrix $\mathbf{H}_1$ instead of $\mathbf{H}$.

The use of the estimator **SRL** [Eq. (7.166)] is identical to the use of the controller **SRL**. A root locus with respect to q is generated, thus yielding sets of optimal estimator poles corresponding more or less to the ratio of process noise intensity to sensor noise intensity. The designer then picks the set of (stable) poles that seems best, considering all aspects of the problem. An important advantage of using the **SRL** technique is that after the process noise input matrix $\mathbf{G}_1$ has been selected, the arbitrariness is reduced to one degree of freedom, the selection q, instead of the many degrees of freedom required to select the poles directly in a higher-order system.

A final comment concerns the reduced-order estimator. Because of the presence of a direct transmission term from y through $\mathbf{L}$ to x_b (see Fig. 7.32), the reduced-order estimator has a much higher bandwidth from sensor to control when compared with the full-order estimator. Therefore, if sensor noise is a significant factor, the reduced-order estimator is less attractive because the potential saving in complexity is more than offset by the increased sensitivity to noise.

EXAMPLE 7.27   *SRL Estimator Design for a Simple Pendulum*

Draw the estimator SRL for the linearized equations of the simple inverted pendulum with $\omega_o = 1$. Take the output to be a noisy measurement of position with noise intensity ratio q.

Solution. We are given the system equations

$$\begin{bmatrix} \dot{x}_1 \\ \dot{x}_2 \end{bmatrix} = \begin{bmatrix} 0 & 1 \\ -\omega_0^2 & 0 \end{bmatrix} \begin{bmatrix} x_1 \\ x_2 \end{bmatrix} + \begin{bmatrix} 0 \\ 1 \end{bmatrix} w,$$

$$y = \begin{bmatrix} 1 & 0 \end{bmatrix} \begin{bmatrix} x_1 \\ x_2 \end{bmatrix} + v.$$

Figure 7.34

Symmetric root locus for the inverted pendulum estimator design

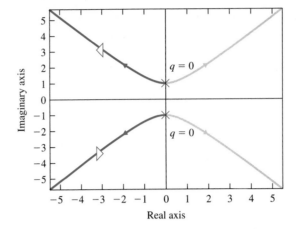

We then calculate from Eq. (7.167) that

$$G_e(s) = \frac{1}{s^2 + \omega_0^2}.$$

The symmetric 180° loci are shown in Fig. 7.34. The MATLAB statements to generate the SRL are (for $\omega_o = 1$)

```
numGG=1;
denGG=conv([1 0 1],[1 0 1]);
sysGG=tf(numGG,denGG);
rlocus(sysGG);
```

We would choose two stable roots for a given value of q, for example $s = -3 \pm j3.18$ for $q = 365$, and use them for estimator pole placement.

7.8 Compensator Design: Combined Control Law and Estimator

Regulator

If we take the control-law design described in Section 7.5, combine it with the estimator design described in Section 7.7, and implement the control law by using the estimated state-variables, the design is complete for a **regulator** that is able to reject disturbances but has no reference input to be tracked. However, because the control law was designed for feedback of the actual (not the estimated) state, you may wonder what effect using $\hat{\mathbf{x}}$ in place of $\mathbf{x}$ has on the system dynamics. In this section we compute this effect. In doing so we will compute the closed-loop characteristic equation and the open-loop compensator transfer function. We will use these results to compare the state-space designs with root-locus and frequency-response designs.

The plant equation with feedback is now

$$\dot{\mathbf{x}} = \mathbf{F}\mathbf{x} - \mathbf{G}\mathbf{K}\hat{\mathbf{x}}, \tag{7.168}$$

which can be rewritten in terms of the state error $\tilde{\mathbf{x}}$ as

$$\dot{\mathbf{x}} = \mathbf{F}\mathbf{x} - \mathbf{G}\mathbf{K}(\mathbf{x} - \tilde{\mathbf{x}}). \tag{7.169}$$

The overall system dynamics in state form are obtained by combining Eq. (7.169) with the estimator error (Eq. 7.135) to get

$$\begin{bmatrix} \dot{\mathbf{x}} \\ \dot{\tilde{\mathbf{x}}} \end{bmatrix} = \begin{bmatrix} \mathbf{F} - \mathbf{GK} & \mathbf{GK} \\ \mathbf{0} & \mathbf{F} - \mathbf{LH} \end{bmatrix} \begin{bmatrix} \mathbf{x} \\ \tilde{\mathbf{x}} \end{bmatrix}. \tag{7.170}$$

The characteristic equation of this closed-loop system is

$$\det \begin{bmatrix} s\mathbf{I} - \mathbf{F} + \mathbf{GK} & -\mathbf{GK} \\ \mathbf{0} & s\mathbf{I} - \mathbf{F} + \mathbf{LH} \end{bmatrix} = 0. \tag{7.171}$$

Because the matrix is block triangular (see Appendix WE), we can rewrite Eq. (7.171) as

$$\det(s\mathbf{I} - \mathbf{F} + \mathbf{GK}) \cdot \det(s\mathbf{I} - \mathbf{F} + \mathbf{LH}) = \alpha_c(s)\alpha_e(s) = 0. \tag{7.172}$$

Poles of the combined control law and estimator

In other words, the set of poles of the combined system consists of the union of the control poles and the estimator poles. This means that the designs of the control law and the estimator can be carried out independently, yet when they are used together in this way, the poles remain unchanged.[9]

To compare the state-variable method of design with the transform methods discussed in Chapters 5 and 6, we note from Fig. 7.35 that the blue shaded portion corresponds to a compensator. The state equation for this compensator is obtained by including the feedback law $u = -\mathbf{K}\hat{\mathbf{x}}$ (because it is part of the compensator) in the estimator Eq. (7.133) to get

$$\dot{\hat{\mathbf{x}}} = (\mathbf{F} - \mathbf{GK} - \mathbf{LH})\hat{\mathbf{x}} + \mathbf{L}y, \tag{7.173a}$$

$$u = -\mathbf{K}\hat{\mathbf{x}}. \tag{7.173b}$$

Note that Eq. (7.173a) has the same structure as Eq. (7.21a), which we repeat here:

$$\dot{\mathbf{x}} = \mathbf{Fx} + \mathbf{G}u. \tag{7.174}$$

Because the characteristic equation of Eq. (7.21a) is

$$\det(s\mathbf{I} - \mathbf{F}) = 0, \tag{7.175}$$

Figure 7.35
Estimator and controller mechanization

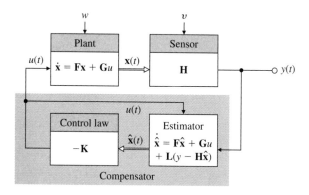

---

[9]This is a special case of the **separation principle** (Gunckel and Franklin, 1963), which holds in much more general contexts and allows us to obtain an overall optimal design by combining the separate designs of control law and estimator in certain stochastic cases.

the characteristic equation of the compensator is found by comparing Eqs. (7.173a) and (7.174) and substituting the equivalent matrices into Eq. (7.175) to get

$$\det(s\mathbf{I} - \mathbf{F} + \mathbf{GK} + \mathbf{LH}) = 0. \tag{7.176}$$

Note that we never specified the roots of Eq. (7.176) nor used them in our discussion of the state-space design technique. (Note also that the compensator is not guaranteed to be stable; the roots of Eq. (7.176) can be in the RHP.) The transfer function from y to u representing the dynamic compensator is obtained by inspecting Eq. (7.48) and substituting in the corresponding matrices from Eq. (7.173):

Compensator transfer function

$$D_c(s) = \frac{U(s)}{Y(s)} = -\mathbf{K}(s\mathbf{I} - \mathbf{F} + \mathbf{GK} + \mathbf{LH})^{-1}\mathbf{L}. \tag{7.177}$$

The same development can be carried out for the reduced-order estimator. Here the control law is

$$u = -[\ K_a \quad \mathbf{K}_b\ ]\begin{bmatrix} x_a \\ \hat{\mathbf{x}}_b \end{bmatrix} = -K_a y - \mathbf{K}_b\hat{\mathbf{x}}_b. \tag{7.178}$$

Substituting Eq. (7.178) into Eq. (7.174) and using Eq. (7.161) and some algebra, we obtain

$$\dot{\mathbf{x}}_c = \mathbf{A}_r\mathbf{x}_c + \mathbf{B}_r y, \tag{7.179a}$$

$$u = \mathbf{C}_r\mathbf{x}_c + D_r y, \tag{7.179b}$$

where

$$\mathbf{A}_r = \mathbf{F}_{bb} - \mathbf{LF}_{ab} - (\mathbf{G}_b - \mathbf{LG}_a)\mathbf{K}_b, \tag{7.180a}$$

$$\mathbf{B}_r = \mathbf{A}_r\mathbf{L} + \mathbf{F}_{ba} - \mathbf{LF}_{aa} - (\mathbf{G}_b - \mathbf{LG}_a)K_a, \tag{7.180b}$$

$$\mathbf{C}_r = -\mathbf{K}_b, \tag{7.180c}$$

$$D_r = -K_a - \mathbf{K}_b\mathbf{L}. \tag{7.180d}$$

Reduced-order compensator transfer function

The dynamic compensator now has the transfer function

$$D_{cr}(s) = \frac{U(s)}{Y(s)} = \mathbf{C}_r(s\mathbf{I} - \mathbf{A}_r)^{-1}\mathbf{B}_r + D_r. \tag{7.181}$$

When we compute $D_c(s)$ or $D_{cr}(s)$ for a specific case, we will find that they are very similar to the classical compensators given in Chapters 5 and 6, in spite of the fact that they are arrived at by entirely different means.

EXAMPLE 7.28   *Full-Order Compensator Design for Satellite Attitude Control*

Design a compensator using pole placement for the satellite plant with transfer function $1/s^2$. Place the control poles at $s = -0.707 \pm 0.707j$ ($\omega_n = 1$ rad/sec, $\zeta = 0.707$) and place the estimator poles at $\omega_n = 5$ rad/sec, $\zeta = 0.5$.

Solution.   A state-variable description for the given transfer function $G(s) = 1/s^2$ is

$$\dot{\mathbf{x}} = \begin{bmatrix} 0 & 1 \\ 0 & 0 \end{bmatrix}\mathbf{x} + \begin{bmatrix} 0 \\ 1 \end{bmatrix}u,$$

$$y = [\ 1 \quad 0\ ]\mathbf{x}.$$

If we place the control roots at $s = -0.707 \pm 0.707j$ ($\omega_n = 1$ rad/sec, $\zeta = 0.7$), then

$$\alpha_c(s) = s^2 + s\sqrt{2} + 1. \tag{7.182}$$

From K = place(F,G,pc), the state feedback gain is found to be

$$\mathbf{K} = \begin{bmatrix} 1 & \sqrt{2} \end{bmatrix}.$$

If the estimator error roots are at $\omega_n = 5$ rad/sec and $\zeta = 0.5$, then the desired estimator characteristic polynomial is

$$\alpha_e(s) = s^2 + 5s + 25 = s + 2.5 \pm 4.3j, \tag{7.183}$$

and, from Lt = place(F',H',pe), the estimator feedback-gain matrix is found to be

$$\mathbf{L} = \begin{bmatrix} 5 \\ 25 \end{bmatrix}.$$

The compensator transfer function given by Eq. (7.177) is

$$D_c(s) = -40.4\frac{(s + 0.619)}{s + 3.21 \pm 4.77j}, \tag{7.184}$$

which looks very much like a lead compensator in that it has a zero on the real axis to the right of its poles; however, rather than one real pole, Eq. (7.184) has two complex poles. The zero provides the derivative feedback with phase lead, and the two poles provide some smoothing of sensor noise.

The effect of the compensation on this system's closed-loop poles can be evaluated in exactly the same way we evaluated compensation in Chapters 5 and 6 using root-locus or frequency-response tools. The gain of 40.4 in Eq. (7.184) is a result of the pole selection inherent in Eqs. (7.182) and (7.183). If we replace this specific value of compensator gain with a variable gain K, then the characteristic equation for the closed-loop system of plant plus compensator becomes

$$1 + K\frac{(s + 0.619)}{(s + 3.21 \pm 4.77j)s^2} = 0. \tag{7.185}$$

The root-locus technique allows us to evaluate the roots of this equation with respect to K, as drawn in Fig. 7.36. Note that the locus goes through the roots selected for Eqs. (7.182) and (7.183), and, when $K = 40.4$, the four roots of the closed-loop system are equal to those specified.

The frequency-response plots given in Fig. 7.37 show that the compensation designed using state-space accomplishes the same results that one would strive for using frequency-response design. Specifically, the uncompensated phase margin of $0°$ increases to $53°$ in the compensated case, and the gain $K = 40.4$ produces a crossover frequency $\omega_c = 1.35$ rad/sec. Both these values are roughly consistent with the controller closed-loop roots, with $\omega_n = 1$ rad/sec and $\zeta = 0.7$, as we would expect, because these slow controller poles are dominant in the system response over the fast estimator poles.

Identical results of state-space and frequency response design methods

Figure 7.36

Root locus for the combined control and estimator, with process gain as the parameter

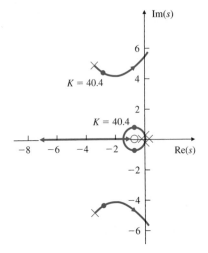

Figure 7.37

Frequency response for $G(s) = \frac{1}{s^2}$

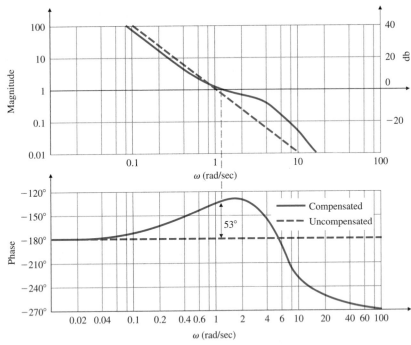

Now we consider a reduced-order estimator for the same system.

EXAMPLE 7.29   *Reduced-Order Compensator Design for a Satellite Attitude Control*

Repeat the design for the $\frac{1}{s^2}$ satellite plant, but use a reduced-order estimator. Place the one estimator pole at -5 rad/sec.

Solution. From Eq. (7.158) we know that the estimator gain is

$$L = 5,$$

Figure 7.38

Simplified block diagram of a reduced-order controller that is a lead network

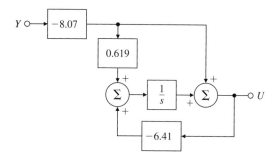

Figure 7.39

Root locus of a reduced-order controller and $1/s^2$ process, root locations at $K = 8.07$ shown by the dots

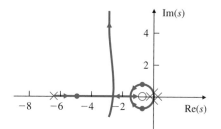

and from Eqs. (7.179a, b) the scalar compensator equations are

$$\dot{x}_c = -6.41x_c - 33.1y,$$
$$u = -1.41x_c - 8.07y,$$

where, from Eq. (7.160),

$$x_c = \hat{x}_2 - 5y.$$

The compensator has the transfer function calculated from Eq. (7.181) to be

$$D_{cr}(s) = -\frac{8.07(s + 0.619)}{s + 6.41}$$

and is shown in Fig. 7.38.

The reduced-order compensator here is precisely a lead network. This is a pleasant discovery, as it shows that transform and state-variable techniques can result in exactly the same type of compensation. The root locus of Fig. 7.39 shows that the closed-loop poles occur at the assigned locations. The frequency response of the compensated system seen in Fig. 7.40 shows a phase margin of about 55°. As with the full-order estimator, analysis by other methods confirms the selected root locations.

More subtle properties of the pole-placement method can be illustrated by a third-order system.

EXAMPLE 7.30   *Full-Order Compensator Design for DC Servo*

Use the state-space pole-placement method to design a compensator for the DC servo system with the transfer function

$$G(s) = \frac{10}{s(s + 2)(s + 8)}.$$

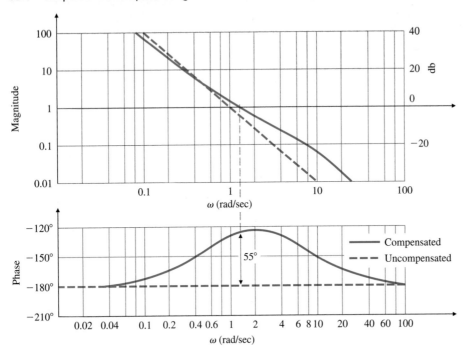

Figure 7.40

Frequency response for $G(s) = 1/s^2$ with a reduced-order estimator

Using a state description in observer canonical form, place the control poles at pc = $[-1.42; -1.04 \pm 2.14j]$ locations and the full-order estimator poles at pe = $[-4.25; -3.13 \pm 6.41j]$.

Solution. A block diagram of this system in observer canonical form is shown in Fig. 7.41. The corresponding state-space matrices are

$$F = \begin{bmatrix} -10 & 1 & 0 \\ -16 & 0 & 1 \\ 0 & 0 & 0 \end{bmatrix}, \quad G = \begin{bmatrix} 0 \\ 0 \\ 10 \end{bmatrix},$$

$$H = \begin{bmatrix} 1 & 0 & 0 \end{bmatrix}, \quad J = 0.$$

The desired poles are

$$pc = \begin{bmatrix} -1.42; -1.04 + 2.14 * j; -1.04 - 2.14 * j \end{bmatrix}.$$

We compute the state feedback gain to be K=(F,G,pc),

$$K = \begin{bmatrix} -46.4 & 5.76 & -0.65 \end{bmatrix}.$$

The estimator error poles are at

$$pe = [-4.25; -3.13 + 6.41 * j; -3.13 - 6.41 * j];$$

We compute the estimator gain to be Lt=place(F',H',pe), L=Lt',

$$L = \begin{bmatrix} 0.5 \\ 61.4 \\ 216 \end{bmatrix}.$$

Figure 7.41

DC servo in observer canonical form

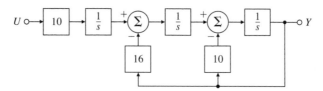

Figure 7.42

Root locus for DC servo pole assignment

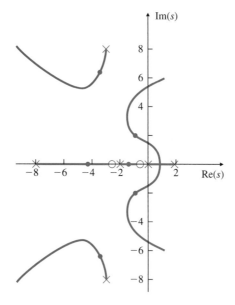

The compensator transfer function, as given by substituting into Eq. (7.177), is

$$D_c(s) = -190\frac{(s + 0.432)(s + 2.10)}{(s - 1.88)(s + 2.94 \pm 8.32j)}.$$

Figure 7.42 shows the root locus of the system of compensator and plant in series, plotted with the compensator gain as the parameter. It verifies that the roots are in the desired locations specified when the gain $K = 190$ in spite of the peculiar (unstable) compensation that has resulted. Even though this compensator has an unstable root at $s = +1.88$, all system closed-loop poles (controller and estimator) are stable.

An unstable compensator is typically not acceptable because of the difficulty in testing either the compensator by itself or the system in open loop during a bench checkout. In some cases, however, better control can be achieved with an unstable compensator; then its inconvenience in checkout may be worthwhile.[10]

Figure 7.33 shows that a direct consequence of the unstable compensator is that the system becomes unstable as the gain is reduced from its nominal value. Such

Conditionally stable compensator

a system is called **conditionally stable** and should be avoided if possible. As we shall see in Chapter 9, actuator saturation in response to large signals has the effect of lowering the effective gain, and in a conditionally stable system, instability can result.

---

[10]There are even systems that cannot be stabilized with a stable compensator.

Also, if the electronics are such that the control amplifier gain rises continuously from zero to the nominal value during startup, such a system would be initially unstable. These considerations lead us to consider alternative designs for this system.

EXAMPLE 7.31   *Redesign of the DC Servo System with a Reduced-Order Estimator*

Design a compensator for the DC servo system of Example 7.30 by using the same control poles but with a reduced-order estimator. Place the estimator poles at $-4.24 \pm 4.24j$ positions with $\omega_n = 6$ and $\zeta = 0.707$.

Solution. The reduced-order estimator corresponds to

$$\text{pe} = \left[-4.24 + 4.24 * j; -4.24 - 4.24 * j \right]$$

After partitioning we have

$$\left[\begin{array}{cc} \mathbf{F}_{aa} & \mathbf{F}_{ab} \\ \mathbf{F}_{ba} & \mathbf{F}_{bb} \end{array} \right] = \left[\begin{array}{ccc} -10 & 1 & 0 \\ -16 & 0 & 1 \\ 0 & 0 & 0 \end{array} \right], \quad \left[\begin{array}{c} G_a \\ G_b \end{array} \right] = \left[\begin{array}{c} 0 \\ 0 \\ 10 \end{array} \right].$$

Solving for the estimator error characteristic polynomial,

$$\det(s\mathbf{I} - \mathbf{F}_{bb} + \mathbf{L}\mathbf{F}_{ab}) = \alpha_e(s),$$

we find (using place) that

$$\mathbf{L} = \left[\begin{array}{c} 8.5 \\ 36 \end{array} \right].$$

The compensator transfer function, given by Eq. (7.181), is computed to be

$$D_{cr}(s) = 20.93 \frac{(s - 0.735)(s + 1.871)}{(s + 0.990 \pm 6.120j)}.$$

A nonminimum-phase compensator

The associated root locus for this system is shown in Fig. 7.43. Note that this time we have a stable but nonminimum-phase compensator and a zero-degree root locus.

Figure 7.43

Root locus for DC servo reduced-order controller

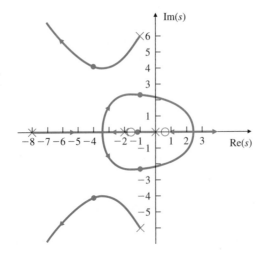

The RHP portion of the locus will not cause difficulties because the gain has to be selected to keep all closed-loop poles in LHP.

As a next pass at the design for this system, we attempt a design with the SRL.

EXAMPLE 7.32   *Redesign of the DC Servo Compensator Using the SRL*

Design a compensator for the DC servo system of Example 7.30 using pole placement based on the SRL. For the control law, let the cost output z be the same as the plant output; for the estimator design, assume that the process noise enters at the same place as the system control signal. Select roots for a control bandwidth of about 2.5 rad/sec, and choose the estimator roots for a bandwidth of about 2.5 times faster than the control bandwidth (6.3 rad/sec). Derive an equivalent discrete controller with a sampling period of $T_s = 0.1$ sec (10 times the fastest pole), and compare the continuous and digital control outputs and control efforts.

Solution. Because the problem has specified that $G_1 = G$ and $H_1 = H$, then the SRL is the same for the control as for the estimator, so we need to generate only one locus based on the plant transfer function. The SRL for the system is shown in Fig. 7.44. From the locus, we select $-2 \pm 1.56j$ and -8.04 as the desired control poles (pc=[−2+1.56*j;−2−1.56*j;−8.04]) and $-4\pm4.9j$ and -9.169 (pe=[−4+4.9*j;−4−4.9*j;−9.169]) as the desired estimator poles. The state feedback gain is K=(F,G,pc), or

$$\mathbf{K} = [\ -0.285 \quad 0.219 \quad 0.204\ ],$$

and the estimator gain is Lt=place(F′,H′,pe), L=Lt′, or

$$\mathbf{L} = \begin{bmatrix} 7.17 \\ 97.4 \\ 367 \end{bmatrix}.$$

Notice that the feedback gains are much smaller than before. The resulting compensator transfer function is computed from Eq. (7.177) to be

$$D_c(s) = -\frac{94.5(s + 7.98)(s + 2.52)}{(s + 4.28 \pm 6.42j)(s + 10.6)}.$$

Figure 7.44

Symmetric root locus

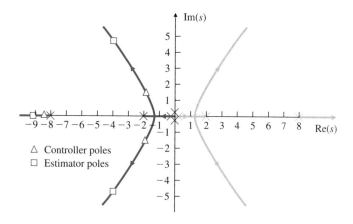

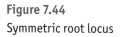
△ Controller poles
□ Estimator poles

Figure 7.45

Root locus for pole assignment from the SRL

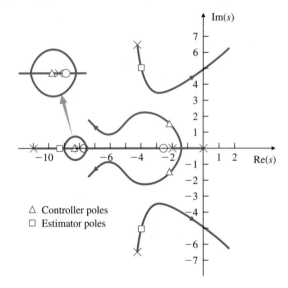

△ Controller poles
☐ Estimator poles

We now take this compensator, put it in series with the plant, and use the compensator gain as the parameter. The resulting ordinary root locus of the closed-loop system is shown in Fig. 7.45. When the root-locus gain equals the nominal gain of 94.5, the roots are at the closed-loop locations selected from the SRL, as they should be.

Note that the compensator is now stable and minimum phase. This improved design comes about in large part because the plant pole at $s = -8$ is virtually unchanged by either controller or estimator. It does not need to be changed for good performance; in fact, the only feature in need of repair in the original $G(s)$ is the pole at $s = 0$. Using the SRL technique, we essentially discovered that the best use of control effort is to shift the two low-frequency poles at $s = 0$ and -2 and to leave the pole at $s = -8$ virtually unchanged. As a result, the control gains are much lower and the compensator design is less radical. This example illustrates why LQR design is typically preferable over pole placement.

The discrete equivalent for the controller is obtained from MATLAB with the c2d command, as in the following code:

```
nc=94.5*conv([1 7.98],[1 2.52]); % form controller numerator
dc=conv([1 8.56 59.5348],[1 10.6]); % form controller denominator
sysDc=tf(nc,dc); % form controller system description
ts=0.1;% sampling time of 0.1 sec
sysDd=c2d(sysDc,ts,'zoh'); % convert controller to discrete time
```

Discrete controller

The resulting controller has the discrete transfer function

$$D_d(z) = \frac{5.9157(z + 0.766)(z + 0.4586)}{(z - 0.522 \pm 0.3903j)(z + 0.3465)}.$$

The equation for the control law (with the sample period suppressed for clarity) is

$$u(k + 1) = 1.3905u(k) - 0.7866u(k - 1) + 0.1472u(k - 2)$$
$$+ e(k) - 7.2445e(k - 2) + 2.0782e(k - 2).$$

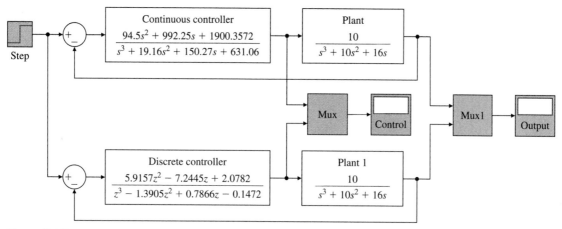

Figure 7.46
SIMULINK® block diagram to compare continuous and discrete controllers

SIMULINK simulation

A SIMULINK® diagram for simulating both the continuous and discrete systems is shown in Fig. 7.46. A comparison of the continuous and discrete step responses and control signals is shown in Fig. 7.47. Better agreement between the two responses can be obtained if the sampling period is reduced.

Armed with the knowledge gained from Example 7.32, let us go back, with a better selection of poles, to investigate the use of pole placement for this example. Initially we used the third-order locations, which produced three poles with a natural frequency of about 2 rad/sec. This design moved the pole at $s = -8$ to $s = -1.4$, thus violating the principle that open-loop poles should not be moved unless they are a problem. Now let us try it again, this time using dominant second-order locations to shift the slow poles, and leaving the fast pole alone at $s = -8$.

EXAMPLE 7.33   *DC Servo System Redesign with Modified Dominant Second-Order Pole Locations*

Design a compensator for the DC servo system of Example 7.30 by using pole placement with control poles given by

$$pc = [-1.41 \pm 1.41 j; -8]$$

and the estimator poles given by

$$pe = [-4.24 \pm 4.24 j; -8]$$

Solution. With these pole locations, we find that the required feedback gain is (using K=place(F,G,pc))

$$\mathbf{K} = [\ -0.469 \quad 0.234 \quad 0.0828 \ ],$$

which has a smaller magnitude than the case where the pole at $s = -8$ was moved.

Figure 7.47

Comparison of step responses and control signals for continuous and discrete controllers: (a) step responses; (b) control signals

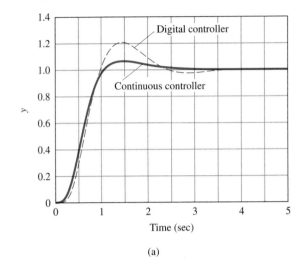

(a)

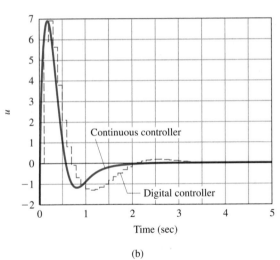

(b)

We find the estimator gain to be (using Lt = place(F',H',pe), L=Lt')

$$\mathbf{L} = \begin{bmatrix} 6.48 \\ 87.8 \\ 288 \end{bmatrix}.$$

The compensator transfer function is

$$D_c(s) = -\frac{414(s+2.78)(s+8)}{(s+4.13 \pm 5.29j)(s+9.05)},$$

which is stable and minimum phase. This example illustrates the value of judicious pole selection and of the SRL technique.

The poor pole selection inherent in the initial use of the poles results in higher control effort and produces an unstable compensator. Both of these undesirable features are eliminated by using the SRL (or LQR), or by improved pole selection. But we really need to use SRL to guide the proper selection of poles. The bottom line is that *SRL (or LQR) is the method of choice!*

As seen from some of the preceding examples, we have shown the use of optimal design via the SRL. However, it is more common in practice to skip that step and use LQR directly.

7.9 Introduction of the Reference Input with the Estimator

The controller obtained by combining the control law studied in Section 7.5 with the estimator discussed in Section 7.8 is essentially a **regulator design**. This means that the characteristic equations of the control and the estimator are chosen for good disturbance rejection—that is, to give satisfactory transients to disturbances such as $w(t)$. However, this design approach does not consider a reference input, nor does it provide for **command following**, which is evidenced by a good transient response of the combined system to command changes. In general, good disturbance rejection and good command following both need to be taken into account in designing a control system. Good command following is done by properly introducing the reference input into the system equations.

Let us repeat the plant and controller equations for the full-order estimator; the reduced-order case is the same in concept, differing only in detail:

$$\text{Plant:} \quad \dot{\mathbf{x}} = \mathbf{F}\mathbf{x} + \mathbf{G}u, \tag{7.186a}$$

$$y = \mathbf{H}\mathbf{x}; \tag{7.186b}$$

$$\text{Controller:} \quad \dot{\hat{\mathbf{x}}} = (\mathbf{F} - \mathbf{G}\mathbf{K} - \mathbf{L}\mathbf{H})\hat{\mathbf{x}} + \mathbf{L}y, \tag{7.187a}$$

$$u = -\mathbf{K}\hat{\mathbf{x}}. \tag{7.187b}$$

Figure 7.48 shows two possibilities for introducing the command input r into the system. This figure illustrates the general issue of whether the compensation should be put in the feedback or feed-forward path. The response of the system to command inputs is different, depending on the configuration, because the zeros of the transfer functions are different. The closed-loop poles are identical, however, as can be easily verified by letting $r = 0$ and noting that the systems are then identical.

The difference in the responses of the two configurations can be seen quite easily. Consider the effect of a step input in r. In Fig. 7.48(a) the step will excite the estimator in precisely the same way that it excites the plant; thus the estimator error will remain zero during and after the step. This means that the estimator dynamics are not excited by the command input, so the transfer function from r to y must have zeros at the estimator pole locations that cancel those poles. As a result, a step command will excite system behavior that is consistent with the control poles alone—that is, with the roots of $\det(s\mathbf{I} - \mathbf{F} + \mathbf{G}\mathbf{K}) = 0$.

In Fig. 7.48(b), a step command in r enters directly only into the estimator, thus causing an estimation error that decays with the estimator dynamic characteristics in addition to the response corresponding to the control poles. Therefore, a step command will excite system behavior consistent with both control roots and estimator

Figure 7.48

Possible locations for introducing the command input: (a) compensation in the feedback path; (b) compensation in the feed-forward path

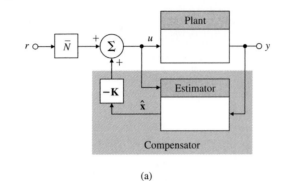

(a)

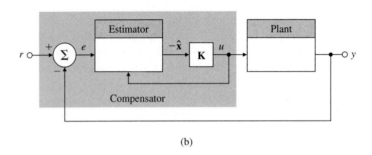

(b)

roots—that is, the roots of

$$\det(s\mathbf{I} - \mathbf{F} + \mathbf{GK}) \cdot \det(s\mathbf{I} - \mathbf{F} + \mathbf{LH}) = 0.$$

For this reason, the configuration shown in Fig. 7.48(a) is typically the superior way to command the system, where $\bar{N}$ is found using Eqs. (7.100)–(7.102).

In Section 7.9.1, we will show a general structure for introducing the reference input with three choices of parameters that implement either the feed-forward or the feedback case. We will analyze the three choices from the point of view of the system zeros and the implications the zeros have for the system transient response. Finally, in Section 7.9.2 we will show how to select the remaining parameter to eliminate constant errors.

7.9.1   A General Structure for the Reference Input

Given a reference input $r(t)$, the most general linear way to introduce r into the system equations is to add terms proportional to it in the controller equations. We can do this by adding $\bar{N}r$ to Eq. (7.187b) and $\mathbf{M}r$ to Eq. (7.187a). Note that in this case, $\bar{N}$ is a scalar and $\mathbf{M}$ is an $n \times 1$ vector. With these additions, the **controller equations** become

Controller equations

$$\dot{\hat{\mathbf{x}}} = (\mathbf{F} - \mathbf{GK} - \mathbf{LH})\hat{\mathbf{x}} + \mathbf{L}y + \mathbf{M}r, \tag{7.188a}$$

$$u = -\mathbf{K}\hat{\mathbf{x}} + \bar{N}r. \tag{7.188b}$$

The block diagram is shown in Fig. 7.49(a). The alternatives shown in Fig. 7.48 correspond to different choices of $\mathbf{M}$ and $\bar{N}$. Because $r(t)$ is an external signal, it is clear that neither $\mathbf{M}$ nor $\bar{N}$ affects the characteristic equation of the combined controller–estimator system. In transfer-function terms, the selection of $\mathbf{M}$ and $\bar{N}$ will affect

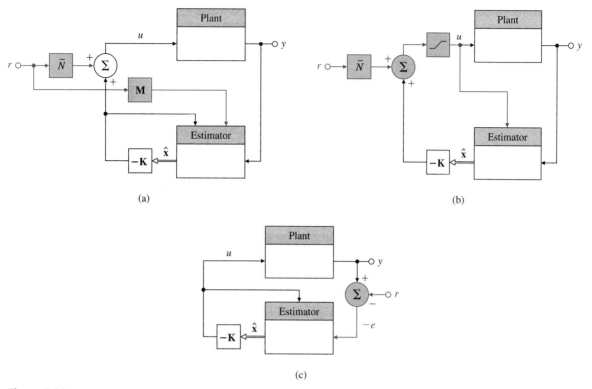

Figure 7.49
Alternative ways to introduce the reference input: (a) general case—zero assignment; (b) standard case—estimator not excited, zeros $= \alpha_e(s)$; (c) error-control case—classical compensation

only the zeros of transmission from r to y and, as a consequence, can significantly affect the transient response but not the stability. How can we choose **M** and $\bar{N}$ to obtain satisfactory transient response? We should point out that we assigned the poles of the system by feedback gains **K** and **L**, and we are now going to assign zeros by feed-forward gains **M** and $\bar{N}$.

There are three strategies for choosing **M** and $\bar{N}$:

Three methods for selecting **M** and *N*

1. *Autonomous estimator:* Select **M** and $\bar{N}$ so that the state estimator error equation is independent of r [Fig. 7.49(b)].
2. *Tracking-error estimator:* Select **M** and $\bar{N}$ so that only the tracking error, $e = (r - y)$, is used in the control [Fig. 7.49(c)].
3. *Zero-assignment estimator:* Select **M** and $\bar{N}$ so that n of the zeros of the overall transfer function are assigned at places of the designer's choice [Fig. 7.49(a)].

CASE 1. From the viewpoint of estimator performance, the first method is quite attractive and the most widely used of the alternatives. If $\hat{\mathbf{x}}$ is to generate a good estimate of $\mathbf{x}$, then surely $\tilde{\mathbf{x}}$ should be as free of external excitation as possible; that is, $\tilde{\mathbf{x}}$ should be uncontrollable from r. The computation of **M** and $\bar{N}$ to bring this about is quite easy. The estimator error equation is found by subtracting Eq. (7.188a) from Eq. (7.186a), with the plant output [Eq. (7.186b)] substituted into the estimator [Eq. (7.187a)] and the control [Eq. (7.187b)] substituted into the plant [Eq. (7.186a)]:

$$\dot{\mathbf{x}} - \dot{\hat{\mathbf{x}}} = \mathbf{Fx} + \mathbf{G}(-\mathbf{K}\hat{\mathbf{x}} + \bar{N}r) - [(\mathbf{F} - \mathbf{GK} - \mathbf{LH})\hat{\mathbf{x}} + \mathbf{L}y + \mathbf{M}r], \quad (7.189a)$$

$$\dot{\tilde{\mathbf{x}}} = (\mathbf{F} - \mathbf{LH})\tilde{\mathbf{x}} + \mathbf{G}\bar{N}r - \mathbf{M}r. \quad (7.189b)$$

If r is not to appear in Eq. (7.189a), then we should choose

$$\mathbf{M} = \mathbf{G}\bar{N}. \quad (7.190)$$

Because $\bar{N}$ is a scalar, $\mathbf{M}$ is fixed to within a constant factor. Note that with this choice of $\mathbf{M}$, we can write the controller equations as

$$u = -\mathbf{K}\hat{\mathbf{x}} + \bar{N}r, \quad (7.191a)$$

$$\dot{\hat{\mathbf{x}}} = (\mathbf{F} - \mathbf{LH})\hat{\mathbf{x}} + \mathbf{G}u + \mathbf{L}y, \quad (7.191b)$$

which matches the configuration in Fig. 7.49(b). The net effect of this choice is that the control is computed from the feedback gain and the reference input *before* it is applied, and then the same control is input to both the plant and the estimator. In this form, if the plant control is subject to saturation (as shown by the inclusion of the saturation nonlinearity in Fig. 7.49(b), and discussed in Chapter 9), the same control limits can be applied in Eq. (7.191) to the control entering the equation for the estimate $\hat{\mathbf{x}}$, and the nonlinearity cancels out of the $\tilde{\mathbf{x}}$ equation. This behavior is essential for proper estimator performance. The block diagram corresponding to this technique is shown in Fig. 7.49(b). We will return to the selection of the gain factor on the reference input, $\bar{N}$, in Section 7.9.2 after discussing the other two methods of selecting $\mathbf{M}$.

CASE 2. The second approach suggested earlier is to use the tracking error. This solution is sometimes forced on the control designer when the sensor measures only the output error. For example, in many thermostats the output is the difference between the temperature to be controlled and the setpoint temperature, and there is no absolute indication of the reference temperature available to the controller. Also, some radar tracking systems have a reading that is proportional to the pointing error, and this error signal alone must be used for feedback control. In these situations, we must select $\mathbf{M}$ and $\bar{N}$ so that Eqs. (7.188) are driven by the error only. This requirement is satisfied if we select

$$\bar{N} = 0 \quad \text{and} \quad \mathbf{M} = -\mathbf{L}. \quad (7.192)$$

Then the estimator equation is

$$\dot{\hat{\mathbf{x}}} = (\mathbf{F} - \mathbf{GK} - \mathbf{LH})\hat{\mathbf{x}} + \mathbf{L}(y - r). \quad (7.193)$$

The compensator in this case, for low-order designs, is a standard lead compensator in the forward path. As we have seen in earlier chapters, this design can have a considerable amount of overshoot because of the zero of the compensator. This design corresponds exactly to the compensators designed by the transform methods given in Chapters 5 and 6.

CASE 3. The third method of selecting $\mathbf{M}$ and $\bar{N}$ is to choose the values so as to assign the system's zeros to arbitrary locations of the designer's choice. This method provides the designer with the maximum flexibility in satisfying transient-response and steady-state gain constraints. The other two methods are special cases of this third method. All three methods depend on the zeros. As we saw in Section 7.5.2, when there is no estimator and the reference input is added to the control, the closed-loop system zeros remain fixed as the zeros of the open-loop plant. We now examine what happens to the zeros when an estimator is present. To do so, we reconsider the controller of Eqs. (7.188). If there is a zero of transmission from r to u, then there is necessarily a zero of transmission from r to y, unless there is a pole at the same location as the zero. It is therefore sufficient to treat the controller alone to determine what effect the choices of $\mathbf{M}$ and $\bar{N}$ will have on the system zeros. The equations for a zero from r to u from Eqs. (7.188) are given by

$$\det \begin{bmatrix} s\mathbf{I} - \mathbf{F} + \mathbf{GK} + \mathbf{LH} & -\mathbf{M} \\ -\mathbf{K} & \bar{N} \end{bmatrix} = 0. \tag{7.194}$$

(We let $y = 0$ because we care only about the effect of r.) If we divide the last column by the (nonzero) scalar $\bar{N}$ and then add to the rest the product of $\mathbf{K}$ times the last column, we find that the feed-forward zeros are at the values of s such that

$$\det \begin{bmatrix} s\mathbf{I} - \mathbf{F} + \mathbf{GK} + \mathbf{LH} - \frac{\mathbf{M}}{\bar{N}}\mathbf{K} & -\frac{\mathbf{M}}{\bar{N}} \\ 0 & 1 \end{bmatrix} = 0,$$

or

$$\det \left(s\mathbf{I} - \mathbf{F} + \mathbf{GK} + \mathbf{LH} - \frac{\mathbf{M}}{\bar{N}}\mathbf{K} \right) = \gamma(s) = 0. \tag{7.195}$$

Now Eq. (7.195) is exactly in the form of Eq. (7.136) for selecting $\mathbf{L}$ to yield desired locations for the estimator poles. Here we have to select $\mathbf{M}/\bar{N}$ for a desired zero polynomial $\gamma(s)$ in the transfer function from the reference input to the control. Thus the selection of $\mathbf{M}$ provides a substantial amount of freedom to influence the transient response. We can add an arbitrary nth-order polynomial to the transfer function from r to u and hence from r to y; that is, we can assign n zeros in addition to all the poles that we assigned previously. If the roots of $\gamma(s)$ are not canceled by the poles of the system, then they will be included in zeros of transmission from r to y.

Two considerations can guide us in the choice of $\mathbf{M}/\bar{N}$—that is, in the location of the zeros. The first is dynamic response. We have seen in Chapter 3 that the zeros influence the transient response significantly, and the heuristic guidelines given there may suggest useful locations for the available zeros. The second consideration, which will connect state-space design to another result from transform techniques, is steady-state error or velocity-constant control. In Chapter 4 we derived the relationship between the steady-state accuracy of a Type 1 system and the closed-loop poles and zeros. If the system is Type 1, then the steady-state error to a step input will be zero and to a unit-ramp input will be

$$e_{\infty} = \frac{1}{K_v}, \tag{7.196}$$

Truxal's formula

where K_v is the velocity constant. Furthermore, it was shown that if the *closed-loop* poles are at $\{p_i\}$ and the *closed-loop* zeros are at $\{z_i\}$, then (for a Type 1 system) **Truxal's formula** gives

$$\frac{1}{K_v} = \sum \frac{1}{z_i} - \sum \frac{1}{p_i}. \tag{7.197}$$

Equation (7.197) forms the basis for a partial selection of $\gamma(s)$, and hence of $\mathbf{M}$ and $\bar{N}$. The choice is based on two observations:

1. If $|z_i - p_i| \ll 1$, then the effect of this pole–zero pair on the dynamic response will be small, because the pole is almost canceled by the zero, and in any transient the residue of the pole at p_i will be very small.
2. Even though $z_i - p_i$ is small, it is possible for $1/z_i - 1/p_i$ to be substantial and thus to have a significant influence on K_v according to Eq. (7.197).

Application of these two guidelines to the selection of $\gamma(s)$, and hence of $\mathbf{M}$ and $\bar{N}$, results in a lag-network design. We illustrate this with an example.

EXAMPLE 7.34

Servomechanism: Increasing the Velocity Constant through Zero Assignment

Lag compensation by a state-space method

Consider the second-order servomechanism system described by

$$G(s) = \frac{1}{s(s+1)}$$

and with state description

$$\dot{x}_1 = x_2,$$

$$\dot{x}_2 = -x_2 + u.$$

Design a controller using pole placement so that both poles are at $s = -2$ and the system has a velocity constant $K_v = 10$. Derive an equivalent discrete controller with a sampling period of $T_s = 0.1$ sec ($20 \times \omega_n = 20 \times 0.05 = 0.1$ sec), and compare the continuous and digital control outputs, as well as the control efforts.

Solution. For this problem, the state feedback gain

$$\mathbf{K} = [\; 8 \quad 3 \;]$$

results in the desired control poles. However, with this gain, $K_v = 2$, and we need $K_v = 10$. What effect will using estimators designed according to the three methods for $\mathbf{M}$ and $\bar{N}$ selection have on our design? Using the first strategy (the autonomous estimator), we find that the value of K_v does not change. If we use the second method (error control), we introduce a zero at a location unknown beforehand, and the effect on K_v will not be under direct design control. However, if we use the third option (zero placement) along with Truxal's formula [Eq. (7.197)], we can satisfy both the dynamic response and the steady-state requirements.

First we must select the estimator pole p_3 and the zero z_3 to satisfy Eq. (7.197) for $K_v = 10$. We want to keep $z_3 - p_3$ small, so that there is little effect on the dynamic response, and yet have $1/z_3 - 1/p_3$ be large enough to increase the value of K_v. To do this, we arbitrarily set p_3 small compared with the control dynamics. For example, we let

$$p_3 = -0.1.$$

Notice that this approach is opposite to the usual philosophy of estimation design, where fast response is the requirement. Now, using Eq. (7.197) to get

$$\frac{1}{K_v} = \frac{1}{z_3} - \frac{1}{p_1} - \frac{1}{p_2} - \frac{1}{p_3},$$

where $p_1 = -2+2j, p_2 = -2-2j$, and $p_3 = -0.1$, we solve for z_3 such that $K_v = 10$ we obtain

$$\frac{1}{K_v} = \frac{4}{8} + \frac{1}{0.1} + \frac{1}{z_3} = \frac{1}{10},$$

or

$$z_3 = -\frac{1}{10.4} = -0.096.$$

We thus design a reduced-order estimator to have a pole at -0.1 and choose $\mathbf{M}/\bar{N}$ such that $\gamma(s)$ has a zero at -0.096. A block diagram of the resulting system is shown in Fig. 7.50(a). You can readily verify that this system has the overall transfer function

$$\frac{Y(s)}{R(s)} = \frac{8.32(s + 0.096)}{(s^2 + 4s + 8)(s + 0.1)}, \tag{7.198}$$

for which $K_v = 10$, as specified.

The compensation shown in Fig. 7.50(a) is nonclassical in the sense that it has two inputs (e and y) and one output. If we resolve the equations to provide pure error compensation by finding the transfer function from e and u, which would give Eq. (7.198), we obtain the system shown in Fig. 7.50(b). This can be seen as follows: The relevant controller equations are

$$\dot{x}_c = 0.8\, e - 3.1\, u,$$

$$u = 8.32\, e + 3.02\, y + x_c,$$

where x_c is the controller state. Taking the Laplace transform of these equations, eliminating $X_c(s)$, and substituting for the output $[Y(s) = G(s)U(s)]$, we find that the compensator is described by

$$\frac{U(s)}{E(s)} = D_c(s) = \frac{(s + 1)(8.32s + 0.8)}{(s + 4.08)(s + 0.0196)}.$$

This compensation is a classical lag–lead network. The root locus of the system in Fig. 7.50(b) is shown in Fig. 7.51. Note the pole–zero pattern near the origin that is

Figure 7.50

Servomechanism with assigned zeros (a lag network): (a) the two-input compensator; (b) equivalent unity-feedback system

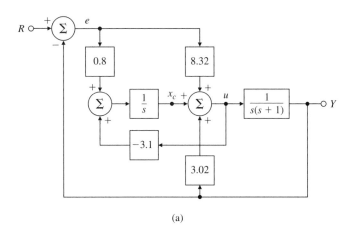

(a)

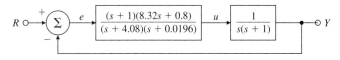

(b)

Figure 7.51

Root locus of lag–lead compensation

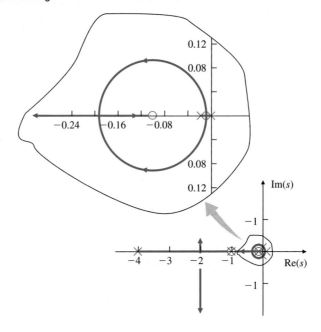

Figure 7.52

Frequency response of lag–lead compensation

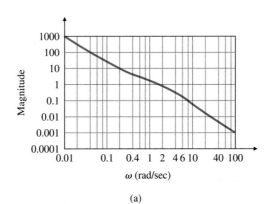

(a)

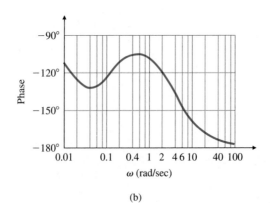

(b)

Figure 7.53

Step response of the system with lag compensation

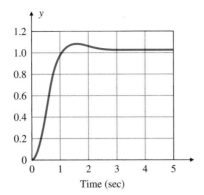

characteristic of a lag network. The Bode plot in Fig. 7.52 shows the phase lag at low frequencies and the phase lead at high frequencies. The step response of the system is shown in Fig. 7.53 and shows the presence of a "tail" on the response due to the slow pole at −0.1. Of course, the system is Type 1 and the system will have zero tracking error eventually.

The discrete equivalent for the controller is obtained from MATLAB by using the c2d command, as in the following code:

MATLAB c2d

```
nc=conv([1 1],[8.32 0.8]); % controller numerator
dc=conv([1 4.08],[1 0.0196]); % controller denominator
sysDc=tf(nc,dc); % form controller system description
ts=0.1; % sampling time of 0.1 sec
sysDd=c2d(sysDc,ts,'zoh'); % convert to discrete time controller
```

The discrete controller has the discrete transfer function

$$D_d(z) = \frac{8.32z^2 - 15.8855z + 7.5721}{z^2 - 1.6630z + 0.6637} = \frac{8.32(z - 0.9903)(z - 0.9191)}{(z - 0.998)(z - 0.6665)}.$$

The equation for the control law (with sample period suppressed for clarity) is

$$u(k + 1) = 1.6630u(k) + 0.6637u(k - 1) + 8.32e(k + 1)$$
$$- 15.8855e(k) + 7.5721e(k - 1).$$

SIMULINK simulation

A SIMULINK diagram for simulating both the continuous and discrete systems is shown in Fig. 7.54. A comparison of the continuous and discrete step responses and control signals is shown in Fig. 7.55. Better agreement between the two responses can be achieved if the sampling period is reduced.

We now reconsider the first two methods for choosing $\mathbf{M}$ and $\bar{N}$, this time to examine their implications in terms of zeros. Under the first rule (for the autonomous estimator), we let $\mathbf{M} = \mathbf{G}\bar{N}$. Substituting this into Eq. (7.195) yields, for the controller feed-forward zeros,

$$\det(s\mathbf{I} - \mathbf{F} + \mathbf{LH}) = 0. \tag{7.199}$$

This is exactly the equation from which $\mathbf{L}$ was selected to make the characteristic polynomial of the estimator equation equal to $\alpha_e(s)$. Thus we have created n zeros in

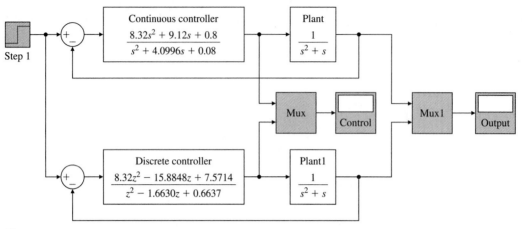

Figure 7.54

SIMULINK® block diagram to compare continuous and discrete controllers

exactly the same locations as the n poles of the estimator. Because of this pole–zero cancellation (which causes "uncontrollability" of the estimator modes), the overall transfer function poles consist only of the state feedback controller poles.

The second rule (for a tracking-error estimator) selects $\mathbf{M} = -\mathbf{L}$ and $\bar{N} = 0$. If these are substituted into Eq. (7.194), then the feed-forward zeros are given by

$$\det \begin{bmatrix} s\mathbf{I} - \mathbf{F} + \mathbf{GK} + \mathbf{LH} & \mathbf{L} \\ -\mathbf{K} & 0 \end{bmatrix} = 0. \tag{7.200}$$

If we postmultiply the last column by $\mathbf{H}$ and subtract the result from the first n columns, and then premultiply the last row by $\mathbf{G}$ and add it to the first n rows, Eq. (7.200) then reduces to

$$\det \begin{bmatrix} s\mathbf{I} - \mathbf{F} & \mathbf{L} \\ -\mathbf{K} & 0 \end{bmatrix} = 0. \tag{7.201}$$

If we compare Eq. (7.201) with the equations for the zeros of a system in a state description, Eq. (7.66), we see that the added zeros are those obtained by replacing the input matrix with $\mathbf{L}$ and the output with $\mathbf{K}$. Thus, if we wish to use error control, we have to accept the presence of these compensator zeros that depend on the choice of $\mathbf{K}$ and $\mathbf{L}$ and over which we have no direct control. For low-order cases this results, as we said before, in a lead compensator as part of a unity feedback topology.

Transfer function for the closed-loop system when reference input is included in controller

Let us now summarize our findings on the effect of introducing the reference input. When the reference input signal is included in the controller, the overall transfer function of the closed-loop system is

$$T(s) = \frac{Y(s)}{R(s)} = \frac{K_s \gamma(s) b(s)}{\alpha_e(s)\alpha_c(s)}, \tag{7.202}$$

where K_s is the total system gain and $\gamma(s)$ and $b(s)$ are monic polynomials. The polynomial $\alpha_c(s)$ results in a control gain $\mathbf{K}$ such that $\det[s\mathbf{I} - \mathbf{F} + \mathbf{GK}] = \alpha_c(s)$. The polynomial $\alpha_e(s)$ results in estimator gains $\mathbf{L}$ such that $\det[s\mathbf{I} - \mathbf{F} + \mathbf{LH}] = \alpha_e(s)$. Because, as designers, we get to choose $\alpha_c(s)$ and $\alpha_e(s)$, we have complete freedom in assigning the poles of the closed-loop system. There are three ways to handle the

Figure 7.55

Comparison of step responses and control signals for continuous and discrete controllers: (a) step responses; (b) control signals

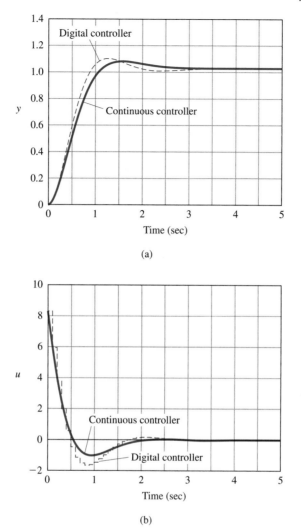

polynomial $\gamma(s)$: We can select it so that $\gamma(s) = \alpha_e(s)$ by using the implementation of Fig. 7.49(b), in which case $\mathbf{M}/\bar{N}$ is given by Eq. (7.190); we may accept $\gamma(s)$ as given by Eq. (7.201), so that error control is used; or we may give $\gamma(s)$ arbitrary coefficients by selecting $\mathbf{M}/\bar{N}$ from Eq. (7.195). It is important to point out that the plant zeros represented by $b(s)$ are not moved by this technique and remain as part of the closed-loop transfer function unless α_c or α_e are selected to cancel some of these zeros.

7.9.2 Selecting the Gain

We now turn to the process of determining the gain $\bar{N}$ for the three methods of selecting $\mathbf{M}$. If we choose method 1, the control is given by Eq. (7.191a) and $\hat{x}_{ss} = x_{ss}$. Therefore, we can use either $\bar{N} = N_u + \mathbf{KN_x}$, as in Eq. (7.102), or $u = N_u r - \mathbf{K}(\hat{\mathbf{x}} - \mathbf{N_x}r)$. *This is the most common choice.* If we use the second method, the result is trivial; recall that $\bar{N} = 0$ for error control. If we use the third method, we pick $\bar{N}$ such that the overall closed-loop DC gain is unity.[11]

The overall system equations then are

$$
\begin{bmatrix} \dot{x} \\ \dot{\tilde{x}} \end{bmatrix} = \begin{bmatrix} F - GK & GK \\ 0 & F - LH \end{bmatrix} \begin{bmatrix} x \\ \tilde{x} \end{bmatrix} + \begin{bmatrix} G \\ G - \bar{M} \end{bmatrix} \bar{N}r, \quad (7.203a)
$$

$$
y = \begin{bmatrix} H & 0 \end{bmatrix} \begin{bmatrix} x \\ \tilde{x} \end{bmatrix}, \quad (7.203b)
$$

where $\bar{M}$ is the outcome of selecting zero locations with either Eq. (7.195) or Eq. (7.190). The closed-loop system has unity DC gain if

$$
-\begin{bmatrix} H & 0 \end{bmatrix} \begin{bmatrix} F - GK & GK \\ 0 & F - LH \end{bmatrix}^{-1} \begin{bmatrix} G \\ G - \bar{M} \end{bmatrix} \bar{N} = 1. \quad (7.204)
$$

If we solve Eq. (7.204) for $\bar{N}$, we get[12]

$$
\bar{N} = -\frac{1}{H(F - GK)^{-1}G[1 - K(F - LH)^{-1}(G - \bar{M})]}. \quad (7.205)
$$

The techniques in this section can be readily extended to reduced-order estimators.

7.10 Integral Control and Robust Tracking

The choices of $\bar{N}$ gain in Section 7.9 will result in zero steady-state error to a step command, but the result is not robust because any change in the plant parameters will cause the error to be nonzero. We need to use integral control to obtain robust tracking.

In the state-space design methods discussed so far, no mention has been made of integral control, and no design examples have produced a compensation containing an integral term. In Section 7.10.1 we show how integral control can be introduced by a direct method of adding the integral of the system error to the equations of motion. Integral control is a special case of tracking a signal that does not go to zero in the steady-state. We introduce (in Section 7.10.2) a general method for robust tracking that will present the internal model principle, which solves an entire class of tracking problems and disturbance-rejection controls. Finally, in Section 7.10.3, we show that if the system has an estimator and also needs to reject a disturbance of known structure, we can include a model of the disturbance in the estimator equations and use the computer estimate of the disturbance to cancel the effects of the real plant disturbance on the output.

---

[11]A reasonable alternative is to select $\bar{N}$ such that, when r and y are both unchanging, the DC gain from r to u is the *negative* of the DC gain from y to u. The consequences of this choice are that our controller can be structured as a combination of error control and generalized derivative control, and if the system is capable of Type 1 behavior, that capability will be realized.

[12]We have used the fact that

$$
\begin{bmatrix} A & C \\ 0 & B \end{bmatrix}^{-1} = \begin{bmatrix} A^{-1} & -A^{-1}CB^{-1} \\ 0 & B^{-1} \end{bmatrix}.
$$

7.10.1 Integral Control

We start with an ad hoc solution to integral control by augmenting the state vector with the desired dynamics. For the system

$$\dot{\mathbf{x}} = \mathbf{Fx} + \mathbf{G}u + \mathbf{G}_1 w, \tag{7.206a}$$

$$y = \mathbf{Hx}, \tag{7.206b}$$

we can feed back the integral of the error,[13] $e = y - r$, as well as the state of the plant, $\mathbf{x}$, by augmenting the plant state with the extra (integral) state x_I, which obeys the differential equation

$$\dot{x}_I = \mathbf{Hx} - r(= e).$$

Thus

$$x_I = \int^t e\, dt.$$

Augmented state equations with integral control

The augmented state equations become

$$\begin{bmatrix} \dot{x}_I \\ \dot{\mathbf{x}} \end{bmatrix} = \begin{bmatrix} 0 & \mathbf{H} \\ \mathbf{0} & \mathbf{F} \end{bmatrix} \begin{bmatrix} x_I \\ \mathbf{x} \end{bmatrix} + \begin{bmatrix} 0 \\ \mathbf{G} \end{bmatrix} u - \begin{bmatrix} 1 \\ \mathbf{0} \end{bmatrix} r + \begin{bmatrix} 0 \\ \mathbf{G}_1 \end{bmatrix} w, \tag{7.207}$$

Feedback law with integral control

and the feedback law is

$$u = -[\begin{array}{cc} K_1 & \mathbf{K}_0 \end{array}] \begin{bmatrix} x_I \\ \mathbf{x} \end{bmatrix},$$

or simply

$$u = -\mathbf{K} \begin{bmatrix} x_I \\ \mathbf{x} \end{bmatrix}.$$

With this revised definition of the system, we can apply the design techniques from Section 7.5 in a similar fashion; they will result in the control structure shown in Fig. 7.56.

EXAMPLE 7.35   *Integral Control of a Motor Speed System*

Consider the motor speed system described by

$$\frac{Y(s)}{U(s)} = \frac{1}{s+3};$$

that is, $F = -3$, $G = 1$, and $H = 1$. Design the system to have integral control and two poles at $s = -5$. Design an estimator with pole at $s = -10$. The disturbance enters at the same place as the control. Evaluate the tracking and disturbance rejection responses.

Figure 7.56

Integral control structure

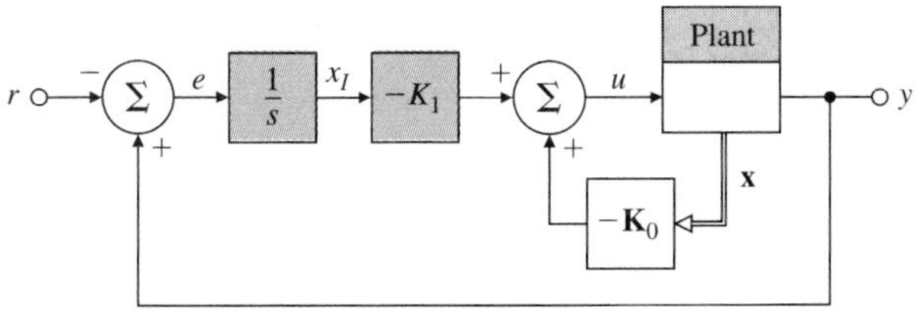

13 Watch out for the sign here; we are using the negative of the usual convention.

Solution. The pole-placement requirement is equivalent to

$$pc = [-5; -5].$$

The augmented system description, including the disturbance w, is

$$\begin{bmatrix} \dot{x}_I \\ \dot{x} \end{bmatrix} = \begin{bmatrix} 0 & 1 \\ 0 & -3 \end{bmatrix} \begin{bmatrix} x_I \\ x \end{bmatrix} + \begin{bmatrix} 0 \\ 1 \end{bmatrix} (u+w) - \begin{bmatrix} 1 \\ 0 \end{bmatrix} r.$$

Therefore, we can find $\mathbf{K}$ from

$$\det \left(s\mathbf{I} - \begin{bmatrix} 0 & 1 \\ 0 & -3 \end{bmatrix} + \begin{bmatrix} 0 \\ 1 \end{bmatrix} \mathbf{K} \right) = s^2 + 10s + 25,$$

or

$$s^2 + (3 + K_0)s + K_1 = s^2 + 10s + 25.$$

Consequently,

$$\mathbf{K} = [\ K_1 \quad K_0\ ] = [\ 25 \quad 7\ ].$$

We may verify this result by using acker. The system is shown with feedbacks in Fig. 7.57, along with a disturbance input w.

The estimator gain $L = 7$ is obtained from

$$\alpha_e(s) = s + 10 = s + 3 + L.$$

The estimator equation is of the form

$$\dot{\hat{x}} = (F - LH)\hat{x} + Gu + Ly$$
$$= -10\hat{x} + u + 7y,$$

and

$$u = -K_0\hat{x} = -7\hat{x}.$$

The step response y_1 due to a step reference input r and the output disturbance response y_2 due to a step disturbance input w are shown in Fig. 7.58(a) and the associated control efforts (u_1 and u_2) are shown in Fig. 7.58(b). As expected, the system is Type 1 and tracks the step reference input and rejects the step disturbance asymptotically.

Figure 7.57

Integral control example

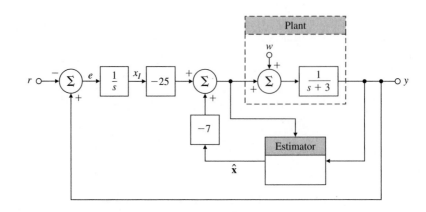

Figure 7.58
Transient response for
motor speed system:
(a) step responses;
(b) control efforts

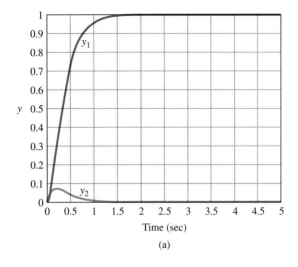

(a)

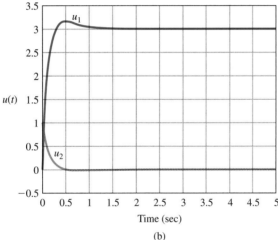

(b)

△ **7.10.2   Robust Tracking Control: The Error-Space Approach**

In Section 7.10.1 we introduced integral control in a direct way and selected the structure of the implementation so as to achieve integral action with respect to reference and disturbance inputs. We now present a more analytical approach to giving a control system the ability to track (with zero steady-state error) a nondecaying input and to reject (with zero steady-state error) a nondecaying disturbance such as a step, ramp, or sinusoidal input. The method is based on including the equations satisfied by these external signals as part of the problem formulation and solving the problem of control in an **error space**, so we are assured that the error approaches zero even if the output is following a nondecaying, or even a growing, command (such as a ramp signal) and even if some parameters change (the robustness property). The method is illustrated in detail for signals that satisfy differential equations of order 2, but the extension to more complex signals is not difficult.

Suppose we have the system state equations

$$\dot{\mathbf{x}} = \mathbf{F}\mathbf{x} + \mathbf{G}u + \mathbf{G}_1 w, \qquad (7.208a)$$

$$y = \mathbf{H}\mathbf{x}, \qquad (7.208b)$$

and a reference signal that is known to satisfy a specific differential equation. The initial conditions on the equation generating the input are unknown. For example, the input could be a ramp whose slope and initial value are unknown. Plant disturbances of the same class may also be present. We wish to design a controller for this system so that the closed-loop system will have specified poles, and can also track input command signals, and reject disturbances of the type described without steady-state error. We will develop the results only for second-order differential equations. We define the reference input to satisfy the relation

$$\ddot{r} + \alpha_1 \dot{r} + \alpha_2 r = 0 \qquad (7.209)$$

and the disturbance to satisfy exactly the same equation:

$$\ddot{w} + \alpha_1 \dot{w} + \alpha_2 w = 0. \qquad (7.210)$$

The (tracking) error is defined as

$$e = y - r. \qquad (7.211)$$

The meaning of robust control

The problem of tracking r and rejecting w can be seen as an exercise in designing a control law to provide *regulation of the error*, which is to say that the error e tends to zero as time gets large. The control must also be **structurally stable** or **robust**, in the sense that regulation of e to zero in the steady-state occurs even in the presence of "small" perturbations of the original system parameters. Note that, in practice, we never have a perfect model of the plant, and the values of parameters are virtually always subject to some change, so robustness is always very important.

We know that the command input satisfies Eq. (7.209), and we would like to eliminate the reference from the equations in favor of the error. We begin by replacing r in Eq. (7.209) with the error of Eq. (7.211). When we do this, the reference cancels because of Eq. (7.209), and we have the formula for the error in terms of the state:

$$\ddot{e} + \alpha_1 \dot{e} + \alpha_2 e = \ddot{y} + \alpha_1 \dot{y} + \alpha_2 y \qquad (7.212a)$$

$$= \mathbf{H}\ddot{\mathbf{x}} + \alpha_1 \mathbf{H}\dot{\mathbf{x}} + \alpha_2 \mathbf{H}\mathbf{x}. \qquad (7.212b)$$

We now replace the plant state vector with the error-space state, defined by

$$\xi \triangleq \ddot{\mathbf{x}} + \alpha_1 \dot{\mathbf{x}} + \alpha_2 \mathbf{x}. \qquad (7.213)$$

Similarly, we replace the control with the control in error space, defined as

$$\mu \triangleq \ddot{u} + \alpha_1 \dot{u} + \alpha_2 u. \qquad (7.214)$$

With these definitions, we can replace Eq. (7.212b) with

$$\ddot{e} + \alpha_1 \dot{e} + \alpha_2 e = \mathbf{H}\xi. \qquad (7.215)$$

Robust control equations in the error space

The state equation for ξ is given by[14]

$$\dot{\xi} = \dddot{\mathbf{x}} + \alpha_1 \ddot{\mathbf{x}} + \alpha_2 \dot{\mathbf{x}} = \mathbf{F}\xi + \mathbf{G}\mu. \qquad (7.216)$$

---

[14]Notice that this concept can be extended to more complex equations in r and to multivariable systems.

Notice that the disturbance, as well as the reference, cancels from Eq. (7.216). Equations (7.215) and (7.216) now describe the overall system in an error space. In standard state-variable form, the equations are

$$\dot{\mathbf{z}} = \mathbf{A}\mathbf{z} + \mathbf{B}\mu, \qquad (7.217)$$

where $\mathbf{z} = [\ e \ \ \dot{e} \ \ \xi^T \ ]^T$ and

$$\mathbf{A} = \begin{bmatrix} 0 & 1 & \mathbf{0} \\ -\alpha_2 & -\alpha_1 & \mathbf{H} \\ \mathbf{0} & \mathbf{0} & \mathbf{F} \end{bmatrix}, \quad \mathbf{B} = \begin{bmatrix} 0 \\ 0 \\ \mathbf{G} \end{bmatrix}. \qquad (7.218)$$

The error system $(\mathbf{A}, \mathbf{B})$ can be given arbitrary dynamics by state feedback if it is controllable. If the plant $(\mathbf{F}, \mathbf{G})$ is controllable and does not have a zero at any of the roots of the reference-signal characteristic equation

$$\alpha_r(s) = s^2 + \alpha_1 s + \alpha_2,$$

then the error system $(\mathbf{A}, \mathbf{B})$ is controllable.[15] We assume these conditions hold; therefore, there exists a control law of the form

$$\mu = -[\ K_2 \ \ K_1 \ \ \mathbf{K_0}\ ] \begin{bmatrix} e \\ \dot{e} \\ \xi \end{bmatrix} = -\mathbf{Kz}, \qquad (7.219)$$

such that the error system has arbitrary dynamics by pole placement. We now need to express this control law in terms of the actual process state $\mathbf{x}$ and the actual control. We combine Eqs. (7.219), (7.213), and (7.214) to get the control law in terms of u and $\mathbf{x}$ (we write $u^{(2)}$ to mean $\frac{d^2u}{dt^2}$):

$$(u + \mathbf{K_0x})^{(2)} + \sum_{i=1}^{2} \alpha_i (u + \mathbf{K_0x})^{(2-i)} = -\sum_{i=1}^{2} K_i e^{(2-i)}. \qquad (7.220)$$

The structure for implementing Eq. (7.220) is very simple for tracking constant inputs. In that case the equation for the reference input is $\dot{r} = 0$. In terms of u and $\mathbf{x}$, the control law [Eq. (7.220)] reduces to

$$\dot{u} + \mathbf{K_0}\dot{\mathbf{x}} = -K_1 e. \qquad (7.221)$$

Here we need only to integrate to reveal the control law and the action of integral control:

$$u = -K_1 \int^t e \, d\tau - \mathbf{K_0x}. \qquad (7.222)$$

A block diagram of the system, shown in Fig. 7.59, clearly shows the presence of a pure integrator in the controller. In this case the only difference between the internal model method of Fig. 7.59 and the ad hoc method of Fig. 7.57 is the relative location of the integrator and the gain.

A more complex problem that clearly shows the power of the error-space approach to robust tracking is posed by requiring that a sinusoid be tracked with zero steady-state error. The problem arises, for instance, in the control of a mass-storage disk-head assembly.

---

[15] For example, it is not possible to add integral control to a plant that has a zero at the origin.

Figure 7.59

Integral control using the internal model approach

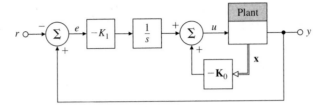

EXAMPLE 7.36

Disk-Drive Servomechanism: Robust Control to Follow a Sinusoid

A simple normalized model of a computer disk-drive servomechanism is given by the equations

$$\mathbf{F} = \begin{bmatrix} 0 & 1 \\ 0 & -1 \end{bmatrix}, \quad \mathbf{G} = \begin{bmatrix} 0 \\ 1 \end{bmatrix},$$

$$\mathbf{G}_1 = \begin{bmatrix} 0 \\ 1 \end{bmatrix}, \qquad \mathbf{H} = \begin{bmatrix} 1 & 0 \end{bmatrix}, \quad J = 0.$$

Because the data on the disk are not exactly on a centered circle, the servo must follow a sinusoid of radian frequency ω_0 determined by the spindle speed.

(a) Give the structure of a controller for this system that will follow the given reference input with zero steady-state error.

(b) Assume $\omega_0 = 1$ and that the desired closed-loop poles are at $-1 \pm j\sqrt{3}$ and $-\sqrt{3} \pm j1$.

(c) Demonstrate the tracking and disturbance rejection properties of the system using MATLAB or SIMULINK.

Solution

(a) The reference input satisfies the differential equation $\ddot{r} = -\omega_0^2 r$ so that $\alpha_1 = 0$ and $\alpha_2 = \omega_0^2$. With these values, the error-state matrices, according to Eq. (7.218), are

$$\mathbf{A} = \begin{bmatrix} 0 & 1 & 0 & 0 \\ -\omega_0^2 & 0 & 1 & 0 \\ 0 & 0 & 0 & 1 \\ 0 & 0 & 0 & -1 \end{bmatrix}, \quad \mathbf{B} = \begin{bmatrix} 0 \\ 0 \\ 0 \\ 1 \end{bmatrix}.$$

The characteristic equation of $\mathbf{A} - \mathbf{BK}$ is

$$s^4 + (1 + K_{02})s^3 + (\omega_0^2 + K_{01})s^2 + [K_1 + \omega_0^2(1 + K_{02})]s + K_{01}\omega_0^2 K_2 = 0,$$

from which the gain may be selected by pole assignment. The compensator implementation from Eq. (7.220) has the structure shown in Fig. 7.60, which clearly shows the presence of the oscillator with frequency ω_0 (known as the **internal model of the input generator**) in the controller.[16]

Internal model principle

---

[16]This is a particular case of the **internal model principle**, which requires that a model of the external or exogenous signal be in the controller for robust tracking and disturbance rejection.

Figure 7.60

Structure of the compensator for the servomechanism to track exactly the sinusoid of frequency ω_0

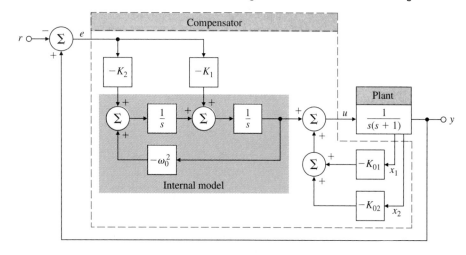

(b) Now assume that $\omega_0 = 1$ rad/sec and the desired closed-loop poles are as given:

$$pc = [-1 + j * \sqrt{3}; -1 - j * \sqrt{3}; -\sqrt{3} + j; -\sqrt{3} - j].$$

Then the feedback gain is

$$\mathbf{K} = [K_2 \ K_1 : \mathbf{K}_o] = [2.0718 \ 16.3923 \ : \ 13.9282 \ 4.4641],$$

which results in the controller

$$\dot{\mathbf{x}}_c = \mathbf{A}_c \mathbf{x}_c + \mathbf{B}_c e,$$
$$u = \mathbf{C}_c \mathbf{x}_c,$$

with

$$\mathbf{A}_c = \begin{bmatrix} 0 & 1 \\ -1 & 0 \end{bmatrix}, \qquad \mathbf{B}_c = \begin{bmatrix} -16.3923 \\ -2.0718 \end{bmatrix},$$

$$\mathbf{C}_c = \begin{bmatrix} 1 & 0 \end{bmatrix}.$$

The relevant MATLAB statements are

```
% plant matrices
F=[0 1; 0 −1];
G=[0;1];
H=[1 0];
J=[0];
% form error space matrices
omega=1;
A=[0 1 0 0;−omega*omega 0 1 0;0 0 0 1;0 0 0 −1];
B=[0;0;G];
```

```
% desired closed-loop poles
pc=[−1+sqrt(3)*j ;−1−sqrt(3)*j;−sqrt(3)+j;−sqrt(3)−j];
K=place(A,B,pc);
% form controller matrices
K1=K(:,1:2);
Ko=K(:,3:4);
Ac=[0 1;−omega*omega 0];
Bc=−[K(2);K(1)];;
Cc=[1 0];
Dc=[0];
```

The controller frequency response is shown in Fig. 7.61 and shows a gain of infinity at the rotation frequency of $\omega_0 = 1$ rad/sec. The frequency response from r to e[i.e., the sensitivity function $\mathcal{S}(s)$], is shown in Fig. 7.62 and reveals a sharp notch at the rotation frequency $\omega_0 = 1$ rad/sec. The same notch is also present in the frequency response of the transfer function from w to y.

(c) Figure 7.63 shows the SIMULINK simulation diagram for the system. Although the simulations can also be done in MATLAB, it is more instructive to use the interactive graphical environment of SIMULINK. SIMULINK also provides the capability to add nonlinearities (see Chapter 9) and carry out robustness studies efficiently.[17] The tracking properties of the system are shown in Fig. 7.64(a), showing the asymptotic tracking property of the system. The associated control effort and the tracking error signal are shown in Fig. 7.64(b) and (c) respectively. The disturbance rejection properties of the system are illustrated in Fig. 7.65(a), displaying asymptotic disturbance rejection of sinusoidal disturbance input. The associated control effort is shown in Fig. 7.65(b). The closed-loop frequency

Figure 7.61

Controller frequency response

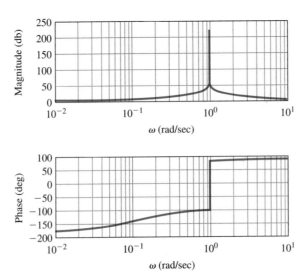

---

[17] In general, the design can be done in MATLAB and (nonlinear) simulations can be carried out in SIMULINK.

Figure 7.62
Sensitivity function
frequency response

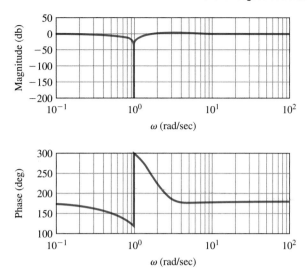

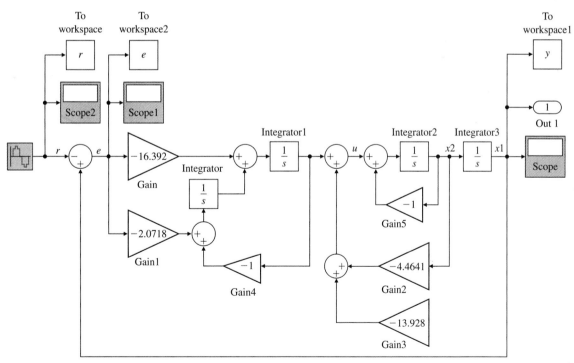

Figure 7.63
SIMULINK® block diagram for robust servomechanism

Figure 7.64

(a) Tracking properties
for robust
servomechanism;
(b) control effort;
(c) tracking error signal

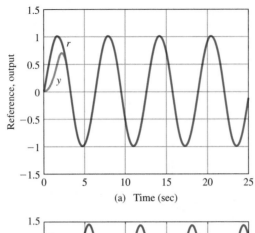

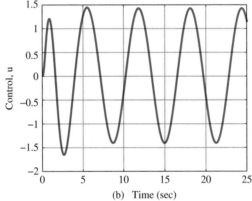

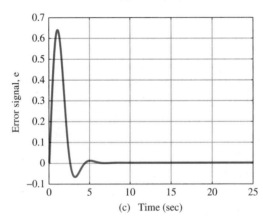

SIMULINK simulation

response [i.e., the complementary transfer function $T(s)$] for the robust servomechanism is shown in Fig. 7.66. As seen from the figure, the frequency response from r to y is unity at $\omega_0 = 1$ rad/sec as expected.

The zeros of the system from r to e are located at $\pm j, -2.7321 \pm j2.5425$. The robust tracking properties are due to the presence of the blocking zeros at

Figure 7.65

(a) Disturbance rejection properties for robust servomechanism; (b) control effort

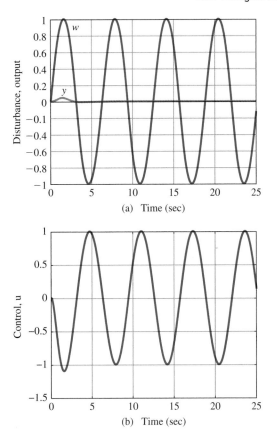

(a) Time (sec)

(b) Time (sec)

Figure 7.66

Closed-loop frequency response for robust servomechanism

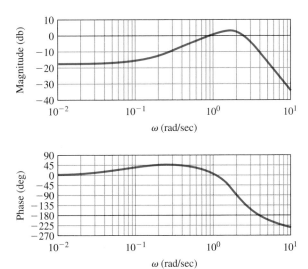

ω (rad/sec)

ω (rad/sec)

$\pm j$. The zeros from w to y, both blocking zeros, are located at $\pm j$. The robust disturbance rejection properties are due to the presence of these blocking zeros.

From the nature of the pole-placement problem, the state z in Eq. (7.217) will tend toward zero for all perturbations in the system parameters as long as $A - BK$ remains stable. Notice that the signals that are rejected are those that satisfy the equations with the values of α_i actually implemented in the model of the external signals. The method assumes that these are known and implemented exactly. If the implemented values are in error, then a steady-state error will result.

Now let us repeat the example of Section 7.10.1 for integral control.

EXAMPLE 7.37

Integral Control Using the Error-Space Design

For the system

$$H(s) = \frac{1}{s+3}$$

with the state-variable description

$$F = -3, \quad G = 1, \quad H = 1,$$

construct a controller with poles at $s = -5$ to track an input that satisfies $\dot{r} = 0$.

Solution. The error system is

$$\begin{bmatrix} \dot{e} \\ \dot{z} \end{bmatrix} = \begin{bmatrix} 0 & 1 \\ 0 & -3 \end{bmatrix} \begin{bmatrix} e \\ z \end{bmatrix} + \begin{bmatrix} 0 \\ 1 \end{bmatrix} \mu.$$

If we take the desired characteristic equation to be

$$\alpha_c(s) = s^2 + 10s + 25,$$

then the pole-placement equation for K is

$$\det[sI - A + BK] = \alpha_c(s). \tag{7.223}$$

In detail, Eq. (7.223) is

$$s^2 + (3 + K_0)s + K_1 = s^2 + 10s + 25,$$

which gives

$$K = [\ 25 \quad 7\ ] = [\ K_1 \quad K_0\ ],$$

and the system is implemented as shown in Fig. 7.67. The transfer function from r to e for this system, the sensitivity function

$$\frac{E(s)}{R(s)} = S(s) = -\frac{s(s+10)}{s^2 + 10s + 25},$$

shows a blocking zero at $s = 0$, which prevents the constant input from affecting the error. The closed-loop transfer function—that is, the complementary sensitivity function—is

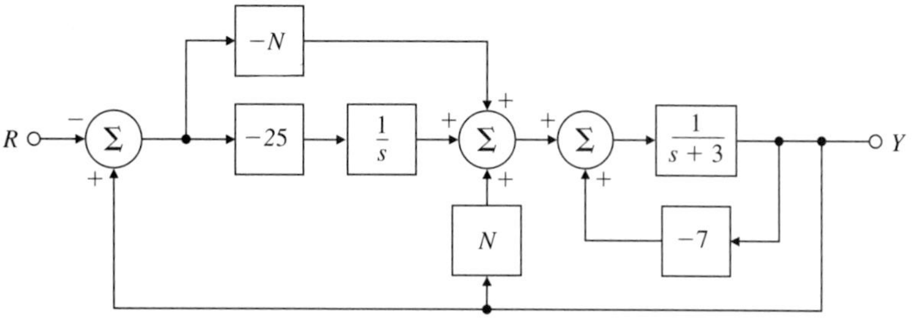

Figure 7.67
Example of internal model with feed-forward

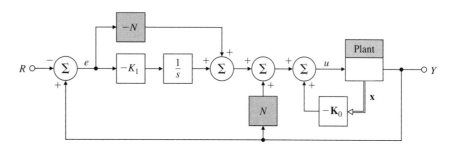

Figure 7.68
Internal model as integral control with feed-forward

$$\frac{Y(s)}{R(s)} = T(s) = \frac{25}{s^2 + 10s + 25}.$$

The structure of Fig. 7.68 permits us to add a feed-forward of the reference input, which provides one extra degree of freedom in zero assignment. If we add a term proportional to r in Eq. (7.222), then

$$u = -K_1 \int^t e(\tau)\, d\tau - \mathbf{K}_0 \mathbf{x} + Nr. \qquad (7.224)$$

This relationship has the effect of creating a zero at $-K_1/N$. The location of this zero can be chosen to improve the transient response of the system. For actual implementation, we can rewrite Eq. (7.224) in terms of e to get

$$u = -K_1 \int^t e(\tau)\, d\tau - \mathbf{K}_0 \mathbf{x} + N(y - e). \qquad (7.225)$$

The block diagram for the system is shown in Fig. 7.68. For our example, the overall transfer function now becomes

$$\frac{Y(s)}{R(s)} = \frac{Ns + 25}{s^2 + 10s + 25}.$$

Notice that the DC gain is unity for any value of N and that, through our choice of N, we can place the zero at any real value to improve the dynamic response. A natural strategy for locating the zero is to have it cancel one of the system poles, in this case at $s = -5$. The step response of the system is shown in Fig. 7.69 for $N = 5$, as well as for $N = 0$ and 8. With the understanding that one pole can be cancelled in integral control designs, we make sure to choose one of the desired control poles such that it is both real and able to be cancelled through the proper choice of N.

Figure 7.69

Step responses with integral control and feed-forward

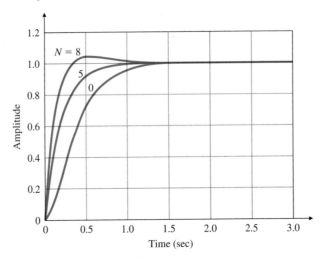

△ 7.10.3 The Extended Estimator

Our discussion of robust control so far has used a control based on full-state feedback. If the state is not available, then as in the regular case, the full-state feedback, $\mathbf{Kx}$, can be replaced by the estimates, $\mathbf{K\hat{x}}$, where the estimator is built as before. As a final look at ways to design control with external inputs, in this section we develop a method for tracking a reference input and rejecting disturbances. The method is based on augmenting the estimator to include estimates from external signals in a way that permits us to cancel out their effects on the system error.

Suppose the plant is described by the equations

$$\dot{\mathbf{x}} = \mathbf{Fx} + \mathbf{G}u + \mathbf{G}w, \qquad (7.226a)$$

$$y = \mathbf{Hx}, \qquad (7.226b)$$

$$e = \mathbf{Hx} - r. \qquad (7.226c)$$

Furthermore, assume that both the reference r and the disturbance w are known to satisfy the equations[18]

$$\alpha_w(s)w = \alpha_\rho(s)w = 0, \qquad (7.227)$$

$$\alpha_r(s)r = \alpha_\rho(s)r = 0, \qquad (7.228)$$

where

$$\alpha_\rho(s) = s^2 + \alpha_1 s + \alpha_2,$$

corresponding to polynomials $\alpha_w(s)$ and $\alpha_r(s)$ in Fig. 7.70(a). In general, we would select the equivalent disturbance polynomial $\alpha_\rho(s)$ in Fig. 7.70(b) to be the *least common multiple* of $\alpha_w(s)$ and $\alpha_r(s)$. The first step is to recognize that, as far as the steady-state response of the output is concerned, there is an input-equivalent signal ρ that satisfies the same equation as r and w and enters the system at the same place as the control signal, as shown in Fig. 7.70(b). As before, we must assume that the

---

[18]Again we develop the results for a second-order equation in the external signals; the discussion can be extended to higher-order equations.

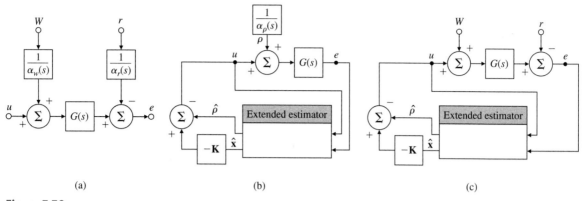

Figure 7.70

Block diagram of a system for tracking and disturbance rejection with extended estimator: (a) equivalent disturbance; (b) block diagram for *design*; (c) block diagram for *implementation*

plant does not have a zero at any of the roots of Eq. (7.227). For our purposes here, we can replace Eqs. (7.226) with

$$\dot{\mathbf{x}} = \mathbf{F}\mathbf{x} + \mathbf{G}(u + \rho), \tag{7.229a}$$

$$e = \mathbf{H}\mathbf{x}. \tag{7.229b}$$

If we can estimate this equivalent input, we can add to the control a term $-\hat{\rho}$ that will cancel out the effects of the real disturbance and reference and cause the output to track r in the steady-state. To do this, we combine Eqs. (7.226) and (7.227) into a state description to get

$$\dot{\mathbf{z}} = \mathbf{A}\mathbf{z} + \mathbf{B}u, \tag{7.230a}$$

$$e = \mathbf{C}\mathbf{z}, \tag{7.230b}$$

where $\mathbf{z} = [\rho \quad \dot{\rho} \quad \mathbf{x}^T]^T$. The matrices are

$$\mathbf{A} = \begin{bmatrix} 0 & 1 & \mathbf{0} \\ -\alpha_2 & -\alpha_1 & \mathbf{0} \\ \mathbf{G} & 0 & \mathbf{F} \end{bmatrix}, \quad \mathbf{B} = \begin{bmatrix} 0 \\ 0 \\ \mathbf{G} \end{bmatrix}, \tag{7.231a}$$

$$\mathbf{C} = [0 \quad 0 \quad \mathbf{H}]. \tag{7.231b}$$

The system given by Eqs. (7.231) is not controllable since we cannot influence ρ from u. However, if $\mathbf{F}$ and $\mathbf{H}$ are observable and if the system $(\mathbf{F}, \mathbf{G}, \mathbf{H})$ does not have a zero that is also a root of Eq. (7.227), then the system of Eq. (7.231) will be observable, and we can construct an observer that will compute estimates of both the state of the plant and of ρ. The estimator equations are standard, but the control is not:

$$\dot{\hat{\mathbf{z}}} = \mathbf{A}\hat{\mathbf{z}} + \mathbf{B}u + \mathbf{L}(e - \mathbf{C}\hat{\mathbf{z}}), \tag{7.232a}$$

$$u = -\mathbf{K}\hat{\mathbf{x}} - \hat{\rho}. \tag{7.232b}$$

In terms of the original variables, the estimator equations are

$$\begin{bmatrix} \dot{\hat{\rho}} \\ \ddot{\hat{\rho}} \\ \dot{\hat{\mathbf{x}}} \end{bmatrix} = \begin{bmatrix} 0 & 1 & \mathbf{0} \\ -\alpha_2 & -\alpha_1 & \mathbf{0} \\ \mathbf{G} & 0 & \mathbf{F} \end{bmatrix} \begin{bmatrix} \hat{\rho} \\ \dot{\hat{\rho}} \\ \hat{\mathbf{x}} \end{bmatrix} + \begin{bmatrix} 0 \\ 0 \\ \mathbf{G} \end{bmatrix} u + \begin{bmatrix} l_1 \\ l_2 \\ \mathbf{L}_3 \end{bmatrix} [e - \mathbf{H}\hat{\mathbf{x}}]. \tag{7.233}$$

The overall block diagram of the system for *design* is shown in Fig. 7.70(b). If we write out the last equation for $\hat{\mathbf{x}}$ in Eq. (7.233) and substitute Eq. (7.232b), a simplification of sorts results because a term in $\hat{\rho}$ cancels out:

$$\dot{\hat{\mathbf{x}}} = \mathbf{G}\hat{\rho} + \mathbf{F}\hat{\mathbf{x}} + \mathbf{G}(-\mathbf{K}\hat{\mathbf{x}} - \hat{\rho}) + \mathbf{L}_3(e - \mathbf{H}\hat{\mathbf{x}})$$
$$= \mathbf{F}\hat{\mathbf{x}} + \mathbf{G}(-\mathbf{K}\hat{\mathbf{x}}) + \mathbf{L}_3(e - \mathbf{H}\hat{\mathbf{x}})$$
$$= \mathbf{F}\hat{\mathbf{x}} + \mathbf{G}\bar{u} + \mathbf{L}_3(e - \mathbf{H}\hat{\mathbf{x}}).$$

With the estimator of Eq. (7.233) and the control of Eq. (7.232b), the state equation is

$$\dot{\mathbf{x}} = \mathbf{F}\mathbf{x} + \mathbf{G}(-\mathbf{K}\hat{\mathbf{x}} - \hat{\rho}) + \mathbf{G}\rho. \tag{7.234}$$

In terms of the estimate errors, Eq. (7.234) can be rewritten as

$$\dot{\mathbf{x}} = (\mathbf{F} - \mathbf{G}\mathbf{K})\mathbf{x} + \mathbf{G}\mathbf{K}\tilde{\mathbf{x}} + \mathbf{G}\tilde{\rho}. \tag{7.235}$$

Because we designed the estimator to be stable, the values of $\tilde{\rho}$ and $\tilde{\mathbf{x}}$ go to zero in the steady state, and the final value of the state is not affected by the external input. The block diagram of the system for *implementation* is drawn in Fig. 7.70(c). A very simple example will illustrate the steps in this process.

EXAMPLE 7.38   *Steady-State Tracking and Disturbance Rejection of Motor Speed by Extended Estimator*

Construct an estimator to control the state and cancel a constant bias at the output and track a constant reference in the motor speed system described by

$$\dot{x} = -3x + u, \tag{7.236a}$$
$$y = x + w, \tag{7.236b}$$
$$\dot{w} = 0, \tag{7.236c}$$
$$\dot{r} = 0. \tag{7.236d}$$

Place the control pole at $s = -5$ and the two extended estimator poles at $s = -15$.

Solution.   To begin, we design the control law by ignoring the equivalent disturbance. Rather, we notice by inspection that a gain of -2 will move the single pole from -3 to the desired -5, Therefore, $K = 2$. The system augmented with equivalent external input ρ, which replaces the actual disturbance w and the reference r, is given by

$$\dot{\rho} = 0,$$
$$\dot{x} = -3x + u + \rho,$$
$$e = x.$$

The extended estimator equations are

$$\dot{\hat{\rho}} = l_1(e - \hat{x}),$$
$$\dot{\hat{x}} = -3\hat{x} + u + \hat{\rho} + l_2(e - \hat{x}).$$

The estimator error gain is found to be $\mathbf{L} = [\ 225 \quad 27\ ]^T$ from the characteristic equation

$$\det \begin{bmatrix} s & l_1 \\ 1 & s + 3 + l_2 \end{bmatrix} = s^2 + 30s + 225.$$

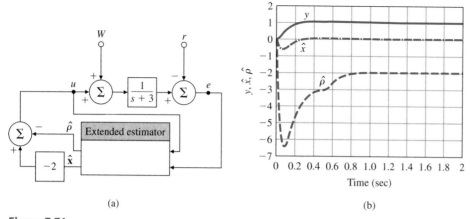

Figure 7.71

Motor speed system with extended estimator: (a) block diagram; (b) command step response and disturbance step response

A block diagram of the system is given in Fig. 7.71(a), and the step responses to input at the command r (applied at $t = 0$ sec) and at the disturbance w (applied at $t = 0.5$ sec) are shown in Fig. 7.71(b).

△ 7.11   Loop Transfer Recovery (LTR)

LTR

The introduction of an estimator in a state feedback controller loop may adversely affect the stability robustness properties of the system [i.e., the phase margin (PM) and gain margin (GM) properties may become arbitrarily poor, as shown by Doyle's famous example (Doyle, 1978)]. However, it is possible to modify the estimator design so as to try to "recover" the LQR stability robustness properties to some extent. This process, called loop transfer recovery (LTR), is especially effective for minimum-phase systems. To achieve the recovery, some of the estimator poles are placed at (or near) the zeros of the plant and the remaining poles are moved (sufficiently far) into the LHP. The idea behind LTR is to redesign the estimator in such a way as to shape the loop gain properties to approximate those of LQR.

The use of LTR means that feedback controllers can be designed to achieve desired sensitivity [$S(s)$] and complementary sensitivity functions [$T(s)$] at critical (loop-breaking) points in the feedback system (e.g., at either the input or output of the plant). Of course, there is a price to be paid for this improvement in stability robustness! The newly designed control system may have worse sensor noise sensitivity properties. Intuitively, one can think of making (some of) the estimator poles arbitrarily fast so that the loop gain is approximately that of LQR. Alternatively, one can think of essentially "inverting" the plant transfer function so that all the LHP poles of the plant are cancelled by the dynamic compensator to achieve the desired loop shape. There are obvious trade-offs, and the designer needs to be careful to make the correct choice for the given problem, depending on the control system specifications.

LTR is a well-known technique now, and specific practical design procedures have been identified (Athans, 1986; Stein and Athans, 1987; Saberi et al., 1993). The same procedures may also be applied to nonminimum phase systems, but there is no guarantee on the extent of possible recovery. The LTR technique may be viewed as a systematic procedure to study design trade-offs for linear quadratic-based compensator design (Doyle and Stein, 1981). We will now formulate the LTR problem.

Consider the linear system

$$\dot{\mathbf{x}} = \mathbf{F}\mathbf{x} + \mathbf{G}u + w, \tag{7.237a}$$

$$y = \mathbf{H}\mathbf{x} + v, \tag{7.237b}$$

where w and v are uncorrelated zero-mean white Gaussian process and sensor noise with covariance matrices $\mathbf{R}_w \geq 0$ and $\mathbf{R}_v \geq 0$. The estimator design yields

$$\dot{\hat{\mathbf{x}}} = \mathbf{F}\hat{\mathbf{x}} + \mathbf{G}u + \mathbf{L}(y - \hat{y}), \tag{7.238a}$$

$$\hat{y} = \mathbf{H}\hat{\mathbf{x}}, \tag{7.238b}$$

resulting in the usual dynamic compensator

$$D_c(s) = -\mathbf{K}(s\mathbf{I} - \mathbf{F} + \mathbf{G}\mathbf{K} + \mathbf{L}\mathbf{H})^{-1}\mathbf{L}. \tag{7.239}$$

We will now treat the noise parameters, $\mathbf{R}_w$ and $\mathbf{R}_v$, as design "knobs" in the dynamic compensator design. Without loss of generality, let us choose $\mathbf{R}_w = \mathbf{\Gamma}\mathbf{\Gamma}^T$ and $\mathbf{R}_v = 1$. For LTR, assume that $\mathbf{\Gamma} = q\mathbf{G}$, where q is a scalar design parameter. The estimator design is then based on the specific design parameters $\mathbf{R}_w$ and $\mathbf{R}_v$. It can be shown that, for a minimum-phase system, as q becomes large (Doyle and Stein, 1979),

$$\lim_{q \to \infty} D_c(s)G(s) = \mathbf{K}(s\mathbf{I} - \mathbf{F})^{-1}\mathbf{G}, \tag{7.240}$$

the convergence is pointwise in s and the degree of recovery can be arbitrarily good. This design procedure in effect "inverts" the plant transfer function in the limit as $q \to \infty$:

Plant inversion

$$\lim_{q \to \infty} D_c(s) = \mathbf{K}(s\mathbf{I} - \mathbf{F})^{-1}\mathbf{G}G^{-1}(s). \tag{7.241}$$

This is precisely the reason that full-loop transfer recovery is not possible for a nonminimum-phase system. This limiting behavior may be explained using the symmetric root loci. As $q \to \infty$, some of the estimator poles approach the zeros of

$$G_e(s) = \mathbf{H}(s\mathbf{I} - \mathbf{F})^{-1}\mathbf{\Gamma}, \tag{7.242}$$

LTR for
nonminimum-phase
systems

and the rest tend to infinity[19] [see Eqs. (7.166) and (7.167)]. In practice, the LTR design procedure can still be applied to a nonminimum-phase plant. The degree of recovery will depend on the specific locations of the nonminimum-phase zeros. Sufficient recovery should be possible at many frequencies if the RHP zeros are located outside the specified closed-loop bandwidth. Limits on achievable performance of feedback systems due to RHP zeros are discussed in Freudenberg and Looze (1985). We will next illustrate the LTR procedure by a simple example.

---

[19] In a Butterworth configuration.

EXAMPLE 7.39   *LTR Design for Satellite Attitude Control*

Consider the satellite system with state-space description

$$\mathbf{F} = \begin{bmatrix} 0 & 1 \\ 0 & 0 \end{bmatrix}, \quad \mathbf{G} = \begin{bmatrix} 0 \\ 1 \end{bmatrix},$$

$$\mathbf{H} = \begin{bmatrix} 1 & 0 \end{bmatrix}, \quad J = 0.$$

(a) Design an LQR controller with $\mathbf{Q} = \rho \mathbf{H}^T \mathbf{H}$ and $R = 1$, $\rho = 1$, and determine the loop gain.
(b) Then design a compensator that recovers the LQR loop gain of part (a) using the LTR technique for $q = 1, 10, 100$.
(c) Compare the different candidate designs in part (b) with respect to the actuator activity due to additive white Gaussian sensor noise.

Solution. Using lqr, the selected LQR weights result in the feedback gain $\mathbf{K} = \begin{bmatrix} 1 & 1.414 \end{bmatrix}$. The loop transfer function is

$$\mathbf{K}(s\mathbf{I} - \mathbf{F})^{-1}\mathbf{G} = \frac{1.414(s + 0.707)}{s^2}.$$

A magnitude frequency response plot of this LQR loop gain is shown in Fig. 7.72. For the estimator design using lqe, let $\Gamma = q\mathbf{G}, \mathbf{R}_w = \Gamma^T\Gamma, R_v = 1$, and choose $q = 10$, resulting in the estimator gain

$$\mathbf{L} = \begin{bmatrix} 14.142 \\ 100 \end{bmatrix}.$$

The compensator transfer function is

$$D_c(s) = \mathbf{K}(s\mathbf{I} - \mathbf{F} + \mathbf{GK} + \mathbf{LH})^{-1}\mathbf{L}$$

$$= \frac{155.56(s + 0.6428)}{(s^2 + 15.556s + 121)} = \frac{155.56(s + 0.6428)}{(s + 7.77 + j7.77)(s + 7.77 - j7.77)},$$

and the loop transfer function is

$$D_c(s)G(s) = \frac{155.56(s + 0.6428)}{s^2(s + 7.77 + j7.77)(s + 7.77 - j7.77)}.$$

Figure 7.72 shows the frequency response of the loop transfer function for several values of q ($q = 1, 10, 100$), along with the ideal LQR loop transfer function frequency response. As seen from this figure, the loop gain tends to approach that of LQR as the value of q increases. As seen in Fig. 7.72, for $q = 10$, the "recovered" gain margin is GM = $11.1 = 20.9$ db and the PM = $55.06°$. Sample MATLAB statements to carry out the preceding LTR design procedure are as follows:

```
F=[0 1; 0 0];
G=[0;1];
H=[1 0];
J=[0];
sys0=ss(F,G,H,J);
H1=[1 0];
sys=ss(F,G,H1,J);
```

Figure 7.72

Frequency response plots for LTR design

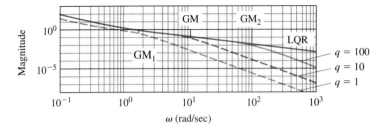

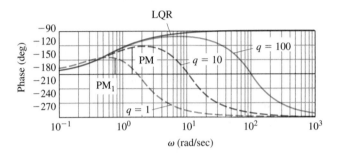

MATLAB lqr

MATLAB lqe

MATLAB bode
MATLAB margin

```
w=logspace(-1,3,1000);
rho=1.0;
Q=rho*H1'*H1;
r=1;
[K]=lqr(F,G,Q,r)
sys1=ss(F,G,K,0);
[maggk1,phasgk1,w]=bode(sys1,w);


q=10;
gam=q*G;
Q1=gam'*gam;
rv=1;
[L]=lqe(F,gam,H,Q1,rv)

aa=F—G*K—L*H;
bb=L;
cc=K;
dd=0;
sysk=ss(aa,bb,cc,dd);
sysgk=series(sys0,sysk);
[maggk,phsgk,w]=bode(sysgk,w);
[gm,phm,wcg,wcp]=margin(maggk,phsgk,w)
loglog(w,[maggk1(:) maggk(:)]);
semilogx(w,[phasgk1(:) phsgk(:)]);
```

To determine the effect of sensor noise, v, on the actuator activity, we determine the transfer function from v to u as shown in Fig. 7.73. For the selected value of LTR

Figure 7.73

Closed-loop system for LTR

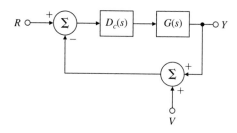

Figure 7.74

SIMULINK block diagram for LTR

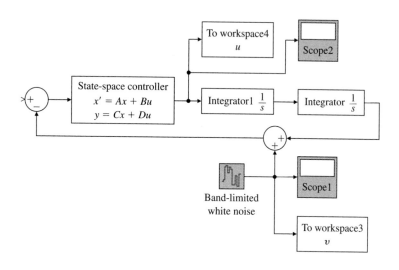

design parameter, $q = 10$, we have

$$\frac{U(s)}{V(s)} = H(s) = \frac{-D_c(s)}{1 + D_c(s)G(s)} = \frac{-155.56s^2(s + 0.6428)}{s^4 + 15.556s^3 + 121s^2 + 155.56s + 99.994}.$$

RMS value

One reasonable measure of the effect of the sensor noise on the actuator activity is the root-mean-square (RMS) value of the control, u, due to the additive noise, v. The RMS value of the control may be computed as

$$\|u\|_{\text{rms}} = \left(\frac{1}{T_0} \int_0^{T_0} u(t)^2 dt \right)^{1/2} \tag{7.243}$$

where T_0 is the signal duration. Assuming white Gaussian noise v, the RMS value of the control can also be determined analytically (Boyd and Barratt, 1991). The closed-loop SIMULINK diagram with band-limited white sensor noise excitation is shown in Fig. 7.74. The values of the RMS control were computed for different values

SIMULINK simulation

of the LTR design parameter q, using the SIMULINK simulations, and are tabulated in Table 7.2. The results suggest increased vulnerability due to actuator wear as q is increased. Refer to MATLAB commands ltry and ltru for the LTR computations.

TABLE 7.2

Computed RMS Control for Various Values of LTR Tuning Parameter q.	
q	$\| u \|_{rms}$
1	0.1454
10	2.8054
100	70.5216

△ **7.12   Direct Design with Rational Transfer Functions**

An alternative to the state-space methods discussed so far is to postulate a general-structure dynamic controller with two inputs (r and y) and one output (u) and to solve for the transfer function of the controller to give a specified overall r-to-y transfer function. A block diagram of the situation is shown in Fig. 7.75. We model the plant as the transfer function

$$\frac{Y(s)}{U(s)} = \frac{b(s)}{a(s)}, \tag{7.244}$$

General controller in polynomial form

rather than by state equations. The controller is also modeled by its transfer function, in this case a transfer function with two inputs and one output:

$$U(s) = -\frac{c_y(s)}{d(s)}Y(s) + \frac{c_r(s)}{d(s)}R(s). \tag{7.245}$$

Here $d(s)$, $c_y(s)$, and $c_r(s)$ are polynomials. In order for the controller of Fig. 7.75 and Eq. (7.245) to be implemented, the orders of the numerator polynomials $c_y(s)$ and $c_r(s)$ must not be higher than the order of the denominator polynomial $d(s)$.

   To carry out the design, we require that the closed-loop transfer function defined by Eqs. (7.244) and (7.245) be matched to the desired transfer function

$$\frac{Y(s)}{R(s)} = \frac{c_r(s)b(s)}{\alpha_c(s)\alpha_e(s)}. \tag{7.246}$$

Equation (7.246) tells us that the zeros of the plant must be zeros of the overall system. The only way to change this is to have factors of $b(s)$ appear in either α_c or α_e. We combine Eqs. (7.244) and (7.245) to get

$$a(s)Y(s) = b(s)\left[-\frac{c_y(s)}{d(s)}Y(s) + \frac{c_r(s)}{d(s)}R(s)\right], \tag{7.247}$$

which can be rewritten as

$$[a(s)d(s) + b(s)c_y(s)]Y(s) = b(s)c_r(s)R(s). \tag{7.248}$$

Diophantine equation

   Comparing Eq. (7.246) with Eq. (7.247) we immediately see that the design can be accomplished if we can solve the **Diophantine equation**

$$a(s)d(s) + b(s)c_y(s) = \alpha_c(s)\alpha_e(s) \tag{7.249}$$

for given arbitrary a, b, α_c, and α_e. Because each transfer function is a ratio of polynomials, we can assume that $a(s)$ and $d(s)$ are **monic polynomials**; that is, the coefficient of the highest power of s in each polynomial is unity. The question is, How

Figure 7.75

Direct transfer-function formulation

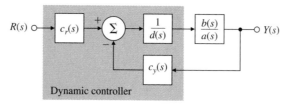

Dynamic controller

many equations and how many unknowns are there if we match coefficients of equal powers of s in Eq. (7.249)? If $a(s)$ is of degree n (given) and $d(s)$ is of degree m (to be selected), then a direct count yields $2m + 1$ unknowns in $d(s)$ and $c_y(s)$ and $n + m$ equations from the coefficients of powers of s. Thus the requirement is that

$$2m + 1 \geq n + m,$$

Dimension of the controller

or

$$m \geq n - 1.$$

One possibility for a solution is to choose $d(s)$ of degree n and $c_y(s)$ of degree $n-1$. In that case, which corresponds to the state-space design for a full-order estimator, there are $2n$ equations and $2n$ unknowns with $\alpha_c\alpha_e$ of degree $2n$. The resulting equations will then have a solution for arbitrary α_i if and only if $a(s)$ and $b(s)$ have no common factors.[20]

EXAMPLE 7.40

Pole Placement for Polynomial Transfer Functions

Using the polynomial method, design a controller of order n for the third-order plant in Example 7.30. Note that if the polynomials $\alpha_c(s)$ and $\alpha_e(s)$ from Example 7.30 are multiplied, the result is the desired closed-loop characteristic equation:

$$\alpha_c(s)\alpha_e(s) = s^6 + 14s^5 + 122.75s^4 + 585.2s^3 + 1505.64s^2 + 2476.8s + 1728. \quad (7.250)$$

Solution. Using Eq. (7.249) with $b(s) = 10$, we find that

$$(d_0s^3 + d_1s^2 + d_2s + d_3)(s^3 + 10s^2 + 16s) + 10(c_0s^2 + c_1s + c_2) \equiv \alpha_c(s)\alpha_e(s). \quad (7.251)$$

We have expanded the polynomial $d(s)$ with coefficients d_i and the polynomial $c_y(s)$ with coefficients c_i.

Now we equate the coefficients of the like powers of s in Eq. (7.251) to find that the parameters must satisfy[21]

$$
\begin{bmatrix}
1 & 0 & 0 & 0 & 0 & 0 & 0 \\
10 & 1 & 0 & 0 & 0 & 0 & 0 \\
16 & 10 & 1 & 0 & 0 & 0 & 0 \\
0 & 16 & 10 & 1 & 0 & 0 & 0 \\
0 & 0 & 16 & 10 & 10 & 0 & 0 \\
0 & 0 & 0 & 16 & 0 & 10 & 0 \\
0 & 0 & 0 & 0 & 0 & 0 & 10
\end{bmatrix}
\begin{bmatrix}
d_0 \\
d_1 \\
d_2 \\
d_3 \\
c_0 \\
c_1 \\
c_2
\end{bmatrix}
=
\begin{bmatrix}
1 \\
14 \\
122.75 \\
585.2 \\
1505.64 \\
2476.8 \\
1728
\end{bmatrix}. \quad (7.252)
$$

----

[20] If they do have a common factor, it will show up on the left side of Eq. (7.249); for there to be a solution, the same factor must be on the right side of Eq. (7.249), and thus a factor of either α_c or α_e.

[21] The matrix on the left side of Eq. (7.252) is called a **Sylvester matrix** and is nonsingular if and only if $a(s)$ and $b(s)$ have no common factor.

The solution to Eq. (7.252) is

$$
\begin{aligned}
d_0 &= 1, & c_0 &= 190.1, \\
d_1 &= 4, & c_1 &= 481.8, \\
d_2 &= 66.75, & c_2 &= 172.8. \\
d_3 &= -146.3,
\end{aligned}
$$

MATLAB a\b

[The solution can be found using x = a\b command in MATLAB, where a is the Sylvester matrix and b is the right-hand side in Eq. (7.252).] Hence the controller transfer function is

$$
\frac{c_y(s)}{d(s)} = \frac{190.1s^2 + 481.8s + 172.8}{s^3 + 4s^2 + 66.75s - 146.3}. \tag{7.253}
$$

Note that the coefficients of Eq. (7.253) are the same as those of the controller $D_c(s)$ (which we obtained using the state-variable techniques), once the factors in $D_c(s)$ are multiplied out.

The reduced-order compensator can also be derived using a polynomial solution.

EXAMPLE 7.41   *Reduced-Order Design for a Polynomial Transfer Function Model*

Design a reduced-order controller for the third-order system in Example 7.30. The desired characteristic equation is

$$
\alpha_c(s)\alpha_e(s) = s^5 + 12s^4 + 74s^3 + 207s^2 + 378s + 288.
$$

Solution. The equations needed to solve this problem are the same as those used to obtain Eq. (7.251), except that we take both $d(s)$ and $c_y(s)$ to be of degree $n - 1$. We need to solve

$$
(d_0s^2 + d_1s + d_2)(s^3 + 10s^2 + 16s) + 10(c_0s^2 + c_1s + c_2) \equiv \alpha_c(s)\alpha_e(s). \tag{7.254}
$$

Equating coefficients of like powers of s in Eq. (7.254), we obtain

$$
\begin{bmatrix}
1 & 0 & 0 & 0 & 0 & 0 \\
10 & 1 & 0 & 0 & 0 & 0 \\
16 & 10 & 1 & 0 & 0 & 0 \\
0 & 16 & 10 & 10 & 0 & 0 \\
0 & 0 & 16 & 0 & 10 & 0 \\
0 & 0 & 0 & 0 & 0 & 10
\end{bmatrix}
\begin{bmatrix}
d_0 \\ d_1 \\ d_2 \\ c_0 \\ c_1 \\ c_2
\end{bmatrix}
=
\begin{bmatrix}
1 \\ 12 \\ 74 \\ 207 \\ 378 \\ 288
\end{bmatrix}. \tag{7.255}
$$

MATLAB a\b

The solution is (again using the x = a\b command in MATLAB)

$$
\begin{aligned}
d_0 &= 1, & c_0 &= -20.8, \\
d_1 &= 2.0, & c_1 &= -23.6, \\
d_2 &= 38, & c_2 &= 28.8,
\end{aligned}
$$

and the resulting controller is

$$\frac{c_y(s)}{d(s)} = \frac{-20.8s^2 - 23.6s + 28.8}{s^2 + 2.0s + 38}. \tag{7.256}$$

Again, Eq. (7.256) is exactly the same as $D_{cr}(s)$ derived using the state-variable techniques in Example 7.31, once the polynomials of $D_{cr}(s)$ are multiplied out and minor numerical differences are considered.

Notice that the reference input polynomial $c_r(s)$ does not enter into the analysis of Examples 7.40 and 7.41. We can select $c_r(s)$ so that it will assign zeros in the transfer function from $R(s)$ to $Y(s)$. This is the same role played by $\gamma(s)$ in Section 7.9. One choice is to select $c_r(s)$ to cancel $\alpha_e(s)$ so that the overall transfer function is

$$\frac{Y(s)}{R(s)} = \frac{K_s b(s)}{\alpha_c(s)}.$$

This corresponds to the first and most common choice of **M** and $\bar{N}$ for introducing the reference input described in Section 7.9.

Adding integral control to the polynomial solution

It is also possible to introduce integral control and, indeed, internal-model-based robust tracking control into the polynomial design method. What is required is that we have error control and that the controller have poles at the internal model locations. To get error control with the structure of Fig. 7.75, we need only let $c_r = c_y$. To get desired poles into the controller, we need to require that a specific factor be part of $d(s)$. For integral control—the most common case—this is almost trivial. The polynomial $d(s)$ will have a root at zero if we set the last term, d_m, to zero. The resulting equations can be solved if $m = n$. For a more general internal model, we define $d(s)$ to be the product of a reduced-degree polynomial and a specified polynomial such as Eq. (7.227), and match coefficients in the Diophantine equation as before. The process is straightforward but tedious. Again we caution that, while the polynomial design method can be effective, the numerical problems of this method are often much worse than those associated with methods based on state equations. For higher-order systems, the state-space methods are preferable.

△ 7.13 Design for Systems with Pure Time Delay

In any linear system consisting of lumped elements, the response of the system appears immediately after an excitation of the system. In some feedback systems—for example, process control systems, whether controlled by a human operator in the loop or by computer—there is a **pure time delay** (also called **transportation lag**) in the system. As a result of the distributed nature of these systems, the response remains identically zero until after a delay of λ seconds. A typical step response is shown in Fig. 7.76(a). The transfer function of a pure transportation lag is $e^{-\lambda s}$. We can represent an overall transfer function of a single-input-single-output (SISO) system with time delay as

Overall transfer function for a time-delayed system

$$G_I(s) = G(s)e^{-\lambda s}, \tag{7.257}$$

where $G(s)$ has no pure time delay. Because $G_I(s)$ does not have a finite state description, standard use of state-variable methods is impossible. However, Smith (1958)

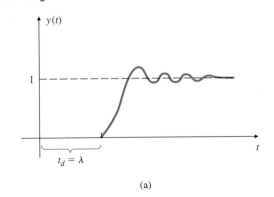

(a)

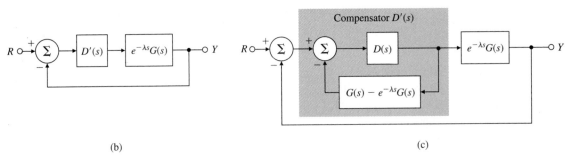

(b)   (c)

Figure 7.76
A Smith regulator for systems with time delay

showed how to construct a feedback structure that effectively takes the delay outside the loop and allows a feedback design based on $G(s)$ alone, which can be
done with standard methods. The result of this method is a design having closed-
loop transfer function with delay λ but otherwise showing the same response as
the closed-loop design based on no delay. To see how the method works, let
us consider the feedback structure shown in Fig. 7.76(b). The overall transfer
function is

$$\frac{Y(s)}{R(s)} = T(s) = \frac{D'(s)G(s)e^{-\lambda s}}{1 + D'(s)G(s)e^{-\lambda s}}. \qquad (7.258)$$

Smith suggested that we solve for $D'(s)$ by setting up a dummy overall transfer function in which the controller transfer function $D(s)$ is in a loop with $G(s)$ with *no loop
delay* but with an overall delay of λ:

$$\frac{Y(s)}{R(s)} = T(s) = \frac{D(s)G(s)}{1 + D(s)G(s)}e^{-\lambda s}. \qquad (7.259)$$

The Smith compensator

We then equate Eqs. (7.258) and (7.259) to solve for $D'(s)$:

$$D'(s) = \frac{D(s)}{1 + D(s)[G(s) - G(s)e^{-\lambda s}]}. \qquad (7.260)$$

If the plant transfer function and the delay are known, $D'(s)$ can be realized
with real components by means of the block diagram shown in Fig. 7.76(c). With
this knowledge, we can design the compensator $D(s)$ in the usual way, based on

Eq. (7.259), as if there were no delay, and then implement it as shown in Fig. 7.76(c). The resulting closed-loop system would exhibit the behavior of a finite closed-loop system except for the time delay λ. This design approach is particularly suitable when the pure delay, λ, is significant as compared to the process time constant, for example, in pulp and paper process applications.

Notice that, conceptually, the Smith compensator is feeding back a simulated plant output to cancel the true plant output and then adding in a simulated plant output without the delay. It can be demonstrated that $D'(s)$ in Fig. 7.76(c) is equivalent to an ordinary regulator in line with a compensator that provides significant phase lead. To implement such compensators in analog systems, it is usually necessary to approximate the delay required in $D'(s)$ by a Padé approximant; with digital compensators the delay can be implemented exactly (see Chapter 8). It is also a fact that the compensator $D'(s)$ is a strong function of $G(s)$, and a small error in the model of the plant used in the controller could lead to large errors in the closed loop, perhaps even to instability. This design is very sensitive. If $D(s)$ is implemented as a PI controller, then one could detune (i.e., reduce the gain) to try to ensure stability and reasonable performance. For automatic tuning of the Smith regulator and a recent application to Stanford's quiet hydraulic precision lathe fluid temperature control, the reader is referred to Huang and DeBra (2000).

EXAMPLE 7.42   *Heat Exchanger: Design with Pure Time Delay*

Figure 7.77 shows the heat exchanger from Example 2.15. The temperature of the product is controlled by controlling the flow rate of steam in the exchanger jacket. The temperature sensor is several meters downstream from the steam control valve, which introduces a transportation lag into the model. A suitable model is given by

$$G(s) = \frac{e^{-5s}}{(10s + 1)(60s + 1)}.$$

Figure 7.77
A heat exchanger

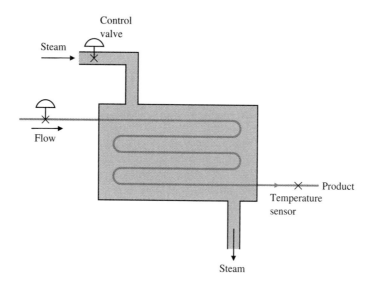

Design a controller for the heat exchanger using the Smith compensator and pole placement. The control poles are to be at

$$p_c = -0.05 \pm 0.087j,$$

and the estimator poles are to be at three times the control poles' natural frequency:

$$p_e = -0.15 \pm 0.26j.$$

Simulate the response of the system with SIMULINK.

Solution. A suitable set of state-space equations is

$$\dot{\mathbf{x}}(t) = \begin{bmatrix} -0.017 & 0.017 \\ 0 & -0.1 \end{bmatrix} \mathbf{x}(t) + \begin{bmatrix} 0 \\ 0.1 \end{bmatrix} u(t-5),$$

$$y = [1 \quad 0]\mathbf{x},$$

$$\lambda = 5.$$

For the specified control pole locations and for the moment ignoring the time delay, we find that the state feedback gain is

$$\mathbf{K} = [5.2 \ -0.17].$$

For the given estimator poles, the estimator gain matrix for a full-order estimator is

$$\mathbf{L} = \begin{bmatrix} 0.18 \\ 4.2 \end{bmatrix}.$$

The resulting controller transfer function is

$$D(s) = \frac{U(s)}{Y(s)} = \frac{-0.25(s+1.8)}{s+0.14 \pm 0.27j}.$$

If we choose to adjust for unity closed-loop DC gain, then

$$\bar{N} = 1.2055.$$

SIMULINK simulation   The SIMULINK diagram for the system is shown in Fig. 7.78. The open-loop and closed-loop step responses of the system and the control effort are shown in Figs. 7.79 and 7.80, and the root locus of the system (without the delay) is shown in Fig. 7.81. Note that the time delay of 5 sec in Figs. 7.79 and 7.80 is quite small compared with the response of the system, and is barely noticeable in this case.

7.14 Historical Perspective

The state-variable approach to solving differential equations in engineering problems was advocated by R. E. Kalman while attending MIT. This was revolutionary and ruffled some feathers as it was going against the grain. The well-established academics, Kalman's teachers, were well-versed in the frequency domain techniques and staunch supporters of it. Beginning in the late 1950s and early 1960s Kalman wrote a series of seminal papers introducing the ideas of state-variables, controllability, observability, the Linear Quadratic (LQ), and the Kalman Filter (LQF). Gunkel and Franklin (1963) and Joseph and Tou (1961) independently showed the separation theorem, which made possible the Linear Quadratic Gaussian (LQG) problem nowadays

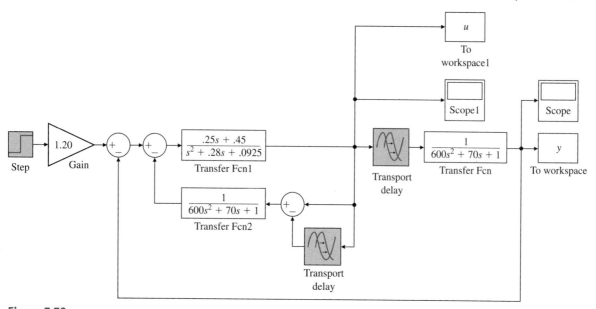

Figure 7.78

Closed-loop SIMULINK® diagram for a heat exchanger

Figure 7.79

Step response for a heat exchanger

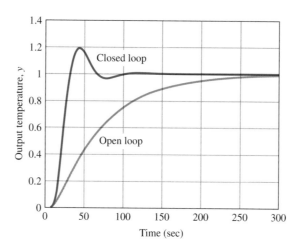

referred to as the H_2 formulation. The separation theorem is a special case of the certainty-equivalence theorem of Simon (1956). The solutions to both LQ and LQG problems can be expressed in an elegant fashion in terms of the solutions to Riccati equations. D. G. Luenberger, who was taking a course with Kalman at Stanford University, derived the observer and reduced-order observer over a weekend after hearing Kalman suggesting the problem in a lecture. Kalman, Bryson, Athans, and others contributed to the field of optimal control theory that was widely employed in aerospace problems including the Apollo program. The book by Zadeh and Desoer published in 1962 was also influential in promoting the state-space method. In the

Figure 7.80

Control effort for a heat exchanger

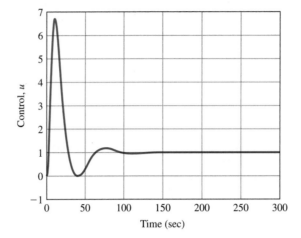

Figure 7.81

Root locus for a heat exchanger

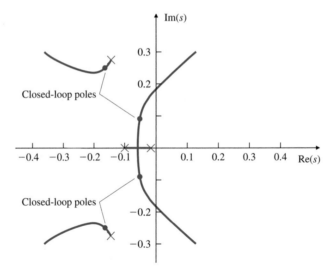

1970s the robustness of LQ and LQG methods were studied resulting in the celebrated and influential paper of Doyle and Stein in 1981. One of the most significant contributions of Doyle and Safonov was to extend the idea of frequency domain gain to multi-input multi-output systems using the singular value decomposition (SVD). Others contributing to this research included G. Zames who introduced the H_∞ methods that were found to be extensions to the H_2 methods. The resulting design techniques are known as H_∞ and μ-synthesis procedures. During the 1980s reliable numerical methods were developed for dealing with state-variable designs and computer-aided software for control design were developed. The invention of MATLAB by Cleve Moler and its wide distribution by The Mathworks has had a huge impact not only in the control design field but on all interactive scientific computations.

While the state-variable methods were gaining momentum particularly in the US, research groups in Europe especially in England led by Rosenbrock, MacFarlane, Munro, and others extended the classical techniques to multi-input multi-output systems. Hence root locus and frequency domain methods such as the (inverse) Nyquist techniques could be used for multi-input multi-output systems. Eventually in the 1980s there was a realization that the power of both frequency domain and state-variable methods should be combined for an eclectic control design method employing the best of both approaches.

We saw in this Chapter 7 that in contrast to frequency response methods of Bode and Nyquist, the state-variable method not only deals with the input and output variables of the system but also with the internal physical variables. The state-variable methods can be used to study linear and nonlinear, as well as time varying systems. Furthermore, the state-variable method handles the multi-input multi-output problems and high-order systems with equal ease. From a computational perspective, the state-variable methods are far superior to the frequency domain techniques that require polynomial manipulations.

SUMMARY

- To every transfer function that has no more zeros than poles, there corresponds a differential equation in state-space form.
- State-space descriptions can be in several canonical forms. Among these are **control**, **observer**, and **modal canonical forms.**
- Open-loop poles and zeros can be computed from the state description matrices $(\mathbf{F}, \mathbf{G}, \mathbf{H}, J)$:

$$\text{Poles: } p = \text{eig}(\mathbf{F}), \quad \det(p\mathbf{I} - \mathbf{F}) = 0,$$

$$\text{Zeros: } \det \begin{bmatrix} z\mathbf{I} - \mathbf{F} & -\mathbf{G} \\ \mathbf{H} & J \end{bmatrix} = 0.$$

- For any controllable system of order n, there exists a state feedback control law that will place the closed-loop poles at the roots of an arbitrary **control characteristic equation** of order n.
- The reference input can be introduced so as to result in zero steady-state error to a step command. This property is not expected to be robust to parameter changes.
- Good closed-loop pole locations depend on the desired transient response, the robustness to parameter changes, and a balance between dynamic performance and control effort.
- Closed-loop pole locations can be selected to result in a dominant second-order response, to match a prototype dynamic response, or to minimize a quadratic performance measure.
- For any observable system of order n, an estimator (or observer) can be constructed with only sensor inputs and a state that estimates the plant state. The n poles of the estimator error system can be placed arbitrarily.
- Every transfer function can be represented by a minimal realization, i.e., a state-space model that is both controllable and observable.
- A single-input single-output system is completely controllable if and only if the input excites all the natural frequencies of the system, i.e., there is no cancellation of the poles in the transfer function.

- The control law and the estimator can be combined into a controller such that the poles of the closed-loop system are the sum of the control-law-only poles and the estimator-only poles.
- With the estimator-based controller, the **reference input** can be introduced in such a way as to permit n arbitrary zeros to be assigned. The most common choice is to assign the zeros to cancel the estimator poles, thus not exciting an estimator error.
- **Integral control** can be introduced to obtain robust steady-state tracking of a step by augmenting the plant state. The design is also robust with respect to rejecting constant disturbances.
- General **robust control** can be realized by combining the equations of the plant and the reference model into an **error space** and designing a control law for the extended system. Implementation of the robust design demonstrates the **internal model principle**. An estimator of the plant state can be added while retaining the robustness properties.
- The estimator can be extended to include estimates of the equivalent control disturbance and so result in robust tracking and disturbance rejection.
- Pole-placement designs, including integral control, can be computed using the polynomials of the plant transfer function in place of the state descriptions. Designs using polynomials frequently have problems with numerical accuracy.
- Controllers for plants that include a pure time delay can be designed as if there were no delay, and then a controller can be implemented for the plant with the delay. The design can be expected to be sensitive to parameter changes.
- Table 7.3 gives the important equations discussed in this chapter. The triangles indicate equations taken from optional sections in the text.
- Determining a model from experimental data, or verifying an analytically based model by experiment, is an important step in system design by state-space analysis, a step that is not necessarily needed for compensator design via frequency-response methods.

REVIEW QUESTIONS

The following questions are based on a system in state-variable form with matrices $\mathbf{F}$, $\mathbf{G}$, $\mathbf{H}$, J, input u, output y, and state $\mathbf{x}$.

1. Why is it convenient to write equations of motion in state-variable form?
2. Give an expression for the transfer function of this system.
3. Give two expressions for the poles of the transfer function of the system.
4. Give an expression for the zeros of the system transfer function.
5. Under what condition will the state of the system be controllable?
6. Under what conditions will the system be observable from the output y?
7. Give an expression for the *closed-loop* poles if state feedback of the form $u = -\mathbf{Kx}$ is used.
8. Under what conditions can the feedback matrix $\mathbf{K}$ be selected so that the roots of $\alpha_c(s)$ are arbitrary?
9. What is the advantage of using the LQR or SRL in designing the feedback matrix $\mathbf{K}$?

TABLE 7.3 **Important Equations in Chapter 7**

Name	Equation	Page
Control canonical form	$$\mathbf{A}_c = \begin{bmatrix} -a_1 & -a_2 & \cdots & \cdots & -a_n \\ 1 & 0 & \cdots & \cdots & 0 \\ 0 & 1 & 0 & \cdots & 0 \\ \vdots & & \ddots & 0 & \vdots \\ 0 & 0 & \cdots & 1 & 0 \end{bmatrix},$$	426
	$$\mathbf{B}_c = \begin{bmatrix} 1 \\ 0 \\ 0 \\ \vdots \\ \vdots \\ 0 \end{bmatrix}, \quad \mathbf{C}_c = [b_1 \quad b_2 \quad \cdots \quad \cdots \quad b_n], \quad D_c = 0.$$	
State description	$\dot{\mathbf{x}} = \mathbf{F}\mathbf{x} + \mathbf{G}u$	428
Output equation	$y = \mathbf{H}\mathbf{x} + Ju$	428
Transformation of state	$\mathbf{A} = \mathbf{T}^{-1}\mathbf{F}\mathbf{T}$	429
	$\mathbf{B} = \mathbf{T}^{-1}\mathbf{G}$	
	$y = \mathbf{H}\mathbf{T}\mathbf{z} + Ju = \mathbf{C}\mathbf{z} + Du,$	
	where $\mathbf{C} = \mathbf{H}\mathbf{T}, D = J$	
Controllability matrix	$\mathcal{C} = [\mathbf{G} \quad \mathbf{F}\mathbf{G} \quad \cdots \quad \mathbf{F}^{n-1}\mathbf{G}]$	430
Transfer function from state equations	$G(s) = \dfrac{Y(s)}{U(s)} = \mathbf{H}(s\mathbf{I} - \mathbf{F})^{-1}\mathbf{G} + J$	437
Transfer-function poles	$\det(p_i\mathbf{I} - \mathbf{F}) = 0$	438
Transfer-function zeros	$\alpha_z(s) = \det\begin{bmatrix} z_i\mathbf{I} - \mathbf{F} & -\mathbf{G} \\ \mathbf{H} & J \end{bmatrix} = 0$	439
Control characteristic equation	$\det[s\mathbf{I} - (\mathbf{F} - \mathbf{G}\mathbf{K})] = 0$	444
Ackermann's control formula for pole placement	$\mathbf{K} = [0 \quad \cdots \quad 0 \quad 1]\mathcal{C}^{-1}\alpha_c(\mathbf{F})$	448
Reference input gains	$\begin{bmatrix} \mathbf{F} & \mathbf{G} \\ \mathbf{H} & J \end{bmatrix}\begin{bmatrix} \mathbf{N_x} \\ \mathbf{N_u} \end{bmatrix} = \begin{bmatrix} \mathbf{0} \\ 1 \end{bmatrix}$	452
Control equation with reference input	$u = N_u r - \mathbf{K}(\mathbf{x} - \mathbf{N_x}r)$	452
	$= -\mathbf{K}\mathbf{x} + (N_u + \mathbf{K}\mathbf{N_x})r$	
	$= -\mathbf{K}\mathbf{x} + \bar{N}r$	
Symmetric root locus	$1 + \rho G_0(-s)G_0(s) = 0$	458
Estimator error characteristic equation	$\alpha_e(s) = \det[s\mathbf{I} - (\mathbf{F} - \mathbf{L}\mathbf{H})] = 0$	467
Observer canonical form	$\dot{\mathbf{x}}_o = \mathbf{F}_o\mathbf{x}_o + \mathbf{G}_o u,$	470
	$y = \mathbf{H}_o\mathbf{x}_o,$	
	where	
	$$\mathbf{F}_o = \begin{bmatrix} -a_1 & 1 & 0 & 0 & \cdots & 0 \\ -a_2 & 0 & 1 & 0 & \cdots & \vdots \\ \vdots & \vdots & & \ddots & & 1 \\ -a_n & 0 & & 0 & & 0 \end{bmatrix},$$	
	$$\mathbf{G}_o = \begin{bmatrix} b_1 \\ b_2 \\ \vdots \\ b_n \end{bmatrix}, \quad \mathbf{H}_o = [1 \quad 0 \quad 0 \quad \cdots \quad 0],$$	

TABLE 7.3	**Continued**		

Name	Equation		Page
Observability matrix	$$\mathcal{O} = \begin{bmatrix} \mathbf{H} \\ \mathbf{HF} \\ \vdots \\ \mathbf{HF}^{n-1} \end{bmatrix},$$		471
Ackermann's estimator formula	$$\mathbf{L} = \alpha_e(\mathbf{F})\mathcal{O}^{-1} \begin{bmatrix} 0 \\ 0 \\ \vdots \\ 1 \end{bmatrix}$$		471
Compensator transfer function	$D_c(s) = \dfrac{U(s)}{Y(s)} = -\mathbf{K}(s\mathbf{I} - \mathbf{F} + \mathbf{GK} + \mathbf{LH})^{-1}\mathbf{L}$		480
Reduced-order compensator transfer function	$D_{cr}(s) = \dfrac{U(s)}{Y(s)} = \mathbf{C}_r(s\mathbf{I} - \mathbf{A}_r)^{-1}\mathbf{B}_r + D_r$		480
Controller equations	$\dot{\hat{\mathbf{x}}} = (\mathbf{F} - \mathbf{GK} - \mathbf{LH})\hat{\mathbf{x}} + \mathbf{L}y + \mathbf{M}r$ $$u = -\mathbf{K}\hat{\mathbf{x}} + \bar{N}r$$		492
Augmented state equations with integral control	$$\begin{bmatrix} \dot{x}_I \\ \dot{\mathbf{x}} \end{bmatrix} = \begin{bmatrix} 0 & \mathbf{H} \\ 0 & \mathbf{F} \end{bmatrix} \begin{bmatrix} x_I \\ \mathbf{x} \end{bmatrix} + \begin{bmatrix} 0 \\ \mathbf{G} \end{bmatrix} u - \begin{bmatrix} 1 \\ 0 \end{bmatrix} r + \begin{bmatrix} 0 \\ \mathbf{G}_1 \end{bmatrix} w$$		503
△ General controller in polynomial form	$U(s) = -\dfrac{c_y(s)}{d(s)} Y(s) + \dfrac{c_r(s)}{d(s)} R(s)$		524
△ Diophantine equation for closed-loop characteristic equation	$a(s)d(s) + b(s)c_y(s) = \alpha_c(s)\alpha_e(s)$		524

10. What is the main reason for using an estimator in feedback control?

11. If the estimator gain $\mathbf{L}$ is used, give an expression for the closed-loop poles due to the estimator.

12. Under what conditions can the estimator gain $\mathbf{L}$ be selected so that the roots of $\alpha_e(s) = 0$ are arbitrary?

13. If the reference input is arranged so that the input to the estimator is identical to the input to the process, what will the overall closed-loop transfer function be?

14. If the reference input is introduced in such a way as to permit the zeros to be assigned as the roots of $\gamma(s)$, what will the overall closed-loop transfer function be?

15. What are the three standard techniques for introducing integral control in the state feedback design method?

PROBLEMS

Problems for Section 7.3: Block Diagrams and State-Space

7.1   Write the dynamic equations describing the circuit in Fig. 7.82. Write the equations as a second-order differential equation in $y(t)$. Assuming a zero input, solve the differential equation for $y(t)$ using Laplace transform methods for the parameter values and

Figure 7.82

Circuit for Problem 7.1

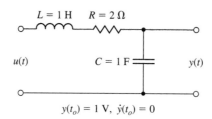

$L = 1\,H \qquad R = 2\,\Omega$

$u(t)$

$C = 1\,F$

$y(t)$

$y(t_o) = 1\ \text{V},\ \dot{y}(t_o) = 0$

initial conditions shown in the figure. Verify your answer using the initial command in MATLAB.

7.2   A schematic for the satellite and scientific probe for the Gravity Probe-B (GP-B) experiment that was launched on April 30, 2004 is sketched in Fig. 7.83. Assume that the mass of the spacecraft plus helium tank, m_1, is 2000 kg and the mass of the probe, m_2, is 1000 kg. A rotor will float inside the probe and will be forced to follow the probe with a capacitive forcing mechanism. The spring constant of the coupling k is 3.2×10^6. The viscous damping b is 4.6×10^3.

(a) Write the equations of motion for the system consisting of masses m_1 and m_2 using the inertial position variables, y_1 and y_2.

(b) The actual disturbance u is a micrometeorite, and the resulting motion is very small. Therefore, rewrite your equations with the scaled variables $z_1 = 10^6 y_1, z_2 = 10^6 y_2$, and $v = 1000u$.

(c) Put the equations in state-variable form using the state $\mathbf{x} = [z_1 \ \dot{z}_1 \ z_2 \ \dot{z}_2]^T$, the output $y = z_2$, and the input an impulse, $u = 10^{-3}\delta(t)$ N·sec on mass m_1.

(d) Using the numerical values, enter the equations of motion into MATLAB in the form

$$\dot{\mathbf{x}} = \mathbf{F}\mathbf{x} + \mathbf{G}v, \tag{7.261}$$

$$y = \mathbf{H}\mathbf{x} + Jv. \tag{7.262}$$

and define the MATLAB system: sysGPB = ss(F,G,H,J). Plot the response of y caused by the impulse with the MATLAB command impulse(sysGPB). This is the signal the rotor must follow.

(e) Use the MATLAB commands p = eig(F) to find the poles (or roots) of the system and z = tzero(F,G,H, J) to find the zeros of the system.

Figure 7.83

Schematic diagram of the GP-B satellite and probe

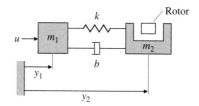

k

Rotor

$u \rightarrow$ m_1

m_2

b

y_1

y_2

Problems for Section 7.4: Analysis of the State Equations

7.3   Give the state description matrices in control-canonical form for the following transfer functions:

(a) $\dfrac{1}{4s + 1}$

(b) $\dfrac{5(s/2 + 1)}{(s/10 + 1)}$

(c) $\dfrac{2s + 1}{s^2 + 3s + 2}$

(d) $\dfrac{s + 3}{s(s^2 + 2s + 2)}$

(e) $\dfrac{(s + 10)(s^2 + s + 25)}{s^2(s + 3)(s^2 + s + 36)}$

7.4 Use the MATLAB function tf2ss to obtain the state matrices called for Problem 7.3.

7.5 Give the state description matrices in normal-mode form for the transfer functions of Problem 7.3. Make sure that all entries in the state matrices are real valued by keeping any pairs of complex conjugate poles together, and realize them as a separate subblock in control canonical form.

7.6 A certain system with state $\mathbf{x}$ is described by the state matrices

$$\mathbf{F} = \begin{bmatrix} -2 & 1 \\ -2 & 0 \end{bmatrix}, \quad \mathbf{G} = \begin{bmatrix} 1 \\ 3 \end{bmatrix},$$

$$\mathbf{H} = \begin{bmatrix} 1 & 0 \end{bmatrix}, \quad J = 0.$$

Find the transformation $\mathbf{T}$ so that if $\mathbf{x} = \mathbf{Tz}$, the state matrices describing the dynamics of $\mathbf{z}$ are in control canonical form. Compute the new matrices $\mathbf{A}$, $\mathbf{B}$, $\mathbf{C}$, and D.

7.7 Show that the transfer function is not changed by a linear transformation of state.

7.8 Use block-diagram reduction or Mason's rule to find the transfer function for the system in observer canonical form depicted by Fig. 7.31.

7.9 Suppose we are given a system with state matrices $\mathbf{F}$, $\mathbf{G}$, $\mathbf{H}$ ($J = 0$ in this case). Find the transformation $\mathbf{T}$ so that, under Eqs. (7.24) and (7.25), the new state description matrices will be in observer canonical form.

7.10 Use the transformation matrix in Eq. (7.41) to explicitly multiply out the equations at the end of Example 7.10.

7.11 Find the state transformation that takes the observer canonical form of Eq. (7.35) to the modal canonical form.

7.12 (a) Find the transformation $\mathbf{T}$ that will keep the description of the tape-drive system of Example 7.11 in modal canonical form but will convert each element of the input matrix $\mathbf{B}_m$ to unity.

(b) Use MATLAB to verify that your transformation does the job.

7.13 (a) Find the state transformation that will keep the description of the tape-drive system of Example 7.11 in modal canonical form but will cause the poles to be displayed in $\mathbf{A}_m$ in order of increasing magnitude.

(b) Use MATLAB to verify your result in part (a), and give the complete new set of state matrices as $\mathbf{A}$, $\mathbf{B}$, $\mathbf{C}$, and D.

7.14 Find the characteristic equation for the modal-form matrix $\mathbf{A}_m$ of Eq. (7.17a) using Eq. (7.58).

7.15 Given the system

$$\dot{\mathbf{x}} = \begin{bmatrix} -4 & 1 \\ -2 & -1 \end{bmatrix} \mathbf{x} + \begin{bmatrix} 0 \\ 1 \end{bmatrix} u$$

with zero initial conditions, find the steady-state value of $\mathbf{x}$ for a step input u.

7.16 Consider the system shown in Fig. 7.84:

   (a) Find the transfer function from U to Y.

   (b) Write state equations for the system using the state-variables indicated.

Figure 7.84

A block diagram for Problem 7.16

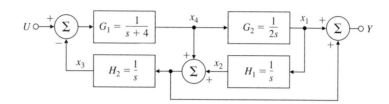

7.17 Using the indicated state-variables, write the state equations for each of the systems shown in Fig. 7.85. Find the transfer function for each system using both block-diagram manipulation and matrix algebra [as in Eq. (7.48)].

Figure 7.85

Block diagrams for Problem 7.17

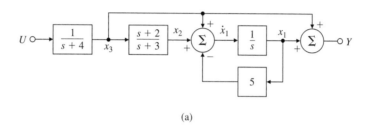

(a)

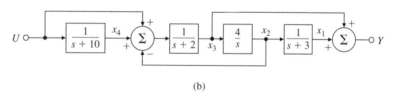

(b)

7.18 For each of the listed transfer functions, write the state equations in both control and observer canonical form. In each case, draw a block diagram and give the appropriate expressions for **F**, **G**, and **H**.

   (a) $\dfrac{s^2 - 2}{s^2(s^2 - 1)}$ (control of an inverted pendulum by a force on the cart).

   (b) $\dfrac{3s + 4}{s^2 + 2s + 2}$.

7.19 Consider the transfer function

$$G(s) = \frac{Y(s)}{U(s)} = \frac{s+1}{s^2 + 5s + 6}. \tag{7.263}$$

   (a) By rewriting Eq. (7.263) in the form

$$G(s) = \frac{1}{s+3}\left(\frac{s+1}{s+2}\right),$$

   find a **series realization** of $G(s)$ as a cascade of two first-order systems.

   (b) Using a partial-fraction expansion of $G(s)$, find a **parallel realization** of $G(s)$.

   (c) Realize $G(s)$ in control canonical form.

Problems for Section 7.5: Control Law Design for Full-State Feedback

7.20 Consider the plant described by

$$\dot{x} = \begin{bmatrix} 0 & 1 \\ 7 & -4 \end{bmatrix} x + \begin{bmatrix} 1 \\ 2 \end{bmatrix} u,$$

$$y = [\ 1 \quad 3 \ ]x.$$

(a) Draw a block diagram for the plant with one integrator for each state-variable.

(b) Find the transfer function using matrix algebra.

(c) Find the closed-loop characteristic equation if the feedback is

(i) $u = -[\ K_1 \quad K_2 \ ]x$;

(ii) $u = -Ky$.

7.21 For the system

$$\dot{x} = \begin{bmatrix} 0 & 1 \\ -6 & -5 \end{bmatrix} x + \begin{bmatrix} 0 \\ 1 \end{bmatrix} u,$$

design a state feedback controller that satisfies the following specifications:

(a) Closed-loop poles have a damping coefficient $\zeta = 0.707$.

(b) Step-response peak time is under 3.14 sec.

Verify your design with MATLAB.

7.22 (a) Design a state feedback controller for the following system so that the closed-loop step response has an overshoot of less than 25% and a 1% settling time under 0.115 sec:

$$\dot{x} = \begin{bmatrix} 0 & 1 \\ 0 & -10 \end{bmatrix} x + \begin{bmatrix} 0 \\ 1 \end{bmatrix} u.$$

(b) Use the step command in MATLAB to verify that your design meets the specifications. If it does not, modify your feedback gains accordingly.

7.23 Consider the system

$$\dot{x} = \begin{bmatrix} -1 & -2 & -2 \\ 0 & -1 & 1 \\ 1 & 0 & -1 \end{bmatrix} x + \begin{bmatrix} 2 \\ 0 \\ 1 \end{bmatrix} u.$$

(a) Design a state feedback controller for the system so that the closed-loop step response has an overshoot of less than 5% and a 1% settling time under 4.6 sec.

(b) Use the step command in MATLAB to verify that your design meets the specifications. If it does not, modify your feedback gains accordingly.

7.24 Consider the system in Fig. 7.86.

(a) Write a set of equations that describes this system in the standard canonical control form as $\dot{x} = Fx + Gu$ and $y = Hx$.

(b) Design a control law of the form

$$u = -[\ K_1 \quad K_2 \ ] \begin{bmatrix} x_1 \\ x_2 \end{bmatrix}$$

that will place the closed-loop poles at $s = -2 \pm 2j$.

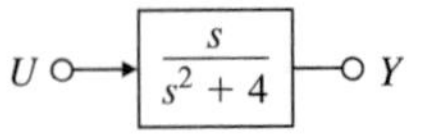

Figure 7.86
System for Problem 7.24

$U \circ\!\!\!\rightarrow \boxed{\dfrac{s}{s^2 + 4}} \!\rightarrow\!\circ Y$

7.25 *Output Controllability.* In many situations a control engineer may be interested in controlling the output y rather than the state $\mathbf{x}$. A system is said to be **output controllable** if at any time you are able to transfer the output from zero to any desired output y^* in a finite time using an appropriate control signal u^*. Derive necessary and sufficient conditions for a continuous system $(\mathbf{F}, \mathbf{G}, \mathbf{H})$ to be output controllable. Are output and state controllability related? If so, how?

7.26 Consider the system

$$
\dot{\mathbf{x}} =
\begin{bmatrix}
0 & 4 & 0 & 0 \\
-1 & -4 & 0 & 0 \\
5 & 7 & 1 & 15 \\
0 & 0 & 3 & -3
\end{bmatrix}
\mathbf{x} +
\begin{bmatrix}
0 \\
0 \\
1 \\
0
\end{bmatrix}
u.
$$

(a) Find the eigenvalues of this system. (*Hint:* Note the block-triangular structure.)

(b) Find the controllable and uncontrollable modes of this system.

(c) For each of the uncontrollable modes, find a vector $\mathbf{v}$ such that

$$
\mathbf{v}^T \mathbf{G} = 0, \quad \mathbf{v}^T \mathbf{F} = \lambda \mathbf{v}^T.
$$

(d) Show that there are an infinite number of feedback gains $\mathbf{K}$ that will relocate the modes of the system to -5, -3, -2, and -2.

(e) Find the unique matrix $\mathbf{K}$ that achieves these pole locations and prevents initial conditions on the uncontrollable part of the system from ever affecting the controllable part.

7.27 Two pendulums, coupled by a spring, are to be controlled by two equal and opposite forces u, which are applied to the pendulum bobs as shown in Fig. 7.87. The equations of motion are

$$
ml^2 \ddot{\theta}_1 = -ka^2(\theta_1 - \theta_2) - mgl\theta_1 - lu,
$$
$$
ml^2 \ddot{\theta}_2 = -ka^2(\theta_2 - \theta_1) - mgl\theta_2 + lu.
$$

(a) Show that the system is uncontrollable. Can you associate a physical meaning with the controllable and uncontrollable modes?

(b) Is there any way that the system can be made controllable?

Figure 7.87

Coupled pendulums for Problem 7.27

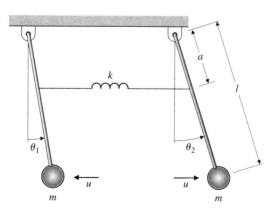

7.28   The state-space model for a certain application has been given to us with the following state description matrices:

$$F = \begin{bmatrix} 0.174 & 0 & 0 & 0 & 0 \\ 0.157 & 0.645 & 0 & 0 & 0 \\ 0 & 1 & 0 & 0 & 0 \\ 0 & 0 & 1 & 0 & 0 \\ 0 & 0 & 0 & 1 & 0 \end{bmatrix}, \quad G = \begin{bmatrix} -0.207 \\ -0.005 \\ 0 \\ 0 \\ 0 \end{bmatrix}, \quad H = [\, 1 \quad 0 \quad 0 \quad 0 \quad 0 \,].$$

(a) Draw a block diagram of the realization with an integrator for each state-variable.

(b) A student has computed $\det \mathcal{C} = 2.3 \times 10^{-7}$ and claims that the system is uncontrollable. Is the student right or wrong? Why?

(c) Is the realization observable?

7.29   *Staircase Algorithm (Van Dooren et al., 1978):* Any realization $(\mathbf{F}, \mathbf{G}, \mathbf{H})$ can be transformed by an **orthogonal similarity transformation** to $(\bar{\mathbf{F}}, \bar{\mathbf{G}}, \bar{\mathbf{H}})$, where $\bar{\mathbf{F}}$ is an **upper Hessenberg matrix** (having one nonzero diagonal above the main diagonal) given by

$$\bar{\mathbf{F}} = \mathbf{T}^T \mathbf{F} \mathbf{T} = \begin{bmatrix} * & \alpha_1 & \mathbf{0} & 0 \\ * & * & \ddots & 0 \\ * & * & \ddots & \alpha_{n-1} \\ * & * & \cdots & * \end{bmatrix}, \quad \bar{\mathbf{G}} = \mathbf{T}^T \mathbf{G} = \begin{bmatrix} 0 \\ \vdots \\ 0 \\ g_1 \end{bmatrix},$$

where $g_1 \neq 0$, and

$$\bar{\mathbf{H}} = \mathbf{H}\mathbf{T} = [\, h_1 \quad h_2 \quad \cdots \quad h_n \,], \quad \mathbf{T}^{-1} = \mathbf{T}^T.$$

Orthogonal transformations correspond to a **rotation** of the vectors (represented by the matrix columns) being transformed with no change in length.

(a) Prove that if $\alpha_i = 0$ and $\alpha_{i+1}, \ldots, \alpha_{n-1} \neq 0$ for some i, then the controllable and uncontrollable modes of the system can be identified after this transformation has been done.

(b) How would you use this technique to identify the observable and unobservable modes of $(\mathbf{F}, \mathbf{G}, \mathbf{H})$?

(c) What advantage does this approach for determining the controllable and uncontrollable modes have over transforming the system to any other form?

(d) How can we use this approach to determine a basis for the controllable and uncontrollable subspaces, as in Problem 7.14?

This algorithm can also be used to design a numerically stable algorithm for pole placement [see Minimis and Paige (1982)]. The name of the algorithm comes from the multi-input version in which the α_i are the blocks that make $\bar{\mathbf{F}}$ resemble a staircase. Refer to ctrbf, obsvf commands in MATLAB.

Problems for Section 7.6: Selection of Pole Locations for Good Design

7.30   The normalized equations of motion for an inverted pendulum at angle θ on a cart are

$$\ddot{\theta} = \theta + u, \quad \ddot{x} = -\beta\theta - u,$$

where x is the cart position, and the control input u is a force acting on the cart.

(a) With the state defined as $\mathbf{x} = \begin{bmatrix} \theta & \dot{\theta} & x & \dot{x} \end{bmatrix}^T$, find the feedback gain $\mathbf{K}$ that places the closed-loop poles at $s = -1, -1, -1 \pm 1j$. For parts (b) through (d), assume that $\beta = 0.5$.

(b) Use the SRL to select poles with a bandwidth as close as possible to those of part (a), and find the control law that will place the closed-loop poles at the points you selected.

(c) Compare the responses of the closed-loop systems in parts (a) and (b) to an initial condition of $\theta = 10°$. You may wish to use the initial command in MATLAB.

(d) Compute $\mathbf{N_x}$ and N_u for zero steady-state error to a constant command input on the cart position, and compare the step responses of each of the two closed-loop systems.

7.31 Consider the feedback system in Fig. 7.88. Find the relationship between K, T, and ξ such that the closed-loop transfer function minimizes the integral of the time multiplied by the absolute value of the error (ITAE) criterion,

$$\mathcal{J} = \int_0^\infty t|e|\, dt,$$

for a step input.

Figure 7.88

Control system for Problem 7.31

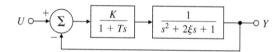

7.32 Prove that the Nyquist plot for LQR design avoids a circle of radius one centered at the -1 point, as shown in Fig. 7.89. Show that this implies that $\frac{1}{2} < \text{GM} < \infty$, the "upward" gain margin is GM$= \infty$, and there is a "downward" GM $= \frac{1}{2}$, and the phase margin is at least PM $= \pm 60°$. Hence the LQR gain matrix, $\mathbf{K}$, can be multiplied by a large scalar or reduced by half with guaranteed closed-loop system stability.

Figure 7.89

Nyquist plot for an optimal regulator

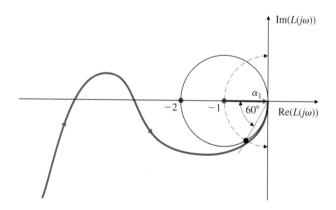

Problems for Section 7.7: Estimator Design

7.33 Consider the system

$$\mathbf{F} = \begin{bmatrix} -2 & 1 \\ 1 & 0 \end{bmatrix}, \mathbf{G} = \begin{bmatrix} 1 \\ 0 \end{bmatrix}, \mathbf{H} = \begin{bmatrix} 1 & 2 \end{bmatrix},$$

and assume that you are using feedback of the form $u = -\mathbf{Kx} + r$, where r is a reference input signal.

(a) Show that $(\mathbf{F}, \mathbf{H})$ is observable.

(b) Show that there exists a $\mathbf{K}$ such that $(\mathbf{F} - \mathbf{GK}, \mathbf{H})$ is unobservable.

(c) Compute a $\mathbf{K}$ of the form $\mathbf{K} = [1, K_2]$ that will make the system unobservable as in part (b); that is, find K_2 so that the closed-loop system is not observable.

(d) Compare the open-loop transfer function with the transfer function of the closed-loop system of part (c). What is the unobservability due to?

7.34 Consider a system with the transfer function

$$G(s) = \frac{9}{s^2 - 9}.$$

(a) Find $(\mathbf{F}_o, \mathbf{G}_o, \mathbf{H}_o)$ for this system in observer canonical form.

(b) Is $(\mathbf{F}_o, \mathbf{G}_o)$ controllable?

(c) Compute $\mathbf{K}$ so that the closed-loop poles are assigned to $s = -3 \pm 3j$.

(d) Is the closed-loop system of part (c) observable?

(e) Design a full-order estimator with estimator error poles at $s = -12 \pm 12j$.

(f) Suppose the system is modified to have a zero:

$$G_1(s) = \frac{9(s+1)}{s^2 - 9}.$$

Prove that if $u = -\mathbf{Kx} + r$, there is a feedback gain $\mathbf{K}$ that makes the closed-loop system unobservable. [Again assume an observer canonical realization for $G_1(s)$.]

7.35 Explain how the controllability, observability, and stability properties of a linear system are related.

7.36 Consider the electric circuit shown in Fig. 7.90.

(a) Write the internal (state) equations for the circuit. The input $u(t)$ is a current, and the output y is a voltage. Let $x_1 = i_L$ and $x_2 = v_c$.

(b) What condition(s) on R, L, and C will guarantee that the system is controllable?

(c) What condition(s) on R, L, and C will guarantee that the system is observable?

Figure 7.90
Electric circuit for
Problem 7.36

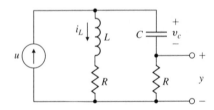

7.37 The block diagram of a feedback system is shown in Fig. 7.91. The system state is

$$\mathbf{x} = \begin{bmatrix} \mathbf{x}_p \\ \mathbf{x}_f \end{bmatrix},$$

Figure 7.91

Block diagram for
Problem 7.37

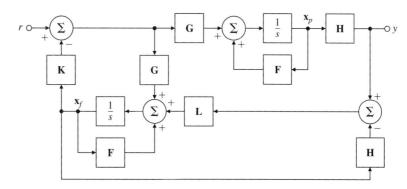

and the dimensions of the matrices are as follows:

$$\mathbf{F} = n \times n, \quad \mathbf{L} = n \times 1,$$
$$\mathbf{G} = n \times 1, \quad \mathbf{x} = 2n \times 1,$$
$$\mathbf{H} = 1 \times n, \quad r = 1 \times 1,$$
$$\mathbf{K} = 1 \times n, \quad y = 1 \times 1.$$

(a) Write state equations for the system.

(b) Let $\mathbf{x} = \mathbf{Tz}$, where

$$\mathbf{T} = \begin{bmatrix} \mathbf{I} & \mathbf{0} \\ \mathbf{I} & -\mathbf{I} \end{bmatrix}.$$

Show that the system is not controllable.

(c) Find the transfer function of the system from r to y.

7.38 This problem is intended to give you more insight into controllability and observability. Consider the circuit in Fig. 7.92, with an input voltage source $u(t)$ and an output current $y(t)$.

(a) Using the capacitor voltage and inductor current as state-variables, write state and output equations for the system.

(b) Find the conditions relating R_1, R_2, C, and L that render the system uncontrollable. Find a similar set of conditions that result in an unobservable system.

(c) Interpret the conditions found in part (b) physically in terms of the time constants of the system.

(d) Find the transfer function of the system. Show that there is a pole–zero cancellation for the conditions derived in part (b) (that is, when the system is uncontrollable or unobservable).

Figure 7.92

Electric circuit for
Problem 7.38

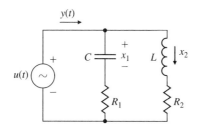

7.39 The linearized equations of motion for a satellite are

$$\dot{x} = Fx + Gu,$$

$$y = Hx,$$

where

$$F = \begin{bmatrix} 0 & 1 & 0 & 0 \\ 3\omega^2 & 0 & 0 & 2\omega \\ 0 & 0 & 0 & 1 \\ 0 & -2\omega & 0 & 0 \end{bmatrix}, \quad G = \begin{bmatrix} 0 & 0 \\ 1 & 0 \\ 0 & 0 \\ 0 & 1 \end{bmatrix}, \quad H = \begin{bmatrix} 1 & 0 & 0 & 0 \\ 0 & 0 & 1 & 0 \end{bmatrix},$$

$$u = \begin{bmatrix} u_1 \\ u_2 \end{bmatrix}, \quad y = \begin{bmatrix} y_1 \\ y_2 \end{bmatrix}.$$

The inputs u_1 and u_2 are the radial and tangential thrusts, the state-variables x_1 and x_3 are the radial and angular deviations from the reference (circular) orbit, and the outputs y_1 and y_2 are the radial and angular measurements, respectively.

(a) Show that the system is controllable using both control inputs.

(b) Show that the system is controllable using only a single input. Which one is it?

(c) Show that the system is observable using both measurements.

(d) Show that the system is observable using only one measurement. Which one is it?

7.40 Consider the system in Fig. 7.93.

(a) Write the state-variable equations for the system, using $[\theta_1 \; \theta_2 \; \dot{\theta}_1 \; \dot{\theta}_2]^T$ as the state vector and F as the single input.

(b) Show that all the state-variables are observable using measurements of θ_1 alone.

(c) Show that the characteristic polynomial for the system is the product of the polynomials for two oscillators. Do so by first writing a new set of system equations involving the state-variables

$$\begin{bmatrix} y_1 \\ y_2 \\ \dot{y}_1 \\ \dot{y}_2 \end{bmatrix} = \begin{bmatrix} \theta_1 + \theta_2 \\ \theta_1 - \theta_2 \\ \dot{\theta}_1 + \dot{\theta}_2 \\ \dot{\theta}_1 - \dot{\theta}_2 \end{bmatrix}.$$

Figure 7.93
Coupled pendulums for
Problem 7.40

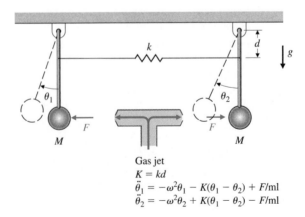

Gas jet
$K = kd$
$\ddot{\theta}_1 = -\omega^2\theta_1 - K(\theta_1 - \theta_2) + F/ml$
$\ddot{\theta}_2 = -\omega^2\theta_2 + K(\theta_1 - \theta_2) - F/ml$

Hint: If **A** and **D** are invertible matrices, then

$$\begin{bmatrix} \mathbf{A} & \mathbf{0} \\ \mathbf{0} & \mathbf{D} \end{bmatrix}^{-1} = \begin{bmatrix} \mathbf{A}^{-1} & \mathbf{0} \\ \mathbf{0} & \mathbf{D}^{-1} \end{bmatrix}.$$

(d) Deduce the fact that the spring mode is controllable with F but the pendulum mode is not.

7.41 A certain fifth-order system is found to have a characteristic equation with roots at 0, -1, -2, and $-1 \pm 1j$. A decomposition into controllable and uncontrollable parts discloses that the controllable part has a characteristic equation with roots 0, and $-1 \pm 1j$. A decomposition into observable and nonobservable parts discloses that the observable modes are at 0, -1, and -2.

(a) Where are the zeros of $b(s) = \mathbf{H}\text{adj}(s\mathbf{I} - \mathbf{F})\mathbf{G}$ for this system?

(b) What are the poles of the reduced-order transfer function that includes only controllable and observable modes?

7.42 Consider the systems shown in Fig. 7.94, employing series, parallel, and feedback configurations.

(a) Suppose we have controllable–observable realizations for each subsystem:

$$\dot{\mathbf{x}}_i = \mathbf{F}_i \mathbf{x}_i + \mathbf{G}_i u_i,$$

$$y_i = \mathbf{H}_i \mathbf{x}_i, \quad \text{where } i = 1, 2.$$

Give a set of state equations for the combined systems in Fig. 7.93.

(b) For each case, determine what condition(s) on the roots of the polynomials N_i and D_i is necessary for each system to be controllable and observable. Give a brief reason for your answer in terms of pole–zero cancellations.

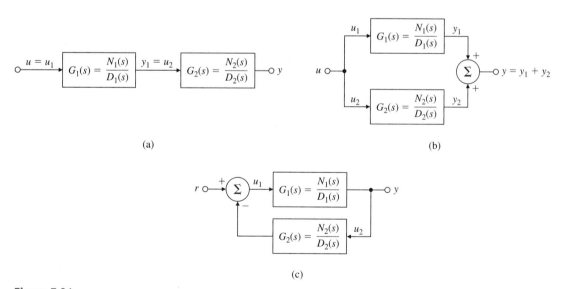

(a)

(b)

(c)

Figure 7.94

Block diagrams for Problem 7.14: (a) series; (b) parallel; (c) feedback

7.43 Consider the system $\ddot{y} + 3\dot{y} + 2y = \dot{u} + u$.

(a) Find the state matrices $\mathbf{F}_c$, $\mathbf{G}_c$, and $\mathbf{H}_c$ in control canonical form that correspond to the given differential equation.

(b) Sketch the eigenvectors of $\mathbf{F}_c$ in the (x_1, x_2) plane, and draw vectors that correspond to the completely observable $(\mathbf{x}_0)$ and the completely unobservable $(\mathbf{x}_{\bar{0}})$ state-variables.

(c) Express $\mathbf{x}_0$ and $\mathbf{x}_{\bar{0}}$ in terms of the observability matrix $\mathcal{O}$.

(d) Give the state matrices in observer canonical form and repeat parts (b) and (c) in terms of controllability instead of observability.

7.44 The equations of motion for a station-keeping satellite (such as a weather satellite) are

$$\ddot{x} - 2\omega\dot{y} - 3\omega^2 x = 0, \quad \ddot{y} + 2\omega\dot{x} = u,$$

where

$$x = \text{radial perturbation,}$$

$$y = \text{longitudinal position perturbation,}$$

$$u = \text{engine thrust in the } y\text{-direction,}$$

as depicted in Fig. 7.95. If the orbit is synchronous with the earth's rotation, then $\omega = 2\pi/(3600 \times 24)$ rad/sec.

(a) Is the state $\mathbf{x} = \begin{bmatrix} x & \dot{x} & y & \dot{y} \end{bmatrix}^T$ observable?

(b) Choose $\mathbf{x} = \begin{bmatrix} x & \dot{x} & y & \dot{y} \end{bmatrix}^T$ as the state vector and y as the measurement, and design a full-order observer with poles placed at $s = -2\omega$, -3ω, and $-3\omega \pm 3\omega j$.

Figure 7.95

Diagram of a station-keeping satellite in orbit for Problem 7.44

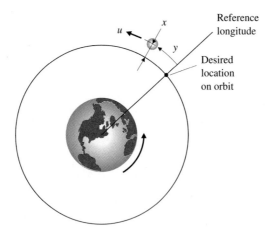

7.45 The linearized equations of motion of the simple pendulum in Fig. 7.96 are

$$\ddot{\theta} + \omega^2 \theta = u.$$

(a) Write the equations of motion in state-space form.

(b) Design an estimator (observer) that reconstructs the state of the pendulum given measurements of $\dot{\theta}$. Assume $\omega = 5$ rad/sec, and pick the estimator roots to be at $s = -10 \pm 10j$.

Figure 7.96

Pendulum diagram for
Problem 7.45

(c) Write the transfer function of the estimator between the measured value of $\dot\theta$ and the estimated value of θ.

(d) Design a controller (that is, determine the state feedback gain **K**) so that the roots of the closed-loop characteristic equation are at $s = -4 \pm 4j$.

7.46 An error analysis of an inertial navigator leads to the set of normalized state equations

$$
\begin{bmatrix} \dot x_1 \\ \dot x_2 \\ \dot x_3 \end{bmatrix} = \begin{bmatrix} 0 & -1 & 0 \\ 1 & 0 & 1 \\ 0 & 0 & 0 \end{bmatrix} \begin{bmatrix} x_1 \\ x_2 \\ x_3 \end{bmatrix} + \begin{bmatrix} 0 \\ 0 \\ 1 \end{bmatrix} u,
$$

where

$$x_1 = \text{east—velocity error,}$$

$$x_2 = \text{platform tilt about the north axis,}$$

$$x_3 = \text{north—gyro drift,}$$

$$u = \text{gyro drift rate of change.}$$

Design a reduced-order estimator with $y = x_1$ as the measurement, and place the observer-error poles at -0.1 and -0.1. Be sure to provide all the relevant estimator equations.

Problems for Section 7.8: Compensator Design: Combined Control Law and Estimator

7.47 A certain process has the transfer function $G(s) = \frac{4}{(s^2-4)}$.

(a) Find **F**, **G**, and **H** for this system in observer canonical form.

(b) If $u = -\mathbf{K}\mathbf{x}$, compute **K** so that the closed-loop control poles are located at $s = -2 \pm 2j$.

(c) Compute **L** so that the estimator error poles are located at $s = -10 \pm 10j$.

(d) Give the transfer function of the resulting controller [for example, using Eq. (7.177)].

(e) What are the gain and phase margins of the controller and the given open-loop system?

7.48 The linearized longitudinal motion of a helicopter near hover (see Fig. 7.97) can be modeled by the normalized third-order system

$$
\begin{bmatrix} \dot q \\ \dot\theta \\ \dot u \end{bmatrix} = \begin{bmatrix} -0.4 & 0 & -0.01 \\ 1 & 0 & 0 \\ -1.4 & 9.8 & -0.02 \end{bmatrix} \begin{bmatrix} q \\ \theta \\ u \end{bmatrix} + \begin{bmatrix} 6.3 \\ 0 \\ 9.8 \end{bmatrix} \delta,
$$

Figure 7.97

Helicopter for
Problem 7.48

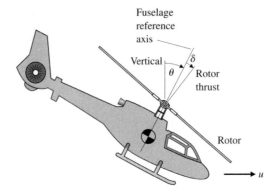

Fuselage
reference
axis

q = pitch rate,

θ = pitch angle of fuselage,

u = horizontal velocity (standard aircraft notation),

δ = rotor tilt angle (control variable).

Suppose our sensor measures the horizontal velocity u as the output; that is, $y = u$.

(a) Find the open-loop pole locations.

(b) Is the system controllable?

(c) Find the feedback gain that places the poles of the system at $s = -1 \pm 1j$ and $s = -2$.

(d) Design a full-order estimator for the system, and place the estimator poles at -8 and $-4 \pm 4\sqrt{3}j$.

(e) Design a reduced-order estimator with both poles at -4. What are the advantages and disadvantages of the reduced-order estimator compared with the full-order case?

(f) Compute the compensator transfer function using the control gain and the full-order estimator designed in part (d), and plot its frequency response using MATLAB. Draw a Bode plot for the closed-loop design, and indicate the corresponding gain and phase margins.

(g) Repeat part (f) with the reduced-order estimator.

(h) Draw the SRL and select roots for a control law that will give a control bandwidth matching the design of part (c), and select roots for a full-order estimator that will result in an estimator error bandwidth comparable to the design of part (d). Draw the corresponding Bode plot and compare the pole placement and SRL designs with respect to bandwidth, stability margins, step response, and control effort for a unit-step rotor-angle input. Use MATLAB for the computations.

7.49 Suppose a DC drive motor with motor current u is connected to the wheels of a cart in order to control the movement of an inverted pendulum mounted on the cart. The linearized and normalized equations of motion corresponding to this system can be put in the form

$$\ddot{\theta} = \theta + v + u,$$
$$\dot{v} = \theta - v - u,$$

where

$$\theta = \text{angle of the pendulum},$$
$$v = \quad \text{velocity of the cart}.$$

(a) We wish to control θ by feedback to u of the form

$$u = -K_1\theta - K_2\dot{\theta} - K_3 v.$$

Find the feedback gains so that the resulting closed-loop poles are located at -1, $-1 \pm j\sqrt{3}$.

(b) Assume that θ and v are measured. Construct an estimator for θ and $\dot{\theta}$ of the form

$$\dot{\hat{\mathbf{x}}} = \mathbf{F}\hat{\mathbf{x}} + \mathbf{L}(y - \hat{y}),$$

where $\mathbf{x} = [\theta \quad \dot{\theta}]^T$ and $y = \theta$. Treat both v and u as known. Select $\mathbf{L}$ so that the estimator poles are at -2 and -2.

(c) Give the transfer function of the controller, and draw the Bode plot of the closed-loop system, indicating the corresponding gain and phase margins.

(d) Using MATLAB, plot the response of the system to an initial condition on θ, and give a physical explanation for the initial motion of the cart.

7.50 Consider the control of

$$G(s) = \frac{Y(s)}{U(s)} = \frac{10}{s(s+1)}.$$

(a) Let $y = x_1$ and $\dot{x}_1 = x_2$, and write state equations for the system.

(b) Find K_1 and K_2 so that $u = -K_1 x_1 - K_2 x_2$ yields closed-loop poles with a natural frequency $\omega_n = 3$ and a damping ratio $\zeta = 0.5$.

(c) Design a state estimator for the system that yields estimator error poles with $\omega_{n1} = 15$ and $\zeta_1 = 0.5$.

(d) What is the transfer function of the controller obtained by combining parts (a) through (c)?

(e) Sketch the root locus of the resulting closed-loop system as plant gain (nominally 10) is varied.

7.51 Unstable equations of motion of the form

$$\ddot{x} = x + u$$

arise in situations where the motion of an upside-down pendulum (such as a rocket) must be controlled.

(a) Let $u = -Kx$ (position feedback alone), and sketch the root locus with respect to the scalar gain K.

(b) Consider a lead compensator of the form

$$U(s) = K\frac{s+a}{s+10}X(s).$$

Select a and K so that the system will display a rise time of about 2 sec and no more than 25% overshoot. Sketch the root locus with respect to K.

(c) Sketch the Bode plot (both magnitude and phase) of the uncompensated plant.

(d) Sketch the Bode plot of the compensated design, and estimate the phase margin.

(e) Design state feedback so that the closed-loop poles are at the same locations as those of the design in part (b).

(f) Design an estimator for x and $\dot{x}$ using the measurement of $x = y$, and select the observer gain **L** so that the equation for $\tilde{x}$ has characteristic roots with a damping ratio $\zeta = 0.5$ and a natural frequency $\omega_n = 8$.

(g) Draw a block diagram of your combined estimator and control law, and indicate where $\hat{x}$ and $\dot{x}$ appear. Draw a Bode plot for the closed-loop system, and compare the resulting bandwidth and stability margins with those obtained using the design of part (b).

7.52 A simplified model for the control of a flexible robotic arm is shown in Fig. 7.98, where

$$k/M = 900 \text{ rad/sec}^2,$$

$$y = \text{output, the mass position,}$$

$$u = \text{input, the position of the end of the spring.}$$

(a) Write the equations of motion in state-space form.

(b) Design an estimator with roots as $s = -100 \pm 100j$.

(c) Could both state-variables of the system be estimated if only a measurement of $\dot{y}$ was available?

(d) Design a full-state feedback controller with roots at $s = -20 \pm 20j$.

(e) Would it be reasonable to design a control law for the system with roots at $s = -200 \pm 200j$? State your reasons.

(f) Write equations for the compensator, including a command input for y. Draw a Bode plot for the closed-loop system and give the gain and phase margins for the design.

Figure 7.98

Simple robotic arm for Problem 7.52

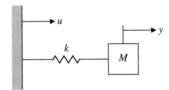

7.53 The linearized differential equations governing the fluid-flow dynamics for the two cascaded tanks in Fig. 7.99 are

$$\delta\dot{h}_1 + \sigma\delta h_1 = \delta u,$$

$$\delta\dot{h}_2 + \sigma\delta h_2 = \sigma\delta h_1,$$

where

$$\delta h_1 = \text{deviation of depth in tank 1 from the nominal level,}$$

$$\delta h_2 = \text{deviation of depth in tank 2 from the nominal level,}$$

$$\delta u = \text{deviation in fluid inflow rate to tank 1 (control).}$$

(a) *Level Controller for Two Cascaded Tanks:* Using state feedback of the form

$$\delta u = -K_1\delta h_1 - K_2\delta h_2,$$

Figure 7.99

Coupled tanks for Problem 7.53

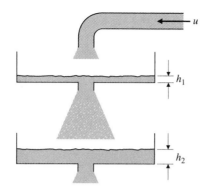

choose values of K_1 and K_2 that will place the closed-loop eigenvalues at

$$s = -2\sigma (1 \pm j).$$

(b) *Level Estimator for Two Cascaded Tanks:* Suppose that only the deviation in the level of tank 2 is measured (that is, $y = \delta h_2$). Using this measurement, design an estimator that will give continuous, smooth estimates of the deviation in levels of tank 1 and tank 2, with estimator error poles at $-8\sigma (1 \pm j)$.

(c) *Estimator/Controller for Two Cascaded Tanks:* Sketch a block diagram (showing individual integrators) of the closed-loop system obtained by combining the estimator of part (b) with the controller of part (a).

(d) Using MATLAB, compute and plot the response at y to an initial offset in δh_1. Assume $\sigma = 1$ for the plot.

7.54 The lateral motions of a ship that is 100 m long, moving at a constant velocity of 10 m/sec, are described by

$$
\begin{bmatrix} \dot{\beta} \\ \dot{r} \\ \dot{\psi} \end{bmatrix} =
\begin{bmatrix} -0.0895 & -0.286 & 0 \\ -0.0439 & -0.272 & 0 \\ 0 & 1 & 0 \end{bmatrix}
\begin{bmatrix} \beta \\ r \\ \psi \end{bmatrix} +
\begin{bmatrix} 0.0145 \\ -0.0122 \\ 0 \end{bmatrix} \delta,
$$

where

$$\beta = \text{sideslip angle (deg)},$$

$$\psi = \text{heading angle (deg)},$$

$$\delta = \text{rudder angle (deg)},$$

$$r = \text{yaw rate (see Fig. 7.99)}.$$

(a) Determine the transfer function from δ to ψ and the characteristic roots of the uncontrolled ship.

(b) Using complete state feedback of the form

$$\delta = -K_1\beta - K_2 r - K_3(\psi - \psi_d),$$

where ψ_d is the desired heading, determine values of K_1, K_2, and K_3 that will place the closed-loop roots at $s = -0.2, -0.2 \pm 0.2j$.

(c) Design a state estimator based on the measurement of ψ (obtained from a gyrocompass, for example). Place the roots of the estimator error equation at $s = -0.8$ and $-0.8 \pm 0.8j$.

(d) Give the state equations and transfer function for the compensator $D_c(s)$ in Fig. 7.100, and plot its frequency response.

(e) Draw the Bode plot for the closed-loop system, and compute the corresponding gain and phase margins.

(f) Compute the feed-forward gains for a reference input, and plot the step response of the system to a change in heading of $5°$.

Figure 7.100

View of ship from above for Problem 7.54

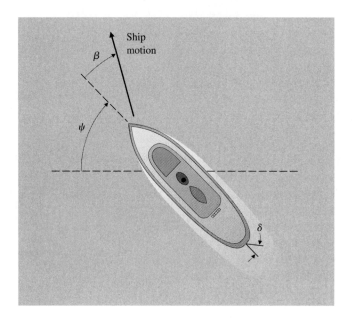

Figure 7.101

Ship control block diagram for Problem 7.54

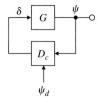

Problems for Section 7.9: Introduction of the Reference Input with the Estimator

△ **7.55** As mentioned in footnote 11 in Section 7.9.2, a reasonable approach for selecting the feed-forward gain in Eq. (7.205) is to choose $\bar{N}$ such that when r and y are both unchanging, the DC gain from r to u is the negative of the DC gain from y to u. Derive a formula for $\bar{N}$ based on this selection rule. Show that if the plant is Type 1, this choice is the same as that given by Eq. (7.205).

Problems for Section 7.10: Integral Control and Robust Tracking

7.56   Assume that the linearized and time-scaled equation of motion for the ball-bearing levitation device is $\ddot{x} - x = u + w$. Here w is a constant bias due to the power amplifier. Introduce integral error control, and select three control gains $\mathbf{K} = [\ K_1 \quad K_2 \quad K_3\ ]$ so that the closed-loop poles are at -1 and $-1 \pm j$ and the steady-state error to w and to a (step) position command will be zero. Let $y = x$ and the reference input $r \overset{\Delta}{=} y_{\text{ref}}$ be a constant. Draw a block diagram of your design showing the locations of the feedback gains K_i. Assume that both $\dot{x}$ and x can be measured. Plot the response of the closed-loop system to a step command input and the response to a step change in the bias input. Verify that the system is Type 1. Use MATLAB (SIMULINK) software to simulate the system responses.

7.57   Consider a system with state matrices

$$\mathbf{F} = \begin{bmatrix} -2 & 1 \\ 0 & -3 \end{bmatrix}, \qquad \mathbf{G} = \begin{bmatrix} 1 \\ 1 \end{bmatrix}, \qquad \mathbf{H} = [\ 1 \quad 3\ ].$$

   (a) Use feedback of the form $u(t) = -\mathbf{Kx}(t) + \bar{N}r(t)$, where $\bar{N}$ is a nonzero scalar, to move the poles to $-3 \pm 3j$.

   (b) Choose $\bar{N}$ so that if r is a constant, the system has zero steady-state error; that is $y(\infty) = r$.

   (c) Show that if $\mathbf{F}$ changes to $\mathbf{F} + \delta\mathbf{F}$, where $\delta\mathbf{F}$ is an arbitrary 2×2 matrix, then your choice of $\bar{N}$ in part (b) will no longer make $y(\infty) = r$. Therefore, the system is not robust under changes to the system parameters in $\mathbf{F}$.

   (d) The system steady-state error performance can be made robust by augmenting the system with an integrator and using unity feedback—that is, by setting $\dot{x}_I = r - y$, where x_I is the state of the integrator. To see this, first use state feedback of the form $u = -\mathbf{Kx} - K_1 x_I$ so that the poles of the augmented system are at $-3, -2 \pm j\sqrt{3}$.

   (e) Show that the resulting system will yield $y(\infty) = r$ no matter how the matrices $\mathbf{F}$ and $\mathbf{G}$ are changed, as long as the closed-loop system remains stable.

   (f) For part (d), use MATLAB (SIMULINK) software to plot the time response of the system to a constant input. Draw Bode plots of the controller, as well as the sensitivity function ($\mathcal{S}$) and the complementary sensitivity function ($\mathcal{T}$).

△   **7.58**   Consider a servomechanism for following the data track on a computer-disk memory system. Because of various unavoidable mechanical imperfections, the data track is not exactly a centered circle, and thus the radial servo must follow a sinusoidal input of radian frequency ω_0 (the spin rate of the disk). The state matrices for a linearized model of such a system are

$$\mathbf{F} = \begin{bmatrix} 0 & 1 \\ 0 & -1 \end{bmatrix}, \qquad \mathbf{G} = \begin{bmatrix} 0 \\ 1 \end{bmatrix}, \qquad \mathbf{H} = [\ 1 \quad 3\ ].$$

The sinusoidal reference input satisfies $\ddot{r} = -\omega_0^2 r$.

   (a) Let $\omega_0 = 1$, and place the poles of the error system for an internal model design at

$$\alpha_c(s) = (s + 2 \pm j2)(s + 1 \pm 1j)$$

and the pole of the reduced-order estimator at

$$\alpha_e(s) = (s + 6).$$

(b) Draw a block diagram of the system, and clearly show the presence of the oscillator with frequency ω_0 (the internal model) in the controller. Also verify the presence of the blocking zeros at $\pm j\omega_0$.

(c) Use MATLAB (SIMULINK) software to plot the time response of the system to a sinusoidal input at frequency $\omega_0 = 1$.

(d) Draw a Bode plot to show how this system will respond to sinusoidal inputs at frequencies different from but near ω_0.

△   **7.59**   Compute the controller transfer function [from $Y(s)$ to $U(s)$] in Example 7.38. What is the prominent feature of the controller that allows tracking and disturbance rejection?

△   **7.60**   Consider the pendulum problem with control torque T_c and disturbance torque T_d:

$$\ddot{\theta} + 4\theta = T_c + T_d.$$

(Here $g/l = 4$.) Assume that there is a potentiometer at the pin that measures the output angle θ, but with a constant unknown bias b. Thus the measurement equation is $y = \theta + b$.

(a) Take the "augmented" state vector to be

$$\begin{bmatrix} \theta \\ \dot{\theta} \\ w \end{bmatrix},$$

where w is the input-equivalent bias. Write the system equations in state-space form. Give values for the matrices $\mathbf{F}$, $\mathbf{G}$, and $\mathbf{H}$.

(b) Using state-variable methods, show that the characteristic equation of the model is $s(s^2 + 4) = 0$.

(c) Show that w is observable if we assume that $y = \theta$, and write the estimator equations for

$$\begin{bmatrix} \hat{\theta} \\ \dot{\hat{\theta}} \\ \hat{w} \end{bmatrix}.$$

Pick estimator gains $[\ l_1 \quad l_2 \quad l_3\ ]^T$ to place all the roots of the estimator error characteristic equation at -10.

(d) Using full-state feedback of the estimated (controllable) state-variables, derive a control law to place the closed-loop poles at $-2 \pm 2j$.

(e) Draw a block diagram of the complete closed-loop system (estimator, plant, and controller) using integrator blocks.

(f) Introduce the estimated bias into the control so as to yield zero steady-state error to the output bias b. Demonstrate the performance of your design by plotting the response of the system to a step change in b; that is, b changes from 0 to some constant value.

Problems for Section 7.13: Design for Systems with Pure Time Delay

△   **7.61**   Consider the system with the transfer function $e^{-Ts}G(s)$, where

$$G(s) = \frac{1}{s(s + 1)(s + 2)}.$$

The Smith compensator for this system is given by

$$D'_c(s) = \frac{D_c}{1 + (1 - e^{-sT})G(s)D_c}.$$

Plot the frequency response of the compensator for $T = 5$ and $D_c = 1$, and draw a Bode plot that shows the gain and phase margins of the system.[22]

---

[22]This problem was given by Åström (1977).

8

Digital Control

A Perspective on Digital Control

Most of the controllers we have studied so far were described by the Laplace transform or differential equations, which, strictly speaking, are assumed to be built using analog electronics, such as those in Figs. 5.31 and 5.35. However, as discussed in Section 4.4, most control systems today use digital computers (usually microprocessors) to implement the controllers. The intent of this chapter is to expand on the design of control systems that will be implemented in a digital computer. The implementation leads to an average delay of half the sample period and to a phenomenon called aliasing, which need to be addressed in the controller design.

Analog electronics can integrate and differentiate signals. In order for a digital computer to accomplish these tasks, the differential equations describing compensation must be approximated by reducing them to algebraic equations involving addition, division, and multiplication, as developed in Section 4.4. This chapter expands on various ways to make these approximations. The resulting design can then be tuned up, if needed, using direct digital analysis and design.

You should be able to design, analyze, and implement a digital control system from the material in this chapter. However, our treatment here is a limited version of a complex subject covered in more detail in *Digital Control of Dynamic Systems* by Franklin et al. (1998 3rd ed.).

Chapter Overview

In Section 8.1 we describe the basic structure of digital control systems and introduce the issues that arise due to the sampling. The digital implementation described in Section 4.4 is sufficient for implementing a feedback control law in a digital control system, which you can then evaluate via SIMULINK® to determine the degradation with respect to the continuous case. However, to fully understand the effect of sampling, it is useful to learn about discrete linear analysis tools. This requires an understanding of the z-transform, which we discuss in Section 8.2. Section 8.3 builds on this understanding to provide a better foundation for design using discrete equivalents that was briefly discussed in Section 4.4. Hardware characteristics and sample rate issues are discussed in Sections 8.4 and 8.5, both of which need to be addressed in order to implement a digital controller.

In contrast to discrete equivalent design, which is an approximate method, optional Section 8.6 explores direct digital design (also called discrete design), which provides an exact method that is independent of whether the sample rate is fast or not.

8.1   Digitization

Figure 8.1(a) shows the topology of the typical continuous system that we have been considering in previous chapters. The computation of the error signal e and the dynamic compensation $D(s)$ can all be accomplished in a digital computer as shown in Fig. 8.1(b). The fundamental differences between the two implementations are that the digital system operates on **samples** of the sensed plant output rather than on the continuous signal and that the control provided by $D(s)$ must be generated by algebraic recursive equations.

We consider first the action of the **analog-to-digital (A/D) converter** on a signal. This device samples a physical variable, most commonly an electrical voltage, and converts it into a binary number that usually consists of 10 to 16 bits. Conversion from the analog signal $y(t)$ to the samples, $y(kT)$, occurs repeatedly at instants of time T seconds apart. T is the **sample period**, and $1/T$ is the **sample rate** in Hertz. The sampled signal is $y(kT)$, where k can take on any integer value. It is often written simply as $y(k)$. We call this type of variable a **discrete signal** to distinguish it from a continuous signal such as $y(t)$, which changes continuously in time. A system having both discrete and continuous signals is called a **sampled data system**.

We make the assumption that the sample period is fixed. In practice, digital control systems sometimes have varying sample periods and/or different periods in different feedback paths. Usually, the computer logic includes a **clock** that supplies a pulse, or **interrupt**, every T seconds, and the A/D converter sends a number to the computer each time the interrupt arrives. An alternative implementation, often referred to as **free running**, is to access the A/D converter after each cycle of code execution has been completed. In the former case the sample period is precisely fixed; in the latter

Sample period

Figure 8.1

Block diagrams for a
basic control system:
(a) continuous system;
(b) with a digital
computer

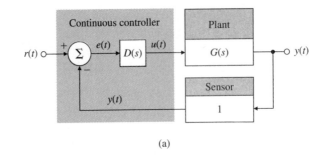

(a)

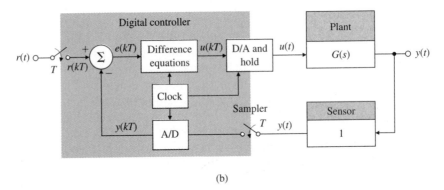

(b)

case the sample period is fixed essentially by the length of the code, provided that no logic branches are present, which could vary the amount of code executed.

There also may be a sampler and an A/D converter for the input command $r(t)$, which produces the discrete $r(kT)$, from which the sensed output $y(kT)$ will be subtracted to arrive at the discrete error signal $e(kT)$. As we saw in Sections 4.4 and 5.4.4, and Example 6.15, the continuous compensation is approximated by difference equations, which are the discrete version of differential equations and can be made to duplicate the dynamic behavior of $D(s)$ if the sample period is short enough. The result of the difference equations is a discrete $u(kT)$ at each sample instant. This signal is converted to a continuous $u(t)$ by the **digital-to-analog (D/A) converter** and the hold: the D/A converter changes the binary number to an analog voltage,

Zero-order hold (ZOH)

and a **zero-order hold** maintains that same voltage throughout the sample period. The resulting $u(t)$ is then applied to the actuator in precisely the same manner as the continuous implementation. There are two basic techniques for finding the difference equations for the digital controller. One technique, called **discrete equivalent**,

Discrete equivalents

consists of designing a continuous compensation $D(s)$ using methods described in the previous chapters, then approximating that $D(s)$ using the method of Section 4.4 (Tustin's Method), or one of the other methods described in Section 8.3. The other technique is **discrete design**, described in Section 8.6. Here the difference equations are found directly without designing $D(s)$ first.

The sample rate required depends on the closed-loop bandwidth of the system. Generally, sample rates should be about 20 times the bandwidth or faster in order

Sample rate selection

to assure that the digital controller will match the performance of the continuous

Figure 8.2

The delay due to the hold operation

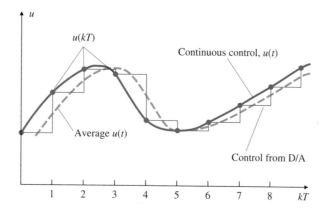

controller. Slower sample rates can be used if some adjustments are made in the digital controller or some performance degradation is acceptable. Use of the discrete design method described in Section 8.6 allows for a much slower sample rate if that is desirable to minimize hardware costs; however, best performance of a digital controller is obtained when the sample rate is greater than 25 times the bandwidth.

It is worth noting that the single most important impact of implementing a control system digitally is the delay associated with the hold. Because each value of $u(kT)$ in Fig. 8.1(b) is held constant until the next value is available from the computer, the continuous value of $u(t)$ consists of steps (see Fig. 8.2) that, on average, are delayed from $u(kT)$ by $T/2$ as shown in the figure. If we simply incorporate this $T/2$ delay into a continuous analysis of the system, an excellent prediction of the effects of sampling results for sample rates much slower than 20 times bandwidth. We will discuss this further in Section 8.3.3.

8.2 Dynamic Analysis of Discrete Systems

The z-transform is the mathematical tool for the analysis of linear discrete systems. It plays the same role for discrete systems that the Laplace transform does for continuous systems. This section will give a short description of the z-transform, describe its use in analyzing discrete systems, and show how it relates to the Laplace transform.

8.2.1 z-Transform

In the analysis of continuous systems, we use the Laplace transform, which is defined by

$$\mathcal{L}\{f(t)\} = F(s) = \int_0^\infty f(t)e^{-st}\, dt,$$

which leads directly to the important property that (with zero initial conditions)

$$\mathcal{L}\{\dot{f}(t)\} = sF(s). \tag{8.1}$$

Relation (8.1) enables us easily to find the transfer function of a linear continuous system, given the differential equation of that system.

Figure 8.3

A continuous, sampled version of signal f

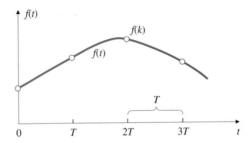

For discrete systems a similar procedure is available. The z-**transform** is defined by

$$\mathcal{Z}\{f(k)\} = F(z) = \sum_{k=0}^{\infty} f(k)z^{-k},\qquad(8.2)$$

where $f(k)$ is the sampled version of $f(t)$, as shown in Fig. 8.3, and $k = 0, 1, 2, 3, \dots$ refers to discrete sample times $t_0, t_1, t_2, t_3, \dots$. This leads directly to a property analogous to Eq. (8.1), specifically, that

$$\mathcal{Z}\{f(k-1)\} = z^{-1}F(z).\qquad(8.3)$$

This relation allows us to easily find the transfer function of a discrete system, given the difference equations of that system. For example, the general second-order difference equation

$$y(k) = -a_1 y(k-1) - a_2 y(k-2) + b_0 u(k) + b_1 u(k-1) + b_2 u(k-2)$$

can be converted from this form to the z-transform of the variables $y(k), u(k), \dots$ by invoking Eq. (8.3) once or twice to arrive at

$$Y(z) = (-a_1 z^{-1} - a_2 z^{-2})Y(z) + (b_0 + b_1 z^{-1} + b_2 z^{-2})U(z).\qquad(8.4)$$

Equation (8.4) then results in the discrete transfer function

$$\frac{Y(z)}{U(z)} = \frac{b_0 + b_1 z^{-1} + b_2 z^{-2}}{1 + a_1 z^{-1} + a_2 z^{-2}}.$$

8.2.2 z-Transform Inversion

Table 8.1 relates simple discrete-time functions to their z-transforms and gives the Laplace transforms for the same time functions.

Given a general z-transform, we could expand it into a sum of elementary terms using partial-fraction expansion (see Appendix A) and find the resulting time series from the table. These procedures are exactly the same as those used for continuous systems; as with the continuous case, most designers would use a numerical evaluation of the discrete equations to obtain a time history rather than inverting the z-transform.

A z-transform inversion technique that has no continuous counterpart is called **long division**. Given the z-transform

$$Y(z) = \frac{N(z)}{D(z)},\qquad(8.5)$$

TABLE 8.1

Laplace Transforms and z-Transforms of Simple Discrete-Time Functions

No.	$F(s)$	$f(kT)$	$F(z)$
1		$1, k = 0; 0, k \neq 0$	1
2		$1, k = k_o; 0, k \neq k_o$	z^{-k_o}
3	$\frac{1}{s}$	$1(kT)$	$\frac{z}{z-1}$
4	$\frac{1}{s^2}$	kT	$\frac{Tz}{(z-1)^2}$
5	$\frac{1}{s^3}$	$\frac{1}{2!}(kT)^2$	$\frac{T^2}{2}\left[\frac{z(z+1)}{(z-1)^3}\right]$
6	$\frac{1}{s^4}$	$\frac{1}{3!}(kT)^3$	$\frac{T^3}{6}\left[\frac{z(z^2+4z+1)}{(z-1)^4}\right]$
7	$\frac{1}{s^m}$	$\lim_{a\to 0}\frac{(-1)^{m-1}}{(m-1)!}\left(\frac{\partial^{m-1}}{\partial a^{m-1}}e^{-akT}\right)$	$\lim_{a\to 0}\frac{(-1)^{m-1}}{(m-1)!}\left(\frac{\partial^{m-1}}{\partial a^{m-1}}\frac{z}{z-e^{-aT}}\right)$
8	$\frac{1}{s+a}$	e^{-akT}	$\frac{z}{z-e^{-aT}}$
9	$\frac{1}{(s+a)^2}$	kTe^{-akT}	$\frac{Tze^{-aT}}{(z-e^{-aT})^2}$
10	$\frac{1}{(s+a)^3}$	$\frac{1}{2}(kT)^2 e^{-akT}$	$\frac{T^2}{2}e^{-aT}z\frac{(z+e^{-aT})}{(z-e^{-aT})^3}$
11	$\frac{1}{(s+a)^m}$	$\frac{(-1)^{m-1}}{(m-1)!}\left(\frac{\partial^{m-1}}{\partial a^{m-1}}e^{-akT}\right)$	$\frac{(-1)^{m-1}}{(m-1)!}\left(\frac{\partial^{m-1}}{\partial a^{m-1}}\frac{z}{z-e^{-aT}}\right)$
12	$\frac{a}{s(s+a)}$	$1 - e^{-akT}$	$\frac{z(1-e^{-aT})}{(z-1)(z-e^{-aT})}$
13	$\frac{a}{s^2(s+a)}$	$\frac{1}{a}(akT - 1 + e^{-akT})$	$\frac{z[(aT-1+e^{-aT})z+(1-e^{-aT}-aTe^{-aT})]}{a(z-1)^2(z-e^{-aT})}$
14	$\frac{b-a}{(s+a)(s+b)}$	$e^{-akT} - e^{-bkT}$	$\frac{(e^{-aT}-e^{-bT})z}{(z-e^{-aT})(z-e^{-bT})}$
15	$\frac{s}{(s+a)^2}$	$(1 - akT)e^{-akT}$	$\frac{z[z-e^{-aT}(1+aT)]}{(z-e^{-aT})^2}$
16	$\frac{a^2}{s(s+a)^2}$	$1 - e^{-akT}(1 + akT)$	$\frac{z[z(1-e^{-aT}-aTe^{-aT})+e^{-2aT}-e^{-aT}+aTe^{-aT}]}{(z-1)(z-e^{-aT})^2}$
17	$\frac{(b-a)s}{(s+a)(s+b)}$	$be^{-bkT} - ae^{-akT}$	$\frac{z[z(b-a)-(be^{-aT}-ae^{-bT})]}{(z-e^{-aT})(z-e^{-bT})}$
18	$\frac{a}{s^2+a^2}$	$\sin akT$	$\frac{z\sin aT}{z^2-(2\cos aT)z+1}$
19	$\frac{s}{s^2+a^2}$	$\cos akT$	$\frac{z(z-\cos aT)}{z^2-(2\cos aT)z+1}$
20	$\frac{s+a}{(s+a)^2+b^2}$	$e^{-akT}\cos bkT$	$\frac{z(z-e^{-aT}\cos bT)}{z^2-2e^{-aT}(\cos bT)z+e^{-2aT}}$
21	$\frac{b}{(s+a)^2+b^2}$	$e^{-akT}\sin bkT$	$\frac{ze^{-aT}\sin bT}{z^2-2e^{-aT}(\cos bT)z+e^{-2aT}}$
22	$\frac{a^2+b^2}{s[(s+a)^2+b^2]}$	$1 - e^{-akT}(\cos bkT + \frac{a}{b}\sin bkT)$	$\frac{z(Az+B)}{(z-1)[z^2-2e^{-aT}(\cos bT)z+e^{-2aT}]}$

$$A = 1 - e^{-aT}\cos bT - \frac{a}{b}e^{-aT}\sin bT$$

$$B = e^{-2aT} + \frac{a}{b}e^{-aT}\sin bT - e^{-aT}\cos bT$$

$F(s)$ is the Laplace transform of $f(t)$, and $F(z)$ is the z-transform of $f(kT)$.
Note: $f(t) = 0$ for $t = 0$.

we simply divide the denominator into the numerator using long division. The result is a series (perhaps with an infinite number of terms) in z^{-1}, from which the time series can be found by using Eq. (8.2).

For example, a first-order system described by the difference equation

$$y(k) = \alpha y(k-1) + u(k)$$

yields the discrete transfer function

$$\frac{Y(z)}{U(z)} = \frac{1}{1 - \alpha z^{-1}}.$$

For a unit-pulse input defined by

$$u(0) = 1,$$

$$u(k) = 0 \quad k \neq 0,$$

the z-transform is then

$$U(z) = 1, \tag{8.6}$$

so

$$Y(z) = \frac{1}{1 - \alpha z^{-1}}. \tag{8.7}$$

Therefore, to find the time series, we divide the numerator of Eq. (8.7) by its denominator using long division:

$$
\require{enclose}
\begin{array}{r}
1 + \alpha z^{-1} + \alpha^2 z^{-2} + \alpha^3 z^{-3} + \cdots \\[4pt]
1 - \alpha z^{-1} \enclose{longdiv}{1 } \\[2pt]
\underline{1 - \alpha z^{-1}} \\[2pt]
\alpha z^{-1} + 0 \\[2pt]
\underline{\alpha z^{-1} - \alpha^2 z^{-2}} \\[2pt]
\alpha^2 z^{-2} + 0 \\[2pt]
\underline{\alpha^2 z^{-2} - \alpha 3 z^{-3}} \\[2pt]
\alpha^3 z^{-3} \\[2pt]
\ddots
\end{array}
$$

This yields the infinite series

$$Y(z) = 1 + \alpha z^{-1} + \alpha^2 z^{-2} + \alpha^3 z^{-3} + \cdots . \tag{8.8}$$

From Eqs. (8.8) and (8.2) we see that the sampled time history of y is

$$y(0) = 1,$$

$$y(1) = \alpha,$$

$$y(2) = \alpha^2,$$

$$\vdots \qquad \vdots$$

$$y(k) = \alpha^k.$$

8.2.3   Relationship between s and z

For continuous systems, we saw in Chapter 3 that certain behaviors result from different pole locations in the s-plane: oscillatory behavior for poles near the imaginary axis, exponential decay for poles on the negative real axis, and unstable behavior for poles with a positive real part. A similar kind of association would also be useful to know when designing discrete systems. Consider the continuous signal

$$f(t) = e^{-at}, \quad t > 0,$$

which has the Laplace transform

$$F(s) = \frac{1}{s + a}$$

and corresponds to a pole at $s = -a$. The z-transform of $f(kT)$ is

$$F(z) = \mathcal{Z}\{e^{-akT}\}. \tag{8.9}$$

From Table 8.1 we can see that Eq. (8.9) is equivalent to

$$F(z) = \frac{z}{z - e^{-aT}},$$

which corresponds to a pole at $z = e^{-aT}$. This means that a pole at $s = -a$ in the s-plane corresponds to a pole at $z = e^{-aT}$ in the discrete domain. This is true in general:

Relationship between z-plane and s-plane characteristics

> The equivalent characteristics in the z-plane are related to those in the s-plane by the expression
> $$z = e^{sT},\tag{8.10}$$
> where T is the sample period.

Table 8.1 also includes the Laplace transforms, which demonstrates the $z = e^{sT}$ relationship for the roots of the denominators of the table entries for $F(s)$ and $F(z)$.

Figure 8.4 shows the mapping of lines of constant damping ζ and natural frequency ω_n from the s-plane to the upper half of the z-plane, using Eq. (8.10). The mapping has several important features (see Problem 8.4):

1. The stability boundary is the unit circle $|z| = 1$.
2. The small vicinity around $z = +1$ in the z-plane is essentially identical to the vicinity around $s = 0$ in the s-plane.
3. The z-plane locations give response information normalized to the sample rate, rather than to time as in the s-plane.
4. The negative real z-axis always represents a frequency of $\omega_s/2$, where $\omega_s = 2\pi/T =$ sample rate in radians per second.
5. Vertical lines in the left half of the s-plane (the constant real part or time constant) map into circles within the unit circle of the z-plane.
6. Horizontal lines in the s-plane (the constant imaginary part of the frequency) map into radial lines in the z-plane.

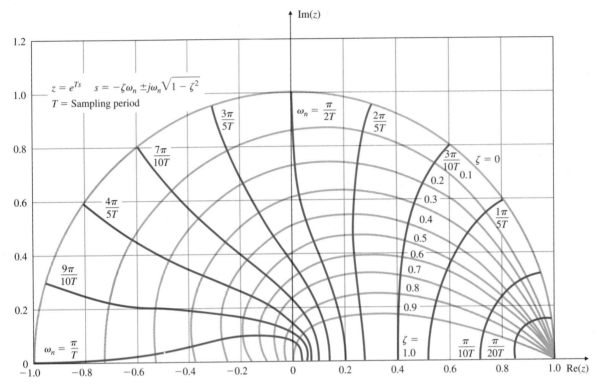

Figure 8.4

Natural frequency (solid color) and damping loci (light color) in the z-plane; the portion below the Re(z)-axis (not shown) is the mirror image of the upper half shown

Nyquist frequency $= \omega_s/2$

7. Frequencies greater than $\omega_s/2$, called the **Nyquist frequency**, appear in the z-plane on top of corresponding lower frequencies because of the circular character of the trigonometric functions imbedded in Eq. (8.10). This overlap is called **aliasing** or **folding**. As a result it is necessary to sample at least twice as fast as a signal's highest frequency component in order to represent that signal with the samples. (We will discuss aliasing in greater detail in Section 8.4.3.)

   To provide insight into the correspondence between z-plane locations and the resulting time sequence, Fig. 8.5 sketches time responses that would result from poles at the indicated locations. This figure is the discrete companion of Fig. 3.15.

8.2.4 Final Value Theorem

The Final Value Theorem for continuous systems, which we discussed in Section 3.1.6, states that

$$\lim_{t \to \infty} x(t) = x_{ss} = \lim_{s \to 0} sX(s), \tag{8.11}$$

as long as all the poles of $sX(s)$ are in the left half-plane (LHP). It is often used to find steady-state system errors and/or steady-state gains of portions of a control

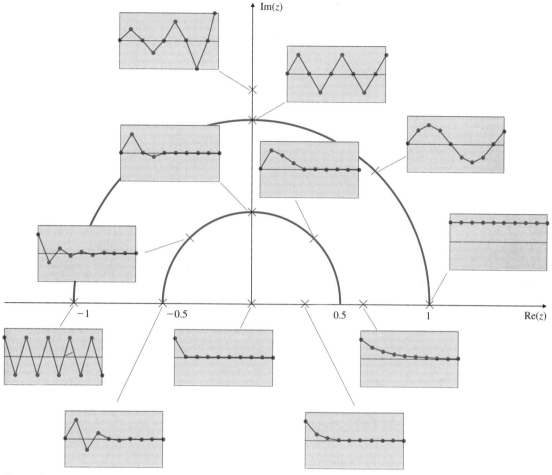

Figure 8.5

Time sequences associated with points in the z-plane

system. We can obtain a similar relationship for discrete systems by noting that a constant continuous steady-state response is denoted by $X(s) = {}^A/_s$ and leads to the multiplication by s in Eq. (8.11). Therefore, because the constant steady-state response for discrete systems is

$$X(z) = \frac{A}{1 - z^{-1}},$$

Final Value Theorem for discrete systems

the discrete Final Value Theorem is

$$\lim_{k \to \infty} x(k) = x_{ss} = \lim_{z \to 1}(1 - z^{-1})X(z) \tag{8.12}$$

if all the poles of $(1 - z^{-1})X(z)$ are inside the unit circle.

For example, to find the DC gain of the transfer function

$$G(z) = \frac{X(z)}{U(z)} = \frac{0.58(1 + z)}{z + 0.16},$$

we let $u(k) = 1$ for $k \geq 0$, so that

$$U(z) = \frac{1}{1 - z^{-1}}$$

and

$$X(z) = \frac{0.58(1 + z)}{(1 - z^{-1})(z + 0.16)}.$$

Applying the Final Value Theorem yields

$$x_{ss} = \lim_{z \to 1} \left[\frac{0.58(1 + z)}{z + 0.16} \right] = 1,$$

DC gain

so the DC gain of $G(z)$ is unity. To find the DC gain of any stable transfer function, we simply substitute $z = 1$ and compute the resulting gain. Because the DC gain of a system should not change whether represented continuously or discretely, this calculation is an excellent aid to check that an equivalent discrete controller matches a continuous controller. It is also a good check on the calculations associated with determining the discrete model of a system.

8.3 Design Using Discrete Equivalents

Stages in design using discrete equivalents

Design by discrete equivalent, sometimes called **emulation**, is partially described in Section 4.4 and proceeds through the following stages:

1. Design a continuous compensation as described in Chapters 1 through 7.
2. Digitize the continuous compensation.
3. Use discrete analysis, simulation, or experimentation to verify the design.

In Section 4.4 we discussed Tustin's method for performing the digitization. Armed with an understanding of the z-transform from Section 8.2, we now develop more digitization procedures and analyze the performance of the digitally controlled system.

Assume that we are given a continuous compensation $D(s)$ as shown in Fig. 8.1(a). We wish to find a set of difference equations or $D(z)$ for the digital implementation of that compensation in Fig. 8.1(b). First we rephrase the problem as one of finding the best $D(z)$ in the digital implementation shown in Fig. 8.6(a) to match the continuous system represented by $D(s)$ in Fig. 8.6(b). In this section we examine and compare three methods for solving this problem.

It is important to remember, as stated earlier, that these methods are approximations; there is no exact solution for all possible inputs because $D(s)$ responds to the complete time history of $e(t)$, whereas $D(z)$ has access to only the samples $e(kT)$. In a sense, the various digitization techniques simply make different assumptions about what happens to $e(t)$ between the sample points.

Figure 8.6

Comparison of
(a) digital and;
(b) continuous
implementation

(a)

(b)

Tustin's Method

As discussed in Section 4.4, one digitization technique is to approach the problem as one of numerical integration. Suppose

$$\frac{U(s)}{E(s)} = D(s) = \frac{1}{s},$$

which is integration. Therefore,

$$u(kT) = \int_0^{kT-T} e(t)\,dt + \int_{kT-T}^{kT} e(t)\,dt, \tag{8.13}$$

which can be rewritten as

$$u(kT) = u(kT - T) + \text{area under } e(t) \text{ over last } T, \tag{8.14}$$

where T is the sample period.

For Tustin's method, the task at each step is to use trapezoidal integration, that is, to approximate $e(t)$ by a straight line between the two samples (Fig. 8.7). Writing $u(kT)$ as $u(k)$ and $u(kT - T)$ as $u(k - 1)$ for short, we convert Eq. (8.14) to

$$u(k) = u(k - 1) + \frac{T}{2}[e(k - 1) + e(k)], \tag{8.15}$$

or, taking the z-transform,

$$\frac{U(z)}{E(z)} = \frac{T}{2}\left(\frac{1 + z^{-1}}{1 - z^{-1}}\right) = \frac{1}{\frac{2}{T}\left(\frac{1 - z^{-1}}{1 + z^{-1}}\right)}. \tag{8.16}$$

For $D(s) = a/(s + a)$, applying the same integration approximation yields

$$D(z) = \frac{a}{\frac{2}{T}\left(\frac{1 - z^{-1}}{1 + z^{-1}}\right) + a}.$$

In fact, substituting

$$s = \frac{2}{T}\left(\frac{1 - z^{-1}}{1 + z^{-1}}\right)$$

Tustin's method or bilinear approximation

for every occurrence of s in any $D(s)$ yields a $D(z)$ based on the trapezoidal integration formula. This is called **Tustin's method** or the **bilinear approximation**. Finding Tustin's approximation by hand for even a simple transfer function requires fairly extensive algebraic manipulations. The c2d function of MATLAB® expedites the process, as shown in the next example.

Figure 8.7
Trapezoidal integration

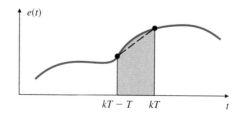

EXAMPLE 8.1   *Digital Controller for Example 6.15 Using Tustin's Approximation*

Determine the difference equations to implement the compensation from Example 6.15,

$$D(s) = 10\frac{s/2 + 1}{s/10 + 1},$$

at a sample rate of 25 times bandwidth using Tustin's approximation. Compare the performance against the continuous system and the discrete implementation done in Example 6.15 at a slower sample rate.

Solution. The bandwidth (ω_{BW}) for Example 6.15 is approximately 10 rad/sec, as can be deduced by observing that the crossover frequency (ω_c) is approximately 5 rad/sec and noting the relationship between ω_c and ω_{BW} in Fig. 6.51. Therefore, the sample frequency should be

$$\omega_s = 25 \times \omega_{BW} = (25)(10) = 250 \text{ rad/sec.}$$

Normally, when a frequency is indicated with the units of cycles per second, or Hz, it is given the symbol f, so with this convention, we have

$$f_s = \omega_s/(2\pi) \simeq 40 \text{ Hz,} \tag{8.17}$$

and the sample period is then

$$T = 1/f_s = 1/40 = 0.025 \text{ sec.}$$

The discrete compensation is computed by the MATLAB statement

```
sysDs = tf(10*[0.5 1],[0.1 1]);
sysDd = c2d(sysDs,0.025,'tustin');
```

which produces

$$D(z) = 10\frac{4.556 - 4.333\,z^{-1}}{1 - 0.7778\,z^{-1}}. \tag{8.18}$$

We can then write the difference equation by inspecting Eq. (8.18) to get

$$u(k) = 0.7778u(k-1) + 45.56e(k) - 43.33e(k-1),$$

or, indexing all time variables by 1, the equivalent is

$$u(k+1) = 0.7778u(k) + 45.56[e(k+1) - 0.9510e(k)]. \tag{8.19}$$

Equation (8.19) computes the new value of the control, $u(k+1)$, given the past value of the control, $u(k)$, and the new and past values of the error signal, $e(k+1)$ and $e(k)$.

In principle, the difference equation is evaluated initially with $k = 0$, then $k = 1, 2, 3, \ldots$ However, there is usually no requirement that values for all times be saved in memory. Therefore, the computer need only have variables defined for the current and past values. The instructions to the computer to implement the feedback loop in Fig. 8.1(b) with the difference equation from Eq. (8.19) would call for a continual looping through the following code:

READ y, r
$\quad e = r - y$
$\quad u = 0.7778u_p + 45.56\left[e - 0.9510e_p\right]$

Figure 8.8

Comparison between the digital and the continuous controller step response with a sample rate 25 times bandwidth: (a) position; (b) control

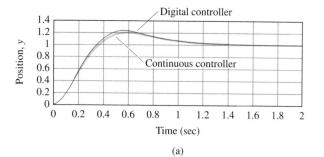

(a)

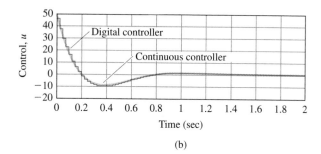

(b)

OUTPUT u

$u_p = u$ (where u_p will be the past value for the next loop through)

$e_p = e$

go back to READ when T sec have elapsed since last READ

Use of SIMULINK to compare the two implementations, in a manner similar to that used for Example 6.15, yields the step responses shown in Fig. 8.8. Note that sampling at 25 times bandwidth causes the digital implementation to match the continuous one quite well. Also note that the same case with half the sampling rate whose step response is shown in Fig. 6.59 contains a noticeable degradation in the overshoot (and damping) compared to the continuous case. Generally speaking, if you want to match a continuous system with a digital approximation of the continuous compensation, it is wise to sample at approximately 25 times bandwidth or faster.

8.3.1 Matched Pole–Zero (MPZ) Method

Another digitization method, called the **matched pole–zero** method, is found by extrapolating from the relationship between the s- and z-planes stated in Eq. (8.10). If we take the z-transform of a sampled function $x(k)$, then the poles of $X(z)$ are related to the poles of $X(s)$ according to the relation $z = e^{sT}$. The MPZ technique applies the relation $z = e^{sT}$ to the poles and zeros of a transfer function, even though, strictly speaking, this relation applies neither to transfer functions nor even to the zeros of a time sequence. Like all transfer-function digitization methods, the MPZ method is an approximation; here the approximation is motivated partly by the fact that $z = e^{sT}$ is the correct s to z transformation for the poles of the transform of a time sequence and partly by the minimal amount of algebra required to determine the digitized transfer function by hand, so as to facilitate checking the computer calculations.

Because physical systems often have more poles than zeros, it is useful to arbitrarily add zeros at $z = -1$, resulting in a $1 + z^{-1}$ term in $D(z)$. This causes an averaging of the current and past input values, as in Tustin's method. We select the low-frequency gain of $D(z)$ so that it equals that of $D(s)$.

MPZ Method Summary

1. Map poles and zeros according to the relation $z = e^{sT}$.
2. If the numerator is of lower order than the denominator, add powers of $(z + 1)$ to the numerator until numerator and denominator are of equal order.
3. Set the DC or low-frequency gain of $D(z)$ equal to that of $D(s)$.

The MPZ approximation of

$$D(s) = K_c \frac{s + a}{s + b} \tag{8.20}$$

is

$$D(z) = K_d \frac{z - e^{-aT}}{z - e^{-bT}}, \tag{8.21}$$

where K_d is found by causing the DC gain of $D(z)$ to equal the DC gain of $D(s)$ using the continuous and discrete versions of the Final Value Theorem. The result is

$$K_c \frac{a}{b} = K_d \frac{1 - e^{-aT}}{1 - e^{-bT}},$$

or

$$K_d = K_c \frac{a}{b} \left(\frac{1 - e^{-bT}}{1 - e^{-aT}} \right). \tag{8.22}$$

For a $D(s)$ with a higher-order denominator, Step 2 in the method calls for adding the $(z + 1)$ term. For example,

$$D(s) = K_c \frac{s + a}{s(s + b)} \Rightarrow D(z) = K_d \frac{(z + 1)(z - e^{-aT})}{(z - 1)(z - e^{-bT})}, \tag{8.23}$$

where, after dropping the poles at $s = 0$ and $z = 1$,

$$K_d = K_c \frac{a}{2b} \left(\frac{1 - e^{-bT}}{1 - e^{-aT}} \right). \tag{8.24}$$

In the digitization methods described so far, the same power of z appears in the numerator and denominator of $D(z)$. This implies that the difference equation output at time k will require a sample of the input at time k. For example, the $D(z)$ in Eq. (8.21) can be written

$$\frac{U(z)}{E(z)} = D(z) = K_d \frac{1 - \alpha z^{-1}}{1 - \beta z^{-1}}, \tag{8.25}$$

where $\alpha = e^{-aT}$ and $\beta = e^{-bT}$. By inspection we can see that Eq. (8.25) results in the difference equation

$$u(k) = \beta u(k - 1) + K_d[e(k) - \alpha e(k - 1)]. \tag{8.26}$$

EXAMPLE 8.2

Design of a Space Station Attitude Digital Controller Using Discrete Equivalents

A very simplified model of the space station attitude control dynamics has the plant transfer function

$$G(s) = \frac{1}{s^2}.$$

Design a digital controller to have a closed-loop natural frequency $\omega_n \cong 0.3$ rad/sec and a damping ratio $\zeta = 0.7$.

Figure 8.9

Continuous-design definition for Example 8.2

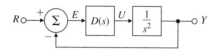

Solution. The first step is to find the proper $D(s)$ for the system defined in Fig. 8.9. After some trial and error, we find that the specifications can be met by the lead compensation

$$D(s) = 0.81 \frac{s + 0.2}{s + 2}. \tag{8.27}$$

The root locus in Fig. 8.10 verifies the appropriateness of using Eq. (8.27).

To digitize this $D(s)$, we first need to select a sample rate. For a system with $\omega_n = 0.3$ rad/sec, the bandwidth will also be about 0.3 rad/sec, and an acceptable sample rate would be about 20 times ω_n. Thus

$$\omega_s = 0.3 \times 20 = 6 \text{ rad/sec.}$$

A sample rate of 6 rad/sec is about 1 Hertz; therefore, the sample period should be $T = 1$ sec. The MPZ digitization of Eq. (8.27), given by Eqs. (8.21) and (8.22), yields

$$D(z) = 0.389 \frac{z - 0.82}{z - 0.135}$$

$$= \frac{0.389 - 0.319z^{-1}}{1 - 0.135z^{-1}}. \tag{8.28}$$

Inspection of Eq. (8.28) gives us the difference equation

$$u(k) = 0.135u(k - 1) + 0.389e(k) - 0.319e(k - 1), \tag{8.29}$$

Figure 8.10

s-Plane locus with respect to K

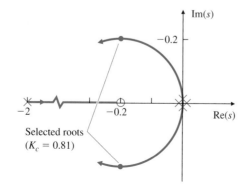

Figure 8.11

A digital control system that is equivalent to Fig. 8.9

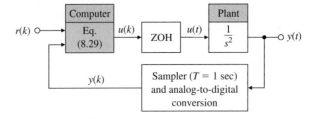

where

$$e(k) = r(k) - y(k),$$

and this completes the digital algorithm design. The complete digital system is shown in Fig. 8.11.

The last step in the design process is to verify the design by implementing it on the computer. Figure 8.12 compares the step response of the digital system using $T = 1$ sec with the step response of the continuous compensation. Note that there is greater overshoot and a longer settling time in the digital system, which suggests a decrease in the damping. The average $T/2$ delay shown in Fig. 8.2 is the cause of the reduced damping. For a better match to the continuous system, it may be prudent to increase the sample rate. Figure 8.12 also shows the response with sampling that is twice as fast and it can be seen that it comes much closer to the continuous system. Note that the discrete compensation needs to be recalculated for this faster sample rate according to Eqs. (8.21) and (8.22).

It is impossible to sample $e(k)$, compute $u(k)$, and then output $u(k)$ all in zero elapsed time; therefore, Eqs. (8.26) and (8.29) are impossible to implement precisely. However, if the equation is simple enough and/or the computer is fast enough, a slight computational delay between the $e(k)$ sample and the $u(k)$ output will have a

Figure 8.12

Step responses of the continuous and digital implementations

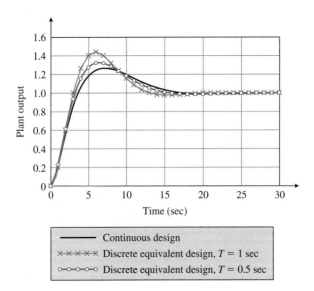

negligible effect on the actual response of the system compared with that expected from the original design. A rule of thumb would be to keep the computational delay on the order of $1/10$ of T. The real-time code and hardware can be structured so that this delay is minimized by making sure that computations between read A/D and write D/A are minimized and that $u(k)$ is sent to the ZOH immediately after its calculation.

8.3.2 Modified Matched Pole–Zero (MMPZ) Method

The $D(z)$ in Eq. (8.23) would also result in $u(k)$ being dependent on $e(k)$, the input at the same time point. If the structure of the computer hardware prohibits this relation or if the computations are particularly lengthy, it may be desirable to derive a $D(z)$ that has one less power of z in the numerator than in the denominator; hence, the computer output $u(k)$ would require only input from the previous time, that is, $e(k-1)$. To do this, we simply modify Step 2 in the matched pole–zero procedure so that the numerator is of lower order than the denominator by 1. For example, if

$$D(s) = K_c \frac{s+a}{s(s+b)},$$

we skip Step 2 to get

$$D(z) = K_d \frac{z - e^{-aT}}{(z-1)(z - e^{-bT})}, \tag{8.30}$$

$$K_d = K_c \frac{a}{b} \left(\frac{1 - e^{-bT}}{1 - e^{-aT}} \right).$$

To find the difference equation, we multiply the top and bottom of Eq. (8.30) by z^{-2} to obtain

$$D(z) = K_d \frac{z^{-1}(1 - e^{-aT}z^{-1})}{1 - z^{-1}(1 + e^{-bT}) + z^{-2}e^{-bT}}. \tag{8.31}$$

By inspecting Eq. (8.31) we can see that the difference equation is

$$u(k) = (1 + e^{-bT})u(k-1) - e^{-bT}u(k-2) + K_d[e(k-1) - e^{-aT}e(k-2)].$$

In this equation an entire sample period is available to perform the calculation and to output $u(k)$, because it depends only on $e(k-1)$. A discrete analysis of this controller would therefore more accurately explain the behavior of the actual system. However, because this controller is using data that are one cycle old, it will typically not perform as well as the MPZ controller in terms of the deviations of the desired system output in the presence of random disturbances.

8.3.3 Comparison of Digital Approximation Methods

A numerical comparison of the magnitude of the frequency response for a first-order lag,

$$D(s) = \frac{5}{s+5},$$

is made in Fig. 8.13 for the three approximation techniques at two different sample rates. The results of the $D(z)$ computations used in Fig. 8.13 are shown in Table 8.2.

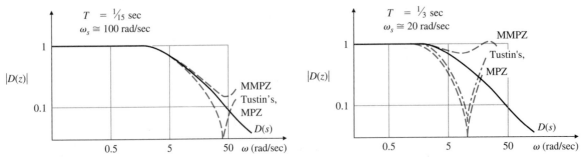

Figure 8.13

A comparison of the frequency response of three discrete approximations

Figure 8.13 shows that all the approximations are quite good at frequencies below about $1/4$ the sample rate, or $\omega_s/4$. If $\omega_s/4$ is sufficiently larger than the filter break-point frequency—that is, if the sampling is fast enough—the break-point characteristics of the lag will be accurately reproduced. Tustin's technique and the MPZ method show a notch at $\omega_s/2$ because of their zero at $z = -1$ from the $z + 1$ term. Other than the large difference at $\omega_s/2$, which is typically outside the range of interest, the three methods have similar accuracies.

8.3.4 Applicability Limits of the Discrete Equivalent Design Method

If we performed an exact discrete analysis or a simulation of a system and determined the digitization for a wide range of sample rates, the system would often be unstable for rates slower than approximately $5\omega_n$, and the damping would be degraded significantly for rates slower than about $10\omega_n$. At sample rates $\gtrsim 20\omega_n$ (or $\gtrsim 20$ times the bandwidth for more complex systems), design by discrete equivalents yields reasonable results, and at sample rates of 30 times the bandwidth or higher, discrete equivalents can be used with confidence.

ZOH transfer function

As shown by Fig. 8.2, the errors come about because the technique ignores the lagging effect of the ZOH which, on the average, is $T/2$. A method to account for

TABLE 8.2

Comparing Digital Approximations of $D(z)$ for $D(s) = 5/(s + 5)$

Method	ω_s	
	100 rad/sec	20 rad/sec
Matched pole–zero (MPZ)	$0.143\dfrac{z + 1}{z - 0.715}$	$0.405\dfrac{z + 1}{z - 0.189}$
Modified MPZ (MMPZ)	$0.285\dfrac{1}{z - 0.715}$	$0.811\dfrac{1}{z - 0.189}$
Tustin's	$0.143\dfrac{z + 1}{z - 0.713}$	$0.454\dfrac{z + 1}{z - 0.0914}$

this is to approximate the $T/2$ delay with Eq. (5.94) by including a transfer function approximation for the ZOH:[1]

$$G_{ZOH}(s) = \frac{2/T}{s + 2/T}.$$

(8.32)

Once an initial design is carried out and the sampling rate has been selected, we could improve on our discrete design by inserting Eq. (8.32) into the original plant model and adjusting the $D(s)$ so that a satisfactory response in the presence of the sampling delay is achieved. Therefore, we see that use of Eq. (8.32) partially alleviates the approximate nature of using discrete equivalents.

For sample rates slower than about $10\omega_n$ it is advisable to analyze the entire system using an exact discrete analysis. If a discrete analysis shows an unacceptable degradation of performance due to the sampling, the design can then be refined using exact discrete methods. We cover this approach in Section 8.6.

8.4   Hardware Characteristics

A digital control system includes several unique components not found in continuous control systems: an **analog-to-digital converter** is a device to sample the continuous signal voltage from the sensor and to convert that signal to a digital word; a **digital-to-analog converter** is a device to convert the digital word from the computer to an analog voltage, an **anti-alias prefilter** is an analog device designed to reduce the effects of aliasing, and the **computer** is the device where the compensation $D(z)$ is programmed and the calculations are carried out. This section provides a brief description of each of these.

8.4.1   Analog-to-Digital (A/D) Converters

As discussed in Section 8.1, A/D converters are devices that convert a voltage level from a sensor to a digital word usable by the computer. At the most basic level, all digital words are binary numbers consisting of many bits that are set to either 1 or 0. Therefore, the task of the A/D converter at each sample time is to convert a voltage level to the correct bit pattern and often to hold that pattern until the next sample time.

Of the many A/D conversion techniques that exist, the most common are based on counting schemes or a successive-approximation technique. In counting methods the input voltage may be converted to a train of pulses whose frequency is proportional to the voltage level. The pulses are then counted over a fixed period using a binary counter, thus resulting in a binary representation of the voltage level. A variation on this scheme is to start the count simultaneously with a voltage that is linear in time and to stop the count when the voltage reaches the magnitude of the input voltage to be converted.

The successive-approximation technique tends to be much faster than the counting methods. It is based on successively comparing the input voltage to reference levels representing the various bits in the digital word. The input voltage is first compared with a reference value that is half the maximum. If the input voltage is greater, the most significant bit is set, and the signal is then compared with a reference level

---

[1] Or other Padé approximate as discussed in Section 5.6.3

that is $3/4$ the maximum to determine the next bit, and so on. One clock cycle is required to set each bit, so an n-bit converter would require n cycles. At the same clock rate a counter-based converter might require as many as 2^n cycles, which would usually be much slower.

With either technique, the greater the number of bits, the longer it will take to perform the conversion. The price of A/D converters generally goes up with both speed and bit size. In 2009, a 14-bit (resolution of 0.006%) converter with a high performance capability of a 10-n sec conversion time (100 million samples per sec) sold for approximately \$25 while a 12-bit (0.025%) converter with a good performance capability of a 1 μ sec conversion time (1 million samples per sec) sold for approximately \$4. An 8-bit (0.4% resolution) with a 1 μ sec conversion time sold for approximately \$1. The performance has been improving considerably every year.

If more than one channel of data needs to be sampled and converted to digital words, it is usually accomplished by use of a multiplexer rather than by multiple A/D converters. The multiplexer sequentially connects the converter into the channel being sampled.

8.4.2   Digital-to-Analog (D/A) Converters

D/A converters, as mentioned in Section 8.1, are used to convert the digital words from the computer to a voltage level and are sometimes referred to as **Sample and Hold** devices. They provide analog outputs from a computer for driving actuators or perhaps a recording device such as an oscilloscope or strip-chart recorder. The basic idea behind their operation is that the binary bits cause switches (electronic gates) to open or close, thus routing the electric current through an appropriate network of resistors to generate the correct voltage level. Because no counting or iteration is required for such converters, they tend to be much faster than A/D converters. In fact, A/D converters that use the successive-approximation method of conversion include D/A converters as components.

8.4.3   Anti-Alias Prefilters

An analog **anti-alias prefilter** is often placed between the sensor and the A/D converter. Its function is to reduce the higher-frequency noise components in the analog signal in order to prevent aliasing, that is, having the noise be modulated to a lower frequency by the sampling process.

An example of aliasing is shown in Fig. 8.14, where a 60 Hertz oscillatory signal is being sampled at 50 Hertz. The figure shows the result from the samples as a 10 Hertz signal and also shows the mechanism by which the frequency of the signal is aliased from 60 to 10 Hertz. Aliasing will occur any time the sample rate is not at least twice as fast as any of the frequencies in the signal being sampled. Therefore, to prevent aliasing of a 60 Hertz signal, the sample rate would have to be faster than 120 Hertz, clearly much higher than the 50 Hertz rate in the figure.

Aliasing is one of the consequences of the **sampling theorem of Nyquist and Shannon**. Their theorem basically states that, for the signal to be accurately reconstructed from the samples, it must have no frequency component greater than half the sample rate ($\omega_s/2$). Another consequence of their theorem is that the highest frequency

Analog prefilters reduce aliasing

Nyquist–Shannon sampling theorem

Figure 8.14

An example of aliasing

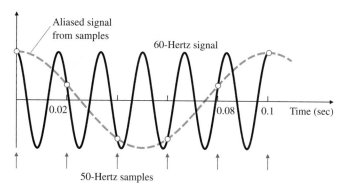

that can be unambiguously represented by discrete samples is the Nyquist rate of $\omega_s/2$, an idea we discussed in Section 8.2.3.

The consequence of aliasing on a digital control system can be substantial. In a continuous system, noise components with a frequency much higher than the control-system bandwidth normally have a small effect because the system will not respond at the high frequency. However, in a digital system, the frequency of the noise can be aliased down to the vicinity of the system bandwidth so that the closed-loop system would respond to the noise. Thus, the noise in a poorly designed digitally controlled system could have a substantially greater effect than if the control had been implemented using analog electronics.

The solution is to place an analog prefilter before the sampler. In many cases a simple first-order low-pass filter will do—that is,

$$H_p(s) = \frac{a}{s + a},$$

where the break point a is selected to be lower than $\omega_s/2$ so that any noise present with frequencies greater than $\omega_s/2$ is attenuated by the prefilter. The lower the break-point frequency selected, the more the noise above $\omega_s/2$ is attenuated. However, too low a break point may force the designer to reduce the control system's bandwidth. The prefilter does not completely eliminate the aliasing; however, through judicious choice of the prefilter break point and the sample rate, the designer has the ability to reduce the magnitude of the aliased noise to some acceptable level.

8.4.4   The Computer

The computer is the unit that does all the computations. Most digital controllers used today are built around a microcontroller that contains both a microprocessor and most of the other functions needed, including the A/D and D/A conversion. For development purposes in a laboratory, a digital controller could be a desktop-sized workstation or a PC. The relatively low cost of microprocessor technology has accounted for the large increase in the use of digital control systems, which started in the 1980s and continues into the 2000s.

The computer consists of a central processor unit (CPU), which does the computations and provides the system logic; a clock to synchronize the system; memory modules for data and instruction storage; and a power supply to provide the various required voltages. The memory modules come in three basic varieties:

1. **Read-only memory (ROM)** is the least expensive, but after its manufacture its contents cannot be changed. Most of the memory in products manufactured in quantity is ROM. It retains its stored values when power is removed.
2. **Random-access memory (RAM)** is the most expensive, but its values can be changed by the CPU. It is required only to store the values that will be changed during the control process and typically represents only a small fraction of the total memory of a developed product. It loses the values in memory when power is removed.
3. **Programmable read-only memory (EPROM)** is a ROM whose values can be changed by a technician using a special device. It is typically used during product development to enable the designer to try different algorithms and parameter values. It retains its stored values when power is removed. In some products, it is useful to have a few of the stored quantities in EPROMs so that individual calibrations can be carried out for each unit.

Microprocessors for control applications generally come with a digital word size of 8, 16, or 32 bits, although some have been available with 12 bits. Larger word sizes give better accuracy, but at an increase in cost. The most economical solution is often to use an 8-bit microprocessor, but to use two digital words to store one value (**double precision**) in the areas of the controller that are critical to the system accuracy. Many digital control systems use computers originally designed for digital signal-processing applications, so-called DSP chips.

8.5   Sample-Rate Selection

The selection of the best sample rate for a digital control system is the result of a compromise of many factors. Sampling too fast can cause a loss of accuracy while the basic motivation for lowering the sample rate ω_s is cost. A decrease in sample rate means more time is available for the control calculations; hence slower computers can be used for a given control function or more control capability can be achieved from a given computer. Either way, the cost per function is lowered. For systems with A/D converters, less demand on conversion speed will also lower cost. These economic arguments indicate that the best engineering choice is the slowest possible sample rate that still meets all performance specifications.

There are several factors that could provide a lower limit on the acceptable sample rate:

1. tracking effectiveness as measured by closed-loop bandwidth or by time-response requirements, such as rise time and settling time;
2. regulation effectiveness as measured by the error response to random plant disturbances;
3. error due to measurement noise and the associated prefilter design methods.

A fictitious limit occurs when using discrete equivalents. The inherent approximation in the method may give rise to decreased performance or even system instabilities as the sample rate is lowered. This can lead the designer to conclude that a faster sample rate is required. However, there are two solutions:

1. sample faster, and
2. recognize that the approximations are invalid and refine the design with a direct digital-design method described in the subsequent sections.

The ease of designing digital control systems with fast sample rates and the low cost of very capable computers often drives the designer to select a sample rate that is $40 \times \omega_{BW}$ or higher. For computers with fixed-point arithmetic, very fast sample rates can lead to multiplication errors that have the potential to produce significant offsets or limit cycles in the control (see Franklin et al., 1998).

8.5.1 Tracking Effectiveness

An absolute lower bound on the sample rate is set by a specification to track a command input with a certain frequency (the system bandwidth). The sampling theorem (see Section 8.4.3 and Franklin et al., 1998) states that in order to reconstruct an unknown, band-limited, continuous signal from samples of that signal, we must sample at least twice as fast as the highest frequency contained in the signal. Therefore, in order for a closed-loop system to track an input at a certain frequency, it must have a sample rate twice as fast; that is, ω_s must be at least twice the system bandwidth ($\omega_s \Rightarrow 2 \times \omega_{BW}$). We also saw from the results of mapping the s-plane into the z-plane ($z = e^{sT}$) that the highest frequency that can be represented by a discrete system is $\omega_s/2$, which supports the conclusion of the theorem.

It is important to note the distinction between the closed-loop bandwidth ω_{BW} and the highest frequency in the open-loop plant dynamics, because the two frequencies can be quite different. For example, closed-loop bandwidths can be an order of magnitude *less* than open-loop modes of resonances for some control problems. Information concerning the state of the plant resonances for purposes of control can be extracted from sampling the output without satisfying the sampling theorem because some a priori knowledge concerning these dynamics (albeit imprecise) is available, and the system is not required to track these frequencies. Thus a priori knowledge of the dynamic model of the plant can be included in the compensation in the form of a notch filter.

The closed-loop-bandwidth limitation provides the fundamental lower bound on the sample rate. In practice, however, the theoretical lower bound of sampling at twice the bandwidth of the reference input signal would not be judged sufficient in terms of the quality of the desired time responses. For a system with a rise time on the order of 1 sec (thus yielding a closed-loop bandwidth on the order of 0.5 Hertz), it is reasonable to insist on a sampling rate of 10 to 20 Hertz, which is a factor of 20 to 40 times ω_{BW}. The purposes of choosing a sample rate much greater than the bandwidth are to reduce the delay between a command and the system response to the command and also to smooth the system output to the control steps coming out of the ZOH.

8.5.2 Disturbance Rejection

Disturbance rejection is an important—if not the most important—aspect of any control system. Disturbances enter a system with various frequency characteristics ranging from steps to white noise. For the purpose of sample-rate selection, the higher-frequency random disturbances are the most influential.

The ability of the control system to reject disturbances with a good continuous controller represents the lower bound on the error response that we can hope for when implementing the controller digitally. In fact, some degradation relative to the continuous design must occur because the sampled values are slightly out of date at all times except precisely at the sampling instants. However, if the sample rate is very fast compared with the frequencies contained in the noisy disturbance, we should expect no appreciable loss from the digital system as compared with the continuous controller. At the other extreme, if the sample rate is very slow compared with the characteristic frequencies of the noise, the response of the system because of noise is essentially the same as the response we would get if the system had no control at all. The selection of a sample rate will place the response somewhere in between these two extremes. Thus, the impact of the sample rate on the ability of the system to reject disturbances may be very important to consider when choosing the sample rate.

Although the best choice of sample rate in terms of the ω_{BW} multiple is dependent on the frequency characteristics of the noise and the degree to which random disturbance rejection is important to the quality of the controller, sample rates on the order of 25 times ω_{BW} or higher are typical.

8.5.3 Effect of Anti-Alias Prefilter

Digital control systems with analog sensors typically include an analog anti-alias prefilter between the sensor and the sampler as described in Section 8.4.3. The prefilters are low-pass, and the simplest transfer function is

$$H_p(s) = \frac{a}{s+a},$$

so that the noise above the prefilter break point a is attenuated. The goal is to provide enough attenuation at half the sample rate ($\omega_s/2$) that the noise above $\omega_s/2$, when aliased into lower frequencies by the sampler, will not be detrimental to control system performance.

A conservative design procedure is to select ω_s and the break point to be sufficiently higher than the system bandwidth that the phase lag from the prefilter does not significantly alter the system stability. This would allow the prefilter to be ignored in the basic control system design. Furthermore, for a good reduction in the high-frequency noise at $\omega_s/2$, we choose a sample rate that is about 5 or 10 times higher than the prefilter break point. The implication of this prefilter design procedure is that sample rates need to be on the order of 30 to 100 times faster than the system bandwidth. Using this conservative design procedure, the prefilter influence will likely provide the lower bound on the selection of the sample rate.

An alternative design procedure is to allow significant phase lag from the prefilter at the system bandwidth. This requires us to include the analog prefilter characteristics in the plant model when carrying out the control design. It allows the use of lower sample rates, but at the possible expense of increased complexity in the compensation because additional phase lead must be provided to counteract the prefilter's phase lag. If this procedure is used and low prefilter break points are allowed, the effect of sample rate on sensor noise is small, and the prefilter essentially has no effect on the sample rate.

It may seem counterintuitive that placing a lag (the analog prefilter) in one portion of the controller and a counteracting lead [extra lead in $D(z)$] in another portion of the controller provides a net positive effect on the overall system. The net gain is a result of the fact that the lag is in the analog part of the system where high frequencies can exist. The counteracting lead is in the digital part of the system where frequencies above the Nyquist rate do not exist. The result is a reduction in the high frequencies before the sampling which are not reamplified by the counteracting digital lead, thus producing net reduction in high frequencies. Furthermore, these high frequencies are particularly insidious with a digital controller because of the aliasing that would result from the sampling.

8.5.4 Asynchronous Sampling

As noted in the previous paragraphs, divorcing the prefilter design from the control-law design may require using a faster sample rate than otherwise. This same result may show up in other types of architecture. For example, a smart sensor with its own computer running asynchronously relative to the primary control computer will not be amenable to direct digital design because the overall system transfer function depends on the phasing between the smart sensor and the primary digital controller. This situation is similar to that of the digitization errors discussed in Section 8.6. Therefore, if asynchronous digital subsystems are present, sample rates on the order of $20 \times \omega_{BW}$ or slower in any module should be used with caution and the system performance checked through simulation or experiment.

△ 8.6 Discrete Design

It is possible to obtain an exact discrete model that relates the samples of the continuous plant $y(k)$ to the input control sequence $u(k)$. This plant model can be used as part of a discrete model of the feedback system including the compensation $D(z)$. Analysis and design using this discrete model is called **discrete design** or, alternatively, **direct digital design**. The following subsections will describe how to find the discrete plant model (Section 8.6.1), what the feedback compensation looks like when designing with a discrete model (Sections 8.6.2 and 8.6.3), and how the design process is carried out (Section 8.6.4).

8.6.1 Analysis Tools

The exact discrete
equivalent

The first step in performing a discrete analysis of a system with some discrete elements is to find the discrete transfer function of the continuous portion. For a system similar to that shown in Fig. 8.1(b), we wish to find the transfer function between $u(kT)$ and $y(kT)$. Unlike the cases discussed in the previous sections, there is an *exact* discrete equivalent for this system, because the ZOH precisely describes what happens between samples of $u(kT)$ and the output $y(kT)$ is dependent only on the input at the sample times $u(kT)$.

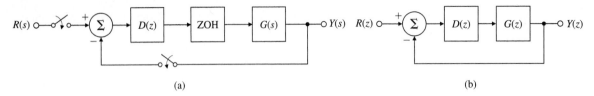

Figure 8.15
Comparison of (a) a mixed control system; and (b) its pure discrete equivalent

For a plant described by $G(s)$ and preceded by a ZOH, the discrete transfer function is

$$G(z) = (1 - z^{-1})\mathcal{Z}\left\{\frac{G(s)}{s}\right\},\qquad(8.33)$$

where $\mathcal{Z}\{F(s)\}$ is the z-transform of the sampled time series whose Laplace transform is the expression for $F(s)$, given on the same line in Table 8.1. Equation (8.33) has the term $G(s)/s$ because the control comes in as a step input from the ZOH during each sample period. The term $1 - z^{-1}$ reflects the fact that a one-sample duration step can be thought of as an infinite duration step followed by a negative step one cycle delayed. For a more complete derivation, see Franklin et al. (1998). Equation (8.33) allows us to replace the mixed (continuous and discrete) system shown in Fig. 8.15(a) with the equivalent pure discrete system shown in Fig. 8.15(b).

The analysis and design of discrete systems is very similar to the analysis and design of continuous systems; in fact, all the same rules apply. The closed-loop transfer function of Fig. 8.15(b) is obtained using the same rules of block-diagram reduction—that is,

$$\frac{Y(z)}{R(z)} = \frac{D(z)G(z)}{1 + D(z)G(z)}.\qquad(8.34)$$

To find the characteristic behavior of the closed-loop system, we need to find the factors in the denominator of Eq. (8.34)—that is, the roots of the discrete characteristic equation

$$1 + D(z)G(z) = 0.$$

The root-locus techniques used in continuous systems to find roots of a polynomial in s apply equally well and without modification to the polynomial in z; however, the interpretation of the results is quite different, as we saw in Fig. 8.4. A major difference is that the stability boundary is now the unit circle instead of the imaginary axis.

EXAMPLE 8.3   *Discrete Root Locus*

For the case in which $G(s)$ in Fig. 8.15(a) is

$$G(s) = \frac{a}{s + a}$$

and $D(z) = K$, draw the root locus with respect to K, and compare your results with a root locus of a continuous version of the system. Discuss the implications of your loci.

Figure 8.16

Root loci for (a) the z-plane; and (b) the s-plane

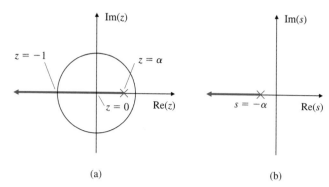

(a)

(b)

Solution. It follows from Eq. (8.33) that

$$G(z) = (1 - z^{-1})\mathcal{Z}\left[\frac{a}{s(s+a)}\right]$$

$$= (1 - z^{-1})\left[\frac{(1 - e^{-aT})z^{-1}}{(1 - z^{-1})(1 - e^{-aT}z^{-1})}\right]$$

$$= \frac{1 - \alpha}{z - \alpha},$$

where

$$\alpha = e^{-aT}.$$

To analyze the performance of the closed-loop system, standard root-locus rules apply. The result is shown in Fig. 8.16(a) for the discrete case and in Fig. 8.16(b) for the continuous case. In contrast to the continuous case, in which the system remains stable for all values of K, in the discrete case the system becomes oscillatory with decreasing damping ratio as z goes from 0 to -1 and eventually becomes unstable. This instability is due to the lagging effect of the ZOH, which is properly accounted for in the discrete analysis.

8.6.2 Feedback Properties

In continuous systems we typically start the design process by using the following basic design elements: proportional, derivative, or integral control laws, or some combination of these, sometimes with a lag included. The same ideas can be used in discrete design. Alternatively, the $D(z)$ resulting from the digitization of a continuously designed $D(s)$ will produce these basic design elements, which will then be used as a starting point in a discrete design. The discrete control laws are as follows:

Discrete control laws

Proportional

$$u(k) = Ke(k) \Rightarrow D(z) = K. \tag{8.35}$$

Derivative

$$u(k) = KT_D[e(k) - e(k-1)], \tag{8.36}$$

for which the transfer function is

$$D(z) = KT_D(1 - z^{-1}) = KT_D \frac{z - 1}{z} = k_D \frac{z - 1}{z}. \tag{8.37}$$

Integral

$$u(k) = u(k - 1) + \frac{K_p}{T_I} e(k), \tag{8.38}$$

for which the transfer function is

$$D(z) = \frac{K}{T_I} \left(\frac{1}{1 - z^{-1}} \right) = \frac{K}{T_I} \left(\frac{z}{z - 1} \right) = k_I \left(\frac{z}{z - 1} \right). \tag{8.39}$$

Lead Compensation

The examples in Section 8.3 showed that a continuous lead compensation leads to difference equations of the form

$$u(k + 1) = \beta u(k) + K[e(k + 1) - \alpha e(k)], \tag{8.40}$$

for which the transfer function is

$$D(z) = K \frac{1 - \alpha z^{-1}}{1 - \beta z^{-1}}. \tag{8.41}$$

8.6.3 Discrete Design Example

Digital control design consists of using the basic feedback elements of Eqs. (8.35) to (8.41) and iterating on the design parameters until all specifications are met.

EXAMPLE 8.4   *Direct Discrete Design of the Space Station Digital Controller*

Design a digital controller to meet the same specifications as in Example 8.2 using discrete design.

Solution. The discrete model of the $1/s^2$ plant, preceded by a ZOH, is found through Eq. (8.33) to be

$$G(z) = \frac{T^2}{2} \left[\frac{z + 1}{(z - 1)^2} \right],$$

which, with $T = 1$ sec, becomes

$$G(z) = \frac{1}{2} \left[\frac{z + 1}{(z - 1)^2} \right].$$

Proportional feedback in the continuous case yields pure oscillatory motion, so in the discrete case we should expect even worse results. The root locus in Fig. 8.17 verifies this. For very low values of K (where the locus represents roots at very low frequencies compared to the sample rate), the locus is tangent to the unit circle ($\zeta \cong 0$ indicating pure oscillatory motion), thus matching the proportional continuous design.

For higher values of K, Fig. 8.17 shows that the locus diverges into the unstable region because of the effect of the ZOH and sampling. To compensate for this, we will add a derivative term to the proportional term so that the control law is

$$U(z) = K[1 + T_D(1 - z^{-1})]E(z), \tag{8.42}$$

Figure 8.17

z-plane root locus for a $1/s^2$ plant with proportional feedback

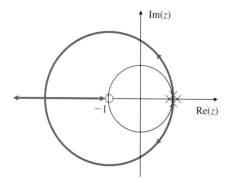

which yields compensation of the form

$$D(z) = K\frac{z - \alpha}{z}, \qquad (8.43)$$

where the new K and α replace the K and T_D in Eq. (8.42). Now the task is to find the values of α and K that yield good performance. The specifications for the design are that $\omega_n = 0.3$ rad/sec and $\zeta = 0.7$. Figure 8.4 indicates that this s-plane root location maps into a desired z-plane location of

$$z = 0.78 \pm 0.18j.$$

Figure 8.18 is the locus with respect to K for $\alpha = 0.85$. The location of the zero (at $z = 0.85$) was determined by trial and error until the locus passed through the desired z-plane location. The value of the gain when the locus passes through $z = 0.78 \pm 0.18j$ is $K = 0.374$. Equation (8.43) now becomes

$$D(z) = 0.374\frac{z - 0.85}{z}. \qquad (8.44)$$

Normally, it is not particularly advantageous to match specific z-plane root locations; rather it is necessary only to pick K and α (or T_D) to obtain acceptable z-plane roots, a much easier task. In this example, we want to match a specific location only so that we can compare the result with the design in Example 8.2.

The control law that results is

$$U(z) = 0.374(1 - 0.85z^{-1})E(z),$$

or

$$u(k) = 0.374e(k) - 0.318e(k - 1), \qquad (8.45)$$

which is similar to the control equation (8.29) obtained previously.

The controller in Eq. (8.45) basically differs from the continuously designed controller [Eq. (8.29)] only in the absence of the $u(k - 1)$ term. The $u(k - 1)$ term in Eq. (8.29) results from the lag term $(s + b)$ in the compensation [Eq. (8.27)]. The lag term is typically included in analog controllers both because it supplies noise attenuation and because pure analog differentiators are difficult to build. Some equivalent lag in discrete design naturally appears as a pole at $z = 0$ (see Fig. 8.18) and represents the one-sample delay in computing the derivative by a first difference. For more noise attenuation, we could move the pole to the right of $z = 0$, thus resulting in less derivative action and more smoothing, the same trade-off that exists in continuous control design.

Figure 8.18

z-plane locus for the $1/s^2$ plant with $D(z) = K(z - 0.85)/z$

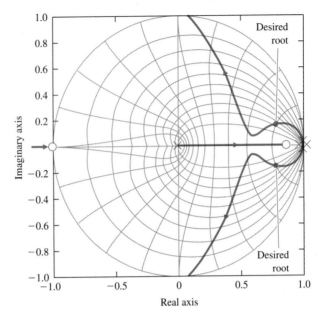

8.6.4 Discrete Analysis of Designs

Any digital controller, whether designed by discrete equivalents or directly in the *z*-plane, can be analyzed using discrete analysis, which consists of the following steps:

1. Find the discrete model of the plant and ZOH using Eq. (8.33).
2. Form the feedback system including $D(z)$.
3. Analyze the resulting discrete system.

We can determine the roots of the system using a root locus, as described in Section 8.6.3, or we can determine the time history (at the sample instants) of the discrete system.

EXAMPLE 8.5   *Damping and Step Response in Digital versus Continuous Design*

Use discrete analysis to determine the equivalent *s*-plane damping and the step responses of the digital designs in Examples 8.2 and 8.4, and compare your results with the damping and step response of the continuous case in Example 8.2.

Solution.   The MATLAB statements to evaluate the damping and step response of the continuous case in Example 8.2 are

```
sysGs = tf(1,[1  0  0]);
sysDs = tf(0.81 * [1  0.2],[1  2]);
sysGDs = series(sysGs,sysDs);
sysCLs = feedback(sysGDs,1,1);
step(sysCLs)
damp(sysCLs)
```

To analyze the digital control cases, the model of the plant preceded by the ZOH is found using the statements

```
T = 1;
sysGz = c2d(sysGs,T,'zoh')
```

Analysis of the digital control designed using the discrete equivalent [Eq. (8.29)] in Example 8.2 is performed by the statements

```
sysDz = tf( [.389 −.319],[1 −.135])
sysDGz = series(sysGz,sysDz)
sysCLz = feedback(sysDGz,1)
step(sysCLz,T)
damp(sysCLz,T)
```

Likewise, the discrete design of $D(z)$ from Eq. (8.44) can be analyzed by the same sequence.

The resulting step responses are shown in Fig. 8.19. The calculated damping ζ and complex root natural frequencies ω_n of the closed-loop systems are

$$\text{Continuous case:} \quad \zeta = 0.705, \quad \omega_n = 0.324;$$

$$\text{Discrete equivalent:} \quad \zeta = 0.645, \quad \omega_n = 0.441;$$

$$\text{Discrete design:} \quad \zeta = 0.733, \quad \omega_n = 0.306.$$

The figure shows increased overshoot for the discrete equivalent method that occurred because of the decreased damping of that case. Very little increased overshoot occurred in the discrete design, because that compensation was adjusted specifically so that the equivalent s-plane damping of the discrete system was approximately at the desired damping value of $\zeta = 0.7$.

Figure 8.19

Step response of the continuous and digital systems in Examples 8.2 and 8.4

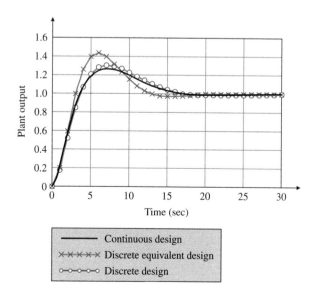

Although the analysis showed some differences between the performance of the digital controllers designed by the two methods, neither the performance nor the control equations [Eqs. (8.29) and (8.45)] are very different. This similarity results because the sample rate is fairly fast compared with ω_n—that is, $\omega_s \cong 20 \times \omega_n$. If we were to decrease the sample rate, the numerical values in the compensations would become increasingly different and the performance would degrade considerably for the discrete equivalent case.

As a general rule, discrete design should be used if the sampling frequency is slower than $10 \times \omega_n$. At the very least, a discrete equivalent design with slow sampling ($\omega_s < 10 \times \omega_n$) should be verified by a discrete analysis or by simulation, as described in Section 4.4, and the compensation adjusted if needed. A simulation of a digital control system is a good idea in any case. If it properly accounts for all delays and possibly asynchronous behavior of different modules, it may expose instabilities that are impossible to detect using continuous or discrete linear analysis. A more complete discussion regarding the effects of sample rate on the design is contained in Section 8.5.

8.7 Historical Perspective

One of the earliest examples of actual control of systems based on sampled data came with the use of search RADAR in WWII. In that case, the position of the target was available only once each revolution of the antenna. The theory of sampled data systems was developed by the mathematician W. Hurewicz[2] and published as a chapter in H. M. James, N. B. Nichols, and R. S. Phillips, *Theory of Servomechanisms*, vol. 25, Rad Lab Series, New York, McGraw Hill, 1947. The historical perspective for Chapter 5 discussed the introduction of computers for engineers performing design activities. The possibility of using computers for direct digital control motivated the continuation of work on sampled data systems during the 1950s, especially at Columbia University under Professor J. R. Ragazzini. That work was published in J. R. Ragazzini and G. F. Franklin, *Sampled-Data Control Systems*, New York, McGraw Hill, 1958. Early applications were in the process control industry where the relatively large and expensive computers available at the time could be justified. Professor Karl Astrom introduced direct digital control of a paper mill in Sweden in the early 1960s.

In 1961, when President Kennedy announced the goal of sending a man to the moon, there were no digital autopilots for aerospace vehicles. In fact, small digital computers suitable for implementing control systems were virtually nonexistent. The team at the MIT Draper Labs (called the Instrumentation Lab at that time) in charge of designing and building the Apollo control systems initially designed the control systems for the lunar and command modules with conventional analog electronics. However, they discovered that those systems would be too heavy and complex for the mission. So the decision was made to design and build the first aerospace digital

---

[2]Hurewicz died in 1956 falling off a ziggurat (a Mexican pyramid) on a conference outing at the International Symposium on algebraic topology in Mexico. It is suggested that he was: "... a paragon of absentmindedness, a failing that probably led to his death."

control system. Bill Widnall, Dick Battin, and Don Fraser were all key players in the successful design and execution of that system for the Apollo flights in late 1960s. The group went on to demonstrate a digital autopilot for NASA's F-8 in the 1970s, and digital autopilots went on to become dominant over the 1980s and beyond. In fact, with the introduction of inexpensive digital signal processors, most control systems of any kind became digital by the turn of the century and, today, very few control systems are being implemented with analog electronics. This evolution has had an effect on the training for controls engineers. In the past, the ability to design and build the specialized circuitry for analog electronic controls caused many controls engineers to have an Electrical Engineering background. Now, with easily programmable digital computers being readily available, the background of controls engineers tends more toward the specialties that are most familiar with the systems being controlled.

SUMMARY

- The simplest and most expedient design technique is to transform a continuous controller design to its discrete form—that is, to use its **discrete equivalent**.
- **Design using discrete equivalents** entails (a) finding the continuous compensation $D(s)$ using the ideas in Chapters 1 to 7, and (b) approximating $D(s)$ with difference equations using Tustin's method or the matched-pole–zero method.
- In order to analyze a discrete controller design, or any discrete system, the z-**transform** is used to determine the system's behavior. The z-transform of a time sequence $f(k)$ is given by

$$\mathcal{Z}\{f(k)\} = F(z) = \sum_{k=0}^{\infty} f(k)z^{-k}$$

and has the key property that

$$\mathcal{Z}\{f(k-1)\} = z^{-1}F(z).$$

This property allows us to find the discrete transfer function of a **difference equation**, which is the digital equivalent of a differential equation for continuous systems. Analysis using z-transforms closely parallels that using Laplace transforms.

- Normally z-transforms are found using the computer (MATLAB) or looking up in Table 8.1.
- The discrete Final Value Theorem is

$$\lim_{k \to \infty} x(k) = \lim_{z \to 1} (1 - z^{-1})X(z),$$

provided that all poles of $(1 - z^{-1})X(z)$ are inside the unit circle.
- For a continuous signal $f(t)$ whose samples are $f(k)$, the poles of $F(s)$ are related to the poles of $F(z)$ by

$$z = e^{sT}.$$

- The following are the most common discrete equivalents:

    1. **Tustin's approximation**:

    $$D(z) = D(s)\big|_{s=\frac{2}{T}\left(\frac{z-1}{z+1}\right)}$$

    2. the **matched pole–zero approximation:**
        - Map poles and zeros by $z = e^{sT}$.
        - Add powers of $z + 1$ to the numerator until numerator and denominator are of equal order or the numerator is one order less than the denominator.
        - Set the low-frequency gain of $D(z)$ equal to that of $D(s)$.

- If designing by discrete equivalents, a minimum **sample rate** of 20 times the bandwidth is recommended. Typically, even faster sampling is useful for best performance.
- Analog **prefilters** are commonly placed before the **sampler** in order to atten-uate the effects of high-frequency measurement noise. A sampler **aliases** all frequencies in the signal that are greater than half the sample frequency to lower frequencies; therefore, prefilter break points should be selected so that no significant frequency content remains above half the sample rate.
- The discrete model of the continuous plant $G(s)$ preceded by a **ZOH** is

$$G(z) = (1 - z^{-1})\mathcal{Z}\left\{\frac{G(s)}{s}\right\}.$$

The discrete plant model plus the discrete controller can be analyzed using the z-**transform** or simulated using SIMULINK.
- **Discrete design** is an exact design method and avoids the approximations inher-ent with discrete equivalents. The design procedure entails (a) finding the discrete model of the plant $G(s)$, and (b) using the discrete model to design the compen-sation directly in its discrete form. The design process is more cumbersome than discrete equivalent design and requires that a sample rate be selected before commencing the design. A practical approach is to commence the design using discrete equivalents, then tune up the result using discrete design.
- Discrete design using $G(z)$ closely parallels continuous design, but the stability boundary and interpretation of z-plane root locations are different. Figure 8.5 summarizes the response characteristics.
- If using discrete design, system stability can theoretically be assured when **sam-pling** at a rate as slow as twice the bandwidth. However, for good transient performance and random disturbance rejection, best results are obtained by sampling at 10 times the closed-loop bandwidth or faster. In some cases with troublesome vibratory modes, it is sometimes useful to sample more than twice as fast as the vibratory mode.

REVIEW QUESTIONS

1. What is the Nyquist rate? What are its characteristics?
2. Describe the discrete equivalent design process.
3. Describe how to arrive at a $D(z)$ if the sample rate is $30 \times \omega_{BW}$.

4. For a system with a 1 rad/sec bandwidth, describe the consequences of various sample rates.

5. Give two advantages for selecting a digital processor rather than analog circuitry to implement a controller.

6. Give two disadvantages for selecting a digital processor rather than analog circuitry to implement a controller.

△ **7.** Describe how to arrive at a $D(z)$ if the sample rate is $5 \times \omega_{BW}$.

PROBLEMS

Problems for Section 8.2: Dynamic Analysis of Discrete Systems

8.1 The z-transform of a discrete-time filter $h(k)$ at a 1 Hertz sample rate is

$$H(z) = \frac{1 + (1/2)z^{-1}}{[1 - (1/2)z^{-1}][1 + (1/3)z^{-1}]}.$$

(a) Let $u(k)$ and $y(k)$ be the discrete input and output of this filter. Find a difference equation relating $u(k)$ and $y(k)$.

(b) Find the natural frequency and damping coefficient of the filter's poles.

(c) Is the filter stable?

8.2 Use the z-transform to solve the difference equation

$$y(k) - 3y(k-1) + 2y(k-2) = 2u(k-1) - 2u(k-2),$$

where

$$u(k) = \begin{cases} k, & k \geq 0, \\ 0, & k < 0, \end{cases}$$

$$y(k) = 0, \qquad k < 0.$$

8.3 The one-sided z-transform is defined as

$$F(z) = \sum_{0}^{\infty} f(k)z^{-k}.$$

(a) Show that the one-sided transform of $f(k+1)$ is $\mathcal{Z}\{f(k+1)\} = zF(z) - zf(0)$.

(b) Use the one-sided transform to solve for the transforms of the Fibonacci numbers generated by the difference equation $u(k+2) = u(k+1) + u(k)$. Let $u(0) = u(1) = 1$. [*Hint:* You will need to find a general expression for the transform of $f(k+2)$ in terms of the transform of $f(k)$.]

(c) Compute the pole locations of the transform of the Fibonacci numbers.

(d) Compute the inverse transform of the Fibonacci numbers.

(e) Show that, if $u(k)$ represents the kth Fibonacci number, then the ratio $u(k+1)/u(k)$ will approach $(\frac{1+\sqrt{5}}{2}$. This is the golden ratio valued so highly by the Greeks.

8.4 Prove the seven properties of the s-plane-to-z-plane mapping listed in Section 8.2.3.

Problems for Section 8.3: Design Using Discrete Equivalents

8.5   A unity feedback system has an open-loop transfer function given by

$$G(s) = \frac{250}{s[(s/10) + 1]}.$$

The following lag compensator added in series with the plant yields a phase margin of $50°$:

$$D(s) = \frac{s/1.25 + 1}{50s + 1}.$$

Using the matched pole–zero approximation, determine an equivalent digital realization of this compensator.

8.6   The following transfer function is a lead network designed to add about $60°$ of phase at $\omega_1 = 3$ rad/sec:

$$H(s) = \frac{s + 1}{0.1s + 1}.$$

(a) Assume a sampling period of $T = 0.25$ sec, and compute and plot in the z-plane the pole and zero locations of the digital implementations of $H(s)$ obtained using (1) Tustin's method and (2) pole–zero mapping. For each case, compute the amount of phase lead provided by the network at $z_1 = e^{j\omega_1 T}$.

(b) Using a log–log scale for the frequency range $\omega = 0.1$ to $\omega = 100$ rad/sec, plot the magnitude Bode plots for each of the equivalent digital systems you found in part (a), and compare with $H(s)$. (*Hint:* Magnitude Bode plots are given by $|H(z)| = |H(e^{j\omega T})|$.)

8.7   The following transfer function is a lag network designed to introduce a gain attenuation of $10(-20 \text{ db})$ at $\omega = 3$ rad/sec:

$$H(s) = \frac{10s + 1}{100s + 1}.$$

(a) Assume a sampling period of $T = 0.25$ sec, and compute and plot in the z-plane the pole and zero locations of the digital implementations of $H(s)$ obtained using (1) Tustin's method and (2) pole–zero mapping. For each case, compute the amount of gain attenuation provided by the network at $z_1 = e^{j\omega_1 T}$.

(b) For each of the equivalent digital systems in part (a), plot the Bode magnitude curves over the frequency range $\omega = 0.01$ to 10 rad/sec.

Problem for Section 8.5: Sample Rate Selection

8.8   For the system shown in Fig. 8.20, find values for K, T_D, and T_I so that the closed-loop poles satisfy $\zeta > 0.5$ and $\omega_n > 1$ rad/sec. Discretize the PID controller using

(a) Tustin's method

(b) the matched pole–zero method

Use MATLAB to simulate the step response of each of these digital implementations for sample times of $T = 1, 0.1$, and 0.01 sec.

Figure 8.20

Control system for Problem 8.8

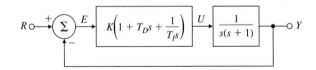

△ *Problems for Section 8.6: Discrete Design*

8.9 Consider the system configuration shown in Fig. 8.21, where

$$G(s) = \frac{40(s+2)}{(s+10)(s^2 - 1.4)}.$$

(a) Find the transfer function $G(z)$ for $T = 1$ assuming the system is preceded by a ZOH.

(b) Use MATLAB to draw the root locus of the system with respect to K.

(c) What is the range of K for which the closed-loop system is stable?

(d) Compare your results of part (c) with the case in which an analog controller is used (that is, where the sampling switch is always closed). Which system has a larger allowable value of K?

(e) Use MATLAB to compute the step response of both the continuous and discrete systems with K chosen to yield a damping factor of $\zeta = 0.5$ for the continuous case.

Figure 8.21

Control system for Problem 8.9

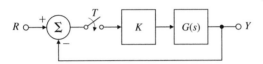

8.10 *Single-Axis Satellite Attitude Control:* Satellites often require attitude control for proper orientation of antennas and sensors with respect to Earth. Figure 2.7 shows a communication satellite with a three-axis attitude-control system. To gain insight into the three-axis problem, we often consider one axis at a time. Figure 8.22 depicts this case, where motion is allowed only about an axis perpendicular to the page. The equations of motion of the system are given by

$$I\ddot{\theta} = M_C + M_D,$$

where

$$I = \text{moment of inertia of the satellite about its mass center,}$$

$$M_C = \text{control torque applied by the thrusters,}$$

$$M_D = \text{disturbance torques,}$$

Figure 8.22

Satellite control schematic for Problem 8.10

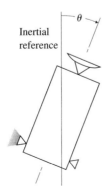

θ = angle of the satellite axis with respect to an inertial reference with no angular acceleration.

We normalize the equations of motion by defining

$$u = \frac{M_C}{I}, \quad w_d = \frac{M_D}{I},$$

and obtain

$$\ddot{\theta} = u + w_d.$$

Taking the Laplace transform yields

$$\theta(s) = \frac{1}{s^2}[u(s) + w_d(s)],$$

which, with no disturbance, becomes

$$\frac{\theta(s)}{u(s)} = \frac{1}{s^2} = G_1(s).$$

In the discrete case in which u is applied through a ZOH, we can use the methods described in this chapter to obtain the discrete transfer function

$$G_1(z) = \frac{\theta(z)}{u(z)} = \frac{T^2}{2}\left[\frac{z+1}{(z-1)^2}\right].$$

(a) Sketch the root locus of this system by hand, assuming proportional control.

(b) Draw the root locus using MATLAB to verify the hand sketch.

(c) Add discrete velocity feedback to your controller so that the dominant poles correspond to $\zeta = 0.5$ and $\omega_n = 3\pi/(10T)$.

(d) What is the feedback gain if $T = 1$ sec? If $T = 2$ sec?

(e) Plot the closed-loop step response and the associated control time history for $T = 1$ sec.

8.11 It is possible to suspend a mass of magnetic material by means of an electromagnet whose current is controlled by the position of the mass (Woodson and Melcher, 1968). The schematic of a possible setup is shown in Fig. 8.23, and a photo of a working system at Stanford University is shown in Fig. 9.2. The equations of motion are

$$m\ddot{x} = -mg + f(x, I),$$

where the force on the ball due to the electromagnet is given by $f(x, I)$. At equilibrium the magnet force balances the gravity force. Suppose we let I_0 represent the current at equilibrium. If we write $I = I_0 + i$, expand f about $x = 0$ and $I = I_0$, and neglect higher-order terms, we obtain the linearized equation

$$m\ddot{x} = k_1 x + k_2 i. \tag{8.46}$$

Reasonable values for the constants in Eq. (8.46) are $m = 0.02$ kg, $k_1 = 20$ N/m, and $k_2 = 0.4$ N/A.

(a) Compute the transfer function from I to x, and draw the (continuous) root locus for the simple feedback $i = -Kx$.

Figure 8.23

Schematic of magnetic
levitation device for
Problems 8.11

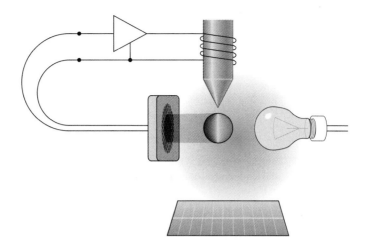

(b) Assume that the input is passed through a ZOH, and let the sampling period be 0.02 sec. Compute the transfer function of the equivalent discrete-time plant.

(c) Design a digital control for the magnetic levitation device so that the closed-loop system meets the following specifications: $t_r \leq 0.1$ sec, $t_s \leq 0.4$ sec, and overshoot $\leq 20\%$.

(d) Plot a root locus with respect to k_1 for your design, and discuss the possibility of using your closed-loop system to balance balls of various masses.

(e) Plot the step response of your design to an initial disturbance displacement on the ball, and show both x and the control current i. If the sensor can measure x only over a range of $\pm 1/4$ cm and the amplifier can provide a current of only 1 A, what is the *maximum* displacement possible for control, neglecting the nonlinear terms in $f(x, I)$?

8.12 Repeat Problem 5.27 in Chapter 5 by constructing discrete root loci and performing the designs directly in the z-plane. Assume that the output y is sampled, the input u is passed through a ZOH as it enters the plant, and the sample rate is 15 Hz.

8.13 Design a digital controller for the antenna servo system shown in Figs. 3.61 and 3.62 and described in Problem 3.31. The design should provide a step response with an overshoot of less than 10% and a rise time of less than 80 sec.

(a) What should the sample rate be?

(b) Use discrete equivalent design with the matched pole–zero method.

(c) Use discrete design and the z-plane root locus.

8.14 The system

$$G(s) = \frac{1}{(s + 0.1)(s + 3)}$$

is to be controlled with a digital controller having a sampling period of $T = 0.1$ sec. Using a z-plane root locus, design compensation that will respond to a step with a rise time $t_r \leq 1$ sec and an overshoot $M_p \leq 5\%$. What can be done to reduce the steady-state error?

8.15 The transfer function for pure derivative control is

$$D(z) = KT_D \frac{z - 1}{Tz},$$

where the pole at $z = 0$ adds some destabilizing phase lag. Can this phase lag be removed by using derivative control of the form

$$D(z) = KT_D \frac{(z - 1)}{T}?$$

Support your answer with the difference equation that would be required and discuss the requirements to implement it.

9

Nonlinear Systems

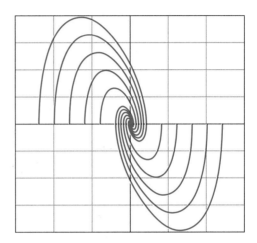

Perspective on Nonlinear Systems

All systems are nonlinear, especially if large signals are considered. On the other hand, almost all physical systems can be well approximated by linear models if the signals are small. For example, if θ is small, then $\sin(\theta) \approx \theta$ and $\cos(\theta) \approx 1$. Similarly, in analog electronic devices such as amplifiers, the operation will be nearly linear if the signals are small with respect to the supply voltage. Finally, as we will consider later in this chapter in an optional section, Lyapunov showed that if the linear approximation of a system is stable near an equilibrium point, then the truly nonlinear system will be stable for some neighborhood of the equilibrium point. For all these reasons, the analysis and design methods presented thus far in this book have considered only the enormously powerful techniques available for linear models. However, if the signals cause a device to saturate or if the system includes nonlinearities that are active for small signals, such as some kinds of friction, then the nonlinear effects must be taken into account to explain the behavior of the system. In this chapter, a few of the tools available for this purpose will be described.

Chapter Overview

Because every nonlinear system is in many ways unique, a vast number of approaches are used in nonlinear control design. The approaches to analysis and design of nonlinear systems that we will describe may be classified under four categories. In Section 9.2 methods of reducing the problem to a linear model are discussed. In most cases, considering the small signal approximation is adequate. In some cases there are nonlinearities for which an inverse can be found, and placing the inverse before the physical nonlinearity results in an overall system that responds linearly. In yet other cases, some nonlinear models can be reduced to an exact linear form by the clever use of feedback, in a technique called "computed torque" in the field of robotics.

The second category is a heuristic approach based on considering the nonlinearity to be a varying gain. In Section 9.3 cases are considered for which the nonlinearity has no memory as, for example, with an amplifier whose output saturates when the signal gets large. The idea is to consider the amplifier as if its gain begins to be reduced as the signal gets large. Because the root locus is based on evaluating the system characteristic roots as gain changes, this point of view leads to a heuristic use of the root locus to predict how such a system will respond to changing input signal sizes. Section 9.4 treats cases in which the nonlinearity has dynamics or memory; then the root locus is not useful. For these cases a technique introduced by Kochenburger in 1950 known as the describing function can be used. To apply this method, a sinusoid is applied to the nonlinear part of the system and the first harmonic of the periodic response is computed. The ratio of the input to the output is taken as if it were a linear but variable frequency response. Thus the Nyquist plot is the natural domain in which to consider the system behavior.

While the heuristic approaches may give very useful insight into the system's behavior, they cannot be used to decide if the system is guaranteed to be stable. For this, one must turn to the analysis of stability as studied in control theory. The most famous of these theories is that of internal stability developed by Lyapunov. As an introduction to the idea of a system response as a trajectory in space, Section 9.5 describes analysis in the phase plane and then presents the stability theory. Examples are given of using the stability theorem to guide design of a controller so the system is guaranteed to be stable if the initial assumptions about the system hold. With these methods, the control engineer is given a start on the path to the effective understanding and design of real control problems. Finally Section 9.6 provides a historical perspective for the material in this chapter.

9.1 Introduction and Motivation: Why Study Nonlinear Systems?

It is intuitively clear that at some level of signal strength any physical system will be nonlinear and that some systems are nonlinear at any and all signal levels. On the other hand, we began our study by developing linear approximate models, and all our design methods thus far have been based on the assumption that the plant can be represented by a linear transfer function. In this chapter we shall give some of the reasons for believing that all the time spent studying linear techniques was not a waste of time, but we shall also try to explain why it is very important to understand how to take nonlinear effects into account in control system design.

We begin by showing that we can combine the root-locus technique, which plots roots of the characteristic equation as a function of various gain values, with the observation that many nonlinear elements can be viewed as a gain that changes as signal level changes. While the method is, at this point, entirely heuristic, the simulation results are very promising. Many properties of systems containing such zero-memory nonlinear elements can be predicted by plotting a root locus versus gain at the point of the nonlinearity. However, the method, as presented, is not on a firm foundation, and the designer is left to wonder if there is some unexplored region of the real state space or the signal spaces where catastrophe awaits. After all, the model is an approximation, and no matter how extensive the simulation, it is not possible to cover every situation.

Following the use of the root locus, we turn to methods based on the frequency response. One of the great advantages of the frequency response is that in many cases one can obtain the transfer function experimentally on the real system. In the most basic approach, a sinusoidal signal is applied to the system and the amplitude and phase of the output sinusoid are measured. However, noise and inevitable nonlinear effects cause the output to be more complicated than a simple sinusoid, so the designer extracts the fundamental component and treats it as if it is the whole story. One gets the same result if a spectrum analyzer is used to compute a transfer function. What has been done is to compute what Kochenburger called the **describing function**. From this point of view, a describing function can be defined for nonlinear elements, including those with memory. Again, simulations are promising and many useful designs are done with this technique but, as with the use of root locus to design nonlinear systems, this method is also on shaky ground.[1]

So what is to be made of this situation? The only possibility is to face up to the facts and take on nonlinear behavior directly. Fortunately, a firm foundation in mathematics was established when A. M. Lyapunov published his work on the stability of motion in 1892. This work was translated into French in 1907 and recovered in a control context by Kalman and Bertram in 1960. Lyapunov gave two methods for the study of stability. For his first method, he considered stability based on the linear approximation, the very thing required to justify our concentration on that approach in this book. He proved the remarkable result that if the linear approximation is strictly stable, having all roots in the left half-plane (LHP), then the nonlinear system will have a region of stability around the equilibrium point where the linear approximation applies. Furthermore, he proved that if the linear approximation has at least one root in the right half-plane (RHP), then the nonlinear system cannot have any region of stability in the neighborhood of the equilibrium. The size of the stable region in the state space is not given by the linear terms but is included in the construction used for the proof. That construction constituted his second method. Lyapunov's second method is based on the mathematical equivalent of finding a scalar function that

---

[1] And as we live in California, we know how dangerous it is to be on shaky ground.

describes the internal energy stored in the system. He proved that if such a function is constructed, and if the derivative of the function is negative on trajectories of the equations of motion, then the function and the state on which it depends will eventually drain away and the state will come to rest at the equilibrium point. A function having these properties is called a *Lyapunov function*. Of course, this simple description omits a great deal of complexity; for example, there are dozens of definitions of stability. However, the concept remains that if a Lyapunov function can be found, then the system on which it is based will be stable. As described, the theory gives a sufficient condition for stability. If a Lyapunov function is not found, the designer does not know if one does not exist or if the search has just been inadequate. A great deal of research has been directed toward finding Lyapunov functions for particular classes of nonlinear systems.

Lyapunov's methods are based on differential equations in normal or state form and are thus concerned with internal stability. Frequency-response methods, on the other hand, are external measures, and there has been interest in developing stability results based on the external response of the system. One such method is the circle criterion, which we will describe in this chapter also. The method can be described as considering the energy seen at the terminal of the system and noting if it is always flowing *into* the terminals. If so, it is reasonable to assume that eventually all energy will be gone and the system will be stable. For a formal proof of the method, researchers have turned to Lyapunov's second method, but the result is expressed in terms of external properties, such as the Nyquist plot of the linear portion of the system that faces the nonlinear elements. Again, this tool gives a basis for setting a firm foundation under a method of design for a particular class of nonlinear systems.

As should be clear at this point, the theory of nonlinear control is a vast and sophisticated topic, and in this book we can give only a brief introduction to a small part of it. However, the foundation of control design rests on this theory, and the more the designer understands of the theory, the better he or she understands both the limits and the opportunities of problems. It is our hope that by considering this material, students will be stimulated to further their profitable study of this fascinating topic.

9.2   Analysis by Linearization

Three methods of reducing some nonlinear systems to a suitable linear model are presented in this section. The differential equations of motion for almost all processes selected for control are nonlinear. On the other hand, both analysis and control design methods we have discussed so far are much easier for linear than for nonlinear models. **Linearization** is the process of finding a linear model that approximates a nonlinear one. Fortunately, as Lyapunov proved over 100 years ago, if a small-signal linear model is valid near an equilibrium and is stable, then there is a region (which may be small, of course) containing the equilibrium within which the nonlinear system is stable. So we can safely make a linear model and design a linear control for it such that, at least in the neighborhood of the equilibrium, our design will be stable. Since a very important role of feedback control is to maintain the process variables near equilibrium, such small-signal linear models are a frequent starting point for control design.

An alternative approach to obtain a linear model for use as the basis of control system design is to use part of the control effort to cancel the nonlinear terms and to design the remainder of the control based on linear theory. This approach—linearization by feedback—is popular in the field of robotics, where it is called the **method of computed torque**. It is also a research topic for control of aircraft. Section 9.2.2 takes a brief look at this method. Finally, some nonlinear functions are such that an **inverse nonlinearity** can be found to be placed in series with the nonlinearity so that the combination is linear. This method is often used to correct mild nonlinear characteristics of sensors or actuators that have small variations in use, as discussed in Section 9.2.3.

9.2.1 Linearization by Small-Signal Analysis

For a system with smooth nonlinearities and a continuous derivative, one can compute a linear model that is valid for small signals. In many cases these models can be used for design. A nonlinear differential equation is an equation for which the derivatives of the state have a nonlinear relationship to the state itself and/or the control. In other words, the differential equations *cannot* be written in the form[2]

$$\dot{\mathbf{x}} = \mathbf{F}\mathbf{x} + \mathbf{G}u,$$

but must be left in the form

$$\dot{\mathbf{x}} = \mathbf{f}(\mathbf{x}, u). \tag{9.1}$$

For small-signal linearization we first determine equilibrium values of $\mathbf{x}_o$, $\mathbf{u}_o$, such that $\dot{\mathbf{x}}_o = \mathbf{0} = \mathbf{f}(\mathbf{x}_o, u_o)$, and we let $\mathbf{x} = \mathbf{x}_o + \delta\mathbf{x}$ and $u = u_o + \delta u$. We then expand the nonlinear equation in terms of the perturbations from these equilibrium values, which yields

$$\dot{\mathbf{x}}_o + \delta\dot{\mathbf{x}} \cong \mathbf{f}(\mathbf{x}_o, u_o) + \mathbf{F}\delta\mathbf{x} + \mathbf{G}\delta u,$$

where $\mathbf{F}$ and $\mathbf{G}$ are the best linear fits to the nonlinear function $\mathbf{f}(\mathbf{x}, u)$ at $\mathbf{x}_o$ and u_o, computed as

$$\mathbf{F} = \left[\frac{\partial \mathbf{f}}{\partial \mathbf{x}}\right]_{x_o, u_o} \quad \text{and} \quad \mathbf{G} = \left[\frac{\partial \mathbf{f}}{\partial u}\right]_{x_o, u_o}. \tag{9.2}$$

Subtracting out the equilibrium solution, this reduces to

$$\delta\dot{\mathbf{x}} = \mathbf{F}\delta\mathbf{x} + \mathbf{G}\delta u, \tag{9.3}$$

which is a linear differential equation approximating the dynamics of the motion *about* the equilibrium point. Normally, the δ notation is dropped and it is understood that $\mathbf{x}$ and u refer to the deviation from the equilibrium.

In developing the models discussed so far in this book, we have encountered nonlinear equations on several occasions: the pendulum in Example 2.5, the hanging crane in Example 2.7, the AC induction motor in Section 2.3, the tank flow in Example 2.16, and the hydraulic actuator in Example 2.17. In each case, we assumed either that the motion was small or that motion from some operating point was small, so that nonlinear functions were approximated by linear functions. The steps followed in those examples essentially involved finding $\mathbf{F}$ and $\mathbf{G}$ in order to linearize the differential equations to the form of Eq. (9.3), as illustrated in the next several examples. The linearization functions in MATLAB® include linmod and linmod2.

---

[2]This equation assumes the system is time invariant. A more general expression would be $\dot{\mathbf{x}} = \mathbf{f}(\mathbf{x}, u, t)$.

EXAMPLE 9.1   *Linearization of Nonlinear Pendulum*

Consider the nonlinear equations of motion of the simple pendulum in Example 2.5. Derive the equilibrium points for the system and determine the corresponding small-signal linear models.

Solution. The equation of motion is

$$\ddot{\theta} + \frac{g}{\ell}\sin\theta = \frac{T_c}{m\ell^2}. \tag{9.4}$$

We can rewrite the equation of motion in state-variable form, with $\mathbf{x} = [\ x_1 \quad x_2\ ]^T = [\ \theta \quad \dot{\theta}\ ]^T$, as

$$\dot{\mathbf{x}} = \begin{bmatrix} x_2 \\ -\omega_o^2 \sin x_1 + u \end{bmatrix} = \begin{bmatrix} f_1(\mathbf{x}, u) \\ f_2(\mathbf{x}, u) \end{bmatrix} = \mathbf{f}(\mathbf{x}, u),$$

where $\omega_o = \sqrt{\dfrac{g}{\ell}}$ and $u = \dfrac{T_c}{m\ell^2}$. To determine the equilibrium state, suppose that the (normalized) input torque has a nominal value of $u_o = 0$. Then

$$\dot{x}_1 = \dot{\theta} = 0,$$

$$\dot{x}_2 = \ddot{\theta} = -\frac{g}{\ell}\sin\theta = 0,$$

so that the equilibrium conditions correspond to $\theta_o = 0, \pi$ (i.e., the downward and the inverted pendulum at rest configurations, respectively). The equilibrium state and the input are $\mathbf{x}_o = [\ \theta_o \quad 0\ ]^T$, $u_o = 0$, and the state-space matrices are given by

$$\mathbf{F} = \begin{bmatrix} \dfrac{\partial f_1}{\partial x_1} & \dfrac{\partial f_1}{\partial x_2} \\ \dfrac{\partial f_2}{\partial x_1} & \dfrac{\partial f_2}{\partial x_2} \end{bmatrix}_{x_o, u_o} = \begin{bmatrix} 0 & 1 \\ -\omega_o^2 \cos\theta_o & 0 \end{bmatrix},$$

$$\mathbf{G} = \begin{bmatrix} \dfrac{\partial f_1}{\partial u} \\ \dfrac{\partial f_2}{\partial u} \end{bmatrix}_{x_o, u_o} = \begin{bmatrix} 0 \\ 1 \end{bmatrix}.$$

The linear system has eigenvalues of $\pm j\omega_o$ and $\pm\omega_o$ corresponding to $\theta_o = 0$ and π, respectively, with the latter inverted case being unstable as expected.

EXAMPLE 9.2   *Linearization of Motion in a Ball Levitator*

Figure 9.1 shows a magnetic bearing used in large turbo machinery. The magnetics are energized using feedback control methods so that the axle is always in the center and never touches the magnets, thus keeping friction to an almost nonexistent level. A simplified version of a magnetic bearing that can be built in a laboratory is shown in Fig. 9.2, where one electromagnet is used to levitate a metal ball. The physical arrangement of the levitator is depicted in Fig. 9.3. The equation of motion of the ball, derived from Newton's law, Eq. (2.1), is

$$m\ddot{x} = f_m(x, i) - mg, \tag{9.5}$$

Figure 9.1

A magnetic bearing

Source: Photo courtesy of Magnetic Bearings, Inc.

Figure 9.2

Magnetic ball levitator used in the laboratory

Source: Photo courtesy of Gene Franklin

where the force $f_m(x, i)$ is caused by the field of the electromagnet. Theoretically, the force from an electromagnet falls off with an inverse square relationship to the distance from the magnet, but the exact relationship for the laboratory levitator is difficult to derive from physical principles because its magnetic field is so complex. However, the forces can be measured with a scale. Figure 9.4 shows the experimental curves for a ball with a 1-cm diameter and a mass of 8.4×10^{-3} kg. At the value for the current of $i_2 = 600$ mA and the displacement x_1 shown in the figure, the magnetic force f_m just cancels the gravity force, $mg = 82 \times 10^{-3}$ N. (The mass of the ball is 8.4×10^{-3} kg, and the acceleration of gravity is 9.8 m/sec^2.) Therefore the point (x_1, i_2) represents an equilibrium. Using the data, find the linearized equations of motion about the equilibrium point.

Figure 9.3

Model for ball levitation

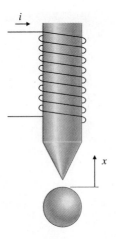

Figure 9.4

Experimentally determined force curves

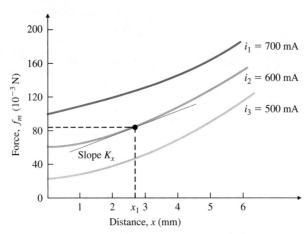

Solution. First we write, in expanded form, the force in terms of deviations from the equilibrium values x_1 and i_2:

$$f_m(x_1 + \delta x, i_2 + \delta i) \cong f_m(x_1, i_2) + K_x \delta x + K_i \delta i. \tag{9.6}$$

The linear gains are found as follows: K_x is the slope of the force versus x along the curve $i = i_2$, as shown in Fig. 9.4, and is found to be about 14 N/m. K_i is the change of force with current for the value of fixed $x = x_1$. We find that for $i = i_1 = 700$ mA at $x = x_1$, the force is about 122×10^{-3} N, and at $i = i_3 = 500$ mA at $x = x_1$, it is about 42×10^{-3} N. Thus

$$K_i \cong \frac{122 \times 10^{-3} - 42 \times 10^{-3}}{700 - 500} = \frac{80 \times 10^{-3} \text{ N}}{200 \text{ mA}}$$

$$\cong 400 \times 10^{-3} \text{ N/A}$$

$$\cong 0.4 \text{ N/A}.$$

Substituting these values into Eq. (9.6) leads to the following linear approximation for the force in the neighborhood of equilibrium:

$$f_m \cong 82 \times 10^{-3} + 14\delta x + 0.4\delta i.$$

Substituting this expression into Eq. (9.5) and using the numerical values for mass and gravity force, we get for the linearized model

$$(8.4 \times 10^{-3})\ddot{x} = 82 \times 10^{-3} + 14\delta x + 0.4\delta i - 82 \times 10^{-3}.$$

Because $x = x_1 + \delta x$, then $\ddot{x} = \delta\ddot{x}$. The equation in terms of δx is thus

$$(8.4 \times 10^{-3})\delta\ddot{x} = 14\delta x + 0.4\delta i,$$

$$\delta\ddot{x} = 1667\delta x + 47.6\delta i, \tag{9.7}$$

which is the desired linearized equation of motion about the equilibrium point. A logical state vector is $\mathbf{x} = [\delta x \ \delta\dot{x}]^T$, which leads to the standard matrices

$$\mathbf{F} = \begin{bmatrix} 0 & 1 \\ 1667 & 0 \end{bmatrix} \quad \text{and} \quad \mathbf{G} = \begin{bmatrix} 0 \\ 47.6 \end{bmatrix}$$

and the control $u = \delta i$.

EXAMPLE 9.3

Linearization of the Water Tank Revisited

Repeat the linearization of Example 2.16 using the concepts presented in this section.
Solution. Equation (2.75) may be written as

$$\dot{x} = f(x, u), \tag{9.8}$$

where $x \triangleq h$, $u \triangleq w_{in}$, and $f(x, u) = -\frac{1}{RA\rho}\sqrt{p_1 - p_a} + \frac{1}{A\rho}w_{in} = -\frac{1}{RA\rho}\sqrt{\rho g h - p_a} + \frac{1}{A\rho}w_{in}$. The linearized equations are of the form

$$\delta\dot{x} = F\delta x + G\delta u, \tag{9.9}$$

where

$$[F]_{x_o, u_o} = \frac{\partial f}{\partial x} = \left[\frac{\partial f}{\partial h}\right]_{h_o, u_o} = \frac{\partial}{\partial h}\left[-\frac{1}{RA\rho}\sqrt{\rho g h - p_a}\right]_{h_o, u_o} \tag{9.10}$$

$$= -\frac{g}{2AR}\frac{1}{\sqrt{\rho g h_o - p_a}} = -\frac{g}{2AR}\frac{1}{\sqrt{p_o - p_a}} \tag{9.11}$$

and

$$[G]_{x_o, u_o} = \frac{\partial f}{\partial u} = \frac{\partial f}{\partial w_{in}} = \frac{1}{A\rho}. \tag{9.12}$$

However, note that some flow is required to maintain the system in equilibrium so that Eq. (9.9) is valid; specifically, we see from Eq. (2.75) that

$$u_o = w_{in_o} = \frac{1}{R}\sqrt{p_o - p_a} \quad \text{for} \quad \dot{h} = 0 \tag{9.13}$$

and the δu in Eq. (9.9) is δw_{in}, where $w_{in} = w_{in_o} + \delta w_{in}$. Therefore, Eq. (9.9) becomes

$$\delta\dot{h} = F\delta h + G\delta w_{in} = F\delta h + Gw_{in} - G\frac{1}{R}\sqrt{p_o - p_a} \tag{9.14}$$

and matches Eq. (2.78) precisely.

9.2.2 Linearization by Nonlinear Feedback

Nonlinear feedback

Linearization by feedback is accomplished by subtracting the nonlinear terms out of the equations of motion and adding them to the control. The result is a linear system, provided that the computer implementing the control has enough capability to compute the nonlinear terms fast enough and the resulting control does not cause the actuator to saturate. A more detailed understanding of the method is best achieved through example.

EXAMPLE 9.4

Linearization of the Nonlinear Pendulum

Consider the equation of a simple pendulum developed in Example 2.5 [Eq. (2.21)]. Linearize the system by using nonlinear feedback.

Solution. The equation of motion is

$$ml^2\ddot{\theta} + mgl \sin \theta = T_c. \tag{9.15}$$

If we compute the torque to be

$$T_c = mgl \sin \theta + u, \tag{9.16}$$

then the motion is described by

$$ml^2\ddot{\theta} = u. \tag{9.17}$$

Eq. (9.17) is a linear equation *no matter how large the angle θ becomes*. We use it as the model for purposes of control design because it enables us to use linear analysis techniques. The resulting linear control will provide the value of u based on measurements of θ; however, the value of the torque actually sent to the equipment would derive from Eq. (9.16). For robots with two or three rigid links, this **computed-torque** approach has led to effective control. It is also being researched for the control of aircraft, where the linear models change considerably in character with the flight regime.

9.2.3 Linearization by Inverse Nonlinearity

Inverse nonlinearity

The simplest case of the introduction of nonlinearities into a control design is that of **inverse nonlinearities**. It is sometimes possible to reverse the effect of some nonlinearities. For example, suppose we have a system whose output is the square of the signal of interest:

$$y = x^2. \tag{9.18}$$

One clever and rather obvious technique is to undo the nonlinearity by preceding the physical nonlinearity with a square root nonlinearity,

$$x = \sqrt{(.)}, \tag{9.19}$$

as shown in the next example. The overall cascaded system would then be linear.

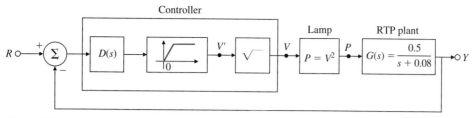

Figure 9.5
Linearization through inverse nonlinearity

EXAMPLE 9.5

Linearization of the Rapid Thermal Processing (RTP) System

Consider the RTP system that uses a nonlinear lamp as an actuator, as shown in Fig. 9.5. Suppose the input to the lamp is voltage V and the output is power P and they are related by

$$P = V^2.$$

Design an inverse nonlinearity to linearize the system.

Solution. We simply precede the lamp input nonlinearity with a square root nonlinearity

$$V = \sqrt{V'}.$$

The overall open-loop cascaded system is now linear for any value of the voltage:

$$Y = G(s)P = G(s)V^2 = G(s)V'.$$

Thus we can use linear control design techniques for the dynamic compensator, $D(s)$. Note that a nonlinear element has been inserted in front of the square root element to ensure that the input to this block remains nonnegative at all times. The controller is then implemented as shown in Fig. 9.5. For a detailed application of this method for control design, we refer the reader to the RTP case study in Section 10.6.

9.3   Equivalent Gain Analysis Using the Root Locus

As we have tried to make clear, every real control system is nonlinear, and the linear analysis and design methods we have described so far use linear approximations to the real models. There is one important category of nonlinear systems for which linearization is not appropriate and for which some significant analysis (and design) can be done. This category comprises the systems in which the nonlinearity has no dynamics and is well approximated as a gain that varies as the size of its input signal varies. Sketches of a few such nonlinear system elements and their common names are shown in Fig. 9.6.

Memoryless nonlinearity

The stability of systems with memoryless nonlinearities can be studied heuristically using the root locus. The technique is to replace the memoryless nonlinearity by an equivalent gain K, and a root locus is plotted versus this gain. For a range of input signal amplitudes, the equivalent gain will take on a range of values, and the

Figure 9.6

Nonlinear elements
with no dynamics:
(a) saturation;
(b) relay; (c) relay with
dead zone; (d) gain
with dead zone;
(e) preloaded spring, or
coulomb plus viscous
friction;
(f) quantization

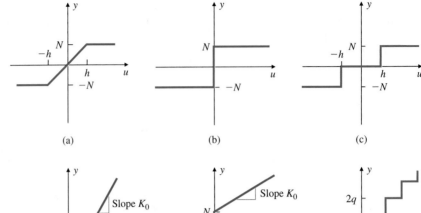

(a)   (b)   (c)

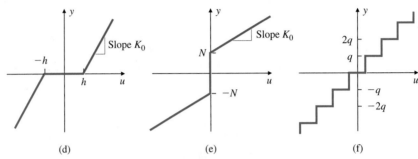

(d)   (e)   (f)

Figure 9.7

Dynamic system with
saturation

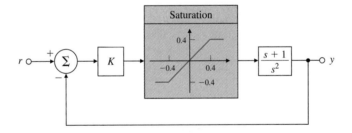

closed-loop roots of the system are examined in this range as if the gain were fixed.
This is illustrated by the next several examples.

EXAMPLE 9.6   *Changing Overshoot and Saturation Nonlinearity*

Consider the system with saturation shown in Fig. 9.7. Determine the stability
properties of the system using the root locus technique.

Solution. The root locus of this system versus K with the saturation removed is
given by Fig. 9.8. At $K = 1$ the damping ratio is $\zeta = 0.5$. As the gain is reduced,
the locus shows that the roots move toward the origin of the s-plane with less and
less damping. Plots of the step responses of this system were obtained using the
SIMULINK® program. A series of step inputs r with magnitudes $r_0 = 2, 4, 6, 8,$
10, and 12 was introduced to the system, and the results are shown in Fig. 9.9. As
long as the signal entering the saturation remains less than 0.4, the system will be
linear and should behave according to the roots at $\zeta = 0.5$. However, notice that
as the input gets larger, the response has more and more overshoot and slower and

Figure 9.8

Root locus of $(s+1)/s^2$, the system in Fig. 9.7 with the saturation removed

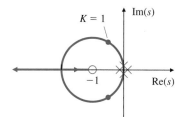

Figure 9.9

Step responses of system in Fig. 9.7 for various input step sizes

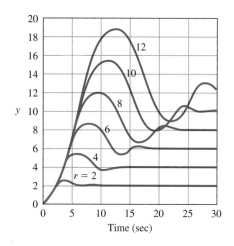

Figure 9.10

General shape of the effective gain of saturation

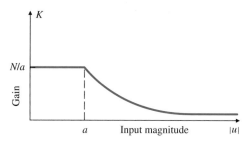

slower recovery. This can be explained by noting that larger and larger input signals correspond to smaller and smaller effective gain K, as seen in Fig. 9.10. From the root-locus plot of Fig. 9.8, we see that as K decreases, the closed-loop poles move closer to the origin and have a smaller damping ζ. This results in the longer rise and settling times, increased overshoot, and greater oscillatory response.

EXAMPLE 9.7

Stability of Conditionally Stable System Using the Root Locus

A nonlinear example: stability depends on input magnitude

As a second example of a nonlinear response described by signal-dependent gain, consider the system with a saturation nonlinearity as shown in Fig. 9.11. Determine whether the system is stable.

Figure 9.11

Block diagram of a conditionally stable system

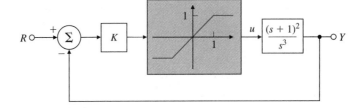

Figure 9.12

Root locus for $G(s) = (s + 1)^2/s^3$ from system in Fig. 9.11

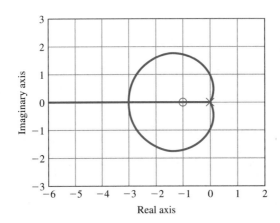

Figure 9.13

Step responses of system in Fig. 9.11

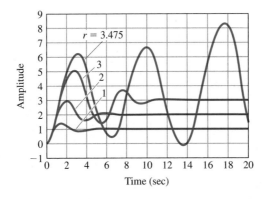

Conditional stability

Solution. The root locus for the system, excluding the saturation, is plotted in Fig. 9.12. From this locus we can readily calculate that the imaginary axis crossing occurs at $\omega_0 = 1$ and $K = \frac{1}{2}$. Systems such as this, which are stable for (relatively) large gains but unstable for smaller gains, are called **conditionally stable systems**. If $K = 2$, which corresponds to $\zeta = 0.5$ on the locus, the system would be expected to show responses consistent with $\zeta = 0.5$ for small reference input signals. However, as the reference input size gets larger, the equivalent gain would get smaller due to the saturation, and the system would be expected to become less well damped. Finally, the system would be expected to become unstable at some point for large inputs. Step responses from nonlinear simulation of the system with $K = 2$ for input steps of size

$r_0 = 1.0, 2.0, 3.0$, and 3.4 are shown in Fig. 9.9. These responses confirm our predictions. Furthermore, the marginally stable case shows oscillations near 1 rad/sec, which is predicted by the frequency at the point at which the root locus crosses into the RHP.

EXAMPLE 9.8

Analysis and Design of System with Limit Cycle Using the Root Locus

A nonlinear example: an oscillatory system with saturation

The final illustration of the use of the root locus to give a qualitative description of the response of a nonlinear system is based on the block diagram in Fig. 9.14. Determine whether the system is stable and find the amplitude and the frequency of the limit-cycle. Modify the controller design to minimize the effect of limit-cycle oscillations.

Solution. This system is typical of electromechanical control problems in which the designer perhaps at first is not aware of the resonant mode corresponding to the denominator term $s^2 + 0.2s + 1, (\omega = 1, \zeta = 0.1)$. The root locus for this system versus K, excluding the saturation, is sketched in Fig. 9.15. The imaginary-axis crossing can be verified to be at $\omega_0 = 1, K = 0.2$; thus a gain of $K = 0.5$ is enough to force the roots of the resonant mode into the RHP, as shown by the dots. If the system gain is set at $K = 0.5$, our analysis predicts a system that is initially unstable but becomes stable as the gain decreases. Thus we would expect the response of the system with the saturation to build up due to the instability until the magnitude is sufficiently large that the effective gain is lowered to $K = 0.2$ *and then stop growing!*

Plots of the step responses with $K = 0.5$ for three steps of size $r_0 = 1, 4$, and 8 are shown in Fig. 9.16, and again our heuristic analysis is exactly correct: The error builds up to a fixed amplitude and then starts to oscillate at a fixed amplitude. The oscillations have a frequency of ≈ 1 rad/sec and hold constant amplitude regardless of the step sizes of the input. In this case, the response always approaches a periodic solution of fixed amplitude known as a **limit cycle**, so-called because the response is cyclic and is approached in the limit as time grows large.

Limit cycle

We can return to Fig. 9.13 and be easily convinced that the first transient to a step of size 3 is nearly a sinusoid. We can predict that the system is just on the border of stability for an equivalent gain corresponding to a root locus gain of $1/2$ when the locus crosses into the RHP. In order to prevent the limit cycle, the locus has to be modified by compensation so that no branches cross into the RHP. One common method of doing this for a lightly damped oscillatory mode is to place compensation zeros near the poles at a frequency such that the angle of departure of the root-locus branch from these poles is toward the LHP, a procedure called phase stabilization earlier. Example 5.8 for collocated mechanical motion demonstrated that a pole–zero

Figure 9.14

Block diagram of a system with an oscillatory mode

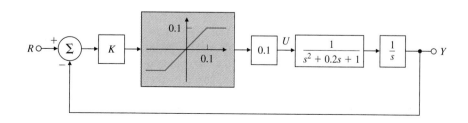

Figure 9.15

Root locus for the system in Fig. 9.14

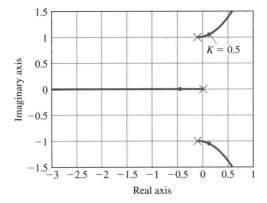

Figure 9.16

Step responses of system in Fig. 9.14

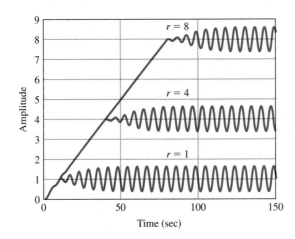

Figure 9.17

Root locus including compensation

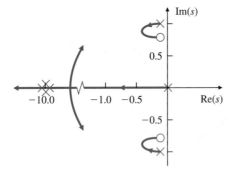

pair located in this manner will often cause a locus branch to go from the pole to the zero, looping to the left, and thus staying away from the RHP. Figure 9.17 shows the root locus for the system, $1/[s(s^2 + 0.2s + 1.0)]$, including a **notch compensation** with zeros located as just discussed. In addition, the compensation also includes two poles to make the compensation physically realizable. In this case, both poles were

Figure 9.18

Block diagram of the system with a notch filter

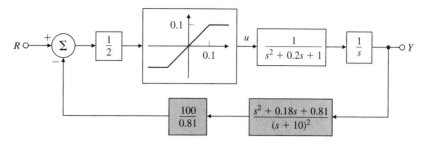

Figure 9.19

Step responses of the system in Fig. 9.18

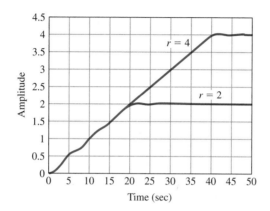

placed at $s = -10$, fast enough to not cause stability problems with the system, yet slow enough that high-frequency noise would not be amplified too much. Thus the compensation used for the root locus is

$$D(s) = 123\frac{s^2 + 0.18s + 0.81}{(s + 10)^2},$$

where the gain of 123 has been selected to make the compensation's DC gain equal to unity. This notch filter compensation attenuates inputs in the vicinity of $\omega_n^2 = 0.81$ or $\omega_n = 0.9$ rad/sec, so that any input from the plant resonance is attenuated and is thus prevented from detracting from the stability of the system. Figure 9.18 shows the system, including the notch filter, and Fig. 9.19 shows the time response for two step inputs. Both inputs, $r_0 = 2$ and 4, are sufficiently large that the nonlinearity is saturated initially; however, because the system is unconditionally stable, the saturation results only in lowering the gain, so the response is slower than predicted by linear analysis but still stable, as also predicted by our piecewise linear analysis. In both cases the nonlinearity eventually becomes unsaturated, and the system stabilizes to its new commanded value of r.

9.3.1 Integrator Antiwindup

In any control system the output of the actuator can saturate because the dynamic range of all real actuators is limited. For example, a valve saturates when it is fully open or closed, the control surfaces on an aircraft cannot be deflected beyond certain angles from their nominal positions, electronic amplifiers can produce only finite voltage outputs, etc. Whenever actuator saturation happens, the control signal to the process stops changing and the feedback path is effectively opened. If the error signal continues to be applied to the integrator input under these conditions, the integrator output will grow (windup) until the sign of the error changes and the integration turns around. The result can be a very large overshoot, as the output must grow to produce the necessary unwinding error, and poor transient response is the result. In effect, the integrator is an unstable element in open loop and must be stabilized when saturation occurs.[3]

Consider the feedback system shown in Fig. 9.20. Suppose a given reference step is more than large enough to cause the actuator to saturate at u_{max}. The integrator continues integrating the error e, and the signal u_c keeps growing. However, the input to the plant is stuck at its maximum value, namely $u = u_{max}$, so the error remains large until the plant output exceeds the reference and the error changes sign. The increase in u_c is not helpful since the input to the plant is not changing, but u_c may become quite large if saturation lasts a long time. It will then take a considerable negative error e and the resulting poor transient response to bring the integrator output back to within the linear band where the control is not saturated.

The solution to this problem is an **integrator antiwindup** circuit, which "turns off" the integral action when the actuator saturates. (This can be done quite easily with logic, if the controller is implemented digitally, by including a statement such as "if $|u| = u_{max}$, $k_I = 0$;" see Chapter 8.) Two equivalent antiwindup schemes are shown in Fig. 9.21(a, b) for a PI controller. The method in Fig. 9.21(a) is somewhat easier to understand, whereas the one in Fig. 9.21(b) is easier to implement, as it does not require a separate nonlinearity but uses the saturation itself.[4] In these schemes, as soon as the actuator saturates, the feedback loop around the integrator becomes active and acts to keep the input to the integrator at e_1 small. During this time the integrator

Figure 9.20

Feedback system with actuator saturation

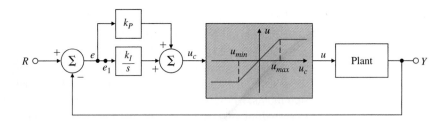

[3] In process control, integral control is usually called **reset control**, and so integrator windup is often called **reset windup**. Without integral control, a given setpoint of, say, 10 results in a response of less value, say, 9.9. The operator must then *reset* to 10.1 to bring the output to the desired value of 10. With integral control the controller automatically brings the output to 10 with a setpoint of 10; hence the integrator does *automatic reset*.

[4] In some cases, especially with mechanical actuators such as an aircraft control surface or a flow control valve, it is not desirable and may cause damage to have the physical device bang against its stops. In such cases it is common practice to include an *electronic* saturation with lower limits than those of the physical device, so that the system hits the electrical stops just before the physical device will saturate.

Figure 9.21

Integrator antiwindup techniques

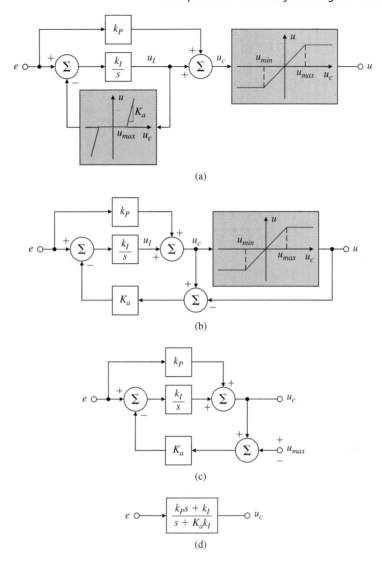

(a)

(b)

(c)

(d)

essentially becomes a fast first-order lag. To see this, note that we can redraw the portion of the block diagram in Fig. 9.21(a) from e to u_c as shown in Fig. 9.21(c). The integrator part then becomes the first-order lag shown in Fig. 9.21(d). The antiwindup gain, K_a, should be chosen to be large enough that the antiwindup circuit keeps the input to the integrator small under all error conditions.

The effect of the antiwindup is to reduce both the overshoot and the control effort in the feedback system. Implementation of such antiwindup schemes is a necessity in any practical application of integral control, and omission of this technique may lead to serious deterioration of the response. From the point of view of stability, the effect of the saturation is to open the feedback loop and leave the open loop plant

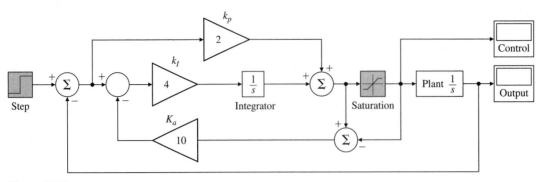

Figure 9.22
SIMULINK® diagram for the antiwindup example

with a constant input and the controller as an open-loop system with the system error as input.

Purpose of antiwindup

> The purpose of the antiwindup is to provide local feedback to make the controller stable alone when the main loop is opened by signal saturation, and any circuit that does this will perform as antiwindup.[5]

EXAMPLE 9.9   *Antiwindup Compensation for a PI Controller*

Consider a plant with the transfer function for small signals,

$$G(s) = \frac{1}{s},$$

and a PI controller,

$$D_c(s) = k_p + \frac{k_I}{s} = 2 + \frac{4}{s},$$

in the unity feedback configuration. The input to the plant is limited to ± 1.0. Study the effect of antiwindup on the response of the system.

Solution. Suppose we use an antiwindup circuit with a feedback gain of $K_a = 10$, as shown in the SIMULINK® block diagram of Fig. 9.22. Figure 9.23(a) shows the step response of the system with and without the antiwindup element. Figure 9.23(b) shows the corresponding control effort. Note that the system with antiwindup has substantially less overshoot and less control effort.

---

[5]A more sophisticated scheme might use an antiwindup feedback at a lower level of saturation than that imposed by the actuator, so PD control continues for a time after integration has been stopped. Any such scheme needs to be analyzed carefully to evaluate its performance and to assure stability.

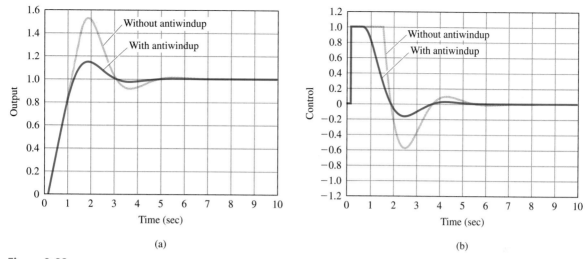

Figure 9.23

Integrator antiwindup: (a) step response; (b) control effort

9.4 Equivalent Gain Analysis Using Frequency Response: Describing Functions

The behavior of systems containing any one of the nonlinearities shown in Fig. 9.6 can be qualitatively described by considering the nonlinear element as a varying signal-dependent gain. For example, with the saturation element (Fig. 9.6a), it is clear that for input signals with magnitudes of less than h, the nonlinearity is linear with the gain N/h. However, for signals larger than h, the output size is bounded by N, while the input size can get much larger than h, so once the input exceeds h, the ratio of output to input goes down. Thus, saturation has the gain characteristics shown in Fig. 9.10. *All* actuators saturate at some level. If they did not, their output would increase to infinity, which is physically impossible. An important aspect of control system design is **sizing the actuator**, which means picking the size, weight, power required, cost, and saturation level of the device. Generally, higher saturation levels require bigger, heavier, and more costly actuators. From the control point of view, the key factor that enters into the sizing is the effect of the saturation on the control system's performance.

A nonlinear analysis method known as **describing functions**, based on the assumption that the input to the nonlinearity is sinusoidal, can be used to predict the behavior of a class of nonlinear systems. A nonlinear element does not have a transfer function. However, for a certain class of nonlinearities, it is possible to replace the nonlinearity by a frequency-dependent equivalent gain for analysis purposes. We can then study the properties of the loop, such as its stability. The describing function method is mostly a heuristic method, and its aim is to try to find something akin to a "transfer function" for a nonlinear element. The idea is that in response to a sinusoidal excitation, most nonlinearities will produce a *periodic* signal (not necessarily sinusoidal) with frequencies being the harmonics of the input frequency. Hence one may view the describing function as an extension of the frequency response to nonlinearities. We can *assume* that in many cases we may approximate the output by the first harmonic alone, and the rest can be neglected. This basic assumption means that the plant behaves approximately as a low-pass filter, and this is luckily a good assumption in most practical situations. The other assumptions behind the describing function are that the nonlinearity is time invariant and that there is a single nonlinear element in the system. Indeed, the describing function is a special case of the more sophisticated harmonic balance analysis. Its roots go back to the early studies in the Soviet Union and elsewhere. The method was introduced by Kochenburger in 1950 in the United States. He proposed that the Fourier series be used to define an equivalent gain, K_{eq} (Truxal, 1955, p. 566). This idea has proved to be very useful in practice. The method is heuristic, but there are attempts at establishing a theoretical justification for the technique (Bergen and Franks, 1973; Khalil, 2002; Sastry, 1999). In fact the method works much better than is warranted by the existing theory!

Consider the nonlinear element $f(u)$ shown in Fig. 9.24. If the input signal $u(t)$ is sinusoidal of amplitude a, or

$$u(t) = a\sin(\omega t), \tag{9.20}$$

then the output $y(t)$ will be *periodic* with a fundamental period equal to that of the input and consequently with a Fourier series described by

$$y(t) = a_0 + \sum_{i=1}^{\infty} a_i \cos(j\omega t) + b_i \sin(j\omega t) \tag{9.21}$$

$$= a_0 + \sum_{i=1}^{\infty} Y_i \sin(j\omega t + \theta_i),$$

where

$$a_i = \frac{2}{\pi} \int_0^{\pi} y(t) \cos(j\omega t)\, d(\omega t), \tag{9.22}$$

$$b_i = \frac{2}{\pi} \int_0^{\pi} y(t) \sin(j\omega t)\, d(\omega t), \tag{9.23}$$

$$Y_i = \sqrt{a_i^2 + b_i^2}, \tag{9.24}$$

$$\theta_i = \tan^{-1}\left(\frac{a_i}{b_i}\right). \tag{9.25}$$

Kochenburger suggested that the nonlinear element could be described by the first fundamental component of this series as if it were a linear system with a gain of Y_1 and phase of θ_1. If the amplitude is varied, the Fourier coefficients and the corresponding phases will vary as a function of the input signal amplitude, due to the nonlinear nature of the element. He called this approximation a **describing function** (*DF*). The describing function is defined as the (complex) quantity that is a ratio of the amplitude of the fundamental component of the output of the nonlinear element to the amplitude

Describing function

Figure 9.24

Nonlinear element

of the sinusoidal input signal and is essentially an "equivalent frequency response" function:

$$DF = K_{eq}(a, \omega) = \frac{b_1 + ja_1}{a} = \frac{Y_1(a, \omega)}{a} e^{j\theta_1} = \frac{Y_1(a, \omega)}{a} \angle\theta_1. \qquad (9.26)$$

Hence the describing function is defined only on the $j\omega$ axis. In the case of memoryless nonlinearities that are also an odd function [i.e., $f(-a) = -f(a)$], then the coefficients of the Fourier series cosine terms are all zeros, and the describing function is simply

$$DF = K_{eq}(a) = \frac{b_1}{a} \qquad (9.27)$$

and is *independent* of the frequency ω. This is the usual case in control, and saturation, relay, and dead-zone nonlinearities all result in such describing functions. Computation of the describing function for the nonlinear characteristics of Fig. 9.6 is generally straightforward, but tedious. It can be done either analytically or numerically and may also be determined by an experiment. We will now focus on computation of several describing functions for some very common nonlinearities.

EXAMPLE 9.10   *Describing Function for a Saturation Nonlinearity*

A saturation nonlinearity is shown in Fig. 9.25(a) and is the most common nonlinearity in control systems. The saturation function (sat) is defined as

$$\text{sat}(x) = \begin{cases} +1, & x > 1, \\ x, & |x| \le 1, \\ -1, & x < -1. \end{cases}$$

If the slope of the linear region is k and the final saturated values are $\pm N$, then the function is

$$y = N \, \text{sat}\left(\frac{k}{N}x\right).$$

Find the describing function for this nonlinearity.

Solution.   Consider the input and output signals of the saturation element shown in Fig. 9.25. For an input sinusoid of $u = a \sin \omega t$ with amplitude $a \le \frac{N}{k}$, the output is such that the DF is just a gain of unity. With $a \ge \frac{N}{k}$, we need to compute the amplitude and phase of the fundamental component of the output. Since saturation is an odd function, all the cosine terms in Eq. (9.21) are zeros and $a_1 = 0$. According to Eq. (9.27),

$$K_{eq}(a) = \frac{b_1}{a},$$

Figure 9.25
(a) Saturation nonlinearity; (b) input and output signals

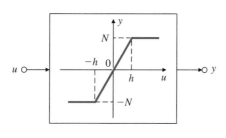

(a)

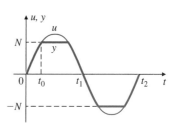

(b)

so that

$$b_1 = \frac{2}{\pi} \int_0^\pi N \operatorname{sat}\left(\frac{k}{N} a \sin \omega t\right) \sin \omega t \, d(\omega t),$$

since the integral for the coefficient b_1 over the interval $\omega t = [0, \pi]$ is simply twice that of the interval $\omega t = [0, \pi/2]$. Then

$$b_1 = \frac{4N}{\pi} \int_0^{\frac{\pi}{2}} \operatorname{sat}\left(\frac{k}{N} a \sin \omega t\right) \sin \omega t \, d(\omega t).$$

We can now divide the integral into two parts corresponding to the linear and saturation parts. Let us define the saturation time t_s as the time when

$$t_s = \frac{1}{\omega} \sin^{-1}\left(\frac{N}{ak}\right) \quad \text{or} \quad \omega t_s = \sin^{-1}\left(\frac{N}{ak}\right). \tag{9.28}$$

Then

$$b_1 = \frac{4N\omega}{\pi a}\left(\int_0^{\omega t_s} \operatorname{sat}\left(\frac{k}{N} a \sin \omega t\right) \sin \omega t \, dt + \int_{\omega t_s}^{\frac{\pi}{2}} \sin \omega t \, dt\right)$$

$$= \frac{4N\omega}{\pi a}\left(\int_0^{\omega t_s} \frac{k}{N} a \sin^2 \omega t \, dt + \int_{\omega t_s}^{\frac{\pi}{2}} \sin \omega t \, dt\right)$$

$$= \frac{4N\omega}{\pi a}\left(\int_0^{\omega t_s} \frac{k}{2N} a(1 - \cos 2\omega t) \, dt + \int_{\omega t_s}^{\frac{\pi}{2}} \sin \omega t \, dt\right)$$

$$= \frac{4N\omega}{\pi a}\left(\frac{k}{2N} at \Big|_0^{t_s} - \frac{k}{2N} a \sin 2\omega t \Big|_0^{t_s} - \frac{1}{\omega}\left(\cos \frac{\pi}{2} - \cos t_s\right)\right).$$

But using Eq. (9.28), we have

$$\sin \omega t_s = \frac{N}{ka} \quad \text{and} \quad \cos \omega t_s = \sqrt{1 - \left(\frac{N}{ka}\right)^2}.$$

We finally obtain

$$K_{eq}(a) = \begin{cases} \frac{2}{\pi}\left(k \sin^{-1}\left(\frac{N}{ak}\right) + \frac{N}{a}\sqrt{1 - \left(\frac{N}{ka}\right)^2}\right), & \frac{ka}{N} > 1, \\ k, & \frac{ka}{N} \le 1. \end{cases} \tag{9.29}$$

Figure 9.26 shows a plot of $K_{eq}(a)$ indicating that it is a real function independent of frequency and results in no phase shifts. It is seen that the describing function is initially a constant and then decays essentially as a function of the reciprocal of the input signal amplitude, a.

Figure 9.26
Describing function for saturation nonlinearity with $k = N = 1$

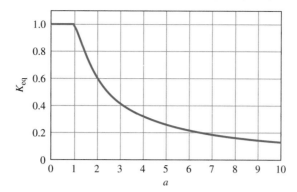

EXAMPLE 9.11

Describing Function for a Relay Nonlinearity

Find the describing function for the relay or sgn function shown in Fig. 9.6(b) and defined as

$$\text{sgn}(x) = 0, \qquad x = 0,$$
$$= \text{sign of } x, \quad \text{otherwise.}$$

Solution. The output is a square wave of amplitude N for *every* size input; thus $Y_1 = \frac{4N}{\pi}$ and $K_{eq} = \frac{4N}{\pi a}$. The solution can also be obtained from Eq. (9.29) if we let $k \to \infty$. For small angles,

$$\sin^{-1}\left(\frac{N}{ak}\right) \cong \frac{N}{ak},$$

and thus, from Eq. (9.27), we have

$$K_{eq}(a) = \frac{2}{\pi}\left(k\left(\frac{N}{ak}\right) + \frac{N}{a}\right) = \frac{4N}{\pi a}. \qquad (9.30)$$

The preceding two nonlinearities were memoryless. Next we consider a nonlinearity with memory. Nonlinearities with memory occur in many applications, including magnetic recording devices, backlash in mechanical systems, and in electronic circuits. Consider the bistable electronic circuit shown in Fig. 9.27 that is called a Schmitt trigger (Sedra and Smith, 1991). This circuit has memory. Referring to Fig. 9.28, if the circuit is in the state where $v_{out} = +N$, then positive values of v_{in} do not change the state. To "trigger" the circuit into the state $v_{out} = -N$, we must make v_{in} negative enough to make v negative. The threshold value is $h = \frac{NR_1}{R_2}$. The Schmitt trigger is employed commonly in spacecraft control (Bryson, 1994). We next find the describing function for a hysteresis nonlinearity.

EXAMPLE 9.12

Describing Function for a Relay with Hysteresis Nonlinearity

Consider the relay function with hysteresis shown in Fig. 9.29(a). Find the describing function for this nonlinearity.

Solution. A system with hysteresis tends to stay in its current state. Until the input to the signum function is past the value h, it is not possible to determine the output

Figure 9.27
Schmitt trigger circuit

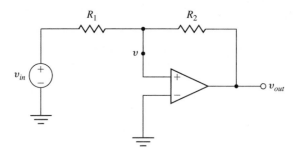

Figure 9.28
Hysteresis nonlinearity
for Schmitt trigger
circuit

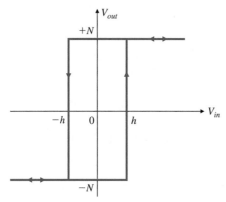

Figure 9.29
(a) Hysteresis
nonlinearity; (b) input
and output to the
nonlinearity

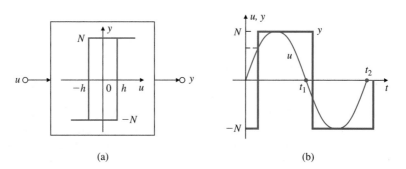

(a) (b)

uniquely without knowing its past history. That implies that we have a nonlinearity
with memory. The output is a square wave with amplitude N as long as the input
amplitude a is greater than the hysteresis level h. From Fig. 9.29(b), it is seen that the
square wave lags the input in time. The lag time can be computed as the time when

$$a \sin \omega t = h \quad \text{or} \quad \omega t = \sin^{-1}\left(\frac{h}{a}\right). \tag{9.31}$$

Because the phase angle is known for all frequencies,

$$K_{eq}(a) = \frac{4N}{\pi a}\angle - \sin^{-1}\left(\frac{h}{a}\right) = \frac{4N}{\pi a}e^{-j\sin^{-1}\left(\frac{h}{a}\right)}, \tag{9.32}$$

$$= \frac{4N}{\pi a}\left(\sqrt{1 - \left(\frac{h}{a}\right)^2} - j\frac{h}{a}\right). \tag{9.33}$$

Figure 9.30

Describing function for the hysteresis nonlinearity for $h = 0.1$ and $N = 1$:
(a) magnitude;
(b) phase

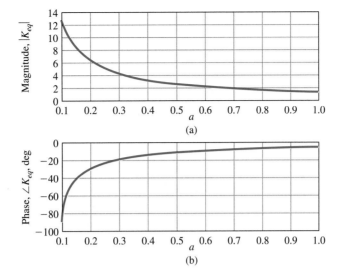

The describing function is then given by

$$K_{eq}(a) = \begin{cases} \frac{4N}{\pi a}\left(\sqrt{1-\left(\frac{h}{a}\right)^2} - j\frac{h}{a}\right), & a \geq h, \\ 0, & a < h. \end{cases} \tag{9.34}$$

The describing function is plotted in Fig. 9.30. The magnitude is proportional to the reciprocal of the input signal amplitude, and the phase varies between $-90°$ and $0°$.

9.4.1 Stability Analysis Using Describing Functions

The Nyquist theorem can be extended to deal with nonlinear systems whose non-linearities have been approximated by describing functions. In the standard linear system analysis the characteristic equation is $1 + KL = 0$, where the loop gain is $L = DG$ and

$$L = -\frac{1}{K}. \tag{9.35}$$

As described in Section 6.3, we look at the encirclements of the $-1/K$ point to determine stability. With a nonlinearity represented by the describing function, $K_{eq}(a)$, the characteristic equation is of the form $1 + K_{eq}(a)L = 0$, and it follows that

$$L = -\frac{1}{K_{eq}(a)}. \tag{9.36}$$

Now we have to look at the intersection of L with a plot of $-1/K_{eq}(a)$. If the curve L intersects $-1/K_{eq}(a)$, then the system will oscillate at the crossing amplitude, a_l, and the corresponding frequency, ω_l, keeping in mind the approximate nature of the describing function. We then look for encirclements to decide whether the system would be stable for that particular value of the gain, as if it were a linear system. If so, we deduce that the nonlinear system is stable. Otherwise, we infer that the nonlinear system is unstable.

Figure 9.31

Closed-loop system with a nonlinearity

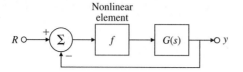

Figure 9.31 shows an example of an otherwise linear system, except for a single nonlinearity. The nonlinear elements may indeed have a beneficial effect and may limit the amplitude of oscillations. The describing function analysis can be used to determine the amplitude and frequency of the limit cycle. Strictly speaking, a limit cycling system can be considered to be unstable. In reality, the trajectory of the limit cycling is confined to a finite region of the state space. If this region is within allowable performance, then the response is tolerable. In some cases, the limit cycling is the beneficial effect (see case study in Section 10.4). The system does not possess asymptotic stability, since the system does not come to a rest at the origin of the state space. The describing function can be beneficial in determining the conditions under which instability results and can even suggest remedies in eliminating instability, as illustrated in the next example, in which the Nyquist plot of the linear loop gain, L, as well as the negative reciprocal of the describing function, $-1/K_{eq}(a)$, are superimposed. The point at which they cross corresponds to the limit cycle. To determine the amplitude and frequency of the limit cycle, we can rewrite Eq. (9.36) as follows:

$$\mathrm{Re}\{L(j\omega)\}\mathrm{Re}\{K_{eq}(a)\} - \mathrm{Im}\{L(j\omega)\}\mathrm{Im}\{K_{eq}(a)\} + 1 = 0, \qquad (9.37)$$

$$\mathrm{Re}\{L(j\omega)\}\mathrm{Im}\{K_{eq}(a)\} + \mathrm{Im}\{L(j\omega)\}\mathrm{Re}\{K_{eq}(a)\} = 0.$$

We can then solve these two equations for the possible two unknown values of the limit-cycle frequency, ω_l, and the corresponding amplitude, a_l, as illustrated in the ensuing examples.

EXAMPLE 9.13   *Conditionally Stable System*

Consider the feedback system in Fig. 9.14. Determine the amplitude and the frequency of the limit cycle using the Nyquist plot.

Solution. The Nyquist plot of the system is superimposed on $-1/K_{eq}(a)$ as shown in Fig. 9.32. Note that the negative of the reciprocal of the describing function, using Eq. (9.29), is

$$-\frac{1}{K_{eq}(a)} = \frac{1}{\frac{2}{\pi}\left(k\sin^{-1}\left(\frac{N}{ak}\right) + \frac{N}{a}\sqrt{1 - \left(\frac{N}{ka}\right)^2}\right)}$$

$$= -\frac{1}{\frac{2}{\pi}\left(\sin^{-1}\left(\frac{0.1}{a}\right) + \frac{0.1}{a}\sqrt{1 - \left(\frac{0.1}{a}\right)^2}\right)},$$

which is a straight line that is coincident with the negative real axis and is parameterized as a function of the input signal amplitude, a. The point of the intersection of the two curves at -0.5 corresponds to the limit-cycle frequency of $\omega_l = 1$. A plot of the

Figure 9.32

Nyquist plot and describing function to determine limit cycle

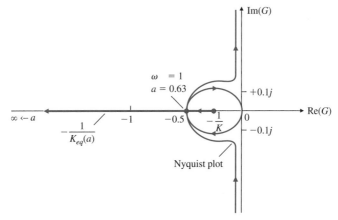

Figure 9.33

Describing function for saturation nonlinearity with $N = 0.1$ and $k = 1$

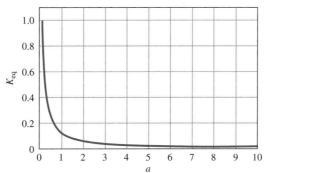

describing function for $k = 1$ and $N = 0.1$ is shown in Fig. 9.33, and a magnitude of $K_{eq} = 0.2$ corresponds to an input amplitude of $a_l = 0.63$.

Alternatively, from the root locus of our example shown in Fig. 9.15, the gain at the imaginary-axis crossover is 0.2; then, from Eq. (9.29), we have

$$K_{eq} = \frac{2}{\pi} \left(\sin^{-1} \left(\frac{0.1}{a} \right) + \frac{0.1}{a} \sqrt{1 - \left(\frac{0.1}{a} \right)^2} \right) = 0.2.$$

If we approximate the arcsine function by its argument as

$$\sin^{-1} \left(\frac{0.1}{a} \right) \approx \frac{0.1}{a},$$

then

$$\frac{2}{\pi} \left(\left(\frac{0.1}{a} \right) + \frac{0.1}{a} \sqrt{1 - \left(\frac{0.1}{a} \right)^2} \right) = 0.2,$$

which leads to the polynomial equation

$$\pi^2 a^4 - 2\pi a^3 + (0.1)^2 = 0$$

and we find the relevant solution to be $a = 0.63$. By measurement on the time history of Fig. 9.16, the amplitude of the oscillation is 0.62, which is in good agreement with our prediction.

For systems with nonlinearities that have memory, we can also use the Nyquist technique, as illustrated in the next example.

EXAMPLE 9.14 *Determination of Stability with a Hysteresis Nonlinearity*

Consider the system with a hysteresis nonlinearity shown in Fig. 9.34. Determine whether the system is stable and find the amplitude and the frequency of the limit cycle.

Solution. The Nyquist plot for the system is shown in Fig. 9.35. The negative reciprocal of the describing function for the hysteresis nonlinearity is

$$-\frac{1}{K_{eq}(a)} = -\frac{1}{\frac{4N}{\pi a}\left(\sqrt{1-\left(\frac{h}{a}\right)^2}-j\frac{h}{a}\right)} = -\frac{\pi}{4N}\left[\sqrt{a^2-h^2}+jh\right].$$

In this case $N = 1$ and $h = 0.1$, and we have

$$-\frac{1}{K_{eq}(a)} = -\frac{\pi}{4}\left[\sqrt{a^2-0.01}+j0.1\right].$$

This is a straight line parallel to the real axis that is parameterized as a function of the input signal amplitude a and is also plotted in Fig. 9.35. The intersection of this curve with the Nyquist plot yields the frequency and the corresponding amplitude of the stable limit cycle. We can also determine the limit-cycle information analytically:

$$-\frac{1}{K_{eq}(a)} = -\frac{\pi}{4}\left[\sqrt{a^2-0.01}+j0.1\right] = G(j\omega) = \frac{1}{j\omega(j\omega+1)}.$$

Clearing the denominator in the preceding equation, we have

$$\frac{\pi}{4}\sqrt{a^2-0.01}\,\omega^2 + \frac{0.1\pi}{4}\omega - 1 + j\left[\frac{0.1\pi}{4}\omega^2 - \frac{\pi}{4}\sqrt{a^2-0.01}\,\omega\right] = 0.$$

Setting the real and imaginary parts equal to zero yields two equations and two unknowns. The relevant solution is $\omega_l = 2.2$ rad/sec and $a_l = 0.24$. A SIMULINK implementation of the closed-loop system is shown in Fig. 9.36. The step response of the system is shown in Fig. 9.37, and the limit cycle has an amplitude of $a_l = 0.24$ and a frequency of $\omega_l = 2.2$ rad/sec and is well predicted by our analysis.

Figure 9.34

Feedback system with hysteresis nonlinearity

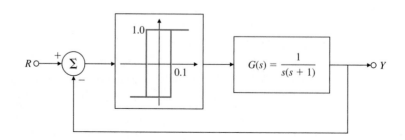

Figure 9.35

Nyquist plot and DF to determine limit-cycle properties

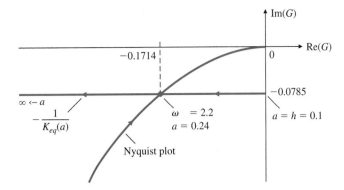

Figure 9.36

SIMULINK® diagram for system with hysteresis

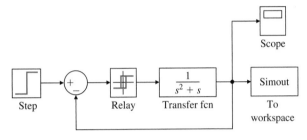

Figure 9.37

Step response displaying limit-cycle oscillations

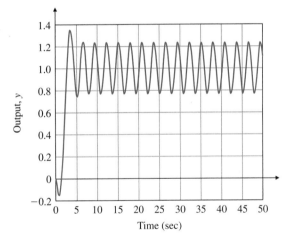

△ **9.5 Analysis and Design Based on Stability**

The central requirement of any control system is stability, and the design methods we have studied are based on this fact. The root locus is a plot of closed-loop poles in the s-plane, and the designer is always aware of the fact that if a root strays into the right half of the plane, the system will be unstable. Designs based on the state representation include pole placement, where the desired locations of the poles are, of course, selected to be well within the stable region. In a similar fashion, Nyquist proved conditions for stability based on the frequency response, and designers are aware of the encirclement requirements of their plots or, equivalently, of the gain and phase margins of stability margins in the Bode plots. Prior to either of these methods, mathematicians studied the stability of ordinary differential equations (ODE), and these and other sophisticated techniques are needed to face the problems of nonlinear systems. We begin with a graphical representation of ODE solutions known as the phase plane and introduce the methods of Lyapunov and others as an introduction to this area of control design.

9.5.1 The Phase Plane

Whereas the root locus and the frequency-response methods consider the system response indirectly via either the poles and zeros of the transfer function or the gain and phase of the frequency response, the phase plane considers the time response directly by plotting the trajectory of the state variables. Although direct visualization restricts the method to second-order systems having only two state variables, the ability of the method to consider nonlinearities, as well as to give new insight into linear systems, makes a quick look at the technique well worthwhile.

To illustrate the ideas of the phase plane, consider a fictional motor system shown in Fig. 9.38 with the open-loop transfer function

$$G(s) = \frac{1}{s(Ts + 1)}.$$

If we assume that $T = 1/6$ and the amplifier is (for the moment) not subject to saturation and has gain K where $K = 5T$, the state equations for the closed-loop system can be written as

$$\dot{x}_1 = x_2, \tag{9.38}$$

$$\dot{x}_2 = -5x_1 - 6x_2, \tag{9.39}$$

$$y = x_1. \tag{9.40}$$

Because these equations are time invariant, the time can be eliminated by dividing Eq. (9.38) into Eq. (9.39), with the result that

$$\frac{dx_2}{dx_1} = \frac{-5x_1 - 6x_2}{x_2}. \tag{9.41}$$

The solution to this equation gives a plot of x_2 versus x_1 or, in other words, a trajectory in the phase plane of coordinates (x_1, x_2).[6] Before plotting Eq. (9.41), it is useful to consider the system equations first in matrix form as $\dot{\mathbf{x}} = \mathbf{F}\mathbf{x}$ for which

$$\mathbf{F} = \begin{bmatrix} 0 & 1 \\ -5 & -6 \end{bmatrix}.$$

Figure 9.38

An elementary position feedback system with a nonlinear actuator

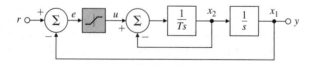

---

[6]If the slope dx_2/dx_1 is set to a constant, the relation between x_2 and x_1 is a straight line. If the known values of slopes are marked along a set of these lines, the trajectories can be readily sketched. For example, along the x_1 axis, where $x_2 = 0$ the slope is ∞ and the trajectories are vertical. This method is called the method of isoclines.

If in this equation we assume that $\mathbf{x} = \mathbf{x}_o e^{st}$, where both s and $\mathbf{x}_o$ are constants, then $\dot{\mathbf{x}} = \mathbf{x}_o s e^{st}$, and the equation can be reduced as follows:

$$\dot{\mathbf{x}} = \mathbf{F}\mathbf{x}, \tag{9.42}$$

$$\mathbf{x}_o s e^{st} = \mathbf{F}\mathbf{x}_o e^{st}, \tag{9.43}$$

$$[s\mathbf{I} - \mathbf{F}]\mathbf{x}_o e^{st} = \mathbf{0}, \tag{9.44}$$

$$[s\mathbf{I} - \mathbf{F}]\mathbf{x}_o = \mathbf{0}. \tag{9.45}$$

Here it should be recognized that Eq. (9.45) is the eigenvector equation for the matrix $\mathbf{F}$, which, in component form, is

$$\begin{bmatrix} s & -1 \\ 5 & s+6 \end{bmatrix} \begin{bmatrix} x_{01} \\ x_{02} \end{bmatrix} = \begin{bmatrix} 0 \\ 0 \end{bmatrix}. \tag{9.46}$$

As described in Appendix WE, Eq. (9.46) has a solution only if the determinant of the coefficient matrix is zero, for which

$$s(s + 6) + 5 = 0, \tag{9.47}$$

$$(s + 1)(s + 5) = 0. \tag{9.48}$$

The two values of s for which the equation has a solution are the eigenvalues $s = -1$ and $s = -5$. If we substitute $s = -1$ into Eq. (9.46), we obtain

$$\begin{bmatrix} -1 & -1 \\ 5 & -1+6 \end{bmatrix} \begin{bmatrix} x_{01} \\ x_{02} \end{bmatrix} = \begin{bmatrix} 0 \\ 0 \end{bmatrix}, \tag{9.49}$$

from which the solution for the initial state vector is $x_{02} = -x_{01}$. This line in the state space is the eigenvector corresponding to the eigenvalue $s = -1$. If we repeat this process with $s = -5$, the result is

$$\begin{bmatrix} -5 & -1 \\ 5 & -5+6 \end{bmatrix} \begin{bmatrix} x_{01} \\ x_{02} \end{bmatrix} = \begin{bmatrix} 0 \\ 0 \end{bmatrix}, \tag{9.50}$$

and in this case the solution for the eigenvector is $x_{02} = -5x_{01}$.

Consider what all this means. We started with the assumption that the time solution for the state is a constant times an exponential. We found that this is possible only if the exponential is either e^{-t} or e^{-5t}. In the first case, the state *must* lie along the vector $x_{02} = -x_{01}$, and in the second, the state *must* lie along the vector $x_{02} = -5x_{01}$. With this knowledge, we compute the solutions to Eq. (9.38) and Eq. (9.39) for different initial conditions and plot $x_1(t)$ vs. $x_2(t)$ in Fig. 9.39. In the figure, the two eigenvectors are identified. When we look at these curves, it is clear that all the paths start parallel to the (fast!) eigenvector corresponding to $s = -5$ and quickly move to the (slow!) one corresponding to $s = -1$. All trajectories approach the equilibrium point at the origin of the state space.

The plot will be substantially changed if the amplifier saturates. For example, if the amplifier saturates at a value of $u = 0.5$, then the velocity, x_2, will rapidly approach this value and will be stuck there until the position reaches a value that brings the amplifier out of saturation. The new plot is shown in Fig. 9.40.

Notice that in the linear region the motion is almost entirely along the slow eigenvector. Finally we note that the phase-plane portrait changes again when the poles are complex. In that case, the motion of the state variable is composed of damped

Figure 9.39

Phase-plane plot of a node with poles at $s = -1$ and $s = -5$

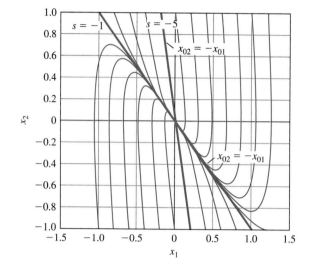

Figure 9.40

Phase-plane plot with saturation

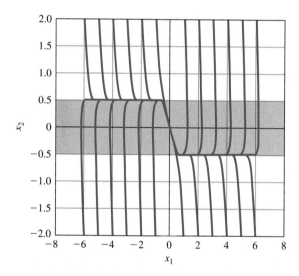

sinusoids and the plot of x_1 versus x_2 is along a spiral. A collection of trajectories for various initial conditions is shown in Fig 9.41.

These few examples just scratch the surface of phase-plane analysis but give some idea of the use of this format in helping a designer to visualize dynamic responses.

Bang-Bang Control

One example of design for a nonlinear system based on the phase plane is that of optimal minimum time control in the face of control saturation. For our purposes here, the simplest version of this widely used technique is introduced: that of the $1/s^2$

Figure 9.41

Phase-plane plot for a system with complex poles

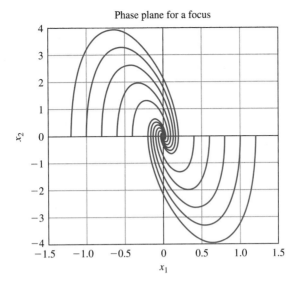

Phase plane for a focus

plant. The equations are

$$\ddot{y} = u, \tag{9.51}$$

$$e = y - y_f, \tag{9.52}$$

where $y(0) = \dot{y}(0) = 0$, y_f is a constant, and the control is constrained to be $|u| \leq 1$. The problem is to drive the error to be identically zero in minimum time. If we define state variables as $x_1 = e$ and $x_2 = \dot{e} = \dot{y}$, the equations reduce to

$$\dot{x}_1 = x_2, \tag{9.53}$$

$$\dot{x}_2 = u, \tag{9.54}$$

$$x_1(0) = -y_f, \tag{9.55}$$

$$x_1(t_f) = x_2(t_f) = 0, \tag{9.56}$$

and the problem is to minimize t_f. Intuitively, this is the problem of the eager driver who wishes to speed from one stop to the next in minimum time. She would put the pedal to the metal for a time and then switch to stand on the brakes as the car skids to a stop at just the right place. A basic result of the theory of optimal control confirms this intuitive idea that the solution to this problem is, if $y_f > 0$, to apply full positive control for a time and then to switch to full negative control at just the right time to cause the error to reach the origin and stop there. To study the case, a plot of the trajectories of the plant in the phase plane for the two cases of $u = 1$ and $u = -1$ is given in Fig. 9.43. For $u = +1$, the trajectories start in the fourth quadrant and move up to the first. For $u = -1$, they start in the second quadrant and move down to the third.

Two segments of this family are of particular interest: those that pass through the origin. Once the path reaches one of these, a constant control will bring the state to the desired final resting place. Therefore, for any initial condition, once the trajectory reaches one of the two curves going through the origin, the correct action is to switch

Figure 9.42

Phase-plane of the $1/s^2$ plant for ± 1 controls

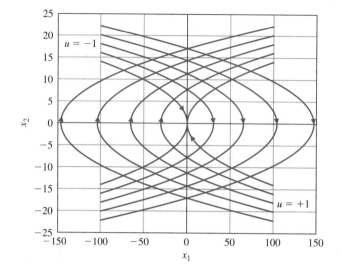

Figure 9.43

Switching curve for the $1/s^2$ plant

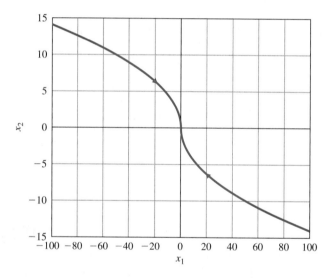

the control ($u = +1$ to -1, or $u = -1$ to $+1$) so that the trajectory will follow that curve to the origin. The "switching curve" is plotted in Fig. 9.44.

For a second-order plant, the switching curve can be found by reversing the time in the equations of motion, setting the initial state to zero and applying the maximum control. The process can be repeated with minimum control to sweep out the other branch of the curve.

For any initial condition above the curve, $u = -1$ is applied and for any initial condition below the curve, $u = +1$ is used. As described, the result will be a minimum-time response. Notice that the curve has vertical slope at the origin; as a result, the implementation is extremely sensitive in this neighborhood. A modified version known as the proximate time-optimal system (PTOS) used in the computer disk drive industry was studied by Workman (1987). The modification consists of shifting the

Figure 9.44

Response of a
time-optimal system

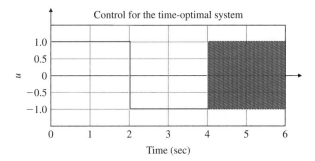

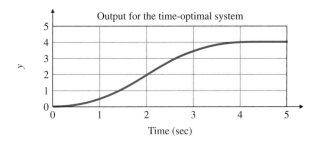

Figure 9.44

Response of a
time-optimal system

Figure 9.45

Response of a
PTOS system

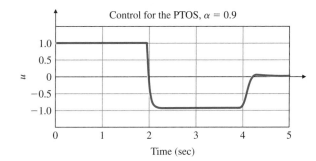

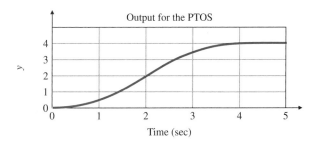

curves a bit and replacing the infinite slope at the origin with a finite-slope, linear
control region. The result has been widely used for hard-disk drives and similar
systems.

Typical responses for a time-optimal system and for a PTOS system generated
with SIMULINK are given in Figs. 9.45 and 9.46. Notice that the response times
are almost exactly the same, but while the time optimal system control has a violent

Figure 9.46

SIMULINK® diagram for position feedback system

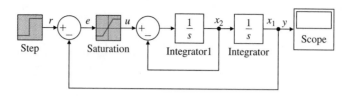

chatter at the end where the switching curve has infinite slope, the output of the PTOS system slips into its final value smoothly. For a more exact study, we need to turn to the nonlinear equations.

9.5.2 Lyapunov Stability Analysis

The stability of motion as studied by Lyapunov involves sophisticated mathematics beyond the scope of this text. Here we will present heuristic arguments giving the flavor of the theory and state a few of the most basic results. Lyapunov presented two methods for the study of stability of motion described by systems of ODEs. His indirect or **first method** is based on linearization of the equations and drawing conclusions about the stability of the nonlinear system by considering the stability of the linear approximation. He proved the results of the first method by use of his direct or **second method**, in which the nonlinear equations are considered directly. A discussion of the indirect method serves to introduce both methods. The problem requires a new definition of stability suitable for the vector-matrix equations. Intuitively, we say a system is stable if initial conditions of moderate size result in a response that remains of moderate size. To express this mathematically, first we need a definition of "size." This is the norm of a vector for which the symbol is $\|\mathbf{x}\|$. Of the many possible definitions, we select here the familiar Euclidean measure defined by its square as $\|\mathbf{x}\|^2 = \mathbf{x}^T\mathbf{x} = \sum_{i=1}^{n} x_i^2$. With this idea, the definition of stability used is that if one is given a sphere of any radius ϵ, one can find a smaller sphere of radius δ such that if the initial state is inside δ, then the trajectory will, for all time, remain inside ϵ. A

Stability in the sense of Lyapunov

bit more formally, the system is **stable** if, for any given $\epsilon > 0$, one can find a $\delta > 0$ such that if $\|\mathbf{x}(0)\| < \delta$, then $\|\mathbf{x}(t)\| < \epsilon$ for all t. If the state is not only stable, but in the limit as $t \to \infty$, $\|\mathbf{x}(t)\| \to 0$, the system is said to be **asymptotically stable**. If, for any ϵ, it is possible to select δ arbitrarily large, then the system is said to be **stable in the large**.

Study of these matters begins with the time-invariant ODE equation

$$\dot{\mathbf{x}} = \mathbf{f}(\mathbf{x}), \tag{9.57}$$

for which the linear approximation is

$$\dot{\mathbf{x}} = \mathbf{F}\mathbf{x} + \mathbf{g}(\mathbf{x}). \tag{9.58}$$

In this equation, it is assumed that all the linear terms are in $\mathbf{F}\mathbf{x}$ and higher-order terms are in $\mathbf{g}(\mathbf{x})$, in the sense that when $\mathbf{x}$ gets small, $\mathbf{g}(\mathbf{x})$ gets small faster, as expressed by

$$\lim_{\|x\| \to 0} \frac{\|\mathbf{g}(\mathbf{x})\|}{\|\mathbf{x}\|} = 0. \tag{9.59}$$

Lyapunov's second method begins with the intuitive notion that one measure of the size of the state of a physical system is the total energy stored in the system at any instant and the observation that when the stored energy is no longer changing, the system must be at rest. For an electric circuit, for example, the electric energy is proportional to the square of the capacitor voltages and the magnetic energy is proportional to the square of the inductor currents. Lyapunov extracted the abstract essence of this idea and defined a scalar function of the state $V(\mathbf{x})$, called a Lyapunov function, having the following properties:

Lyapunov function

1. $V(\mathbf{0}) = 0$;
2. $V(\mathbf{x}) > 0, \|\mathbf{x}\| \neq 0$;
3. V is continuous and has continuous derivatives with respect to all components of $\mathbf{x}$;
4. $\dot{V}(\mathbf{x}) = \frac{\partial V}{\partial \mathbf{x}}\dot{\mathbf{x}} = \frac{\partial V}{\partial \mathbf{x}}\mathbf{f}(\mathbf{x}) \leq 0$ along trajectories of the equation.

The first three conditions ensure that in a neighborhood of the origin the function is like a smooth bowl sitting at the origin of the state space. The fourth condition, which obviously depends on the equations of motion, ensures that if δ is selected so that the initial conditions are deeper in the bowl than any part of the ball defined by ϵ, the trajectory never climbs higher on the bowl than it was at the start and so remains within ϵ, and the system will be stable. Furthermore, if condition 4 is strengthened to be $V(x) < 0$, then the value of the function must fall to zero and, by condition 1, the state also goes to zero. The stability theorem, which is the basis for Lyapunov's second method, states that

Lyapunov's second method

> If a Lyapunov function can be found for a system, then the motion is stable and, furthermore, if $V(x) < 0$, the motion is asymptotically stable. The second method is to search for a Lyapunov function.

The hard part for the application of this theory is the statement "If a Lyapunov function can be found." Only in the linear case is a prescription given for finding a Lyapunov function; otherwise the theory only gives the engineer a hunting license to look for such a function. We are now in a position to consider the indirect method for stability of Eq. (9.58).

Perhaps because energy in simple systems is a sum of the squares of the variables, for this problem Lyapunov considers a quadratic candidate for V by supposing that a symmetric positive definite matrix $\mathbf{P}$ can be found and the function defined as $V(\mathbf{x}) = \mathbf{x}^T\mathbf{P}\mathbf{x}$. Clearly the first three conditions are satisfied by this function; the fourth condition must be tested before it can be concluded that we have a Lyapunov function. The calculation of $\dot{V}$ is

$$V(\mathbf{x}) = \mathbf{x}^T\mathbf{P}\mathbf{x}, \tag{9.60}$$

$$\dot{V}(\mathbf{x}) = \dot{\mathbf{x}}^T\mathbf{P}\mathbf{x} + \mathbf{x}^T\mathbf{P}\dot{\mathbf{x}} \tag{9.61}$$

$$= (\mathbf{F}\mathbf{x} + \mathbf{g}(\mathbf{x}))^T\mathbf{P}\mathbf{x} + \mathbf{x}^T\mathbf{P}(\mathbf{F}\mathbf{x} + \mathbf{g}(\mathbf{x})) \tag{9.62}$$

$$= \mathbf{x}^T(\mathbf{F}^T\mathbf{P} + \mathbf{P}\mathbf{F})\mathbf{x} + 2\mathbf{x}^T\mathbf{P}\mathbf{g}(\mathbf{x}). \tag{9.63}$$

A basic matrix result, known as a Lyapunov equation, is

$$\mathbf{F}^T\mathbf{P} + \mathbf{P}\mathbf{F} = -\mathbf{Q}, \tag{9.64}$$

and he showed that if $\mathbf{F}$ is a stability matrix having all its eigenvalues in the LHP, then for *any* positive definite matrix $\mathbf{Q}$, the solution $\mathbf{P}$ of this equation will also be positive definite. The argument from here is to select $\mathbf{Q}$ and solve for $\mathbf{P}$. Then, if the eigenvalues of $\mathbf{F}$ are in the LHP, $\mathbf{P}$ will be positive definite, so $V(\mathbf{x})$ is a possible Lyapunov function and

$$\dot{V}(\mathbf{x}) = -\mathbf{x}^T \mathbf{Q}\mathbf{x} + 2\mathbf{x}^T \mathbf{P}\mathbf{g}. \tag{9.65}$$

The final part of the argument is to note that, by Eq. (9.59), if $\mathbf{x}$ is small enough, then the first term of Eq. (9.65) will dominate, the fourth condition is satisfied, V is a Lyapunov function, and the system has been proven to be stable. Note that the requirement that $\mathbf{x}$ be small enough guarantees only that there is a *neighborhood* of the origin which is stable. Further conditions are needed to show that the bowl defined by V extends to ∞ in all directions as $\|\mathbf{x}\|$ tends to ∞ (and not before!), so that stability holds for all states and is "in the large."

There is also an instability theorem which shows that if *any* eigenvalue of $\mathbf{F}$ is in the RHP, then the origin will be unstable. If all the poles of $\mathbf{F}$ are in the LHP except for some simple poles on the imaginary axis, then stability depends on further properties of the nonlinear terms, $\mathbf{g}(\mathbf{x})$. With this result in hand, the first or indirect method of Lyapunov can be stated:

Lyapunov's first method

1. Find the linear approximation and compute the eigenvalues of $\mathbf{F}$.
2. If all the eigenvalues are in the LHP, then there is a region of stability about the origin.
3. If at least one of the eigenvalues is in the RHP, then the origin is unstable.
4. If there are simple eigenvalues on the imaginary axis and all other values are in the LHP, then no statement about stability can be made based on this method.

EXAMPLE 9.15   *Lyapunov Stability for a Second-Order System*

Use Lyapunov's method to find conditions for the stability of a second-order linear system described by the state matrix

$$\mathbf{F} = \begin{bmatrix} -\alpha & \beta \\ -\beta & -\alpha \end{bmatrix}.$$

Solution. For the linear case we can take any positive definite $\mathbf{Q}$ we like; the simplest is $\mathbf{Q} = \mathbf{I}$. The corresponding Lyapunov equation is

$$\begin{bmatrix} -\alpha & -\beta \\ \beta & -\alpha \end{bmatrix}\begin{bmatrix} p & q \\ q & r \end{bmatrix} + \begin{bmatrix} p & q \\ q & r \end{bmatrix}\begin{bmatrix} -\alpha & \beta \\ -\beta & -\alpha \end{bmatrix} = \begin{bmatrix} -1 & 0 \\ 0 & -1 \end{bmatrix}. \tag{9.66}$$

Multiplying out Eq. (9.66) and equating coefficients, we get

$$-\alpha p - \beta q - \alpha p - \beta q = -1, \tag{9.67}$$

$$-\alpha q - \beta r + \beta p - \alpha q = 0, \tag{9.68}$$

$$\beta q - \alpha r + \beta q - \alpha r = -1. \tag{9.69}$$

Equations (9.67) to (9.69) are readily solved to get $p = r = 1/2\alpha$, $q = 0$, so that

$$\mathbf{P} = \begin{bmatrix} \frac{1}{2\alpha} & 0 \\ 0 & \frac{1}{2\alpha} \end{bmatrix},$$

and the determinants are $1/2\alpha > 0$ and $1/4\alpha^2 > 0$. Thus $\mathbf{P} > 0$, so we conclude that the system is stable if $\alpha > 0$.

For systems with many state variables and nonnumeric parameters, solution of the Lyapunov equation can be burdensome, but the result is an equivalent alternative to Routh's method for computing the conditions for stability in a system with literal parameters.

EXAMPLE 9.16   *Lyapunov's Direct Method for a Position Feedback System*

Consider the position feedback system modeled in Fig. 9.38. Illustrate the use of the direct method on this nonlinear system. Simulate the system using SIMULINK, assuming that $T = 1$, and evaluate the step response of the system.

Solution.   We assume that the actuator, which is perhaps only an amplifier in this case, has a significant nonlinearity, which is shown in the figure as a saturation but is possibly more complex. We will assume only that $u = f(e)$, where the function lies in the first and third quadrants so that $\int_0^e f(e)\,de > 0$. We also assume that $f(e) = 0$ implies that $e = 0$, and we will assume that $T > 0$, so the system is open-loop stable. The equations of motion are

$$\dot{e} = -x_2, \tag{9.70a}$$

$$\dot{x}_2 = -\frac{1}{T}x_2 + \frac{f(e)}{T}. \tag{9.70b}$$

For a Lyapunov function, consider something like kinetic plus potential energy:

$$V = \frac{T}{2}x_2^2 + \int_0^e f(\sigma)\,d\sigma. \tag{9.71}$$

Clearly, $V = 0$ if $x_2 = e = 0$ and, because of the assumptions about f, $V > 0$ if $x_2^2 + e^2 \neq 0$. To see whether the V in Eq. (9.71) is a Lyapunov function, we compute $\dot{V}$ as follows:

$$\dot{V} = Tx_2\dot{x}_2 + f(e)\dot{e}$$

$$= Tx_2\left[-\frac{1}{T}x_2 + \frac{f(e)}{T}\right] + f(e)(-x_2)$$

$$= -x_2^2.$$

Hence $\dot{V} \leq 0$ and the origin is Lyapunov stable. Moreover, $\dot{V}$ is always decreasing if $x_2 \neq 0$, and Eq. (9.70b) indicates that the system has no trajectory with $x_2 \equiv 0$, except $x_2 = 0$. Thus we can conclude that the system is asymptotically stable for every f that satisfies two conditions: (1) $\int f\,d\sigma > 0$ and (2) $f(e) = 0$ implies that

Figure 9.47

Step response for position control system

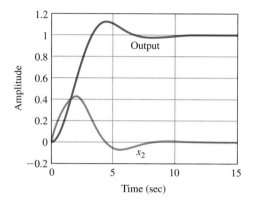

$e = 0$. The SIMULINK diagram for the system is shown in Fig. 9.46 for $T = 1$. The step response of the system is shown in Fig. 9.47.

As we mentioned earlier, the study of the stability of nonlinear systems is vast, so we have only touched here on some important points and methods. Further material for study can be found in LaSalle and Lefschetz (1961), Kalman and Bertram (1960), Vidyasagar (1993), Khalil (2002), and Sastry (1999).

Lyapunov Redesign of Adaptive Control

One of the classical applications of Lyapunov stability theory to control is a technique known as Lyapunov redesign. The idea is to construct the system with some key control parameters unspecified, propose a candidate Lyapunov function and then select the available components to force the candidate to succeed and be an actual Lyapunov function from which stability can be concluded. The method was applied in an early paper by Parks (1966) to a model reference adaptive control system. A block diagram of the simple system first considered is drawn in Fig. 9.48.

In this system, the model and the plant have the same dynamics but different gains. The objective is to adjust the control gain, K_c, so that $K_c K_p = K_m$, and the plant output, y_p, will equal the model output, y_m. A proposed heuristic rule, known as the "MIT" rule, was based on the idea that if we define the cost as the square of the instantaneous error and move K_c so as to make this cost smaller, the result should drive K_c to the right value. If the gradient of the cost is positive (pointing uphill, so to speak), the gain should be reduced, and if the gradient is negative, the gain should be increased. Thus the time derivative of the gain should be proportional to the *negative*

Figure 9.48

Block diagram of a simple model reference adaptive system

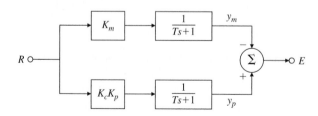

of the gradient. In equation form,

$$J = e^2, \tag{9.72}$$

$$\frac{\partial J}{\partial K_c} = 2e \frac{\partial e}{\partial K_c}, \tag{9.73}$$

$$\frac{dK_c}{dt} = -Be \frac{\partial e}{\partial K_c}, \tag{9.74}$$

where B is the "adaptive gain" to be chosen. From the block diagram,

$$E(s) = \frac{K_c K_p - K_m}{Ts + 1} R(s), \tag{9.75}$$

$$\frac{\partial E}{\partial K_c} = \frac{K_p}{Ts + 1} R \tag{9.76}$$

$$= \frac{K_p}{K_m} Y_m. \tag{9.77}$$

If we substitute the result of Eq. (9.77) into Eq. (9.74), the result is the MIT rule,

$$\frac{dK_c}{dt} = -B' e Y_m, \tag{9.78}$$

where there is a new adaptive gain, B'. Unfortunately, the stability of this rule is not established, and some analysis showed that it could be unstable under reasonable circumstances, such as if there are unmodeled dynamics or disturbances. Parks proposed that Lyapunov redesign would be a better idea and also proposed that, rather than taking $\dot{K}_c$ given by Eq. (9.74), this choice be made in a way that guarantees stability. His idea begins with the differential equations where $r = r_o$ a constant:

$$T\dot{e} + e = (K_c K_p - K_m) r_o, \tag{9.79}$$

$$\dot{K}_c = -B' e Y_m. \tag{9.80}$$

To simplify things, the definition is made $x = (K_c K_p - K_m)$ and $\dot{x}$ is to be found. Parks selects $V = e^2 + \lambda x^2$ as a candidate Lyapunov function and computes

$$\dot{V} = 2e\dot{e} + 2\lambda x \dot{x} \tag{9.81}$$

$$= 2e \left(\frac{x r_o}{T} - \frac{e}{T} \right) + 2\lambda x \dot{x}. \tag{9.82}$$

If $\dot{x}$ in the last equation is selected to be $\dot{x} = -\frac{e r_o}{\lambda T}$, then $\dot{V} = -2\frac{e^2}{T}$, the conditions for a Lyapunov function are satisfied, and stability is assured for the given assumptions. Working back, we find that the new algorithm is

$$\dot{K}_c = -B'' e r_o. \tag{9.83}$$

Obviously, this result does not answer the questions of unmodeled dynamics or disturbances, but the principle is clear: Leaving key control equations to be defined so as to obtain a Lyapunov function can put the stability of a system on a firm foundation.

As a second example of Lyapunov redesign, consider the adaptive control of a motor shown in Fig. 9.49. Defining the model output as y_m and the plant output as y_p, the equations are

$$\ddot{y}_m + 2\zeta \omega_n \dot{y}_m + \omega_n^2 y_m = \omega_n^2 r, \tag{9.84}$$

$$\ddot{y}_p + 2\zeta \omega_n \dot{y}_p + \omega_n^2 y_p = K_c K_p \omega_n^2 (r - y_p) + \omega_n^2 y_p. \tag{9.85}$$

Figure 9.49

Block diagram of adaptive control of a motor

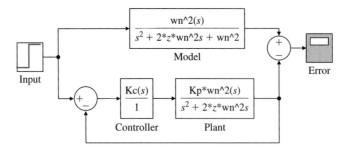

(In the equation for y_p the term $\omega_n^2 y_p$ as been added to both sides to make the error equation simpler.) The error is defined as $e = y_m - y_p$ and an equation for error can be obtained by subtracting the equation for y_p from that for y_m. The result is

$$\ddot{e} + 2\zeta\omega_n\dot{e} + \omega_n^2 ye = \omega_n^2(1 - K_cK_p)(r - y_p). \qquad (9.86)$$

The idea now is to find an equation for K_c that will result in a Lyapunov function for the error equation. To simplify the calculation, we define the parameter as $x = 1 - K_cK_p$, for which $\dot{x} = -K_p\dot{K_c}$ and in terms of which the error equation is

$$\ddot{e}_p + 2\zeta\omega_n\dot{e} + \omega_n^2 e = \omega_n^2 x(r - y_p). \qquad (9.87)$$

At this point, Parks suggests consideration of $V = e^2 + \alpha\dot{e}^2 + \beta x^2$ as a possible function. We need to find $\dot{x}$ so this V will be a Lyapunov function. The equation for the derivative is

$$\dot{V} = 2e\dot{e} + 2\alpha\dot{e}\ddot{e} + 2\beta x\dot{x} \qquad (9.88)$$

$$= 2e\dot{e} + 2\alpha\dot{e}\{-2\zeta\omega_n\dot{e} - \omega_n^2 e + \omega_n^2 x(r - y_p)\} + 2\beta x\dot{x} \qquad (9.89)$$

$$= -4\alpha\zeta\omega_n\dot{e}^2 + 2e\dot{e}(1 - \alpha\omega_n^2) + x\{2\alpha\dot{e}\omega_n^2(r - y_p) + 2\beta\dot{x}\}. \qquad (9.90)$$

If we take $1 - \alpha\omega_n^2 = 0$ and $2\alpha\dot{e}\omega_n^2(r - y_p) + 2\beta\dot{x} = 0$, then the equation for $\dot{V}$ simplifies to $\dot{V} = -4\alpha\zeta\omega_n\dot{e}^2$, which is always negative, and V is a Lyapunov function and the system is stable. Substituting for x, we get the adaptive control law

$$\dot{K_c} = -\beta'\dot{e}(r - y_p), \qquad (9.91)$$

where β' is a new constant equal to $\frac{\alpha\omega_n^2}{K_p\beta}$.

Clearly, we have only touched on Lyapunov's theory of stability, and our examples of redesign are ancient history from 1966, but they illustrate the principle very well and give a good start to further study of this important area.

9.5.3 The Circle Criterion

A nonlinear system with only one single-input-single-output nonlinearity may be represented as shown in Fig. 9.50 by drawing the block diagram from the points of the input and output of the nonlinearity. In the literature this is referred to as the Lur'e problem after the Soviet scientist who first studied it.

We assume that the system is unforced and thus $r \equiv 0$. It is possible to derive a graphical sufficient condition for stability of such systems. Even though this method is practical, it may lead to conservative results in some cases, although extensions exist that yield less conservative results (see Safonov, et al., 1987). First we define sector conditions for memoryless nonlinearities.

Figure 9.50

Block diagram of a nonlinear system

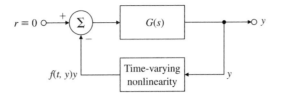

Figure 9.51

Output of the nonlinearity confined in a sector

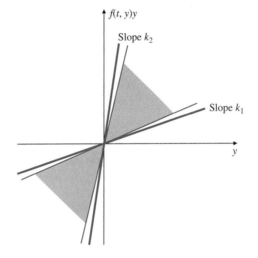

Sector Conditions

A function $f(x)$ with a scalar input and a scalar output belongs to the sector $[k_1, k_2]$ if, for all inputs x,

$$k_1 x^2 \leq f(x)x \leq k_2 x^2. \tag{9.92}$$

This relationship may be rewritten as

$$k_1 \leq \frac{f(x)}{x} \leq k_2 \quad x \neq 0. \tag{9.93}$$

Basically, the definition says that the graph of $f(x)$ lies between two straight lines of slopes k_1 and k_2 going through the origin, as shown in Fig. 9.51. In this definition, k_1 and k_2 are allowed to be $-\infty$ or $+\infty$. Note that the sector conditions place no limits on the incremental gain or *slope* of the function $f(x)$. The ensuing examples illustrate how k_1 and k_2 are determined.

EXAMPLE 9.17   *Computation of a Sector for Signum Nonlinearity*

Determine a sector that contains the signum function $y = f(u)$ shown in Fig. 9.6(b).

Solution.   Since $\text{sgn}(0) = 0$, we know that the only line going through the origin that bounds the signum function from above is the y-axis, corresponding to a slope of $k_2 = \infty$. Similarly the line going through the origin that bounds the signum function from below has a slope of zero and corresponds to the x-axis; therefore, $k_1 = 0$. Hence the sector for the signum function is $[0, \infty]$.

Figure 9.52

Sector for saturation

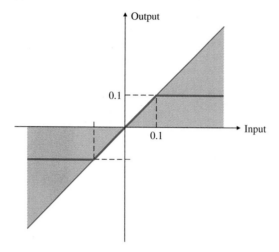

EXAMPLE 9.18

Sector for a Saturation Nonlinearity

Consider the saturation nonlinearity shown Fig. 9.52. Determine a sector for this function.

Solution. The function is bounded above by a line of slope 1, $k_2 = 1$, and is bounded below by the x-axis, $k_1 = 0$, as shown in the figure. Therefore, the sector for this function is $[0, 1]$.

Circle Criterion

In 1949 the Russian scientist Aizermann conjectured that if a Lur'e system is stable with f replaced by any linear gain between the limits $k_1 < k < k_2$, then the system will be stable, with the gain replaced by a nonlinearity in the sector $[k_1, k_2]$. That means that if a single-loop (strictly proper) continuous-time feedback system as shown in Fig. 9.50 with a linear forward path $(\mathbf{F}, \mathbf{G}, \mathbf{H})$ is stable for all linear fixed feedback gains k in the range $k_1 < k < k_2$, such that the resultant closed-loop system matrix $\mathbf{F} + k\mathbf{GH}$ is stable, then the nonlinear system having a memoryless nonlinear time-varying feedback term $f(t, y)$ in the sector $[k_1, k_2]$, shown in Fig. 9.50, is also stable. Unfortunately, this conjecture is *not* true as counterexamples exist.[7] However, a variation of Aizermann's conjecture *is* true and is known as the circle criterion.

Rather than give a rigorous proof of the criterion, we describe a heuristic argument that gives insight into the problem and motivates the proof. An electric circuit with a linear impedance, $Z(j\omega) = R(\omega) + jX(\omega)$, is described by Ohm's law as $V = IZ(s)$. We assume that Z is composed of real components, which means that the real part R

---

[7] Aizermann's conjecture spurred a lot of research in this area and led to the development of the Kalman–Yakubovich–Popov lemma, giving state-space conditions for a passive system. The lemma is used in a proof of the circle criterion.

is even and the imaginary part X is odd; that is $R(-\omega) = R(\omega)$ and $X(-\omega) = -X(\omega)$. If $R(\omega) \geqslant \delta > 0$ for all ω, the impedance is called strictly passive. It will dissipate energy. The instantaneous power *into* the circuit is $p = v(t)i(t)$, and the total energy absorbed by the circuit is $e = \int_0^\infty v(t)i(t)\,dt$. Referring to the figure, Ohm's law is equivalent to the plant equation $Y = UG(s)$ with Y as voltage, U as current, and $G(s) = R + jX$ as the impedance. Applying the expression for energy to the plant equation and using the theorem by Parseval[8] to convert this to the frequency domain yields

$$\int_0^\infty y(t)u(t)\,dt = \frac{1}{2\pi}\int_{-\infty}^\infty U(j\omega)Y(-j\omega)\,d\omega \tag{9.94}$$

$$= \frac{1}{2\pi}\int_{-\infty}^\infty U(j\omega)U(-j\omega)G(-j\omega)\,d\omega \tag{9.95}$$

$$= \frac{1}{2\pi}\int_{-\infty}^\infty |U(j\omega)|^2\,(R - jX)\,d\omega \tag{9.96}$$

$$= \frac{1}{2\pi}\int_{-\infty}^\infty |U(j\omega)|^2\,R(\omega)\,d\omega. \tag{9.97}$$

In the last step, the fact the X is odd was used. At this point, the use of more or less conventional notation will simplify the equations substantially. We define inner products and norms as

$$\int_0^\infty y(t)u(t)\,dt = \; <y,u>, \tag{9.98}$$

$$\|u\|^2 = \int_0^\infty [u(t)]^2\,dt = <u,u>. \tag{9.99}$$

With this notation, and with the assumption that $R \geq \delta > 0$, Eq. (9.97) is reduced to

$$<y,u> \geqq \delta\,\|u\|^2. \tag{9.100}$$

Turning now to the nonlinear component, using the same concept of "energy" and assuming that f is in the sector $[0, K]$, we have

$$\int_0^\infty y(t)f(y,t)\,dt = \int_0^\infty \frac{[f(y,t)]^2}{\dfrac{f(y,t)}{y(t)}}\,dt \tag{9.101}$$

$$\geqq \frac{\|f(y,t)\|^2}{K} \tag{9.102}$$

$$\geqq \frac{\|u(t)\|^2}{K}. \tag{9.103}$$

The assumption now is that if the total energy given by the sum of Eq. (9.100) and Eq. (9.102) is positive, then the system *must be stable*, as energy is being steadily lost. The actual value of the energy lost would be equal to the initial energy stored in the

---

[8] See Appendix A.

elements of the system. From this one would conclude that if $\delta\|u\|^2 + \frac{\|u(t)\|^2}{K} > 0$, then the system is stable. Thus the criterion is

$$\delta\|u\|^2 + \frac{\|u(t)\|^2}{K} > 0, \qquad (9.104)$$

$$\left[\text{Re}\{G(j\omega)\} + \frac{1}{K}\right]\|u(t)\|^2 > 0, \qquad (9.105)$$

$$\text{Re}\{KG(j\omega) + 1\} > 0. \qquad (9.106)$$

In deriving Eq. (9.106), the assumption was made that the nonlinearity was in a zero sector, $[0, K]$. If the function is actually in the sector $[k_1, k_2]$, it can be reduced to a zero sector by adding and subtracting k_1 in the block diagram as shown in Fig. 9.53. With this change, the dynamic system is replaced by $H = \frac{G}{1+k_1 G}$ and the function by $f' = f - k_1$, which is in the sector $[k_2 - k_1, 0]$. With these changes, the stability criterion is transformed to

$$\text{Re}\left\{1 + (k_2 - k_1)\frac{G}{1 + k_1 G}\right\} > 0, \qquad (9.107)$$

$$\text{Re}\left\{\frac{1 + k_1 G + (k_2 - k_1)G}{1 + k_1 G}\right\} > 0, \qquad (9.108)$$

$$\text{Re}\left\{\frac{1 + k_2 G(j\omega)}{1 + k_1 G(j\omega)}\right\} > 0. \qquad (9.109)$$

It is a fact that a bilinear function such as $F = \frac{1+k_2 G(j\omega)}{1+k_1 G(j\omega)}$ in Eq. (9.109) will map a circle in the F plane into another circle in the G plane (see Appendix WD). In this case, the acceptable region is $\text{Re}\{F\} > 0$, of which the boundary is the imaginary axis, so the map is from the *imaginary* axis, a circle of infinite radius, into a finite circle. Because the functions are real, the circle must be centered on the real axis and we need only locate the two points on the real axis. For example, when $F = 0$, we have $1 + k_2 G = 0$ or $G = -\frac{1}{k_2}$. The other point on the real axis is when the function is infinite, at which point $1 + k_1 G = 0$ or $G = -\frac{1}{k_1}$. Thus the circle in the G plane is centered on the real axis and goes through the points $\left[-\frac{1}{k_2}, -\frac{1}{k_1}\right]$ as plotted in

Figure 9.53

Block diagram manipulation for sector

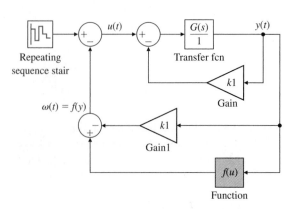

Figure 9.54

Illustration of circle criterion

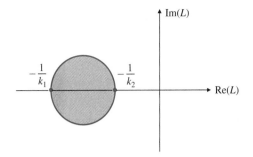

Fig. 9.54. Since F had to avoid the LHP, if we set $F = -1$, which is in the forbidden region, and solve, we find that $G = \frac{-2}{k_1+k_2}$, which is *inside* the circle, from which we conclude that the system will be stable if the plot of $G(j\omega)$ *avoids* this circle.

The actual theorem is as follows:

The nonlinear system described is asymptotically stable given that

1. $f(t, y)$ lies in the sector $[k_1, k_2]$ with $0 \le k_1 < k_2$ and
2. the Nyquist plot of the transfer function $G(j\omega) = \mathbf{H}(j\omega\mathbf{I} - \mathbf{F})^{-1}\mathbf{G}$ does not intersect or encircle the "critical circle," which is centered on the real axis and passes through the two points $-1/k_1$ and $-1/k_2$, as shown in Fig. 9.54.

Circle criterion

In effect, the usual Nyquist "-1" point is replaced by the critical disk. This result is known as the **circle criterion** or circle theorem and is due to Sandberg (1964) and Zames (1966). Note that these conditions are *sufficient* but not necessary, because intersection of the transfer function $G(s)$ with the circle as defined does not prove instability. The critical circle is centered at

$$c = \frac{1}{2}\left[-\frac{1}{k_1} - \frac{1}{k_2} \right] = -\frac{k_1 + k_2}{2k_1 k_2},$$

and has a radius of

$$\frac{k_2 - k_1}{2k_1 k_2}.$$

If $k_1 = 0$, then the critical circle degenerates into a half plane defined by $\mathrm{Re}\,\{G\} \ge -1/k_2$.

The circle criterion and the describing function are related. In fact, for the case of time-invariant odd nonlinearities that are within a sector and whose describing functions are real, the describing function satisfies the relationship

$$k_1 \le K_{eq}(a) \le k_2 \quad \text{for all } a, \tag{9.110}$$

so that

$$-\frac{1}{k_1} \le -\frac{1}{K_{eq}(a)} \le -\frac{1}{k_2}, \tag{9.111}$$

and the plot of the negative reciprocal of the describing function will lie inside the critical circle. This can be seen by the following lower and upper bounds:

$$K_{eq}(a) = \frac{2}{\pi a}\int_0^\pi f(a\sin(\omega t))\sin(\omega t)\,d(\omega t) \ge \frac{2k_1}{\pi}\int_0^\pi \sin^2(\omega t)\,d(\omega t) = k_1, \tag{9.112}$$

Figure 9.55

Nyquist plot and circle criterion

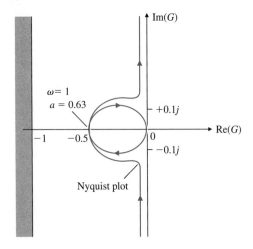

$$K_{eq}(a) = \frac{2}{\pi a} \int_0^\pi f(a \sin(\omega t)) \sin(\omega t) \, d(\omega t) \le \frac{2k_2}{\pi} \int_0^\pi \sin^2(\omega t) \, d(\omega t) = k_2.$$

(9.113)

The equivalent gain analysis and describing functions yield the same results. If we take the gain of the describing function, then the amplitude of the limit cycle can be predicted as with the describing functions. Both equivalent gain techniques can be used to determine stability, but as we have seen, the circle criterion allows for time-varying nonlinearities.

EXAMPLE 9.19   *Determination of Stability Using the Circle Criterion*

Consider the system in Example 9.7. Determine the stability properties of the system using the circle criterion.

Solution.   The related sector is the same one found in Example 9.18. The critical circle degenerates into a half plane defined by $\text{Re}(G) \le -1$, as shown in Fig. 9.55. Since the Nyquist plot lies entirely to the right of the critical circle, the system is stable.

9.6 Historical Perspective

Almost all physical dynamic systems are nonlinear; hence it is not surprising that the study of nonlinear systems has a long and rich history. The study of nonlinear systems goes back to astronomy and the study of the stability of the solar system dating back to Torricelli (1608–1647), Laplace, and Lagrange. The field got a jolt of "energy" with the doctoral dissertation of A. M. Lyapunov in Russia in 1892. He was trying to solve the stability of rotating bodies of fluids posed by Poincaré and recognized that if he could show that system stored energy was always decreasing, then the system would be stable and eventually come to rest. The study of Lyapunov functions was introduced to the control field in 1960 by Kalman and Bertram and has evolved rapidly since then.

Maxwell was the first to study stability by linearization about an equilibrium point by the derivation of the linear model for the Watt's fly-ball governor and stating that the system will be stable if the characteristic roots have negative real parts. Kochenberger derived the describing function method in an attempt to handle nonlinearities in 1950 based on frequency-response ideas. Lur'e proposed the absolute stability problem in 1944 and in 1961 Popov developed the circle criterion for nonlinear stability analysis. Yakubovich (1962) and Kalman (1963) later established connections between Lur'e and Popov's results.

The study of adaptive control received a lot of attention during the three decades of the 1960s, 1970s, and 1980s. Adaptive controllers are both time varying and nonlinear in general. During the 1960s sensitivity methods and the MIT rule for adaptive adjustments were developed by Draper and others. Methods to study adaptive systems based on Lyapunov's methods and passivity were developed in the 1970s. Robust adaptive control methods were studied in the 1980s. Also, there has been a lot of research on systems, such as the weather, where a minute change in initial conditions or parameters can cause drastic changes in the response of the system. Such systems are said to be chaotic. In all recent studies of nonlinear systems, the availability of powerful computers to solve the equations and to graph the results has been critical. Development of a general theory of nonlinear control continues to be a dream of control theorists and is an ongoing quest.

SUMMARY

- The nonlinear equations of motion may be approximated by linear ones by considering a small-signal linear model that is accurate near an equilibrium.
- In many cases, the inverse of a nonlinearity may be used to linearize a system.
- Nonlinearities with no dynamics, such as saturation, can be analyzed using the root locus by considering the nonlinearity to be a variable gain.
- The root-locus technique can be used to determine the limit-cycle properties for memoryless nonlinearities, and yields the same results as the describing function.
- The describing function is essentially a heuristic method with the goal of finding a frequency-response function for a nonlinear element.
- The stability of systems with a single nonlinearity can be studied using the describing function method.
- The describing function can be used to predict periodic solutions in feedback systems.
- The Nyquist plot together with the describing function can be used to determine limit-cycle properties.
- The stability of a nonlinear system in state-space description can be studied by the methods of **Lyapunov**.
- The circle criterion provides a sufficient condition for stability.

REVIEW QUESTIONS

1. Why do we approximate a physical model of the plant (which is *always* nonlinear) with a linear model?

2. How would you linearize the nonlinear system equation for radiation heat transfer $\dot{T} = T^4 + T + u$?

3. A lamp used as a thermal actuator has a nonlinearity such that the experimentally measured output power is related to the input voltage by $P = V^{1.6}$. How would you deal with such a nonlinearity in feedback control design?

4. What is integrator windup?

5. Why is an antiwindup circuit important?

6. Using the nonlinear saturation function having gain 1 and limits ± 1, sketch the block diagram of saturation for an actuator that has gain 7 and limits of ± 20.

7. What is a describing function and how is it related to a transfer function?

8. What are the assumptions behind the use of the describing function?

9. What is a limit cycle in a nonlinear system?

10. How can one determine the describing function for a nonlinear system in the laboratory?

11. What is the minimum time control strategy for a satellite attitude control with bounded controls?

12. How are the two Lyapunov methods used?

PROBLEMS

Problems for Section 9.2: Analysis by Linearization

9.1 Fig. 9.56 shows a simple pendulum system in which a cord is wrapped around a fixed cylinder. The motion of the system that results is described by the differential equation

$$(l + R\theta)\ddot{\theta} + g \sin\theta + R\dot{\theta}^2 = 0,$$

where

$$l = \text{length of the cord in the vertical (down) position,}$$
$$R = \text{radius of the cylinder.}$$

(a) Write the state-variable equations for this system.

Figure 9.56

Motion of cord wrapped around a fixed cylinder

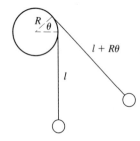

(b) Linearize the equation around the point $\theta = 0$, and show that for small values of θ, the system equation reduces to an equation for a simple pendulum—that is,

$$\ddot{\theta} + (g/l)\theta = 0.$$

9.2 The circuit shown in Fig. 9.57 has a nonlinear conductance G such that $i_G = g(v_G)$ $= v_G(v_G - 1)(v_G - 4)$. The state differential equations are

$$\frac{di}{dt} = -i + v,$$

$$\frac{dv}{dt} = -i + g(u - v),$$

where i and v are the state variables and u is the input.

(a) One equilibrium state occurs when $u = 1$, yielding $i_1 = v_1 = 0$. Find the other two pairs of v and i that will produce equilibrium.

(b) Find the linearized model of the system about the equilibrium point $u = 1$, $i_1 = v_1 = 0$.

(c) Find the linearized models about the other two equilibrium points.

Figure 9.57

Nonlinear circuit for Problem 9.2

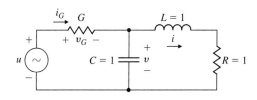

9.3 Consider the circuit shown in Fig. 9.58; u_1 and u_2 are voltage and current sources, respectively, and R_1 and R_2 are nonlinear resistors with the following characteristics:

$$\text{Resistor 1:} \quad i_1 = G(v_1) = v_1^3,$$

$$\text{Resistor 2:} \quad v_2 = r(i_2).$$

Figure 9.58

A nonlinear circuit

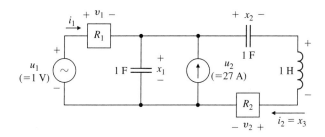

Here the function r is defined in Fig. 9.59.

(a) Show that the circuit equations can be written as

$$\dot{x}_1 = G(u_1 - x_1) + u_2 - x_3,$$

$$\dot{x}_2 = x_3,$$

$$\dot{x}_3 = x_1 - x_2 - r(x_3).$$

Figure 9.59

Nonlinear resistance

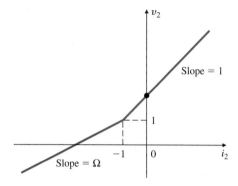

Suppose we have a constant voltage source of 1 volt at u_1 and a constant current source of 27 amps (i.e., $u_1^o = 1$, $u_2^o = 27$). Find the *equilibrium state* $\mathbf{x}^o = \left[x_1^o, x_2^o, x_3^o\right]^T$ for the circuit. For a particular input u^o, an equilibrium state of the system is defined to be any constant state vector whose elements satisfy the relation

$$\dot{x}_1 = \dot{x}_2 = \dot{x}_3 = 0.$$

Consequently, any system started in one of its equilibrium states will remain there indefinitely until a different input is applied.

(b) Due to disturbances, the initial state (capacitance, voltages, and inductor current) is slightly different from the equilibrium and so are the independent sources; that is,

$$u(t) = u^o + \delta u(t),$$

$$x(t_0) = x^o(t_0) + \delta x(t_0).$$

Do a small-signal analysis of the network about the equilibrium found in (a), displaying the equations in the form

$$\delta \dot{x}_1 = f_{11}\delta x_1 + f_{12}\delta x_2 + f_{13}\delta x_3 + g_1\delta u_1 + g_2\delta u_2.$$

(c) Draw the circuit diagram that corresponds to the linearized model. Give the values of the elements.

9.4   Consider the nonlinear system

$$\dot{x} = -x^2 e^{-\frac{1}{x}} + \sin u, \quad x(0) = 1.$$

(a) Assume $u^o = 0$ and solve for $x^o(t)$.

(b) Find the linearized model about the nominal solution in part (a).

9.5   *Linearizing effect of feedback:* We have seen that feedback can reduce the sensitivity of the input–output transfer function with respect to changes in the plant transfer function, and reduce the effects of a disturbance acting on the plant. In this problem we explore another beneficial property of feedback: It can make the input–output response *more linear* than the open-loop response of the plant alone. For simplicity, let us ignore all the dynamics of the plant and assume that the plant is described by the static nonlinearity

$$y(t) = \begin{cases} u, & u \leq 1, \\ \frac{u+1}{2}, & u > 1. \end{cases}$$

(a) Suppose we use proportional feedback

$$u(t) = r(t) + \alpha(r(t) - y(t)),$$

where $\alpha \geq 0$ is the feedback gain. Find an expression for $y(t)$ as a function of $r(t)$ for the closed-loop system. (This function is called the *nonlinear characteristic* of the system.) Sketch the nonlinear transfer characteristic for $\alpha = 0$ (which is really open loop), $\alpha = 1$, and $\alpha = 2$.

(b) Suppose we use integral control:

$$u(t) = r(t) + \int_0^t (r(\tau) - y(\tau)) \, d\tau.$$

The closed-loop system is therefore nonlinear and dynamic. Show that if $r(t)$ is a constant, say r, then $\lim_{t \to \infty} y(t) = r$. Thus, the integral control makes the steady-state transfer characteristic of the closed-loop system *exactly linear*. Can the closed-loop system be described by a transfer function from r to y?

9.6 This problem shows that linearization does not always work. Consider the system

$$\dot{x} = \alpha x^3, \quad x(0) \neq 0.$$

(a) Find the equilibrium point and solve for $x(t)$.

(b) Assume that $\alpha = 1$. Is the linearized model a valid representation of the system?

(c) Assume that $\alpha = -1$. Is the linearized model a valid representation of the system?

9.7 Consider the object moving in a straight line with constant velocity shown in Fig. 9.60. The only available measurement is the range to the object. The system equations are

$$\begin{bmatrix} \dot{x} \\ \dot{v} \\ \dot{z} \end{bmatrix} = \begin{bmatrix} 0 & 1 & 0 \\ 0 & 0 & 0 \\ 0 & 0 & 0 \end{bmatrix} \begin{bmatrix} x \\ v \\ z \end{bmatrix},$$

where

$$z = \text{constant},$$
$$\dot{x} = \text{constant} = v_0,$$
$$r = \sqrt{x^2 + z^2}.$$

Derive a linear model for this system.

Figure 9.60

Diagram of moving object for Problem 9.7

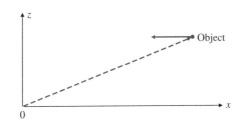

Object

Problems for Section 9.3: Equivalent Gain Analysis Using the Root Locus

9.8   Consider the third-order system shown in Fig. 9.61.

(a) Sketch the root locus for this system with respect to K, showing your calculations for the asymptote angles, departure angles, and so on.

(b) Using graphical techniques, locate carefully the point at which the locus crosses the imaginary axis. What is the value of K at that point?

(c) Assume that, due to some unknown mechanism, the amplifier output is given by the following saturation non-linearity (instead of by a proportional gain K):

$$u = \begin{cases} e, & |e| \le 1, \\ 1, & e > 1, \\ -1, & e < -1. \end{cases}$$

Qualitatively describe how you would expect the system to respond to a unit-step input.

Figure 9.61
Control system for
Problem 9.8

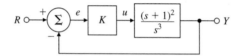

Problems for Section 9.4: Equivalent Gain Analysis

Using Frequency Response: Describing Functions

9.9   Compute the describing function for the relay with dead-zone nonlinearity shown in Fig. 9.6(c).

9.10 Compute the describing function for gain with dead-zone nonlinearity shown in Fig. 9.6(d).

9.11 Compute the describing function for the preloaded spring or Coulomb plus viscous friction nonlinearity shown in Fig. 9.6(e).

9.12 Consider the quantizer function shown in Fig. 9.62 that resembles a staircase. Find the describing function for this nonlinearity and write a MATLAB .m function to generate it.

9.13 Derive the describing function for the ideal contactor controller shown in Fig. 9.63. Is it frequency dependent? Would it be frequency dependent if it had a time delay or hysteresis? Graphically sketch the time histories of the output for several amplitudes of the input and determine the describing function values for those inputs.

9.14 A contactor controller of an inertial platform is shown in Fig. 9.64, where

$$I = 0.1 \text{ kg} \cdot \text{m}^2,$$

$$\frac{I}{B} = 10 \text{ sec},$$

$$\frac{h}{c} = 1,$$

$$\frac{J}{c} = 0.01 \text{ sec},$$

Figure 9.62
Quantizer nonlinearity
for Problem 9.12

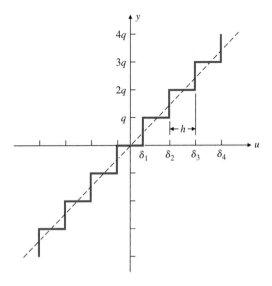

Figure 9.63
Contactor for
Problem 9.13

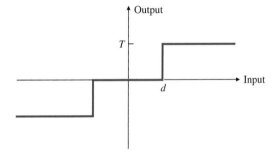

Figure 9.64
Block diagram of the
system for Problem 9.14

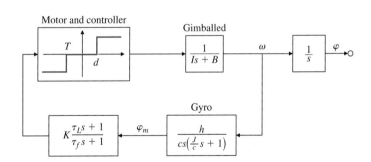

$$\tau_L = 0.1 \text{ sec},$$

$$\tau_f = 0.01 \text{ sec},$$

$$d = 10^{-5} \text{ rad},$$

$$T = 1 \text{ N} \cdot \text{m}.$$

The required stabilization resolution is approximately 10^{-6} rad:

$$K\varphi_m > d \quad \text{for} \quad \varphi_m > 10^{-6} \text{ rad.}$$

Discuss the existence, amplitude, and frequency of possible limit cycles as a function of the gain K and the DF of the controller. Repeat the problem for a deadband with hysteresis.

9.15 *Nonlinear Clegg Integrator:* Here there have been some attempts over the years to improve upon the linear integrator. A linear integrator has the disadvantage of having a phase lag of 90° at all frequencies. In 1958, J. C. Clegg suggested that we modify the linear integrator to reset its state, x, to zero whenever the input to the integrator, e, crosses zero (i.e., changes sign). The Clegg integrator has the property that it acts like a linear integrator whenever its input and output have the same sign. Otherwise, it *resets* its output to zero. The Clegg integrator can be described by

$$\begin{aligned} x(t) &= e(t), \quad \text{if } e(t) \neq 0, \\ x(t+) &= 0, \quad \text{if } e(t) = 0, \end{aligned}$$

where the latter equation implies that the state of the integrator, x, is reset to zero immediately after e changes sign. It can be implemented with op-amps and diodes. A potential disadvantage of the Clegg integrator is that it may induce oscillations.

(a) Sketch the output of the Clegg integrator if the input is $e = a\sin(\omega t)$.

(b) Prove that the DF for the Clegg integrator is

$$N(a, \omega) = \frac{4}{\pi \omega} - j\frac{1}{\omega}.$$

and this amounts to a phase lag of only 38°.

△ *Problems for Section 9.5: Analysis and Design Based on Stability*

9.16 Compute and sketch the optimal reversal curve and optimal control for the minimal time control of the plant

$$\begin{aligned} \dot{x}_1 &= x_2, \\ \dot{x}_2 &= -x_2 + u, \\ |u| &\leq 1. \end{aligned}$$

Use the reverse-time method and eliminate the time.

9.17 Sketch the optimal reversal curve for the minimal time control with $|u| \leq 1$ of the linear plant

$$\begin{aligned} \dot{x}_1 &= x_2, \\ \dot{x}_2 &= -2x_1 - 3x_2 + u. \end{aligned}$$

9.18 Sketch the time-optimal control law for

$$\begin{aligned} \dot{x}_1 &= x_2, \\ \dot{x}_2 &= -x_1 + u, \\ |u| &\leq 1, \end{aligned}$$

and show a trajectory for $x_1(0) = 3$ and $x_2(0) = 0$.

9.19 Consider the thermal control system shown in Fig. 9.65. The physical plant can be a room, an oven, etc.

  (a) What is the limit-cycle period?

  (b) If T_r is commanded as a slowly increasing function, sketch the output of the system, T. Show the solution for T_r "large."

Figure 9.65

Thermal system for Problem 9.19

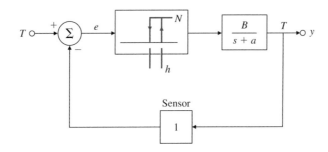

9.20 Several systems, such as a spacecraft, a spring-mass system with resonant frequency well below the frequency of switching, a large motor-driven load with very small friction, etc., can be modeled as just an inertia. For an ideal switching curve, sketch the phase portraits of the system. The switching function is $e = \theta + \tau \omega$. Assume that $\tau = 10$ sec and the control signal $= 10^{-3}$ rad/sec^2. Now sketch the results with

  (a) deadband,

  (b) deadband plus hysteresis,

  (c) deadband plus time delay T,

  (d) deadband plus a constant disturbance.

9.21 Compute the amplitude of the limit cycle in the case of satellite attitude control with delay

$$I\ddot{\theta} = N\,u(t - \Delta),$$

using

$$u = -\text{sgn}(\tau\dot{\theta} + \theta).$$

Sketch the phase-plane trajectory of the limit cycle and time history of θ giving the maximum value of θ.

9.22 Consider the point mass pendulum with zero friction as shown in Fig. 9.66. Using the method of isoclines as a guide, sketch the phase-plane portrait of the motion. Pay particular attention to the vicinity of $\theta = \pi$. Indicate a trajectory corresponding to spinning of the bob around and around rather than oscillating back and forth.

Figure 9.66

Pendulum for Problem 9.22

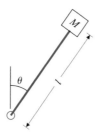

9.23 Draw the phase trajectory for a system

$$\ddot{x} = 10^{-6} \text{ m/sec}^2$$

between $\dot{x}(0) = 0$, $x(0) = 0$, and $x(t) = 1$ mm. Find the transition time t_f by graphical means from the parabolic curve by comparing your solution with two different interval sizes and the exact solution.

9.24 Consider the system with equations of motion

$$\ddot{\theta} + \dot{\theta} + \sin \theta = 0.$$

(a) What physical system does this correspond to?

(b) Draw the phase portraits for this system.

(c) Show a specific trajectory for $\theta_0 = 0.5$ rad and $\dot{\theta} = 0$.

9.25 Consider the nonlinear upright pendulum with a motor at its base as an actuator. Design a feedback controller to stabilize this system.

9.26 Consider the system

$$\dot{x} = -\sin x.$$

Prove that the origin is an asymptotically stable equilibrium point.

9.27 A first-order nonlinear system is described by the equation $\dot{x} = -f(x)$, where $f(x)$ is a continuous and differentiable nonlinear function that satisfies the following:

$$f(0) = 0,$$
$$f(x) > 0, \quad \text{for} \quad x > 0,$$
$$f(x) < 0, \quad \text{for} \quad x < 0.$$

Use the Lyapunov function $V(x) = x^2/2$ to show that the system is stable near the origin $(x = 0)$.

9.28 Use the Lyapunov equation

$$\mathbf{F}^T \mathbf{P} + \mathbf{P}\mathbf{F} = -\mathbf{Q} = -\mathbf{I}$$

to find the range of K for which the system in Fig. 9.67 will be stable. Compare your answer with the stable values for K obtained using Routh's stability criterion.

Figure 9.67
Control system for
Problem 9.28

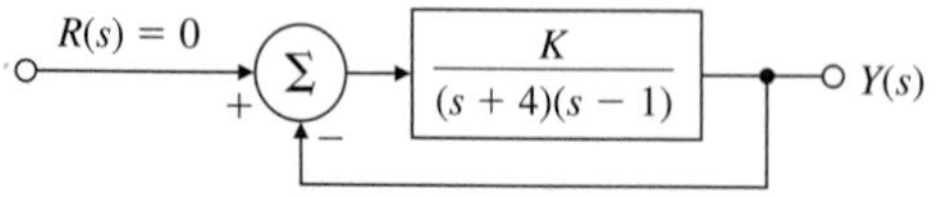

9.29 Consider the system

$$\frac{d}{dt}\begin{bmatrix} x_1 \\ x_2 \end{bmatrix} = \begin{bmatrix} x_1 + x_2 u \\ x_2(x_2 + u) \end{bmatrix}, \quad y = x_1.$$

Find all values of α and β for which the input $u(t) = \alpha y(t) + \beta$ will achieve the goal of maintaining the output $y(t)$ near 1.

9.30 Consider the nonlinear autonomous system

$$\frac{d}{dt}\begin{bmatrix} x_1 \\ x_2 \\ x_3 \end{bmatrix} = \begin{bmatrix} x_2(x_3 - x_1) \\ x_1^2 - 1 \\ -x_1 x_3 \end{bmatrix}.$$

(a) Find the equilibrium point(s).

(b) Find the linearized system about each equilibrium point.

(c) For each case in part (b), what does Lyapunov theory tell us about the stability of the nonlinear system near the equilibrium point?

9.31 *Van der Pol's equation:* Consider the system described by the nonlinear equation

$$\ddot{x} + \varepsilon(1 + x^2)\dot{x} + x = 0$$

with the constant $\varepsilon > 0$.

(a) Show that the equations can be put in the form [Liénard or (x, y) plane]

$$\dot{x} = y + \varepsilon\left(\frac{x^3}{3} - x\right),$$

$$\dot{y} = -x.$$

(b) Use the Lyapunov function $V = \frac{1}{2}(x^2 + \dot{x}^2)$, and sketch the *region* of stability as predicted by this V in the Liénard plane.

(c) Plot the trajectories of part (b) and show the initial conditions that tend to the origin. Simulate the system in SIMULINK using various initial conditions on $x(0)$ and $\dot{x}(0)$. Consider two cases, with $\varepsilon = 0.5$ and $\varepsilon = 1.0$.

10

Control System Design: Principles and Case Studies

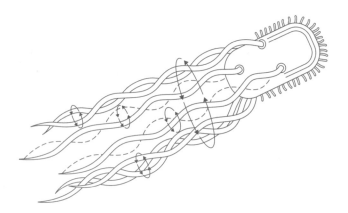

A Perspective on Design Principles

In Chapters 5, 6, and 7 we presented techniques for analyzing and designing feedback systems based on the root-locus, frequency-response, and state-variable methods. Thus far we have had to consider somewhat isolated, idealized aspects of larger systems and to focus on applying one analysis method at a time. In this chapter we return to the theme of Chapter 4—the advantages of feedback control—to reconsider the overall problem of control systems design with the sophisticated tools developed in Chapters 5 to 7 and 9 in hand. We will apply these tools to several complex, real-world applications in a case study-type format.

Having an overarching, step-by-step design approach serves two purposes: It provides a useful starting point for any real-world controls problem, and it provides meaningful checkpoints once the design process is underway. This chapter develops just such a general approach, which will be applied in the case studies.

Chapter Overview

Section 10.1 opens the chapter with a step-by-step design process that is sufficiently general to apply to any control design process, but which also provides useful definitions and directions. We then apply the design process to four practical, complex applications: design of an attitude control system for a satellite (Section 10.2), lateral and longitudinal control of a Boeing 747 (Section 10.3), and control of the fuel–air ratio in an automotive engine (Section 10.4), control of a disk drive (Section 10.5), and control of rapid thermal processing (RTP) system (Section 10.6). The satellite case study is representative of the control of geosynchronous communications satellite systems. The study addresses the design of robust control systems in which the physical parameters are known to vary within a given range. In this system the control system needs to meet specifications from the "beginning-of-life" (BOL) to the "end-of-life" (EOL) that spans a period of 12–15 years. The satellite's moment of inertia and mass will vary as fuel is expended for attitude control, and by deployment and re-orientation of satellite antennas. The satellite case study illustrates the use of a notch compensation for a system with lightly damped resonance. We will see from this case study that collocated actuator and sensor systems are much easier to control than the noncollocated systems. The Boeing 747 case study addresses the familiar flight control system of commercial passenger aircraft. The nonlinear equations of motion are given and are linearized about a particular flight condition. The rigid body dynamics, longitudinal and lateral-directional, are each fourth order. Of course, the flexible modes need also be considered for a more accurate model. The Boeing 747 lateral-stabilization case study will illustrate the use of feedback as an inner-loop designed to aid the pilot, who will provide the primary outer-loop control. The altitude control will show how to combine inner-loop feedback with outer-loop compensation to design a complete control system. The air-to-fuel ratio automotive case study is a real-world example that includes a nonlinear sensor and a pure time delay. We will use the describing function method of Chapter 9 to analyze the behavior of this system. Another familiar problem to every PC user is the control of data stored on a disk drive. This case study is about position control and bandwidth will be a key performance parameter. The RTP case study from semiconductor wafer fabrication is remarkably close to the industrial application. The problem concerns temperature tracking and disturbance rejection for a highly nonlinear thermal system. The actuator (lamp) is also nonlinear and we will use the technique from Chapter 9 to try to cancel the effect of this nonlinearity. Another key aspect of this system is actuator saturation and the fact that the control signal cannot go negative. In all these case studies the designer needs to be able to use multiple tools from previous chapters, including the root locus, the frequency response, pole placement by state feedback, and (nonlinear) simulation of time responses to obtain a satisfactory design. In Section 10.7 we present a case study from the emerging field of systems biology and describe chemotaxis or how Escherichia coli (*E. coli*) swims away from trouble. Section 10.8 provides a historical perspective on applications of feedback control.

10.1 An Outline of Control Systems Design

Control engineering is an important part of the design process of many dynamic systems. As suggested in Chapter 4, the deliberate use of feedback can stabilize an otherwise unstable system, reduce the error due to disturbance inputs, reduce the tracking error while following a command input, and reduce the sensitivity of a closed-loop transfer function to small variations in internal system parameters. In those situations for which feedback control is required, it is possible to outline an approach to control systems design that often leads to a satisfactory solution.

Before describing this approach, we wish to emphasize that the purpose of control is to aid the product or process—the mechanism, the robot, the chemical plant, the aircraft, or whatever—to do its job. Engineers engaged in other areas of the design process are increasingly taking the contribution of control into account early in their plans. As a result, more and more systems have been designed so that they will not work at all without feedback. This is especially significant in the design of high-performance aircraft, where control has taken its place along with structures and aerodynamics as essential to assuring that the craft will even fly at all. It is impossible to give a description of such overall design in this book, but recognizing the existence of such cases places in perspective not only the specific task of control system design but also the central role this task can play in an enterprise.

Control system design begins with a proposed product or process whose satisfactory dynamic performance depends on feedback for stability, disturbance regulation, tracking accuracy, or reduction of the effects of parameter variations. We will give an outline of the design process that is general enough to be useful whether the product is an electronic amplifier or a large structure to be placed in earth orbit. Obviously, to be so widely applicable, our outline has to be vague with respect to physical details and specific only with respect to the feedback-control problem. To present our results, we will divide the control design problem into a sequence of characteristic steps.

STEP 1. *Understand the process and translate dynamic performance requirements into time, frequency, or pole–zero specifications.* The importance of understanding the process, what it is intended to do, how much system error is permissible, how to describe the class of command and disturbance signals to be expected, and what the physical capabilities and limitations are can hardly be overemphasized. Regrettably, in a book such as this, it is easy to view the process as a linear, time-invariant transfer function capable of responding to inputs of arbitrary size, and we tend to overlook the fact that the linear model is a *very* limited representation of the real system, valid only for small signals, short times, and particular environmental conditions. Do not confuse the approximation with the reality. You must be able to use the simplified model for its intended purpose, and to return to an accurate model or the actual physical system to really verify the design performance.

Typical results of this step are specifications that the system have a step response inside some constraint boundaries (as shown in Fig. 10.1a), an open-loop frequency response satisfying certain constraints (Fig. 10.1b), or closed-loop poles to the left of some constraint boundary (Fig. 10.1c).

STEP 2. *Select sensors.* In **sensor** selection, consider which variables are important to control and which variables can physically be measured. For example, in a jet engine there are critical internal temperatures that must be controlled, but that cannot

Specifications

Sensors

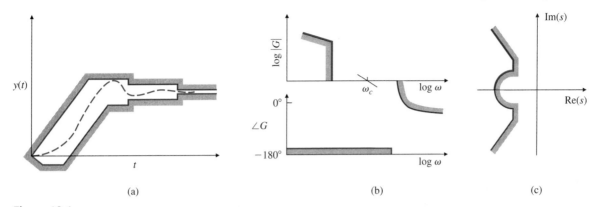

Figure 10.1
Examples of (a) time-response; (b) frequency-response; and (c) pole–zero specifications resulting from Step 10.1

be measured directly in an operational engine. Select sensors that indirectly allow a good estimate to be made of these critical variables. It is important to consider sensors for the disturbance. Sometimes, especially in chemical processes, it is beneficial to sense a load disturbance directly, because improved performance can be obtained if this information is fed forward to the controller.

Following are some factors that influence sensor selection:

Number of sensors and locations:	Select minimum required number of sensors and their optimal locations
Technology:	Electric or magnetic, mechanical, electromechanical, electro-optical, piezoelectric
Functional performance:	Linearity, bias, accuracy, bandwidth, resolution, dynamic range, noise
Physical properties:	Weight, size, strength
Quality factors:	Reliability, durability, maintainability
Cost:	Expense, availability, facilities for testing and maintenance

Actuators

STEP 3. *Select actuators.* In order to control a dynamic system, obviously you must be able to influence the response. The device that does this is the **actuator**. Before choosing a specific actuator, consider which variables can be influenced. For example, in a flight vehicle many configurations of movable surfaces are possible, and the influence these have on the performance and controllability of the craft can be profound. The locations of jets or other torque devices are also a major part of the control design of spacecraft.

Having selected a particular variable to control, you may need to consider other factors:

Number of actuators and locations:	Select required actuators and their optimal locations
Technology:	Electric, hydraulic, pneumatic, thermal, other
Functional performance:	Maximum force possible, extent of the linear range, maximum speed possible, power, efficiency
Physical properties:	Weight, size, strength
Quality factors:	Reliability, durability, maintainability
Cost:	Expense, availability, facilities for testing and maintenance

Linearization

STEP 4. *Make a linear model.* Here you take the best choice for process, actuator, and sensor; identify the equilibrium point of interest; and construct a small-signal dynamic model valid over the range of frequencies included in the specifications of Step 1. You should also validate the model with experimental data where possible. To be able to make use of all the available tools, express the model in state-variable and pole–zero form as well as in frequency-response form. As we have seen, MATLAB® and other computer-aided control systems design software packages have the means to perform the transformations among these forms. Simplify and reduce the order of the model if necessary. Quantify model uncertainty.

Simple compensation
PID/lead–lag design

STEP 5. *Try a simple proportional–integral–derivative (PID) or lead–lag design.* To form an initial estimate of the complexity of the design problem, sketch a frequency response (Bode plot) and a root locus with respect to plant gain. If the plant–actuator–sensor model is stable and minimum phase, the Bode plot will probably be the most useful; otherwise, the root locus shows very important information with respect to behavior in the right half-plane (RHP). In any case, try to meet the specifications with a simple controller of the lead–lag variety, including integral control if steady-state error response requires it. Do not overlook feed-forward of the disturbances if the necessary sensor information is available. Consider the effect of sensor noise, and compare a lead network to a direct sensor of velocity to see which gives a better design.

STEP 6. *Evaluate/modify plant.* Based on the simple control design, evaluate the source of the undesirable characteristics of the system performance. Reevaluate the specifications, the physical configuration of the process, and the actuator and sensor selections in light of the preliminary design, and return to Step 1 if improvement seems necessary or feasible. For example, in many motion-control problems, after testing the first-pass design, you might find vibrational modes that prevent the design from meeting the initial specifications of the problem. It may be much easier to meet the specifications by altering the structure of the plant through the addition of stiffening members or by passive damping than to meet them by control strategies alone. An alternative solution may be to move a sensor so it is at a node of a vibration mode, thus providing no feedback of the motion. Also, some actuator technologies (such as hydraulic) have many more low-frequency vibrations than others (such as electric) do and changing the actuator technology may be indicated. In a digital implementation, it may be possible to revise the sensor–controller–actuator system structure so as to reduce time delay, which is always a destabilizing element. In thermal systems, it is often possible to change heat capacity or conductivities by material substitution that will enhance the control design. It is important to consider all parts of the design, not only the control logic, to meet the specifications in the most cost-effective way. If the plant is modified, go back to Step 1. If the design now seems satisfactory, go to Step 8; otherwise try Step 7.

Optimal design

STEP 7. *Try an optimal design.* If the trial-and-error compensators do not give entirely satisfactory performance, consider a design based on optimal control. The symmetric root locus (SRL) will show possible root locations from which to select locations for the control poles that meet the response specifications; you can select locations for the estimator poles that represent a compromise between sensor and process noise. Plot the corresponding open-loop frequency response and the root locus to evaluate the stability margins of this design and its robustness to parameter changes. You can modify the pole locations until a best compromise results. Returning to the SRL with different cost measures is often a part of this step, or computations via the direct functions lqr and lqe can be used. Another variation on optimal control is to propose a fixed structure controller with unknown parameters, formulate a performance cost function, and use parameter optimization to find a good set of parameter values.

Compare the optimal design yielding the most satisfactory frequency response with the transform-method design you derived in Step 5. Select the better of the two before proceeding to Step 8.

STEP 8. *Build a computer model, and compute (simulate) the performance of the design.* After reaching the best compromise among process modification, actuator and sensor selection, and controller design choice, run a computer model of the system. This model should include important nonlinearities, such as actuator saturation, realistic noise sources, and parameter variations you expect to find during operation of the system. The simulation will often identify sensitivities that may lead to going back to Step 5 or even Step 2. Design iterations should continue until the simulation confirms acceptable stability and robustness. As part of this simulation you can often include parameter optimization, in which the computer tunes the free parameters for best performance. In the early stages of design the model you simulate will be relatively simple; as the design progresses, you will study more complete and detailed models. At this step it is also possible to compute a digital equivalent of the analog controller as described in Chapters 4 and 8. Some refinement of the controller parameters may be required to account for the effects of digitization. This allows the final design to be implemented with digital processor logic.

If the results of the simulation prove the design satisfactory, go to Step 9, otherwise return to Step 1.

Prototype
Prototype testing

STEP 9. *Build a prototype.* As the final test before production, it is common to build and test a prototype. At this point you verify the quality of the model, discover unsuspected vibration and other modes, and consider ways to improve the design. Implement the controller using embedded software/hardware. Tune the controller if necessary. After these tests, you may want to reconsider the sensor, actuator, and process and return to Step 1—unless time, money, or ideas have run out.

This outline is an approximation of good practice; other engineers will have variations on these themes. In some cases you may wish to carry out the steps in a different order, to omit a step, or to add one. The stages of simulation and prototype construction vary widely, depending on the nature of the system. For systems for which a prototype is difficult to test and rework (for example, a satellite) or where a failure is dangerous (for example, active stabilization of a high-speed centrifuge or landing a human on the moon), most of the design verification is done through simulation of some sort. It may take the form of a digital numerical simulation, a laboratory scale model, or a full-size laboratory model with a simulated environment. For systems that are easy to build and modify (for example, feedback control for an automotive fuel system), the simulation step is often skipped entirely; design verification and refinement are instead accomplished by working with prototypes.

One of the issues raised in the preceding discussion (Step 6) was the important consideration for **changing the plant itself**. In many cases, proper plant modifications can provide additional damping or increase in stiffness, change in mode shapes, reduction of system response to disturbances, reduction of Coulomb friction, change in thermal capacity or conductivity, etc. It is worth elaborating on this by way of specific examples from the authors' experiences. In a semiconductor wafer-processing example, the edge ring holding the wafer was identified as a limiting factor in closed-loop control. Modifying the thickness of the edge ring and using a different coating material reduced the heat losses and, together with relocating one of the temperature sensors closer to the edge ring, resulted in significant improvement in control performance. In another application, thin film processing, simply changing the order of the two incoming flows resulted in significant improvement in the mixing of the precursor and oxidizer materials, and led to improvement in uniformity of the film. In an application on physical vapor deposition using RF-plasma, the shape of the target was modified to be curved to counter the geometry effects of the chamber, and yielded substantial improvements in deposition uniformity. As the last example, in a hydraulic spindle control problem, adding oil temperature control with ceramic insulation and a temperature sink for the bell housing resulted in several orders of magnitude reduction in disturbances not achievable by feedback control alone.[1] One can also mention aerospace applications for which the control was an afterthought, and the feedback control problem became exceedingly difficult and resulted in poor closed-loop performance. The moral of this discussion is that one must not forget the option of modifying the plant itself to make the control problem easier and provide maximum closed-loop performance.

The usual approach of designing the system and "throwing it over the fence" to the control group has proved to be inefficient and flawed. A better approach that is gaining momentum is to get the control engineer involved from the onset of a project to provide early feedback on how hard it is to control the system. The control engineer can provide valuable feedback on choice of actuators and sensors and can even suggest modifications to the plant. It is often much more efficient to change the plant design while it is on the drawing board before "any metal has been bent." Closed-loop performance studies can then be performed on a simple model of the system early on.

Implicit in the process of design is the well-known fact that designs within a given category often draw on experience gained from earlier models. Thus, good designs evolve rather than appear in their best form after the first pass. We will illustrate the method with several cases (Sections 10.2 to 10.6). For easy reference, we summarize the steps here.

Summary of Control Design Steps

1. Understand the process and its performance requirements.
2. Select the types and number of sensors considering location, technology, and noise.

---

[1] Our colleague Prof. Daniel DeBra strongly believes in considering modifying the plant itself an option for improved control. He cites this particular application to make the point. Of course, we agree with him!

3. Select the types and number of actuators considering location, technology, noise, and power.
4. Make a linear model of the process, actuator, and sensor.
5. Make a simple trial design based on the concepts of lead–lag compensation or PID control. If satisfied, go to Step 8.
6. Consider modifying the plant itself for improved closed-loop control.
7. Make a trial pole-placement design based on optimal control or other criteria.
8. Simulate the design, including the effects of nonlinearities, noise, and parameter variations. If the performance is not satisfactory, return to Step 1 and repeat. Consider modifying the plant itself for improved closed-loop control.
9. Build a prototype and test it. If not satisfied, return to Step 1 and repeat.

10.2   Design of a Satellite's Attitude Control

Our first example, taken from the space program, is suggested by the need to control the pointing direction, or attitude, of a satellite in orbit about the earth. Figure 10.2(a) shows a picture of a geosynchronous communications satellite. We will go through each step in our design outline and touch on some of the factors that might be considered for the control of such a system.

STEP 1. *Understand the process and its performance specifications.* A satellite is sketched in Fig. 10.2(b). We imagine that the vehicle has an astronomical survey mission requiring accurate pointing of a scientific sensor package. This package must be maintained in the quietest possible environment, which entails isolating it

Figure 10.2

(a) Picture of the geosynchronous communications satellite IPSTAR; (b) diagram of a satellite and its two-body model

Source: Courtesy Thaicom plc and Space Systems/Loral

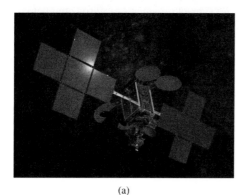

(a)

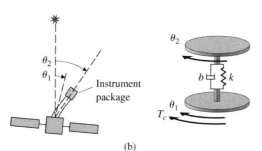

(b)

from the vibrations and electrical noise of the main service body and from its power supplies, thrusters, and communication gear. We model the resulting structure as two masses connected by a flexible boom. In Fig. 10.2(b), the satellite attitude θ_2 is the angle between the star sensor and the instrument package, and θ_1 is the angle of the main satellite with respect to the star. Figure 10.2(b) shows the equivalent mechanical system diagram for the satellite, where the sensor is mounted to the disk associated with θ_2. Disturbance torques due to solar pressure, micrometeorites, and orbit perturbations are computed to be negligible. The pointing requirement arises when it is necessary to point the unit in another direction. It can be met by dynamics with a transient settling time of 20 sec and an overshoot of no more than 15%. The dynamics of the satellite include parameters that can vary. The control must be satisfactory for any parameter values in a prespecified range to be given when the equations are written.

STEP 2. *Select sensors.* In order to orient the scientific package, it is necessary to measure the attitude angles of the package. For this purpose we propose to use a **star tracker**, a system based on gathering an image of a specific star and keeping it centered on the focal plane of a telescope. This sensor gives a relatively noisy but very accurate (on the average) reading proportional to θ_2, the angle of deviation of the instrument package from the desired angle. To stabilize the control, we include a rate gyro to give a clean reading of $\dot{\theta}_2$, because a lead network on the star-tracker signal would amplify the noise too much. Furthermore, the rate gyro can stabilize large motions before the star tracker has acquired the target star image.

STEP 3. *Select actuators.* Major considerations in selecting the actuator are precision, reliability, weight, power requirements, and lifetime. Alternatives for applying torque are cold-gas jets, reaction wheels or gyros, magnetic torquers, and a gravity gradient. The jets have the most power and are the least accurate. Reaction wheels are precise but can transfer only momentum, so jets or magnetic torquers are required to "dump" momentum from time to time. Magnetic torquers provide relatively low levels of torque and are suitable only for some low-altitude satellite missions. A gravity gradient also provides a very small torque that limits the speed of response and places severe restrictions on the shape of the satellite. For purposes of this mission, we select cold-gas jets as being fast and adequately accurate.

STEP 4. *Make a linear model.* For the satellite, we assume two masses connected by a spring with torque constant k and viscous-damping constant b as shown in Fig. 10.2. The equations of motion are

$$J_1\ddot{\theta}_1 + b(\dot{\theta}_1 - \dot{\theta}_2) + k(\theta_1 - \theta_2) = T_c, \tag{10.1a}$$

$$J_2\ddot{\theta}_2 + b(\dot{\theta}_2 - \dot{\theta}_1) + k(\theta_2 - \theta_1) = 0, \tag{10.1b}$$

where T_c is the control torque on the main body. With inertias $J_1 = 1$ and $J_2 = 0.1$, the transfer function is

$$G(s) = \frac{10bs + 10k}{s^2(s^2 + 11bs + 11k)}. \tag{10.2}$$

If we choose

$$\mathbf{x} = [\ \theta_2 \quad \dot{\theta}_2 \quad \theta_1 \quad \dot{\theta}_1\ ]^T$$

as the state vector, then, using Eq. (10.1a) and assuming $T_c \equiv u$, we find that the equations of motion in state-variable form are

$$\dot{\mathbf{x}} = \begin{bmatrix} 0 & 1 & 0 & 0 \\ -\dfrac{k}{J_2} & -\dfrac{b}{J_2} & \dfrac{k}{J_2} & \dfrac{b}{J_2} \\ 0 & 0 & 0 & 1 \\ \dfrac{k}{J_1} & \dfrac{b}{J_1} & -\dfrac{k}{J_1} & -\dfrac{b}{J_1} \end{bmatrix} \mathbf{x} + \begin{bmatrix} 0 \\ 0 \\ 0 \\ \dfrac{1}{J_1} \end{bmatrix} u, \qquad (10.3a)$$

$$y = [\ 1 \quad 0 \quad 0 \quad 0\ ] \mathbf{x}. \qquad (10.3b)$$

Physical analysis of the boom leads us to assume that the parameters k and b vary as a result of temperature fluctuations but are bounded by

$$0.09 \le k \le 0.4, \qquad (10.4a)$$

$$0.038 \sqrt{\frac{k}{10}} \le b \le 0.2 \sqrt{\frac{k}{10}}. \qquad (10.4b)$$

As a result, the vehicle's natural resonance frequency ω_n can vary between 1 and 2 rad/sec, and the damping ratio ζ varies between 0.02 and 0.1.

Selecting nominal values for varying parameters

One approach to control design when parameters are subject to variation is to select nominal values for the parameters, construct the design for this model, and then test the controller performance with other parameter values. In the present case we choose nominal values of $\omega_n = 1$ and $\zeta = 0.02$. The choice is somewhat arbitrary, being based on experience and heuristic analysis. However, note that these are the lowest values in their respective ranges and thus correspond to the plant that is probably the most difficult to control so as to meet the specifications. We assume that a design for this model has a good chance to meet the specifications for other parameter values as well. (Another choice would be to select a model with average values for each parameter.) The selected parameter values are $k = 0.091$ and $b = 0.0036$; with $J_1 = 1$ and $J_2 = 0.1$, the nominal equations become

$$\dot{\mathbf{x}} = \begin{bmatrix} 0 & 1 & 0 & 0 \\ -0.91 & -0.036 & 0.91 & 0.036 \\ 0 & 0 & 0 & 1 \\ 0.091 & 0.0036 & -0.091 & -0.0036 \end{bmatrix} \mathbf{x} + \begin{bmatrix} 0 \\ 0 \\ 0 \\ 1 \end{bmatrix} u, \qquad (10.5a)$$

$$y = [\ 1 \quad 0 \quad 0 \quad 0\ ] \mathbf{x}. \qquad (10.5b)$$

The corresponding transfer function, using the MATLAB ss2tf function, is then

$$G(s) = \frac{0.036(s + 25)}{s^2(s^2 + 0.04s + 1)}. \qquad (10.6)$$

When a trial design is completed, the computer simulation should be run with a range of possible parameter values to ensure that the design has adequate robustness to tolerate these changes. Equations (3.66)–(3.68) tell us that the dynamic performance specifications will be met if the closed-loop poles have a natural frequency of 0.5 rad/sec and a closed-loop damping ratio of 0.5; these correspond to an open-loop

Figure 10.3

Root locus of $KG(s)$

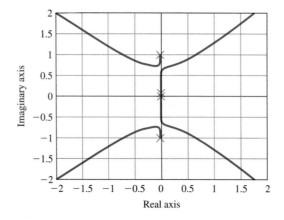

Figure 10.4

Open-loop Bode plot of $KG(s)$ for $K = 0.5$

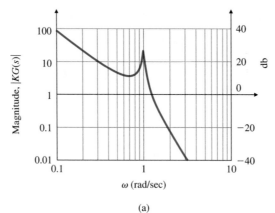

(a)

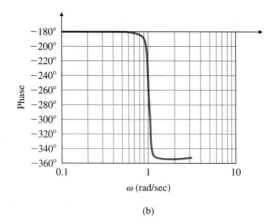

(b)

crossover frequency of $\omega_c \cong 0.5$ rad/sec and a phase margin of about PM $= 50°$. We will try to meet these design criteria.

STEP 5. *Try a lead–lag or PID controller.* The proportional-gain root locus for the nominal plant is drawn in Fig. 10.3, and the Bode plot is given in Fig. 10.4. We can see

Figure 10.5

Root locus of $KD_1(s)G(s)$

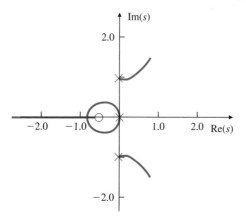

from Fig. 10.4 that this may be a difficult design problem because the frequency of the lightly damped resonance is greater than the crossover-frequency design point by only a factor of 2. This situation will require that the compensation can correct for the phase lag of the plant at the resonance frequency. Such correction is very dependent on knowing the resonance frequency, which is subject to change in this case. There may be trouble ahead.

In order to illustrate some important aspects of compensation design, we will at first ignore the resonance and generate a design that would be acceptable for the rigid body alone. We take the process transfer function to be $1/s^2$, the feedback to be position plus derivative (star tracker plus rate gyro) or PD control with the transfer function $D(s) = K(sT_D + 1)$, and the response objective to be $\omega_n = 0.5$ rad/sec and $\zeta = 0.5$. A suitable controller would be

$$D_1(s) = 0.25(2s + 1). \tag{10.7}$$

The root locus for the actual plant with D_1 is shown in Fig. 10.5 and the Bode plot in Fig. 10.6. From these plots we can see that the low-frequency poles are reasonable but that the system will be unstable because of the resonance.[2] At this point we take the simple actions of reducing our expectation with respect to bandwidth and of slowing the system down by lowering the gain until the system is stable. With so little damping, we must really go slowly. A bit of experimentation leads to

$$D_2(s) = 0.001(30s + 1), \tag{10.8}$$

for which the root locus is drawn in Fig. 10.7 and the Bode plot given in Fig. 10.8. The Bode plot shows that we have a phase margin of 50° but a crossover frequency of only $\omega_c = 0.04$ rad/sec. While this is too low to meet the settling-time specification, a low crossover frequency is unavoidable if we expect to keep the gain at the resonance frequency below unity so that it is gain stabilized.

An alternative approach to the problem is to place zeros near the lightly damped poles and use them to hold these poles back from the RHP. Such a compensation has

---

[2]If this system were built, the actuator jets would saturate as the response grew. We could analyze the response using the method described in Section 9.3 for nonlinear systems. From the analysis we would expect the signal to grow and the equivalent gain of the actuator to fall until the roots return to the imaginary axis near ω_n. The resulting limit cycle would rapidly deplete the control gas supply.

Figure 10.6

Bode plot of $KD_1(s)G(s)$

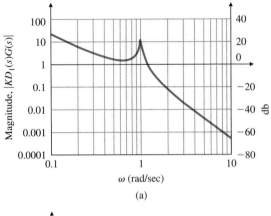

(a)

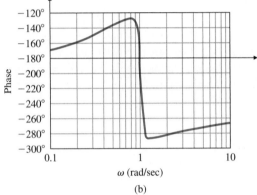

(b)

Figure 10.7

Root locus of $KD_2(s)G(s)$

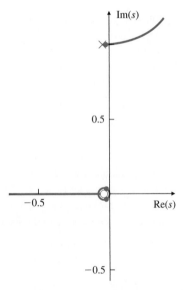

Figure 10.8

Bode plot of $D_2(s)G(s)$

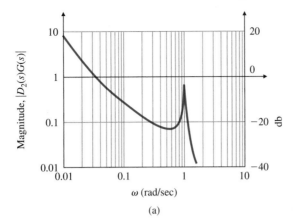

(a)

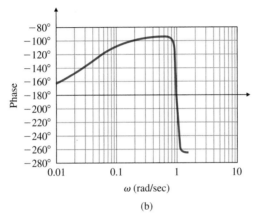

(b)

Figure 10.9

Twin-tee realization of a notch filter

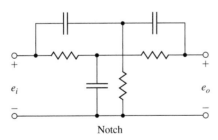

Notch

Notch filter

a frequency response with a very low gain near the frequency of the offending poles and a reasonable gain elsewhere. Because the frequency response seems to have a dent or notch in it, the device is called a **notch filter**. (It is also called a **band reject filter** in electric network theory.) An RC circuit with a notch characteristic is shown in Fig. 10.9, its pole–zero pattern in Fig. 10.10, and its frequency response in Fig. 10.11. The $+180°$ phase lead of the notch can be used to correct for the $180°$ phase lag of the resonance; if the notch frequency is *lower* than the plant's resonance frequency, the system phase is kept above $180°$ near resonance.

Figure 10.10

Notch filter pole–zero pattern

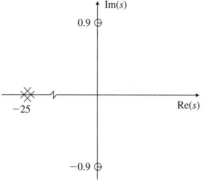

Figure 10.11

Bode plot of a notch filter

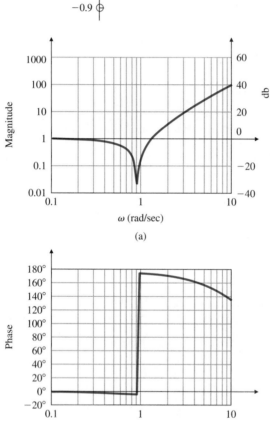

With this idea we return to the compensation given by Eq. (10.7) and add the notch, producing the revised compensator transfer function

$$D_3(s) = 0.25(2s + 1)\frac{(s/0.9)^2 + 1}{[(s/25) + 1]^2}. \tag{10.9}$$

The Bode plot for this case is shown in Fig. 10.12, the root locus in Fig. 10.13, and the unit step response in Fig. 10.14. The settling time of the design is too long for the

Figure 10.12

Bode plot of $KD_3(s)G(s)$

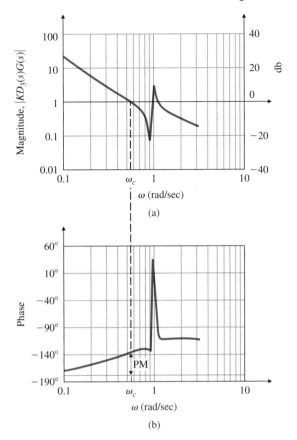

(a)

(b)

specification and the overshoot is too large, but this design approach seems promising; with iteration it could lead to a satisfactory compensator.

We now recall that the compensator is expected to provide adequate performance as the parameters vary over the ranges given by Eq. (10.3a). An examination of the robustness of the design can be made by looking at the root locus shown in Fig. 10.15, which is drawn using the compensator of Eq. (10.9) and the plant with $\omega_n = 2$, rather than 1, such that

$$\hat{G}(s) = \frac{(s/50 + 1)}{s^2(s^2/4 + 0.02s + 1)}. \qquad (10.10)$$

This assumes that the boom is as stiff as possible. Notice that now the low-frequency poles have a damping ratio of only 0.02. Combining the various parameter values, we get the frequency response and transient response shown in Figs. 10.16 and 10.17. We could make a few more trial-and-error iterations with the notch filter and rate feedback, but the system is complex enough that a look at state-space designs now seems reasonable. We go to Step 7.

STEP 6. *Evaluate/modify plant.* Refer to the collocated control discussion after Step 8.

Figure 10.13

Root locus of $KD_3(s)G(s)$

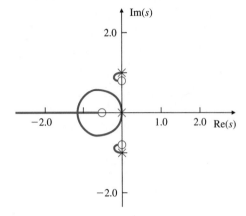

Figure 10.14

Closed-loop step response of $D_3(s)G(s)$ where $\theta_2(0) = 0.2$ rad

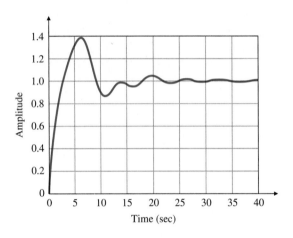

Figure 10.15

Root locus of $KD_3(s)\hat{G}(s)$

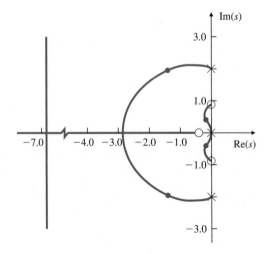

Figure 10.16

Bode plot of $KD_3(s)\hat{G}(s)$

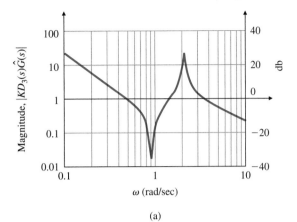

(a)

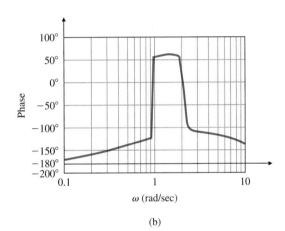

(b)

Figure 10.17

Closed-loop step response of $D_3(s)\hat{G}(s)$

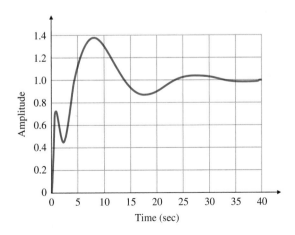

STEP 7. *Try an optimal design using pole placement.* Using the state-variable formulation of the equations of motion in Eq. (10.4a), we devise a controller that will place the closed-loop poles in arbitrary locations. Of course, used without thought, the method of pole placement can also result in a design that requires unreasonable levels of control effort or is very sensitive to changes in the plant transfer function. Guidelines for pole placement are given in Chapter 7; an often successful approach is to derive optimal pole locations using the SRL. Figure 10.18 shows the SRL for the problem at hand. To obtain a bandwidth of about 0.5 rad/sec, we select closed-loop control poles from this locus at $-0.45 \pm 0.34j$ and $-0.15 \pm 1.05j$.

If we select $\alpha_c(s)$, as discussed earlier, from the SRL, the control gain using the MATLAB function place is

$$\mathbf{K} = \begin{bmatrix} -0.2788 & 0.0546 & 0.6814 & 1.1655 \end{bmatrix}. \tag{10.11}$$

Figure 10.19 shows the step responses for the nominal plant parameters and stiff-spring plant models. The Bode plot of the SRL controller design with the nominal plant parameters can be computed from the loop transfer function (by breaking the

Figure 10.18

Symmetric root locus of the satellite

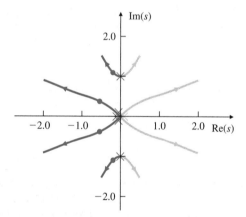

Figure 10.19

Closed-loop step response of the SRL design

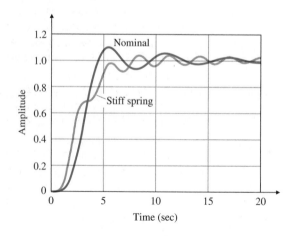

loop at u)

$$\frac{\mathbf{KX}(s)}{U(s)} = \mathbf{K}(s\mathbf{I} - \mathbf{F})^{-1}\mathbf{G}$$

and results in a phase margin of about 60°, as shown in Fig. 10.20. While the speed of response of the design meets the specifications with the nominal plant, the settling time when the plant has the stiff spring is a bit longer than the specifications call for. We might be able to get a better compromise between the nominal and the stiff-spring cases by selecting another point on the SRL; at this point we do not know. The designer must face alternatives such as these and select the best compromise for the problem at hand.

The design of Fig. 10.19 is based on full-state feedback. To complete the optimal design, we need an estimator. We select the closed-loop estimator error poles to be about eight times faster than the control poles. The reason for this is to keep the error poles from reducing the robustness of the design; a fast estimator will have almost the same effect on the response as no estimator at all. We choose the error poles from the SRL at $-7.7 \pm 3.12j$ and $-3.32 \pm 7.85j$. Pole placement with these values leads

Figure 10.20

Frequency response of the SRL design from u to Kx

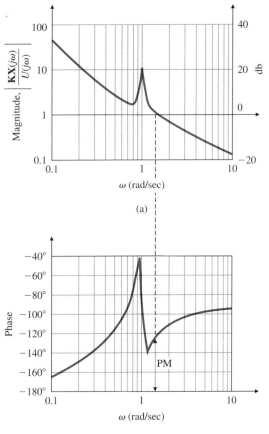

to an estimator (filter) gain, using the MATLAB function place:

$$L = \begin{bmatrix} 22 \\ 242.3 \\ 1515.4 \\ 5503.9 \end{bmatrix}. \tag{10.12}$$

After we combine the control gain and estimator, as described in Section 7.8, the compensator transfer function that results from Eq. (7.177) is

$$D_4(s) = \frac{-745(s+0.3217)(s+0.0996 \pm 0.9137 j)}{(s+3.1195 \pm 8.3438 j)(s+8.4905 \pm 3.6333 j)}. \tag{10.13}$$

The frequency response of this compensator (Fig. 10.21) shows that pole placement has introduced a notch directly. The frequency response and the root locus of the combined system $D_4(s)G(s)$ are given in Figs. 10.22 and 10.23, while Fig. 10.24 shows the step response for both the nominal and the stiff-spring plants. Notice that the design almost meets the specifications.

STEP 8. *Simulate the design, and compare the alternatives.* At this point we have two designs, with differing complexities and different robustness properties. The

Figure 10.21

Bode plot of optimal compensator $D_4(s)$

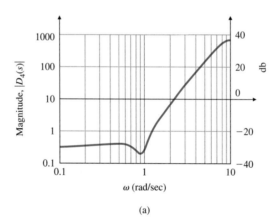

(a)

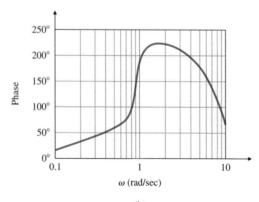

(b)

Figure 10.22
Bode plot of the compensated system $D_4(s)G(s)$

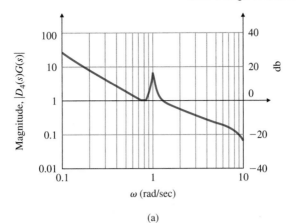

(a)

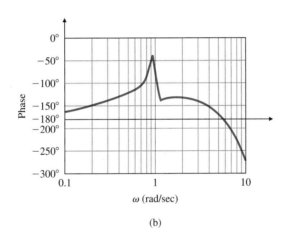

(b)

Figure 10.23
Root locus of $D_4(s)G(s)$

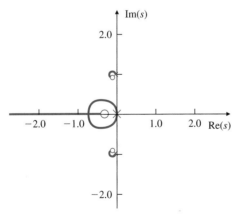

notch-filter design might be improved with further iterations or by starting with a different nominal case. The SRL design meets the specifications for the nominal plant but is too slow for the stiff-spring case, although alternative selections for the pole locations might lead to a better design. In either case, much more extensive studies

Figure 10.24

Closed-loop step response of $D_4(s)G(s)$

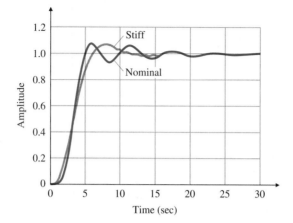

need to be made to explore the robustness and noise-response properties. Rather than follow any of these paths, we consider some aspects of the physical system.

Both designs are strongly influenced by the presence of the lightly damped resonant mode caused by the coupled masses. However, the transfer function of this system is strongly dependent on the fact that the actuator is on one body and the sensor is on the other (that is, not collocated). Suppose that, rather than considering pointing the star tracker on the small mass, we have the mission of pointing the main mass, perhaps toward an Earth station for communications purposes. For this purpose we can put the sensor on the *same* mass that holds the actuator—to give control with a collocated actuator and sensor. Due to the physics of the situation, the system's transfer function now has zeros close to the flexible modes, so control can be achieved by using PD feedback alone, because the plant already has the effect of a notch compensator. Consider the transfer function of the satellite with collocated actuator and sensor (to measure θ_1) for which the state matrices are

Collocated actuator and sensor

$$\mathbf{F} = \begin{bmatrix} 0 & 1 & 0 & 0 \\ -0.91 & -0.036 & 0.91 & 0.036 \\ 0 & 0 & 0 & 1 \\ 0.091 & 0.0036 & -0.091 & -0.0036 \end{bmatrix}, \quad \mathbf{G} = \begin{bmatrix} 0 \\ 0 \\ 0 \\ 1 \end{bmatrix},$$

$$\mathbf{H} = [\, 0 \quad 0 \quad 1 \quad 0 \,].$$

The transfer function of the system using the MATLAB ss2tf function is

$$G_{\text{co}}(s) = \mathbf{H}(s\mathbf{I} - \mathbf{F})^{-1}\mathbf{G} = \frac{(s + 0.018 \pm 0.954j)}{s^2(s + 0.02 \pm j)}. \qquad (10.14)$$

Notice the presence of the zeros in the vicinity of the complex conjugate poles. If we now use the same PD feedback as before, namely,

$$D_5(s) = 0.25(2s + 1), \qquad (10.15)$$

then the system will not only be stabilized, but will also have a satisfactory response (if we consider θ_1 as the output), because the resonant poles tend to be cancelled by the complex conjugate zeros.

Figures 10.25–10.27 show the frequency response, the root locus, and the step response, respectively, for this system. Note from Fig. 10.27 that the step response

Figure 10.25

Bode plot of $D_5(s)G_{co}(s)$

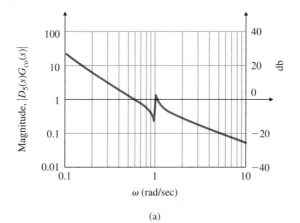

(a)

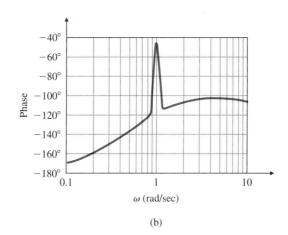

(b)

Figure 10.26

Root locus for $D_5(s)G_{co}(s)$

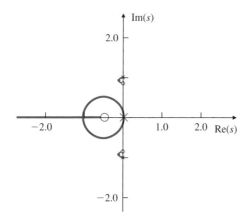

Figure 10.27

Closed-loop step response of the system with collocated control, $D_5(s)G_{co}(s)$ and $D_5(s)\hat{G}_{co}(s)$

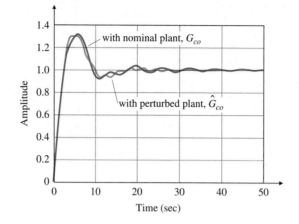

Figure 10.28

Response at θ_2 of the collocated design

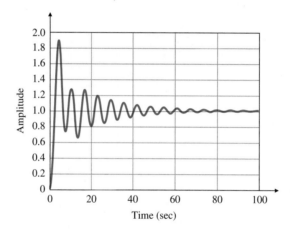

has the excess overshoot associated with the zero of the compensator in the forward path of the transfer function.

The result is a very simple robust design achieved by moving the sensor from a noncollocated position to one collocated with the actuator. The result illustrates that, to achieve good feedback control, it is very important to consider sensor location and other features of the physical problem. However, this last control design will *not* do for pointing the star tracker. This is evident from plotting the output θ_2 corresponding to the nice step response of Fig. 10.27. The result is shown in Fig. 10.28.

An architecture suggested by the results is to place a coarse star tracker on the satellite body to be used for search and initial settling. Then switch to a star tracker on the instrument package with longer settling time for fine control.

10.3 Lateral and Longitudinal Control of a Boeing 747

The Boeing 747 (Fig. 10.29) is a large wide-body transport jet. A schematic with the relevant coordinates that move with the airplane is shown in Fig. 10.30. The linearized

Figure 10.29

Boeing 747

Source: Courtesy Boeing Commercial Airplane Co.

Figure 10.30

Definition of aircraft coordinates

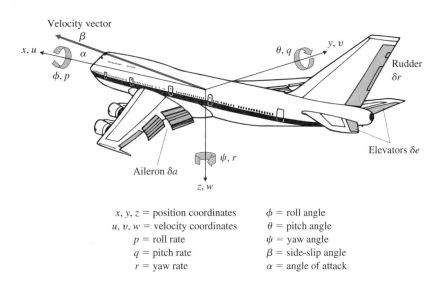

x, y, z = position coordinates
u, v, w = velocity coordinates
p = roll rate
q = pitch rate
r = yaw rate

ϕ = roll angle
θ = pitch angle
ψ = yaw angle
β = side-slip angle
α = angle of attack

equations of (rigid-body) motion[3] for the Boeing 747 are of eighth order but are separated into two fourth-order sets representing the perturbations in longitudinal (U, W, θ, and q in Fig. 10.30) and lateral (ϕ, β, r, and p) motion. The longitudinal motion consists of axial (X), vertical (Z), and pitching (θ, q) motion, while the lateral motion consists of rolling (ϕ, p) and yawing (r, β) movement. The side-slip angle β is a measure of the direction of forward velocity relative to the direction of the nose of the airplane. The elevator control surfaces and the throttle affect the longitudinal motion, whereas the aileron and rudder primarily affect lateral motion. Although there is a small amount of coupling of lateral motion into longitudinal motion, this is usually ignored, so the equations of motion are treated as two decoupled fourth-order sets for designing the control, or **stability augmentation**, for the aircraft.

---

[3] For derivation of equations of motion for an aircraft, the reader is referred to Bryson (1994), Etkin and Reid (1996), and McRuer et al. (1973).

The nonlinear rigid body equations of motion in body-axis coordinates, under proper assumptions,[4] can be derived as (Bryson, 1994)

$$m(\dot{U} + qW - rV) = X - mg \sin\theta + \kappa T \cos\theta, \qquad (10.16)$$

$$m(\dot{V} + rU - pW) = Y + mg \cos\theta \sin\phi,$$

$$m(\dot{W} + pV - qU) = Z + mg \cos\theta \cos\phi - \kappa T \sin\theta,$$

$$I_x\dot{p} + I_{xz}\dot{r} + (I_z - I_y)qr + I_{xz}qp = L, \qquad (10.17)$$

$$I_y\dot{q} + (I_x - I_z)pr + I_{xz}(r^2 - p^2) = M,$$

$$I_z\dot{r} + I_{xz}\dot{p} + (I_y - I_x)qp - I_{xz}qr = N,$$

where

m = mass of the aircraft,

$[U, V, W]$ = body-axis components of the velocity of the center of mass (c.m.),

$$\beta = \tan^{-1}\left(\frac{V}{U}\right),$$

$[U_o, V_o, W_o]$ = reference velocities,

$[p, q, r]$ = the body-axis components of the angular velocity of the aircraft (roll, pitch, and yaw respectively)

$[X, Y, Z]$ = the body-axis aerodynamic forces about the c.m.,

$[L, M, N]$ = the body-axis aerodynamic torques about the c.m.,

g_o = the gravitational force per unit mass,

I_i = the inertias in body axes,

(θ, ϕ) = the Euler pitch and roll angles of the aircraft body axes with respect to horizontal,

V_{ref} = reference flight speed,

T = the propulsive thrust resultant, and

κ = the angle between thrust and body x-axis.

The linearization of these equations can be carried out as follows: In the steady-state straight, level, and constant speed flight condition, $\dot{U} = \dot{V} = \dot{W} = \dot{p} = \dot{q} = \dot{r} = 0$. Furthermore, there is no turning in any axis so that $p_o = q_o = r_o = 0$, and the wings will be level so that $\phi = 0$. However, there will be an angle of attack in order to provide some lift from the wings to counteract the aircraft's weight, so θ_o and $W_o \neq 0$, where

$$U = U_o + u, \qquad (10.18)$$

$$V = V_o + v,$$

$$W = W_o + w.$$

---

[4]x–z is the body-axis plane of mass symmetry.

The steady-state velocity body axis components will be

$$U_o = V_{ref} \cos(\theta_o), \qquad (10.19)$$

$$V_o = 0 \ (\beta_o = 0),$$

$$W_o = V_{ref} \sin(\theta_o),$$

as depicted in Fig. 10.31. With these conditions, the *equilibrium* (see Chapter 9) equations are

$$0 = X_0 - mg_o \sin\theta_0 + \kappa T \cos\theta_0, \qquad (10.20)$$

$$0 = Y_0,$$

$$0 = Z_0 + mg_o \cos\theta_0 - \kappa T \sin\theta_0,$$

$$0 = L_0,$$

$$0 = M_0,$$

$$0 = N_0.$$

Figure 10.31

Steady-state flight condition

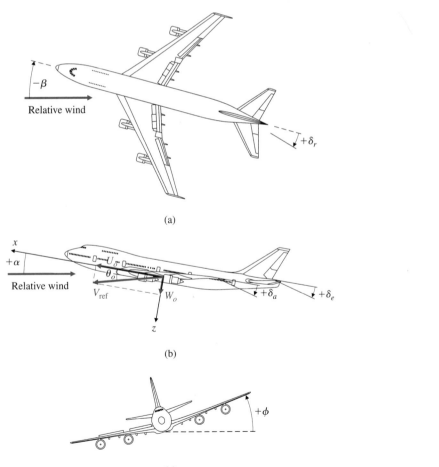

(a)

(b)

(c)

With the assumptions (Bryson, 1994)

$$(v^2, w^2) \ll u^2, \tag{10.21}$$

$$(\phi^2, \theta^2) \ll 1,$$

$$(p^2, q^2, r^2) \ll \frac{u^2}{b^2},$$

where b denotes the wingspan, many of the nonlinear terms in Eqs. (10.16) and (10.17) can be neglected. Substitution of Eq. (10.20) in the nonlinear equations of motion leads to a set of linear perturbational equations that describe small deviations from constant speed, straight and level flight. The equations of motion then divide into two uncoupled sets of *longitudinal* and *lateral* equations of motion.

For linearized longitudinal motion, the results are

$$\begin{bmatrix} \dot{u} \\ \dot{w} \\ \dot{q} \\ \dot{\theta} \end{bmatrix} = \begin{bmatrix} X_u & X_w & -W_o & -g_o \cos\theta_o \\ Z_u & Z_w & U_o & -g_o \sin\theta_o \\ M_u & M_w & M_q & 0 \\ 0 & 0 & 1 & 0 \end{bmatrix} \begin{bmatrix} u \\ w \\ q \\ \theta \end{bmatrix} + \begin{bmatrix} X_{\delta e} \\ Z_{\delta e} \\ M_{\delta e} \\ 0 \end{bmatrix} \delta e, \tag{10.22}$$

where

$u =$ forward velocity perturbation in the aircraft in x direction (Fig. 10.30),

$w =$ velocity perturbation in the z direction (also proportional to perturbations in the angle of attack, $\alpha = \frac{w}{U_0}$),

$q =$ angular rate about the positive y-axis, or pitch rate,

$\theta =$ pitch-angle perturbation from the reference θ_o value,

$X_{u,w,\delta e} =$ partial derivative of the aerodynamic force in x direction with respect to perturbations in u, w, and δe,[5]

$Z_{u,w,\delta e} =$ partial derivative of the aerodynamic force in z direction with respect to perturbations in u, w, and δe,

$M_{u,w,q,\delta e} =$ partial derivative of the aerodynamic (pitching) moment with respect to perturbations in u, w, q, and δe,

$\delta e =$ movable tail-section, or "elevator," angle for pitch control.

$W_o q$, $U_o q$ terms in the equations are due to the angular velocity of the body fixed (rotating) reference frame and arise directly from the left-hand side of Eq. (10.16). To determine altitude changes, we need to add the following equation to the longitudinal equations of motion:

$$\dot{h} = V_{ref} \sin\theta - w \cos\theta. \tag{10.23}$$

This equation will result in the linearized altitude equation

$$\dot{h} = V_{ref}\,\theta - w, \tag{10.24}$$

which is to be augmented with Eq. (10.22).

---

[5]X, Z, M are stability derivatives and are identified from wind tunnel and flight tests.

For linearized lateral motion, the results are

$$
\begin{bmatrix} \dot{\beta} \\ \dot{r} \\ \dot{p} \\ \dot{\phi} \end{bmatrix} = \begin{bmatrix} Y_v & -U_o & V_o & g_o \cos\theta_o \\ N_v & N_r & N_p & 0 \\ L_v & L_r & L_p & 0 \\ 0 & \tan\theta_o & 1 & 0 \end{bmatrix} \begin{bmatrix} \beta \\ r \\ p \\ \phi \end{bmatrix} + \begin{bmatrix} Y_{\delta r} & Y_{\delta a} \\ N_{\delta r} & N_{\delta a} \\ L_{\delta r} & L_{\delta a} \\ 0 & 0 \end{bmatrix} \begin{bmatrix} \delta r \\ \delta a \end{bmatrix},
$$

$$(10.25)$$

where

$$\beta = \text{side-slip angle, defined to be } \frac{v}{U_o},$$

$$r = \text{yaw rate},$$

$$p = \text{roll rate},$$

$$\phi = \text{roll angle},$$

$Y_{v,\delta r,\delta a}$ = partial derivative of the aerodynamic force in the y direction with respect to perturbations in $\beta, \delta r$, and δa,

$N_{v,r,p,\delta r,\delta a}$ = aerodynamic (yawing) moment stability derivatives,

$L_{v,r,p,\delta r,\delta a}$ = aerodynamic (rolling) moment stability derivatives,

$$\delta r = \text{rudder deflection},$$

$$\delta a = \text{aileron deflection}.$$

We will next discuss the design of a stability-augmentation system for the lateral dynamics, called a **yaw damper**, and the autopilot affecting the longitudinal behavior.

10.3.1 Yaw Damper

STEP 1. *Understand the process and its performance specifications.* Swept-wing aircraft have a natural tendency to be lightly damped in the lateral modes of motion. At typical commercial-aircraft cruising speeds and altitudes, this dynamic mode is sufficiently difficult to control that virtually every swept-wing aircraft has a feedback system to help the pilot. Therefore, the goal of our control system is to modify the natural dynamics so that the plane is acceptable for the pilot to fly.[6] Studies have shown that pilots like natural frequencies $\omega_n \lesssim 0.5$ and damping ratio of $\zeta \geq 0.5$. Aircraft with dynamics that violate these guidelines are generally considered fatiguing to fly and highly undesirable. Thus our system specifications are to achieve lateral dynamics that meet these constraints.

STEP 2. *Select sensors.* The easiest measurement of aircraft motion to take is the angular rate. The side-slip angle can be measured with a wind-vane device, but it is noisier and less reliable for stabilization. Two angular rates—roll and yaw—partake in the lateral motion. Study of the lightly damped lateral mode indicates that it is primarily a yawing phenomenon, so measurement of the yaw rate is a logical starting point for the design. Until the early 1980s the measurement was made with a **gyroscope** with a small, fast-spinning rotor that can yield an electric output proportional to the angular yaw rate of the aircraft. Since the early 1980s most new aircraft systems

---

[6]The mode is sufficiently difficult to control manually that, if the yaw damper fails in cruise, the pilot is instructed to descend and slow down where the mode is more manageable.

have relied on a laser device (called a **ring-laser gyroscope**) for the measurement. Here, two laser beams traverse a closed path (often a triangle) in opposite directions. As the triangular device rotates, the detected frequencies of the two beams appear to shift, and this frequency shift is measured, producing a measure of rotational rate. These devices have fewer moving parts and are more reliable at less cost than the spinning-rotor variety of gyroscope.

STEP 3. *Select actuators.* Two aerodynamic surfaces typically influence the lateral aircraft motion: the rudder and the ailerons (see Fig. 10.30). The lightly damped yaw mode that will be stabilized by the yaw damper is most affected by the rudder. Therefore, use of that single control input is a logical starting point for the design. Hence, it is best to choose the rudder as our actuator. Hydraulic devices are universally employed in large aircraft to provide the force that moves the aerodynamic surfaces. No other kind of device has been developed to provide the combination of high force, high speed, and light weight desirable for the actuation of the controlling aerodynamic surfaces. On the other hand, the low-speed flaps, which are extended slowly prior to landing, are typically actuated by an electric motor with a worm gear. For small aircraft with no autopilot, no actuator is required at all; the pilot yoke is directly connected to the aerodynamic surface by means of wire cables, and all the force required to move the surfaces is provided by the pilot.

STEP 4. *Make a linear model.* The lateral-perturbation equations of motion for Boeing 747 in horizontal flight at 40,000 ft and nominal forward speed $U_0 = 774$ ft/sec (Mach 0.8) (Heffley and Jewell, 1972), with the rudder chosen as the actuator (Step 3), are

$$\begin{bmatrix} \dot{\beta} \\ \dot{r} \\ \dot{p} \\ \dot{\phi} \end{bmatrix} = \begin{bmatrix} -0.0558 & -0.9968 & 0.0802 & 0.0415 \\ 0.598 & -0.115 & -0.0318 & 0 \\ -3.05 & 0.388 & -0.4650 & 0 \\ 0 & 0.0805 & 1 & 0 \end{bmatrix} \begin{bmatrix} \beta \\ r \\ p \\ \phi \end{bmatrix}$$

$$+ \begin{bmatrix} 0.00729 \\ -0.475 \\ 0.153 \\ 0 \end{bmatrix} \delta r,$$

$$y = \begin{bmatrix} 0 & 1 & 0 & 0 \end{bmatrix} \begin{bmatrix} \beta \\ r \\ p \\ \phi \end{bmatrix},$$

where β and ϕ are in radians and r and p are in radians per second. The transfer function, using the MATLAB ss2tf function, is

$$G(s) = \frac{r(s)}{\delta r(s)} = \frac{-0.475(s + 0.498)(s + 0.012 \pm 0.488j)}{(s + 0.0073)(s + 0.563)(s + 0.033 \pm 0.947j)}, \tag{10.26}$$

so that the system has two stable real poles and a pair of stable complex poles. Notice first that the low-frequency gain is negative, corresponding to the simple physical fact that a positive or clockwise rudder motion causes a negative or counterclockwise yaw rate. In other words, turning the rudder left (clockwise) causes the front of the aircraft to rotate left (counterclockwise). The natural motion corresponding to the complex poles is referred to as the **Dutch roll**; the name comes from the motions of a person skating on the frozen canals of Holland. The motion corresponding to the stable real poles is referred to as the **spiral mode** ($s_1 = -0.0073$) and the **roll mode** ($s_2 = -0.563$). From looking at the system poles, we see that the offending mode that needs repair for good pilot handling is the Dutch roll, with the poles at $s = -0.033 \pm 0.95j$. The roots have an acceptable frequency, but their damping ratio $\zeta \cong 0.03$ is far short of the desired value $\zeta \cong 0.5$.

STEP 5. *Try a lead–lag or PID design.* As a first try at the design, we will consider proportional feedback of the yaw rate to the rudder. The root locus with respect to the gain of this feedback is shown in Fig. 10.32, and its frequency response is shown in Fig. 10.33. The figures show that $\zeta \cong 0.45$ is achievable and can be computed to occur at a gain of about 3.0.

This feedback, however, creates an objectionable situation during a steady turn when the yaw rate is constant: Because the feedback produces a steady rudder input opposite the yaw rate, the pilot must introduce a much larger steady command for the same yaw rate than is necessary in the open-loop case. This dilemma is solved by attenuating the feedback at DC (i.e., "washing out" the feedback). This is accomplished by inserting

$$H(s) = \frac{s}{s + 1/\tau}$$

in the feedback, which passes the yaw rate feedback at frequencies above $1/\tau$ and provides no feedback at DC. Therefore, in a steady turn, the damper will provide no correction. Figure 10.34 shows a block diagram of the yaw damper with the washout.

For a more complete model, we include the rudder servo, which represents the actuator dynamics and has the transfer function

$$A(s) = \frac{\delta r(s)}{e_{\delta r}(s)} = \frac{10}{s + 10},$$

which is fast compared with the dynamics of the rest of the system and is not expected to change the response very much. The root locus, including actuator dynamics and a washout circuit with $\tau = 3$, is shown in Fig. 10.35. As seen from the root locus, the addition of the yaw rate feedback, including the washout, allows the damping ratio to be increased from 0.03 to about 0.35. The associated frequency response of the system

Figure 10.32

Root locus for yaw damper with proportional feedback

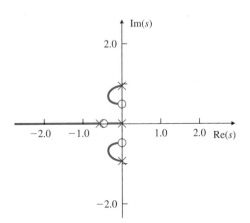

Figure 10.33

Bode plot of yaw
damper with
proportional feedback

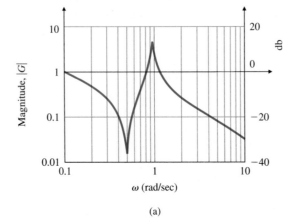

(a)

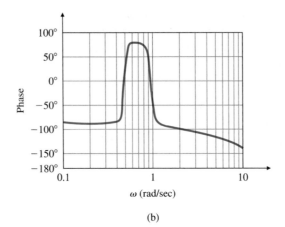

(b)

Figure 10.34

Yaw damper:
(a) functional block
diagram; (b) block
diagram for analysis

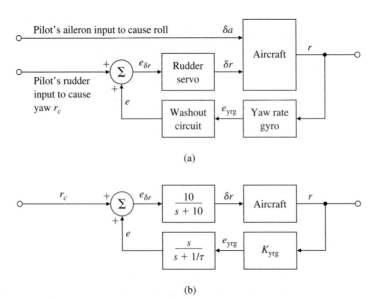

(a)

(b)

Figure 10.35

Root locus with washout circuit, $\tau = 3$

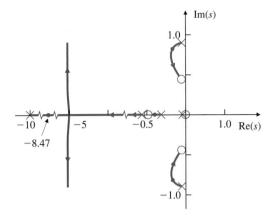

is shown in Fig. 10.36. The response of the closed-loop system to an initial condition of $\beta_0 = 1°$ is shown in Fig. 10.37 for a root-locus gain of 2.6. For reference, the response of yaw rate with no feedback is also given. Although feedback of yaw rate through the washout circuit results in a considerable improvement over the original aircraft control, the response is not as good as originally specified. Further iterations, not included here, could include other gain values or more complex compensations.

STEP 6. *Evaluate/modify plant.* The solution would be to unsweep the wings, which would cause a large drag penalty.

STEP 7. *Try an optimal design using pole placement.* If we augment the dynamic model of the system by adding the actuator and washout, we obtain the state-variable model

$$
\begin{bmatrix} \dot{x}_A \\ \dot{\beta} \\ \dot{r} \\ \dot{p} \\ \dot{\phi} \\ \dot{x}_{wo} \end{bmatrix} = \begin{bmatrix} -10 & 0 & 0 & 0 & 0 & 0 \\ 0.0729 & -0.0558 & -0.997 & 0.0802 & 0.0415 & 0 \\ -4.75 & 0.598 & -0.1150 & -0.0318 & 0 & 0 \\ 1.53 & -3.05 & 0.388 & -0.465 & 0 & 0 \\ 0 & 0 & 0.0805 & 1 & 0 & 0 \\ 0 & 0 & 1 & 0 & 0 & -0.333 \end{bmatrix}
$$

$$
\times \begin{bmatrix} x_A \\ \beta \\ r \\ p \\ \phi \\ x_{wo} \end{bmatrix} + \begin{bmatrix} 10 \\ 0 \\ 0 \\ 0 \\ 0 \\ 0 \end{bmatrix} e_{\delta r},
$$

$$
e = \begin{bmatrix} 0 & 0 & 1 & 0 & 0 & -0.333 \end{bmatrix} \begin{bmatrix} x_A \\ \beta \\ r \\ p \\ \phi \\ x_{wo} \end{bmatrix},
$$

where $e_{\delta r}$ is the input to the actuator and e is the output of the washout. The SRL for the augmented system is as shown in Fig. 10.38. If we select the state-feedback

Figure 10.36

Bode plot of yaw damper, including washout and actuator

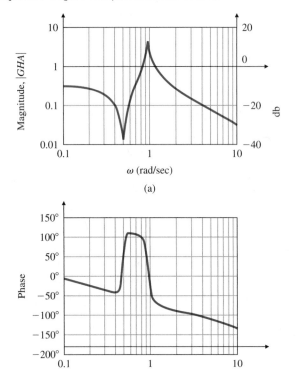

(a)

(b)

Figure 10.37

Initial condition response with yaw damper and washout, and SRL design, for $\beta_0 = 1°$

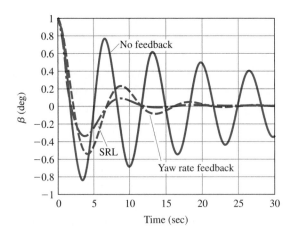

poles from the SRL so that the complex roots have maximum damping ($\zeta = 0.4$), we find that

$$pc = [-0.0051; -0.468; 0.279 + 0.628 * j; 0.279 - 0.628 * j; -1.106; -9.89]$$

Then we can compute the state-feedback gain, using the MATLAB function place, to be

$$\mathbf{K} = [\ 1.059 \quad -0.191 \quad -2.32 \quad 0.0992 \quad 0.0370 \quad 0.486\ ].$$

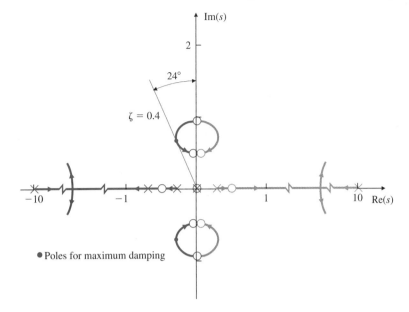

Note that the third entry in **K** is larger than the others, so the feedback of all six state variables is essentially the same as proportional feedback of r. This is also evident from the similarity of the root locus in Fig. 10.31 and the SRL of Fig. 10.38. If we select the estimator poles to be five times faster than the controller poles, then

$$\mathrm{pe} = [-0.0253; -2.34; -1.39 + 3.14*\mathrm{j}; -1.39 - 3.14*\mathrm{j}; -5.53; -49.5]$$

and the estimator gain, again using the MATLAB function place, is found to be

$$\mathbf{L} = \begin{bmatrix} 25.0 \\ -2{,}044 \\ -5{,}158 \\ -24{,}843 \\ -40{,}113 \\ -15{,}624 \end{bmatrix}.$$

The compensator transfer function from Eq. (7.177) is

$$D_c(s) = \frac{-844(s+10.0)(s-1.04)(s+0.974 \pm 0.559j)(s+0.0230)}{(s+0.0272)(s+0.837 \pm 0.671j)(s+4.07 \pm 10.1j)(s+51.3)}. \quad (10.27)$$

Figure 10.37 also shows the response of the yaw rate to an initial condition of $\beta_0 = 1°$. It is clear from the root locus that the damping can be improved by the SRL approach, and this is borne out by the reduced oscillatory behavior in the transient response of the system. However, this improvement has come at a considerable price. Note that the order of the compensator has increased from one in the original design (Fig. 10.33) to six and washout in the design obtained using the controller-estimator-SRL approach.

Design trade-off: system response vs. system complexity

Aircraft yaw dampers in use today generally employ a proportional feedback of yaw rate to rudder through a washout or through minor modifications to this design. The improved performance achievable with an optimal design approach utilizing full-state feedback and estimation is not judged to be worth the increase in complexity.

Perhaps a more fruitful approach to improving the design would be to add the aileron surface as a control variable along with the rudder.

STEP 8 and 9. *Verify the design.* Linear models of aircraft motion are reasonably accurate as long as the motion is small enough that the actuators and surfaces do not saturate. Because actuators are sized for safety in order to handle large transients, such saturation is very rare. Therefore, the linear-analysis-based design is reasonably accurate, and we will not pursue a nonlinear simulation or further design verification. However, aircraft manufacturers do carry out extensive nonlinear simulations and flight testing under all possible flight conditions before obtaining Federal Aviation Administration (FAA) certification to carry passengers.

10.3.2 Altitude-Hold Autopilot

STEP 1. *Understand the process and its performance specifications.* One of the pilot's many tasks is to hold a specific altitude. As an aid to keeping aircraft from colliding, those craft on an easterly path are required to be on an odd multiple of 1000 ft and those on a westerly path on an even multiple of 1000 ft. Therefore, the pilot needs to be able to hold the altitude to less than a hundred feet. A well-trained, attentive pilot can easily accomplish this task manually to within ± 50 ft, and air-traffic controllers expect pilots to maintain this kind of tolerance. However, since this task requires the pilot to be fairly diligent, sophisticated aircraft often have an altitude-hold autopilot to lessen the pilot's work. This system differs fundamentally from the yaw damper because its role is to replace the pilot for certain periods of time, while the yaw damper's role is to help the pilot fly. Dynamic specifications, therefore, need not require that pilots like the craft's "feel" (how it responds to their handling of the controls); instead, the design should provide the kind of ride that pilots and passengers like. The damping ratio should still be in the vicinity of $\zeta \cong 0.5$, but for a smooth ride the natural frequency should be much slower than $\omega_n = 1$ rad/sec.

STEP 2. *Select sensors.* Clearly needed is a device to measure altitude, a task most easily done by measuring the atmospheric pressure. Almost from the time of the first Wright brothers' flight, this basic idea has been used in a device called a **barometric altimeter**. Before autopilots, the device consisted of a bellows whose free end was connected to a needle that directly indicated altitude on a dial. The same bellows concept is used today for the altitude display, but the pressure is sensed electrically for the autopilot.

Because the transfer function from the controlling elevator input to the altitude control consists of five poles [see Eq. (10.30)], stabilization of the feedback loop cannot be accomplished by simple proportional feedback. Therefore the pitch rate q is also used as a stabilizing feedback; it is measured by a gyroscope or ring-laser gyro identical to that used for yaw-rate measurement. Further stabilization using pitch-angle feedback is also helpful. It is obtained either from an inertial reference system based on a ring-laser gyro or from a rate-integrating gyro. The latter is a device similar to the rate gyro, but structured differently so that its outputs are proportional to the angles of the aircraft's pitch θ and roll ϕ.

STEP 3. *Select actuators.* The only aerodynamic surface typically used for pitch control on most aircraft is the elevator δe. It is located on the horizontal tail, well removed from the aircraft's center of gravity, so that its force produces an angular pitch rate and thus a pitch angle, which acts to change the lift from the wing. In some high-performance aircraft there are direct-lift control devices on the wing or perhaps small canard surfaces, which are like tiny wings forward of the main wing, which produce vertical forces on the aircraft that are much faster than elevators on the tail are able to generate. However, for purposes of our altitude hold, we will consider only the typical case of an elevator surface on the tail.

As for the rudder, hydraulic actuators are the preferred devices to move the elevator surface, mainly because of their favorable force-to-weight ratio.

STEP 4. *Make a linear model.* The longitudinal perturbation equations of motion for the Boeing 747 in horizontal flight at a nominal speed $U_0 = 830$ ft/sec at 20,000 ft (Mach 0.8) with a weight of 637,000 lb are

$$\dot{\mathbf{x}} = \mathbf{F}\mathbf{x} + \mathbf{G}\delta e, \tag{10.28}$$

$$
\begin{bmatrix} \dot{u} \\ \dot{w} \\ \dot{q} \\ \dot{\theta} \\ \dot{h} \end{bmatrix} =
\begin{bmatrix}
-0.00643 & 0.0263 & 0 & -32.2 & 0 \\
-0.0941 & -0.624 & 820 & 0 & 0 \\
-0.000222 & -0.00153 & -0.668 & 0 & 0 \\
0 & 0 & 1 & 0 & 0 \\
0 & -1 & 0 & 830 & 0
\end{bmatrix}
\begin{bmatrix} u \\ w \\ q \\ \theta \\ h \end{bmatrix}
$$

$$
+ \begin{bmatrix} 0 \\ -32.7 \\ -2.08 \\ 0 \\ 0 \end{bmatrix} \delta e,
$$

where the desired output for an altitude-hold autopilot is

$$h = \mathbf{H}\mathbf{x},$$

$$
h = \begin{bmatrix} 0 & 0 & 0 & 0 & 1 \end{bmatrix}
\begin{bmatrix} u \\ w \\ q \\ \theta \\ h \end{bmatrix}, \tag{10.29}
$$

and

$$\frac{h(s)}{\delta e(s)} = \frac{32.7(s + 0.0045)(s + 5.645)(s - 5.61)}{s(s + 0.003 \pm 0.0098j)(s + 0.6463 \pm 1.1211j)}. \tag{10.30}$$

Phugoid mode
Short-period modes

The system has two pairs of stable complex poles and a pole at $s = 0$. The complex pair at $-0.003 \pm 0.0098j$ are referred to as the **phugoid mode**,[7] and the poles at -0.6463 ± 1.1211 are the **short-period modes**, as computed using the MATLAB eig command.

Inner-loop design

STEP 5. *Try a lead–lag or PID controller.* As a first step in the design, it is typically helpful to use an inner-loop feedback of pitch rate q to δe so as to improve the damping

---

[7] The name was adopted by F. W. Lanchester (1908), who was the first to study the dynamic stability of aircraft analytically. It is apparently an incorrect version of a Greek word.

of the short-period mode of the aircraft (see Fig. 10.39). The transfer function from δe to q, using the MATLAB ss2tf function, is

$$\frac{q(s)}{\delta e(s)} = -\frac{2.08s(s + 0.0105)(s + 0.596)}{(s + 0.003 \pm 0.0098j)(s + 0.646 \pm 1.21j)}. \tag{10.31}$$

The inner loop root locus for q feedback using Eq. (10.31) is as shown in Fig. 10.40. Because k_q is the root-locus parameter, the system matrix [Eq. (10.28)] is now modified to

$$\mathbf{F}_q = \mathbf{F} + k_q \mathbf{G} \mathbf{H}_q, \tag{10.32}$$

where $\mathbf{F}$ and $\mathbf{G}$ are defined in Eq. (10.28) and $\mathbf{H}_q = [\ 0\ \ 0\ \ 1\ \ 0\ \ 0\ ]$. The process of picking a suitable gain k_q is an iterative one. The selection procedure is the same one discussed in Chapter 5. (Recall the tachometer feedback example in Section 5.6.2.) If we choose $k_q = 1$, then the closed-loop poles will be located at $-0.0039 \pm 0.0067j$, $-1.683 \pm 0.277j$ on the root locus, and

$$\mathbf{F}_q = \begin{bmatrix} -0.00643 & 0.0263 & 0 & -32.2 & 0 \\ -0.0941 & -0.624 & 787.3 & 0 & 0 \\ -0.000222 & -0.00153 & -2.75 & 0 & 0 \\ 0 & 0 & 1 & 0 & 0 \\ 0 & -1 & 0 & 830 & 0 \end{bmatrix}. \tag{10.33}$$

Figure 10.39

Altitude-hold feedback system

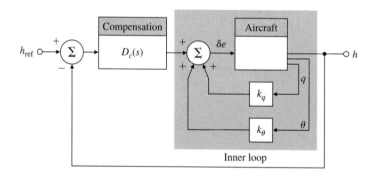

Figure 10.40

Inner-loop root locus for altitude-hold dynamics with q feedback

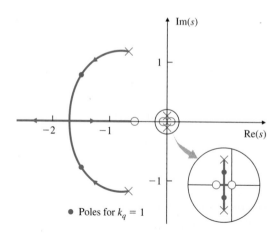

• Poles for $k_q = 1$

Note that only the third column of $\mathbf{F}_q$ is different from $\mathbf{F}$. To further improve the damping, it is useful to feed back the pitch angle of the aircraft. By trial and error, we select

$$\mathbf{K}_{\theta q} = [\, 0 \quad 0 \quad -0.8 \quad -6 \quad 0 \,]$$

in order to feed back θ and q, and the system matrix becomes

$$\mathbf{F}_{\theta q} = \mathbf{F}_q - \mathbf{G}\mathbf{K}_{\theta q},$$

$$= \begin{bmatrix} -0.0064 & 0.0263 & 0 & -32.2 & 0 \\ -0.0941 & -0.624 & 761 & -196.2 & 0 \\ -0.0002 & -0.0015 & -4.41 & -12.48 & 0 \\ 0 & 0 & 1 & 0 & 0 \\ 0 & -1 & 0 & 830 & 0 \end{bmatrix},$$

with poles at $s = 0, -2.25 \pm 2.99j, -0.531, -0.0105$.

So far, the inner loop of the aircraft has been stabilized significantly. The uncontrolled aircraft has a natural tendency to return to equilibrium in level flight, as evidenced by the open-loop roots in the LHP. The inner-loop stabilization is necessary to enable an outer-loop feedback of h and $\dot{h}$ to be successful; furthermore, the feedbacks of θ and q can be used by themselves in an attitude-hold mode of the autopilot, when a pilot wishes to control θ directly through input command. Figure 10.41 shows the response of the inner loop to a $2°$ (0.035-rad) step command in θ. With the inner loop in place the transfer function of the system from elevator angle to altitude is now

$$\frac{h(s)}{\delta e(s)} = \frac{32.7(s + 0.0045)(s + 5.645)(s - 5.61)}{s(s + 2.25 \pm 2.99j)(s + 0.0105)(s + 0.0531)}. \tag{10.34}$$

The root locus for this system, given in Fig. 10.42, shows that proportional feedback of altitude by itself does not yield an acceptable design. For stabilization we may also feed back the rate of change in the altitude in a PD controller. The root locus of the system with feedback of both h and $\dot{h}$ is shown in Fig. 10.43. After some iteration we find that the best ratio of $\dot{h}$ to h is 10:1, that is,

$$D_e(s) = K_h(s + 0.1).$$

Figure 10.41

Response of altitude-hold autopilot to a command in θ

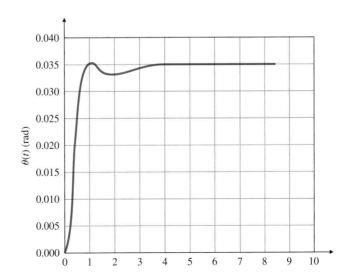

Figure 10.42

0° root locus with feedback of h only

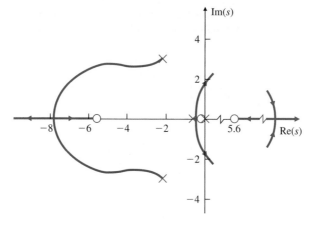

Figure 10.43

0° root locus with feedback of h and $\dot{h}$

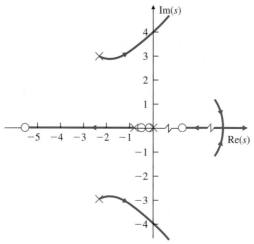

The final design is the result of iterations between the q, θ, $\dot{h}$, and h feedback gains, obviously a lengthy process. Although this trial design was successful, use of the SRL approach promises to expedite the process.

STEP 6. *Evaluate/modify plant.* Not applicable here.

STEP 7. *Do an optimal design.* The SRL of the system is shown in Fig. 10.44. If we choose the closed-loop poles at

$$pc = [-0.0045; -0.145; -0.513; -2.25 - 2.98*j; -2.25 + 2.98*j]$$

then the required feedback gain, using the MATLAB function place, is

$$\mathbf{K} = [\ -0.0009 \quad 0.0016 \quad -1.883 \quad -7.603 \quad -0.001 \ ].$$

The step response of the system to a 100-ft step command in h is shown in Fig. 10.45, and the associated control effort is shown in Fig. 10.46.

This design has been carried out with the assumption that the linear model is valid for the altitude changes under consideration. We should perform simulations to verify this or to determine the range of validity of the linear model.

Figure 10.44

SRL for altitude-hold design

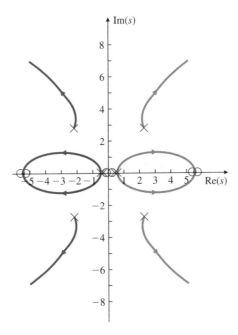

Figure 10.45

Step response of altitude-hold autopilot to a 100-ft step command

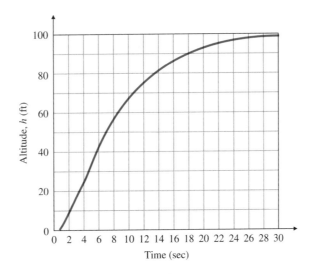

STEP 8. and 9. *Verify the design.* The comments in Steps 7 and 8 of Section 10.3.1 apply to this design as well.

For small airplane autopilots now in production, such as the one described in Chapter 5, it is interesting to note that, for the inner loop, some manufacturers employ only θ feedback while others use q feedback. The use of θ enables faster response, but use of q is less costly. Both, of course, use the altimeter for h feedback.

Figure 10.46

Control effort for 100-ft step command in altitude

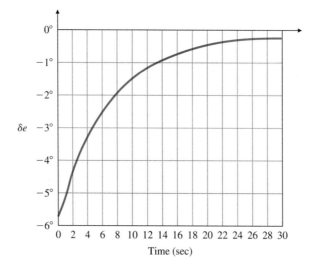

10.4 Control of the Fuel–Air Ratio in an Automotive Engine

Until the 1980s most automobile engines had a carburetor to meter the fuel so that the ratio of the gasoline-mass flow to air-mass flow, or fuel-to-air ratio (F/A), remained in the vicinity of 1:15. This device metered the fuel by relying on a pressure drop produced by the air flowing through a venturi. The device performed adequately in terms of keeping the engine running satisfactorily, but it historically allowed excursions of up to 20% in the F/A. After the implementation of federal exhaust-pollution regulations, this level of inaccuracy in the F/A was unacceptable because neither excess hydrocarbons (HCs) nor excess oxygen could be accepted. During the 1970s, automobile companies improved the design and manufacturing process of the carburetors so that they became more accurate and delivered a F/A accuracy in the vicinity of 3% to 5%.[8] Through a combination of factors, this improved F/A accuracy helped lower the exhaust pollution levels. However, the carburetors were still open-loop devices because the system did not measure the F/A of the mixture entering the engine for subsequent feedback into the carburetor. During the 1980s almost all manufacturers turned to feedback control systems to provide a much-improved level of F/A accuracy, an action made necessary by the decreasing levels of allowable exhaust pollutants.

We now turn to the design of a typical feedback system for engine control, again using the step-by-step design outline given in Section 10.1.

STEP 1. *Understand the process and its performance.* The method chosen to meet the exhaust-pollution standards has been to use a catalytic converter that simultaneously oxidizes excess levels of exhaust carbon monoxide (CO) and unburned HCs and reduces excess levels of the oxides of nitrogen (NO and NO_2, or NO_x). This device is usually referred to as a three-way catalyst because of its effect on all three pollutants. This catalyst is ineffective when the F/A is much different from the stoichiometric

---

[8]A review of automotive engine control is contained in a book by Alexander Stotsky, *Automotive Engine: Control Estimation, Statistical Detection.*

level of 1:14.7; therefore, a feedback control system is required to maintain the F/A within ±1% of that desired level. The system is depicted in Fig. 10.47.

The dynamic phenomena that affect the relationship between the sensed F/A output from the exhaust and the fuel-metering command in the intake manifold are (1) intake fuel and air mixing, (2) cycle delays due to the piston strokes in the engine, and (3) the time required for the exhaust to travel from the engine to the sensor. All these effects are strongly dependent on the speed and load of the engine. For example, engine speeds typically vary from 600 to 6000 rpm. The result of these variations is that the time delays in the system that will affect the feedback control-system behavior will also vary by at least 10:1, depending on the operating condition. The system undergoes transients as the driver demands more or less power through changes in the accelerator pedal, with the changes taking place over fractions of a second. Ideally, the feedback control system should be able to keep up with these transients.

STEP 2. *Select sensors.* The discovery and development of the exhaust sensor was the key technological step that made possible this concept of exhaust-emission reduction by feedback control. The active element in the device, zirconium oxide, is placed in the exhaust stream, where it yields a voltage that is a monotonic function of the oxygen content of the exhaust gas. The F/A is uniquely related to the oxygen level. The voltage of the sensor is highly nonlinear with respect to F/A (Fig. 10.48); almost all the change in voltage occurs precisely at the F/A value at which the feedback system must operate for effective performance of the catalyst. Therefore, the gain of

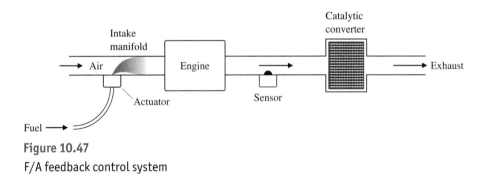

Figure 10.47
F/A feedback control system

Figure 10.48
Exhaust sensor output

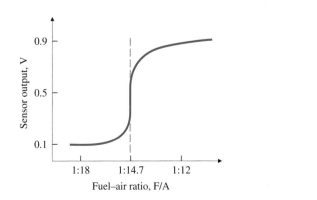

Nonlinear sensor

the sensor will be very high when the F/A is at the desired point (1:14.7) but will fall off considerably for F/A excursions away from 1:14.7.

Although other sensors have been under development for possible use in F/A feedback control, no other cost-effective sensor has so far demonstrated the capability to perform adequately. All manufacturers of production-line automobiles currently use zirconium oxide sensors in their feedback control systems.

STEP 3. *Select actuators*. Fuel metering can be accomplished by a carburetor or by fuel injection. Implementing a feedback F/A system requires the capability of adjusting the fuel metering electrically, because the sensor used provides an electric output. Initially, carburetors were designed to provide this capability by including adjustable orifices that modify the primary fuel flow in response to the electric error signal. However, today manufacturers accomplish the metering by use of fuel injection. Fuel-injection systems are typically electrical by nature, so they can be used to perform the fuel adjustment for F/A feedback simply by including the capability of using the feedback signal from the sensor. Today, fuel injectors are placed at the inlet to every cylinder (called **multipoint injection**); in the past, there was one large injector upstream from all the cylinders (called **single-point** or **throttle body injection**). Multipoint injection offers improved performance because the fuel is introduced much closer to the engine, with better distribution to the cylinders. Being closer reduces the time delays and thus yields better engine response and enables lower exhaust pollution.

STEP 4. *Make a linear model*. The sensor nonlinearity shown in Fig. 10.48 is severe enough that any design effort based on a linearized model of it should be used with caution. Figure 10.49 shows a block diagram of the system, with the sensor shown to have a gain K_s. The time constants τ_1 and τ_2 indicated for the inlet manifold dynamics represent, respectively, fast fuel flow in the form of vapor or droplets and slow fuel flow in the form of a liquid film on the manifold walls. The time delay is the sum of (1) the time it takes the pistons to move through the four strokes from the intake process until

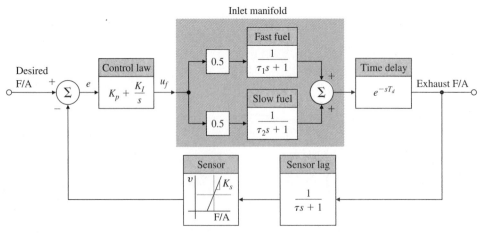

Figure 10.49
Block diagram of an F/A control system

the exhaust process and (2) the time required for the exhaust to travel from the engine to the sensor located roughly 1 ft away. A sensor lag with time constant τ is also included in the process to account for the mixing that occurs in the exhaust manifold. Although the time constants and the delay time change considerably, primarily as a function of engine load and speed, we will examine the design at a specific point where the values are

$$\tau_1 = 0.02 \text{ sec}, \qquad T_d = 0.2 \text{ sec},$$

$$\tau_2 = 1 \text{ sec}, \qquad \tau = 0.1 \text{ sec}.$$

In an actual engine, designs would be carried out for all speed loads.

STEP 5. *Try a lead–lag or PID controller.* Given the tight error specifications and the wide variations in the required fuel command u_f due to varying engine-operating conditions, an integral control term is mandatory. With integral control, any required steady state u_f can be provided when the error signal $e = 0$. The addition of a proportional term, although not often used, allows for an increase (doubling) in the bandwidth without degrading steady-state characteristics. In this example we use a control law that is proportional plus integral (PI). The output from the control law is a voltage that drives the injector's pulse former to give a fuel pulse whose duration is proportional to the voltage. The controller transfer function can be written as

$$D_c(s) = K_p + \frac{K_I}{s} = \frac{K_p}{s}(s + z), \tag{10.35}$$

where

$$z = \frac{K_I}{K_p}$$

and z can be chosen as desired.

First, let us assume that the sensor is linear and can be represented by a gain K_s. Then we can choose z for good stability and good response of the system. Figure 10.50 shows the frequency response of the system for $K_s K_p = 1.0$ and $z = 0.3$, while Fig. 10.51 shows a root locus of the system with respect to $K_s K_p$ with $z = 0.3$. Both analyses show that the system becomes unstable for $K_s K_p \cong 2.8$. Figure 10.50 shows that to achieve a phase margin of approximately 60°, the gain $K_s K_p$ should be ~2.2. Figure 10.50 also shows that this produces a crossover frequency of 6.0 rad/sec (~1 Hz). The root locus in Fig. 10.51 verifies that this candidate design will achieve acceptable damping ($\zeta \cong 0.5$).

Although this linear analysis shows that acceptable stability at a reasonable bandwidth (~1 Hz) can be achieved with a PI controller, a look at the nonlinear sensor characteristics (Fig. 10.48) shows that this indeed may not be achievable. Note that the slope of the sensor output is extremely high near the desired setpoint, thus producing a very high value of K_s. Therefore, lower values of the controller gain K_p need to be used to maintain the overall $K_s K_p$ value of 2.2 when including the effect of the high sensor gain. On the other hand, a value of K_p low enough to yield a stable system at F/A = 1 : 14.7 (= 0.068) will yield a very sluggish response to transient errors that deviate much from the setpoint, because the effective sensor gain will be reduced substantially. It is therefore necessary to account for the sensor nonlinearity in order to obtain satisfactory response characteristics of the system for anything other than minute disturbances about the setpoint. A first approximation to the sensor is shown

Complications of nonlinearity

Figure 10.50

Bode plot of a PI F/A controller

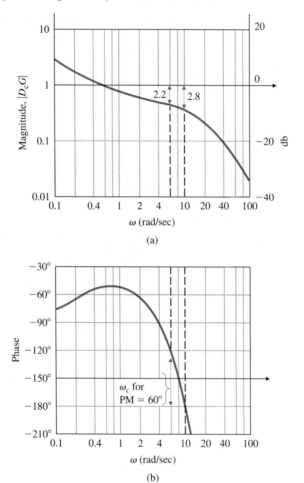

(a)

(b)

Figure 10.51

Root locus of a PI F/A controller

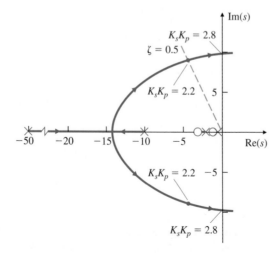

in Fig. 10.52. Because the actual sensor gain at the setpoint is still quite different from its approximation, this approximation will yield erroneous conclusions regarding stability about the setpoint; however, it will be useful in a simulation to determine the response to initial conditions.

STEP 6. *Evaluate/modify plant.* The nonlinear sensor is undesirable; however, no suitable linear sensor has been found.

STEP 7. *Try an optimal controller.* The response of this system is dominated by the sensor nonlinearity, and any fine tuning of the control needs to account for that feature. Furthermore, the system dynamics are relatively simple, and it is unlikely that an optimal design approach will yield any improvement over the PI controller used. We will thus omit this step.

STEP 8. *Simulate design with nonlinearities.* The nonlinear closed-loop simulation of the system implemented in SIMULINK® is shown in Fig. 10.53. The MATLAB function (fas) implements the approximate nonlinear sensor characteristics of Fig. 10.53,

```
function y = fas(u)
if u < 0.0606,
```

Figure 10.52

Sensor approximation

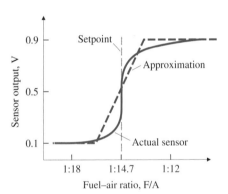

Figure 10.53

Closed-loop nonlinear simulation implemented in SIMULINK®

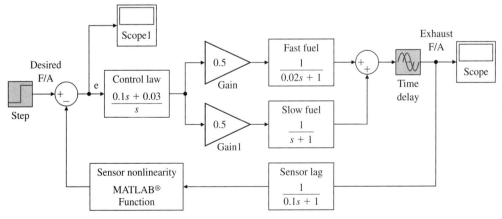

```
    y = 0.1 ;
elseif u < 0.0741,
    y = 0.1 + (u − 0.0606) * 20;
else y = 0.9;
end
```

Figure 10.54(a) is a plot of the system error using the approximate sensor of Fig. 10.52 and $K_p K_s = 2.0$. The slow response is apparent with 12.5 sec before the error comes out of saturation and a time constant of almost 5 sec once the linear region is reached. In real automobiles these systems are operated with much higher gains. To show these effects, a simulation with $K_p K_s = 6.0$ is plotted in Fig. 10.54(b, c). At this gain the linear system is unstable and up until about 5 sec the signals grow. The growth halts after 5 sec due to the fact that, as the input to the sensor nonlinearity gets large,

Figure 10.54

System response with nonlinear sensor approximation

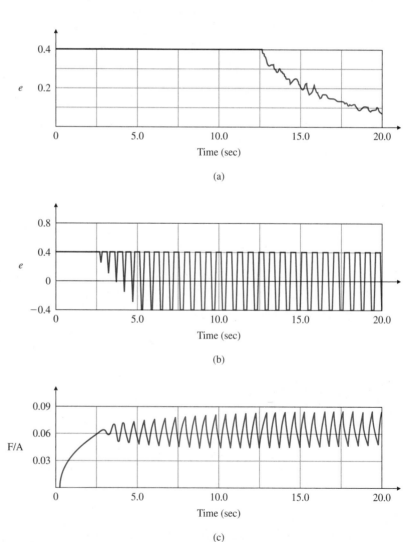

(a)

(b)

(c)

the *effective* gain of the sensor decreases due to the saturation, and eventually, a limit cycle is reached. The frequency of this limit cycle corresponds to the point at which the root locus crosses the imaginary axis and has an amplitude such that the total effective gain of $K_p K_{s,eq} = 2.8$. As described in Section 9.3, the effective gain of a saturation for moderately large inputs can be computed and is given by the describing function to be approximately $4N/\pi a$, where N is the saturation level and a is the amplitude of the input signal. Here $N = 0.4$, and if $K_p = 0.1$, then $K_{s,eq} = 28$. Thus we predict an input signal amplitude of $a = 4(0.4)/28\pi = 0.018$. This value is closely verified by the plot of Fig. 10.54(c), the input to the nonlinearity in this case. The frequency of oscillation is also nearly 10.1 rad/sec, as predicted by the root locus in Fig. 10.51.

SIMULINK nonlinear
simulation

In the actual implementation of F/A feedback controllers in automobile engines, sensor degradation over thousands of miles of use is of primary concern, because the federal government mandates that the engines meet the exhaust-pollution standards for the first 50,000 mi. In order to reduce the sensitivity of the average setpoint to changes in the sensor output characteristics, manufacturers typically modify the design discussed here. One approach is to feed the sensor output into a relay function [see Fig. 9.6(b)], thus completely eliminating any dependency on the sensor gain at the setpoint. The frequency of the limit cycle is then solely determined by controller constants and engine characteristics. Average steady-state F/A accuracy is also improved. The oscillations in the F/A are acceptable because they are not noticeable to the car's occupants. In fact, the F/A excursions are beneficial to the catalyst operation in reducing pollutants.

10.5 Control of the Read/Write Head Assembly of a Hard Disk

The first mass storage device based on recording data on hard disks was introduced by IBM in 1956 as the model 350 RAMAC.[9] It consisted of a stack of fifty 24-inch diameter aluminum disks that were coated with a magnetic material, and the data were recorded in concentric tracks at 100 bytes per inch with 20 tracks per inch. The disks were rotated at 1200 rpm. There was a single read/write head assembly mounted on an arm that could be moved vertically from disk to disk and horizontally across the chosen disk to reach a desired data track. The heads were held above the disk surface by an air bearing generated by blowing air through holes in the fixture holding the heads. The assembly was held on a particular disk by a detent on the elevator mechanism and held on a particular track by an arm detent. The entire head assembly was driven by a single electric motor. The system held 5 MB of data, and consideration had to be given to be sure that the final device could be passed through a door 36 in. wide. The technical advances in this field have been such that in the year 2000 Seagate introduced a hard drive magnetic memory consisting of three disks, each 2.5 inches in diameter, rotating at 15,000 rpm designed to be included in a portable laptop computer. This device could hold 18,350 megabytes of data. The read/write assembly consisted of a single arm moving a comb of heads, one per surface, in a rotary motion to move the heads from track to track. The heads are mountable on a gimbal at the end of the

---

[9]Random Access Method of Accounting and Control

arm and fly above the surfaces of the disks. To follow a track, the assembly is under active feedback control using samples of position data recorded between the sectors of user data around each track. An economic measure of the progress in the field is that while the cost of the RAMAC data was about $10,000 per megabyte, that of a modern drive is less than 1 cent per megabyte. A brief summary of this remarkable history, with many references is given in Abramovitch and Franklin (2002), and a table of a few disk parameters over time is presented in Table 10.1. A large number of people from both industrial and academic institutions have contributed to the many technologies involved in the advances in hard-disk memory devices made over the past 50 years, and one of the enabling technologies has been feedback control. A picture of a Seagate 1000-GB disk drive is shown in Fig. 10.55. In this brief case study we will point out a number of issues involving control, but the design example will be concerned only with the issue of track following. We will follow the outline given in Section 10.1 in presenting the case.

STEP 1. *Understand the process.* An exploded view of the track-following servo problem is given in Fig. 10.56. The mechanism consists of a rotary voice-coil motor moving an assembly of a light arm supporting gimbal-mounted sliders that include the magnetoresistive read heads and the light, thin-film inductive write heads. The slider flies above the disk surface on an air bearing produced by the disk rotation. The power amplifier is usually connected as a current amplifier so that the basic motion can be modeled as simple inertia, described by

$$G_o(s) = \frac{A}{Js^2},\qquad (10.36)$$

where J is the total inertia and A includes both the motor torque constant and the amplifier gain. The structure is flexible, however, and the detailed motion is very complex, with many lightly damped modes. It is also subject to buffeting from the air flow and from vibration caused by housing motion. For purposes of control design, a single resonant mode will be included according to the model

$$G(s) = \frac{A}{Js^2} \frac{\left(2\zeta\dfrac{s}{\omega_1}+1\right)}{\left(\dfrac{s^2}{\omega_1}+2\zeta\dfrac{s}{\omega_1}+1\right)},\qquad (10.37)$$

where the vibration frequency, ω_1, and the damping ratio, ζ, are known only within bounds.

The motion control of the head assembly is in two modes: the seek motion to move the head from track to track and the track-follow motion to maintain the heads over the center of the selected track. In the seek mode the criterion is minimum time, and theory would call for "on–off"or "bang-bang"[10] control. In order to use the same controller for many units, which differ in the maximum torque available and other critical parameters, the method used in disk drives is a bang-curve-follow technique in which the assembly is accelerated under full torque until the velocity reaches a torque reversal curve based on the distance to the desired track and deceleration

---

[10]Common names for the case in which the control is saturated with one polarity for half the time, then reversed for the remaining half.

TABLE 10.1 Disk Drive Parameters Over Time

No.	Year	Unit	Capacity	Size (N/d)	tpi	bpi	rpm	Fly Height	Head Type	Sensor Type	Actuator Type	Seek Time	Comment
1	1956	IBM RAMAC	5 MB	50/24"	20	100	1200	20 μ	Air bearing	Detent	dc motor		The first hard disk
2	1962	IBM 1301	28 MB	25/24"	50	520	1800		Flying head	Detent	Hydraulic piston	165 ms	
3	1971	IBM 3330	100 MB	11/14"	192	4040		1.2 μ	Ferrite,	Dedicated surface	Linear voice coil flying	30 ms	The first feedback
4	1973	3340 Winchester	70 MB	4/14"	270	5600		0.5 μ	Ferrite, flying	Dedicated surface	Linear voice coil		Low-mass heads
5	1979	IBM 3370	571 MB	7/14"	635	12,134	2964	0.324 μ	Thin film	Dedicated surface	Linear voice coil		
6	1979	IBM 3310	64.5 MB	6/8"	450	8530				Hybrid, sector servo	Rotary voice coil	27 ms	
7	1980	SeagateST506	5 MB	4/5.25"	255	7690				Open loop	Stepper motor	170 ms	5.25" disk for PCs
8	1983	MaxtorXT1140	126 MB	8/5.25"						Sector servo	Rotary voice coil		In-hub spindle motor
9	1991	IBM Corsair	1 GB	8/3.5"	2238	58,874			MR head	Sector servo	Rotary voice coil		
10	1993	Seagate 12550	2.19 GB	10/3.5"			7200			Sector servo	Rotary voice coil		
11	1997	IBMTravelstar	4 GB	3/2.5"	12,500	211,000				Sector servo	Rotary voice coil		
12	2000	Seagate ST318451	18.3 GB	5/2.5"	21.5k	343k	15,000		Thin-film/ GMR	Sector servo	Rotary voice coil	3.9(r), 4.5(w)	First 15,000 RPM disk drive
13	2003	Seagate ST300007	300 GB	4/3.3"	105k	658k	10,000		Thin-film/ GMR	Sector servo	Micro-actuator	4.9(r), 5.4(w)	First micro-actuators
14	2006	Seagate ST300655	300 GB	4/2.75"	125k	890k	15,000		Thin-film/ GMR	Sector servo	Rotary voice coil	3.5(r), 4.0(w)	First perpendicular recording drive
15	2007	Barracuda ES.2	1000 GB	4/3.75"	150k	1090k	7,200		Thin-film/ TMR	Sector servo	Rotary voice coil	7.4(r), 8.5(w)	First SAS drive

GMR = giant magnetoresitive head; TMR = tunneling magnetoresistive head.

Figure 10.55

Picture of 1000-GB disk drive

Source: Courtesy Seagate Technology LLC

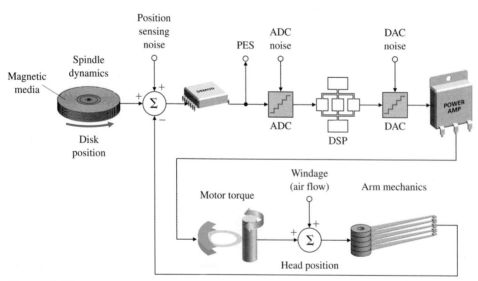

Figure 10.56

Generalized view of track-following model

is under feedback control to follow this curve to reach the desired track with zero velocity. The curve approximates the optimal minimum time switching curve with torque discounted to the extent that the weakest motor will have a reserve of torque adequate to follow the curve. When the selected track is reached, the control transfers to track-following mode. A scheme to avoid mode switching when the selected track is approached and to cause the servo to move seamlessly into track-follow mode has been called the Proximate Time Optimal Servo or PTOS (see Chapter 9).[11]

As a mature technology, many trends have influenced the nature of the control problem over the years. For example, as the table shows, disks have become smaller and thus stiffer and smoother. As the arm assembly has become smaller, it has less inertia to the extent that for very small motions as in a one- or two-track transfer, friction is more important than inertia. For recent drives, the width of a track is on the order of 0.2 micron (μ), a value comparable to the feature dimensions on a modern integrated circuit chip! To counter this trend, research is exploring ways to add a second actuator, either on the arm or on the gimbal, to make small moves much as the wrist acts on the end of a robot arm. Because of the difficulty of controlling a very lightly damped flexibility, consideration is also given to adding a coating to the arm to increase the damping of the principal modes of vibration. Other proposals include adding sensors on the arm to allow extra feedback to control the flexibility. In this case study, we will assume a single voice-coil actuator and that the flexibility is described as in Eq. (10.37), where $\omega_1 \geq 2\pi \times 2.500$ and $\zeta \geq 0.05$. Because the details of the actual resonance are not well known, the resonance will need to be gain stabilized.

STEP 2. *Select sensors.* The earliest drives were controlled open loop with one mechanical detent to hold the assembly on a disk and another detent to hold the heads on a track. Feedback control was introduced in 1971 using position information recorded on a special disk surface dedicated to the servo data. The entire comb of heads was positioned by the servo surface information. If the comb were to tilt or otherwise be misaligned, the data would be that much more difficult to read. Such issues limited the number of disks and the track density possible with this arrangement. The track position information in modern disks is recorded on each track in a gap between the sectors of user data. Controls based on this information are called sector servos, and the data are sampled of necessity. There is a conflict between the desire to record large amounts of data, which calls for fewer and larger sectors, and the control requirement to have a high sample rate, which calls for smaller sectors. Each case is a compromise between these conflicting demands. Because the position data are sampled, the controllers are digital devices to make the best possible use of the position data. Theoretical study has been given to using a multirate control to apply more than one control correction for each sensor reading, but the method has not been found to be cost effective yet. For the case study here, we will design an analog controller.

The position information extracted from data recorded on the disk is subject to errors caused by run-out in the track path, which means that the radius of the track is not constant. In general, there is a repeatable component in each trip around the track, and this element can be estimated, often harmonic by harmonic, and a signal used as

---

[11]Workman (1987), Franklin, Powell, and Workman (1998).

feed-forward to the motor to cancel it out. The position error signal (PES) also contains random noise from many sources. These include the buffeting by the airflow over the slider, wobble and vibration of the disks, noise in the signal-processing electronics used to decode the position information, noise from the power amplifier used to provide torque to the motor, and errors caused by the analog-to-digital converters needed in the process.

STEP 3. *Select actuators.* The RAMAC used a DC motor as actuator, and later drives used hydraulic actuators. When the 5.25-in. drive was introduced by Seagate in 1980, the actuator was a stepping motor. Each of these were used in open loop. The first feedback control of the head position was on the IBM 3330 in 1971, and the actuator was a linear-motion voice-coil motor. In 1979 a rotary voice-coil motor was introduced, and today almost all hard disk drives use a rotary motion actuator. The power amplifier is usually connected as a current amplifier to simplify the dynamics. The feedback from the current-sensing resistor to the amplifier constitutes a "torque loop" that is designed separately and carefully, so that the dynamics of the motor can be ignored most of the time in considering the outer loop position control in track following.

STEP 4. *Make a linear model.* As mentioned in the discussion of the process, the linear model has one flexible mode, namely

$$G(s) = \frac{1}{s^2} \frac{(2\zeta s/\omega_1 + 1)}{\left(\dfrac{s^2}{\omega_1} + 2\zeta\dfrac{s}{\omega_1} + 1\right)}, \tag{10.38}$$

where we take $\zeta = 0.05$ and $\omega_1 = 2.5$, corresponding to measuring time in milliseconds rather than seconds. The gain A and the inertia J will be absorbed in the gain of the compensator. The power amplifier is thus assumed to be an ideal current amplifier. Also we are considering only track following, and not seek.

STEP 5. *Try a PID or lead–lag design.* Because the nominal model is so simple, the first design will be a lead compensation with the objective of achieving the greatest possible bandwidth subject to having a phase margin of 50° and such that it will gain stabilize the resonance with a gain margin of at least 4. This approach was already published by R. K. Oswald (1974). We will try two designs and compare them for bandwidth and the quality of their step responses. In the first case, we will use a simple lead compensation, selected to give 50° phase margin and a factor of 4 gain margin. To get the phase margin, the lead will be designed with an α of 0.1, and the crossover frequency will be placed as high as possible while keeping a gain margin of 4 at the resonance, which rises by a factor of $1/2\zeta = 10$ above the Bode asymptote. Thus the crossover must be located so that the asymptote is a factor of $10 \times 4 = 40$ below 1 at $\omega_1 = 5\pi$. The resulting lead transfer function is

$$D(s) = 0.617 \frac{(2.22s + 1)}{(0.222s + 1)}, \tag{10.39}$$

and the Bode plot of the lead design is shown in Fig. 10.57.

The gain crossover frequency for this design is $\omega_c = 1.39$ rad/msec and the step response is plotted in Fig. 10.58, which shows a rise time of about $t_r = 0.8$ msec with an overshoot of about 25%. We have shown before that a phase margin of 50° should correspond to a damping of 0.5 and thus an overshoot of about 17%. However,

Figure 10.57

The Bode plot of the design with a single lead

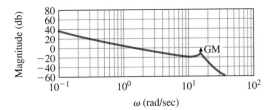

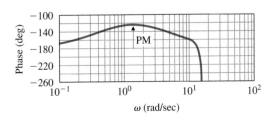

Figure 10.58

Step response of disk drive control with PM = 50°

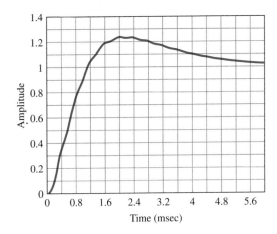

because the zero of the lead is in the forward path, we get the extra overshoot that goes with such a zero.

As a second design, a roll-off filter is to be added to try to suppress the resonance peak in order to gain a bit in speed of response and bandwidth. The idea is to put the filter cutoff frequency between the crossover frequency and the resonance frequency and to give it a damping ratio low enough that it does not reduce the phase margin too much but high enough that it does not interfere with the gain margin. After some experimentation, the trial design

$$D(s) = 1.44 \frac{(1.48s + 1)}{(0.148s + 1)} \tag{10.40}$$

is tested with a filter of

$$F(s) = \frac{1}{\dfrac{s^2}{(10.3)^2} + 0.6\dfrac{s}{10.3} + 1}. \tag{10.41}$$

For this case the Bode plot is given in Fig. 10.59 and the step response in Fig. 10.60.

In this case the crossover frequency is 2.13, a 35% increase, and the rise time is 0.3 msec, a 60% reduction from the case without the roll-off filter. The overshoot is a bit higher in this case. Although not presented here, further possibilities for the control compensation might include a notch filter rather than the low pass filter designed here. A notch might be able to further suppress the resonance and permit further increase in the bandwidth. A great deal depends on the degree of understanding of the resonance and how much uncertainty surrounds its behavior. In some cases, it is possible to phase-stabilize the resonance and to raise the crossover to be higher than the resonance frequency.

STEP 6. *Evaluate/modify plant*. Possible changes to the process that involve major design changes were introduced in the discussion concerning understanding the process in Step 1 above. Once the major parameters of the design have been selected, the remaining possibilities for improvement might include a change in the fabrication of the arm to add stiffness, which will raise the frequency of the vibration, and to add a damping coating to the arm to increase the damping ratio of the flexibility. Other

Figure 10.59

Bode plot of system with lead plus roll-off filter

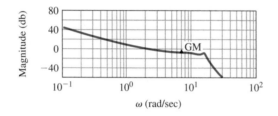
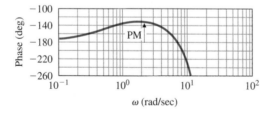

Figure 10.60

Step response of system with lead plus roll-off filter

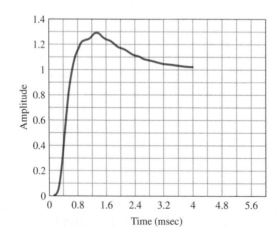

Figure 10.61

Step response for LQR design

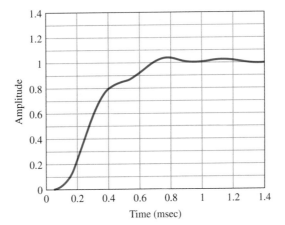

possibilities for improvement concern changes in the PES decoding methodology to reduce the noise.

STEP 7. *Try an optimal controller or adaptive control.* A design was done with the linear quadratic performance measure with the performance index (loss function) selected to obtain a rise time of about 0.3 msec to match the classical design. The result is shown in Fig. 10.61. Although further effort might produce an acceptable design, the clearly oscillatory response tolerated by this particular technique does not look promising. In particular, a design that includes a cost on $\dot{y}$ as well as y should be considered. Such extensions are considered in more advanced courses.

STEP 8. *Simulate the design, and compare the alternatives.* Usually done in parallel with the design.

STEP 9. *Build a prototype.* Done early in the design process as a bench model so trial schemes can be tested on hardware as designed.

For digital control design and implementation of disk drive servos, the reader is referred to Franklin, Powell, and Workman (1998).

10.6   Control of RTP Systems in Semiconductor Wafer Manufacturing

Figure 10.62 diagrams the major steps in the manufacture of an ultra-large-scale integrated circuit such as a microprocessor and some of the associated control aspects. Many of the steps described in this process, such as chemical vapor deposition or etching, must be done at closely controlled and timed temperature sequences (Sze, 1988). The standard practice for many years has been to perform these steps in batches on many wafers at a time to produce large numbers of identical chips. In response to the demand for ever smaller critical dimensions of the devices on the chip, and to give more flexibility in the variety and number of chips to be produced, the makers of the tools for fabrication of integrated circuits are asked to provide more and more precise control of temperature and time profiles during thermal processing. In response to these demands, an important trend is to perform the thermal steps on one wafer at a

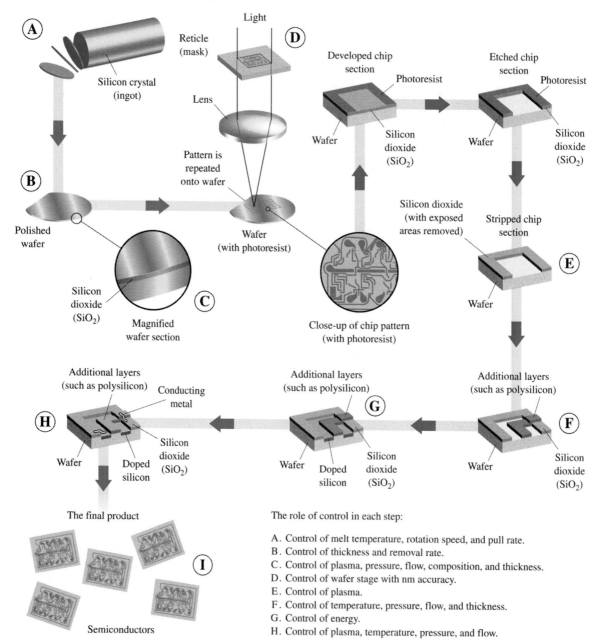

Figure 10.62

Steps in making an integrated circuit

Source: Courtesy International Sematech

time in a chamber with cold walls and a flexible heat source called a rapid thermal processor (RTP) as shown in Fig. 10.63.

The demands on an RTP system are illustrated by the requirement that it cause the wafer temperature to follow a profile such as that shown in Fig. 10.64, where the ramp-up speeds are at rates of 25° to 150°C/sec, and the soak temperatures range from 600°C to 1100°C and last from a few to as many as 120 sec. The ramp-up rates

Figure 10.63

Applied Materials' Radiance RTP system

Source: Courtesy Applied Materials

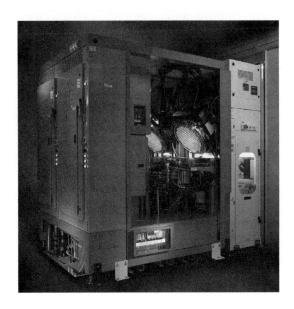

Figure 10.64

Typical RTP temperature trajectory

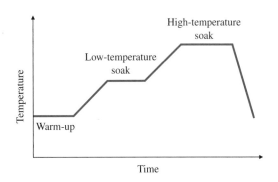

are limited by the danger of causing damage to the crystal structure if the temperature gradients become too large. The ability of the RTP to change temperature rapidly permits fabrication of devices with very small critical lengths by being able to stop the processes such as deposition or etching quickly and accurately.

Figure 10.65 shows a generic RTP reactor with tungsten halogen lamps, stainless steel walls that are water cooled, and quartz windows. Temperature measurement can be done by a variety of methods, including thermocouples, RTDs, and pyrometers. For various reasons (particle generation, minimal disturbance, etc.), it is desirable to use noncontact temperature sensing; therefore, pyrometric techniques are the most commonly employed. A pyrometer is a noncontact temperature sensor that measures infrared (IR) radiation, which is directly a function of the temperature. It is known that objects emit radiant energy proportional to T^4, where T is the temperature of the object. Among the advantages of pyrometers are that they have very fast response time, and can be used to measure the temperature of moving objects (e.g., a rotating semiconductor wafer), and in vacuum for semiconductor manufacturing.

The selection of the actuator depends on the choice of techniques for supplying power (tungsten halogen lamps, arc lamps, hot susceptor, etc.) to heat the wafer. Tungsten halogen lamps are now commonly used in RTP in semiconductor manufacturing (Emami-Naeini et al., 2003). Figure 10.66(a) shows a system with two-sided heating by linear tungsten halogen lamps (typical of systems produced by Mattson). The lamp arrays on the top and bottom are at right angles to provide more of an axisymmetry. Fig. 10.66(b) shows one-sided heating with lamps in a honeycomb configuration (typical of the Applied Materials systems). Finally Fig. 10.66(c) shows a configuration of lamps arranged in concentric rings (typical of the Stanford–TI MMST chamber, Gyugyi et al., 1993). The lamps do saturate and, for practical reasons, it is desired to operate them within 5%–95% of power settings.

To illustrate the design of an RTP system, we give the results of a specific design carried out at SC Solutions as a laboratory model constructed to study problems associated with RTP design and operation. The laboratory model is shown schematically in Fig. 10.67. It is made of aluminum. It consists of three standard 35-W 12-V tungsten halogen lamps heating a rectangular plate that simulates the wafer. The plate

Pyrometer

Tungsten halogen lamp

Figure 10.65

Generic RTP system

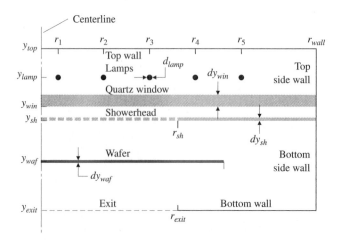

Figure 10.66

Various lamp geometries for RTP

Source: Norman, 1992

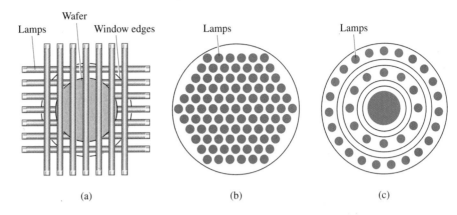

(a)   (b)   (c)

Figure 10.67

Block diagram of the RTP laboratory model

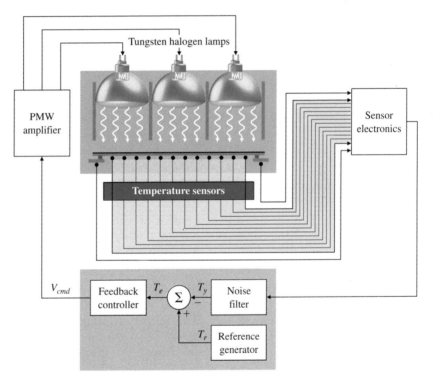

measures 4 in. $\times$ $1\frac{3}{4}$ in. and is blackened to increase its radiation absorption. The plate is mounted parallel to the lamps. The lamps are mounted in the lamp housing. The lamp assembly is mounted on a railing so that the distance from the lamps to the plate is adjustable. As the lamps are moved out the gain of the system decreases, but the radiation cross talk (coupling) increases. On the other hand, as the lamps are moved closer to the plate, the gain of the system increases and the coupling is reduced. The nominal distance from the lamps to the plate is 1 in., but it is adjustable to several inches. The lamps are driven by a pulse-width modulated (PWM) amplifier driver. There is a separate power supply unit. There are three dials mounted on the side for open-loop and manual system operations. There are 14 resistive temperature detector (RTD) strips mounted vertically behind the back of the plate: 12 on the plate and 2 on each support on either side. There is a noise source filter that generates periodic sensor noise at 1.5 Hertz so as to represent noise seen in real RTP systems. All electronics (i.e., sensor signal processing and PWM amplifier) reside in the enclosure at the bottom of the unit. Because there is exposure to the outside, the surrounding environment provides sources of disturbance.

RTP laboratory model

STEP 1. *Understand the process and its performance specifications.* RTP is an inherently dynamic and nonlinear process. Among interesting properties of the system are multiple time scales (time constants for lamps, wafer, showerhead, and quartz window are different); nonlinear (radiation dominant) behavior; nonlinear lamps; effects of power supplies; number and placement of sensors; number, placement, and grouping of lamps; and large temperature variations. The DC gain in the system (δ temperature/δ power) decreases with increasing temperature due to the nonlinear increase in radiative losses. Various types of physical models are needed. Detailed physical models are required for equipment design, but reduced-order models are needed for fast evaluation of geometry changes, recipe development, and for feedback control design. Smooth transition between manual and automatic control is also required.

STEP 2. *Select sensors.* This was discussed earlier. For the laboratory model, the sensors were a set of 14 RTDs, but three (located at the center and the support edges of the plate) can be used for feedback and the rest can be used for temperature monitoring purposes. In our case, we will use *only the center temperature* for feedback control. (Another alternative would be to sum the three temperatures into one signal and control the *average* temperature.)

STEP 3. *Select actuators.* This was also discussed earlier. For the laboratory model, the actuators were composed of three standard tungsten halogen lamps previously described. In our case, we shall tie up all three lamps *into a single actuator* by applying the same input command to each lamp.

STEP 4. *Make a linear model.* The laboratory model was built (see Step 9). The nonlinear system equations involve both conduction (Chapter 2) and radiation terms (see Emami-Naeini et al., 2003). Nonlinear system identification approaches were used to derive a model for the system. Specifically, the three lamps were stepped up, held constant, and then stepped down sequentially, and the three output temperatures were recorded. System identification studies[12] resulted in the following nonlinear model for the system that contains the radiation and conduction terms ($\mathbf{A}_r$ and $\mathbf{A}_{con}$ respectively):

Nonlinear radiation heat transfer

$$\mathbf{M}\,\dot{\mathbf{T}} = \mathbf{A}_r \left[\begin{array}{c} \mathbf{T} \\ T_\infty \end{array}\right]^4 + \mathbf{A}_{con} \left[\begin{array}{c} \mathbf{T} \\ T_\infty \end{array}\right] + \mathbf{B}\,\mathbf{u}. \qquad (10.42)$$

Here $\mathbf{T} = [T_1\ T_2\ T_3]^T$ denote the temperatures, T_∞ = constant ambient temperature ($\dot{T}_\infty = 0$), $\mathbf{u} = [v_{cmd1}\ v_{cmd2}\ v_{cmd3}]^T$ are the voltage commands, and the system matrices are

---

[12]Performed by Dr. G. van der Linden.

$$\mathbf{M}^{-1} = \begin{bmatrix} 1.000040 & 0 & 0 \\ 0 & 5.557443 & 0 \\ 0 & 0 & 13.638218 \end{bmatrix},$$

$$\mathbf{A}_r = \begin{bmatrix} 5.4762e-2 & -8.5706e-3 & -8.2961e-4 & -4.5361e-2 \\ -8.5706e-3 & 8.5709e-3 & -1.6213e-7 & -8.9134e-8 \\ -8.2961e-4 & -1.6213e-7 & 8.2998e-4 & 2.0976e-7 \end{bmatrix},$$

$$\mathbf{A}_{con} = \begin{bmatrix} 3.5599e-7 & -1.1136e-7 & -1.1976e-7 & -4.7011e-8 \\ -1.1136e-7 & 1.1602e-2 & -2.5027e-3 & -9.0992e-3 \\ -1.9761e-7 & -2.5027e-3 & 6.3736e-3 & -3.8707e-3 \end{bmatrix},$$

$$\mathbf{B} = \begin{bmatrix} 3.4600e-1 & 1.1772e-1 & 2.8380e-2 \\ 3.8803e-11 & 8.0249e-2 & 1.8072e-2 \\ 8.0041e-9 & 2.7216e-3 & 3.1713e-2 \end{bmatrix}.$$

A linear model for the system was derived as

$$\dot{\mathbf{T}} = \mathbf{F}_3\mathbf{T} + \mathbf{G}_3\mathbf{u}, \qquad (10.43)$$

$$\mathbf{y} = \mathbf{H}_3\mathbf{T} + \mathbf{J}_3\mathbf{u},$$

RTP linear model

where $\mathbf{y} = [T_{y1}\ T_{y2}\ T_{y3}]^T$ and

$$\mathbf{F}_3 = \begin{bmatrix} -0.0682 & 0.0149 & 0.0000 \\ 0.0458 & -0.1181 & 0.0218 \\ 0.0000 & 0.04683 & -0.1008 \end{bmatrix}, \quad \mathbf{G}_3 = \begin{bmatrix} 0.3787 & 0.1105 & 0.0229 \\ 0.0000 & 0.4490 & 0.0735 \\ 0.0000 & 0.0007 & 0.4177 \end{bmatrix},$$

$$\mathbf{H}_3 = \begin{bmatrix} 1 & 0 & 0 \\ 0 & 1 & 0 \\ 0 & 0 & 1 \end{bmatrix}, \qquad\qquad \mathbf{J}_3 = \begin{bmatrix} 0 & 0 & 0 \\ 0 & 0 & 0 \\ 0 & 0 & 0 \end{bmatrix}.$$

The three open-loop poles are computed from MATLAB and are located at -0.0527, -0.0863, and -0.1482. For our case, because we tied the three lamps into one actuator and are using only the center temperature for feedback, the linear model is then

$$\mathbf{F} = \begin{bmatrix} -0.0682 & 0.0149 & 0.0000 \\ 0.0458 & -0.1181 & 0.0218 \\ 0.0000 & 0.04683 & -0.1008 \end{bmatrix}, \quad \mathbf{G} = \begin{bmatrix} 0.5122 \\ 0.5226 \\ 0.4185 \end{bmatrix},$$

$$\mathbf{H} = \begin{bmatrix} 0 & 1 & 0 \end{bmatrix}, \qquad\qquad \mathbf{J} = [0],$$

resulting in the transfer function

$$G(s) = \frac{T_{y2}(s)}{V_{cmd}(s)} = \frac{0.5226(s+0.0876)(s+0.1438)}{(s+0.1482)(s+0.0527)(s+0.0863)}.$$

STEP 5. *Try a lead–lag or PID controller.* We may try a simple PI controller of the form

$$D_c(s) = \frac{(s+0.0527)}{s},$$

so as to cancel the effect of one of the slower poles. The linear closed-loop response is shown in Fig. 10.68(a) and the associated control effort is shown in Fig. 10.68(b). The system response follows the commanded trajectory with a time delay of

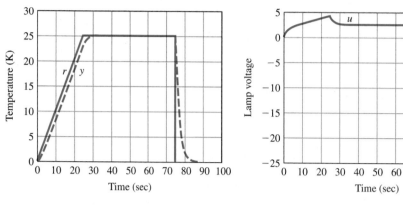

(a) Temperature tracking response

(b) Control effort

Figure 10.68

Linear closed-loop RTP response for PI controller

approximately 2 sec and no overshoot. The lamp has its normal response until 75 sec and goes negative (shown dashed) to try to follow the sharp drop in commanded temperature. This behavior is not possible in the system, as there is no means of active cooling and the lamps do saturate low. Note that there is no explicit means of controlling the temperature nonuniformity here.

STEP 6. *Evaluate/modify plant.* This was discussed already in connection with actuator and sensor selection.

STEP 7. *Try an optimal design.* We use the error-space approach for inclusion of integral control and employ the linear quadratic Gaussian technique of Chapter 7. The error system is

$$\begin{bmatrix} \dot{e} \\ \dot{\xi} \end{bmatrix} = \begin{bmatrix} 0 & \mathbf{H} \\ 0 & \mathbf{F} \end{bmatrix} \begin{bmatrix} e \\ \xi \end{bmatrix} + \begin{bmatrix} J \\ \mathbf{G} \end{bmatrix} \mu, \tag{10.44}$$

where

$$\mathbf{A} = \begin{bmatrix} 0 & \mathbf{H} \\ 0 & \mathbf{F} \end{bmatrix}, \quad \mathbf{B} = \begin{bmatrix} J \\ \mathbf{G} \end{bmatrix},$$

$e = y - r, \xi = \dot{T}$, and $\mu = \dot{u}$. For state feedback design, the LQR formulation of Chapter 7 is used; that is,

$$\mathcal{J} = \int_0^\infty \{\mathbf{z}^T \mathbf{Q} \, \mathbf{z} + \rho \mu^2\} \, dt,$$

where $\mathbf{z} = [e \; \xi^T]^T$. Note that $\mathcal{J}$ needs to be chosen in such a way as to penalize the tracking error e and the control u, as well as the differences in the three temperatures.

Temperature uniformity

Therefore, the performance index should include a term of the form

$$10 \left\{ (T_1 - T_2)^2 + (T_1 - T_3)^2 + (T_2 - T_3)^2 \right\},$$

which minimizes the *temperature nonuniformity*. The factor of 10 was determined by trial and error as the relative weighting between the error state and the plant state.

The state and control weighting matrices, $\mathbf{Q}$ and R, respectively, are then

$$\mathbf{Q} = \begin{bmatrix} 1 & 0 & 0 & 0 \\ 0 & 20 & -10 & -10 \\ 0 & -10 & 20 & -10 \\ 0 & -10 & -10 & 20 \end{bmatrix}, \quad R = \rho = 1.$$

The following MATLAB command is used to design the feedback gain:

[K] = lqr(A,B,Q,R).

The resulting feedback gain matrix computed from MATLAB is

$$\mathbf{K} = [K_1 : \mathbf{K}_0],$$

where

$$K_1 = 1, \quad \mathbf{K}_0 = \begin{bmatrix} 0.1221 & 2.0788 & -0.2140 \end{bmatrix},$$

which results in the internal model controller of the form

$$\dot{x}_c = B_c e, \tag{10.45}$$

$$u = C_c x_c - \mathbf{K}_0 \mathbf{T},$$

with x_c denoting the controller state and

$$B_c = -K_1 = -1, C_c = 1.$$

The resulting state-feedback closed-loop poles computed from MATLAB are at $-0.5574 \pm 0.4584j, -0.1442$, and -0.0877. The full-order estimator was designed with the process and sensor noise intensities selected as the estimator design knobs:

$$R_w = 1, \quad R_v = 0.001.$$

The following MATLAB command is used to design the estimator:

[L] = lqe(F,G,H,Rw,Rv).

The resulting estimator gain matrix is

$$\mathbf{L} = \begin{bmatrix} 16.1461 \\ 16.4710 \\ 13.2001 \end{bmatrix},$$

with estimator error poles at $-16.5268, -0.1438$, and -0.0876. The estimator equation is

$$\dot{\hat{\mathbf{T}}} = \mathbf{F}\hat{\mathbf{T}} + Gu + \mathbf{L}(y - \mathbf{H}\hat{\mathbf{T}}). \tag{10.46}$$

With the estimator, the internal model controller equation is modified as

$$\dot{x}_c = B_c e, \tag{10.47}$$

$$u = C_c x_c - \mathbf{K}_0 \hat{\mathbf{T}}.$$

The closed-loop system equations are given by

$$\dot{\mathbf{x}}_{cl} = \mathbf{A}_{cl}\mathbf{x}_{cl} + \mathbf{B}_{cl}r, \tag{10.48}$$

$$y = \mathbf{C}_{cl}\mathbf{x}_{cl} + \mathbf{D}_{cl}r,$$

where r is the reference input temperature trajectory, the closed-loop state vector is $\mathbf{x}_{cl} = [\mathbf{T}^T \ x_c^T \ \hat{\mathbf{T}}^T]^T$ and the system matrices are

$$\mathbf{A}_{cl} = \begin{bmatrix} \mathbf{F} & \mathbf{GC}_c & -\mathbf{GK}_0 \\ \mathbf{B}_c\mathbf{H} & 0 & 0 \\ \mathbf{LH} & \mathbf{GC}_c & \mathbf{F}-\mathbf{GK}_0-\mathbf{LH} \end{bmatrix}, \quad \mathbf{B}_{cl} = \begin{bmatrix} 0 \\ -\mathbf{B}_c \\ 0 \end{bmatrix},$$

$$\mathbf{C}_{cl} = \begin{bmatrix} \mathbf{H} & 0 & 0 \end{bmatrix}, \quad \mathbf{D}_{cl} = [0],$$

with closed-loop poles (computed with MATLAB) located at $-0.5574 \pm 0.4584j$, $-0.1442, -0.0877, -16.5268, -0.1438$ and -0.0876 as expected. The closed-loop control structure is shown in Fig. 10.69.

The closed-loop control system diagram implemented in SIMULINK is shown in Fig. 10.70. The linear closed-loop response is shown in Fig. 10.71(a), and the associated control effort is shown in Fig. 10.71(b). The commanded temperature trajectory, r, is a ramp from 0°C to 25°C, with a 1°C/sec slope followed by 50-sec soak time and a drop back to 0°C. (Note that the ramp rate is very slow here because we have only three lamps for our RTP laboratory model, whereas a real RTP system would

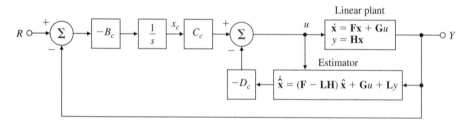

Figure 10.69
Closed-loop control structure diagram

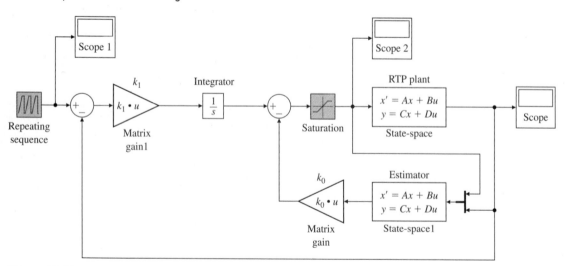

Figure 10.70
SIMULINK® block diagram for RTP closed-loop control

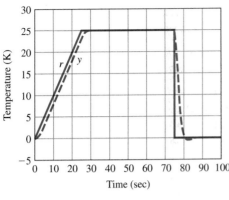

(a) Temperature tracking response

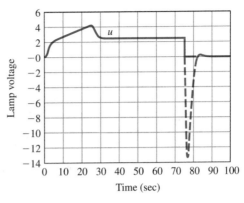

(b) Control effort

Figure 10.71

Linear closed-loop RTP response

have hundreds of lamps, and the much faster ramp rates mentioned earlier would be relevant.) The system tracks the commanded temperature trajectory—albeit with a time delay of approximately 2 sec for the ramp and a maximum of 0.089°C overshoot. As expected, the system tracks a constant input asymptotically, with zero steady-state error. The lamp command increases as expected to allow for tracking the ramp input, reaches a maximum value at 25 sec, and then drops to a steady-state value around 35 sec. The normal response of the lamp is seen from 0 to 75 sec, followed by a negative commanded voltage for a few seconds corresponding to fast cooling. Again, the negative control effort voltage (shown in dashed lines) is physically impossible as there is no active cooling in the system. Hence, in the nonlinear simulations, commanded lamp power must be constrained to be strictly nonnegative (Step 8). Note that the response from 75 to 100 sec is that of the (negative) step response of the system.

STEP 8. *Simulate the design with nonlinearities.* The nonlinear closed-loop system was simulated in SIMULINK as shown in Fig. 10.72a. The model was implemented in temperature units of degrees Kelvin and the ambient temperature is 301K.[13] The nonlinear plant model is the implementation of Eq. (10.42). There is a prefilter following the reference temperature trajectory (to smoothen the sharp corners) with the transfer function

$$G_{pf}(s) = \frac{0.2}{s + 0.2}. \tag{10.49}$$

Note that conversion from voltage to power was determined experimentally to be given by

$$P = V^{1.6} \tag{10.50}$$

and is implemented as a nonlinear block (named VtoPower) in the SIMULINK diagram accordingly. The inverse of the static nonlinear lamp model is also included as a block

Temperature trajectory following

SIMULINK nonlinear simulation

Prefilter

Lamp nonlinearity

---

[13]$[K] = [°C] + 273.$

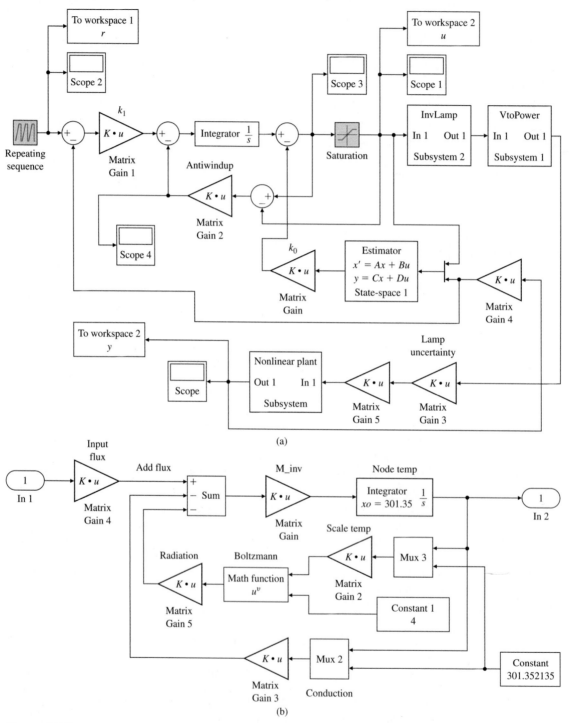

(a)

(b)

Figure 10.72a

SIMULINK® diagram for nonlinear closed-loop RTP system: (a) nonlinear closed-loop; (b) nonlinear plant

Figure 10.72b

SIMULINK® diagram for nonlinear closed-loop RTP system: (c) subsystem to convert voltage to power; (d) subsystem for lamp model inversion

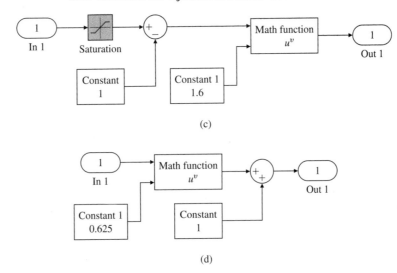

(c)

(d)

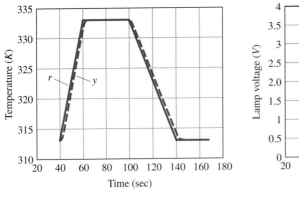

(a) Temperature tracking response

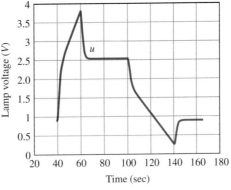

(b) Control effort

Figure 10.73
Nonlinear closed-loop response

(named InvLamp):

$$V = P^{0.625}. \tag{10.51}$$

This will cancel the lamp nonlinearity. The voltage range for system operation is between 1 and 4 volts, as seen from the diagram. A saturation nonlinearity is included for the lamp as well as integrator antiwindup logic to deal with lamp saturation. The nonlinear dynamic response is shown in Fig. 10.73(a) and the control effort is shown in Fig. 10.73(b). Note that the nonlinear response is in general agreement with the linear response.

Figure 10.74

RTP temperature control laboratory model

Source: Photo courtesy of Abbas Emami-Naeini

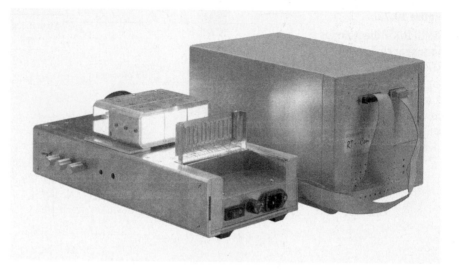

A prototype of the RTP laboratory model was designed, built,[14] and was demonstrated at the Sematech AEC/APC'98 Conference, in Vail, Colorado. Figure 10.74 shows a photograph of the operational system. This system is really multivariable in nature. The three-input–three-output multivariable controller used on the prototype system was designed using the same approach discussed in Step 7, and was implemented on an embedded controller platform that uses a real-time operating system.

The continuous controller (i.e., the combined internal model controller and the estimator) is of the form

$$\dot{\mathbf{x}}^c = \mathbf{A}^c \mathbf{x}^c + \mathbf{B}^c \mathbf{e}, \qquad (10.52)$$

$$\mathbf{u} = \mathbf{C}^c \mathbf{x}^c,$$

where $\mathbf{x}^c = [\mathbf{x}_c^T \ \hat{\mathbf{T}}^T]^T$,

$$\mathbf{A}^c = \begin{bmatrix} \mathbf{0} & \mathbf{0} \\ \mathbf{G}\mathbf{C}_c & \mathbf{F} - \mathbf{G}\mathbf{K}_0 - \mathbf{L}\mathbf{H} \end{bmatrix}, \quad \mathbf{B}^c = \begin{bmatrix} \mathbf{B}_c \\ \mathbf{L} \end{bmatrix}, \qquad (10.53)$$

and

$$\mathbf{C}^c = \begin{bmatrix} \mathbf{C}_c & -\mathbf{K}_0 \end{bmatrix}.$$

The controller was discretized (see Chapter 8) with a sampling period of $T_s = 0.1$ sec and implemented digitally (with appropriate antiwindup logic) as

$$\mathbf{x}_{k+1}^c = \mathbf{\Phi}^c \mathbf{x}_k^c + \mathbf{\Gamma}^c \mathbf{e}_k, \qquad (10.54)$$

$$\mathbf{u}_k = \mathbf{C}^c \mathbf{x}_k^c.$$

The response of the actual system to the reference temperature trajectory, along with the three lamp voltages, is shown in Fig. 10.75. It is in good agreement with the nonlinear closed-loop simulation of the system (once noise is accounted for).

For further information on modeling and control of RTP systems, the reader is referred to Emami-Naeini et al. (2003), Ebert et al. (1995a,b), de Roover et al. (1998), and Gyugyi et al. (1993).

---

[14]By Dr. J. L. Ebert.

Figure 10.75

Response of the RTP temperature control laboratory model

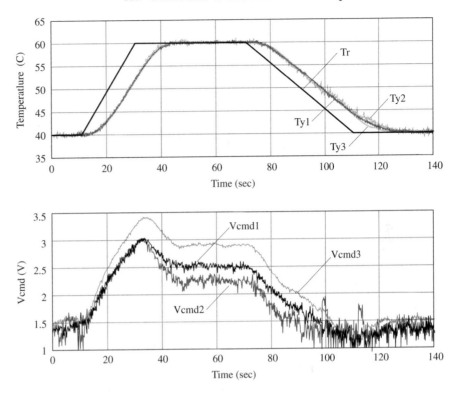

10.7   Chemotaxis or How *E. Coli* Swims Away from Trouble

Background

The *cell* is the basic structural and physiological subsystem of all living organisms and most of the biochemical activities necessary for life are performed in cells. Some organisms, such as bacteria, consist of only a single cell. A prokaryotic cell is shown in Fig. 10.76. Escherichia coli (*E. coli*), photographed in Fig. 10.77, is one of these single-cell organisms that has been extensively studied and whose interesting motion and control will be described in a highly simplified way in this case study. The technical results for the study come from the field of systems biology. *Systems biology* is an emerging field with the goal of creating dynamic models to describe the incredibly complex processes in many biological systems. The aim is to determine how shifting variables in one part impacts the whole. In this case study, a highly simplified model is presented to suggest how ideas from control can contribute to this effort. In preparing the study, we have tried to minimize the use of technical terms from biology and to define clearly those found useful and necessary for the presentation. It is hoped that this simple introduction will inspire control engineers to conduct direct study of this important field. First, a bit of background.

E. coli was discovered by German pediatrician and bacteriologist Theodor Escherich in 1885. The bacterium is a cylindrical organism with hemispherical end-caps similar to the sketch shown in Fig. 10.76. A photograph of *E. coli* is shown

Systems Biology

E.coli

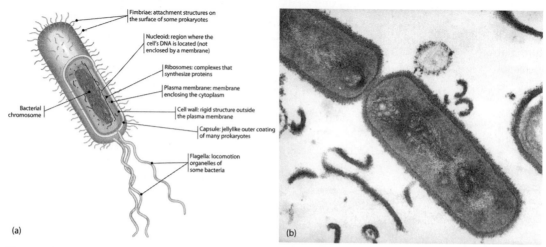

(a)

(b)

Figure 10.76

A cell structure: (a) a typical bacterium; (b) TEM of bacterium *Bacillus coagulans*

Source: (a) Campbell and Reece, page 98, 2008. © Pearson Education; (b) © Stanley C. Holt/Biological Photo Service. All rights reserved

Figure 10.77

Photograph of
Escherichia coli (E. coli)
bacteria

*Source: United States
Department of Health and
Human Services, National
Institutes of Health*

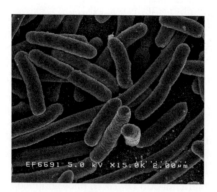

in Fig. 10.77. It is approximately 1 micron (μ) in diameter and 2 microns (μ) in length and weighs about 1 picogram (pg). *E. coli* has been studied extensively by geneticists because of its rather small genome size and the ease of growth in the laboratory. The entire genome, or the "library" of inherited genetic information, has been sequenced: it contains 4,639,221 of the adenosine (A), cytosine (C), guanine (G), and thymine (T) nitrogenous bases arranged into a total of 4288 genes. These genes serve as instructions for the synthesis of specific proteins, and are transcribed and eventually translated into the primary structure, or amino acid sequence, of a protein. *E. coli* grows longer and divides by binary fission to create two genetically identical "daughter" bacteria. It is a "cell division machine" and divides continuously such that under optimal conditions, a population of *E. coli* can double every 20 minutes. In 2003 researchers demonstrated that solitary *E. coli* cells exhibit positive *chemotaxis*, which means that they are attracted to like cells enabling formation of *E. coli* colonies. *E. coli* lives in the lower intestine of warm blooded animals including humans and feed on amino acids. The bacterium helps in maintaining the balance of normal intestinal flora (bacteria) against harmful bacteria and synthesizing or producing some vitamins. Most *E. coli* strains are harmless but a particular strain (*E. coli* O157:H7) can cause food poisoning in humans.

Escherischia coli has a set of 6 to 10 rotary motors each driving a thin *helical* filament about 10 μm long through a short, flexible and proximal hook that acts as a universal joint. This entire assembly is called a flagellum (Berg, 2003). The motor runs either clockwise (CW), as seen by an observer outside the cell looking down at the hook, or counterclockwise (CCW). When all the motors rotate CCW, the flagella filaments bundle together and the cell swims steadily forward in a "run" as suggested in Fig. 10.78. When one or more motors switches to CW rotation, the corresponding flagella unbundle and reorient the cell in a "tumble" resulting in little displacement as shown in Fig. 10.79. The two modes of motion alternate and in a state of equilibrium with its environment the runs last about 1 sec and the tumbles about 0.1 sec resulting in a 3-D random walk. Through control of tumbling frequency, the bacteria can direct their motion toward a concentration of attractant molecules or away from a concentration of repellent molecules as suggested in Fig. 10.80.

The Problem

Chemotaxis is the name given to the process by which a motile bacterium tastes the changes in its environment and moves toward places with a more favorable environment. Chemotaxis is important for proper functioning of the cell. An *E. coli* bacterium compares the current attractant concentration with the past attractant concentration. If it detects a positive change in the attractant concentration, it should move up the gradient. To do so, the probability of a tumble and hence its tumbling frequency is reduced and the runs are correspondingly longer. In contrast, if it detects an increase in repellent concentration, the assumption seems to be that it must have been swimming in a bad direction so it increases its tumbling frequency and tries to change

Figure 10.78

Flagella motors turning CCW resulting in a run

Source: Courtesy Nima Cyrus Emami

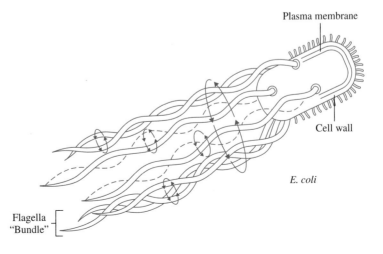

Figure 10.79

Flagella motors turning CW resulting in a tumble

Source: Courtesy Nima Cyrus Emami

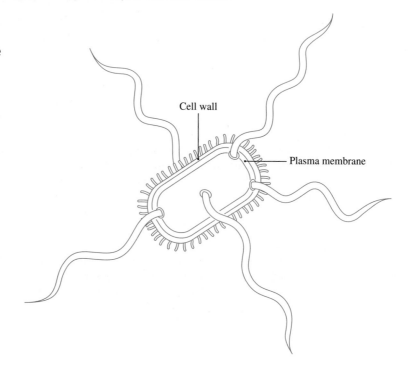

Cell wall

Plasma membrane

Figure 10.80

Escherschia coli movements resembling a biased random walk

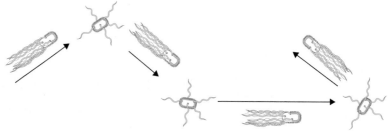

Ligand

direction so as to swim away from the repellents. The dynamics of this chemotaxis is the subject of our case study.

Several different models of bacterial chemotaxis have been developed by researchers in systems biology. Our discussion is based on two of these (Barkai & Libler, 1997; Yi et al, 2000). The different proteins involved in chemotactic response have been well studied and their interactions have been characterized in some detail as shown in Fig. 10.81. Biologists have named the proteins involved in Chemotaxis by letters of the alphabet prefixed by "Che." Thus we have CheA, CheB, and so on to CheZ. On the surface of the bacterium are receptor complexes, which include the CheW and CheA, to which the attractant or repellent molecules may bind. These chemicals constitute the input to the system and are called collectively *ligands*. The system is set up to control the frequency of tumbling, which is done by control of the activity of CheY, the protein that acts directly on the motor of the flagella.

Figure 10.81

The chemotaxis signal transconduction pathway in *E. coli*

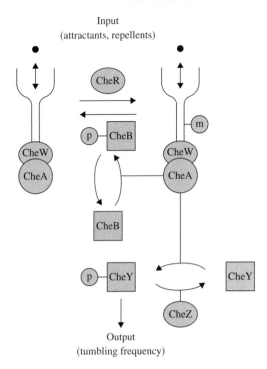

Input
(attractants, repellents)

Output
(tumbling frequency)

The receptors are either active and awaiting a ligand or are inactive and not accepting any ligand. A receptor complex becomes active if a methyl group (–CH3) is added to it by CheR and inactive if the group is removed by CheB. The CheR level is mainly fixed while the CheB level is controlled by the activity of the receptor via CheA. As part of the steady-state dynamics of chemotaxis, methyl groups are regularly being added by CheR and equally removed by CheB. This balance is upset when a ligand binds to an active receptor. If the ligand is an attractant, the activity of CheA is reduced and consequently the action of CheB in demethylation is reduced, more receptors are made active and the activity of CheA slowly returns to the steady state. This is the feedback loop in chemotaxis. At the same time as it reduces the rate of activation of CheB, CheA reduces its rate of activating CheY and this causes the tumbling frequency to be *reduced*. As a consequence, the bacteria swim more and presumably swim toward the attractant concentration. Now, if the ligand is a repellent, the activity of CheA is increased, which causes increased rate of CheY activity and *increased* frequency of tumbling. The bacteria swims less while it "looks" for a new direction in order to escape the concentration of repellents. At the same time, in the feedback loop, CheB is also more active, receptors are made inactive at a greater rate and again CheA and the tumble frequency return to their steady-state values. The fact that the activity and the tumble frequency return to exactly the same value after a change in ligand concentration is a remarkable property called by system biologists exact adaptation. As we will see, to a control engineer, this is a very common control method. An experimental plot of chemotaxis is reproduced in Fig. 10.82.

Figure 10.82

Experimental data of
E. coli chemotaxis
(Berg, 1972). The plots
are planar projections
of 3-D paths

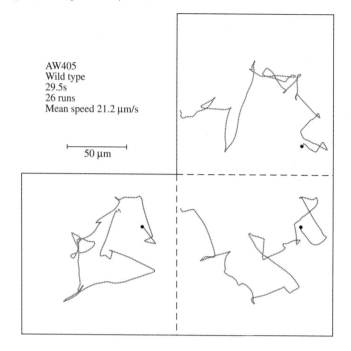

AW405
Wild type
29.5s
26 runs
Mean speed 21.2 μm/s

50 μm

The Model

The problem, then, is to develop a model as a control system block diagram that will describe the average motion of this chemotaxis situation. We represent the averages as if they were one receptor complex with the related proteins acting on the flagella. As the research shows, the equations are complex and highly nonlinear. Also, the surface of the bacterium contains hundreds of receptor complexes and these interact as suggested already in Fig. 10.81. For our study, the variables for the block diagram are selected as linear, small signal deviations of the averages of the several quantities away from their equilibrium values. The input is taken to be the concentration of ligand, with attractors being positive and repellents being negative. The outputs of the system are the activity of CheA–P and resulting motion in the single x direction. The parameters of our model were selected so the responses matched the curves given in Fig. 10 of (Mello et al. 2004). The mechanics of one-dimensional motion assume that the viscous friction dominates the mass so the dynamics are a single integrator. The model is based on the following facts.

- It is observed that when a ligand binds to an active receptor site, the changes in concentrations of CheA–P and resulting CheB and CheY–P are almost instantaneous.

- However, the changed CheB concentration only changes the rate of de-methylation, not the extent of de-methyl itself. The changes in methyl level take place much more slowly than the changes in tumble rate.

- Upon insertion of a concentration of attractants, the "activity" as measured by the concentration of CheA drops quickly, then slowly recovers to exactly the same steady-state level. This property is called *adaptation* of activity.

Adaptation

Figure 10.83

Simplified block diagram of *E. coli* chemotaxis. ℓ represents ligand, m the methylation, CheR the steady-state rate of methylation, and w the steady-state random walk motion

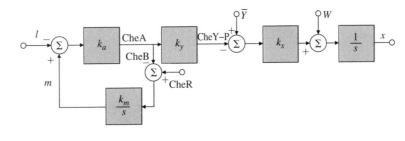

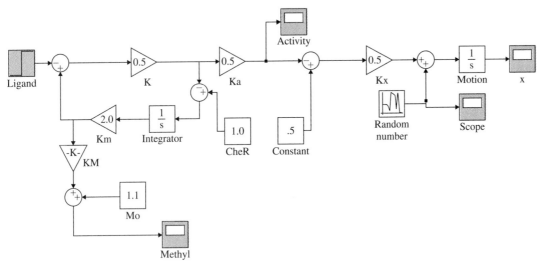

Figure 10.84

A SIMULINK schematic diagram for simulating *E. coli* chemotaxis

A control block diagram shown in Fig. 10.83 implements these facts, including the adaptation. As seen, the adaptation result is accomplished by the standard control scheme of integral control. A SIMULINK schematic is shown in Fig. 10.84 and the responses in Figures 10.85, 10.86 and 10.87 for fixed concentrations of CheR. If the value of CheR is changed, the steady-state intensity of the activity changes and the time constant of the methylation also changes.

In the end, we leave this case study with more questions than answers. For example, one should be able to derive the model by a small signal analysis from the basic chemical and physical equations of the processes. The model as presented could be modified to account for changes in the concentration of CheR, for example. Finally, how would the model be extended to describe the motion in three dimensions? We hope someone using this book is inspired to find the answers to these questions.

Summary and Recap

For years biologists had been focusing on studying various parts of living organisms. Recently, the focus has shifted to studying the whole organism's behavior as a system

Figure 10.85

Simulated tumble frequency of the chemotaxis model following insertion of attractant at $t = 20$ sec

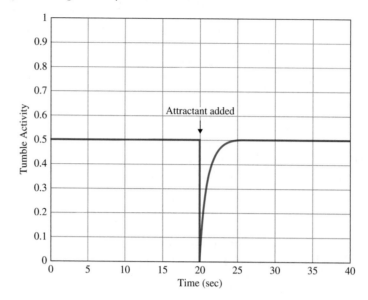

Figure 10.86

Methylation of the chemotaxis model following insertion of attractant at $t = 20$ sec

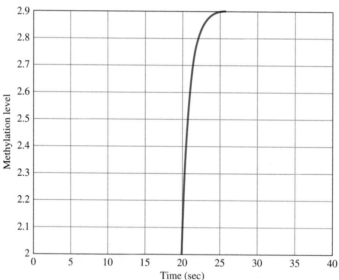

of interconnected parts. Since the 1970s, it had been known experimentally that many biological systems adjust to their environment in an adaptive way. Recently, analytical models have been developed to explain this phenomenon as we discussed in this case study. The new analytical models can explain the inherent properties of the biological system such as robust perfect adaptation as given by the integral control of the active sites. Control theory methods and interpretations have proved helpful in increasing the level of our understanding of the behavior and properties of biological systems. We hope that this simple example helps stimulate interest in this exciting field.

Figure 10.87

Motion response of the chemotaxis model following insertion of attractant at $t = 20$ sec

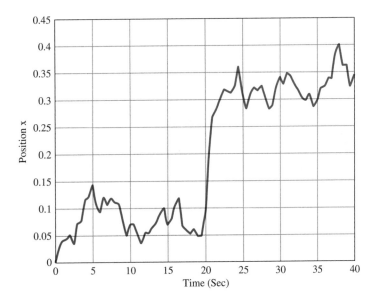

10.8 Historical Perspective

The first autopilot was tested on a Curtis flying boat in 1912, just 9 years after the first Wright brothers' flight. It consisted of a gyroscope to measure the attitude and servo motors to activate the control surfaces and was designed by Elmer Sperry. In part, it was a result of the Wright brothers design to intentionally make the aircraft slightly unstable in order to make it more controllable by the pilot. This system gained fame in 1914 when it won a prize in France by demonstrating its capabilities by flying close to the ground with the mechanic walking back and forth along the wing with the pilot, Lawrence Sperry, standing in the cockpit with his hands in the air.[15]

Autopilot development went underground in 1915 due to military security for WWI. The next public display was an adaptation of the Sperry system for Wiley Post in his 1933 flight around the world in "Winnie Mae." The flight would have been near impossible without the autopilot because it allowed Post to doze off on occasion. It has been reported that Post had a system consisting of a wrench and string tied to his finger that would wake him up if he slept too soundly. The success of this flight led to the development of an autopilot that included some navigational capabilities as well as attitude control and, in 1947, the Air Force demonstrated an automatic trans-Atlantic flight in a DC-3 type airplane from take-off to landing.

Subsequently, airplanes developed swept wings and higher speeds which required stability augmentation systems to help the pilot control the aircraft even when not on the autopilot. These systems are on all high-performance military and commercial airliners today. In 1974, the F-16 became the first airplane to have aerodynamically unstable regimes and was, therefore, highly dependent on the stability augmentation for sustained flight. This was implemented in order to make the airplane more

---

[15]No mechanical connection from the stick to the control surfaces.

maneuverable, but required a "fly-by-wire" and quad redundancy for acceptable reliability.

The first spacecraft in the late 1950s had no attitude control since their only mission was to take measurements and broadcast the information back to earth. However, they were followed in the early 1960s with the Corona spacecraft whose mission was to take photographs of the earth, which required that the camera be pointed and stabilized very accurately. At the time, these missions were classified for military purposes and called Discoverer for public consumption, but since then have been declassified and described in some detail.[16]

The first digital autopilots were in the Apollo program lunar module and the command module in the late 1960s. They were developed primarily by MIT's Instrumentation Lab under the direction of Bill Widnall, Don Fraser, and Dick Battin. The decision to take the bold step of using digital technology for the first time rather than the traditional analog implementation was made by NASA in order to handle the complexity required at a reasonable weight.

Prior to 1980, automobile engine control systems consisted of a mechanical arrangement in the distributor to vary the spark timing and a fluidic system in the carburetor that varied fuel flow in response to the airflow rate or sudden changes in the accelerator pedal position. These were open-loop systems that essentially programmed the proper control setting based on the operating condition of the engine. In 1980, cars were required to improve their polluting characteristics; therefore, it was essential to improve the controls by using feedback as described in Section 10.4. These systems still exist today along with variable valve timing, variable fuel injection timing, and variable valve opening levels.

Application of control to semiconductor wafer manufacturing automation is gaining momentum. Many important process steps such as RTP, chemical–mechanical planarization, and lithography use advanced real-time controllers. It is anticipated that during the next decade many more of the semiconductor fabrication equipment will employ sophisticated *in-situ* feedback control as new sensors become available. This adoption of sophisticated closed-loop control systems by the semiconductor industry presents new challenges and opportunities for control system engineers especially for the upcoming 450-mm diameter wafers. Application of control to magnetic resonance force microscopy (MRFM) for imaging atomic structure of materials (de Roover et al., 2008) can fundamentally change our understanding of atomic structures of devices and enable imaging of biological subsystems.

The emerging field of systems biology marks the coming of age of the life sciences. The usual approach of studying individual components is being replaced by a new approach focused on understanding the behavior of the whole biological system. Among the admirable goals are understanding the behavior of biological systems and discovering cure for diseases such as cancer, as well as developing novel approaches to discovery of new drugs, production of antibiotics, and vaccines.

The applications of control theory have never been more exciting than they are today. Applications of feedback control ideas to biological systems, network congestion control, and new aerospace systems are emerging. Applications in genomics

---

[16]Taubman (2003).

treating the human body as a dynamic system are underway. The Internet has attracted the attention of many control systems researchers eager to understand the tremendous success of this technology and how to improve it. Network design and control including Internet modeling, and development of congestion routing and control are under study. A number of our colleagues are enthusiastic about application of control theory to the financial field. If you like that, you will love the real-estate bubble!

SUMMARY

• In this chapter we have laid out a basic outline of control systems design and applied it to six typical case studies. The design outline calls for a number of explicit steps.

1. *Make a system model and determine the required performance specifications.* The purpose of this step is to answer the question, What is the system, and what is it supposed to do?
2. *Select sensors.* A basic rule of control is that if you can't observe it, you can't control it. Following are some factors to consider in the selection of sensors:
   (a) Number and location of sensors;
   (b) Technology to be used;
   (c) Performance of the sensor, such as its accuracy;
   (d) Physical size and weight;
   (e) Quality of the sensor, such as lifetime and robustness to environment changes;
   (f) Cost.
3. *Select actuators.* The actuators must be capable of driving the system so as to meet the required performance specifications. The selection is governed by the same factors that apply to sensor selection.
4. *Make a linear model.* All our design methods are based on linear models. Both small-signal perturbation models and feedback-linearization methods can be used.
5. *Try a simple PID controller.* An effort to meet the specifications with a PID or its cousin, the lead–lag compensator, may succeed; in any case such an effort will expose the nature of the control problem.
6. *Evaluate/modify plant.* Evaluate whether plant modifications enhance closed-loop performance; if so, return to Step 1 or 4.
7. *Try an optimal design.* The SRL method for control-law selection and estimator design based on state equations is guaranteed to produce a stable control system and can be structured to show a trade-off between error reduction and control effort. A related alternative is arbitrary pole placement, which gives the designer direct control over the dynamic response. Both the SRL and the pole-placement methods may result in designs that are not robust to parameter changes.
8. *Simulate the design, and verify its performance.* All the tools of analysis should be used here, including the root locus, the frequency response, *GM*

and *PM* measurements, and transient responses. Also, the performance of the design can be tested in simulation against changes in model parameters and the effects of approximating the compensator with a discrete model if digital control is to be used.

9. *Build a prototype, and measure the performance with typical input signals.* The proof of the pudding is in the eating, and no control design is acceptable until it has been tested. No model can include all the features of a real physical device; so the final step before fixing the design is to try it out on a physical prototype if time and budget permit.

- The satellite case study illustrated particularly the use of a notch compensation for a system with lightly damped resonance. It was also shown that collocated actuator and sensor systems are much easier to control than the noncollocated systems.
- The Boeing 747 lateral-stabilization case study illustrated the use of feedback as an inner-loop designed to aid the pilot, who provides the primary outer-loop control.
- The Boeing 747 altitude control showed how to combine inner-loop feedback with outer-loop compensation to design a complete control system.
- The automobile fuel–air ratio control illustrated the use of the Bode plot to design a system that includes time delay. Simulation of the design with the nonlinear sensor verified our heuristic analysis of limit cycles using the concept of equivalent gain with a root locus.
- The disk-drive case study illustrated control in an uncertain environment, where bandwidth is very important.
- The RTP case study illustrated modeling and control of a nonlinear thermal system.
- The *E. coli* chemotaxis case study illustrated a simple example of the application of ideas from control theory to the emerging field of systems biology.
- In all cases the designer needs to be able to use multiple tools, including the root locus, the frequency response, pole placement by state feedback, and simulation of time responses to get a good design. We promised an understanding of these tools at the beginning of the text, and we trust you are now ready to practice the art of control engineering.

REVIEW QUESTIONS

1. Why is a collocated actuator and sensor arrangement for a lightly damped structure such as a robot arm easier to design than a noncollocated setup?

2. Why should the control engineer be involved in the design of the process to be controlled?

3. Give examples of an actuator and a sensor for the following control problems:

   (a) Attitude control of a geosynchronous communication satellite.

   (b) Pitch control of a Boeing 747 airliner.

   (c) Track-following control of a CD player.

   (d) Fuel–air ratio control of a spark-ignited automobile engine.

(e) Position control for an arm of a robot used to paint automobiles.

(f) Heading control of a ship.

(g) Attitude control of a helicopter.

PROBLEMS

10.1 Of the three types of PID control (proportional, integral, or derivative), which one is the most effective in reducing the error resulting from a constant disturbance? Explain.

10.2 Is there a greater chance of instability when the sensor in a feedback control system for a mechanical structure is not collocated with the actuator? Explain.

10.3 Consider the plant $G(s) = 1/s^3$. Determine whether it is possible to stabilize this plant by adding the lead compensator

$$D_C(s) = K\frac{s+a}{s+b}, \quad (a < b).$$

(a) What is the maximum phase margin of the resulting feedback system?

(b) Can a system with this plant, together with any number of lead compensators, be made unconditionally stable? Explain why or why not.

10.4 Consider the closed-loop system shown in Fig. 10.88.

(a) What is the phase margin if $K = 70{,}000$?

(b) What is the gain margin if $K = 70{,}000$?

(c) What value of K will yield a phase margin of $\sim 70°$?

(d) What value of K will yield a phase margin of $\sim 0°$?

(e) Sketch the root locus with respect to K for the system, and determine what value of K causes the system to be on the verge of instability.

(f) If the disturbance w is a constant and $K = 10{,}000$, what is the maximum allowable value for w if $y(\infty)$ is to remain less than 0.1? (Assume $r = 0$.)

(g) Suppose the specifications require you to allow larger values of w than the value you obtained in part (f) but with the same error constraint $[|y(\infty)| < 0.1]$. Discuss what steps you could take to alleviate the problem.

Figure 10.88
Control system for
Problem 10.4

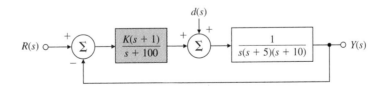

10.5 Consider the system shown in Fig. 10.89, which represents the attitude rate control for a certain aircraft.

(a) Design a compensator so that the dominant poles are at $-2 \pm 2j$.

(b) Sketch the Bode plot for your design, and select the compensation so that the crossover frequency is at least $2\sqrt{2}$ rad/sec and PM $\geq 50°$.

(c) Sketch the root locus for your design, and find the velocity constant when $\omega_n > 2\sqrt{2}$ and $\zeta \geq 0.5$.

Figure 10.89

Block diagram for aircraft-attitude rate control

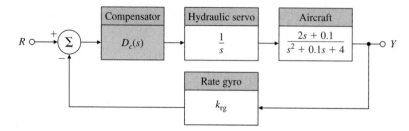

10.6 Consider the block diagram for the servomechanism drawn in Fig. 10.90. Which of the following claims are true?

(a) The actuator dynamics (the pole at 1000 rad/sec) must be included in an analysis to evaluate a usable maximum gain for which the control system is stable.

(b) The gain K must be negative for the system to be stable.

(c) There exists a value of K for which the control system will oscillate at a frequency between 4 and 6 rad/sec.

(d) The system is unstable if $|K| > 10$.

(e) If K must be negative for stability, the control system cannot counteract a positive disturbance.

(f) A positive constant disturbance will speed up the load, thereby making the final value of e negative.

(g) With only a positive constant command input r, the error signal e must have a final value greater than zero.

(h) For $K = -1$ the closed-loop system is stable, and the disturbance results in a speed error whose steady-state magnitude is less than 5 rad/sec.

Figure 10.90

Servomechanism for Problem 10.6

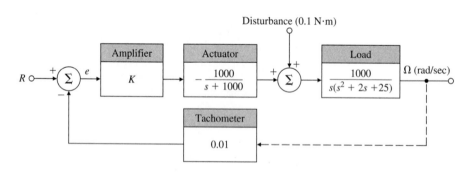

10.7 A stick balancer and its corresponding control block diagram are shown in Fig. 10.91. The control is a torque applied about the pivot.

(a) Using root-locus techniques, design a compensator $D(s)$ that will place the dominant roots at $s = -5 \pm 5j$ (corresponding to $\omega_n = 7$ rad/sec, $\zeta = 0.707$).

(b) Use Bode plotting techniques to design a compensator $D(s)$ to meet the following specifications:

- Steady-state θ displacement of less than 0.001 for a constant input torque $T_d = 1$,
- Phase margin $\geq 50°$,
- Closed-loop bandwidth $\cong 7$ rad/sec.

Figure 10.91

Servomechanism for
Problem 10.7

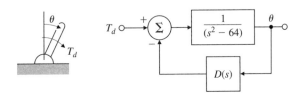

10.8   Consider the standard feedback system drawn in Fig. 10.92.

(a) Suppose

$$G(s) = \frac{2500\,K}{s(s + 25)}.$$

Design a lead compensator so that the phase margin of the system is more than
45°; the steady-state error due to a ramp should be less than or equal to 0.01.

(b) Using the plant transfer function from part (a), design a lead compensator so that
the overshoot is less than 25% and the 1% settling time is less than 0.1 sec.

(c) Suppose

$$G(s) = \frac{K}{s(1 + 0.1s)(1 + 0.2s)}$$

and let the performance specifications now be $K_v = 100$ and PM $\geq 40°$. Is the
lead compensation effective for this system? Find a lag compensator, and plot the
root locus of the compensated system.

(d) Using $G(s)$ from part (c), design a lag compensator such that the peak overshoot
is less than 20% and $K_v = 100$.

(e) Repeat part (c) using a lead–lag compensator.

(f) Find the root locus of the compensated system in part (e), and compare your
findings with those from part (c).

Figure 10.92

Block diagram of a
standard feedback
control system

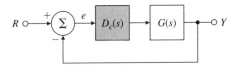

10.9   Consider the system in Fig. 10.92, where

$$G(s) = \frac{300}{s(s + 0.225)(s + 4)(s + 180)}.$$

The compensator $D_C(s)$ is to be designed so that the closed-loop system satisfies the
following specifications:

•   Zero steady-state error for step inputs,
•   PM $= 55°$, GM ≥ 6 db,
•   Gain crossover frequency is not smaller than that of the uncompensated plant.

(a) What kind of compensation should be used and why?

(b) Design a suitable compensator $D_C(s)$ to meet the specifications.

10.10   We have discussed three design methods: the root-locus method of Evans, the
frequency-response method of Bode, and the state-variable pole-assignment method.
Explain which of these methods is *best* described by the following statements (if you
feel more than one method fits a given statement equally well, say so and explain why):

(a) This method is the one most commonly used when the plant description must be obtained from experimental data.

(b) This method provides the most direct control over dynamic response characteristics such as rise time, percent overshoot, and settling time.

(c) This method lends itself most easily to an automated (computer) implementation.

(d) This method provides the most direct control over the steady-state error constants K_p and K_v.

(e) This method is most likely to lead to the *least complex* controller capable of meeting the dynamic and static accuracy specifications.

(f) This method allows the designer to guarantee that the final design will be unconditionally stable.

(g) This method can be used without modification for plants that include transportation lag terms—for example,

$$G(s) = \frac{e^{-2s}}{(s+3)^2}.$$

10.11 Lead and lag networks are typically employed in designs based on frequency-response (Bode) methods. Assuming a Type 1 system, indicate the effect of these compensation networks on each of the listed performance specifications. In each case, indicate the effect as "an increase," "substantially unchanged," or "a decrease." Use the second-order plant $G(s) = K/[s(s+1)]$ to illustrate your conclusions.

(a) K_v

(b) Phase margin

(c) Closed-loop bandwidth

(d) Percent overshoot

(e) Settling time

10.12 *Altitude Control of a Hot-Air Balloon*: American solo balloonist Steve Fossett landed in the Australian outback aboard *Spirit of Freedom* on July 3rd, 2002, becoming the first solo balloonist to circumnavigate the globe (see Fig. 10.93). The equations of vertical motion for a hot-air balloon (Fig. 10.94), linearized about vertical equilibrium, are

$$\delta \dot{T} + \frac{1}{\tau_1}\delta T = \delta q,$$

$$\tau_2 \ddot{z} \;\; + \dot{z} = \;\;\; a\delta T + w,$$

where

$\delta T = $ deviation of the hot-air temperature from the equilibrium temperature where buoyant force equals weight,

$z = $ altitude of the balloon,

$\delta q = $ deviation in the burner heating rate from the equilibrium rate (normalized by the thermal capacity of the hot air),

$w = $ vertical component of wind velocity,

$\tau_1, \tau_2, a = $ parameters of the equations.

An altitude-hold autopilot is to be designed for a balloon whose parameters are

$$\tau_1 = 250 \text{ sec} \quad \tau_2 = 25 \text{ sec} \quad a = 0.3 \text{ m/(sec·°C)}.$$

Figure 10.93
Spirit of Freedom
balloon
*Source: French
Navy/Tahitipresse*

Figure 10.94
Hot-air balloon

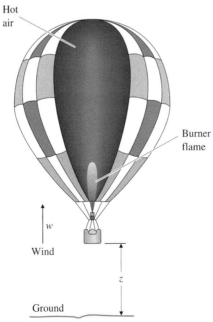

Only altitude is sensed, so a control law of the form

$$\delta q(s) = D(s)[z_d(s) - z(s)]$$

will be used, where z_d is the desired (commanded) altitude.

(a) Sketch a root locus of the closed-loop eigenvalues with respect to the gain K for a proportional feedback controller, $\delta q = -K(z - z_d)$. Use Routh's criterion (or let $s = j\omega$ and find the roots of the characteristic polynomial) to determine the value of the gain and the associated frequency at which the system is marginally stable.

(b) Our intuition and the results of part (a) indicate that a relatively large amount of lead compensation is required to produce a satisfactory autopilot. Because Steve Fossett was a millionaire, he could afford a more complex controller implementation. Sketch a root locus of the closed-loop eigenvalues with respect to the gain K for a double-lead compensator, $\delta q = D(s)(z_d - z)$, where

$$D(s) = K \left(\frac{s + 0.03}{s + 0.12} \right)^2.$$

(c) Sketch the magnitude portions of the Bode plots (straight-line asymptotes only) for the open-loop transfer functions of the proportional feedback and lead-compensated systems.

(d) Select a gain K for the lead-compensated system to give a crossover frequency of 0.06 rad/sec.

(e) With the gain selected in part (d), what is the steady-state error in altitude for a steady vertical wind of 1 m/sec? (Be careful: First find the closed-loop transfer function from w to the error.)

(f) If the error in part (e) is too large, how would you modify the compensation to give higher low-frequency gain? (Give a qualitative answer only.)

10.13 Satellite-attitude control systems often use a reaction wheel to provide angular motion. The equations of motion for such a system are

$$\text{Satellite}: \quad I\ddot{\phi} = T_c + T_{ex},$$
$$\text{Wheel}: \quad J\dot{r} = -T_c,$$
$$\text{Measurement}: \quad \dot{Z} = \dot{\phi} - aZ,$$
$$\text{Control}: \quad T_c = -D(s)(Z - Z_d),$$

where

$$J = \text{moment of inertia of the wheel},$$
$$r = \text{wheel speed},$$
$$T_c = \text{control torque},$$
$$T_{ex} = \text{disturbance torque},$$
$$\phi = \text{angle to be controlled},$$
$$Z = \text{measurement from the sensor},$$
$$Z_d = \text{reference angle},$$
$$I = \text{satellite inertia (1000 kg/m}^2\text{)},$$

$$a = \text{sensor constant (1 rad/sec)},$$

$$D(s) = \text{compensation}.$$

(a) Suppose $D(s) = K_0$, a constant. Draw the root locus with respect to K_0 for the resulting closed-loop system.

(b) For what range of K_0 is the closed-loop system stable?

(c) Add a lead network with a pole at $s = -1$ so that the closed-loop system has a bandwidth $\omega_{BW} = 0.04$ rad/sec, a damping ratio $\zeta = 0.5$, and compensation given by

$$D(s) = K_1 \frac{s+z}{s+1}.$$

Where should the zero of the lead network be located? Draw the root locus of the compensated system, and give the value of K_1 that allows the specifications to be met.

(d) For what range of K_1 is the system stable?

(e) What is the steady-state error (the difference between Z and some reference input Z_d) to a constant disturbance torque T_{ex} for the design of part (c)?

(f) What is the type of this system with respect to rejection of T_{ex}?

(g) Draw the Bode plot asymptotes of the *open-loop* system, with the gain adjusted for the value of K_1 computed in part (c). Add the compensation of part (c), and compute the phase margin of the closed-loop system.

(h) Write state equations for the open-loop system, using the state variables ϕ, $\dot{\phi}$, and Z. Select the gains of a state-feedback controller $T_c = -K_\phi \phi - K_{\dot\phi} \dot\phi$ to locate the closed-loop poles at $s = -0.02 \pm 0.02j\sqrt{3}$.

10.14 Three alternative designs are sketched in Fig. 10.95 for the closed-loop control of a system with the plant transfer function $G(s) = 1/s(s + 1)$. The signal w is the plant noise and may be analyzed as if it were a step; the signal v is the sensor noise and may be analyzed as if it contained power to very high frequencies.

(a) Compute values for the parameters K_1, a, K_2, K_T, K_3, d, and K_D so that in each case (assuming $w = 0$ and $v = 0$),

$$\frac{Y}{R} = \frac{16}{s^2 + 4s + 16}.$$

Note that in system III, a pole is to be placed at $s = -4$.

(b) Complete the following table, expressing the last entries as A/s^k to show how fast noise from v is attenuated at high frequencies:

System	K_v	$\dfrac{y}{w}\Big|_{s=0}$	$\dfrac{y}{v}\Big|_{s\to\infty}$
I			
II			
III			

(c) Rank the three designs according to the following characteristics (the best as "1," the poorest as "3"):

Figure 10.95

Alternative feedback
structures for
Problem 10.14

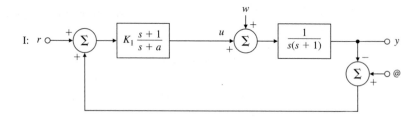

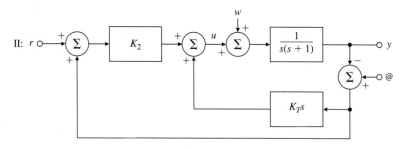

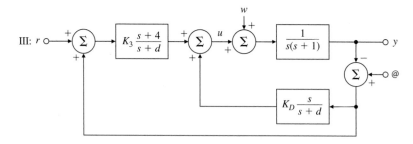

	I	II	III
Tracking			
Plant-noise rejection			
Sensor-noise rejection			

10.15 The equations of motion for a cart–stick balancer with state variables of stick angle, stick angular velocity, and cart velocity are

$$\dot{x} = \begin{bmatrix} 0 & 1 & 0 \\ 31.33 & 0 & 0.016 \\ -31.33 & 0 & -0.216 \end{bmatrix} x + \begin{bmatrix} 0 \\ -0.649 \\ 8.649 \end{bmatrix} u,$$

$$y = [\; 10 \quad 0 \quad 0 \;]x,$$

where the output is stick angle, and the control input is voltage on the motor that drives the cart wheels.

(a) Compute the transfer function from u to y, and determine the poles and zeros.

(b) Determine the feedback gain **K** necessary to move the poles of the system to the locations -2.832 and $-0.521 \pm 1.068j$, with $\omega_n = 4$ rad/sec.

(c) Determine the estimator gain **L** needed to place the three estimator poles at -10.

(d) Determine the transfer function of the estimated-state-feedback compensator defined by the gains computed in parts (b) and (c).

(e) Suppose we use a reduced-order estimator with poles at -10 and -10. What is the required estimator gain?

(f) Repeat part (d) using the reduced-order estimator.

(g) Compute the frequency response of the two compensators.

10.16 A 282-ton Boeing 747 is approaching land at sea level. If we use the state given in the case study (Section 10.3) and assume a velocity of 221 ft/sec (Mach 0.198), then the lateral-direction perturbation equations are

$$
\begin{bmatrix} \dot{\beta} \\ \dot{r} \\ \dot{p} \\ \dot{\phi} \end{bmatrix} = \begin{bmatrix} -0.0890 & -0.989 & 0.1478 & 0.1441 \\ 0.168 & -0.217 & -0.166 & 0 \\ -1.33 & 0.327 & -0.975 & 0 \\ 0 & 0.149 & 1 & 0 \end{bmatrix} \begin{bmatrix} \beta \\ r \\ p \\ \phi \end{bmatrix} + \begin{bmatrix} 0.0148 \\ -0.151 \\ 0.0636 \\ 0 \end{bmatrix} \delta r,
$$

$$
y = \begin{bmatrix} 0 & 1 & 0 & 0 \end{bmatrix} \begin{bmatrix} \beta \\ r \\ p \\ \phi \end{bmatrix}.
$$

The corresponding transfer function is

$$
G(s) = \frac{r(s)}{\delta r(s)} = \frac{-0.151(s+1.05)(s+0.0328\pm 0.414j)}{(s+1.109)(s+0.0425)(s+0.0646\pm 0.731j)}.
$$

(a) Draw the uncompensated root locus [for $1 + KG(s)$] and the frequency response of the system. What type of classical controller could be used for this system?

(b) Try a state-variable design approach by drawing a SRL for the system. Choose the closed-loop poles of the system on the SRL to be

$$
\alpha_c(s) = (s+1.12)(s+0.165)(s+0.162\pm 0.681j),
$$

and choose the estimator poles to be five times faster at

$$
\alpha_e(s) = (s+5.58)(s+0.825)(s+0.812\pm 3.40j).
$$

(c) Compute the transfer function of the SRL compensator.

(d) Discuss the robustness properties of the system with respect to parameter variations and unmodeled dynamics.

(e) Note the similarity of this design to the one developed for different flight conditions earlier in the chapter. What does this suggest about providing a continuous (nonlinear) control throughout the operating envelope?

10.17 (Contributed by Prof. L. Swindlehurst) The feedback control system shown in Fig. 10.96 is proposed as a position control system. A key component of this system is an armature-controlled DC motor. The input potentiometer produces a voltage E_i that is proportional to the desired shaft position: $E_i = K_p\theta_i$. Similarly, the output potentiometer produces a voltage E_0 that is proportional to the actual shaft position: $E_0 = K_p\theta_0$. Note that we have assumed that both potentiometers have the same proportionality constant. The error signal $E_i - E_0$ drives a compensator, which in turn produces an armature voltage that drives the motor. The motor has an armature resistance R_a, an armature inductance L_a, a torque constant K_t, and a back emf constant K_e. The moment of inertia of the motor shaft is J_m, and the rotational damping due to

Figure 10.96

A servomechanism with gears on the motor shaft and potentiometer sensors

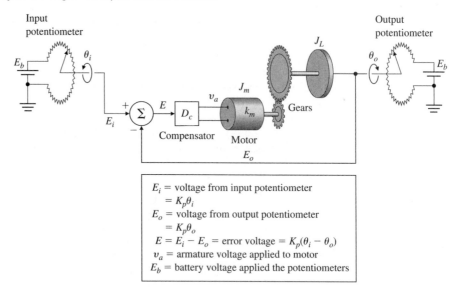

E_i = voltage from input potentiometer
  = $K_p\theta_i$
E_o = voltage from output potentiometer
  = $K_p\theta_o$
$E = E_i - E_o$ = error voltage = $K_p(\theta_i - \theta_o)$
v_a = armature voltage applied to motor
E_b = battery voltage applied the potentiometers

bearing friction is B_m. Finally, the gear ratio is $N : 1$, the moment of inertia of the load is J_L, and the load damping is B_L.

(a) Write the differential equations that describe the operation of this feedback system.

(b) Find the transfer function relating $\theta_0(s)$ and $\theta_i(s)$ for a general compensator $D_C(s)$.

(c) The open-loop frequency-response data shown in Table 10.2 were taken using the armature voltage v_a of the motor as an input and the output potentiometer voltage E_0 as the output. Assuming that the motor is linear and minimum-phase, make an estimate of the transfer function of the motor,

$$G(s) = \frac{\theta_m(s)}{V_a(s)},$$

where θ_m is the angular position of the motor shaft.

TABLE 10.2

Frequency-Response Data for Problem 10.8

Frequency (rad/sec)	$\left\|\dfrac{E_0(s)}{V_a(s)}\right\|$ (db)	Frequency (rad/sec)	$\left\|\dfrac{E_0(s)}{V_a(s)}\right\|$ (db)
0.1	60.0	10.0	14.0
0.2	54.0	20.0	2.0
0.3	50.0	40.0	−10.0
0.5	46.0	60.0	−20.0
0.8	42.0	65.0	−21.0
1.0	40.0	80.0	−24.0
2.0	34.0	100.0	−30.0
3.0	30.5	200.0	−48.0
4.0	27.0	300.0	−59.0
5.0	23.0	500.0	−72.0
7.0	19.5		

(d) Determine a set of performance specifications that are appropriate for a position control system and will yield good performance. Design $D_c(s)$ to meet these specifications.

(e) Verify your design through analysis and simulation using MATLAB.

10.18 Design and construct a device to keep a ball centered on a freely swinging beam. An example of such a device is shown in Fig. 10.97. It uses coils surrounding permanent magnets as the actuator to move the beam, solar cells to sense the ball position, and a hall-effect device to sense the beam position. Research other possible actuators and sensors as part of your design effort. Compare the quality of the control achievable for ball-position feedback only with that of multiple-loop feedback of both ball and beam position.

Figure 10.97

Ball-balancer design example

Source: Photo courtesy of David Powell

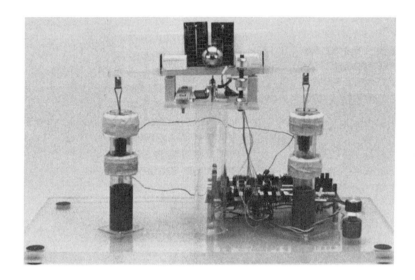

10.19 Design and construct the magnetic levitation device shown in Fig. 9.2. You may wish to use LEGO components in your design.

10.20 *Run-to-Run Control*: Consider the RTP system shown in Fig. 10.98. We wish to heat up a semiconductor wafer, and control the wafer surface temperature accurately using rings of tungsten halogen lamps. The output of the system is temperature T as a function of time: $y = T(t)$. The system reference input R is a desired step in temperature (700° C), and the control input is lamp power. A pyrometer is used to measure the wafer center temperature. The model of the system is first order, and an integral controller is used as shown in Fig. 10.98. Normally, there is not a sensor bias ($b = 0$).

Figure 10.98

RTP system

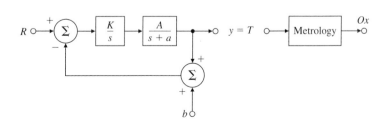

(a) Suppose the system suddenly develops a sensor bias $b \neq 0$, where b is known. What can be done to ensure zero steady-state tracking of temperature command R, despite the presence of the sensor bias?

(b) Now assume $b = 0$. In reality, we are trying to control the thickness of the oxide film grown (Ox) on the wafer and not the temperature. At present, no sensor can measure Ox in real time. The semiconductor process engineer must use off-line equipment (called *metrology*) to measure the thickness of the oxide film grown on the wafer. The relationship between the system output temperature and Ox is nonlinear and given by

$$\text{Oxide thickness} = \int_0^{t_f} p e^{-\frac{c}{T(t)}} \, dt,$$

where t_f is the process duration, and p and c are known constants. Suggest a scheme in which the center wafer oxide thickness Ox can be controlled to a desired value (say, $Ox = 5000$ Å) by employing the temperature controller and the output of the metrology.

10.21 Develop a nonlinear model for a tungsten halogen lamp and simulate it in SIMULINK.

10.22 Develop a nonlinear model for a pyrometer. Show how temperature can be deduced from the model.

10.23 Repeat the RTP case study design by summing the three sensors to form a single signal to control the average temperature. Demonstrate the performance of the linear design, and validate the performance on the nonlinear SIMULINK simulation.

10.24 One of the steps in semiconductor wafer manufacturing during photolithography is performed by placement of the wafer on a heated plate for a certain period of time. Laboratory experiments have shown that the transfer function from the heater power, u, to the wafer temperature, y, is given by

$$\frac{y(s)}{u(s)} = G(s) = \frac{0.09}{(s+0.19)(s+0.78)(s+0.00018)}.$$

(a) Sketch the $180°$ root locus for the uncompensated system.

(b) Using the root-locus design techniques, design a *dynamic* compensator, $D(s)$, such that the system meets the following time-domain specifications

  i. $M_p \leq 5\%$

  ii. $t_r \leq 20$ sec

  iii. $t_s \leq 60$ sec

  iv Steady-state error to a $1°$C step input command $< 0.1°$C.

   Draw the $180°$ root locus for the compensated system.

10.25 *Excitation-Inhibition Model from Systems Biology* (Yang and Iglesias, 2005): In *Dictyostelium* cells, the activation of key signaling molecules involved in chemoattractant sensing can be modeled by the following third order linearized model. The external disturbance to the output transfer function is:

$$\frac{y(s)}{w(s)} = S(s) = \frac{(1-\alpha)s}{(s+\alpha)(s+1)(s+\gamma)}$$

where, w is the external disturbance signal proportional to chemoattractant concentration, and y is the output which is the fraction of active response regulators. Show that there is an alternate representation of the system with the "plant" transfer function

$$G(s) = \frac{(1-\alpha)}{s^2 + (1+\alpha+\gamma)s + (\alpha+\gamma+\alpha\gamma)}$$

and the "feedback regulator"

$$D(s) = \frac{\alpha\gamma}{(1-\alpha)s}.$$

It is known that $\alpha \neq 1$ for this version of the model. Draw the feedback block diagram of the system showing the locations of the disturbance input and the output. What is the significance of this particular representation of the system? What hidden system property does it reveal? Is the disturbance rejection a robust property for this system? Plot the disturbance rejection response of the system for a unit step disturbance input. Assume the system parameter values are $\alpha = 0.5$ and $\gamma = 0.2$.

Appendix A   Laplace Transforms

A.1 The $\mathcal{L}_-$ Laplace Transform

Laplace transforms can be used to study the complete response characteristics of feedback systems, including the transient response. This is in contrast to Fourier transforms, in which the steady-state response is the main concern. In many applications it is useful to define the Laplace transform of $f(t)$, denoted by $\mathcal{L}_-\{f(t)\} = F(s)$, as a function of the complex variable $s = \sigma + j\omega$, where

$$F(s) \triangleq \int_{0^-}^{\infty} f(t)e^{-st}\, dt, \tag{A.1}$$

which uses 0^- (that is, a value just before $t = 0$) as the lower limit of integration and is referred to as the **unilateral** (or **one-sided**) **Laplace transform**.[1] A function $f(t)$ will have a Laplace transform if it is of **exponential order**, which means that there exists a real number σ such that

$$\lim_{t \to \infty} |f(t)e^{-\sigma t}| = 0. \tag{A.2}$$

The decaying exponential term in the integrand in effect provides a built-in convergence factor. This means that even if $f(t)$ does not vanish as $t \to \infty$, the integrand will vanish for sufficiently large values of σ if f does not grow at a faster-than-exponential rate. For example, ae^{bt} is of exponential order, whereas e^{t^2} is not. If $F(s)$ exists for some $s_0 = \sigma_0 + j\omega_0$, then it exists for all values of s such that

$$\mathrm{Re}(s) \geq \sigma_0. \tag{A.3}$$

The smallest value of σ_0 for which $F(s)$ exists is called the **abscissa of convergence**, and the region to the right of $\mathrm{Re}(s) \geq \sigma_0$ is called the **region of convergence**. Typically, two-sided Laplace transforms exist for a specified range

$$\alpha < \mathrm{Re}(s) < \beta, \tag{A.4}$$

which defines the strip of convergence. Table A.2 gives some Laplace transform pairs. Each entry in the table follows from direct application of the transform definition.[2]

A.1.1 Properties of Laplace Transforms

In this section we will address and prove each of the significant properties of the Laplace transform as discussed in Chapter 3 and listed in Table A.1. In addition we show how these properties can be used through examples.

---

[1]**Bilateral** (or **two-sided**) **Laplace transforms** and the so-called $\mathcal{L}_+$ transforms, in which the lower value of integral is 0^+, also arise elsewhere.

[2]As for the one-sided Laplace transform, an astute reader would wonder what happens to the validity of the Laplace transform for the rest of the s-plane, namely, the region where $\mathrm{Re}(s) < \sigma_0$. Indeed it would be disappointing if $F(s)$ was only valid for $\mathrm{Re}(s) \geq \sigma_0$ and not elsewhere in the s-plane. Fortunately, except for some pathological cases (which do not arise in practice), one can invoke an important result from the theory of complex variables known as the **Analytic Continuation Theorem** to extend the region of the validity of $F(s)$ to the whole s-plane excluding the locations of the poles.

TABLE A.1 **Properties of Laplace Transforms**

Number	Laplace Transform	Time Function	Comment
—	$F(s)$	$f(t)$	Transform pair
1	$\alpha F_1(s) + \beta F_2(s)$	$\alpha f_1(t) + \beta f_2(t)$	Superposition
2	$F(s)e^{-s\lambda}$	$f(t - \lambda)$	Time delay ($\lambda \geq 0$)
3	$\dfrac{1}{|a|}F\left(\dfrac{s}{a}\right)$	$f(at)$	Time scaling
4	$F(s + a)$	$e^{-at}f(t)$	Shift in frequency
5	$s^m F(s) - s^{m-1}f(0)$ $-s^{m-2}\dot{f}(0) - \cdots - f^{(m-1)}(0)$	$f^{(m)}(t)$	Differentiation
6	$\dfrac{1}{s}F(s)$	$\displaystyle\int_0^t f(\zeta)\,d\zeta$	Integration
7	$F_1(s)F_2(s)$	$f_1(t) * f_2(t)$	Convolution
8	$\displaystyle\lim_{s\to\infty} sF(s)$	$f(0^+)$	Initial Value Theorem
9	$\displaystyle\lim_{s\to 0} sF(s)$	$\displaystyle\lim_{t\to\infty} f(t)$	Final Value Theorem
10	$\dfrac{1}{2\pi j}\displaystyle\int_{\sigma_c-j\infty}^{\sigma_c+j\infty} F_1(\zeta)F_2(s - \zeta)\,d\zeta$	$f_1(t)f_2(t)$	Time product
11	$\dfrac{1}{2\pi}\displaystyle\int_{-j\infty}^{+j\infty} Y(-j\omega)U(j\omega)\,d\omega$	$\int_0^\infty y(t)u(t)\,dt$	Parseval's Theorem
12	$-\dfrac{d}{ds}F(s)$	$tf(t)$	Multiplication by time

1. Superposition

One of the more important properties of the Laplace transform is that it is linear. We can prove this as follows:

$$\mathcal{L}\{\alpha f_1(t) + \beta f_2(t)\} = \int_0^\infty [\alpha f_1(t) + \beta f_2(t)]e^{-st}\,dt \tag{A.5}$$

$$= \alpha \int_0^\infty f_1(t)e^{-st}\,dt + \beta \int_0^\infty f_2(t)e^{-st}\,dt$$

$$= \alpha F_1(s) + \beta F_2(s).$$

The scaling property is a special case of linearity; that is,

$$\mathcal{L}\{\alpha f(t)\} = \alpha F(s). \tag{A.6}$$

EXAMPLE A.1   *Sinusoidal Signal*

Find the Laplace transform of $f(t) = 1 + 2\sin(\omega t)$.

Solution. The Laplace transform of $\sin(\omega t)$ is

$$\mathcal{L}\{\sin(\omega t)\} = \frac{\omega}{s^2 + \omega^2}.$$

TABLE A.2

Table of Laplace Transforms

Number	$F(s)$	$f(t), t \geq 0$
1	1	$\delta(t)$
2	$1/s$	$1(t)$
3	$1/s^2$	t
4	$2!/s^3$	t^2
5	$3!/s^4$	t^3
6	$m!/s^{m+1}$	t^m
7	$\dfrac{1}{s+a}$	e^{-at}
8	$\dfrac{1}{(s+a)^2}$	te^{-at}
9	$\dfrac{1}{(s+a)^3}$	$\dfrac{1}{2!}t^2 e^{-at}$
10	$\dfrac{1}{(s+a)^m}$	$\dfrac{1}{(m-1)!}t^{m-1}e^{-at}$
11	$\dfrac{a}{s(s+a)}$	$1 - e^{-at}$
12	$\dfrac{a}{s^2(s+a)}$	$\dfrac{1}{a}(at - 1 + e^{-at})$
13	$\dfrac{b-a}{(s+a)(s+b)}$	$e^{-at} - e^{-bt}$
14	$\dfrac{s}{(s+a)^2}$	$(1 - at)e^{-at}$
15	$\dfrac{a^2}{s(s+a)^2}$	$1 - e^{-at}(1 + at)$
16	$\dfrac{(b-a)s}{(s+a)(s+b)}$	$be^{-bt} - ae^{-at}$
17	$\dfrac{a}{s^2+a^2}$	$\sin at$
18	$\dfrac{s}{s^2+a^2}$	$\cos at$
19	$\dfrac{s+a}{(s+a)^2+b^2}$	$e^{-at}\cos bt$
20	$\dfrac{b}{(s+a)^2+b^2}$	$e^{-at}\sin bt$
21	$\dfrac{a^2+b^2}{s[(s+a)^2+b^2]}$	$1 - e^{-at}\left(\cos bt + \dfrac{a}{b}\sin bt\right)$

Therefore, using Eq. (A.5) we obtain

$$F(s) = \frac{1}{s} + \frac{2\omega}{s^2 + \omega^2} = \frac{s^2 + 2\omega s + \omega^2}{s^3 + \omega^2 s}.$$

The following commands in MATLAB® yields the same result,

```
syms s t w
laplace(1+2*sin(w*t)).
```

2. Time Delay

Suppose a function $f(t)$ is delayed by $\lambda > 0$ units of time. Its Laplace transform is

$$F_1(s) = \int_0^\infty f(t - \lambda)e^{-st}\,dt.$$

Let us define $t' = t - \lambda$. Then $dt' = dt$, because λ is a constant and $f(t) = 0$ for $t < 0$. Thus

$$F_1(s) = \int_{-\lambda}^\infty f(t')e^{-s(t'+\lambda)}\,dt' = \int_0^\infty f(t')e^{-s(t'+\lambda)}\,dt'.$$

Because $e^{-s\lambda}$ is independent of time, it can be taken out of the integrand, so

$$F_1(s) = e^{-s\lambda} \int_0^\infty f(t')e^{-st'}\,dt' = e^{-s\lambda}F(s). \qquad \text{(A.7)}$$

From this result we see that a time delay of λ corresponds to multiplication of the transform by $e^{-s\lambda}$.

EXAMPLE A.2

Delayed Sinusoidal Signal

Find the Laplace transform of $f(t) = A\sin(t - t_d)$.

Solution.   The Laplace transform of $\sin(t)$ is

$$\mathcal{L}\{\sin(t)\} = \frac{1}{s^2 + 1}.$$

Therefore, using Eq. (A.7) we obtain

$$F(s) = \frac{A}{s^2 + 1}\,e^{-st_d}.$$

---

3. Time Scaling

If the time t is scaled by a factor a, then the Laplace transform of the time-scaled signal is

$$F_1(s) = \int_0^\infty f(at)e^{-st}\,dt.$$

Again, we define $t' = at$. As before, $dt' = a\,dt$ and

$$F_1(s) = \int_0^\infty f(t')\frac{e^{-st'/a}}{|a|}\,dt' = \frac{1}{|a|}F\left(\frac{s}{a}\right). \qquad \text{(A.8)}$$

EXAMPLE A.3

Sinusoid with Frequency ω

Find the Laplace transform of $f(t) = A\sin(\omega t)$.

Solution.   The Laplace transform of $\sin(t)$ is

$$\mathcal{L}\{\sin(t)\} = \frac{1}{s^2 + 1}.$$

Therefore, using Eq. (A.8) we obtain

$$F(s) = \frac{1}{|\omega|} \frac{1}{\left(\frac{s}{\omega}\right)^2 + 1}$$

$$= \frac{A\omega}{s^2 + \omega^2},$$

as expected. The following commands in MATLAB yields the same result,

```
syms s t w A
laplace(A*sin(w*t)).
```

4. Shift in Frequency

Multiplication (modulation) of $f(t)$ by an exponential expression in the time domain corresponds to a shift in frequency:

$$F_1(s) = \int_0^\infty e^{-at} f(t) e^{-st}\, dt = \int_0^\infty f(t) e^{-(s+a)t}\, dt = F(s+a). \tag{A.9}$$

EXAMPLE A.4   *Exponentially Decaying Sinusoid*

Find the Laplace transform of $f(t) = A\sin(\omega t)e^{-at}$.

Solution.   The Laplace transform of $\sin(\omega t)$ is

$$\mathcal{L}\{\sin(\omega t)\} = \frac{\omega}{s^2 + \omega^2}.$$

Therefore, using Eq. (A.9) we obtain

$$F(s) = \frac{A\omega}{(s+a)^2 + \omega^2}.$$

5. Differentiation

The transform of the derivative of a signal is related to its Laplace transform and its initial condition as follows:

$$\mathcal{L}\left\{\frac{df}{dt}\right\} = \int_{0^-}^\infty \left(\frac{df}{dt}\right) e^{-st}\, dt = e^{-st} f(t)|_{0^-}^\infty + s\int_{0^-}^\infty f(t) e^{-st}\, dt. \tag{A.10}$$

Because $f(t)$ is assumed to have a Laplace transform, $e^{-st}f(t) \to 0$ as $t \to \infty$. Thus

$$\mathcal{L}[\dot{f}] = -f(0^-) + sF(s). \tag{A.11}$$

Another application of Eq. (A.11) leads to

$$\mathcal{L}\{\ddot{f}\} = s^2 F(s) - sf(0^-) - \dot{f}(0^-). \tag{A.12}$$

Repeated application of Eq. (A.11) leads to

$$\mathcal{L}\{f^m(t)\} = s^m F(s) - s^{m-1}f(0^-) - s^{m-2}\dot{f}(0^-) - \cdots - f^{(m-1)}(0^-), \quad \text{(A.13)}$$

where $f^m(t)$ denotes the mth derivative of $f(t)$ with respect to time.

EXAMPLE A.5   *Derivative of Cosine Signal*

Find the Laplace transform of $g(t) = \frac{d}{dt}f(t)$, where $f(t) = \cos(\omega t)$.
Solution.   The Laplace transform of $\cos(\omega t)$ is

$$F(s) = \mathcal{L}\{\cos(\omega t)\} = \frac{s}{s^2 + \omega^2}.$$

Using Eq. (A.11) with $f(0^-) = 1$, we have

$$G(s) = \mathcal{L}\{g(t)\} = s \cdot \frac{s}{s^2 + \omega^2} - 1 = -\frac{\omega^2}{s^2 + \omega^2}.$$

---

6. Integration

Let us assume that we wish to determine the Laplace transform of the integral of a time function—that is, to find

$$F_1(s) = \mathcal{L}\left\{\int_0^t f(\xi)\, d\xi\right\} = \int_0^\infty \left[\int_0^t f(\xi)\, d\xi\right] e^{-st}\, dt.$$

Employing integration by parts, where

$$u = \int_0^t f(\xi)\, d\xi \quad \text{and} \quad dv = e^{-st}\, dt,$$

we get

$$F_1(s) = \left[-\frac{1}{s} e^{-st}\left(\int_0^t f(\xi)\, d\xi\right)\right]_0^\infty - \int_0^\infty -\frac{1}{s} e^{-st}f(t)\, dt = \frac{1}{s} F(s). \quad \text{(A.14)}$$

EXAMPLE A.6   *Time Integral of Sinusoidal Signal*

Find the Laplace transform of $f(t) = \int_0^t \sin \omega \tau\, d\tau$.
Solution.   The Laplace transform of $\sin(\omega t)$ is

$$\mathcal{L}\{\sin(\omega t)\} = \frac{\omega}{s^2 + \omega^2}.$$

Therefore, using Eq. (A.14), then

$$F(s) = \frac{\omega}{s^3 + \omega^2 s}.$$

7. Convolution

Convolution in the time domain corresponds to multiplication in the frequency domain. Assume that $\mathcal{L}\{f_1(t)\} = F_1(s)$ and $\mathcal{L}\{f_2(t)\} = F_2(s)$. Then

$$\mathcal{L}\{f_1(t) * f_2(t)\} = \int_0^\infty f_1(t) * f_2(t)e^{-st}\, dt = \int_0^\infty \left[\int_0^t f_1(\tau)f_2(t-\tau)\, d\tau\right] e^{-st}\, dt.$$

We see that t varies from zero to infinity and τ varies from zero to t. With the aid of Fig. A.1, we reverse the order of integration and change the limits of integration accordingly so that τ varies from zero to infinity and $\infty \geqslant t \geqslant \tau$, to yield

$$\mathcal{L}\{f_1(t) * f_2(t)\} = \int_0^\infty \int_\tau^\infty f_1(\tau)f_2(t-\tau)e^{-st}\, dt\, d\tau.$$

Multiplying by $e^{-s\tau}e^{s\tau}$ results in

$$\mathcal{L}\{f_1(t) * f_2(t)\} = \int_0^\infty f_1(\tau)e^{-s\tau}\left[\int_\tau^\infty f_2(t-\tau)e^{-s(t-\tau)}\, dt\right] d\tau.$$

If we change variables $t' \triangleq t - \tau$, then

$$\mathcal{L}\{f_1(t) * f_2(t)\} = \int_0^\infty f_1(\tau)e^{-s\tau}\, d\tau \int_0^\infty f_2(t')e^{-st'}\, dt',$$

$$\mathcal{L}\{f_1(t) * f_2(t)\} = F_1(s)F_2(s).$$

This implies that

$$\mathcal{L}^{-1}\{F_1(s)F_2(s)\} = f_1(t) * f_2(t). \tag{A.15}$$

EXAMPLE A.7

Ramp Response of a First-Order System

Find the ramp response of a first-order system with a pole at $+a$.

Solution. Let $f_1(t) = t$ be the ramp input and $f_2(t) = e^{at}$ be the impulse response of the first-order system. Then, using Eq. (A.15) we find that

$$\mathcal{L}^{-1}\left\{\frac{1}{s^2}\frac{1}{s-a}\right\} = f_1(t) * f_2(t)$$

$$= \int_0^t f_1(\tau)f_2(t-\tau)\, d\tau$$

$$= \int_0^t \tau e^{a(t-\tau)}\, d\tau$$

$$= \frac{1}{a^2}(e^{at} - at - 1).$$

The following commands in MATLAB yields the same result,

```
syms s t a
ilaplace(1/(s^3-a*s^2)).
```

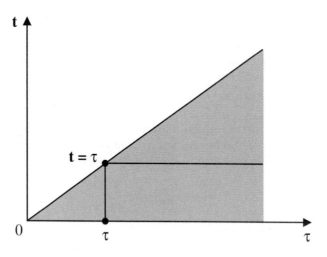

8. Time Product

Multiplication in the time domain corresponds to convolution in the frequency domain:

$$\mathcal{L}\{f_1(t)f_2(t)\} = \frac{1}{2\pi j} \int_{\sigma_c-j\infty}^{\sigma_c+j\infty} F_1(\xi)F_2(s-\xi)\,d\xi.$$

To see this, consider the relation

$$\mathcal{L}\{f_1(t)f_2(t)\} = \int_0^\infty f_1(t)f_2(t)e^{-st}\,dt.$$

Substituting the expression for $f_1(t)$ given by Eq. (3.25) yields

$$\mathcal{L}\{f_1(t)f_2(t)\} = \int_0^\infty \left[\frac{1}{2\pi j} \int_{\sigma_c-j\infty}^{\sigma_c+j\infty} F_1(\xi)e^{\xi t}\,d\xi \right] f_2(t)e^{-st}\,dt.$$

Changing the order of integration results in

$$\mathcal{L}\{f_1(t)f_2(t)\} = \frac{1}{2\pi j} \int_{\sigma_c-j\infty}^{\sigma_c+j\infty} F_1(\xi) \int_0^\infty f_2(t)e^{-(s-\xi)t}\,dt\,d\xi.$$

Using Eq. (A.9), we get

$$\mathcal{L}\{f_1(t)f_2(t)\} = \frac{1}{2\pi j} \int_{\sigma_c-j\infty}^{\sigma_c+j\infty} F_1(\xi)F_2(s-\xi)\,d\xi = \frac{1}{2\pi j}F_1(s) * F_2(s). \quad \text{(A.16)}$$

9. Parseval's Theorem

Parseval's famous theorem is used to compute the "energy" in a signal or "correlation" between two signals. It tells us that mentioned quantities can be computed either in the time domain or in the frequency domain. If

$$\int_0^\infty |y(t)|^2 dt < 1 \quad \text{and} \quad \int_0^\infty |u(t)|^2\, dt < 1 \quad \text{(A.17)}$$

(i.e., $y(t)$ and $u(t)$ are square integrable), then

$$\int_0^\infty y(t)u(t)dt = \frac{1}{2\pi} \int_{-\infty}^\infty Y(-j\omega)U(j\omega)\,d\omega. \quad \text{(A.18)}$$

Parseval's result involves only a substitution of the transform for the time functions and an exchange of integration:

$$\int_0^\infty y(t)u(t)\,dt = \int_0^\infty y(t) \left[\frac{1}{2\pi} \int_{-\infty}^\infty U(j\omega)e^{j\omega t}\,d\omega \right] dt \quad \text{(A.19)}$$

$$= \frac{1}{2\pi} \int_{-\infty}^\infty U(j\omega) \left[\int_0^\infty y(t)e^{j\omega t}\,dt \right] d\omega \quad \text{(A.20)}$$

$$= \frac{1}{2\pi} \int_{-\infty}^\infty U(j\omega)Y(-j\omega)\,d\omega. \quad \text{(A.21)}$$

10. Multiplication by Time

Multiplication by time corresponds to differentiation in the frequency domain. Let us consider

$$\frac{d}{ds}F(s) = \frac{d}{ds} \int_0^\infty e^{-st}f(t)\,dt$$

$$= \int_0^\infty -te^{-st}f(t)\,dt$$

$$= -\int_0^\infty e^{-st}[tf(t)]\,dt$$

$$= -\mathcal{L}\{tf(t)\}.$$

Then

$$\mathcal{L}\{tf(t)\} = -\frac{d}{ds}F(s), \quad \text{(A.22)}$$

which is the desired result.

EXAMPLE A.8   *Time Product of Sinusoidal Signal*

Find the Laplace transform of $f(t) = t \sin \omega t$.

Solution. The Laplace transform of $\sin \omega t$ is

$$\mathcal{L}\{\sin(\omega t)\} = \frac{\omega}{s^2 + \omega^2}.$$

Hence, using Eq. (A.22), we obtain

$$F(s) = -\frac{d}{ds}\left[\frac{\omega}{s^2 + \omega^2}\right] = \frac{2\omega s}{(s^2 + \omega^2)^2}.$$

The following commands in MATLAB yields the same result,

```
syms s t w
laplace(t*sin(w*t)).
```

A.1.2 Inverse Laplace Transform by Partial-Fraction Expansion

As we saw in Chapter 3, the easiest way to find $f(t)$ from its Laplace transform $F(s)$, if $F(s)$ is rational, is to expand $F(s)$ as a sum of simpler terms that can be found in the tables via partial-fraction expansion. We have already discussed this method in connection with simple roots in Section 3.1.5. In this section, we discuss partial-fraction expansion for cases of complex and repeated roots.

Complex Poles In the case of quadratic factors in the denominator, the numerator of the quadratic factor is chosen to be first order as shown in Example A.9. Whenever there exists a complex conjugate pair of poles such as

$$F(s) = \frac{C_1}{s - p_1} + \frac{C_2}{s - p_1^*},$$

we can show that

$$C_2 = C_1^*$$

(see Problem 3.1) and that

$$f(t) = C_1 e^{p_1 t} + C_1^* e^{p_1^* t} = 2\mathrm{Re}(C_1 e^{p_1 t}).$$

Assuming that $p_1 = \alpha + j\beta$, we may rewrite $f(t)$ in a more compact form as

$$f(t) = 2\mathrm{Re}\{C_1 e^{p_1 t}\} = 2\mathrm{Re}\{|C_1| e^{j \arg(C_1)} e^{(\alpha + j\beta)t}\} \tag{A.23}$$

$$= 2|C_1| e^{\alpha t} \cos[\beta t + \arg(C_1)].$$

EXAMPLE A.9

Partial-Fraction Expansion: Distinct Complex Roots

Find the function $f(t)$ for which the Laplace transform is

$$F(s) = \frac{1}{s(s^2 + s + 1)}.$$

Solution. We rewrite $F(s)$ as

$$F(s) = \frac{C_1}{s} + \frac{C_2 s + C_3}{s^2 + s + 1}.$$

Using the cover-up method, we find that

$$C_1 = sF(s)|_{s=0} = 1.$$

Setting $C_1 = 1$ and then equating the numerators in the partial-fraction expansion relation, we obtain

$$(s^2 + s + 1) + (C_2 s + C_3)s = 1.$$

After solving for C_2 and C_3, we find that $C_2 = -1$ and $C_3 = -1$. To make it more suitable for using the Laplace transform tables, we rewrite the partial fraction as

$$F(s) = \frac{1}{s} - \frac{s + \frac{1}{2} + \frac{1}{2}}{\left(s + \frac{1}{2}\right)^2 + \frac{3}{4}}.$$

From the tables we have,

$$f(t) = \left(1 - e^{-t/2} \cos \sqrt{\frac{3}{4}}\, t - \frac{1}{\sqrt{3}} e^{-t/2} \sin \sqrt{\frac{3}{4}}\, t\right) 1(t)$$

$$= \left(1 - \frac{2}{\sqrt{3}} e^{-t/2} \cos\left(\frac{\sqrt{3}}{2} t - \frac{\pi}{6}\right)\right) 1(t).$$

Alternatively, we may write $F(s)$ as

$$F(s) = \frac{C_1}{s} + \frac{C_2}{s - p_1} + \frac{C_2^*}{s - p_1^*}, \tag{A.24}$$

where $p_1 = -\frac{1}{2} + j\frac{\sqrt{3}}{2}$. $C_1 = 1$, as before, and now

$$C_2 = (s - p_1)F(s)|_{s=p_1} = -\frac{1}{2} + j\frac{1}{2\sqrt{3}},$$

$$C_2^* = -\frac{1}{2} - j\frac{1}{2\sqrt{3}},$$

and

$$f(t) = (1 + 2|C_2|e^{\alpha t} \cos[\beta t + \arg(C_2)]) 1(t)$$

$$= \left(1 + \frac{2}{\sqrt{3}} e^{-t/2} \cos\left[\frac{\sqrt{3}}{2} t + \frac{5\pi}{6}\right]\right) 1(t)$$

$$= \left(1 - \frac{2}{\sqrt{3}} e^{-t/2} \cos\left[\frac{\sqrt{3}}{2} t - \frac{5\pi}{6}\right]\right) 1(t).$$

The latter partial-fraction expansion can be readily computed using the MATLAB statements

$$\text{num} = 1; \qquad \text{\% form numerator}$$

$$\text{den} = \text{conv}([1\ 0],[1\ 1\ 1]); \qquad \text{\% form denominator}$$

$$[r,p,k] = \text{residue(num,den)} \qquad \text{\% compute residues}$$

which yield the results

$$r = [-0.5000 + 0.2887i - 0.5000 - 0.2887i\ 1.0000]';$$
$$p = [-0.5000 + 0.8660i - 0.5000 - 0.8660i\ 0]';\ k = []$$

and agrees with the previous hand calculations. Note that if we are using the tables, the first method is preferable, while the second method is preferable for checking MATLAB results.

The following commands in MATLAB yield the same result for the inverse Laplace transform,

```
syms s t
ilaplace(1/(s*(s^2+s+1))).
```

Repeated poles   For the case in which $F(s)$ has repeated roots, the procedure to compute the partial-fraction expansion must be modified. If p_1 is repeated three times, we write the partial fraction as

$$F(s) = \frac{C_1}{s - p_1} + \frac{C_2}{(s - p_1)^2} + \frac{C_3}{(s - p_1)^3} + \frac{C_4}{s - p_4} + \cdots + \frac{C_n}{s - p_n}.$$

We determine the constants C_4 through C_n as discussed previously. If we multiply both sides of the preceding equation by $(s - p_1)^3$, we obtain

$$(s - p_1)^3 F(s) = C_1 (s - p_1)^2 + C_2 (s - p_1) + C_3 + \cdots + \frac{C_n (s - p_1)^3}{s - p_n}. \quad \text{(A.25)}$$

If we then set $s = p_1$, all the factors on the right side of Eq. (A.25) will go to zero except C_3, which is

$$C_3 = (s - p_1)^3 F(s)|_{s=p_1},$$

as before. To determine the other factors, we differentiate Eq. (A.25) with respect to the Laplace variable s:

$$\frac{d}{ds}[(s - p_1)^3 F(s)] = 2C_1 (s - p_1) + C_2 + \cdots + \frac{d}{ds}\left[\frac{C_n (s - p_1)^3}{s - p_n}\right]. \quad \text{(A.26)}$$

Again, if we set $s = p_1$, we have

$$C_2 = \frac{d}{ds}[(s - p_1)^3 F(s)]_{s=p_1}.$$

Similarly, if we differentiate Eq. (A.26) again and set $s = p_1$ a second time, we get

$$C_1 = \frac{1}{2}\frac{d^2}{ds^2}[(s - p_1)^3 F(s)]_{s=p_1}.$$

In general, we may compute C_i for a factor with multiplicity k as

$$C_{k-i} = \frac{1}{i!}\left[\frac{d^i}{ds^i}[(s - p_1)^k F(s)]\right]_{s=p_1}, \quad i = 0, \ldots, k - 1.$$

EXAMPLE A.10

Partial-Fraction Expansion: Repeated Real Roots

Find the function $f(t)$ that has the Laplace transform

$$F(s) = \frac{s + 3}{(s + 1)(s + 2)^2}.$$

Solution.   We write the partial fraction as

$$F(s) = \frac{C_1}{s + 1} + \frac{C_2}{s + 2} + \frac{C_3}{(s + 2)^2}.$$

Then

$$C_1 = (s+1)F(s)|_{s=-1} = \frac{s+3}{(s+2)^2}|_{s=-1} = 2,$$

$$C_2 = \frac{d}{ds}\left[(s+2)^2F(s)\right]|_{s=-2} = -2,$$

$$C_3 = (s+2)^2F(s)|_{s=-2} = \frac{s+3}{s+1}|_{s=-2} = -1.$$

The function $f(t)$ is

$$f(t) = (2e^{-t} - 2e^{-2t} - te^{-2t})1(t).$$

The partial fraction computation can also be carried out using MATLAB's residue function,

```
num = [1 3];              % form numerator
den = conv([1 1],[1 4 4]);  % form denominator
[r,p,k] = residue(num,den)   % compute residues
```

which yields the result

```
r = [−2 −1 2]', p = [−2 −2 −1]', and k = [ ];
```

and agrees with the hand calculations.

The following commands in MATLAB yields the same result for the inverse Laplace transform,

```
syms s t
ilaplace((s+3)/((s+1)*(s+2)^2)).
```

A.1.3 The Initial Value Theorem

We discussed the Final Value Theorem in Chapter 3. A second valuable Laplace transform theorem is the **Initial Value Theorem**, which states that it is always possible to determine the initial value of the time function $f(t)$ from its Laplace transform. We may also state the theorem in this way:

The Initial Value Theorem

For any Laplace transform pair,

$$\lim_{s \to \infty} sF(s) = f(0^+). \tag{A.27}$$

We may show this as follows.

Using Eq. (A.11), we get

$$\mathcal{L}\left\{\frac{df}{dt}\right\} = sF(s) - f(0^-) = \int_{0^-}^{\infty} e^{-st}\frac{df}{dt}\,dt. \tag{A.28}$$

Let us consider the case in which $s \to \infty$ and rewrite the integral as

$$\int_{0^-}^{\infty} e^{-st}\frac{df(t)}{dt}\,dt = \int_{0^+}^{\infty} e^{-st}\frac{df(t)}{dt}\,dt + \int_{0^-}^{0^+} e^{-st}\frac{df(t)}{dt}\,dt.$$

Taking the limit of Eq. (A.28) as $s \to \infty$, we get

$$\lim_{s \to \infty} [sF(s) - f(0^-)] = \lim_{s \to \infty} \left[\int_{0^-}^{0^+} e^0 \frac{df(t)}{dt} \, dt + \int_{0^+}^{\infty} e^{-st} \frac{df(t)}{dt} \, dt \right].$$

The second term on the right side of the preceding equation approaches zero as $s \to \infty$, because $e^{-st} \to 0$. Hence

$$\lim_{s \to \infty} [sF(s) - f(0^-)] = \lim_{s \to \infty} [f(0^+) - f(0^-)] = f(0^+) - f(0^-)$$

or

$$\lim_{s \to \infty} sF(s) = f(0^+).$$

In contrast with the Final Value Theorem, the Initial Value Theorem can be applied to any function $F(s)$.

EXAMPLE A.11   *Initial Value Theorem*

Find the initial value of the signal in Example 3.11.

Solution.   From the Initial Value Theorem, we get

$$y(0^+) = \lim_{s \to \infty} sY(s) = \lim_{s \to \infty} s \frac{3}{s(s-2)} = 0,$$

which checks with the expression for $y(t)$ computed in Example 3.11.

A.1.4   Final Value Theorem

The Final Value Theorem

If all poles of $sY(s)$ are in the left half of the s-plane, then

$$\lim_{t \to \infty} y(t) = \lim_{s \to 0} sY(s). \qquad (3.46)$$

Proof of the Final Value Theorem

We may prove this result as follows.

The derivative relationship developed in Eq. (3.33) is

$$\mathcal{L}\left\{ \frac{dy}{dt} \right\} = sY(s) - y(0^-) = \int_{0^-}^{\infty} e^{-st} \frac{dy}{dt} \, dt.$$

We assume we are interested in the case where $s \to 0$. Then

$$\lim_{s \to 0} [sY(s) - y(0)] = \lim_{s \to 0} \left(\int_{0}^{\infty} e^{-st} \frac{dy}{dt} \, dt \right) = \lim_{t \to \infty} [y(t) - y(0)],$$

and we have

$$\lim_{t \to \infty} y(t) = \lim_{s \to 0} sY(s).$$

Another way to see this same result is to note that the partial-fraction expansion of $Y(s)$ [Eq. (3.43)] is

$$Y(s) = \frac{C_1}{s - p_1} + \frac{C_2}{s - p_2} + \cdots + \frac{C_n}{s - p_n}.$$

Let us say that $p_1 = 0$ and all other p_i are in the LHP so that C_1 is the steady-state value of $y(t)$. Using Eq. (3.45), we see that

$$C_1 = \lim_{t \to \infty} y(t) = sY(s)|_{s=0},$$

which is the same as the previous result.

For a thorough study of Laplace transforms and extensive tables, see Churchill (1972) and Campbell and Foster (1948); for the two-sided transform, see Van der Pol and Bremmer (1955).

Appendix B Solutions to the Review Questions

Chapter 1

1. What are the main components of a feedback control system?
   The process, the actuator, the sensor, and the controller.

2. What is the purpose of the sensor?
   To measure the output variable and, usually, to convert it to an electrical voltage.

3. Give three important properties of a good sensor.
   A good sensor is linear (the output is proportional to the input signal) over a large range of amplitudes and a large range of frequencies at its input, has low noise, is unbiased, is easy to calibrate, and has low cost. The relative values of these properties varies with the particular application.

4. What is the purpose of the actuator?
   The actuator takes an input, usually electrical, and converts it to a signal such as a force or torque that causes the process output to move or change over the required range.

5. Give three important properties of a good actuator.
   A good actuator has fast response, and adequate power, energy, speed, torque, and so on, to be able to cause the process output to meet the design specifications and is efficient, lightweight, small, cheap, and so on. As with sensors, the relative value of these properties varies with the application.

6. What is the purpose of the controller? Give the input(s) and output(s) of the controller.
   The controller is to take the sensor output (the input to the controller) and compute the control signal (the output of the controller) to be sent to the actuator.

7. What physical variable(s) of a process can be directly measured by a Hall effect sensor?
   A Hall-effect device measures the strength of a magnetic field and can be most easily configured to measure relative positions of two bodies or relative angles.

8. What physical variable is measured by a tachometer?
   A tachometer measures speed of rotation or angular velocity.

9. Describe three different techniques for measuring temperature.
   In each of the following cases, it is important to realize that the devices mentioned need to be calibrated and often corrected for nonlinearity in order to give a reliable, accurate measure of temperature.

(a) *An early technique still used in many home thermostats is based on the bimetallic strip composed of two strips of different metals that expand with different coefficients with temperature. As a result, the strip bends with temperature and the resulting motion can be used as a measure of temperature. This principle was introduced in the 18th century to maintain a constant length to a clock pendulum for precision time keeping.*

(b) *A technique related to the bimetallic strip is based on the fact that metals with different work functions placed in contact will produce a voltage that is proportional to temperature. Such a device is called a thermocouple and is the basis of a standard laboratory technique for measuring temperature.*

(c) *A number of materials have electrical resistance that is dependent in a monotonic way on temperature, and a resistance bridge can be used with one of these to indicate temperature. Such devices are called thermistors.*

(d) *For high temperatures, it is well known that the color of the radiation due to heat depends on temperature. A piece of iron placed in a fire will glow orange, then red, and finally become white hot at high temperatures. An instrument for measuring the frequency of the radiation, and thus the temperature, is a pyrometer.*

(e) *In ceramic kilns, cones of different materials that melt at different and known temperatures are placed near the products in the kiln to indicate when the design temperature has been reached. The potter watches until the cone of importance begins to sag and then knows that the products should be removed. These give a quantized measure of temperature.*

10. Why do most sensors have an electrical output, regardless of the physical nature of the variable being measured?
Electrical signals are the most easily manipulated; therefore, most controllers are electrical devices, either analog or digital. To provide the signal input to such a device, the sensor needs to produce an electrical output.

Chapter 2

1. What is a "free-body" diagram?
To write the equations of motion of a system of connected bodies, it is useful to draw each body in turn with the influence of all other bodies represented by forces and torques on the body in question. A drawing of the collection of such isolated bodies is called a "free-body diagram."

2. What are the two forms of Newton's law?
Translational motion is described by $F = ma$. Rotational motion is described by $M = I\alpha$.

3. For a structural process to be controlled, such as a robot arm, what is the meaning of "collocated control"? "Noncollocated control"?
When the actuator and the sensor are located on the same rigid body, the control is said to be "collocated." When they are on different bodies that are connected by springs, the control is "noncollocated."

4. State Kirchhoff's current law.

   The algebraic sum of all currents entering a junction or circuit is zero.

5. State Kirchhoff's voltage law.

   The algebraic sum of voltages around a closed path in an electric circuit is zero.

6. When, why, and by whom was the device named an "operational amplifier"?

   In a paper in 1947, Ragazzini, Randall, and Russell named the high-gain, wide-bandwidth amplifier used in feedback to realize operational calculus "operations" the operational amplifier.

7. What is the major benefit of having zero input current to an operational amplifier?

   With zero input current the amplifier does not load the input circuit; thus, the transfer function of the device is not dependent of the amplifier characteristics. Also, the analysis of the circuit is simplified in this case.

8. Why is it important to have a small value for the armature resistance R_a of an electric motor?

   The armature resistance causes power loss when the armature current flows and thus reduces the efficiency of the motor.

9. What are the definition and units of the electric constant of a motor?

   A rotating motor produces a voltage (called the back emf) in its armature proportional to the rotational speed. The electric constant K_e is the ratio of this voltage to the speed, so that $e = K_e \dot{\theta}$. The units are volt-sec/radians.

10. What are the definition and units of the torque constant of an electric motor?

    When current i_a flows in the armature of an electrical motor, a torque τ is produced that is proportional to the current. The torque constant K_t is the constant of proportionality, so that $\tau = K_t i_a$. The units are Newton-meters/amp.

11. Why do we approximate a physical model of the plant (which is *always* nonlinear) with a linear model?

    Analysis and design of linear models is vastly simpler than with nonlinear models. Furthermore, it has been shown (by Lyapunov) that, if the linear approximation is stable, then there is at least some region of stability for the nonlinear model.

△ 12. Give the relationships for (a) heat flow across a substance, and (b) heat storage in a substance.

   (a) *Heat flow is proportional to the temperature difference divided by the thermal resistance; that is,*

   $$q = \frac{1}{R}(T_1 - T_2).$$

   (b) *The differential equation describing the heat storage is*

   $$\dot{T} = \frac{1}{C}q,$$

   where C is the thermal capacity of the material.

$\triangle$ 13. Name and give the equations for the three relationships governing fluid flow.

$$\textit{Continuity:} \qquad \dot{m} = w_{in} - w_{out}.$$

$$\textit{Force equilibrium:} \qquad f = pA.$$

$$\textit{Resistance:} \quad w = \frac{1}{R}(p_1 - p_2)^{1/\alpha}.$$

Chapter 3

1. What is the definition of "transfer function"?
   The Laplace transform of the output of a linear, time-invariant system, $Y(s)$, is proportional to the transform of its input, $U(s)$. The function of proportionality is the transfer function $F(s)$, so that $Y(s) = F(s)U(s)$. It is assumed that all initial conditions are zero.

2. What are the properties of systems whose responses can be described by transfer functions?
   The system must be both linear (superposition applies) and time invariant (the parameters do not vary with time).

3. What is the Laplace transform of $f(t - \lambda)1(t - \lambda)$ if the transform of $f(t)$ is $F(s)$?
   $$\mathcal{L}\{f(t - \lambda)1(t - \lambda)\} = e^{-s\lambda}F(s).$$

4. State the Final Value Theorem.
   If all the poles of $sF(s)$ are in the LHP, then the final value of $f(t)$ is given by
   $$\lim_{t \to \infty} f(t) = \lim_{s \to 0} sF(s).$$

5. What is the most common use of the Final Value Theorem in control?
   A standard test of a control system is the step response, and the FVT is used to determine the steady-state error to such an input.

6. Given a second-order transfer function with damping ratio ζ and natural frequency ω_n, what is the estimate of the step response rise time? What is the estimate of the percent overshoot in the step response? What is the estimate of the settling time?
   These are given by $t_r \cong 1.8/\omega_n$, M_p is set by the damping ratio (see the curve in Fig. 3.23) and $t_s \cong 4.6/\sigma$.

7. What is the major effect of an extra zero in the LHP on the second-order step response?
   Such a zero causes additional overshoot, and the closer the zero is to the imaginary axis, the higher the overshoot. If the zero is more than six times the real part of the complex poles, the effect is negligible.

8. What is the most noticeable effect of a zero in the RHP on the step response of the second-order system?
   Such a zero often causes an initial undershoot of the response.

9. What is the main effect of an extra real pole on the second-order step response?
   A pole slows down the response and makes the rise time longer. The closer the pole is to the imaginary axis, the more pronounced is the effect. If the pole is more than six times the real part of the complex poles, the effect is negligible.

10. Why is stability an important consideration in control system design?
    Almost any useful dynamic system must be stable to perform its function. Feedback around a system that is normally stable can actually introduce instability, so control designers must be able to assure the stability of their designs.

11. What is the main use of Routh's criterion?
    With this method, we can find (symbolically) the range of a parameter such as the loop gain for which the system will be stable.

12. Under what conditions might it be important to know how to estimate a transfer function from experimental data?
    In many cases, the equations of motion are either extremely complex or not known at all. Chemical processes such as a paper-making machine are often of this kind. In these cases, if one wishes to build a good control, it is very useful to be able to take transient data or steady-state frequency-response data and to estimate a transfer function from these.

Chapter 4

1. Give three advantages of feedback in control.

   (a) *Feedback can reduce the steady-state error in response to disturbances.*
   (b) *Feedback can reduce the steady-state error in tracking a reference.*
   (c) *Feedback can reduce the sensitivity of a transfer function to parameter changes.*
   (d) *Feedback can stabilize an unstable process.*

2. Give two disadvantages of feedback in control.

   (a) *Feedback requires a sensor, which can be very expensive and may introduce additional noise.*
   (b) *Feedback systems are often more difficult to design and operate than open-loop systems.*

3. A temperature control system is found to have zero error to a constant tracking input and an error of 0.5°C to a tracking input that is linear in time, rising at the rate of 40°C/sec. What is the system type of this control system and what is the relevant error constant (K_p or K_v or K_a)?
   The system is Type 1 and the K_v is the ratio of input rate to error or $K_v = 40/0.5 = 80/\text{sec}$.

4. What are the units of K_p, K_v, and K_a?
   K_p is dimensionless, K_v is $\sec^{-1}$, and K_a is $\sec^{-2}$.

5. What is the definition of system type with respect to reference inputs?
   With only a polynomial of degree k reference input (no disturbances), the type is the largest value of k for which the steady-state error is a constant.

6. What is the definition of system type with respect to disturbance inputs?
   With only a polynomial of degree k disturbance input (no reference), the type is the largest value of k for which the steady-state error is a constant.

7. Why does system type depend on where the external signal enters the system?
   Because the error depends on where the input enters, so does the type.

8. What is the main objective of introducing integral control?
   Integral control will make the error to a constant input go to zero. It removes the effects of process noise bias. It cannot remove the effects of sensor bias.

9. What is the major objective of adding derivative control?
   Derivative control typically makes the system better damped and more stable.

10. Why might a designer wish to put the derivative term in the feedback rather than in the error path?
    When a reference input might include sudden changes, including it in the derivative action might cause unnecessary large controls.

11. What is the advantage of having a 'tuning rule' for PID controllers?
    PID controllers are typically packaged as a unit with knobs on the front for the several gain constants. These devices are widely installed in factories and operated by technicians with modest knowledge of control theory. A tuning rule permits such a person to measure parameters of the process experimentally and use this data to set the parameters in such a way as to give good response.

12. Give two reasons to use a digital controller rather than an analog controller.

    (a) *The control law is easier to change if the controller is digital.*
    (b) *A digital controller can perform logic and other nonlinear operations much easier than an analog controller.*
    (c) *The hardware of a digital controller can be fixed in the design before the details of the actual control design are finished.*

13. Give two disadvantages to using a digital controller.

    (a) *The bandwidth of a digital controller is limited by the possible sample frequency.*
    (b) *The digital controller introduces noise by the quantization process.*

14. Give the substitution in the discrete operator z for the Laplace operator s if the approximation to the integral in Eq. (4.98) is taken to be the rectangle of height $e(kT_s)$ and base T_s.

$$s = \frac{z - 1}{T_s}$$

Chapter 5

1. Give two definitions for the root locus.

    (a) *The root locus is the locus of points in the s-plane where the equation $a(s) + Kb(s) = 0$ has a solution.*
    (b) *The root locus is the locus of points in the s-plane where the angle of $G(s) = b(s)/a(s)$ is 180°.*

2. Define the negative root locus.
    The negative root locus is the locus of points where the equation $a(s) - Kb(s) = 0$ has a solution or where the angle of $G(s) = b(s)/a(s)$ is 0°.

3. Where are the sections of the (positive) root locus on the real axis?
    Segments of the real axis to the left of an odd number of zeros and poles are on the root locus.

4. What are the angles of departure from two coincident poles at $s = -a$ on the real axis? Assume there are no poles or zeros to the right of $-a$.
    The loci depart at $\pm 90°$.

5. What are the angles of departure from *three* coincident poles at $s = -a$ on the real axis? Assume there are no poles or zeros to the right of $-a$.
    The loci depart at $\pm 60°$ and 180°.

6. What is the principal effect of a lead compensation on a root locus?
    The lead compensation generally causes the locus to bend toward the LHP, moving the dominant roots to a place of higher damping.

7. What is the principal effect of a lag compensation on a root locus in the vicinity of the dominant closed-loop roots?
    The lag compensation is normally placed so near the origin that it has negligible effect on the root locus in the vicinity of the dominant closed-loop roots.

8. What is the principal effect of a lag compensation on the steady-state error to a polynomial reference input?
    A lag compensation normally raises the gain at $s = 0$ and thus increases the velocity constant of a Type 1 system and lowers the error to polynomial inputs.

9. Why is the angle of departure from a pole near the imaginary axis especially important?
    If the locus starts toward the RHP, then feedback will make the system less stable. On the other hand if the locus departs toward the LHP, then feedback is going to make the system more stable.

10. Define a conditionally stable system.
    *A system that becomes **unstable** as gain is reduced is considered to be conditionally stable. That is, its stability is conditioned on having an operating compensator with at least a minimum value of gain.*

11. Show, with a root locus argument, that a system having three poles at the origin *must* be conditionally stable.

    With three poles at the origin, the angles of departure ensure that two poles leave the origin at 180°, ±60°, or, if there are poles on the real axis in the RHP, they may leave at 0°, ±120° which is to say that at least one pole begins by moving into the RHP. As gain is reduced from the operating level, at least one root must pass into the RHP for gain low enough; therefore, the system must be conditionally stable.

Chapter 6

1. Why did Bode suggest plotting the magnitude of a frequency response on log–log coordinates?

   In log–log coordinates, the plot for a rational transfer function can be well guided by linear asymptotes and thus easily plotted and visualized.

2. Define a decibel.

   If a power ratio is P_1/P_2, then the measure in decibels is $10 \log(P_1/P_2)$. Because power is proportional to voltage squared, and a transfer function would give a ratio of voltages, then the gain of a transfer function $G(j\omega)$ in decibels is $G_{db} = 20 \log |G(j\omega)|$.

3. What is the transfer function magnitude if the gain is listed as 14 db?

   $14 = 20 \log M$, *therefore* $M = 5.01$.

4. Define gain crossover.

   The gain crossover ω_c is the value of frequency where the magnitude gain is 1 (or 0 db).

5. Define phase crossover.

   The phase crossover ω_{cp} is the value of the frequency where the phase crosses $-180°$.

6. Define phase margin, PM.

   The phase margin PM is a measure of how far in phase the Nyquist plot is from instability. In the typical case, if the phase of the system at gain crossover is ϕ, then the phase margin is $180° + \phi$. For example, if $\phi = -150°$, then the phase margin is 30°.

7. Define gain margin, GM.

   The gain margin is a measure of how far the system is from instability by changes in gain alone. If the gain at phase crossover, where the system phase is 180°, is $|G(j\omega_{cp})|$, then the gain margin is $GM * |G(j\omega_{cp})| = 1.0$ or $GM = 1/|G(j\omega_{cp})|$.

8. What Bode plot characteristic is the best indicator of the closed-loop step response overshoot?

The phase margin is related to the equivalent closed-loop damping ratio approximately by $\zeta_{eq} = PM/100$. As we saw in Chapter 3, the step response overshoot is monotonically related to the damping ratio.

9. What Bode plot characteristic is the best indicator of the closed-loop step response rise time?

   The rise time is measured by the closed-loop natural frequency, which in turn is adequately approximated by the gain crossover. Thus the best indicator of rise time is ω_{cg}.

10. What is the principal effect of a lead compensation on Bode plot performance measures?

    The lead compensation usually is used to raise the phase margin at a desired gain crossover frequency.

11. What is the principal effect of a lag compensation on Bode plot performance measures?

    The lag compensation is usually used to raise the low-frequency gain to reduce the steady-state error to polynomial or low-frequency sinusoidal inputs. It can also be used to lower the crossover frequency ω_c, where a more favorable phase exists.

12. How do you find the K_v of a type 1 system from its Bode plot?

    The K_v is determined by the low-frequency asymptote, which has a slope of -1 for a type 1 system and is given by the expression K_v/ω. The value of the constant may be found either from the frequency where the asymptote reaches 1.0 (or 0 db) or else as the value of the asymptote at the frequency of $\omega = 1$.

13. Why do we need to know beforehand the number of open-loop unstable poles in order to tell stability from the Nyquist plot?

    The Nyquist plot encirclements counts the difference in the number of zeros and the number of poles in the RHP of $1 + KDG$. In order to know the number of zeros of this function (which are closed-loop poles and thus unstable poles of the closed loop), we must know the number of unstable open-loop poles for the plot.

14. What is the main advantage in control design of counting the encirclements of $-1/K$ of $D(j\omega)G(j\omega)$ rather than encirclements of -1 of $KD(j\omega)G(j\omega)$?

    If we plot DG alone, then the stability depends on the encirclements of $-1/K$. The designer can thus easily look at the entire range of real K and determine the best value of gain for the design without having to make any more plots.

15. Define a conditionally stable feedback system. How can you identify one on a Bode plot?

    A conditionally stable system becomes unstable as gain is reduced. If the low-frequency phase drops below $-180°$ then a reduction in gain until gain crossover occurs where there is no phase margin, then the system is almost surely unstable. A look at the Nyquist plot is necessary to be certain. This condition can also be *seen easily from a root locus; the locus will have segments in the RHP for low values of gain.*

△ 16. A certain control system is required to follow sinusoids, which may be any frequency in the range $0 \leq \omega_\ell \leq 450$ rad/sec and have amplitudes up to 5 units with (sinusoidal) steady-state error to be never more than 0.01. Sketch (or describe) the corresponding performance function $W_1(\omega)$.
The magnitude of W_1 is given by the ratio $|R|/e_b = 5/0.01 = 500$. The performance function would then have the value 500 for frequencies up to 450 rad/sec. The Bode magnitude plot would be required to be above this curve for these frequencies.

Chapter 7

The following questions are based on a system in state-variable form with matrices **F, G, H,** and J, input u, output y, and state **x**:

1. Why is it convenient to write equations of motion in state-variable form?
It provides a standard way to describe the differential equations for any dynamic system so that computer-aided analysis can be carried out more conveniently. It is also more convenient to analyze linear systems in terms of the standard description matrices.

2. Give an expression for the transfer function of this system.

$$G(s) = \mathbf{H}(s\mathbf{I} - \mathbf{F})^{-1}\mathbf{G} + J.$$

3. Give two expressions for the poles of the transfer function of the system.

(a) $p = \mathsf{eig}(\mathbf{F})$.
(b) $p = $ *roots of* $\det[s\mathbf{I} - \mathbf{F}] = a(s) = 0$.

4. Give an expression for the zeros of the system transfer function.

$$z = \text{roots of } \det \begin{bmatrix} s\mathbf{I} - \mathbf{F} & -\mathbf{G} \\ \mathbf{H} & J \end{bmatrix} = b(s) = 0.$$

5. Under what condition will the state of the system be controllable?

(a) *If the pair* $(\mathbf{F}, \mathbf{G})$ *is controllable—that is, if the matrix*

$$\mathcal{C} = \begin{bmatrix} \mathbf{G} & \mathbf{FG} & \cdots & \mathbf{F}^{n-1} \end{bmatrix}$$

is full rank.
(b) *If the system can be put into control canonical form.*

6. Under what conditions will the system be observable from the output y?

(a) *If the matrices* $(\mathbf{F}, \mathbf{H})$ *are observable—that is, if the matrix*

$$\mathcal{O} = \begin{bmatrix} \mathbf{H} \\ \mathbf{HF} \\ \vdots \\ \mathbf{HF}^{(n-1)} \end{bmatrix}$$

has full rank.

(b) *If the system can be put into observable canonical form.*

7. Give an expression for the *closed-loop* poles if state feedback of the form $u = -\mathbf{Kx}$ is used.

(a) $p_c = \text{eig}(\ \mathsf{F} - \mathsf{G} * \mathsf{K})$.
(b) $p_c = $ *roots* of $\det(s\mathbf{I} - \mathbf{F} + \mathbf{GK}) = \alpha_c(s) = 0$.

8. Under what conditions can the feedback matrix $\mathbf{K}$ be selected so that the roots of $\alpha_c(s)$ are placed arbitrarily?
If the system is controllable.

9. What is the advantage of using the LQR or symmetrical root locus in designing the feedback matrix $\mathbf{K}$?
With LQR, the closed-loop system will be more robust to parameter changes, and the designer has some control over the control effort used by the closed-loop system.

10. What is the main reason for using an estimator in feedback control?
When the state is not available (usually because it is too expensive or impractical to put sensors on each state variable), then an estimator using only the output y can give an estimate that can be used in place of the actual state.

11. If the estimator gain $\mathbf{L}$ is used, give an expression for the closed-loop poles due to the estimator.

(a) $p_e = \text{eig}(\mathsf{F} - \mathsf{L} * \mathsf{H})$.
(b) $p_e = $ *roots of* $\det(s\mathbf{I} - \mathbf{F} + \mathbf{LH}) = \alpha_e(s) = 0$.

12. Under what conditions can the estimator gain $\mathbf{L}$ be selected so that the roots of $\alpha_e(s) = 0$ are placed arbitrarily?
If the system is observable.

13. If the reference input is arranged so that the input to the estimator is identical to the input to the process, what will be the overall closed-loop transfer function?

$$T(s) = N\frac{b(s)}{\alpha_c(s)}.$$

14. If the reference input is introduced in such a way as to permit the zeros to be assigned as the roots of $\gamma(s)$, what will the overall closed-loop transfer function be?

$$T(s) = N \frac{\gamma(s)b(s)}{\alpha_e(s)\alpha_c(s)},$$

usually $\gamma(s) = \alpha_e(s)$.

15. What are the three standard techniques for introducing integral control in the state-feedback design method?

   (a) *By augmenting the process state to include an integrator state variable.*
   (b) *By the internal model approach.*
   (c) *By using the extended estimator approach.*

Chapter 8

1. What is the Nyquist rate? What are its characteristics?
   The Nyquist rate is half the sample rate, or $= \omega_s/2$. Above this rate, no frequencies can be represented by a sampled signal.

2. Describe the discrete equivalent design process.
   The controller for a system is designed as if the controller will be analog. The resulting controller is then approximated by a digital equivalent.

3. Describe how to arrive at a $D(z)$ if the sample rate is $30 \times \omega_{BW}$.
   Use the discrete equivalent design method. It typically yields satisfactory results for such a high sample rate. But after using the discrete equivalent, check the result using a simulation that includes the effect of sampling or else perform an exact discrete linear analysis. It is best to use a simulation that includes all known sampling effects and system delays.

4. For a system with a 1 rad/sec bandwidth, describe the consequences of various sample rates.
   An absolute minimum sample rate is 2 rad/sec (or 0.32 Hz and $T = 3$ sec). From 2 rad/sec to 10 or 20 rad/sec, the control will be jerky with noticeable steps in the control and the design needs to be done very carefully. Between 20 and 30 rad/sec, the magnitude of the control steps become progressively smaller and design using discrete equivalents works reasonably well. Above 30 rad/sec, the control steps are hardly noticeable and the discrete equivalent can be used with confidence.

5. Give two advantages for selecting a digital processor rather than analog circuitry to implement a controller.

   (a) *The physical layout of a digital controller can be done before the final design is complete, often resulting in completing the hardware implementation in* much less time than required to get an analog controller specified and constructed.

(b) *A digital processor is more flexible in making design changes as software is easier to reprogram than rewiring and/or adding op-amps to a printed circuit board.*

(c) *A digital processor can much more easily include nonlinear terms and logic decision steps in the overall controller design to permit adaptive control or gain scheduling, for example.*

(d) *Many models of the same basic controller can be accommodated by simply using different PROMS with the same hardware design. For example, an automobile manufacturer might have one engine controller hardware design for its entire product line; but have a different PROM for each engine/vehicle combination.*

(e) *Digital controllers are less sensitive to temperature variations than analog controllers.*

6. Give two disadvantages of selecting a digital processor rather than analog circuitry to implement a controller.

(a) *The finite sampling rate of the A/D and D/A converters and the finite compute speed of the processor limit the bandwidth of the controller to about 1/10 of the sample frequency.*

(b) *The finite accuracy or bit length of the converters introduce extra noise or offsets into the control loop if using low-end controllers.*

(c) *Cost. For simple controllers, a digital implementation will typically be more expensive than an analog implementation.*

△ 7. Describe how to arrive at a $D(z)$ if the sample rate is $5 \times \omega_{BW}$.
   Start by using the discrete equivalent, but include an approximation of the effect of the delay in the plant model when carrying out the analog design. Then check the result via an exact discrete analysis by converting the plant to its discrete equivalent and combining that with the discrete controller. If performance is degraded from that desired, modify the discrete controller using discrete design methods. Finish by using a simulation that includes all known sampling effects and system delays.

Chapter 9

1. Why do we approximate a physical model of the plant (which is *always* nonlinear) with a linear model?
   Analysis and design of linear models is vastly simpler than with nonlinear models. Furthermore, it has been shown (by Lyapunov) that if the linear approximation is stable, then there is at least some region of stability for the nonlinear model.

2. How would you linearize the nonlinear system equation for radiation heat transfer $\dot{T} = T^4 + T + u$?

$$\delta\dot{T} = (4T_o^3 + 1)\delta T + \delta u,$$

where T_o is the nominal operating temperature. (See the RTP case study in Chapter 10.)

3. A lamp used as a thermal actuator has a nonlinearity such that the experimentally measured output power is related to the input voltage by $P = V^{1.6}$. How would you deal with such a nonlinearity in feedback control design?

We precede the lamp with an inverse nonlinearity—that is, $V = P^{0.625}$—so as to linearize the cascaded system (see the RTP case study in Chapter 10).

4. What is integrator windup?

If the plant actuator output signal saturates, then it may take a long time for the error to be brought back to zero from an initial upset and during this time the integrator output may grow or "windup" much more than it would if the system were linear. Special "antiwindup" circuits are designed to prevent windup.

5. Why is an antiwindup circuit important?

When a control includes integral action and is subject to saturation, large inputs can cause large overshoots and slow recovering unless an antiwindup circuit is included.

6. Using the nonlinear saturation function having gain 1 and limits ± 1, sketch the block diagram of saturation for an actuator that has gain 7 and limits ± 20.

If the output of the actuator is u_{out} and its input is u_{in}, the control is given by

$$u_{out} = 20\,\text{sat}\left(\frac{7u_{in}}{20}\right).$$

7. What is a describing function and how is it related to a transfer function?

The goal of the describing function approach is to find something like a "transfer function" for a nonlinear element. One may view the describing function as an extension of the frequency response to nonlinearities.

8. What are the assumptions behind the use of the describing function?

The basic assumption is that the plant behaves approximately as a low-pass filter. The other assumptions are that the nonlinearity is time invariant, and there is a single nonlinear element in the system.

9. What is a limit cycle in a nonlinear system?

In some nonlinear systems the error builds up and the response approaches a periodic solution of fixed amplitude, the limit cycle, as time grows large.

10. How can one determine the describing function for a nonlinear system in the laboratory?

We can inject sinusoidal signals into the system and place a low-pass filter with a sharp cutoff at the output of the system to measure the fundamental component of the output. The describing function is then computed as the ratio of the amplitude of the fundamental component of the output of the nonlinear system over the amplitude of the sinusoidal input signal.

11. What is the minimum time-control strategy for a satellite attitude control with bounded controls?
    Bang-bang.

12. How are the two Lyapunov methods used?
    His indirect or first method is based on linearization of the equations of motion and drawing conclusions about the stability of the nonlinear system by considering the stability of the linear approximation. In his direct or second method, the nonlinear equations are considered directly.

Chapter 10

1. Why is a collocated actuator and sensor arrangement for a lightly damped structure such as a robot arm easier to design than a noncollocated setup?
   In the collocated case, the process naturally has zeros near the lightly damped poles, which keep the root locus in the LHP.

2. Why should the control engineer be involved in the design of the process to be controlled?
   In many cases, the characteristics and locations of the actuators and sensors can have a major impact on the complexity and difficulty in design of the controller. If the needs of control are included in the process design, the final systems are often more effective (better closed-loop performance) and less expensive.

3. Give examples of an actuator and a sensor for the following control problems:

   (a) Attitude control of a geosynchronous communication satellite.
       Actuators: Cold gas-jet thrusters, momentum wheels, magnetic torquers (coils, torque rod), plasma thruster.
       Sensors: Earth sensor (roll, pitch), digital integrated rate assembly (DIRA) gyro (for rates), star tracker.
   (b) Pitch control of a Boeing 747 airliner.
       Actuators: Elevator.
       Sensors: Pitch rate and/or pitch angle is measured using a gyro or a ring-laser gyro.
   (c) Track-following control of a CD player.
       Actuators: DC motor to move the (dual stage sledge) arm mechanism, magnetic coils (two) for focusing on tracks.
       Sensors: Array of photodiodes.
   (d) Fuel–air ratio control of a spark-ignited automobile engine.
       Actuators: Fuel injection.
       Sensors: Zirconium oxide sensor.
   (e) Position control for an arm of a robot used to paint automobiles.
       Actuators: Hydraulic actuators or electric motors.
       Sensors: Encoders to measure arm rotations, pressure sensors, and force sensors.

(f) Heading control of a ship.
Actuators: Rudder.
Sensors: Gyrocompass.

(g) Attitude control of a helicopter.
Actuators: Moving swash plate (either via direct link or servo) rotates main blade angle of attack.
Sensors: Same as aircraft (pitot tube, accelerometers, rate gyros).

Appendix C MATLAB® Commands

MATLAB function (.m file) or Variable	Description	Page (s)
angle	Phase Angle	
ans	Most recent answer	
abs	Absolute value	
acker	Ackermann's formula for pole placement	448,456,469,472,504
atan2	Four quadrant inverse tangent	
axis	Control axis scaling	329,332
bilin	Bilinear transform	
bode	Bode frequency response	84,300,315,380,522
bodemag	Bode magnitude frequency response	
c2d	Continuous-to-discrete conversion	203,353,488,499,570
canon	State-space canonical forms	435
clear	Clear variables and functions	
clf	Clear current figure	
close	Close figure	
close all	Close all figures	
conj	Complex conjugate	
conv	Polynomial multiplication	92,459,461,767
cos	Cosine	
ctrb	Controllability matrix	
ctrbf	Staircase canonical form, controllability	430
damp	Damping and natural frequency	588–589
dcgain	Computes DC gain of LTI system	
deconv	Division of polynomials	
det	Determinant of a matrix	
diag	Diagonal matrix, diagonals of a matrix	
diary	Save text of MATLAB session	
dstep	Step response of a discrete system	
eig	Eigenvalues and eigenvectors	
exp	Exponential	432
expm	Matrix exponential	
eye	Identity matrix	
ezplot	Easy-to-use function plotter	
feedback	Feedback connection of two systems	262–263

MATLAB function (.m file) or Variable	Description	Page (s)
figure	Create figure window	
figure(i)	Make i the current figure	
find	Find indices of nonzero elements	
format	Set output format	
freqresp	Frequency response of LTI systems	
gram	Controllability/observability Gramian	
grid	Grid lines	
hold	Hold current plot	
i	$\sqrt{-1}$	
ilaplace	Inverse Laplace transform	
imag	Complex imaginary part	
impulse	Impulse response of LTI system	111,115,127
inf	Infinity	
initial	Initial condition response of state-space system	469,475
inv	Matrix inverse	434,436
j	$\sqrt{-1}$	
laplace	Laplace transform	86
linmod	Linearization	620
linmod2	Linearization (advanced)	603
line	Create a line	
linspace	Linearly spaced vector	
load	Load in workspace variables	
log	Natural logarithm	
log10	Logarithm to the base 10	
loglog	Log–log plot	84,300,315,522
logspace	Logarithmically spaced frequency points	84
lqe	Linear Quadratic Estimator design	521
lqr	Linear Quadratic Regulator design	463,464,725
lsim	Simulation of LTI system with arbitrary input	101,102
ltiview	Opens the LTI viewer GUI	
ltru	Loop transfer recovery (LTR)	
ltry	Loop transfer recovery (LTR)	
margin	Gain and phase margins	357,398,522
max	Largest component	380
mean	Average or mean value	
min	Smallest component	
nan	Not-a-number	
nichols	Nichol's chart	383
norm	Matrix or vector norm	
nyquist	Nyquist plot	326,329,333
obsv	Observability matrix	471
obsvf	Staircase canonical form, observability	471
ones	Array of ones	101

MATLAB function (.m file) or Variable	Description	Page (s)
pade	Pade approximation for time delay	273
parallel	Parallel connection of two LTI systems	107
place	Pole placement	451,472,484,490,510
pi	3.141592653589793	
plot	Plot function	33,99,101
pole	Poles of LTI system	
poly	Form polynomial from its roots	96
polyval	Evaluate polynomial	
printsys	Print system in pretty format	98
pzmap	Pole–zero map	109
rand	Uniformly distributed random numbers	
randn	Normally distributed random numbers	
rank	Matrix rank	
real	Complex real part	
residue	Residues in partial-fraction expansion	92,96,767
rlocfind	Find Root-Locus gain	246,399
rlocus	Root locus	225,399,459,461
rltool	Interactive Root-Locus tool	282
roots	Roots of a polynomial	136,239,441,451
save	Save workspace variables	
semilogx	Semi-log plot	84,300,522
semilogy	Semi-log plot	380
series	Series connection of two LTI systems	107,522,588–589
sgrid	s-plane grid lines	
sin	Sine	
sim	Simulate a SIMULINK model	
sisotool	SISO design tool	
size	Size of an array	
sort	Sort in ascending or descending order	
sqrt	Square root	510
squeeze	Remove singleton dimensions	
ss2ss	State-space similarity transformation	
ss2tf	State space to transfer function conversion	438,440,441
ss2zp	State space to pole–zero conversion	423
ss	Conversion to state-space	419,435,440,442
ssdata	Create a state-space model	435
std	Standard deviation	
step	Step response	24,137,138
subplot	Multiple plots on the same window	
sum	Sum of Elements	
svd	Singular-value decomposition	
syms	Declaration of symbolic variables	759
text	Text annotation	

MATLAB function (.m file) or Variable	Description	Page (s)
tf2ss	Transfer function to state-space conversion	426
tf2zp	Transfer function to pole–zero conversion	98,100
tf	Creation or conversion to transfer function	24,84,107,111,127,201,459
tfdata	Transfer function data	
title	Plot title	
tzero	Transmission zeros	468,479
var	Variance	
who	List of current variables	
why	Answers any question you may have	
whos	List of current variables, long form	
xlabel	x-axis label	
xlsread	Get data from an Excel spreadsheet	
ylabel	y-axis label	
zero	Transmission zeros	
zeros	Array of zeros	101
zgrid	z-plane grid lines	
zpk	Zero–pole–gain	
zp2tf	Zero–pole to transfer function conversion	100

Bibliography

Abramovitch, D. and G. F. Franklin, "A brief history of disk drive control," *IEEE Control System Magazine,* Vol. 22, pp. 28–42, June 2002.

Ackermann, J., "Der entwurf linearer regelungssysteme im zustandsraum," *Regelungstech. Prozess-Datenverarb.*, Vol. 7, pp. 297–300, 1972.

Airy, G. B., "On the regulator of the clock-work for effecting uniform movement of equatorials," *Mem. R. Astron. Soc.*, Vol. 11, pp. 249–267, 1840.

Alon, U., *An Introduction to Systems Biology*. Chapman & Hall/CRC, 2007.

Alon, U., M. G. Surette, N. Barkai, and S. Leibler, "Robustness in bacterial chemotaxis," *Nature*, Vol. 397, pp. 168–171, January 1999.

Anderson, B. D. O. and J. B. Moore, *Optimal Control: Linear Quadratic Methods.* Upper Saddle River, NJ: Prentice Hall, 1990.

Anderson, E., et al., *LAPACK User's Guide,* 3rd ed., Philadelphia, PA: SIAM, 1999.

Åström, K. J., "Frequency domain properties of Otto Smith regulators," *Int. J. Control*, Vol. 26, No. 2, pp. 307–314, 1977.

Åström, K. J. and T. Hägglund, *PID Controllers: Theory, Design, and Tuning,* 2nd ed., Research Triangle, NC: International Society for Measurement and Control, 1995.

Åström, K. J. and T. Hägglund, *Advanced PID Control.* Research Triangle, NC: International Society for Measurement and Control, 2006.

Athans, M., "A tutorial on the LQG/LTR method," *Proc. American Control Conf,* pp. 1289–1296, June 1986.

Barkai, N. and S. Leibler, "Robustness in simple biochemical networks," *Nature*, Vol. 387, pp. 913–917, 1997.

Bellman, R. and R. Kalaba, eds., *Mathematical Trends in Control Theory*. New York: Dover, 1964.

Berg, H. C., *E. coli in Motion*. New York: Springer-Verlag, 2004.

Bergen, A. R. and R. L. Franks, "Justification of the describing function method," *SIAM J. Control,* Vol. 9, pp. 568–589, 1971.

Blakelock, J. H., *Automatic Control of Aircraft and Missiles.* 2nd ed., New York: John Wiley, 1991.

Bodanis, D., $E = MC^2$: *A Biography of the World's Most Famous Equation.* New York: Walker and Co., 2000.

Bode, H. W., "Feedback: The history of an idea," *Conference on Circuits and Systems*. New York (1960): Reprinted in Bellman and Kalaba, 1964.

—— *Network Analysis and Feedback Amplifier Design.* New York: Van Nostrand, 1945.

Boyd, S. P. and C. H. Barratt, *Linear Controller Design: Limits of Performance.* Upper Saddle River, NJ: Prentice Hall, 1991.

Brennan, R. P., *Heisenberg Probably Slept Here*. New York: John Wiley & Sons, 1997.

Brown, J. W. and R. V. Churchill, *Complex Variables and Applications*. 6th ed., New York: McGraw-Hill, 1996.

Bryson, A. E., Jr. and W. F. Denham, "A steepest-ascent method for solving optimum programming problems," *J. Appl. Mech.*, June 1962.

Bryson, A. E., Jr. and Y. C. Ho, *Applied Optimal Control*. Waltham, MA: Blaisdell, 1969.

Bryson, A. E., Jr., *Control of Spacecraft and Aircraft*. Princeton, NJ: Princeton University Press, 1994.

Callender, A., D. R. Hartree, and A. Porter, "Time lag in a control system," *Philos. Trans. R. Soc. London A*, London: Cambridge University Press, 1936.

Campbell, G. A. and R. N. Foster, *Fourier Integrals for Practical Applications*. New York: Van Nostrand, 1948.

Campbell, N. A. and J. B. Reece, *Biology*, 8th ed., Benjamin Cummings, 2008.

Cannon, R. H., Jr., *Dynamics of Physical Systems*. New York: McGraw-Hill, 1967.

Churchill, R. V., *Operational Mathematics*, 3rd ed., New York: McGraw-Hill, 1972.

Clark, R. N., *Introduction to Automatic Control Systems*. New York: John Wiley, 1962.

Clegg, J. C., "A nonlinear integrator for servomechanisms," *Trans. AIEE*, Pt. II, Vol. 77, pp. 41–42, 1958.

de Roover, D., L. Porter, A. Emami-Naeini, J. A. Marohn, S. Kuehn, S. Garner, and D. Smith, "An all-digital cantilever controller for MRFM and scanned probe microscopy using a combined DSP/FPGA design," *Am. Lab.*, Vol. 40, No. 8, pp. 12–17, 2008.

de Roover, D., A. Emami-Naeini, and J. L. Ebert, "Model-based control of fast-ramp RTP systems," *Sixth International Conference on Advanced Thermal Processing of Semiconductors*, pp. 177–186, Kyoto, Japan, September 1998.

de Vries, G., T. Hillen, M. Lewis, J. Muller, and B. Schonfisch, *A Course in Mathematical Biology*. SIAM, 2006.

Dorato, P., *Analytic Feedback System Design: An Interpolation Approach, Pacific Grove,* CA: Brooks/Cole, 2000.

Doyle, J. C., B. A. Francis, and A. Tannenbaum, *Feedback Control Theory*. New York: Macmillan, 1992.

Doyle, J. C. and G. Stein, "Multivariable feedback design: Concepts for a classical/ modern synthesis," *IEEE Trans. Autom. Control*, Vol. AC-26, No. 1, pp. 4–16, February 1981.

Doyle, J. C., "Guaranteed margins for LQG regulators," *IEEE Trans. Autom. Control*, Vol. AC-23, pp. 756–757, 1978.

Doyle, J. C. and G. Stein, "Robustness with observers," *IEEE Trans. Autom. Control*, Vol. AC-24, pp. 607–611, August 1979.

Ebert, J. L., A. Emami-Naeini, H. Aling, and R. L. Kosut, "Thermal modeling of rapid thermal processing systems," *Third International Rapid Thermal Processing Conference*, pp. 343–355, Amsterdam, August 1995a.

Ebert, J. L., A. Emami-Naeini, and R. L. Kosut, "Thermal Modeling and control of rapid thermal processing systems," *Proceeding 34th IEEE Conference Decision and Control*, pp. 1304–1309, December 1995b.

Elgerd, O. I., *Electric Energy Systems Theory*. New York: McGraw-Hill, 1982.

——— and W. C. Stephens, "Effect of closed-loop transfer function pole and zero locations on the transient response of linear control systems," *Trans. Am. Inst. Electr. Eng. Part 1*, Vol. 42, pp. 121–127, 1959.

Emami-Naeini, A., "The shapes of Nyquist plots: Connections with classical plane curves," *IEEE Control Systems Magazine*, October 2009.

Emami-Naeini, A., J. L. Ebert, D. de Roover, R. L. Kosut, M. Dettori, L. Porter, and S. Ghosal, "Modeling and control of distributed thermal systems," *Proc. IEEE Trans. Control Systems Technol.*, Vol. 11, No. 5, pp. 668–683, September 2003.

Emami-Naeini, A. and G. F. Franklin, "Zero assignment in the multivariable robust servomechanism," *Proc. IEEE Conf. Dec. Control*, pp. 891–893, December, 1982.

Emami-Naeini, A., and P. Van Dooren, "Computation of zeros of linear multivariable systems," *Automatica*, Vol. 18, No. 4, pp. 415–430, 1982.

——— "On computation of transmission zeros and transfer functions," *Proc. IEEE Conf. Dec. Control*, pp. 51–55, December 1982.

Etkin, B. and L. D. Reid, *Dynamics of Flight: Stability and Control*, 3rd ed., New York: John Wiley, 1996.

Evans, G. W., "Bringing root locus to the classroom," *IEEE Control Systems Magazine*, Vol. 24, pp. 74–81, 2004.

Freudenberg, J. S. and D. P. Looze, "Right half plane zeros and design tradeoffs in feedback systems," *IEEE Trans. Autom. Control*, Vol. AC-30, pp. 555–561, June 1985.

Fuller, A. T., "The Early development of control theory," *J. Dyn. Syst. Meas. Control*, Vol. 98, pp. 109–118 and 224–235, 1976.

Gardner, M. F. and J. L. Barnes, *Transients in Linear Systems*. New York: John Wiley, 1942.

Gunckel, T. L., III and G. F. Franklin, "A general solution for linear sampled data control," *J. Basic Eng.*, Vol. 85-D, pp. 197–201, 1963.

Gyugyi, P., Y. Cho, G. F. Franklin, and T. Kailath, "Control of rapid thermal processing: A system theoretic approach," *Proceedings of IFAC World Congress*, 1993.

Hanselman, D. C. and B. C. Littlefield, *Mastering MATLAB 7*, Upper Saddle River. NJ: Prentice Hall, 2005.

Heffley, R. K. and W. F. Jewell, *Aircraft Handling Qualities*, Technical Report 1004-1, System Technology, Inc., Hawthorne, CA, May 1972.

Higham, D. J. and N. J. Higham, *MATLAB Guide*, 2nd ed., Philadelphia: SIAM, 2005.

Ho, M.-T., A. Datta, and S. P. Bhattacharyya, "An elementary derivation of the Routh-Hurwitz criterion," *IEEE Trans. Autom. Control*, Vol. 43, No. 3, pp. 405–409, 1998.

Huang, J-J. and D. B. DeBra, "Automatic tuning of Smith-predictor design using optimal parameter mismatch," *Proc. IEEE Conf. Dec. Contr.*, pp. 3307–3312, December 2000.

Hubbard, M., Jr., and J. D. Powell, "Closed-loop control of internal combustion engine exhaust emissions," SUDAAR No. 473, Department of Aero/Astro, Stanford University, Stanford, CA, February 1974.

James, H. M., N. B. Nichols, and R. S. Phillips, *Theory of Servomechanisms*, Radiation Lab. Series, 25. New York: McGraw-Hill, 1947.

Johnson, R. C., Jr., A. S. Foss, G. F. Franklin, R. V. Monopoli, and G. Stein, "Toward development of a practical benchmark example for adaptive control," *IEEE Control System Magazine*, Vol. 1, No. 4, pp. 25–28, December 1981.

Joseph, P. D. and J. T. Tou., "On linear control theory," *AIEE Transactions*, Vol. 80, pp. 193–196, 1961.

Kailath, T., *Linear Systems*. Upper Saddle River, NJ: Prentice Hall, 1980.

Kalman, R. E., "A new approach to linear filtering and prediction problems," *J. Basic Eng.*, Vol. 85, pp. 34–45, 1960a.

Kalman, R. E. and J. E. Bertram, "Control system analysis and design via the second method of Lyapunov. II. Discrete Systems," *J. Basic Eng.*, Vol. 82, pp. 394–400, 1960.

Kalman, R. E., Y. C. Ho, and K. S. Narendra, "Controllability of linear dynamical systems," *Contributions to Differential Equations*, Vol. 1. New York: John Wiley, 1962.

Khalil, H. K., *Nonlinear Systems*, 3rd ed., Upper Saddle River, NJ: Prentice Hall, 2002.

Kharitonov, V. L., "Asymptotic stability of an equilibrium position of a family of systems of linear differential equations," *Differential'nye Uraveniya*, Vol. 14, pp. 1483–1485, 1978.

Kochenburger, R. J., "A frequency response method for analyzing and synthesizing contactor servomechanisms," *Trans. Am. Inst. Electr. Eng.*, Vol. 69, pp. 270–283, 1950.

Kuo, B. C., ed., *Incremental Motion Control, Vol. 2: Step Motors and Control Systems*, Champaign, IL: SRL Publishing, 1980.

—— *Proceedings of the Symposium Incremental Motion Control Systems and Devices, Part. 1: Step Motors and Controls*, Champaign-Urbana, IL: University of Illinois, 1972.

Lanchester, F. W., *Aerodonetics*. London: Archibald Constable, 1908.

LaSalle, L. P. and S. Lefschetz, *Stability by Lyapunov's Direct Method*. New York: Academic Press, 1961.

Lyapunov, A. M., "Problème général de la stabilité du mouvement," *Ann. Fac. Sci. Univ. Toulouse Sci. Math. Sci. Phys.*, Vol. 9, pp. 203–474, 1907; original paper published in 1892 in *Commun. Soc. Math. Kharkow*, 1892; reprinted as Vol. 17 in *Annals of Math Studies*. Princeton, NJ: Princeton University Press, 1949.

Ljüng, L., *System Identification: Theory for the User*, 2nd ed., Upper Saddle River, NJ: Prentice Hall, 1999.

Luenberger, D. G., "Observing the state of a linear system," *IEEE Trans. Mil. Electron.*, Vol. MIL-8, pp. 74–80, 1964.

Mahon, B., *The Man who Changed Everything: The Life of James Clerk Maxwell*, UK: Wiley, 2003.

Marsden, J. E. and M. J. Hoffman, *Basic Complex Analysis,* 3rd ed., Freeman, 1999.

Mason, S. J., "Feedback theory: Some properties of signal flow graphs," *Proc. IRE*, Vol. 41, pp. 1144–1156, 1953.

—— "Feedback theory: Further properties of signal flow graphs," *Proc. IRE*, Vol. 44, pp. 920–926, 1956.

Maxwell, J. C., "On governors," *Proc. R. Soc. Lond.*, Vol. 16, pp. 270–283, 1868.

Mayr, O., *The Origins of Feedback Control*. Cambridge, MA: MIT Press, 1970.

McRuer, D. T., I. Askenas, and D. Graham, *Aircraft Dynamics and Automatic Control*, Princeton, NJ: Princeton University Press, 1973.

Mello, B. A., L. Shaw, and Y. Tu, "Effects of receptor interaction in bacterial chemotaxis," *Biophysical Journal*, Vol. 87, pp. 1578–1595, September 2004.

Messner, W. C. and D. M. Tilburry, *Control Tutorials for MATLAB and Simulink: A Web-Based Approach,* Upper Saddle River, NJ: Prentice Hall, 1999.

Minimis, G. S. and C. C. Paige, "An algorithm for pole assignment of time invariant systems," *Int. J. Control*, Vol. 35, No. 2, pp. 341–354, 1982.

Moler, C. B., "Nineteen dubious ways to compute the exponential of a matrix, twenty-five years later," *SIAM Rev.*, Vol. 45, No. 1, pp. 3–49, 2003.

—— *Numerical Computing with MATLAB*, Philadelphia, PA: SIAM, 2004.

Norman, S. A., "Wafer temperature control in rapid thermal processing," Ph.D. Dissertation, Stanford University, Stanford, CA, 1992.

Nyquist, H., "Regeneration theory," *Bell Systmes Technical. J.*, Vol. 11, pp. 126–147, 1932.

Oswald, R. K., "Design of a disk file head positioning servo," *IBM J. Res. Dev.*, Vol. 18, pp. 506–512, November 1974.

Parks, P., "Lyapunov redesign of model reference adaptive control systems," *IEEE Trans. Autom. Control*, AC-11, No. 3, 1966.

Perkins, W. R., P. V. Kokotovic, T. Boureret, and J. L. Schiano, "Sensitivity function methods in control system education," *IFAC Conference on Control Education*, June 1991.

Ragazzini, J. R., R. H. Randall, and F. A. Russell, "Analysis of problems in dynamics by electronic circuits," *Proc. IRE*, Vol. 35, No. 5, pp. 442–452, May 1947.

Ragazzini, J. R. and G. F. Franklin, *Sampled-Data Control Systems*, New York: McGraw-Hill, 1958.

Reliance Motion Control Corp., *DC Motor Speed Controls Servo Systems,* 5th ed., Eden Prairie, MN: Reliance Motion Control Corp., 1980.

Routh, E. J., *Dynamics of a System of Rigid Bodies*. London: MacMillan, 1905.

Saberi, A., B. M. Chen, P. Sannuti, *Loop Transfer Recovery: Analysis and Design.* New York: Springer-Verlag, 1993.

Safonov, M. G. and G. Wyetzner, "Computer-aided stability analysis renders Popov criterion obsolete," *IEEE Trans. Autom. Control*, Vol. AC-32, pp. 1128–1131, 1987.

Sandberg, I. W., "A frequency domain condition for stability of feedback systems containing a single time varying nonlinear element," *Bell Systems Technical J.*, Vol. 43, pp. 1581–1599, 1964.

Sastry, S. S., *Nonlinear Systems; Analysis, Stability, and Control*. New York: Springer-Verlag, 1999.

Schmitz, E., "Robotic arm control," Ph.D. Dissertation, Stanford University, Stanford, CA, 1985.

Sedra, A. S., and K. C. Smith, *Microelectronics Circuits*, 3rd ed., New York: Oxford University Press, 1991.

Simon, H. A., "Dynamic programming under uncertainty with a quadratic function," *Econometrica*, Vol. 24, pp. 74–81, 1956.

Sinha, N. K. and B. Kuszta, *Modeling and Identification of Dynamic Systems*. New York: Van Nostrand, 1983.

Smith, O. J. M., *Feedback Control Systems*. New York: McGraw-Hill, 1958.

Sobel, D., *Galileo's Daughter*. New York: Penguin Books, 2000.

Stein, G. and M. Athans, "The LQG/LTR procedure for multivariable feedback control design," *IEEE Trans. Autom. Control*, Vol. AC-32, pp. 105–114, February 1987.

Strang, G., *Linear Algebra and Its Applications*, 3rd ed., New York: Harcourt Brace, 1988.

Swift, J., *On Poetry: A Rhapsody*, 1973, J. Bartlett, ed., *Familiar Quotations*, 15th ed., Boston: Little Brown, 1980.

Sze, S. M., ed. *VLSI Technology*, 2nd ed., New York: McGraw-Hill, 1988.

Taubman, P., *Secret Empire: Eisenhower, the CIA, and the Hidden Story of America's Space Espionage*, New York: Simon and Schuster, 2003.

Thomson, W. T. and M. D. Dahleh, *Theory of Vibration with Applications*, 5th ed., Upper Saddle River, NJ: Prentice Hall, 1998.

Trankle, T. L., "Development of WMEC Tampa maneuvering model from sea trial data," Report MA-RD-760-87201. Palo Alto, CA: Systems Control Technology, March 1987.

Truxal, J. G., *Control System Synthesis*. New York: McGraw-Hill, 1955.

van der Linden, G., J. L. Ebert, A. Emami-Naeini, and R. L. Kosut "RTP robust control design: Part II: controller synthesis," Fourth *International Rapid Thermal Processing Conference*, pp. 263–271, September 1996.

Van der Pol, B., and H. Bremmer, *Operational Calculus*. New York: Cambridge University Press, 1955.

Van Dooren, P., A. Emami-Naeini, and L. Silverman, "Stable extraction of the kronecker structure of Pencils," *Proc. IEEE Conf. Dec. Control*, San Diego, CA, pp. 521–524, December 1978.

Vidyasagar, M., *Nonlinear Systems Analysis*, 2nd ed., Upper Saddle River, NJ: Prentice Hall, 1993.

Wiener, N., "Generalized harmonic analysis," *Acta Math.*, Vol. 55, pp. 117, 1930.

Woodson, H. H. and J. R. Melcher, *Electromechanical Dynamics, Part I: Discrete Systems*. New York: John Wiley, 1968.

Workman, M. L., "Adaptive proximate time-optimal servomechanisms," Ph.D. Dissertation, Stanford University, Stanford, CA, 1987.

Yi, T.-M., Y. Huang, M. I. Simon, and J. C. Doyle, "Robust perfect adaptation in bacterial chemotaxis through integral feedback control," *PNAS*, Vol. 97, No. 9, pp. 4649–4653, April 2000.

Zames, G., "On the input-output stability of time-varying nonlinear feedback systems—Part I: Conditions derived using concepts of loop gain, conicity and positivity," *IEEE Trans. Autom. Control*, Vol. AC-11, pp. 465–476, 1966.

—— "On the input-output stability of time-varying nonlinear feedback systems—Part II: Conditions involving circles in the frequency plane and sector nonlinearities," *IEEE Trans. Autom. Control*, Vol. AC-11, pp. 228–238, 1966.

Ziegler, J. G. and N. B. Nichols, "Optimum settings for automatic controllers," *Trans. ASME*, Vol. 64, pp. 759–768, 1942.

—— "Process lags in automatic control circuits," *Trans. ASME*, Vol. 65, No. 5, pp. 433–444, July 1943.

Index

801